ISBN 978-0-365-87321-1
PIBN 11346147

1 MONTH OF
FREE
READING

at

www.ForgottenBooks.com

By purchasing this book you are eligible for one month membership to ForgottenBooks.com, giving you unlimited access to our entire collection of over 1,000,000 titles via our web site and mobile apps.

To claim your free month visit:

www.forgottenbooks.com/free1346147

1

CATSKILL MOUNTAIN HOUSE.

71st SEASON.

Eight miles west of the Hudson River. By way of Catskill, within FOU:
HOURS of New York City,. Also accessible by way of Rhinebeck,
Rondout, and Kingston. Passengers by either of the above
routes may purchase tickets to CATSKILL MOUN-
TAIN STATION, which is only 300 feet from
the Hotel, and is the terminus of all
rail routes to this region.

Open June 20th.

THIS well-known Summer Hotel is situated on one of the eastern ledges of the sum
of the Catskill Mountains, 2,250 feet above tide water, and by reason of its peculia
advantageous location, on the front of the range, is the only hotel that commands
famous view of the Hudson Valley, which stretches out from the base of the mountains
low, to the Adirondacks in the north, the Green Mountain and Berkshire Hills in the ea
and the Highlands in the south, embracing an area of 12,000 square miles, with sixty mi
of the Hudson River in the foreground.

THE MOUNTAIN-HOUSE PARK

Has a valley frontage of over three miles in extent, and consists of 3,000 acres, or about :
square miles, of magnificent forest and farming lands, traversed in all directions by mi
miles of carriage-roads and paths leading to various noted places of interest.

The Crest, Newman's Ledge, Bears' Den, and Prospect Rock on North Mountain, and Ea
Rock and Palenville Overlook on South Mountain, from which the grandest views of
region are obtained, are included in the property. It also includes within its bounda
North and South Lakes, both plentifully stocked with various kinds of fish, and well supp
with boats.

The principal drives include Kaaterskill Falls, Haines's Falls, Kaaterskill Clove, Pa
ville, Tannersville, and Hunter Village. The atmosphere is delightful, invigorating,
pure, the great elevation and surrounding forest rendering it absolutely free from mala
It affords relief to sufferers from Chills and Fever, Asthma, Hay-Fever, Loss of Appe
and General Debility. The temperature is always fifteen to twenty degrees lower that
Catskill Village, New York City, or Philadelphia. The location and surroundings are in
respects the most desirable in the entire range of the Catskills, and no hotel similarly situ
is so easy of access, or so near in time to New York City. As a resort for transient visi
to the Mountains it has many great attractions over other localities.

For Circular, containing rates, etc., address

PLAZA HOTEL.

CURTIS HOTEL,

LENOX, MASS.

In center of the Berkshire Hills. 1,270 feet above tide water. Pure air, pure water, fine drives, and walks. Good rooms in hotel or cottages.

✽— *OPEN ALL THE YEAR.* —✽

Copley Square Hotel.

American and

European Plans.

HUNTINGTON AVENUE AND EXETER STREET, BOSTON.

Located in the Fashionable and Beautiful Back Bay District.

CONTAINING 300 ROOMS, SINGLE AND EN SUITE, RICHLY FURNISHED.

IT IS BUT SIX MINUTES' RIDE BY HORSE OR ELECTRIC CARS TO THE SHOPPING AND AMUSEMENT CENTRES.
FIVE MINUTES TO PROVIDENCE DEPOT, THE TERMINUS OF THE SHORE LINE R. R.,
FALL RIVER, STONINGTON, AND PROVIDENCE BOAT LINES.
PASSENGERS VIA BOSTON & ALBANY R. R. MAY LEAVE THE TRAIN AT HUNTINGTON AVENUE STATION, WITHIN
ONE MINUTE'S WALK OF HOTEL.
HOTEL PORTER WILL BE IN ATTENDANCE AT TRAINS ARRIVING FROM NEW YORK AND THE WEST

F. S. RISTEEN & CO., PROPRIETORS.

THE HOTEL CHAMPLAIN,

On the west shore of Lake Champlain, three miles south of Plattsburg. Delaware & Hudson R. R. station and steamboat landing in hotel grounds. All trains and boats stop. The natural stopping-over point for tourists to and from Montreal and the Adirondacks. Extensive grounds, unrivaled scenery.

O. D. SEAVEY, Manager,

P. O. address, HOTEL CHAMPLAIN, Clinton Co., N. Y.

7

12

Hotel Florence,
QUEBEC, CANADA.

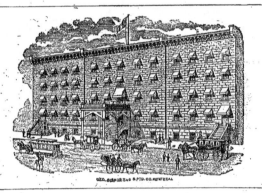

"THE FLORENCE" is the most pleasant, attractive, and comfortable house for tourists that can be found on this continent. Its location is unequaled, and the panoramic view to be had from the Balcony is not even surpassed by the world-renowned Dufferin Terrace, as it commands a full view of the River St. Lawrence, the St. Charles Valley, Montmorency Falls, Laurentian Range of Mountains, and overlooks the largest part of the city.

Rooms with bath, and *en suite*, elegantly furnished and well ventilated, and the CUISINE FIRST CLASS.

Street-cars pass the door every five minutes.

Telephone communication Electric light and bell in every room.

Iron balconies and iron stairs from every floor. Perfect safety assured.

"The Florence" Hotel Observation Cars run every 15 minutes up to 10.30 P. M. to the Basilica, the Post-Office, Grand Battery, Dufferin Terrace, Court-House, Governor's Garden, Ursuline Convent, House where Montgomery was laid, City Hall, Union Club, Esplanade, Garrison Club, the foot of the Citadel, and the Parliament Buildings.

"The Florence" being the centre of all these interesting points, the round trip from the hotel and back is made in 30 minutes. Passengers are landed and taken at any point. Fare, five cents.

Tourists staying at other hotels especially invited to visit "The Florence," and also enjoy the magnificent view to be had from its veranda.

BENJ. TRUDEL, Proprietor.

The Cambridge,

Fifth Ave. and Thirty-third St. - New York City.

The only English, palatial, model Hotel
in America.

CUISINE UNSURPASSED. Henry Walter, Proprietor,

ay be kept out

UNITED STATES AND
CANADA

Illustrated

ITH RAILWAY MAPS, PLANS OF CITIES, SPECIAL
ITINERARIES, TABLE OF RAILWAY AND
STEAMBOAT FARES,
AND AN APPENDIX DESCRIBING THE
COLUMBIAN EXPOSITION

REVISED EACH YEAR TO DATE OF ISSUE

NEW YORK
D. APPLETON AND COMPANY
1893

910
.Ap5

29797

PREFACE.

DURING the past two years the editor of the GENERAL GUIDE has made a trip over the entire United States and Canada, and also special side trips to important centers, including Chicago. The information gathered by him has been incorporated in the present edition.

Among the new features will be found :

1. Descriptions of routes, resulting from increased railroad facilities.

2. Descriptions of resorts, notably those on the Pacific coast.

3. The leading cities have been visited, and the latest information concerning each has been gathered for this work by some special expert.

4. Itineraries of the larger cities will be found at the proper places, describing how the salient features may be seen in the shortest space of time.

5. New plans and new maps of the environs of the cities have been specially prepared.

6. The old illustrations give place to new ones, including several of the buildings of the Chicago Exposition.

7. A specially prepared APPENDIX containing a description of the characteristic features of the great World's Fair.

The leading idea which has governed the preparation of this work has been to produce an American Guide-Book prepared with the special knowledge of an American. Each year finds an increasing number of our citizens who desire to know more about their own country, and each year brings an increasing influx of foreign tourists who desire to see those features which are most significantly American. For both of these classes this book is designed.

All the important cities and great routes of travel in the United States and Canada are carefully and minutely described in it, and also every locality which is sufficiently visited for its own sake to entitle it to a place in such a work. At the same time, it is believed that the book will be not less useful for what it excludes than for what it includes. In the present work the gazetteer plan has been deliberately discarded, and mention is made only of those places, facts, and items which are considered in some way interesting and worthy of attention. Small stations *en route* are often mentioned in order to indicate distances and rate of progress—in itself frequently a highly interesting item of information; but, as a general rule, not only are merely local lines of travel and off-route places (unless attractive for special reasons) omitted entirely, but the tourist's attention is invited only to such things as are really worth attention; and the editor has been much more anxious in describing a route to indicate the characteristic features of the country traversed, and where fine views may be obtained, than to enumerate and describe all the little stations at which the train may happen to pause.

As much aid as possible is afforded to the eye by printing the names of places and objects either in *italics*, or, where they are of sufficient importance, in **large-faced type.** Objects worthy of special attention are further distinguished by asterisks (*).

The Plans of Cities also follow the excellent system of numbered and lettered squares, with figures corresponding to similar figures prefixed to lists of the principal public buildings, hotels, churches, and objects of interest. The illustrations afford a trustworthy idea of American architecture, and, to some extent, of American scenery.

The tourist will find in APPLETONS' CANADIAN GUIDE-BOOK, Part I, *Eastern Canada*, and Part II, *Western Canada*, full information concerning that important division of America which now demands more space than could be given to it in the GENERAL GUIDE. Both the HAND-BOOK OF SUMMER RESORTS and HAND-BOOK OF WINTER RESORTS, as implied by their titles, treat very fully of places frequented during the summer and winter seasons.

In dealing with so many and diverse facts it is probable that some errors have crept in and that there are some omissions. The book will continue to receive a thorough annual revision, and the editor will be grateful for any corrections and suggestions.

CONTENTS.

WESTERN AND SOUTHERN STATES.

LIST OF ILLUSTRATIONS.

MAPS.

PLANS OF CITIES.

KOUNTZE BROTHERS,

Bankers,

120 BROADWAY, NEW YORK.

* * * * * * *

Deposits received subject to check, and interest allowed on balances.

Government Bonds and other Securities bought and sold for the usual commission.

State and Municipal Bonds negotiated. Information solicited in regard to any new or proposed issues of Bonds.

Advances made to correspondents against available collateral ; also approved business paper discounted or received as security for loans.

Collections made throughout the United States and Territories, the British Provinces and Europe.

Letters of Credit, both Foreign and Domestic, also Circular Notes issued for the use of travelers, available in all parts of the world.

Cable Transfers made to, and Bills drawn on, Great Britain, Ireland and the Continent. Also Telegraphic Transfers to various places in the United States.

INTRODUCTION.

I. Passports, Customs Duties, etc.

PASSPORTS are not required in the United States. The examinations of baggage at the ocean ports and the Canadian frontier are usually conducted in a courteous manner, but are at times very rigid; and the visitor from abroad will do well to include in his luggage only such articles as can be strictly regarded as of necessary personal use. The articles most watched for and guarded against by the customs authorities are clothing (new and in undue quantity), silks, linens, laces, cigars, watches, jewelry, and precious stones. In case of any portion of the luggage being found "dutiable," it is best to pay the charges promptly (under protest), and forward complaint to the Treasury Department at Washington.

II. Currency.

The present currency of the United States consists of gold and silver coin, and of United States Treasury notes (called "greenbacks") and national-bank bills redeemable in coin at par. The fractional currency (which includes all sums below a dollar) is of silver, with nickel five-cent pieces, and copper pieces of the value of one and two cents. In Canada the currency is coin, or the notes of the local banks, which are at par. Foreign money is not current in the United States, but may be exchanged for the usual currency at the brokers' offices at fixed rates. For practical purposes, a pound sterling may be rated as equivalent to five dollars of American money, and a shilling as equivalent to twenty-five cents, or a "quarter." A franc is equivalent to about twenty cents of American money; five francs to a dollar.

III. Hotels.

The hotels of the United States are commonly well-equipped and conducted. In the larger cities there are two kinds: those conducted on what is called the American plan, by which a fixed charge includes lodgings and the usual meals at *table d'hôte;* and those conducted on the European plan, where the charge is made for lodgings alone, and the meals are taken *à la carte* in the hotel or elsewhere. At some hotels the two plans are combined, and the traveler has his choice between them. The charge at first-class hotels (on the American plan) is from $3 to $5 a day; but good accommodations may be had at houses of the second class for $2 to $3 a day. A considerable reduction is usually made on board by the week. The charge for rooms at hotels on the European plan ranges from $1 to $3 a day. The "extras" and "sundries" which make European hotel-bills so exasperating are unknown in America; and the practice of feeing servants is not so general as it is in Europe. The best hotels at the various points are designated at their proper places in the body of the GUIDE; they are named usually in alphabetical order. At the larger hotels, besides a reading-room for the use of guests, there will nearly always be found a letter-box, a telegraph-office, and offices for the sale of railroad and theatre tickets.

IV. Conveyances.

The average cost of travel by *Railroad* is two to three cents a mile in the Middle States and New England, and from three to five cents in the Western and Southern States. Children between the ages of five and twelve are generally charged half price; those under five are passed free. Between distant places which may be reached by competing lines there are usually what are called "through tickets," costing much less than regular mileage rates. These tickets are good only for the day and train for which they were purchased, and, if the traveler wishes to stop at any intermediate point, he must notify the conductor and get a "stop-over check." Attached to all "through trains" on the longer routes are drawing-room cars, which are richly finished and furnished, provided with easy-chairs, tables, mirrors, etc., and, being mounted on twelve wheels, run much easier than the ordinary coaches. Those attached to the night-trains are so arranged as to be ingeniously converted into sleeping-berths, and are provided with lavatories in addition to the usual conveniences. From $2 to $3 a day in addition to the regular fare is charged for a seat or berth in these parlor cars, or a whole "section" may be secured at double rates. On a few of the more important lines have been placed what are called "hotel or dining cars," and on these lines the "limited" trains are formed of vestibuled cars, consisting of parlor and sleeping cars, and buffet, smoking, and library car, and a dining-car.

Travel by *Steamboats* is somewhat less expensive and less expeditious than by rail. The ticket (in case of a night-passage) gives the right to a sleeping-berth in the lower saloon; but the extra cost of a state-room (usually $2 a night) is more than compensated by the greater comfort and privacy. On the much-traveled lines, state-rooms should be secured a day or two in advance, and, if possible, in the outside tier. Meals are usually an extra on steamboats, and will cost about $1 each when the service is not *à la carte*.

The vast extension of the railway system has nearly superseded the old *Stages* and *Coaches*, but a few lines still run among the mountains and in remote rural districts. Where the object is not merely to get quickly from point to point, this is perhaps the most enjoyable mode of travel, and, in pleasant weather, the traveler should try to get an outside seat. The charges for stage-travel are relatively high—often as much as 10c. or 15c. a mile.

In all the cities and larger towns there are *Omnibuses* at the station on the arrival of every train, which connect directly with the principal hotels; a small charge (usually 50c.) is made for this conveyance.

V. Baggage—The Check System.

It is the custom in America to deliver baggage to a person known as the baggage-master, who will in return give a small numbered brass plate (called a "check") for each piece, on presentation of which the baggage is delivered. Baggage may be "checked" over long routes in this way, and the traveler, no matter how many times he changes cars or vehicles, has no concern about it. The railroad company are responsible if the baggage should be injured or lost, the "check" being evidence of delivery into their hands. The traveler, arrived at the station or depot, should first procure his ticket at the ticket-office, and then, proceeding to the baggage-room or proper station of the baggage-master, have his trunks checked to the point to which he wishes them sent. (The baggage-master usually requires the traveler to exhibit his ticket before he will check the trunks.) Arriving at his destination, the checks may be handed to the hotel-porter, always in waiting, who will procure the various articles and have them sent to the hotel. Should the owner be delayed on the route, the baggage is stored safely at its destined station until he calls or sends for it (of course presenting the check). Beyond a certain weight (from 100 to 150 lbs.) for each ticket bought, baggage is charged for extra; and this may become a serious item where the distances are great. Before arriving at the principal cities, a baggage or express man generally passes through the cars and gives receipts (in exchange for checks) for delivering baggage at any point desired.

VI. Round-trip Excursions.

Every summer the leading railway companies issue excursion-tickets at greatly reduced prices. These excursions embrace the principal places of interest throughout the country, and are arranged in a graded series, so that the tourist may have choice of a number of round trips of a day or two to popular resorts near by, or may make one of the grand tours to distant points affording thousands of miles of travel. As the tickets are good for thirty, sixty, and ninety days, the traveler can consult his convenience *en route*, lingering or hastening on as he may happen to choose. Lists of these excursions and such information about them as may be required can be obtained at the central offices of the various companies in the larger cities, either by personal application or by letter.

VII. Climate and Dress.

Of course, in a country so extensive as the United States, the differences of climate are very great, New England and the Middle States being frequently buried in snow at the very moment when the Southern States are enjoying their most genial season, while California has but two seasons (the wet and the dry) instead of the four seasons of the temperate zone. It is true of the country as a whole, however, that the summers are hotter and the winters colder than those of Europe; and that there is greater liability to sudden changes from heat to cold, or from cold to heat. For this reason it is highly important that the traveler should be dressed with sufficient warmth; it will be better for him to suffer at noonday from too much clothing than to expose himself at night, in storms, or to sudden changes of temperature, with too little. Woolen underclothing should be worn both summer and winter, and a shawl or extra wrap should always be on hand. At the same time, exposure to the vertical rays of the sun in summer must be carefully avoided; sunstroke being by no means unusual even in the Northern cities.

II

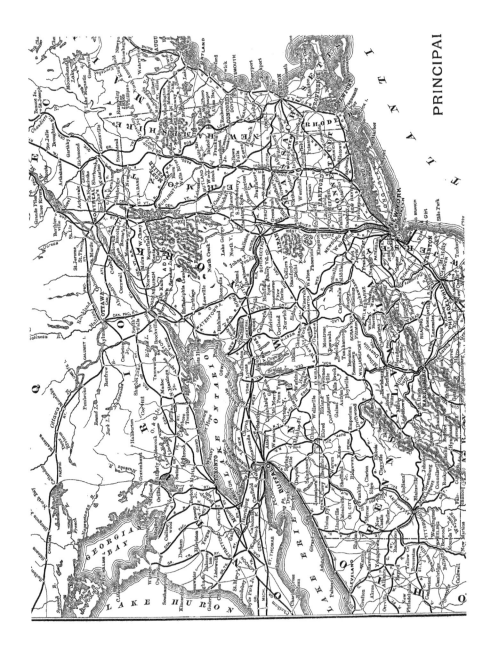

NEW ENGLAND AND MIDDLE STATES AND CANADA.

1. New York City.

Hotels.—The best hotels include the *Fifth Avenue*, the *Hoffman House*, the *Holland House*, the *Waldorf*, the *Windsor*, the *Buckingham*, the *Plaza*, the *Savoy*, and the *New Netherland*. A fuller list is as follows : On the American plan, beginning down town, the *Metropolitan*, 584 Broadway ; the *Broadway Central*, 671 Broadway, opposite Bond St.; the *Berkeley*, cor. 5th Ave. and 9th St.; the *Gramercy Park*, 35 Gramercy Park ; the *Fifth Avenue*, cor. Broadway and 23d St.; the *Cambridge*, 334 5th Ave.; the *Hotel Bristol*, cor. 5th Ave. and 42d St.; the *Vendome*, cor. Broadway and 41st St.; the *Sherwood*, cor. 5th Ave. and 44th St.; the *Windsor*, cor. 5th Ave. and 46th St.; the *Madison Avenue*, cor. Madison Ave. and 58th St.; and the *Winthrop*, cor. 7th Ave. and 125th St. On the European plan are the *Astor House*, cor. Broadway and Vesey St.; the *St. Denis*, cor. Broadway and 11th St.; the *Brevoort House*, cor. 5th Ave. and Clinton Pl.; the *Albert*, 37 University Pl.; the *Union Square*, E. Union Square and 15th St.; the *Westminster*, Irving Pl. and 16th St.; the *Everett House*, 4th Ave. and 17th St.; the *Hotel de Logerot*, 126 5th Ave.; the *Belvedere*, 4th Ave. and 18th St.; the *New Amsterdam*, cor. 4th Ave. and 21st St.; the *Bartholdi*, cor. Broadway and 23d St.; the *Hoffman House*, cor. Broadway and 25th St.; the *Albemarle*, cor. Broadway and 24th St.; the *St. James*, cor. Broadway and 26th St.; the *Holland House*, cor. 5th Ave. and 30th St.; the *Gilsey House*, cor. Broadway and 30th St.; the *Grand*, cor. Broadway and 31st St.; the *Imperial*, cor. Broadway and 32d St.; the *Normandie*, cor. Broadway and 38th St.; the *St. Marc*, cor. 5th Ave. and 39th St.; the *Grand Union*, cor. 4th Ave. and 42d St.; the *Metropole*, cor. Broadway and 42d St.; the *Barrett House*, cor. Broadway and 43d St.; the *Buckingham*, cor. 5th Ave. and 50th St.; the *Lincoln*, cor. Broadway and 52d St.; and the *Grenoble*, cor. 7th Ave. and 56th St. Both plans are combined in the following : the *Clarendon*, cor. 4th Ave. and 18th St.; the *Brunswick*, cor. 5th Ave. and 26th St.; the *Victoria*, cor. 5th Ave. and 27th St.; the *Sturtevant*, cor. Broadway and 29th St.; the *Park Avenue*, cor. Park Ave. and 33d St.; the *Waldorf*, cor. 5th Ave. and 33d St.; the *Marlborough*, cor. Broadway and 36th St.; the *Oriental*, cor. Broadway and 39th St.; the *Murray Hill*, cor. Park Ave. and 40th St.; the *Plaza*, cor. 5th Ave. and 59th St.; the *Savoy*, cor. 5th Ave. and 59th St.; and the *New Netherland*, cor. 5th Ave. and 59th St.

Restaurants.—*Delmonico's* (cor. 5th Ave. and 26th St.), the *Holland House Café* (Fifth Ave. and 30th St.), the *Café Brunswick* (also at the cor. of 5th Ave. and 26th St.), and *Sherry's* (cor. Fifth Ave. and 37th St.), are among the best. The *St. Denis* (ccr. Broadway and 11th St.), *Clarke* (22 W. 23d St.), *Purssell's* (914 Broadway), and the *Vienna Bakery* (cor. Broadway and 10th St.), are of excellent repute, and places where ladies or families may lunch or dine. The *cafés* and restaurants attached to the large hotels on the European plan are generally well kept ; among the best of these are the *Hoffman House*, cor. Broadway and 24th St.; the *St. James*, cor. Broadway and 26th St.; the *Coleman House*, Broadway, between 26th and 27th Sts.; and the *Clifton*, 6th Ave. and 35th St. *Delmonico's*, 22 Broad St. and at junction of Beaver and William Sts. ; *Cable's*, 130 Broadway ; the *Hoffmann House Café*, in the Consolidated Stock and Petroleum Exchange, 7 Beaver and 23 New Sts. ; *Sutherland's*, 64 Liberty St. ; the *Café Savarin*, in the Equitable Building, 120 Broadway ; the *Astor House*, in Broadway, are first-class restaurants. There are a number of restaurants where *table-d'hôte* dinners may be got from 5 to 8 P. M.,

1

for from 75c. to $1.50, usually including wine ; of these may be mentioned the *Brunswick*, cor. 5th Ave. and 26th St. ; the *Murray·Hill*, cor. Park Ave. and 40th St. ; and *Morello's*, 4 W 29th St. *Ricadonna's* (42 Union Square) and *Moretti's* (22 E. 21st St.) have the Italian *cuisine*, on the *table-d'hôte* plan. There are also English chop-houses ; of these, *Farrish's* (64 John St.), *Browne's* (31 W. 27th St.), and *The Studio* (332 6th Ave.), are noted.

Modes of Conveyance.—Elevated R. R.—Four lines extend lengthwise from South Ferry east side of the Battery. The *3d Ave.* line runs to Chatham Sq. (where passengers may transfer to 2d Ave. line), thence by Bowery and 3d Ave. to Harlem River. This line has a branch from City Hall to Chatham Sq., one at 42d St. to Grand Central Depot, and one at 34th St. for ferry to Long Island City. The Chatham Sq. branch connects with Brooklyn Bridge cars. The stations of 3d Ave. line are : Battery, foot Whitehall St.; in Pearl St., at Hanover Sq., at cor. of Fulton St.; at Franklin Sq.; and at Chatham Sq.; in Bowery, cor. Canal, Grand, and Houston Sts.; in 3d Ave., cor. 9th, 14th, 18th, 23d, 28th, 34th, 42d, 47th, 53d, 59th, 67th, 76th, 84th, 86th, 98th, 106th, 116th, 125th, and 129th Sts. (Harlem River). The *6th Ave.* line starts from Battery, foot of Whitehall St., and has stations at Battery Place cor. Greenwich St., in New Church St., cor. of Rector St., and Cortlandt St.; in Church St., cor. Park Pl.; in W. Broadway, cor. Chambers St., and Franklin St.; in S. 5th Ave., cor. Grand St., and Bleecker St.; in 6th Ave., cor. 8th, 14th, 18th, 23d, 28th, 33d, 42d, 50th, and 58th Sts. ; in 8th Ave., cor. 53d St.; in 9th Ave., cor. 59th, 66th, 72d, 81st, 93d, and 104th Sts. ; and in 8th Ave. again, cor. 116th, 125th, 135th, 145th, and 155th Sts. (Harlem River). ·Certain trains run only to 58th St., cor. 6th Ave. The line connects at cor. 42d St. and 6th Ave. with Grand Central Depot by surface cars. The *9th Ave.* line starts from Battery, foot Whitehall St.; stations at Battery Place, in Greenwich St., cor. Rector, Cortlandt, Barclay, Warren, Franklin, Desbrosses, Houston, and Christopher Sts.; and in 9th Ave., cor. 14th, 23d, 30th, 34th, 42d, 50th, and 59th Sts., at the last-named station connecting with the 6th Ave. line. The *2d Ave.* line starts at Battery, running on the same track as the 3d Ave. line to Chatham Sq. It has stations in Allen St., cor. Canal, Grand, and Rivington Sts.; in 1st Ave., cor. 1st, 8th, 14th, and 19th Sts.; in 23d St., bet. 1st and 2d Aves. ; and in 2d Ave., cor. 34th, 42d, 50th, 57th, 65th, 80th, 86th, 92d, 99th, 111th, 117th, 121st, 127th, and 129th Sts. (Harlem River), where it connects with the Suburban line to 166th St. The trains run at intervals of about 4 minutes. On Sundays the trains run on all the lines, but at somewhat longer intervals ; and at night from 12.30 A. M. to 5.30 A. M. trains run on the 6th Ave. road every half-hour, and the 3d Ave. line every quarter-hour. Fares are 5 cents on all the lines. Express trains morning and evening make but 4 stops bet. 23d and 155th Sts., on 9th Ave. line, and make sharp connection with express trains on the New York & Northern R. R.

Street-Cars, etc.—*Broadway Line*, from South Ferry through Whitehall St. to Broadway, thence along Broadway to 44th St., thence through *7th Ave.* to Central Park. The *4th Ave.* to 138th St. (Mott Haven); and the *3d Ave.* to Harlem, start from the Post-Office; the *6th Ave.* line to Central Park starts from Broadway, cor. Vesey St.; while the *8th Ave.* line, starting from the same place, runs to 145th St. There are cross-town lines along Broadway at Chambers St., at Canal St., at Grand St., at Prince St., at Houston St., at 8th St., at 14th St., at 23d St., at 34th St., at 42d St., at 59th St., and at 125th St.; and the *Bleecker St.* cars from Fulton Ferry by Broadway and Bleecker St. to W. 23d St. Ferry. The *Belt Line* runs from South Ferry to Central Park, the E. branch passing all ferries on East River ; the W. branch, all ferries on North River. The *Boulevard Line* runs from E. 34th St. to 42d St., to Boulevard at 72d St., to Fort Lee Ferry, W. 129th St.: with a branch from 92d St., East River, to 110th St., to St. Nicholas Ave., to foot of W. 129th St. Fare on all the lines 5c. A CABLE line runs from foot of E. 125th St. west to 10th Ave., to 187th St., past High Bridge. Stages run from Bleecker St. through S. 5th Ave. to 5th Ave., to 86th St. *Hackney-coaches* have stands in different parts of the city, and attend the arrival of every train and steamboat. A tariff of fares is or ought to be hung in each carriage. Disputed questions as to time, distance, or price, must be settled at the Mayor's office (City Hall). The legal rates for cabs are, for one or more passengers for a distance of 1 m. or less, 50c., and 25c. each half-mile additional ; coaches, $1 for 1 m., and 40c. additional for each half-mile. The principal hotels have carriages in waiting for the use of guests.

for from 75c. to $1.50, usually including wine ; of these may be mentioned the *Brunswick*, cor. 5th Ave. and 26th St. ; the *Murray Hill*, cor. Park Ave. and 40th St. ; and *Morello's*, 4 W. 29th St. *Ricadonna's* (42 Union Square) and *Moretti's* (22 E. 21st St.) have the Italian *cuisine*, on the *table-d'hôte* plan. There are also English chop-houses ; of these, *Farrish's* (64 John St.), *Browne's* (31 W. 27th St.), and *The Studio* (382 6th Ave.), are noted.

· **Modes of Conveyance.—Elevated R. R.**—Four lines extend lengthwise from South Ferry east side of the Battery. The 3*d Ave.* line runs to Chatham Sq. (where passengers may transfer to 2d Ave. line), thence by Bowery and 3d Ave. to Harlem River. This line has a branch from City Hall to Chatham Sq., one at 42d St. to Grand Central Depot, and one at 84th St. for ferry to Long Island City. The Chatham Sq. branch connects with Brooklyn Bridge cars. The stations of 3d Ave. line are : Battery, foot Whitehall St.; in Pearl St., at Hanover Sq., at cor. of Fulton St., at Franklin Sq. ; and at Chatham Sq.; in Bowery, cor. Canal, Grand, and Houston Sts. ; in 3d Ave., cor. 9th, 14th, 18th, 23d, 28th, 34th, 42d. 47th, 53d, 59th, 67th, 76th, 84th, 89th, 98th, 106th, 116th, 125th, and 129th Sts. (Harlem River). The 6*th Ave.* line starts from Battery, foot of Whitehall St., and has stations at Battery Place cor. Greenwich St., in New Church St., cor. of Rector St., and Cortlandt St.; in Church St., cor. Park Pl.; in W. Broadway, cor. Chambers St., and Franklin St.; in S. 5th Ave., cor. Grand St., and Bleecker St.; in 6th Ave., cor. 8th, 14th, 18th, 23d, 28th, 33d, 42d, 50th, and 58th Sts. ; in 8th Ave., cor. 53d St.; in 9th Ave., cor. 59th, 66th, 72d, 81st, 93d, and 104th Sts. ; and in 8th Ave. again, cor. 116th, 125th, 135th, 145th, and 155th Sts. (Harlem River). ·Certain trains run only to 58th St., cor. 6th Ave. The line connects at cor. 42d St. and 6th Ave. with Grand Central Depot by surface cars. The 9*th Ave.* line starts from Battery, foot Whitehall St.; stations at Battery Place, in Greenwich St., cor. Rector, Cortlandt, Barclay, Warren, Franklin, Desbrosses, Houston, and Christopher Sts.; and in 9th Ave., cor. 14th, 23d, 30th, 34th, 42d, 50th, and 59th Sts., at the last-named station connecting with the 6th Ave. line. The 2*d Ave.* line starts at Battery, running on the same track as the 3d Ave. line to Chatham Sq. It has stations in Allen St., cor. Canal, Grand, and Rivington Sts. ; in 1st Ave., cor. 1st, 8th, 14th, and 19th Sts.; in 23d St., bet. 1st and 2d Aves. ; and in 2d Ave., cor. 34th, 42d, 50th, 57th, 65th, 80th, 86th, 92d, 99th, 111th, 117th, 121st, 127th, and 129th Sts. (Harlem River), where it connects with the Suburban line to 166th St. The trains run at intervals of about 4 minutes. On Sundays the trains run on all the lines, but at somewhat longer intervals ; and at night from 12.30 A. M. to 5.30 A. M. trains run on the 6th Ave. road every half-hour, and the 3d Ave. line every quarter-hour. Fares are 5 cents on all the lines. Express trains morning and evening make but 4 stops bet. 23d and 155th Sts., on 9th Ave. line, and make sharp connection with express trains on the New York & Northern R. R.

Street-Cars, etc.—*Broadway Line*, from South Ferry through Whitehall St. to Broadway, thence along Broadway to 44th St., thence through 7*th Ave.* to Central Park. The 4*th Ave.* to 188th St. (Mott Haven); and the 3*d Ave.* to Harlem, start from the Post-Office ; the 6*th Ave.* line to Central Park starts from Broadway, cor. Vesey St.·; while the 8*th Ave.* line, starting from the same place, runs to 145th St. There are cross-town lines along Broadway at Chambers St., at Canal St., at Grand St., at Prince St., at Houston St., at 8th St., at 14th St., at 23d St., at 34th St., at 42d St., at 59th St., and at 125th St.; and the *Bleecker St.* cars from Fulton Ferry by Broadway and Bleecker St. to W. 23d St. Ferry. The *Belt Line* runs from South Ferry to Central Park, the E. branch passing all ferries on East River ; the W. branch, all ferries on North River. The *Boulevard Line* runs from E. 34th St. to 42d St., to Boulevard at 72d St., to Fort Lee Ferry, W. 129th St.; with a branch from 92d St., East River, to 110th St., to St. Nicholas Ave., to foot of W. 129th St. Fare on all the lines 5c. A CABLE line runs from foot of E. 125th St. west to 10th Ave., to 187th St., past High Bridge. Stages run from Bleecker St. through S. 5th Ave. to 5th Ave., to 86th St. *Hackney-coaches.* have stands in different parts of the city, and attend the arrival of every train and steamboat. A tariff of fares is or ought to be hung in each carriage. Disputed questions as to time, distance, or price, must be settled at the Mayor's office (City Hall). The legal rates for cabs are, for one or more passengers for a distance of 1 m. or less, 50c., and 25c. each half-mile additional ; coaches, $1 for 1 m., and 40c. additional for each half-mile. The principal hotels have carriages in waiting for the use of guests.

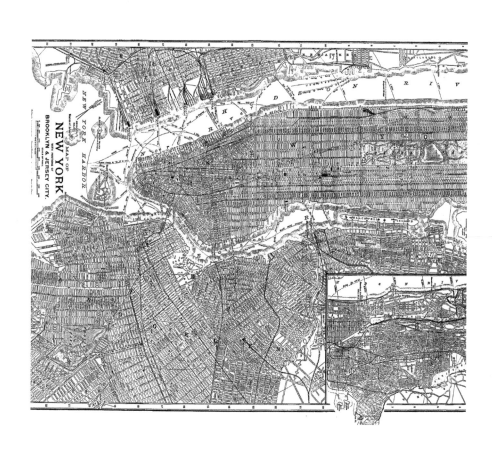

MAP OF
NEW YORK
WITH PORTIONS OF
BROOKLYN & JERSEY CITY.

Railroad Stations.—The *Grand Central Station*, in 42d St., between Lexington and Madison Avenues, is used by most of the passenger trains of the New York Central & Hudson River R. R., the New York & Harlem Div., and New York, New Haven & Hartford R. R. The Harlem River branch to New Rochelle starts from 122d St. and 2d Ave. Local trains to Spuyten Duyvel (Hudson River R. R.) leave the station at 10th Ave. and 30th St. The depot of the *Pennsylvania R. R.* (in Jersey City), the *New York, Susquehanna & Western R. R.*, *Northern Railroad of New Jersey*, *New York & Greenwood Lake R. R.*, and *New Jersey & New York R. R.* is reached by ferries from foot of Desbrosses and Cortlandt Sts.; the *West Shore R. R.* and *New York, Ontario & Western R. R.*, same as Penn. R. R., Jersey City; also at Weehawken, by ferry from foot of W. 42d St.; the *Lehigh Valley R. R.* and the *Central of New Jersey* from foot of Liberty St.; the *Delaware, Lackawanna & Western* from foot of Barclay and Christopher Sts.; the *Long Island R. R.*, from James Slip and foot E. 34th St.

Ferries.—There are ferries to *Brooklyn* from foot of Whitehall St., Wall St., Fulton St., and Catherine St.; to East Brooklyn (Williamsburgh) from foot of Roosevelt St., Grand St., E. Houston St., and E. 23d St.; to *Greenpoint* from foot of 10th and E. 23d Sts.; to *Long Island City* from James Slip and foot of E. 34th St. To *Jersey City* from foot of Liberty St., Cortlandt St., Desbrosses St., Chambers St., and W. 23d St. To *Hoboken* from foot of Barclay St., foot of Christopher St., and foot of W. 14th St. To *Weehawken* from foot of Jay St. and W. 42d St. To *Fort Lee* from foot of Canal (touching at W. 22d St.) and W. 129th St. To *Staten Island*, to *St. George's*, from foot of Whitehall St., east side of the Battery. To *Astoria* from foot of E. 92d St. and foot of E. 34th St. (to Long Island City, thence by cars); and by boat from Pier 22, East River. To *Blackwell's*, *Randall's*, and *Ward's Islands* from foot of E. 26th St.

Churches.—There are nearly 500 churches of all denominations in the city, and at any of them the visitor is sure of a polite reception. The following are the principal of those whose Sunday services are most attended by strangers: Trinity Church (Episcopal), in Broadway, opposite Wall St., with cathedral choral service; Trinity Chapel (Episcopal), 25th St., near Broadway; St. George's (Episcopal), in Stuyvesant Square, E. 16th St.; Grace Church (Episcopal), Broadway, near 10th St., fine music; and St. Mary the Virgin (Ritualistic), 223 W. 45th St. Of the Roman Catholic churches, the Cathedral of St. Patrick (5th Ave. between 50th and 51st Sts.), St. Leo's (11 E. 28th St.), and St. Stephen's (149 E. 28th St., famed for its musical services) are most attended. The Presbyterian churches of Dr. John Hall (cor. 5th Ave. and 55th St.) and the Brick Church (5th Ave. and 37th St.) are very popular; also the Methodist Madison Avenue Church (cor. 60th St. and Madison Ave.); the Unitarian Church of All Souls (cor. 4th Ave. and 20th St.), and the Church of the Messiah (Dr. Robert Collyer, cor. Park Ave. and 34th St.); the Universalist Church of the Divine Paternity (cor. 5th Ave. and 45th St.). Among the Baptist churches are Fifth Avenue (6 W. 46th St.) and Judson Memorial (S. Washington Square); the Congregational Tabernacle (cor. Broadway and 34th St.); the Reformed Dutch Collegiate Churches (cor. 5th Ave. and 29th St. and cor. 5th Ave. and 48th St.); the Swedenborgian Church (114 E. 35th St.); the Moravian (cor. Lexington Ave. and 30th St.); and the Church of the Strangers (259 Mercer St., near 8th St.). The Sabbath (Saturday) services of the Jewish Temple Emanuel (5th Ave. cor. 43d St.) are very impressive, and the interior decorations of the building remarkably rich. The newspapers on Saturday and Sunday give the place and time of the most important services of the ensuing Sunday.

Theatres and Amusements.—The *Metropolitan Opera-House*, in Broadway, between 39th and 40th Sts. (destroyed by fire, but to be reconstructed), is the home of the grand opera during the winter season, and many of the large balls are given there, as well as of the Vaudeville Club. Other places of amusement include the *Academy of Music*, Irving Place, cor. 14th St.; *Amberg Theatre*, Irving Place and 14th St.; *Bijou Theatre*, Broadway, between 30th and 31st Sts.; *Broadway Theatre*, cor. Broadway and 41st St.; *Casino*, cor. Broadway and 39th St.; *Columbus Theatre*, 112 E. 125th St.; *Daly's Theatre*, cor. Broadway and 30th St.; *Empire Theatre*, cor. Broadway and 40th St.; *Fifth Avenue Theatre*, cor. Broadway and 28th St.; *Fourteenth St. Theatre*, W. 14th St., near 6th Ave.; *Garden Theatre*, cor. Madison Ave. and 27th St.; *Grand Opera-House*, cor. 8th Ave. and 23d St.; *Harlem Opera-House*, 125th St., west of 7th Ave.; *Harlem Theatre*, E. 125th St., near 3d Ave.; *Harrigan's Theatre*, 35th St. and 6th Ave. ; *Herrman's Theatre*, cor. Broadway and 29th St. ; *London Theatre*. 235 Bowery: the *Lyceum Opera-House*, 160 E. 34th St.; *Lyceum Theatre*, 4th Ave. near 23d St.; *Madison Square Theatre*, 24th. St., west of Broadway ; *Manhattan Opera-House*, E. 34th St., near Broadway; *Niblo's Garden*, 570 Broadway, near Prince St. ; *Palmer's Theatre*, Broadway and 30th St.; *Park Theatre*, cor. Broadway and 35th St.; *Proctor's Theatre*, 139 W. 23d St.; *Standard Theatre*, Broadway and 33d St.; *Star Theatre*, Broadway and 13th St. ; and *Union Square Theatre*, 14th St., near Broadway. *Chickering Hall*, cor. 5th Ave. and 18th St., and *Music Hall*, cor. 57th St. and 7th Ave., are concert and music halls. The *Cyclorama Building* is at the corner of 18th St. and 4th Ave. The *Eden Musée*, in 23d St., between 5th and 6th Aves., is devoted to wax-works. Summer-night concerts are given at the *Lenox Lyceum*, cor. Madison Ave. and 59th St., and at *Madison Square Garden*, Madison Ave. and 26th St. *Horse-races* at Morris Park, under the auspices of the N. Y. Jockey Club ; and at Sheepshead Bay near Coney Island, under the auspices of the Coney Island Jockey Club, near Brooklyn.

Reading-Rooms.—The *Astor Library*, Lafayette Place, near 8th St., contains 270,000 volumes ; open from 9 A. M. to 5 P. M. (in winter, 4 P. M.). The *Cooper Institute*, cor. 4th Ave. and 8th St. (32,000 volumes), is open to all from 8 A. M. to 10 P.M., on Sunday, 10 A. M. to 9 P. M. in winter. *Young Men's Christian Association* has free reading-rooms at 4th Ave. cor. 23d St., and at 5 W. 125th St.—both open from 8 A. M. to 10 P. M. The *Mercantile Library*, Astor Place near Broadway (240,000 volumes), has an excellent reading-room, to which strangers are admitted on introduction by a member. The libraries at *Columbia College* and the *College of the City of New York* are large. The *New York Free Circulating Library*, with about 80,000 volumes, 49 Bond St., and branches at 135 2d Ave., 226 W. 42d St., and 251 W. 13th St., are open to all. The *Society Library* (90,000 volumes), 67 University Place, founded 1757, is the oldest in the city. The library of the *Historical Society*, 2d Ave., opposite St. Mark's Church, is rich in documents relating to Revolutionary and colonial history ; and the *Geographical Society* has a valuable series of maps, etc., in its rooms, No. 11 W. 29th St. These are accessible by introduction of a member. The *Harlem Library*, 2238 3d Ave. and 123d St., is one of the oldest in the city.

Art Collections.—At the *National Academy of Design* (cor. 4th Ave. and 23d St.) there are annual exhibitions of recent works of American artists (entrance, 25c.). The *Metropolitan Museum of Art* (5th Ave. and 82d St.) has a fine collection of paintings by old and modern masters, and usually has on exhibition paintings loaned by the wealthy *virtuosi* of the city, including pictures by American artists, statuary, pottery and porcelain-ware, arms and armor, coins and medals, antiques, and various articles of *vertu*. It also contains the famous Cesnola Collection of Cypriote Antiquities. It is open daily from 10 A. M. till sunset, and on Sundays from 1 P. M.; also on Tuesday and Saturday evenings, from 8 to 10 P. M. At the *Historical Society* (cor. 11th St. and 2d Ave.) is a gallery of paintings with many old portraits, the Abbott Collection of Egyptian Antiquities, the Lenox Collection of Nineveh Sculptures, etc., admission by card from member. The *Lenox Library* (cor. 5th Ave. and 70th St.) contains a fine collection of paintings: admission free. The *American Art Gallery*, 6 E. 23d St., has frequent exhibitions of pictures and *bric-à-brac*. There are usually pictures on exhibition at the sales-galleries of *Knoedler*, cor. 5th Ave. and 22d St.; *Schaus*, 5th Ave., near 25th St.; *Avery*, 368 Fifth Ave.; and *Cottier*, 144 5th Ave. Many artists' studios may be found in the Young Men's Christian Association Building, cor. 4th Ave. and 23d St.; Studio Building, cor. 4th Ave. and 25th St. ; the *Sherwood*, 57 W. 57th St.; the *Rembrandt*, 152 W. 57th St.; and the Studio Building, 51 W. 10th St. The best private collections in the city are those of August Belmont, John Hoey, John Wolfe, Henry G. Marquand, Thomas B. Clark, William Rockefeller, Robert Hoe, and Mrs. W. H. Vanderbilt. Admission to these may be obtained by sending a letter (inclosing card) to their owners.

Clubs.—The principal are the *Century*, 7 W. 43d St. ; the *Knickerbocker*, 319 5th Ave. ; the *Manhattan*, 5th Ave. cor. 34th St.; the *Union*, 1 W. 21st St. ; the *Union League*, 1 E. 39th St.: the *Lotos*, 149 5th Ave.; the *Players*, 16 Gramercy Park ; the *New York*, 2 W. 35th St.; the *St. Nicholas*, 386 5th Ave. ; the *University*, Madison Ave. and 26th St. ; the *Calumet*, 267 Fifth Ave. ; the *Prog-*

Academy of Design

Madison Square Garden

Printing-House Square.

Views in New York.

ress, 5th Ave. cor. 63d St.; the *Democratic*, 617 5th Ave.; and the *New York Athletic*, 104 W. 55th St. Admission to these is obtained only through introduction by a member.

Post-Office.—The General Post-Office, at the southern end of City Hall Park, is open continuously, except Sundays, when it is open only from 9 to 11 A. M. There are also 20 sub-post-offices in the city, called "Stations," and alphabetically named; these are open from 7 A. M. to 8 P. M.; on Sundays, from 9 to 11 A. M. Letters may also be mailed in the lamp-post boxes (of which there are 700), or at any hotel.

NEW YORK CITY, the commercial metropolis of the United States, and largest city of the Western Hemisphere, is situated on New York Bay, in latitude about 41° N. and longitude 71° W. from Greenwich (3° 1′ 13″ E. from Washington), at the junction of the Hudson or North River, which washes its western shore, and of the East River, as the narrower portion of Long Island Sound is named, which separates it from Brooklyn It occupies the entire surface of Manhattan Island; Randall's, Ward's, and Blackwell's Islands in the East River; and a portion of the mainland, annexed from Westchester County, north of Manhattan Island and separated from it by Harlem River and Spuyten Duyvel Creek. Governor's, Bedloe's, and Ellis's Islands, in the Bay, are used by the United States Government. Ellis's Island has recently been selected as the landing-place of immigrants. The extreme length north from the Battery is 16 miles; greatest width from the Hudson to the mouth of Bronx River, 4¼ miles; area, nearly 41½ square miles, or 26,500 acres, of which 12,100 acres are on the mainland. Manhattan Island, on which the city proper stands, is 13½ miles long, and varies in breadth from a few hundred yards to 2¼ miles, having an area of nearly 22 square miles. The older portion of the city below 14th St. is somewhat irregularly laid out. The plan of the upper part includes avenues running N. to the boundary of the island, and streets running across them at right angles from river to river. The avenues are numbered from the east to 12th Ave.; east of 1st Ave. in the widest part of the city are Aves. A, B, C, and D. Above 21st, between 3d and 4th Aves., is Lexington Ave., and above 23d St., between 4th and 5th Aves., is Madison Ave.; 6th and 7th Aves. are intersected by Central Park. Above 59th St., on the west side, 8th Ave. is known as Central Park, west, 9th Ave. as Columbus Ave., 10th Ave. as Amsterdam Ave., 11th Ave. as West End Ave., while extending along the line of the river is Riverside Ave. St. Nicholas Ave. starts from 110th St. and Lenox, which is the name given to 6th Ave. above Central Park, and extends irregularly northward. Manhattan Ave. extends northward from 100th St. to 123d St., and thence westward to 130th St. and North River. Morningside Ave., east and west, are on either side of Morningside Park, between 110th and 123d St. Above 125th St. and in the annexed district the avenues and streets are still somewhat irregular. The streets are numbered consecutively N. to 225th St., at the end of the island; 21 blocks, including streets, average a mile. The house-numbers on the avenues run N.; those on the streets E. and W. from 5th Ave. The city is compactly built to Harlem, about 8½ miles from the Battery. Distances are usually calculated from the City Hall.

The harbor of New York is one of the finest and most picturesque in the world. The outer bar is at Sandy Hook, 18 miles from the Battery, and is crossed by two ship-channels, either of which admits vessels of the heaviest draught. On the steamers from Europe the American coast is usually first sighted at the line of the Navesink Highlands, or off Fire Island Light, and the bar is crossed soon after. As the steamer enters the Bay and sails through the Narrows, between the villa-crowned shores of Staten and Long Islands, on the left are seen the massive battlements of *Fort Wadsworth* and *Fort Tompkins;* while opposite, on the Long Island shore, are *Fort Hamilton* and old *Fort Lafayette*, the latter more famous as a political prison than as a fortress. Passing amid these fortifications, the panorama of city and harbor rapidly unfolds itself. To the left is Bedloe's Island, the site of the colossal statue of Liberty, by Bartholdi (see p. 19); Ellis's Island, with the immigration bureau, stands still farther toward the Jersey shore; and to the right is Governor's Island, with *Castle William* and old *Fort Columbus*. Directly ahead, the city opens to view, with Brooklyn on the right and Jersey City on the left.

The site of New York is said to have been discovered by Giovanni de Verrazzano, a Florentine mariner, in 1524; but authentic history begins with the visit of Henry Hudson, an Englishman in the service of the Dutch East India Company, who arrived there Sept. 3, 1609. Hudson afterward ascended the river as far as the site of Albany, and claimed the land by right of discovery as an appanage of Holland. In 1614 a Dutch colony came over and began a settlement. At the close of that year the future metropolis consisted of a small fort (on the site of the present Bowling Green) and four houses, and was known as New Amsterdam. As late as 1648 it contained but 1,000 inhabitants. In 1664 it was surrendered to the British, and, passing into the hands of the Duke of York, was thenceforward called New York. In 1667 the city contained 384 houses. In 1700 the population had increased to about 6,000. In 1696 Trinity Church was founded. In 1711 a slave-market was established in Wall Street; and in 1725 the *New York Gazette* was started. The American army under Washington occupied the city in 1776; but after the battles of Long Island and Harlem Heights, it was captured by the British forces, and remained their headquarters for 7 years. The British troops evacuated the city Nov. 25, 1783. Within ten years after the War of Independence, New York had doubled its population. In 1807 the first steamboat was put on the Hudson; the completion of the Erie Canal followed in 1825; and since that time the growth of the city has been rapid. Its population in 1800 was 60,489; it was 123,706 in 1820, 515,-847 in 1850, 812,869 in 1860, 942,377 in 1870, 1,206,590 in 1880, 1,515,301 in 1890, and 1,801,739 (State census) in 1892. Commerce and industry have kept pace with the population. More than half the foreign commerce of the United States is carried on through the customs district of which this is the port, and about two thirds of the duties are here collected. In 1890 the exports from this port were of the value of $347,500,252, and the imports $542,366,800. The manufactures of New York, though secondary in importance to its commercial and mercantile interests, are varied and extensive. In the value of products, in 1890, it was the first city in the Union, the whole number of manufacturing establishments being over 14,000, employing 351,757 hands, and producing goods valued at $763,-833,923.

The * **Battery** is a pretty little park at the southern extremity of the city, looking out upon the Bay, and protected by a massive granite sea-wall. It was the site of a fort in the early years of the city, and later was the fashionable quarter. At the S. W. end is *Castle Garden*, where an aquarium is soon to be built. At the S. end is the U. S. Revenue barge-office. From the pier adjoining boats start for *The Statue of Liberty* on Bedloe's Island every hour from 6 A. M. to 7.30 P. M.

(round trip, 25c.). Next to it is the ferry to Staten Island. Just E. of the Battery is *Whitehall St.*, at the foot of which are the South, Hamilton, and 39th St. Ferries to Brooklyn, where the lines of the elevated railways converge. *South St.*, beginning here, follows the East River shore for over 2 miles, passing the East River piers and the Long Island ferries, while *West St.* skirts the shore of the Hudson (or North) River for 2 miles, passing the North River piers, and the ferries to the Jersey shore. A little higher up Whitehall St. is the *U. S. Army Building.* Just N. of the Battery, at the foot of Broadway, is **Bowling Green,** the cradle of New York, and in Revolutionary times the residential end of the town. The row of 6 buildings facing the Green on the S. side covers the location of the Dutch and English fort. No. 1 Broadway is the site of the house (now the Washington Building) which was built in 1760 by the Hon. Archibald Kennedy, then collector of the port, and successively the headquarters of Lords Cornwallis and Howe and Sir Henry Clinton. Talleyrand also lived there. Benedict Arnold occupied No. 5, long since pulled down. Robert Fulton died at No. 1 Marketfield St., now covered by the **Produce Exchange,** in Whitehall St. To the N. are the *Welles* (No. 18) and *Standard Oil* (No. 24) buildings, and on the W. of Broadway is *Columbia* building (No. 29), *Aldrich Court* (No. 45), and nearly opposite is the *Consolidated Stock and Petroleum Exchange* (No. 58); at No. 50 is the tall *Tower* building, and at No. 80 is the *Union Trust Co.'s* building.

Passing up **Broadway** from the Green, between continuous rows of large offices, in a short time ***Trinity Church** towers up on the left, with its beautiful spire 284 ft. high. It is in the Gothic style, of solid brownstone, and is 192 ft. long, 80 wide, and 60 high. It has rich stained-glass windows, and the finest chime of bells in America. The **Astor Memorial Reredos*, in the chancel, is one of the richest and costliest in the world; it is 33 ft. wide and nearly 20 high, its materials being marble, glass, and precious stones, with statuary, the most delicate and elaborate carving, and the richest mosaics. It was erected in 1878 at a cost of upward of $100,000. The Trinity Parish is the oldest in the city; its first church was built in 1696 and destroyed by fire in 1776; its present edifice was begun in 1839 and consecrated in 1846. The church is open all day; there are prayers twice daily (at 9 A. M. and 3 P. M.), and imposing choral services on Sunday. The graveyard surrounding the church is one of the most picturesque spots in the city. It occupies nearly two acres of ground, is embowered in trees, and contains many venerated tombs—among them those of Alexander Hamilton, Captain Lawrence (the hero of the "Chesapeake"), Robert Fulton, and the unfortunate Charlotte Temple. In the N. E. corner is a stately Gothic monument erected to the memory of the patriots who died in British prisons at New York during the Revolution.

Beginning directly opposite Trinity Church, **Wall St.,** the monetary center of the country and resort of bankers and brokers, runs to the East River. One block down (at the corner of Nassau St.) is the *** U. S. Sub-Treasury,** a stately white-marble building in the Doric style, 200 ft. long, 80 wide, and 80 high. The main entrance in Wall

St. is reached by a flight of 18 marble steps, and in the interior is a lofty Rotunda, 60 ft. in diameter and supported by 16 Corinthian columns (visitors admitted from 10 to 3 o'clock)   The old Federal Hall stood here, and from its balcony Washington delivered his first address as President. A bronze statue of him, by J. Q. A. Ward, was unveiled here November 26, 1883. Next to it is the *U. S. Assay Office*, open from 10 to 3 o'clock on Wednesdays. At the opposite corner is the *Drexel Building*, of white marble, and just below it, in Broad St., the *Mills Building*, an immense brick pile for offices; and nearly opposite, in Broad St., is the * **Stock Exchange.** A visit to the Stock Exchange is well worth making. In Wall St. below the Treasury (at the cor. of William St.) is the * **U. S. Custom-House,** built in 1835 as the Merchants' Exchange, and famous for the great granite plinths of the columns that support the pediment of the front elevation. It is of massive Quincy granite, with a depth of 200 ft., a frontage of 144 ft., and a rear breadth of 171 ft. Its height to the top of the central dome is 124 ft. Beneath this dome, in the interior of the building, is the Rotunda, around which are eight lofty columns of Italian marble, the superb Corinthian capitals of which were carved in Italy. They support the base of the dome and are probably the largest marble columns in the country (open to visitors from 10 to 3 o'clock). Many new buildings of great height and beauty have been erected in Wall St., conspicuous among which are the *Wilkes* building, *Bank of America, Royal Insurance* building, etc., while to the south, on William St., near Hanover Square, is the *Cotton Exchange.* From the foot of Wall St. a ferry runs to Montague St., Brooklyn. *Pearl St.*, crossing Wall just beyond the Custom-House, is the seat of a heavy wholesale trade in cotton and other staples. *Nassau St.*, one of the busiest in the city, extends from Wall St. to Printing-House Square. In this street, between Cedar and Liberty Sts., on the site of the Middle Dutch Church, long occupied as Post-Office, is the spacious structure of the *Mutual Life Insurance Co.* It is of the Renaissance style, and one of the most notable specimens of architecture in New York.

Continuing up Broadway from Wall St., the *United Bank* building is passed on the corner, and the massive building of the * *Equitable Life Ins. Co.* (No. 120), extending between Cedar and Pine Sts., next attracts attention on the right. From the U. S. Signal-Service Station, at the top, an excellent view of the city, bay, and neighborhood is obtained. Just above is the six-story *Mutual Life* building (No. 142); and above is the *Williamsburgh Insurance Co.* building. On the other side of Broadway is the *Boreel* building (No. 115), and the building of the *Western Union Telegraph Co.*, on the cor. of Dey St. The *Mail and Express* building is in the same block, near Fulton St. The junction of Broadway and Fulton St. is the place of all others to see what Dr. Johnson calls "the full tide of human life"; from morning to night it presents a struggling throng of vehicles and pedestrians. To the E. *Fulton St.* runs through an active business quarter to *Fulton Ferry*, where the *Fulton Market* is, noted for its fish; to the W. it leads to *Washington Market*, the principal distributing market of the city, where may be

Madison Square, New York.

seen an unequaled display of fruits, vegetables, meats, fish, etc. At the S. E. cor. of Fulton St. is the *Evening Post* building, and on the next block (adjoining each other on the east side of Broadway) are the *Park Bank* and *New York Herald* buildings, both of white marble. **St. Paul's Church** (chapel of Trinity Church), on the west side, is a venerable structure, built in 1776, and standing in the midst of a graveyard in which are monuments of great interest. The pediment of the façade contains a white-marble statue of St. Paul, and under the rear portico is a monument to Gen. Richard Montgomery. Immediately above (on the left) is the long and severely simple front of the historic *Astor House*, opposite and on each side of which most of the horse-car lines have their termini. Opposite the Astor House, at the S. end of the City Hall Park, is the *** Post-Office,** an imposing granite building of Doric and Renaissance architecture, four stories high, besides a Mansard roof, with a front of 279 ft. toward the Park and of 144 ft. toward the south, and two equal façades of 262½ ft. on Broadway and Park Row. It is fire-proof, and cost $7,000,000. The upper floors are for U. S. Courts. The *City Hall,* in the Park, N. of the Post-Office, is a pleasing structure in the Italian style, 3 stories high, with front and ends of white marble and rear of brown-stone. It is 216 ft. long by 105 ft. deep, with Ionic, Corinthian, and composite pilasters lining its front, and surmounted by a cupola containing a four-dial clock which is illuminated at night by gas. It was erected from 1803 to 1812, at a cost of $500,000, and is occupied by the Mayor, Common Council, and other public officers. The Governor's Room, in the second story, contains the writing-desk on which Washington wrote his first message to Congress, the chairs used by the first Congress, the chair in which Washington was inaugurated first President, and a number of portraits of American worthies, mostly by eminent artists. It has also a very fine portrait of Columbus. N. of the City Hall is the *** Court-House,** which was begun in 1861, and has been occupied since 1867; the dome of it is not yet completed. It is of white marble, in the Corinthian style, 3 stories high, 250 ft. long and 150 wide, and the crown of the dome is to be 210 ft. above the sidewalk; the walls are of marble; the beams, staircases, etc., are of iron; while black walnut and pine are employed in the interior decoration. The main entrance in Chambers St. is reached by a flight of 30 broad steps, which are flanked by marble columns. The cost of the building and furniture was over $12,000,000, the result of the notorious "Ring frauds," of which it was the instrument.

On the E. side of the City Hall Park are *Printing-House Square* and *Park Row,* where are the offices of most of the daily and many of the weekly newspapers. Fronting the Square on the E. is the *** Tribune Building,** a very lofty structure. It is built of red pressed brick, granite, and iron, is absolutely fire-proof, and has a clock-tower 285 ft. high, with four dials. In front of the building is a bronze statue of Horace Greeley, by J. Q. A. Ward. The building of **The World,** 11 stories high, built of brown-stone, is on the cor. of Frankfort St. It is 309 ft. high, and the highest of its kind in the world. On the N. is the stately granite building of the **Staats-Zeitung,** with statues of Guten-berg and Franklin above the portal; and on the S. is the granite Roman-esque building of **The Times,** towering above the huge *Potter Build-ing* on the cor. of Beekman. In the Square stands a bronze statue of Franklin, of heroic size. Leading northward from Printing-House Square is *Centre St.*, which 4 squares above passes the city prison called *The Tombs*, a vast granite building in a gloomy Egyptian style, covering an entire block, and on the block above is the *City Building*, where the law courts will be held; and a part of *Park Row* (late *Chatham St.*), the habitat of Jew tradesmen, old-clothes dealers, and low concert-saloons. At the N. end of Park Row is *Chatham Square*, running N. from which about a mile is the **Bowery,** a broad and crowded thoroughfare, de-voted principally to retail-shops of every kind, with numerous beer-rooms and cheap shows. The Third Ave. line of the Elevated Railway be-gins in Park Row at the Brooklyn Bridge, and runs up Park Row, the Bowery, and 3d Ave. to Harlem River. The * great **East River** or **Brooklyn Bridge,** the largest suspension-bridge in the world, has its New York terminus in Park Row, opposite the City Hall Park, in direct connection with the City Hall branch of the Third Ave. Elevated road, and was opened for travel and traffic on May 24, 1883. The whole length of the bridge is 5,989 ft. Its width is 85 ft., which in-cludes a promenade for foot-passengers, 2 railroad-tracks on which run passenger-cars propelled by a stationary engine from the Brooklyn side, and two roadways for vehicles. The distance from high-water mark to the floor of the bridge is 135 ft. The central span of the bridge is sus-pended to 4 cables of steel wire, each 15¾ inches in diameter, which have a deflection of 128 ft. The towers at each end of the bridge are 140 ft. long by 50 ft. wide at the water's edge and are 278 ft. in height above high water. At the anchorages each of the four cables, after passing over the towers, enters the anchor-walls at an elevation of nearly 80 ft. above high water, and passes through the masonry a distance of 20 ft., at which point a connection is formed with the anchor-chains. The ap-proach on the New York side from Park Row to the foot of Roosevelt St. is 1,562 ft. On the Brooklyn side the approach is 930 ft., the ter-minus being in Sands St. of that city. This triumph of engineering was planned by Col. John A. Roebling and built by his son, Wash-ington Roebling. It was thirteen years constructing, and cost about $15,000,000, including $4,000,000 for real-estate at the termini. The visitor should not fail crossing it, as it affords splendid views of the river, the distant bay, and the cities of New York and Brooklyn.

Above City Hall Park on Broadway (cor. Chambers St.) is the mar-ble Stewart building occupied by offices. It stands on the site of a fort erected by the British during the Revolution. Farther up (on the corner of Leonard St.) is the beautiful building of the * **N. Y. Life Insurance Co.,** of pure white marble, in the Ionic style; and a num-ber of other fine buildings line the roadway on either side. *Canal St.*, once the bed of a rivulet, is one of the chief thoroughfares running across the city from E. to W. Above Canal St. a succession of fine buildings present themselves, among them the *Metropolitan Hotel*, a noble brown-stone building, occupying more than half the square on the right hand of Broadway above Prince St.   This building is also the location of Niblo's Garden Theatre.   Opposite Bond St., in Broadway, is the lofty marble façade of the *Broadway Central Hotel.*   At 49 Bond St. is the *Free Circulating Library,* which has branches at 135 2d Ave., 226 W. 42d St., and 251 W. 13th St., near 7th Ave.   One block from Broadway, in Astor Place, is *Clinton Hall,* containing the *Mercantile Library.* On the little opening is Miss Louise Lawton's statue of Samuel S. Cox. Half a block S., in Lafayette Place, is the **Astor Library,** occupying a spacious brick building in the Romanesque style.   It was founded by John Jacob Astor, who endowed it with $400,000, to which additions were made by his son, Wm. B. Astor.   It contains over 250,000 volumes, and is complete in many special departments of study.   It is open to the public daily, free.   At the end of Astor Place (2 blocks from Broadway) is the **Cooper Institute,* a large brown-stone building, occupying the entire square bounded by 3d and 4th Avenues and 7th and 8th Sts. It was founded and endowed by Peter Cooper, a wealthy and philanthropic merchant; and contains a free library, a free reading-room, free schools of art and telegraphy for women, a free night-school of art for men, a free night-school of science for both sexes, and free lectures.   The reading-room is open to all from 8 A. M. to 10 P. M., on Sundays, 10 A. M. to 9 P. M.   Opposite is the *Bible House,* a large brick structure, containing the headquarters of the American Bible Society, next to the British the largest in the world.

Returning to Broadway and passing N., the spacious iron building occupied by the successors of A. T. Stewart & Co. is seen on the right. It is 5 stories high, occupying the entire block between 9th and 10th Sts. and Broadway and 4th Ave.   At 10th St. Broadway turns slightly toward the left, and **Grace Church* (Episcopal), with its fine marble façade, seems to project into the middle of the highway.   The interior of Grace Church is extremely rich, and the music is very fine.   Passing the *Star Theatre* (near 13th St.), and the lofty and florid building of the *Domestic Sewing-Machine Co.* (cor. 14th St.), we enter **Union Square,* a pretty little park, oval in shape, 3½ acres in extent, and filled with trees, shrubbery, and green lawns.   At its southern end, on the E., are the **bronze equestrian statue of Washington by H. K. Browne, and Bartholdi's bronze statue of Lafayette; and on the W. is a bronze statue of Lincoln, while just above is a fine bronze fountain. The Square is surrounded by hotels and shops, chief among which are the jewelry-store of Tiffany & Co. (cor. W. 15th St.).   On 15th St., east of Broadway, is the attractive building of the *Young Women's Christian Association.*   On the N. side is a Plaza or parade-ground, the tall *Jackson Building,* and the *Everett House,* one of the oldest of the better uptown hotels.   *Fourteenth St.,* a leading cross-town thoroughfare, runs E. from Union Square, past the *Academy of Music* and *Tammany Hall;* and to the W. it passes for several blocks through a line of handsome retail stores.   Just W. of 6th Ave. is the *Fourteenth St. Theatre,* and beyond are private residences.

Above Union Square, Broadway contains a number of large dry-goods and carpet warehouses, furniture, *bric-à-brac,* and fancy-goods dealers. At * **Madison Square** is another beautiful little park, from 23d St. to 26th St. On the N. W. of the park, facing Delmonico's, is St. Gaudens' bronze statue of Admiral Farragut, and nearly opposite, at the junction of Broadway and 5th Ave., is a monument to General Worth; near the S. W. corner is Randolph Rogers's bronze statue of Seward. On the S. E. corner of 23d St. and Broadway is the *Bartholdi Hotel.* Overlooking the Square, on the W. side, are the buildings of the *Fifth Avenue Hotel* and the *Hoffman House.* In 23d St., running *west* to the river and a fashionable shopping-street, is the *Eden Musée,* a granite building with a very ornately-carved front, and, at the cor. of 6th Ave., the * **Masonic Temple,** of granite, 100 by 140 feet, 5 stories high, and with a dome 50 feet square, rising 155 feet above the pavement. It contains fine rooms, and the Grand Lodge Hall, 84 by 90 feet, and 30 feet high, will seat 1,200 persons. In 23d St., between 7th and 8th Aves., the immense building, 12 stories in height, known as the *Chelsea Apartment-House,* will attract the eye of the stranger. Half a block W. (cor. of 8th Ave.) is the * **Grand Opera-House.** On the Square, between 20th and 21st. Sts. and 9th and 10th Aves., are the buildings and chapel of the General (P. E.) Theological Seminary.

Returning eastward to *Broadway,* which we have been following as our main route, we see the marble *Metropolitan* office building; while one block *east* of Madison Square (in 23d St. cor. 4th Ave.) is the * **National Academy of Design,** built of gray and white marbles and blue-stone, copied from a famous palace in Venice. It has an imposing entrance and stairway leading to extensive galleries, where every spring and fall are held exhibitions of recent works of American artists (admission 25c.). Opposite is the building of the *Young Men's Christian Association,* constructed in the Renaissance style. Besides a library, free reading-room (open from 8 A. M. to 10 P. M.), gymnasium, etc., it contains a lecture-hall capable of seating 1,500 persons. Above the academy in 4th Ave. is the *Lyceum Theatre.* On the S. E. corner is the large building of the *Society for the Prevention of Cruelty to Children,* and adjacent to it on 4th Ave. is the home of the *Charity Organization Society.* The **Madison Square Garden,** designed by Stanford White, occupying the whole square bounded by 26th and 27th Sts. and 4th and Madison Aves., is one of the grandest structures in the city, and has the largest auditorium in America; connected with it is the *Garden Theatre.* At the cor. of 21st St. is *Calvary Church* (Episcopal), a brown-stone Gothic building. At the cor. of 20th St. is the *Church of All Souls* (formerly Dr. Bellows's), a curious structure in the Italian style. A short distance to the E. on 20th St. is the aristocratic *Gramercy Park,* with the *Players' Club* at No. 16 and the *Tilden House* at No. 14, on the south side. Taking 16th St. to the E. from 4th Ave., Stuyvesant Square is soon reached, in which stands * **St. George's** (Episcopal), one of the largest churches in the city. It is of brown-stone, in the Byzantine style, and the interior is magnificent. The *Florence,* an apartment-house is on the cor. of 18th St. and 4th Ave., and opposite is the *Clarendon Hotel.* On the corner above is the *Cyclorama Building.*

Broadway runs from Madison Square 2 miles N. to Central Park, passing a number of theatres and hotels, among which the most noteworthy are the *Victoria Hotel* (cor. 27th St.), *Fifth Avenue Theatre* (cor. 28th St.), *Daly's Theatre* (cor. 30th St.), *Palmer's Theatre* (cor. 30th St.), the *Hotel Normandie* (cor. 38th St.), the *Casino Theatre* (cor. 39th St.), the *Broadway Theatre* (cor. of 41st St.), the *St. Cloud* and the *Rossmore* (cor. 42d St.), and the Albany, Newport, Saratoga, Rockingham, and Fennimore apartment-houses. One of the most striking buildings in this part is the *** Metropolitan Opera-House** (occupying the whole of the square between 39th and 40th Sts.). It is built of yellow brick, and makes but little pretension to exterior decoration. The theatre was destroyed by fire in August, 1892, but will be rebuilt during the present year. The continuation of Broadway above 59th St. is known as the *** Boulevard,** a grand avenue 150 ft wide, divided in the center by trees and grass, and extending N. to 167th St. By it may be reached what were formerly the villages of *Manhattanville* (125th–132d St.) and *Carmansville* (1 mile beyond); still N. of which is *** Fort Washington** (or Washington Heights), the chief summit on Manhattan Island (238 ft. high), and commanding a noble view of the city, the Hudson, and the opposite Jersey shore. It is now occupied by elegant villa residences. (Fort Washington is easily reached from the lower part of the city by taking the Sixth Ave. Elevated Railway to 145th St.)

Fifth Avenue begins at *Washington Square* (a pleasant park of 9½ acres laid out on the site of the old Potter's Field, where over 100,000 bodies were buried) and runs N. for 6 miles to Harlem River. Within the square are bronze monuments to Garibaldi and A. L. Holley. At the S. end of 5th Ave. is the **Washington Memorial Arch.** As far as Central Park it is lined with compact rows of houses; between 59th and 110th Sts. it has the Park on the left, and houses at greater or less intervals on the right; and from Mt. Morris to Harlem River (124th to 135th Sts.) it is lined with residences.

Washington Square has fine old residences on the north side, and on the E. side is the *University of the City of New York*, a marble building in the Gothic style, 200 by 100 ft. The Chapel, with its spacious window 50 ft. high and 24 ft. wide, is a noble room. The University was founded in 1831, and has about 50 instructors and 500 students. Adjoining it is a handsome church (Methodist) of granite in the Gothic style. On the south side is the large *Judson Memorial Church* (Methodist), erected in memory of the great missionary. Passing up Fifth Ave. from Washington Square, the *Church of the Ascension* (Episcopal) is seen at the cor. of 10th St., and the *First Presbyterian* at the cor. of 11th. At the cor. of 16th St. is the *Judge* building; and a short distance to the left (in 15th St.) are the Italian-Gothic buildings and church of the *College of St. Francis Xavier*, the headquarters of the Order of Jesus in North America. Near by is the spacious building of the *New York Hospital*. Among the large business houses, the *Methodist Book Concern*, cor. of 19th St., is conspicuous. At the cor. of 18th St. is *Chickering Hall*, and at the cor. of 21st St. is the house of the wealthy *Union Club*, while opposite is the *Lotos Club*. At

the cor. of 21st St. is the *Mohawk Building*, and at the cor. of 22d
St. **Knoedler's* art-gallery; beyond which the avenue leads past Madi-·
son Square on the right and a line of hotels, among which are the
Fifth Avenue and *Hoffman House*, on the left. From Madison Square
to Central Park, Fifth Ave. is the aristocratic street of America, lined
with handsome residences, and presenting a brilliant spectacle, especially on Sunday mornings. At all times it is thronged with the
equipages of the wealthy and richly dressed pedestrians, and a succession of costly churches challenges the attention of the passer-by. Just
off the avenue in W. 25th St. is *Trinity Chapel* (Episcopal), with its
richly decorated interior and impressive choral services; and on opposite
corners of 26th St. are *Delmonico's* world-famous restaurant and the
Café Brunswick, whose reputation is scarcely inferior. At the cor. of
28th St. is the *Knickerbocker* apartment-house, eleven stories high. To
the E. in 28th St. at No. 11 is *St. Leo's Church* (Roman Catholic), which
is perhaps the most aristocratic church of its denomination in the city;
beyond at No. 149 is * *St. Stephen's Church* (Roman Catholic), unattractive as a building, but containing some excellent paintings and the
most expensive and elegant altar-piece in the country. Its music is
famous and attracts many visitors. At the foot of E. 26th St. is **Bellevue Hospital,** the largest in the city, with accommodations for 1,200
patients. At the cor. of 5th Ave. and 29th St. is the *Collegiate Church*
(Dutch Reformed); and in 29th St. just E. of the avenue is the picturesque *Church of the Transfiguration* (Episcopal), known familiarly as
"the little church round the corner." A block beyond is the *Holland
House*, a family hotel for the wealthy, while at the cor. of 33d St. is the
Waldorf Hotel, opened in the spring of 1893. At the cor. of 34th St.
is the *Manhattan Club* building, formerly the residence of A. T. Stewart,
a large white-marble structure, 8 stories high with a Mansard roof; and
splendidly decorated and furnished. Passing W. along 34th St. the spacious *Congregational Tabernacle* is seen at the cor. of 6th Ave.; and the
vast marble buildings of the *N. Y Institution for the Blind*, with turrets
and battlements, at 9th Ave. At the cor. of 35th St. and 7th Ave. is
the brick and gray-stone structure of the *State Arsenal*, the headquarters of the Ordnance Department of the State; and at the cor. of 36th
St. and 9th Ave. is the Gothic edifice of the *Northwestern Dispensary*.
One block E. of 5th Ave., 34th St. emerges into * **Park Avenue,** a
beautiful street 140 ft. wide, bordered by handsome private residences,
and divided in the center by a row of beautiful little parks, surrounding openings in the railroad tunnel which runs underneath. In
Park Ave., the *Murray Hill Hotel* is at the cor. of 40th St.; at the
cor. of 35th St. is the *Church of the Covenant* (Presbyterian), of gray-·
stone in the Lombardo-Gothic style; at the cor. of 34th St. is the
Church of the Messiah (Unitarian); and just below (cor. 4th Ave. and
32d St.) is the vast iron building erected by A. T. Stewart as a Work-
ing-women's Home, but now the *Park Hotel*. In 5th Ave., at the
cor. of 35th St., is the *New York Club*, formerly the Caswell home;
and at the cor. of 37th St. is the *Brick Church* (Presbyterian). Farther
up, at the cor. of 39th St., the *Union League Club* has a spacious build-

St. Patrick's Cathedral, New York.

ing. Occupying the left side from 40th to 42d St. is the disused *Distributing Reservoir* of the Croton Aqueduct, covering 4 acres. West of it is the pretty little Bryant Square, with a bust of Washington Irving, and opposite is the tall building of the *Columbia Bank*. Two squares E. in 42d St. is the *Grand Central Depot, built of brick, stone, and iron, 692 ft. long and 240 ft. wide. An addition to the east is now used for incoming trains. In 43d St., a few doors west of the avenue, is the beautiful home of the *Century Club*, the building was designed by Stanford White, and opposite is the *Hotel Renaissance*, while adjacent to it is the *Racket and Tennis Club*. At the cor. of 5th Ave. and 43d St. is the Jewish *Temple Emanuel, the chief synagogue of the city, and the finest specimen cf Saracenic architecture in America. The interior is gorgeously decorated. In W. 44th St. are the *Academy of Medicine* and the *Berkeley Lyceum*. At the cor. of 45th St. is the Universalist Church of the *Divine Paternity;* at 46th St. is the *Windsor Hotel;* and at the cor. of 48th St. is the costly *Collegiate Church* (Dutch Reformed). Passing E. along 50th St. to Madison Ave., we reach *Columbia College, the buildings of which are very handsome. It is the oldest college in the State, having been chartered in 1754, and is richly endowed. Occupying the square on 5th Ave. between 50th and 51st Sts., is the * *Cathedral of St. Patrick* (Roman Catholic), the largest church in the city, and one of the largest and finest on the continent. It is of white marble in the decorated Gothic style, and is 332 ft. long, with a general breadth of 132 ft., and at the transept of 174 ft. At the front are two spires, and each 328 ft. high, flanking a central gable 156 ft. high. Between 51st and 52d Sts. are the *Vanderbilt* houses. That at the N. W. cor. of 53d St. is the house of William K. Vanderbilt, and is considered next to Trinity Church, Boston, the finest piece of architecture in the United States, while on the cor. of 57th St. is the house of Cornelius Vanderbilt At the cor. of 53d St. is the handsome church of *St. Thomas* (Episcopal); and at 54th St. is **St. Luke's Hospital,** one of the most notable objects on the avenue. It is in charge of the Episcopal Sisters of the Holy Communion. On the S. W. corner is the residence of William C. Whitney, and opposite on the east side of the avenue is the residence of C. P. Huntington. At 55th St. is the **Fifth Avenue Presbyterian Church** (Dr. John Hall's), the largest of that sect in the world; and at 59th St. Central Park is reached. The opening at the entrance of Central Park is called the *Plaza.* At its S. end, facing 5th Ave., is the *Veteran Club*, while on the W. side is the *Plaza Hotel*, and on the E. side are the *Savoy* and the *New Netherlands.* In 59th St., facing Central Park, west of 5th Ave., are large apartment-houses, the elegant homes of the *Deutsche Verein* and the *Catholic Club*, while beyond, the huge "Spanish" apartment-houses attract the attention for their beauty and spaciousness. From 59th St. to 72d, facing the Park, are many handsome residences of attractive architecture, and on the N. E. cor. of 63d St. is the *Progress Club*, built in the Italian Renaissance style, one of the finest structures of its kind in the city. The * **Lenox Library,** of Lockport limestone, extends from 70th to 71st St. It was founded and erected by the late

James Lenox. It possesses, besides other valuable donations, "the collection of MSS., printed books, engravings and maps, statuary, paintings, drawings, and other works of art," made by the founder, and is particularly rich in early American history, biblical bibliography, and Elizabethan literature. Close by it is the *Presbyterian Hospital* (founded by Mr. Lenox), a pleasing brick and stone structure with graceful spires. A short distance E. (cor. 4th Ave. and 69th St.) is the *Normal College, a beautiful building in the secular Gothic style, 300 ft. long, 125 ft. wide, and 70 ft. high, with a lofty and massive Victoria tower. It is part of the common-school system, and is free. On Park Ave., between 69th and 70th Sts., is the *Union Theological Seminary*, while at the cor. of 72d St. is the handsome building of the *Freundschaft Club*. Beyond 72d St. the residences are fewer, but on 4th Ave. and 94th St. is the imposing armory of the *Eighth Regiment*. Between 120th and 124th Sts. is *Mount Morris Square*, a park of 20 acres, with a rocky hill in the center 101 ft. high, commanding picturesque views. Beyond this the avenue passes amid tasteful residences to the Harlem River, at the end of the island. (Fifth Ave., up to 72d St., may be advantageously seen, on week-days, by taking a 5th Ave. stage. In the afternoon it is the fashionable promenade.)

Among the institutions and buildings not yet mentioned but worthy of notice are the following: The *Five Points House of Industry* (155 Worth St.) and the *Five Points Mission*, facing each other on what was once the vilest and most dangerous part of the city. The *Howard Mission*, 204 5th St., supports day and Sunday schools and a home for needy children, and distributes food, clothing, and fuel to the deserving poor. The *Deaf and Dumb Institution is located on Washington Heights (see p. 13); the buildings, which are the largest and finest of the kind in the world, cover 2 acres and stand in a park of 28 acres (visitors admitted from 1.30 to 4 daily). The *Convent of the Sacred Heart*, in Manhattanville (see p. 13), is beautifully situated on a hill surrounded by park-like grounds. The *Bloomingdale Asylum for the Insane occupies a commanding site in 117th St. near 10th Ave.; the buildings, 3 in number, can accommodate 170 patients, and are always full. Columbia College has purchased this site for its new buildings. At Manhattanville is *Manhattan College* (Roman Catholic), with stately buildings and 700 students. . The *Sheltering Arms*, at 10th Ave. and 129th St., receives children between 2 and 10 years of age, for whom no other institution provides.

**Central Park is reached from the lower part of the city by the 6th Ave. Elevated Railway; by the street-cars of the Broadway, 6th, 7th, and 8th Ave. lines; or by stage up 5th Ave. It is one of the finest parks in the world, embracing a rectangular area of 843 acres, extending from 59th to 110th St. and from 5th to 8th Ave. It has 18 entrances (4 at each end and 5 at each side), and four streets (65th, 79th, 85th, and 97th) cross it, to afford opportunity for traffic, passing under the park walks and drives. The original surface was exceedingly rough and unattractive, consisting chiefly of rock and marsh; but by engineering skill the very defects that once seemed fatal have been converted into its most attractive features. Between 79th and 96th streets a large portion of the Park is occupied by the two Croton reservoirs, the smaller one comprising 35 and the larger 107 acres. The Lakes, five in number, occupy 43¼ acres more. There are 10 miles of carriage-roads, 6 miles of bridle-paths, and 30 miles of footpaths, with numerous bridges, arches, and other architectural monuments, together with many statues. The **Mall*, near the 5th Ave. entrance, is the principal promenade; it is a wide esplanade, nearly a quarter of a mile long, and bordered by double rows of stately elms. At various points are bronze statues of *Shakespeare, Walter Scott, Goethe, Burns, Halleck, and *Daniel Webster; and also the Puritan and 7th Regiment statues. Particularly worthy of notice are the bronze groups of "The Indian Hunter and his Dog" (near the S. end) and "The Falconer" (near the upper end). In the Music Pavilion, in the upper part of the Mall, concerts are given on Saturday and Sunday afternoons in the summer. The Mall is terminated on the N. by **The Terrace*, a sumptuous pile of masonry, richly carved and decorated. Descending the Terrace by a flight of broad stone stairs, *Central Lake* is reached, the prettiest piece of water in the Park. Between the Terrace and the Lake is a costly fountain with large granite basins and a colossal statue of the Angel of Bethesda. The *Ramble*, covering 36 acres of sloping hills, and abounding in pleasant shady paths, lies N. of Central Lake. On the highest point of the Ramble stands the **Belvedere*, a tower in the Norman style of architecture. From the top attractive views in all directions may be had. Just above the Belvedere is the Old Croton Reservoir (holding 150,000,000 gallons), and above this the New Reservoir (holding 1,000,000,000 gallons). Still above this is the Upper Park, less embellished by art than the lower portion, but richer in natural beauties. About the Old State Arsenal, at the S. E. end, are the **Zoölogical Gardens*, or *Menagerie*, with an interesting collection of animals, birds, reptiles, etc.; and at 82d St. on the 5th Ave. side is the spacious building of the **Metropolitan Museum of Art* (see "Art Collections" on p. 4). The Egyptian **Obelisk* (Cleopatra's Needle) stands on an eminence just W. of the Museum, and is one of the most striking objects in the Park. This obelisk is one of the most ancient of the world's monuments. Originally hewn and inscribed by Thothmes III, one of the sides is also inscribed with the victories of Rameses II (a contemporary of Moses), who lived three centuries afterward. It was presented to the city of New York by Ismail Pasha, and brought to this country at the expense of William H. Vanderbilt. Hackney-coaches may be hired at the entrances for $2 per hour, and the circuit can be made in an hour. Park carriages run to Mount St. Vincent and back from 5th and 6th Aves. (fare 25c.). In Manhattan Square, which adjoins Central Park on the W. between 77th and 81st Sts., is the **American Museum of Natural History*, in a large brick building, containing Indian antiquities, minerals, shells, and stuffed and mounted specimens of birds, fishes, quadrupeds, insects, etc. Open daily to the public from 9 A. M. till sunset, and on Sundays from 1 P. M. till sunset; also on Tuesday and Saturday evenings from 8 to 10 P. M. *Riverside*

Park extends from 72d to 130th Sts. along the Hudson River, and can be reached most easily by the 6th Ave. Elevated Railway to 125th St. Near 125th St., on Claremont Hill, is the site of the Grant Memorial, and there the remains of the great hero are buried.

In 1873 Morrisania, West Farms, and Kingsbridge, with adjacent territory aggregating some 13,000 acres, were annexed to the city. This section is very rapidly being built up, but as yet has no special attractive features worthy of description. From the terminus of the 6th Ave Elevated Railroad the New York & Northern R. R. passes along the N. bank of the Harlem River, stopping at the stations of High Bridge, Morris Heights, Fordam Heights, and Kingsbridge, and thence northward through Van Cortlandt, Mosholu, Lowerre, and Park Hill to Yonkers. On the east side the 3d Ave. extension of the elevated system runs from 129th St. and 2d Ave. to E. 170th St., a distance of 2½ miles. This road is to be extended to New Rochelle, a distance of 16 miles. On the N. an elaborate series of parks is being laid out, beginning with Pelham Bay Park (1,700 acres), on Long Island Sound, which is connected by the Bronx and Pelham Parkway with Bronx Park (653 acres), where is the proposed site of the Botanical Gardens; thence by the Mosholu Parkway to the Van Cortlandt Park (1,069 acres), used as a parade-ground by the National Guard.

Itineraries.

The following series of excursions have been prepared so as to enable the visitor, whose time is limited, to see as much of the city as possible in the least amount of time. Each excursion is planned to occupy a single day, but the visitor can readily spend more time as special features crowd upon his attention. In making these various excursions, the visitor should provide himself with APPLETONS' DICTIONARY OF NEW YORK AND ITS VICINITY. in which he will find a detailed description of the places referred to in these itineraries.

1. An excellent idea of the magnitude of the city may be gained by starting from Madison Square and proceeding eastward along 23d St. until 3d Ave. is reached, where the elevated railroad may be taken to the Battery; then, without descending to the street, take either the 9th Ave. or the 6th Ave. train to 155th St.; whence proceed by the New York and Northern R. R. to Yonkers, and then return by the train of the New York Central & Hudson R. R. to 125th St.; thence across to 3d Ave., where the elevated railroad may be taken running southward, or north to Tremont by the Suburban Rapid Transit R. R., a distance of 2½ miles, returning by the same route. This excursion may be shortened by crossing the city at 125th St., and foregoing a visit to the newer portion of the city. It may be lengthened by visiting the new parks and parkways in the upper part of the city.

2. Starting from the vicinity of Madison Square, take the Broadway cars down town, passing on the left Union Square, with the statues of Lafayette, Lincoln, and Washington; then past Grace Church, at 10th St., and the many large business houses that line the great thoroughfare. Alight at Chambers St.; visit the Court-House, City Hall, and other public buildings, including the Post-Office. The great East River Bridge and the newspaper-offices are in Printing-House Square, which is to the

east of City Hall Park. The buildings of the *World*, *Tribune*, and *Times* are the most conspicuous. Then proceed along Park Row to Broadway, past the *Herald* Building, on the corner of Ann St.; cross over to St. Paul's Church. On the block beyond is the *Mail and Express* Building; and on the corner of Dey St. is the Western Union Building, where visitors are admitted to see the operating-room, in which messages from all parts of the country are being received. A few blocks farther, at 120 Broadway, is the Equitable Building; and then, at the head of Wall St., is Trinity Church, with its historic grave-yard. Continuing down Broadway, visit the Consolidated Stock and Petroleum Exchange, on the corner of Exchange Place. From the visitors' gallery a view of the floor can be obtained. Passing the many tall office-buildings in the vicinity, Bowling Green is reached, and to the left is the Produce Exchange, from the tall tower of which a fine view can be obtained. Continue down Whitehall St. to the Battery, with the Barge-Office on the water's edge; then pass along South St., in front of the shipping, till Wall St. is reached, by which return to Broadway, calling at the U. S. Custom-House, the U. S. Assay-Office, the U. S. Sub-treasury, and the Stock Exchange (entrance to visitors' gallery is on Wall St.). Walk up Nassau St. to Printing-House Square, where the elevated railway may be taken, or preferably the 4th Ave. street-cars, which pass the Tombs in Centre St., and then up the Bowery, one of the unique streets in New York; turning into 4th Ave. at 8th St., where the Cooper Union is. Return to Madison Square at 23d St., passing the Young Men's Christian Association and the Academy of Design at 4th Ave. The large building of the College of the City of New York is at the corner of Lexington Ave., one block to the east of 4th Ave.

3. Proceed westward through 23d St. to 6th Ave. The Masonic Temple is on the N. E. corner. Then down 6th Ave. past the large retail dry-goods stores to 9th St., where the Jefferson Market Court-House, one of the really fine pieces of architecture, is situated. A few minutes may be spent with interest in the court-rooms. Adjoining is the Jefferson Market. In the immediate vicinity are many of the old houses of New York, and a ramble to the westward and southward brings one in the midst of the old village of Greenwich, and later the residential quarter of the wealthy. Passing from 6th Ave. through 10th St. the Studio Building is seen on the left, and at times the studios are open to the public. At the corner of 5th Ave. is the Church of the Ascension, with a fine painting by John La Farge over the altar, and numerous memorial windows of considerable merit. Turning S., the Memorial Arch is seen, and in a few minutes Washington Square is reached. In the square on the west side is J. Q. A. Ward's bust of A. L. Holley, and on the east is Turini's Garibaldi. Beyond, in University Place, is the New York University, with many historic associations; to the S. of which is the Benedict Apartment House, where many artists have their studios. At the S. W. corner of the square is the Judson Memorial Church, with its tall tower. A block below the square is Bleecker St., which is about the centre of the French quarter of New York. To those who desire to study the peculiarities

of a foreign people in a large city, an hour or so may be spent in visiting this district. Returning to the square, a stage may be taken up 5th Ave. to 86th St., passing the many churches, club-houses, and private residences. At the end of the stage line, cross to the E., and return by way of the Madison Ave. cars. The Tiffany house at 72d St., and the Villard houses at 50th St. are passed, and at 49th St. the buildings of Columbia College are reached. These, with their museums, deserve careful inspection. Then, again taking the cars, the Manhattan Athletic Club building, at 45th St., the Grand Central Station at 42d St., are passed; then through the tunnel, and down 4th. Ave. to 23d St.

4. Madison Ave., between 23d and 42d Sts., is worthy of a visit. The University Club is on the S. E. corner of 26th St., on the upper corner is the Madison Square Garden. Visitors can ascend the tower. Several churches are on the route, among which are the Church of Transfiguration at No. 5 East 29th St., and the Church of the Incarnation at the cor. of 35th St. Residences of prominent citizens, with pleasing architecture, line the avenue till 42d St. is reached. Then, by either the Madison Ave. or 3d Ave. cars, proceed to 67th St. This vicinity, once city property, is now the site of numerous public institutions. In 67th St., near 3d Ave., is the headquarters of the Fire Department. On the S. W. cor. of Lexington Ave. is the Mount Sinai Hospital, while adjoining, in 66th St., is the Chapin Home. Between 66th and 67th Sts. is the Seventh Regiment Armory, which is best seen during the winter months in the evening, as then the militia are drilling. Passing up 4th Ave. the Hahnemann Hospital is reached, and at 68th St. is the Female Normal College, occupying the entire block. A technical school for the deaf and dumb is in Lexington Ave., between 67th and 68th Sts., and also in Lexington Ave., directly opposite the College, is the Foundling Asylum of the Roman Catholic Church. Returning to 4th Ave., the Union Theological Seminary of the Presbyterian Church occupies, with its fine building, the block between 69th and 70th Sts. In 70th St. is the Lenox Presbyterian Hospital, and beyond, in 5th Ave., is the Lenox Library, while on the cor. of 72d St. and 4th Ave. is the Freundschaft Club-House. In returning to 23d St. the visitor may take the street cars on Madison or 3d Ave., or the 5th Ave. stages.

5. An entire day may be given up to Central Park, which should be entered at 59th St. and 5th St. and beyond Ave., and at either of which gates there are carriages in waiting, which will take the visitor around the park (fare 25c.). The old Arsenal and zoölogical collection are on the east side, near 64th St., and should be visited independently of the carriage-ride. Stops are made at the Metropolitan Museum of Art, near 82d St., in the vicinity of which is also the Obelisk, at Mount St. Vincent, and on the west side at the American Museum of Natural History, and at the Terrace Bridge, whence the Mall and the Belvedere may be visited. From the Museum of Natural History the visitor may take the street cars and ride down 8th Ave., past the large St. Remo and Dakota apartment-houses to 59th St.; thence eastward past the Spanish apartment-house, and club-houses to 6th Ave., where the cars may be taken to 23d St.

The Accepted Design for the Tomb of General Grant.

(COPYRIGHTED BY THE GRANT MONUMENT ASSOCIATION.)

6. Take the Broadway cars up Broadway from 23d St., past the theatres, hotels, Metropolitan Opera-House to 40th St., to 42d St., then by a Boulevard line car to 72d St., whence proceed westward to River-side Park. At the upper end of the Park is Grant's tomb, where the great memorial is now being built to the deceased hero. To the E. of the Park is West End Ave., which is fast being built up with fine residences. At the cor. of 110th St. and Amsterdam Ave. is the accepted site of the Episcopal Cathedral of St. John the Divine, and at 117th St. is the site to which Columbia College will remove, while still to the E. is Morningside Park. At 125th St. and 10th Ave. take the cable-cars to their extreme end at Fort George, where during Revolutionary times a fort existed, then return a few blocks and cross the Harlem River over the Washington Bridge, from where a fine view is obtained. Pass down on the N. bank of the river to High Bridge, over which return to Manhattan Island, and visit the Water Works. Descend to the river and take the ferry down the stream to 130th St. and 3d Ave., where connection is made with the east side elevated rail-road, or, if there is time, a boat may be taken for the lower portion of the city, affording the visitor a view of Hell Gate, the islands, and shipping.

7. A pleasant walk from Madison Square is up 5th Ave. to 26th St., passing the statue of Farragut ; then eastward past the University Club, Madison Square Garden to Bellevue Hospital, which with its various departments is worthy of careful inspection. Then by ferry to Blackwell, Ward's, and Randall Islands, visiting the public institutions there (a pass is required). Returning, the remainder of the day may be spent in visiting the treasures of the Astor Library, reached by cross-town car from foot of E. 23d St. The Mercantile Library, Cooper Union, and Bible House are in the immediate vicinity. Thence pass up 4th Ave. to 14th St., then E. to Irving Place, and N. past the Academy of Music, Amberg's Theatre, to Gramercy Park, with its notable residences on the north side, and the Players' Club and Tilden house on the south side. A few steps to the W. is 4th Ave., thence proceed N. to 23d St.

Several days may be agreeably spent in visiting the various suburban resorts near New York, including Brooklyn.

An excursion which no visitor should fail to make is that to *** High Bridge** (reached by 6th Ave. Elevated road to 155th St.), connecting with N. Y. & Northern R. R. trains ; by horse-cars or 3d Ave. Elevated to 125th St., and thence by Cable-Road to 10th Ave. and 175th St., or by 3d Ave. Elevated to Harlem, and thence by steamboat. This noble structure, by which the Croton Aqueduct is carried across Harlem River, is of granite throughout, and spans the entire width of valley and river. It is 1,450 ft. long, 114 ft. high, and supported on 14 massive piers, and has been well called "a structure worthy of the Roman Empire." On the lofty bank at its S. end is a capacious reservoir. At 181st St., reached by cable-car from 125th St., is the *Washington Bridge* across the Harlem, connecting Edgecombe road and Boscobel Ave. Passing thence along the roadway, a short distance beyond the Berkeley Oval is reached, where the great intercollegiate foot-ball matches are held.

The public institutions on the East River islands are places of special interest. Opposite the foot of E. 46th St. is *Blackwell's Island*, 120 acres in extent; upon it are located the Almshouse, Lunatic Asylum (for females), Penitentiary, Workhouse, Blind Asylum, Charity, Small-pox, and Typhus-Fever Hospitals, Hospital for Incurables, and Convalescent Hospital, all built of granite, quarried on the island by the convicts. North of the island, between the village of Astoria (reached by 92d St. Ferry) and New York, on the opposite shore is *Hell-Gate*, long the dread of all vessels passing through the Sound, but now largely shorn of its terrors, while the U. S. engineers are engaged in removing the remaining rocks. To the left is *Ward's Island* (200 acres), which divides the Harlem from the East River; upon it are the Lunatic Asylum (for males), the Emigrant Hospital, and the Inebriate Asylum. *Randall's Island*, separated from Ward's Island by a narrow channel, is the site of the Idiot Asylum, the House of Refuge, the Infant Hospital, Nurseries, and other charities provided by the city for destitute children. (Permits for visiting any of these islands must be procured at the office of the Commissioners of Public Charities, cor. 3d Ave. and 11th St.)

Governor's Island (reached by ferry from pier adjoining Staten Island ferry) is a national military station, with two forts (Fort Columbus and Castle William) and some attractive officers' quarters.

* Staten Island is reached by ferry-boats from the foot of Whitehall St. every 20 minutes to St. George, whence trains of the Staten Island Rapid Transit R. R. run E. to Tompkinsville, Stapleton, and Clifton, and W. to New Brighton, Sailors' Snug Harbor, Livingston (Cricket Club Station), West Brighton, Port Richmond, Sharp Ave., Elm Park, Erastina, and Arlington. (Last boat from St. George 12 P. M.) It is the largest island in the harbor, having an area of 58½ square miles, and separated from New Jersey by Staten Island Sound and the Kill Van Kull, and from Long Island by the Narrows. It is connected with the New Jersey shore by the Arthur Kill bridge, and has become the freight terminus of the Baltimore & Ohio R. R. The drives about the upper part are attractive, especially those on Vanderbilt Ave., Richmond Terrace, the Serpentine and Clove Roads. From the heights there are broad views over harbor and ocean. *New Brighton* is the largest village on the island, and contains the *Castleton* and other fine summer hotels, a number of churches, and many handsome villas. From Port Richmond an electric railway connects with *Prohibition Park*, where during the summer the American Institute of Christian Philosophy holds its meetings. Horse-cars and the *Rapid Transit R. R.* traverse the North Shore, and the Staten Island R. R. runs from St. George to Tottenville (14 miles). One mile S. E. of Clifton is ** Fort Wadsworth*, commanding a fine view of the upper and lower bays and vicinity.

Bedloe's Island, with Bartholdi's colossal statue of *Liberty Enlightening the World*, is reached by ferry from pier adjoining Staten Island Ferry, every hour from 6 A. M. to 7.30 P. M. (fare, 25c., round-trip). The statue was presented by the French nation to the American people, the cost being defrayed by public subscription, and the artist,

Auguste Bartholdi, taking no remuneration. The pedestal was built by public subscription collected in the United States. The weight of the statue is 450,000 lbs. of copper and iron; height from base to torch, 151 ft. 1 in.; from foundation to torch, 305 ft. 6 in. Total cost, nearly $1,000,000. On the 28th of October, 1886, the French delegates handed it over to the President of the United States.

Coney Island.

Coney Island lies just outside the Bay, about 10 miles from the city (by water), and consists of a very narrow island 4½ miles long It is separated from the mainland by Gravesend Bay on the west, Sheepshead Bay on the east, and Sheepshead Bay and Coney Island Creek on. the north, and has the broad Atlantic for its southern boundary. The island is divided into four parts, known as *Coney Island Point*, or *Norton's*, at the west end, *West Brighton Beach*, *Brighton Beach*, and *Manhattan Beach* at the east end. Brighton and Manhattan have extensive hotels, complete opportunity for bathing, and are the preferred resorts of the better class. *Norton's* possesses no attractions. At * **West Brighton** there are two iron piers extending 1,300 feet into the ocean, with restaurants, bath-houses, promenades; and the huge *Elephant Hotel*, built in the shape of an elephant, a camera, shows, etc., make a stirring but rude scene. At this point is the end of the Ocean Parkway drive from Brooklyn. The Concourse which leads to Brighton Beach is a wide drive and promenade nearly a mile long. Coaches connect West Brighton with * **Brighton Beach,** where there is a good hotel, and a pavilion for parties with lunch-baskets. The Marine Railway, laid on piles, connects Brighton with * **Manhattan Beach,** less than a mile distant, which is the most fashionable resort on the island. There are two hotels here: the *Manhattan*, with wide piazzas, large dining-rooms, and garden, on the European plan, and the *Oriental*, a family hotel, on the American plan. The bathing establishment has separate sections for men and women, an amphitheatre with 3,500 seats for those who prefer to look on, and an inclosed beach for the bathers. There is music by an excellent band (admission 25c. and 10c.) every day at 2 and 7.30 P. M., and a pyrotechnic display every evening (50c. for reserved seats). A drive has been laid out for the guests of the Manhattan Beach Hotel, connecting the Oriental with the Ocean Park Boulevard. It is known as the *Manhattan Boulevard*, and is 60 ft. wide, skirting the shores of *Sheepshead Bay*.

To reach Coney Island.—(1) *Manhattan Beach* is reached by trains from Long Island City (ferries from James Slip and E. 34th St.); also by steamboat from the Battery, east side (in connection with the elevated trains), to Bay Ridge, and thence by rail. Fare, 25c.; excursion tickets, 40c. (2) *Brighton Beach* is reached by Brighton Beach R. R. from Prospect Park station, Bedford station, Bergen, St. station, and Butler St. station in Brooklyn, reached by cars from the ferries. Fare from Brooklyn, 25c.; excursion tickets, 40c. (3) *West Brighton Beach* is reached by steamboat direct to the Iron Pier from Pier No. 1, North River, at Battery, and foot of W. 23d St.; also by boat from east end of Battery to Bay Ridge, and thence by the New York and Sea Beach R. R. Fare, 35c.; excursion, 50c. From Brooklyn *via* Prospect Park & Coney Island R. R. (depot cor. 20th St. and 9th Ave., reached by horse-cars from ferries); also by horse-cars from Fulton Ferry. Round-trip ticket, 25c. (4) From Brooklyn by the Brooklyn, Bath & West End R. R. from main entrance to Greenwood Cemetery; also *via* Prospect Park & Coney Island R. R. (depot cor. 9th Ave. & 20th St.).

Rockaway Beach has been hitherto a popular but not fashionable resort, and the scene on the beach is like that at *West Brighton.* The best hotel is the *East End Hotel*, situated on the inlet, where good accommodation can be had. The vast hotel that stood so long unopen has been pulled down, and the grounds transformed into a seaside resort, like Asbury Park. The beach is reached by railway from Long Island City (34th St. ferry; round-trip ticket, 50c.), and from Brooklyn by the Long Island R. R. (depot cor. Flatbush and Atlantic Aves.), and by the New York & Rockaway Beach R. R. from Flatbush Ave. The favorite way of reaching it, however, is by the large steamers from New York and Brooklyn. By this route a delightful sail of 20 miles each way is obtained (round-trip ticket, 50c.). A colossal tubular *Iron Pier*, 1,200 ft. long, affords safe landing for steamboats.

Long Beach is a favorite point for summer excursions; it has an excellent hotel, numerous cottages, good bathing, music, etc., and is a quiet family resort, attracting a high class of visitors. It is reached by railway from Long Island City (fare, 70c., round-trip).

Long Branch.

There are three routes from New York to Long Branch : (1) An all-rail route by the New York & Long Branch R. R., operated in common by the Central R. R. of New Jersey, and the Pennsylvania R. R., foot of Liberty St. (fare, $1; round trip, $1.50). (2) By steamer leaving Pier 8 North River, 4 times daily in summer, to Sandy Hook (20 miles), and thence *via* Sandy Hook Div. of Central R. R. of New Jersey (11 miles). Time, under 2 hours ; fare, $1. From Philadelphia Long Branch is reached *via* the Pennsylvania R. R. (Amboy Div.) ; distance, 79 miles ; fare, $2.25.

Hotels.—The *West End*, located at the W. end of the Beach, is very popular. The *Elberon*, at the W. end of the beach, is also a very large hotel. *Howland's Hotel*, N. of the West End ; the *Ocean House* near the R. R. depot ; the *Hotel Brighton* on the site of the old "Metropolitan"; and the *Hollywood*, open all the year round, are the larger hotels. The charges are from $3 to $5 per day. These are the principal hotels, and are provided with ballrooms, billiard-rooms, brass and string bands, bowling-alleys, shooting-galleries, and the like. Good hotels on a smaller scale are the *Atlantic, Iauch's, Scarboro, United States*, and others. Boarding-houses charge $10 to $18 per week.

Long Branch, the other great summer resort in the vicinity of New York, is situated on the Jersey shore of the Atlantic, where a long beach affords admirable facilities for bathing. The old village of Long Branch lies back from the shore about a mile, but the great summer hotels and cottages occupy a broad plateau 20 ft. above the sea. *Ocean Avenue*, on which are the leading hotels, runs directly along the bluff, beneath which is the beach. The regular time for bathing is near high tide, when white flags are displayed over the hotels, and boats are stationed outside the surf-line to aid persons who get into too deep water. The finest private cottages lie S. of the West End Hotel. The *Monmouth Park Race-Course* is 4 miles from Long Branch, on the line of the New York & Long Branch R. R. The *Iron Pier*, running from shore to deep water, is worth attention. The drives in the vicinity of Long Branch are very attractive. One excellent road extends S. to old

Long Branch, Oceanport, and Red Bank (8 miles), and another leads to Atlanticville, Seabright, and the Highlands (8 miles). At *Elberon*, a continuation of Long Branch on the south, is the the Francklyn cottage where President Garfield died. The Elberon Casino is worthy of note. Here also is the cottage where President Grant formerly passed his summer. *Deal* is a quaint old village on the shore, 5 miles S. of Long Branch, and near by is the great Methodist camp-meeting ground of **Ocean Grove.** Beyond is **Asbury Park,** containing several hundred cottages and numerous hotels, of which the *Sheldon House* and *Arlington* at Ocean Grove are the principal. *Shark River*, just S. of Deal, is a favorite resort for picnickers from Long Branch, and is noted for its oysters and crabs. **Pleasure Bay,** on the Shrewsbury River, about a mile N. of the Branch, is another favorite picnic resort, also famous for its oysters. Here are several hotels, and yachts and boats may be hired.

The **Highlands of Navesink** are a series of bold and picturesque bluffs on the Shrewsbury River, extending S. E. from Sandy Hook Bay, which are passed on the way to Long Branch. The highest point, Mt. Mitchell, is 282 ft. above the sea-level, and from its summit extensive views may be obtained. These highlands are usually the first land seen on approaching New York from the ocean, and the last to sink below the horizon on leaving. There are two lighthouses about 100 ft. apart on Beacon Hill, at the mouth of the Shrewsbury; the southern one, a revolving "Fresnel," 248 ft. above the water, being one of the most powerful on the Atlantic coast. On the river, a short distance from Beacon Hill, is the little village of *Highlands*, an attractive resort, with fine bathing and fishing, and pleasing scenery. The Red Bank boat from New York touches at Highlands daily, and it is also reached *via* Sandy Hook Div. of the Central R. R. of New Jersey. **Red Bank** is a remarkably pretty town of 4,145 inhabitants, at the head of navigation on Shrewsbury River. It possesses among other attractions sailing, fishing, and bathing, and being only 8 miles from Long Branch by an excellent driveway (9 by railway), many summer visitors, who wish to be within easy reach of that fashionable resort yet away from its excitement, pass the season here. A branch of the Central R. R. of New Jersey runs from the *Atlantic Highlands* to Raritan Bay, opening the country intervening to travel, and is a convenient route for reaching Red Bank. Red Bank is reached from New York by the New York & Long Branch R. R. (fare, $1; excursion ticket, $1.50); also by steamer from Pier 35, North River (fare, 50c.).

2. Brooklyn.

Brooklyn is conveniently reached from the City Hall Park by the *East River Bridge*, either on foot or by cable-cars to Sands, near Fulton St., where the street-car lines converge.

Ferries.—The principal lines are as follows: Fulton St. to Fulton St.; Whitehall St. to Atlantic Ave., to Hamilton Ave.; and 39th St., S. B.; and from Wall St. to Montague St. There is a line of "Annex" boats from Fulton St., Brooklyn, to Jersey City, connecting with the Pennsylvania R. R., the Lehigh Valley R. R.. and *Fall River* steamers, and during the summer with *Albany* boats.

Other ferries from Brooklyn (E. D.), are : Grand St. to Grand and Houston Sts.; Broadway to Grand, Roosevelt, and 23d Sts.; Main St. to Catherine St.; Long Island City to 34th St.; and Greenpoint Ave. to 10th and 23d Sts.

Hotels.—The *Pierrepont*, cor. Montague and Hicks Sts.; the *St. George*, Clark and Hicks Sts.; the *Clarendon*, cor. Washington and Johnson Sts.; and the *Mansion House* (family hotel), Hicks St., are the only first-class houses.

Restaurants.—The leading ones are : *Hubel's*, 301 Washington St.; *Clarendon*, Washington St. and Johnson ; and *Marisi's*, in Clinton St.

Clubs.—The leading clubs are the *Union League*, Bedford Ave. cor. Dean St. ; *Montauk*, 8th Ave. and Lincoln Pl. ; the *Hamilton*, cor. Clinton and Remsen Sts.; the *Brooklyn*, cor. Pierrepont and Clinton Sts.; the *Lincoln*, 65—7 Putnam Ave ; the *Oxford*, cor. S. Oxford St. and Lafayette Ave.; the *Crescent Athletic Club*, with its home at 166 Montague St., and grounds on the Shore Road between 83d and 85th Sts.; and the *Hanover*, in the Eastern district, cor. Bedford Ave. and Rodney St.

Modes of Conveyance.—*Brooklyn Elevated Railway.* The *Broadway Line* runs from Broadway Ferry to Van Sicklen Ave., a distance of 4·80 miles in 20 minutes. The stations are : Ferry, Driggs St., Marcy Ave., Hewes St., Lorimer St., Flushing Ave., Park Ave., Stuyvesant Ave. (Myrtle Ave.), Kosciusko St. (De Kalb Ave.), Gates Ave., Halsey St., Chauncey St., Manhattan Beach R. R., Alabama Ave., Van Sicklen Ave. Last train leaves ferry at 12.34 A. M. Passengers are transferred to and from Broadway and main-line trains at Gates Ave.. either way. Connection is made with the Long Island R. R. at Manhattan Junction, East New York, and with the Canarsie & Rockaway Beach R. R. at Alabama Ave. The *Brooklyn Bridge, Grand, and Lexington Ave. Line* runs from Brooklyn Bridge to Van Sicklen Ave., a distance of 6·41 miles, in 30 minutes. The stations are : Sands and Washington Sts., Myrtle Ave. and Adams St., Bridge St., Navy St., Vanderbilt Ave., Washington Ave., Myrtle and Grand Aves., De Kalb Ave., Greene Ave., Franklin Ave., Nostrand Ave., Tompkins Ave., Sumner Ave., Reid Ave. Gates Ave., Halsey St., Chauncey St., Manhattan Beach R. R., Alabama Ave., Van Sicklen Ave. Trains run at intervals of 30 minutes during the night. The *5th Ave. Line* runs from the Brooklyn Bridge to 5th Ave. and 36th St., a distance of 4·25 miles, in 20 minutes. The stations are : Brooklyn Bridge, City Hall, Bridge St., Fulton St., Flatbush and Atlantic Aves. (L. I R. R. station), St. Mark's Pl., Union St., 3d St., 9th St., 16th St., 20th St., 25th St., 36th St. Last train leaves the bridge at 1.25 A. M. The *Fulton Ferry and Myrtle Ave. Line* runs from Fulton ferry to Wyckoff Ave., Ridgewood, a distance of 4·91 miles, in 24 minutes. The stations are : Fulton Ferry, Washington and York Sts., Bridge and York Sts., on Myrtle Ave., Navy St., Vanderbilt Ave., Washington Ave., Myrtle and Grand Aves , Franklin Ave., Nostrand Ave., Tompkins Ave., Sumner Ave., Broadway and Myrtle Ave., Evergreen Ave., De Kalb Ave., Knickerbocker Ave., Wyckoff Ave. Last train leaves ferry at 12.48 A. M. The *Fulton Elevated Railway*, with its road under construction, is built and operated under lease to the Kings County Elevated Railway Company as far as Schenck Ave. Its route is by way of Fulton and Sackman Sts. to Jamaica Ave., to Williams Pl., to Snediker Ave., to Eastern Parkway, to Market St., to Liberty Ave., to city line. The length completed is 1·5 miles, and the stations are Manhattan Crossing, Atlantic Ave., Eastern Park, Pennsylvania Ave., and Van Sicklen Ave. The *Kings County Elevated Railway* runs from Fulton Ferry and Brooklyn Bridge to Montauk Ave., a distance of 7¼ miles, in 30 minutes. The stations are : Fulton Ferry, Brooklyn Bridge, Clark St., Tillary St., Court St., Myrtle Ave., Boerum Pl., Elm Pl., Duffield St., Flatbush Ave., Lafayette Ave., Cumberland St , Vanderbilt Ave., Grand Ave., Franklin Ave., Nostrand Ave., Brooklyn Ave., Tompkins Ave., Albany Ave., Sumner Ave., Utica Ave., Ralph Ave., Saratoga Ave., Rockaway Ave., Manhattan Crossing, Atlantic Ave., Eastern Park, Pennsylvania Ave., Van Sicklen Ave., Linwood St., and Montauk Ave. Up stations only are Clark St., Court St., Elm Pl., and Brooklyn Ave. ; down stations only are Sumner Ave., Tompkins Ave., Duffield St., Myrtle Ave., and Tillary St. Trains for bridge and ferry display no signals ; for bridge only, white signals ; for ferry only, green signals. Trains on this line run continuously day and night, and from 12.30 A. M to 5 A. M. every 45 minutes. Fare, 5c.

Street-car Lines connect with Atlantic, Fulton, Broadway, and other ferries, and with the Bridge, and there are numerous cross-town roads. Fare, 5c.

Surface Railways (steam) are: *Brooklyn & Brighton Beach R. R.* (station at Franklin and Atlantic Ave.), to Brighton Beach; *Long Island Ry.* (same place, main station at Flatbush Ave.), to Long Island City and Sag Harbor, and intermediate points, and to Manhattan Beach; *Prospect Park & Coney Island Ry.* (station, 9th Ave. and 20th St.), to West Brighton Beach; *New York & Sea Beach R. R.* (station, 3d Ave. and 65th St.), to West Brighton Beach; *Brooklyn, Bath & West End Ry.* (Fifth Ave. and 25th St.), to West End, Coney Island.

Hackney-coaches and Cabs have stands at Fulton Ferry, the Bridge (Sands St.), and the City Hall. Each carriage contains a card of the legal rates, but the charge should be settled before starting.

BROOKLYN, the fourth largest city in the United States, lies just across East River from New York, at the W. end of Long Island. Its extreme length from N. to S. is 7¾ miles, and its average breadth 3½, embracing an area of 20·84 square miles. The surface is elevated and diversified. Brooklyn was settled in 1623, near Wallabout Bay, by a band of Walloons, and during the Revolutionary War was the scene of events that give great interest to some of its localities. On the Heights back of the city the battle of Long Island was fought (Aug. 26, 1776), and the Americans defeated with a loss of 2,000 out of 5,000 men. The population of Brooklyn was 3,298 in 1800, 566,689 in 1880, and 806,343 in 1890. The main business street is **Fulton Street,** from Fulton Ferry to East New York (5½ miles). *Atlantic Ave.* runs nearly parallel with Fulton St. from South Ferry to East New York; it is an active business street in its lower part, and from Flatbush Ave. to East New York is occupied by the tracks of the "rapid transit" railroad. *** Clinton Ave.** is the handsomest street in the city, being embowered with trees and lined with fine residences surrounded by ornamental grounds. *St. Mark's Place* is scarcely less attractive. Remsen and Montague Sts., on the Heights, contain many fine residences; and from ** Montague Terrace,* on the latter, is obtained a magnificent view of New York city and harbor. The favorite drive is through Prospect Park and along the *** Ocean Parkway,** a splendid boulevard 210 ft. wide, extending from the S. W. cor. of the Park to the sea-shore at Coney Island (5½ miles). The *Eastern Parkway,* also a popular drive, extends from the Park entrance to East New York (2¼ miles). Still another attractive drive is to Bay Ridge and Fort Hamilton.

The *City Hall* (reached from Fulton Ferry *via* Fulton St. in ½ mile) is within easy walking distance of nearly all the public buildings in Brooklyn that are worth attention. It is of white marble in the Ionic style, surmounted by a belfry with a four-dial clock, and stands in an open square. Just E. of the Hall, fronting toward Fulton St., is the *** County Court-House,** a large building with white-marble front, with a Corinthian portico, and an iron dome 104 ft. high. On the W. of the Court-House stands the *Municipal Building,* of marble, 4 stories high, with a tower at each of the 4 corners, and on the E. the *Hall of Records.* On the plaza in front of the City Hall is J. Q. A. Ward's statue of Henry Ward Beecher. A short distance W. of the City Hall, cor. Clinton and Pierrepont Sts., is the *Long Island Historical Society* building, containing a valuable reference library of 50,000 volumes and many curious relics (free). In Fulton St., opposite the City Hall, is the

Park Theatre. Other theatres are the *Columbia Theatre* (cor. Tillary and Washington Sts.), *Grand Opera-House* (Elm Place, near Fulton St.), *Star Theatre* (Jay St., near Fulton), *Hyde and Behman's Theatre* (Adams St., near Myrtle Ave.), *Amphion Academy* (Bedford Ave., near Division Ave., E. D.) *Lee Avenue Academy* (Lee Ave., near Division Ave., E. D.), and the *Bedford Avenue Theatre* (South 6th St., near Broadway, E. D.). The *Post-Office*, at the intersection of Washington, Adams, and Johnson Sts., is the finest public building in the city, built of granite and iron, at a cost of $5,000,000. In Montague St., W. of the City Hall, is the *Academy of Music*, a brick building of slight architectural merit, but with fine interior decorations. Adjoining it is the **Art Association** building, with highly ornate front. Opposite is the **Brooklyn Library,** a handsome structure in the Gothic style, containing 100,000 volumes and two reading-rooms. At the cor. of Clinton St. is the beautiful * **Church of the Holy Trinity** (Episcopal), in the decorated Gothic style, with stained windows, and a spire 275 ft. high. To the left, down Clinton St. (cor. Livingston), is the church of *St. Ann* (Episcopal), in the pointed Gothic style, with ornate interior. To the right, in Pierrepont St., is the *Dutch Reformed Church*, of brown-stone in the Roman Corinthian style, with a Corinthian portico, and a rich interior. Near by (cor. Pierrepont St. and Monroe Place) is the Unitarian *Church of the Saviour*, a structure in the pointed Gothic style. Other noteworthy churches in this vicinity are *Grace* (Episcopal), cor. Grace Court and Hicks St. ; *Christ* (Episcopal), cor. Clinton and Harrison Sts.; and the *Church of the Pilgrims* (Congregational), cor. Remsen and Henry Sts. *Plymouth Church* (Lyman Abbott) is a large, plain building in Orange St. near Hicks. Other churches are the *Lafayette Ave. Presbyterian*, the *Clinton Ave. Congregational*, the "*New*" *Church*, Monroe Place, and the *Tabernacle* (or Dr. Talmage's church), cor. Clinton and Greene Aves. The church of **St. Charles Borromeo* (R. C.), in Sidney Place, is famous for its music. *Pratt Institute*, on Ryerson St., between De Kalb and Willoughby Aves., founded by Charles Pratt, comprises classes for thorough instruction in trades and useful arts, for both sexes. It has a large library, maintains lecture courses, and is one of the most complete and extensive institutions of its kind in this country. The *Brooklyn Polytechnic Institution*, on Livingston St., directly S. of the City Hall, is the only college in Brooklyn. The *Long Island College Hospital* has a large and imposing building, in extensive grounds, in Henry St., near Pacific. The *County Jail*, in Raymond St., is a castellated Gothic edifice of red sandstone ; the *Penitentiary* is an immense stone pile in Nostrand Ave. near the city limits. The *Young Men's Christian Association* has a fine building in Fulton St., cor. Bond St., with library (21,000 vols.) and reading-room (free).

In crossing Fulton Ferry to or from New York the massive towers and ponderous cables of the * **East River Bridge** are conspicuous objects. This bridge has already been described in section on New York City. The *United States Navy-Yard* (reached by horse-cars from Fulton Ferry), on the S. shore of Wallabout Bay, is the chief naval station of the Republic. It contains 45 acres, inclosed by a high brick wall, within which are numerous foundries, workshops, and storehouses.

Representative vessels of every kind used in the U. S. navy may usually be seen at the Yard. The * **Atlantic Dock**, at the south end of the city, a mile below South Ferry, has a basin which covers an area of 42½ acres, and surrounding it are piers of solid granite, on which are spacious warehouses.

* **Prospect Park** (reached by several lines of cars from Fulton or Broadway Ferries) is one of the most beautiful in America. It contains 516⅛ acres, is situated on an elevated ridge, and commands magnificent views of the two cities, of the inner and outer harbor, Long Island, the Jersey shore, and the Atlantic. It is beautifully shaded in many parts by old woods which have been skillfully improved, and its combination of broad meadows, grassy slopes, and wooded hills, is unequaled elsewhere. It contains 8 miles of drives, 4 miles of bridle-paths, and 11 miles of walks. The main entrance on Flatbush Ave., known as the Plaza, is paved with stone and bordered by grassy mounds; in the center are a fine fountain and a bronze statue of President Lincoln. A memorial arch to the soldiers and sailors stands at the entrance to the Park. Park carriages, starting from the entrance, make the circuit of the leading points of interest (fare, 25c.). *Washington Park* (30 acres) is an elevated plateau ⅓ mile E. of City Hall, between Myrtle and De Kalb Aves., commanding extensive views. During the Revolutionary War it was the site of extensive fortifications, of which Fort Greene was the principal.

* **Greenwood Cemetery** (reached by cars from Fulton, Wall, and Hamilton Ferries, and by 5th Ave. Branch of Union Elevated R. R.) is situated on Gowanus Heights in the S. portion of the city. It contains 474 acres, and more than 270,000 interments have been made in it since its opening in 1843. The main entrance, near 5th Ave. and 23d St., is an elegant monumental structure in the pointed Gothic style, ornamented with sculptures representing scenes from the Gospels; and the later entrance on the E. side is of scarcely inferior beauty. The grounds have a varied surface of hill, valley, and plain, and are traversed by 19 miles of carriage-roads and 17 miles of footpaths. The elevations afford extensive views. There are many beautiful monuments, chief among which are the Pilots' and Firemen's, Charlotte Canda's, and that to the "mad poet" McDonald Clark. By keeping in the main avenue called *The Tour*, as indicated by finger-posts, visitors will obtain the best general view of the cemetery, and will be able to regain the entrance without difficulty. About 4 miles E. of Greenwood are the cemeteries of *The Evergreens*, covering 360 acres, and *Cypress Hills*, 340 acres.

3. New York to Philadelphia.

a. Via Pennsylvania R. R., 90 miles. Limited Express of parlor-cars, 10 A. M. There are 23 other trains on week days and 16 on Sundays. Time, 1¾ to 2¾ hours. Fare, $2.50. Round-trip ticket, $4.

FERRY-BOATS run from foot of Desbrosses and Cortlandt Sts. every ten minutes during the day; also from Fulton St., Brooklyn (Pennsylvania "Annex" boat), every half-hour, to Jersey City depot. **Jer-**

sey City is on the Hudson River, opposite New York, of which it is practically a portion. It is a place of much commercial and industrial activity, is agreeably situated and well built, and had in 1890 a population of 163,003; but, except for the fact that it contains the stations of several of the most important railways leading south and west from New York, and the docks of leading transatlantic steamers, it possesses no interest for the tourist. The route after leaving Jersey City is across broad meadows to **Newark** (*Continental* and *Park*) (9 miles), a large manufacturing city with 181,830 inhabitants, but, like its rival Jersey City, offering little of interest to the tourist. The city is on an elevated plain upon the right bank of the Passaic River, 4 miles from Newark Bay, and is regularly laid out in wide streets crossing each other at right angles. Broad St. is the main business thoroughfare, and runs N. and S. through the heart of the city. The principal E. and W. street is Market St., on which are some of the finest buildings, including the *Court-House*, an imposing stone edifice in the Egyptian style. Other noteworthy public buildings are the *City Hall* (cor. Broad and William Sts.), the *Custom-House and Post-Office* (cor. Broad and Academy), and many handsome churches. The building of the *Mutual Benefit Life Ins. Co.* is said to be the finest in the State. Of the literary institutions the most noteworthy are the *Library Association* (20,000 volumes), the *State Historical Society*, and the *Newark Academy*. From the grounds of the latter (on High St.) an extensive view of the Passaic Valley is had. Newark is distinguished for its manufactures of jewelry, carriages, paper, and leather; and its lager-beer is excellent.

Six miles beyond Newark is **Elizabeth** (*Sheridan House*), one of the attractive cities in New Jersey, with 37,764 inhabitants, and many fine residences, a few of which are visible from the cars. **New Brunswick** (*New Brunswick*) at the head of navigation on Raritan River (32 miles from New York), has a population of some 18,603, with extensive manufactures of India-rubber, harness, and hosiery. There are fine residences in the upper part of the city, but the "institution" of New Brunswick is *Rutgers College*, an old, richly-endowed, and flourishing establishment. The buildings are visible from the cars. *Princeton Junction* (48 miles) is 2½ miles from Princeton, noted as the seat of **Princeton College** (*Nassau Hotel*). The college buildings (especially the Library, Nassau Hall, and Dickinson Hall) are remarkably fine, and stand in a green and shady campus. A branch line conveys passengers from the junction to the town. **Trenton** (*American, National, Windsor*) (58 miles) is the capital of New Jersey, and is pleasantly situated at the head of navigation on the Delaware. It had in 1890 a population of 57,458, with important manufacturing interests (chief among which are the potteries), and is a remarkably well-built, cleanly, and attractive town. State St. is the principal thoroughfare, and next to this is Greene St., which crosses State at right angles. The leading event in the past history of Trenton is the famous victory over the Hessians won by Washington, Dec. 26, 1776; and its chief present attractions are the public buildings. The *State-House* (in State St.) is a stone structure, beautifully situated on the Delaware, and overlooking the river and vicinity. The *Post-Office*, also in State St., is a massive stone building in the Renaissance style; and the vast *State Penitentiary* (in Federal St.), the *State Arsenal* (near the Penitentiary), and the *State Lunatic Asylum* (1½ mile N. of the city) are all worth visiting. The only place between Trenton and Philadelphia requiring mention is *Bristol* (67 miles), a pretty town of 6,553 inhabitants, on the Delaware nearly opposite Burlington.

b. Via " Bound Brook Route," 88 miles.   Time, 1¾ to 2¾ hours.
Drawing-room cars with all trains.

The depot in Jersey City, a splendid structure of iron, is reached by ferry from foot of Liberty St.   The country along this route is very similar in character to that along the preceding route, but there are fewer large towns and a scantier population.   Highly cultivated farms and smiling orchards stretch away on every side, and the prospect in summer is very pleasing.   The first station that will attract the attention of the traveler is *Elizabeth* (13 miles), which has been described on page 30, and the only important town on the route is **Plainfield** (24 miles), containing some 11,267 inhabitants, and pleasantly situated near the foot of Orange Mountain.   Washington's Rock (seen from the train on the right) is on the mountain 2 miles W. of Plainfield, and is noted as the place whence Washington watched the movements of the enemy during the campaign in this vicinity.   At *Bound Brook* (31 miles) the Americans were defeated in 1777 by Lord Cornwallis.   A short branch road diverges from the regular route and runs to Trenton.

c. Via Amboy Division of the Penn. R. R., 92 miles.   Time, 5 hours.

This route was formerly much used, but with the increased facilities for rapid transit, it is now seldom taken, except by tourists anxious to see the country.

From pier 6 North River, foot of Rector St., a steamer runs daily, at 2.30 P. M., to South Amboy (27 miles), and the sail past the shores of Staten Island, and up the Raritan River, is very pleasant, particularly in summer.   *South Amboy* is situated on Raritan Bay, at the mouth of Raritan River, across which is **Perth Amboy** (*Hotel Central*), a port of entry, and one of the oldest cities in New Jersey, much frequented in summer.   At South Amboy the cars are taken, and the route leads through a barren and uninteresting country to the Delaware River at Bordentown (64 miles).   **Bordentown** (*Bordentown House*) is a flourishing town of 4,232 inhabitants, situated on the E. bank of the Delaware, with extensive foundries and machine-shops, and the terminal basins of the Delaware and Raritan Canal.   The principal object of interest is the mansion and park occupied for 26 years by Joseph Bonaparte, once King of Spain.   **Burlington** (73 miles) is a city of 7,264 inhabitants on the Delaware, 19 miles above Philadelphia, whence it is much visited in summer by steamboat.   *Burlington College* (Episcopal) is located here, and there are handsome churches and school-buildings. **Camden** (*Hobkirk Inn, West Jersey House ; 92 miles*) is a flourishing

city of 58,313 inhabitants on the Delaware opposite Philadelphia, with
which it is connected by 4 ferries. It is the terminus also of the *West
Jersey* and *Camden & Atlantic* and the *Philadelphia & Reading Railways ;* and there are extensive ship-yards, besides manufactures of iron,
glass, chemicals, etc.

4. Philadelphia.

Hotels.—The leading hotels on the American plan are the *Continental*, cor.
Chestnut and 9th Sts.; the *Girard House*, on Chestnut St., at the corner of 9th
St.; the *Colonnade*, cor. Chestnut and 15th Sts.; the *Stratford*, cor. Broad and
Walnut Sts.; the *Aldine*, 1914 Chestnut St.; the *Bellevue*, cor. of Walnut and S.
Broad; the *Bingham*, cor. 11th and Market; *Guy's Hotel*, 7th and Chestnut
Sts.; and the *Windsor Hotel.* 1219 Filbert St., are much patronized. The *Hôtel
Lafayette*, 108 S. Broad St.; the *Stratford*, cor. Broad and Walnut; and *Green's*;
Chestnut and 8th Sts., are conducted on both the American and European plans.
The *Stenton*, cor. Broad and Spruce Sts., is on the European plan. The rates
on the American plan are from $2.50 to $5 a day ; on the European plan the
charges are $1 to $3 a day for rooms.
    Restaurants.—The *Bellevue*, cor. of Broad and Walnut ; the *Colonade*, cor.
Chestnut and 15th Sts.; *Dooner's*, in 10th St., north of Chestnut ; *Green's*, cor.
Chestnut and 8th Sts.; *Boldt's*, on 4th St. below Chestnut ; and *Vendig's*, cor.
12th and Market Sts., are first class. *Partridge's*, 19 S. 8th St. and 15 N. 8th
St., and *Cabadi's*, 5 S. 8th St., are much frequented by ladies. Other first-class
restaurants are those attached to the hotels on the European plan.
    Modes of Conveyance.—The streets are traversed by *horse* and *cable-
cars* between the Delaware and 42d St., on Market and Chestnut Sts.; by *cable-
cars* from Columbia Ave. entrance to Fairmount Park, to Delaware Ave. and
Market Sts., and N. on 9th St., and S. on 7th St.; and by *horse-cars* on almost
every principal street, north, east, south, and west, to any point of interest in
the city. The fare is 5c., and points on any connecting line may be reached by
transfer. There is also a line of *omnibuses* on Broad St., conducted on the English plan. *Carriages* are found at all the depots, and at various stands. The
fares are regulated by law, and a card containing them should be in every carriage. There are also *Hansom cabs*, carrying two people, rate 65c. per hour.
In case of dispute, call a policeman, or apply at the Mayor's office.
    Railroad Stations.—The station of the *Pennsylvania R. R.* for Wilming-
ton and Baltimore, New York and Pittsburg, and of the *Schuylkill Valley, R. R.*
(division of the same), is at Broad and Market Sts.; of the *Amboy Division* (for
New York) by ferry from foot of Market St. to Camden ; of the *Philadelphia
& Reading R. R.* (main line), and of its *North Pennsylvania Division*, cor. 12th
and Market Sts.; of the *West Chester & Philadelphia*, cor. Broad and Market
Sts.; of the *Bound Brook Route* (Philadelphia & Reading) to New York, and
the *Germantown & Norristown*, cor. 12th and Market Sts.; of the *Baltimore &
Ohio* at Chestnut St. Bridge. The *Camden & Atlantic* is reached by ferry from
foot of Vine St.; the *West Jersey*, by ferry from foot of Market St.; the *Phila-
delphia & Atlantic City*, by ferry from foot of Walnut St.
    Ferries.—To *Camden* (fare 3c.) from foot of Market St., Vine St., Walnut
St., South St., in the lower part of the city, and from Shackamaxon St. in
Kensington. To *Gloucester, N. J.*, from foot of South St. (fare 10c.).
    Churches.—Among the 550 churches the following are those most vis-
ited : The *Cathedral of St. Peter and St. Paul* (Roman Catholic), in Logan
Square, 18th St.; *St. John's*, 13th St. above Chestnut, noted for its fine music ;
St. Peter's (Episcopal), 3d and Pine Sts., a relic of the early days of the city ;
the *Holy Trinity* (Episcopal), cor. 19th and Walnut; *St. Stephen's* (Episcopal), in 10th St. near Market, with the beautiful Burd monument ; *St. Andrew's* (Episcopal), in 8th St. near Spruce ; the *First Baptist*, cor. Broad and
Arch Sts.; the *West Arch St. Presbyterian*, in Arch St.; *Bethany* (Presbyterian), 22d and Bainbridge Sts., with the largest Sunday-school in the country ;
the *Second Presbyterian*, cor. 21st and Walnut; the *Washington Square Presbyterian ;* the *Arch St. Methodist*, cor. Broad and Arch Sts.; the *Lutheran
Church*, cor. Broad and Arch ; the *Gloria Dei* (Old Swede), Swanson below
Christian St. (1698); *Christ Church*, 2d above Market St. (1695); and the Jewish

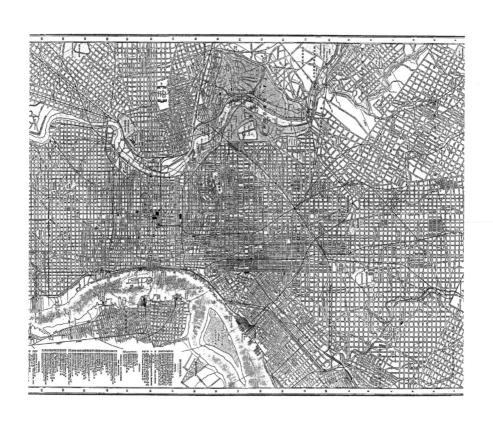

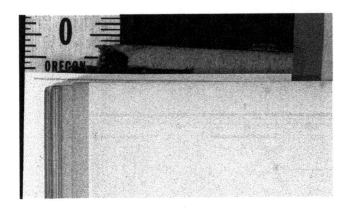

Synagogue, in Broad St. near Green. Very fine new churches on N. Broad St. are the *Universalist Church of the Messiah*, cor. Montgomery Ave.; the *Memorial Baptist* and *Grace Methodist*, cor. Master St. Among the Friends' meeting-houses those at the cor. of Arch and 4th and Race and 15th Sts. are best worth a visit. The *Tabernacle* (Presbyterian), 37th and Chestnut Sts., *St. James* (Catholic), at 38th and Chestnut Sts., and *Christ Memorial*, 43d and Chestnut Sts., are handsome structures.

Theatres and Amusements.—The *Academy of Music*, cor. of Broad and Locust Sts., is one of the largest houses in America, with sittings for 3,000 persons. It is used for operas, concerts, lectures, balls, etc. The *Opera-House* is at the cor. of Broad St. and Montgomery Av. The *Arch St. Theatre* (Mrs. John Drew's) is in Arch St., near 6th. The *Chestnut St. Theatre* is in Chestnut St. above 12th; Chestnut St. *Opera-House*, Chestnut above 10th; *Broad St. Theatre* is in Broad St. near Locust; *Eleventh St. Opera-House* in 11th St. above Chestnut; *National*, 10th and Callowhill Sts.; *Lyceum*, Race below 8th St.; *Forepaugh's*, 8th below Vine St. Other theatres are the *Central*, in Walnut St. above 8th; *Bijou*, in 8th above Race; *New Park*, cor. Broad and Fairmount Ave.; *Kensington*, Franklin Ave. and Morris St.; *Germania*, 532 N. 3d; *Empire*, Broad and Locust Sts.; *Girard Ave.*, Girard Ave., above 15th St.; *People's*, Kensington Ave. and Cumberland St.; *Keller's Egyptian Hall*, Chestnut, above 12th St.; and *Standard*, in South St. above 11th. Musical entertainments are given at the *Drexel Institute*, Chestnut and 32d Sts.; *Musical Fund Hall*, in Locust St. below 9th; at *Association Hall*, cor. Chestnut and 15th; at *St. George's Hall*, cor. Arch and 13th Sts.; and *Industrial Hall*, Broad and Wood Sts. *Horticultural Hall*, cor. Broad and Locust Sts., is the scene of the annual floral displays of the Horticultural Society.

Reading-Rooms.—The *Mercantile Library*, in 10th St. near Chestnut, contains 165,000 volumes and a well-supplied reading-room (open from 9 A. M. to 10 P. M.). The *Philadelphia Library*, in Locust St. east of Broad, 155,000 volumes, is free from 10 o'clock till sunset; and the "Ridgway Branch," with two reading-rooms, is at the cor. of Broad and Christian. The *Philosophical Society* (founded by Franklin in 1743) is located in its original building in 5th St. below Chestnut, with a library of 60,000 volumes. The *Academy of Natural Sciences*, in Logan Square, has a valuable collection of books and specimens (admission from 9 A. M. to 5 P. M., daily). The *Athenæum*, cor. 6th and Adelphi Sts., has a library of 25,000 volumes (introduction by a member). The *Apprentices' Library*, at the cor. of 5th and Arch Sts., is also open daily. The *Young Men's Christian Association*, cor. 15th and Chestnut Sts., has a free reading-room (open from 9 A. M. to 10 P. M.). The *Historical Society of Pennsylvania*, 13th and Locust Sts., has a rich library (open from 10 A. M. to 5 P. M.). The *Franklin Institute*, in 7th St. above Chestnut, has a free library, rich in scientific and technical books, and patent reports, and reading-room.

Art Collections.—At the *Academy of Fine Arts*, cor. Broad and Cherry Sts., is one of the best collections of paintings, statuary, casts, and prints in America (entrance 25c., free on Sundays and Mondays). Fine pictures may usually be seen (free) at the sales-galleries of *Earle*, 816 Chestnut St., and *Haseltine*, 1516 Chestnut St. Among the richest private collections in the country are those of Henry C. Gibson (1612 Walnut St.), W. B. Bement (1812 Spring Garden St.), the late James L. Claghorn (on W. Logan Square), the late Joseph Harrison, Jr. (in Rittenhouse Square), Artemas Partridge, and A. E. Borie. Admission to these may be obtained by application.

Clubs.—The *Philadelphia* (Walnut and 13th Sts.) is the oldest in the city. The *Union League Club* has a handsome building cor. Broad and Sansom Sts. The *Manufacturers'* occupies an imposing home in Walnut St. above Broad, and the *Columbia*, in Broad above Master, has one of the finest houses in the city. A member's introduction will secure the visitor the privileges of the Club for one month. The *Penn* (8th and Locust Sts.), the *University* (1316 Walnut St.), and the *Catholic Club*, in Broad St. below Walnut, are literary and social. The *New Century Club* for women, occupies its own house on 12th St., near Walnut. The *Union Republican*, at 11th and Chestnut Sts., and the *Americus*, at Broad and Chestnut Sts., are the principal political associations. The *Rittenhouse Club* occupies a fine marble building in Walnut St. near 18th. The *Philadelphia Art Club* is in Broad St. above Locust. The *Athletic Club* of the Schuylkill Navy, in Arch St. below 17th, is a massive structure.

3

Post-Office.—The general Post-Office is at the cor. of Chestnut and 9th Sts., extending on 9th St. to Market. Letters may be mailed in the lamp-post boxes in all parts of the city. There are also several sub-stations in different parts of the city. The courts of the United States are held in the same building.

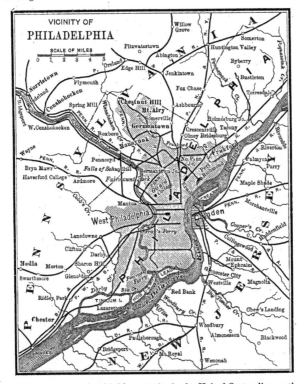

PHILADELPHIA, the third largest city in the United States, lies on the west bank of the Delaware River, 90 miles from the Atlantic Ocean. Its latitude is 39° 57′ N. and longitude 75° 10′ W. from Greenwich. It is 22 miles long from N. to S., with a breadth of 5 to 8 miles, and an area of 1,294 square miles. The city, as originally incorporated, was bounded by the rivers Delaware and Schuylkill and Vine and South Sts., and this area was not enlarged until 1854, when the corporation was extended over the the entire county. Within its present area there are over 1,000 miles of paved streets. The city is regularly laid out, the streets running N. and S. being numbered in succession from the Delaware to the Schuylkill, which is reached at 23d St., the first street on the W. side of that river being 30th. These are crossed at right angles by named streets. A few irregular avenues, formerly country-roads, stretch away from the original town-plot. The houses on the streets running E. and W. are numbered toward the W., all between 1st and 2d streets being between 100 and 200, and all between 2d and 3d streets between 200 and 300, and so on; so that the number of the house indicates the number of the street as well. Hence, if the number of the house be 836, it is between 8th St. and 9th St. In like manner, the streets running N. and S. are allowed 100 numbers for every square they are distant from Market St., either N. or S. Thus, whenever one can see a number, he can calculate his exact distance from Market St. or the Delaware. The great business thoroughfare is *Market St.;* it runs E. and W., is 100 ft. wide, and contains the principal wholesale stores. *Broad St.*, the dividing street running N. and S., is 113 ft. wide, and is lined with churches and elegant private residences. *Chestnut St.*, parallel with Market on the S., is the fashionable promenade, containing the finest hotels and retail stores. Walnut, Spruce, and Pine, farther S., and Arch, Race, and Vine, N. of Market, west of Broad, are leading streets. Third St., between Market and Walnut, Walnut St. from 2d to 7th, and Chestnut St. from 2d to Broad, are now the banking and financial centers. The principal drives are through Fairmount Park, out Broad St. toward Germantown, down Broad to Point Breeze Park, and Woodland Ave. to Suffolk Park.

Philadelphia was founded by William Penn, who came over from England in 1682, accompanied by a colony of Quakers, and purchased the site from the Indians. The emigration thither was very rapid, and in 1684 the population was estimated at 2,500. Penn presented the city with a charter in 1701. It prospered greatly, and was the most important city in the country during the colonial period and for more than a quarter of a century after the Revolution. The first Continental Congress assembled here (in 1774), as did also the subsequent Congresses during the war. The Declaration of Independence was made and issued here, July 4, 1776. The convention which formed the Constitution of the Republic assembled here in May, 1787. Here resided the first President of the United States, and here Congress continued to meet until 1797. Until 1799 it was the capital of the colony and State of Pennsylvania, and from 1790 to 1800 was the seat of the government of the United States. The city was in possession of the British from September, 1777, to June, 1778, a result of the unfortunate battles of Brandywine and Germantown. Since the Revolution the city has grown steadily and rapidly. The population, which in 1800 was 67,811, had increased to 132,376 in 1820, to 565,529 in 1860, to 846,984 in 1880, and to 1,046,964 in 1890. The commerce of Philadelphia is large and increasing, but manufactures are its chief source of wealth. In heavy manufactures Philadelphia is only approached by Pittsburg. The leading industries are the manufacture of locomotives and all kinds of iron-ware, ships, carpets, woolen and cotton goods, shoes, umbrellas, and books.

Chestnut St. begins at the Delaware River and runs W. to the city limits, crossing the Schuylkill at 24th St. by a bridge 1,528 ft. long, at the E. end of which is the depot of the *Baltimore & Ohio R. R.* In 2d St. N. of Chestnut is the *Commercial Exchange*, a large brown-

stone building standing on the site of the old "Slate-roof House," once
the residence of William Penn, and later the home of John Adams,
John Hancock, Baron De Kalb, and Benedict Arnold. Opposite are
the massive buildings used as U. S. Appraisers' Stores, extending westward to Dock St. At the corner of Walnut and 2d Sts. is the Coal Exchange. In 2d St. near Market St. is **Christ Church** (Episcopal), one
of the most venerable of the antiquarian relics of the city, begun in
1695, and still a fine building. Its steeple is 196 ft. high, and contains
the oldest (1754) bells in America. At the cor. of Market and Front Sts.
is a small brick house, now used as a tobacco-shop; it was built in 1702,
and a hundred years ago was the famous *London Coffee-House*, frequented by the magnates of the city. A few steps from this (in Leti-
tia St., S. of Market) was *Penn's Cottage*, the first brick building erected
in Philadelphia. It has been removed to Fairmount Park.

At the corner of *3d* and *Walnut Sts.* is the * **Merchants' Exchange,** a fine marble building, with an ornamented front on Dock
St., a semicircular colonnade of 8 pillars, and a spacious rotunda within
on that side. The reading-room in the rotunda of the second story is
handsomely frescoed. Opposite, in Walnut St., is the fine building of
the *North American Insurance Co.* Near by is the *Girard National
Bank*, a stately edifice with handsome portico, originally built for the
first United States Bank, and occupied by Stephen Girard until his
death. It was copied from the Dublin Exchange. To the S. (cor. 3d
and Pine Sts.) is the church of *St. Peter's* (Episcopal), begun in 1758
and finished in 1761. On the S. E. corner of Walnut and 4th Sts. is
the *Manhattan Life Insurance Co. of New York*, with a front of 50 ft.
on Walnut and 101 ft. on 4th St. In 4th St. above Walnut is the *Insurance Co. of the State of Pennsylvania*. Midway between 4th and 5th
Sts., on the S. side of Walnut, towers the building of the *Commercial
Union Assurance Co. of London*. On the E. side of 4th St. below Walnut are the main office buildings of the *Philadelphia & Reading* and
Pennsylvania Railroad Companies.

Above 3d St., in Chestnut, are the brown-stone *Bank of North America*, and the costly buildings of the *Fidelity Safe Deposit Co.*, the *First
National Bank*, the *National Bank of the Republic*, and the *Guarantee
Trust and Safe Deposit Co.* On the S. side of Chestnut St. between 3d
and 4th a narrow court leads to **Carpenters' Hall**, where assembled
(in 1774) the first Congress of the United Colonies. It is a plain two-
story brick building, carefully preserved. On the S. E. corner of 4th
and Chestnut is the *Banking House* of the Brown Brothers. On
the N. W. corner is the massive addition to the *Provident Life and
Trust Co.*, and on the S. W. corner the *Wood Building*, adjoining which
is the *Western Bank*, beside the * **U. S. Custom-House,** with imposing fronts on Chestnut and Library Sts., originally the *United States
Bank*. Just above, on the opposite side, is a fine series of commercial
buildings, including the *Provident Life and Trust Co.* (granite), on the
corner, the *Philadelphia Trust Co.*, the *Farmers and Mechanics' Bank*,
the *Philadelphia Bank* (granite), the *Pennsylvania Life Ins. Co.*, and the
Girard Building. On the S. side is the *Independence National Bank*,

Independence Hall, Philadelphia.

and on either side of it, towering above it, is the 10-storied *Drexel Building*, within which is the *Stock Exchange*, the *Board of Trade*, many offices, and on the cor. of 5th St., the bank of Drexel and Co. Between 5th and 6th Sts. stands *Independence Hall,** the most interesting object in Philadelphia. It was begun in 1729 and completed in 1735, at a cost of £5,600. In the E. room (Independence Hall proper) the Continental Congress met, and here on July 4, 1776, the Declaration of Independence was adopted, and publicly proclaimed from the steps on the same day. The room presents the same appearance now as it did at that time; the furniture is that used by Congress; there are a statue of Washington and numerous portraits and pictures, including Benjamin West's "Penn's Treaty with the Indians." The W. room is a depository of many curious Revolutionary relics. In it is preserved the old "Liberty Bell," the first bell rung in the United States after the passage of the Declaration. In Congress Hall, in the 2d story, Washington delivered his farewell address. Visitors are admitted from 9 A. M. to 4 P. M. daily. The wings and adjacent buildings are at present occupied by the law courts, and the mayor's and other municipal and county offices. On the sidewalk in front of the Hall stands Bailey's statue of Washington, erected by public-school children; and in the rear is **Independence Square,** an open space of four acres. Diagonally opposite Independence Square (on the S. W.) is **Washington Square,** celebrated for containing nearly every variety of tree that will grow in this climate. There is a map of the Square showing the position of each tree. Opposite, on the S. W. cor. of Walnut and 7th, is the *Philadelphia Saving-Fund Building* (granite), the oldest institution of the kind in the country. Fronting the Square, in 6th St., is the *Athenæum*, with a library of 25,000 volumes. In 5th St. near Chestnut is the building of the Philosophical Society, called **Philosophical Hall,** completed in 1791. The Society was founded in 1743, through the influence of Benjamin Franklin. Farther along 5th St. (at the S. E. cor. of Arch St., in Christ Church cemetery) is *Franklin's Grave*, which may be seen through iron railings in the brick wall of the cemetery.

At the cor. of Chestnut and 6th Sts. is the *Ledger Building*, of brown-stone, 5 stories high. Adjoining is the *Land, Title, and Trust Co.'s* building. The office of the *German Democrat* is at 614 Chestnut St., that of the *Press* at 700 Chestnut St., and that of the *North American* at 701 Chestnut St., while on 7th St. above Chestnut are located a number of other journals; and near by, also in 7th St., is the *Franklin Institute*, provided with a library (33,000 volumes), a reading-room, and free courses of scientific lectures. Directly opposite is the *Master Builders' Exchange*. On Chestnut St. above 7th is the massive *Singerly Building* for the Union Trust Co. and the Chestnut St. National Bank. Just above 7th in Market St. is the 6-story publishing-house of the *J. B. Lippincott Co.* On the S. W. corner of 7th and Market is the *Penn National Bank*, on the site of the house where Jefferson wrote the Declaration of Independence. To the S. from Chestnut, 8th St. leads past the **Pennsylvania Hospital,** standing in ample grounds shaded by venerable trees, and containing a medical library and anatomical museum. At 13th and Locust Sts. is the building of the *Pennsylvania Historical Society*, containing a large library, rich in local and family histories, and interesting historical relics (open from 10 A. M. to 5 P. M.). At the corner of Chestnut and 8th Sts. is the handsome *Times Building.*

At the cor. of 9th St. is the *Continental Hotel*, and opposite is the *Girard House*. At the N. W. cor. is the * **Post-Office,** in the Renaissance style, 4 stories high, with an iron dome, and costing $4,000,-000, containing also the *United States Courts* and Federal offices. Adjoining the post-office, on Chestnut St., is the imposing tower of the *Record* office, 8 stories high, and this in turn is surpassed by the tower of the *Penn Mutual Life Insurance Building*, next to which is the 8-story structure of the *City Trust, Safe Deposit, and Security Co.* At the N. W. cor. of 10th St. is the building of the *New York Mutual Life Insurance Co.* On the N. E. corner is the office of the *Inquirer*, one of the oldest daily journals, and on the S. W. corner is the *Assembly Buildings*, on the first floor of which is the *Western Union Telegraph Co's* offices. To the right, in 10th St., is the * **Mercantile Library,** with 165,000 volumes and a spacious reading-room (open from 9 A. M. to 10 P. M.). In *St. Stephen's Church* (Episcopal), opposite the library, are some fine monuments. In 10th St., S. of Chestnut, is *Jefferson Medical College*. At the corner of 12th St. is the elegant white-marble jewelry-store of *Bailey & Co.*, and at the corner of Chestnut and 13th Sts. is the retail store of *John Wanamaker*. Just above is the * **U. S. Mint,** a white-marble building in the Ionic style, with a graceful portico. The collection of coins preserved here is the largest and most valuable in America. Visitors are admitted from 9 to 12 o'clock. Adjoining the Mint is the structure of the *Girard Trust Co.* Adjacent, in Broad St., is the ornate *Betz Building*, 16 stories high. The *terminal station* of the Philadelphia & Reading R. R. and its branches is at 12th and Market Sts. ; and the great Broad St. station of the Pennsylvania R. R. and its branches is at 15th and Market Sts., facing the Public Buildings.

Crossing Broad (14th) St., with its imposing hotels and churches, Chestnut St. passes in sight of the *Public Buildings* (to the right), and the massive and spacious * building of the *Young Men's Christian Association* and the old *Church of the Epiphany*, which, with the *Colonnade Hotel*, occupy opposite corners of 15th and Chestnut Sts. The Y. M. C. A. building is of sand-stone and marble, 230 by 72 ft., 4 stories high, with a tower, and containing a library, reading-room, etc. (open from 9 A. M. to 10 P. M.). We have now entered the residence quarter. Up 18th St., to the right, is **Logan Square,** a pretty little park of 7 acres, neatly laid out and delightfully shaded. Fronting the square on the E. side is the Roman Catholic * **Cathedral of St. Peter and St. Paul,** the largest church edifice in the city. It is of red sandstone in the Roman-Corinthian style, 136 by 216 ft., with a dome 210 ft. high. The façade consists of a classic pediment, upheld by 4 lofty Corinthian columns, flanked by pilastered wings. The interior is cruciform and adorned with frescoes ; the altar-piece, by Brumidi, is conspicuous for its fine coloring. Also fronting on the square

(at the cor. of 19th and Race Sts.) is the handsome building of the
*Academy of Natural Sciences,** of serpentine stone trimmed with
Ohio sandstone, in the Collegiate Gothic style. Its library contains
30,000 volumes, and there are extensive collections in zoölogy, ornithol-
ogy, geology, mineralogy, conchology, ethnology, archæology, and botany. The museum contains upward of 250,000 specimens; and Louis
Agassiz pronounced it one of the finest natural science collections in
the world (open daily from 9 A. M. to 5 P. M.; admission, 10c.). Facing
the square on the S. is *Wills's Hospital* for the treatment of diseases of
the eye; and at the cor. of Race and 20th Sts. is the *Institution for the
Blind,* who are instructed in useful trades, in music, and in the usual
branches taught in schools. Farther along 20th (at the cor. of Spring
Garden St.) is a lying-in hospital, the *Preston Retreat.* To the left (S.)
from Chestnut St., 18th St. leads in one block to the aristocratic *Rittenhouse Square,** surrounded by costly private residences. In Walnut St., W. of 18th, stands the *Rittenhouse Club.* At the cor. Chestnut
and 24th Sts. is the *Baltimore & Ohio R. R. Station,* from which trains
leave for Baltimore and the South and West, and for New York by the
Reading route. Above 24th St., Chestnut St. crosses the Schuylkill on
a massive iron bridge (completed in 1866), and leads for a mile or so
amid the beautiful residences of West Philadelphia. (Street-cars traverse Chestnut St. from Front to 42d Sts., but the points we have described are not beyond the limits of a morning or afternoon stroll.)

Broad St. is a noble thoroughfare, 113 ft. wide, extending N. from
the Delaware for 15 miles through the heart of the city. At the foot
of Broad St. is **League Island** (600 acres), on which is the *U. S.
Navy-Yard,* and which is being converted into a naval depot. The
site was presented by the city to the U. S. Government in 1862, and
the material from the old navy-yard transferred to it in 1875. For 3
miles after leaving the river, Broad St. passes across flats occupied by
truck-farms. The first building requiring notice is *St. Agnes Hospital,* a
Roman Catholic charity, at the cor. of Mifflin St., running along to
McKean St. At the cor. Christian St., the splendid *Ridgway Library** (a branch of the Philadelphia Library), an elegant granite structure 220 by 105 ft., contains the Loganian Library. It was a bequest
of Dr. Benjamin Rush, and cost $1,500,000 (open from 9 A. M. to 5 P. M.).
In Locust St., E. of Broad, is the home of the Philadelphia Library. At
the cor. Pine St. is the long granite building of the *Deaf and Dumb
Asylum* (tickets at *Ledger* office); and one square above stands the superb **Beth-Eden Baptist Church.** Just beyond is *Horticultural
Hall,* where are held the annual floral displays of the Horticultural Society; and next door is the *Academy of Music,** one of the largest
opera-houses in America, with seats for 3,000 persons. Opposite Horticultural Hall is the *Broad St. Theatre,* and just beyond is the *Empire
Theatre,* while between Locust and Walnut Sts. is the handsome building of the *Philadelphia Art Club;* at the cor. Walnut St. is the *Stratford Hotel,* and opposite the *Bellevue;* and at the cor. Sansom St. is
the **Union League Club** (*see* p. 33). Also at the cor. Sansom St. is
the lofty *Hôtel Lafayette,* a short distance beyond which Broad St. is

crossed by Chestnut St. On *Penn Square*, at the intersection of Broad and Market Sts., are being erected the vast * **Public Buildings** (for law-courts and public offices), of white marble, 486½ ft. long by 470 wide, 4 stories high, and covering an area of nearly 4½ acres, not including a court-yard in the center 200 ft. square, containing 520 rooms. The central tower will be 537½ ft. high, and the total cost of the building over $15,000,000. At the cor. of Filbert St., the * **Masonic Temple** lifts its front, 250 ft. long by 150 wide, with a tower 230 ft. high. At the intersection of Broad and Arch Sts. is a cluster of fine churches: * *Arch St. Methodist*, of white marble ; the * *Holy Communion* (Lutheran), of green serpentine, in the Gothic style ; and the *First Baptist*, of brown-stone. Beyond, at the cor. of Cherry St., is the * **Academy of Fine Arts,** in the Byzantine style, 260 by 100 ft., and containing an excellent collection of pictures, etc. (entrance, 25c. ; free on Sunday afternoon and Mondays), and diagonally opposite is a circular building (cyclorama) for the exhibition of panoramas, etc. On the E. side, below Race, is the *Armory of the State Fencibles*, and on the W. side of the street, above Race, is the *Hahnemann Medical College*. On the E. side, on the cor. of Vine St., is the *Catholic Boys' High-School*. At the cor. of Callowhill St. stands the armory of the First Regiment of State militia. At the opposite corner is the old depot of the Reading R. R., and just above are the *Baldwin Locomotive-Works*. N. of the Baldwin Works, Broad is crossed by * *Spring-Garden St.*, leading toward Fairmount Park, on the N. E. cor. of which and Broad St. is the *Spring-Garden Institute* and *School of Design for Young Men and Boys*. At the cor. of 17th St. is the *Girls' Normal School*. At the cor. of Broad and Green Sts. is a Presbyterian church, beside which stands the * **Synagogue Rodef Shalom,** and near by the *Central High-School;* the Widener mansion, on the cor. of Broad St. and Girard Ave., and that of W. L. Elkins, nearly opposite, are among the finest residences in the country ; and at the cor. of Master St. is the *School of Design for Women*, in the house once occupied by Edwin Forrest, the actor. Broad St. now traverses for about two miles a residence quarter, with a number of fine churches, forming a popular promenade (with the *Monument Cemetery*), and then runs N. to **Germantown** (22d Ward, 6 miles from Chestnut St.), a pretty suburb, with fine villas and churches, inhabited chiefly by the business men of Philadelphia. Here was fought the battle of Germantown (Oct. 4, 1777), in which Washington was defeated by Lord Howe. (Germantown may be reached from Philadelphia by railroad from Pennsylvania R. R. depot, Broad St. ; or from Reading depot at 12th and Market Sts. ; or by horse-cars from 8th and Dauphin Sts., 4th and 8th St. line).

Other places of interest are as follows : * **Girard College** (2 miles N. W. of the State-House by Ridge Ave. cars) was founded by Stephen Girard, a native of France, who died in Philadelphia in 1831, leaving an immense fortune. He bequeathed $2,000,000 to erect suitable buildings "for the gratuitous instruction and support of destitute orphans," and the institution is supported by the income of the residue of the estate after the payment of certain legacies. The estate now is estimated at about $15,000,000. The site of the college grounds comprises 42 acres. The college building is a noble marble structure of the Corinthian order, 218 ft. long, 160 wide, and 97 high. The roof commands a wide * view over the city. In the building are interesting relics òf Girard, and in the grounds is a monument to the graduates of the college who fell in the civil war. (Permits to visit the college may be obtained at the principal hotels, of the Secretary, or of the Directors; clergymen are not admitted.) The *German Hospital* and the *Mary J. Drexel Home* are opposite its main entrance; and within one block, in the opposite direction, are the *Woman's Medical College* and *Woman's Hospital.* At Chestnut and 32d St., in West Philadelphia, is the beautiful building of the * **Drexel Institute,** constructed of light buff brick with terra-cotta ornamentations. It has a library, museum of art and technical products, class-roòms, laboratories, and a large auditorium (seating 1,500), where organ-recitals and other entertainments are held at times. The * **University of Pennsylvania** occupies a group of stone buildings on Woodland Ave., bet. 34th and 36th Sts. (reached by Spruce St. cars *via* Market and Walnut Sts.). It has a library of 85,000 volumes, a fine museum and cabinets, and a hospital and medical college. Near by (on 34th St.) is the *Blockley Hospital* and *Almshouse,* with four buildings 500 ft. long, and grounds of 187 acres (tickets of admission at 42 N. 7th St.). The * **Pennsylvania Hospital for the Insane** (42d to 50th Sts.), Haverford Road, W. Philadelphia (take traction cars in Market St. to 41st St.; tickets at *Ledger* office), is worth a visit to see Benjamin West's picture of "Christ Healing the Sick." (Admittance every day except Saturday and Sunday.) The *Presbyterian Hospital* is at Powelton Ave. and 39th St., opposite which is the *Old Man's Home,* the grounds of which occupy nearly an entire square. The * *Episcopal Hospital* is at 2649 N. Front St. The * **U. S. Naval Asylum** (in Gray's Ferry Road, below Washington Ave.; take Spruce and Pine St. cars) is an immense building, standing in highly-cultivated grounds. The portico, with 8 columns, the trophy cannon, and the official residences, are worthy of notice. There are two *U. S. Arsenals,* one a short distance S. E. of the Naval Asylum, and the other at Bridesburg, near Frankford (reached by steam-cars from the Broad St. or Kensington stations, or by the red cars of the 2d and 3d St. line). The former is devoted to the manufacture of clothing for the army; the latter to the manufacture of fixed ammunition. The * **Eastern Penitentiary,** in Fairmount Ave. above 21st St., covers about 10 acres of ground, and in architecture resembles a baronial castle. The separate (*not* solitary) system is adopted here, and furnished Charles Dickens with a pathetic passage in his "American Notes." Each prisoner is furnished with work, and is allowed to see and converse with the officials, but not with any of his fellow-prisoners. (Tickets of admission are obtained at the *Ledger* office.) The *Moyamensing Prison,* 10th St. and Passyunk Road, is a vast granite building, holding 900 prisoners, appropriated to persons awaiting trial or sentenced for short periods.

* **Fairmount Park,** the largest city park in the world, extends along both banks of the Schuylkill River for more than 7 miles, and along

both banks of Wissahickon Creek for more than 6 miles, commencing at Fairmount, an elevation on the Schuylkill from which the Park derives its name, and extending to Chestnut Hill on the Wissahickon, a distance of nearly 14 miles, embracing an area of over 3,000 acres. The total length of driveways is 32½ miles. It possesses much natural beauty, being well wooded and having a great variety of surface; but art, other than that of landscape-gardening, has as yet done little for it. The main entrances are at 25th and Green Sts. and by the Girard Ave. bridge, and are reached by horse-cars from all parts. Just inside, on the right, is Fairmount Hill, on the summit of which are 4 reservoirs of the Schuylkill Water-Works, covering 6 acres, and surrounded by a graveled walk. The *East Park Reservoir* is reached by Ridge Ave. cars. The buildings containing the water-works machinery lie just in front of the visitor as he enters the Park; and in the grounds adjoining them are several fountains and statues. Beyond the buildings is an open plaza, surrounded by flower-beds and shrubbery, and containing Randolph Rogers's colossal bronze statue of Abraham Lincoln; and beyond this still is * *Lemon Hill*, on the summit of which is the mansion (now used as a restaurant) in which Robert Morris lived during the Revolutionary War. The principal points of interest in the Park, besides those we have mentioned, are *Sedgeley Hill*, above Lemon Hill, on the carriage-road; the *Solitude*, a villa built in 1785 by John Penn, grandson of William Penn; the * *Zoölogical Gardens*, containing a fine collection, reached most easily by Pennsylvania R. R., Broad St. station (admission: adults, 25c.; children, 10c.); *George's Hill* and the *Belmont Mansion*, from both of which there are noble views; *Belmont Glen*, a picturesque ravine; the various bridges across the Schuylkill River; and the romantic drive up the *Wissahickon*. The grounds on which the Centennial Exhibition of 1876 was held are located in the Park, commencing on the N. side of Elm Ave., between 41st and 52d Sts., and extended to a line with George's Hill, and may be reached by several lines of street-cars. Many of the buildings which then crowded the space have been removed; but enough are still standing to make the spot worth a visit. Several of them, indeed, were planned for permanent use, and are well fitted to add to the attractions of the Park, being large and of striking design. * *Memorial Hall*, erected by the State and city at a cost of $1,500,000, stands on an elevated terrace just N. of Elm Ave., and is a splendid stone edifice 365 ft. long, 210 wide, and 150 high. It was built for the Exhibition art-gallery, and now contains a permanent art and industrial collection similar to the famous South Kensington Museum in London. Just N. of Memorial Hall stands the * *Horticultural Building*, a charming structure in the Moresque style, with polychromatic frescoes and arabesques. It is a conservatory, filled with tropical and other plants, and around it are 35 acres of ground devoted to horticultural purposes.

　　* **Laurel Hill** adjoins the upper part of East Fairmount Park, and is one of the most beautiful cemeteries in the country. It embraces nearly 200 acres, and is divided into North, South, and Central Laurel Hill. Many fine monuments adorn it; but the distinctive feature of the cemetery is its unique garden landscape, and the profusion of beau-

tiful trees, shrubs, and flowers. (Admission every day except Sunday from 9 o'clock till sunset.) *Mount Vernon Cemetery* is nearly opposite Laurel Hill; *Glenwood* is prettily situated near by (reached by Ridge Ave. cars); and *Woodland Cemetery* is in West Philadelphia (reached by Spruce St. cars). The latter contains the Drexel Mausoleum, the costliest in America.

Itineraries.

The following series of excursions has been prepared so as to enable the visitor whose time is limited to see as much of the city as possible in the least amount of time. Each excursion is planned to occupy a single day, but the visitor can readily spend more time as special features crowd upon his attention.

1. Visit the City Hall, from the tower of which an extended view of the city may be had; walk to Broad and Arch Sts., and take cars out Arch to 19th St. One block N. is Logan Square; see the Academy of Natural Sciences and the Cathedral, both on the square; next take car going N. to Fairmount Ave.; walk to Corinthian Ave., and see the *Eastern Penitentiary. A few squares N. on Corinthian Ave. are the German Hospital and Girard College.[1] Take Girard Ave. car W., crossing the Schuylkill on the wide bridge, and visit the Zoölogical Gardens (getting there, if possible, before feeding-time, 3.30 P. M.); visit Penn's house in the Park across Girard Ave., and then take the train at Zoological Station on the Pennsylvania R. R. for Broad St. Station.

2. Starting from Broad and Chestnut Sts., notice the Betz and Girard Trust Buildings in Broad St.; visit the U. S. Mint (open from 9 A. M. to 12 M., free); see Wanamaker's Bazaar, and walk down Chestnut St., passing the various trust, safe-deposit, insurance, and newspaper buildings, and the U. S. Post-Office; visit Independence Hall, with its collections; back of this is Independence Square, and across Walnut St. is Washington Square; visit the Drexel Building (from the roof of which there is a fine view of the Delaware River), and note the many imposing buildings in the neighborhood of 4th, Chestnut, and Walnut Sts.; visit Carpenter's Hall and the old Christ Church, and then take car S. on 2d St. to Christian; visit old Swedes Church in Swanson St., and the sugar-refineries, returning along the river front to Walnut and take car W. to Broad St.

3. Starting from City Hall northward in Broad St., visit the Masonic Temple, and then the Academy of Fine Arts (Broad and Cherry Sts.), and then the Cyclorama of the Battle of Gettysburg on the opposite corner. Continuing northward, fine buildings are passed, and at Broad and Callowhill Sts. the extended plant of the Baldwin Locomotive Works should be visited. On the opposite corner is the Spring Garden Institute of Mechanic Arts, and in Spring Garden St., just off from Broad, the Textile School of the Pennsylvania School of Industrial Art. Take omnibus and ride out Broad St. to terminus of line, passing many fine residences and churches. Return by same route, or go a few blocks farther and take train at Germantown Junction for Broad St. Station.

[1] Tickets to be had at the Ledger Office, 6th and Chestnut Sts. .

4. Take car at cor. Broad and Arch Sts. going westward, and get out at Park entrance (eastern end of Spring Garden St. bridge): visit the Fairmount Water Works, and the Lincoln-Monument in the Park; take Schuylkill boat at the landing for Wissahickon; then follow stream by path for a time (or take carriage for Wissahickon drive), and returning cross by boat to Belmont Landing; return on foot through the West Park, visiting Horticultural and Memorial Halls, Sunken Gardens and George's Hill, taking train at 52d St., or Park stations of the Pennsylvania R. R. for Broad St. Station.

5. Take Woodland Ave. (or Spruce St.) car in front of Broad St. Station to cor. of 34th St. and Woodland Ave.; visit the University of Pennsylvania buildings, especially the Egyptian and Archæological collections in the Library Building; walk back three blocks to the Drexel Institute to see its library and museum; take green car passing the front of the building going westward, passing several fine churches and numerous residences; walk from car terminus to 44th and Chestnut; see Reformed Episcopal Church and Seminary; then, after passing fine residences on 42d St., take car for city at 41st and Spruce Sts.; take Broad St. omnibus going S. to Ridgway Library, seeing thus the Union League Club-house, Academy of Music, and other fine buildings on South Broad St.; return by same route.

6. Take train at Broad St. Station for Wissahickon Heights; visit Wissahickon Inn and St. Martin's-in-the-Fields; take train back as far as Tulpehocken Station and walk up Walnut Lane to Main St.; visit the old Chew Mansion (scene of the battle of Germantown in Revolutionary times) and the beautiful residences of Germantown, taking train for the city at Walnut Lane Station of the Philadelphia & Reading R. R.; walk S. from 9th and Green Station to intersection of Ridge Ave.; take car on Ridge Ave. westward, and continue to Diamond St.; enter East Park, and walk by Reservoir and down to the River Drive, passing the Dairy; follow the River Drive around past the boat-houses, and see Lemon Hill; ascend the Lemon Hill Observatory; take car for city at Spring Garden St. entrance to the Park.

7. A ride along the Delaware River front, going from foot of Chestnut St. to Port Richmond coal-wharves, and Cramp's Ship-yards on the N. to Greenwich, and the Girard Point elevators on the S., passing Spreckel's Sugar Refinery, and the works of the Pennsylvania Salt Co.

Cape May.

From Philadelphia Cape May is reached *via* West Jersey R. R. (ferry from foot of Market St.), in 2¼ hrs.; fare, $2.50 (distance, 81 miles). The road traverses an uninteresting and thinly-populated section of New Jersey, the only important station being *Vineland* (34 miles). There are also daily steamers in summer to and from Philadelphia. From New York *via* Pennsylvania and West Jersey R. Rs. (distance, 172 miles; fare, $4 25); also *via* New Jersey Southern Div. of the Central R. R. of New Jersey (distance, 141 miles; fare, $4.50).

Hotels.—The leading hotels are the *Congress Hall, Hotel Lafayette, Stockton Hotel,* and *Windsor Hotel.* Other good houses are the *Arctic, Arlington, Chalfonte, Marine Villa,* and the *West End,* besides many smaller ones. The charges are from $3 to $4 per day, according to the rank of hotel. The "cottage system" is growing in favor, and there are boarding-houses where board may be had at $10 to $18 a week.

Cape May is the extreme southern point of New Jersey, fronting the Atlantic at the entrance of Delaware Bay. Its beach is over 5 miles long, and, being hard and smooth, affords a splendid drive, which has been artificially improved. The bathing is unsurpassed, the surf being especially fine, and the water (so it is claimed) less chilling than elsewhere on the coast. The fashionable hours for bathing are from 11 A. M. to 1 P. M., and the spectacle is then very brilliant. A long promenade extends along the water-front, and is generally thronged in the forenoon and late afternoon. Cape May was at one time the favorite resort of Southern and Western people, besides being the place of all places for Philadelphians. The hotels and cottages are built on a small piece of land, about 250 acres in extent, known as Cape Island, having formerly been separated from the mainland by a small creek. The village contains 6 churches, 2,136 permanent residents, and many fine villas.

The most popular resorts in the vicinity of Cape May are *Cape May Point, Sewell's Point,* and *Schellinger's Landing ;* they are on the Atlantic, and are reached by steam and horse cars from Cape May. Six miles from Cape May are *Holly Beach, Wildwood,* and *Anglesea,* where there are good boating and fishing. *Cold Spring* is on the line of the railroad, about 2 miles N. of the beach. The steamboat-landing is on Delaware Bay, about 2 miles from the village, and a lighthouse, with powerful revolving light, is down the beach to the W. The favorite drive is on the beach, which may be traversed from Poverty Beach to Diamond Beach, a distance of 10 miles; but the roads inland have lately been much improved.

Atlantic City.

From Philadelphia, Atlantic City is reached in 75 to 90 minutes *via* Camden & Atlantic R. R. (distance, 59 miles), and Atlantic City Div. of the Philadelphia & Reading R. R. (distance, 57 miles). From New York *via* Pennsylvania R. R. to Camden, and thence *via* Camden & Atlantic R. R. in 3¼ hours (distance, 146 miles ; fare, $3.25); also *via* Central R. R. of New Jersey to *Winslow,* and thence *via* Camden & Atlantic (distance, 126 miles ; fare, $3.30).

Hotels.—The principal are the *Albion, Brighton, Chalfonte, Colonnade, Congress Hall, Dennis, Haddon Hall, Hoffman, Irvington, Mansion, Normandie, Sea-side, Shelburne, Traymore, Windsor,* and *United States.* The charges at the above-mentioned hotels vary, according to the excellence of accommodations, from $3 a day upward.

Atlantic City is now the favorite resort of the citizens of Philadelphia, but during the season it draws thousands of visitors from all parts of the country. The hotels and larger cottages are located on an island, just off the mainland, and the beach is one of the best and safest on the coast. The regular bathing-hour is 11 o'clock A. M. The city proper contains some 13,055 inhabitants, and is laid out in broad and pleasant avenues. The island on which the city is built is cut off from the mainland by a series of wide-stretching salt-marshes, which are said to contribute materially to the superior healthfulness of this favorite resort. Boating and fishing in the vicinity are excellent, and game can generally be found by the persistent sportsman. The Atlantic City Div. of the Philadelphia and Reading R. R. and the Southern

Div. of the Central R. R. of New Jersey (connecting with the Camden
& Atlantic) place Atlantic City in easy connection with the famous
hunting-grounds of *Barnegat*, *Waretown*, *West Creek*, and *Tuckerton.*
A short distance N. of Atlantic City is the beautiful but ill-omened
Brigantine Beach, called by the sailors "the graveyard," on account of
the number of fatal wrecks that have occurred there. Also near by
is the famous *Long Beach*, favorite of fishermen and hunters.

5. Philadelphia to Baltimore.

Via Philadelphia, Wilmington & Baltimore R. R. (Broad St. Station). Distance, 96 miles; time, 2 hrs.; fare, $2.80. Also *via* Baltimore & Ohio R. R.
(station, 24th and Chestnut Sts.). Distance, 96 miles; time, 2 hrs. Fare, $2.80.
The through-trains of the Pennsylvania R. R. system, from New York to Baltimore (Union Station), make the entire distance (188 miles) in about 4 hrs.;
fare, $5.30. Through-trains of the Baltimore & Ohio R. R. from New York
from station of Central R. R. of New Jersey (Liberty St. ferry), 188 miles; time,
about 4 hrs.; fare, $5.30.

THE country traversed on this route has few scenic attractions,
though the highly-cultivated farms and clustering towns indicate a
populous and long-settled region. *Chester* (14 miles) is the oldest town
in Pennsylvania, having been settled by the Swedes in 1643. It now
has 20,226 inhabitants, and is notable for its extensive ship-yards.
The *Brandywine* (crossed 4 miles beyond Chester) is famous for the battle fought on its banks in September, 1777. **Wilmington** (*Clayton
House*; 28 miles) is the chief city of the State of Delaware. It has
61,431 inhabitants, and its manufactures are very extensive and various,
embracing ship-building, car-manufactories, cotton and woolen factories,
flour-mills, powder-mills, and shoe and leather factories. The city is
regularly laid out, with streets at right angles, the principal ones paved
with stone, and all lined with brick sidewalks. The buildings are uniformly of brick, of which an excellent quality is made in the vicinity.
The public buildings are the *City Hall*, the county *Almshouse*, the *Custom-House and Post-Office* (cor. King and 6th Sts.), the *Wilmington
Institute and Public Library*, and the *Opera-House*. There are several
handsome churches, including the Central and West Presbyterian, the
Grace (Methodist), and the Church of the Sacred Heart (Roman Catholic). The * *Old Swedes' Church*, of stone, erected in 1698, is still in
good condition. There is a restaurant in the depot, and the trains usually stop from 5 to 10 minutes.

Newark (40 miles) is an academic town, seat of several excellent
educational institutions, and 4 miles beyond the train crosses the celebrated *Mason & Dixon's Line* (long the boundary between the Northern and Southern States), and enters Maryland. At *Havre de Grace* (62
miles), the Susquehanna River is crossed by the Baltimore & Ohio trains
on a lofty iron bridge nearly a mile long. In entering Baltimore a view
of the Patapsco River and Fort McHenry may be obtained by this route
from the car-window on the left.

6. Baltimore.

Hotels.—The *Hotel Rennert* (cor. Saratoga and Liberty Sts.), the *St. James* (cor. Charles and Centre Sts.), *Mount Vernon* (in Monument St., near Mount Vernon Pl.), the *Altamont* (Eutaw Pl.), and the *Albion* (Cathedral and Richmond Sts.), are on the European plan. The *Carrollton Hotel* (cor. German and Light Sts.), the *Eutaw House* (cor. Baltimore and Eutaw Sts.), the *Howard House* (Howard near Baltimore St.), are on the American plan. The *Maltby House* (in W. Pratt St. between Light and S. Charles Sts.) is conducted on both the American and European plans. The *Brexton* (Park Ave.), the *Langham* (Charles and Center Sts.), and the *Shirley* (Madison St.), are family hotels.

Restaurants.—For ladies and gentlemen : the *Woman's Industrial Exchange*, cor. Charles and Pleasant Sts.; *Rennert's*, cor. Saratoga and Liberty Sts.; *St. James*, cor. Charles and Centre Sts.; and *Painter's*, on Lexington St., near St. Charles St. For gentlemen : *Rennert's*, cor. Calvert and German Sts.; *Sheehan's*, in Light St., opposite the Carrollton ; and *Pepper's*, in Holliday St.

Modes of Conveyance.—*Street-cars*, fare 5c., afford easy access everywhere. *Public carriages* wait at the depots and at stands in various parts of the city. Tariffs of fares are placed inside the carriages ; in case of disagreement with the driver, apply to a policeman. *One-horse cabs and hansoms* run to and from depots and boats, 25c. a passenger, and by the hour 75c.; also, *street-cars* to Franklin and Powhatan ; and street and steam cars to Catonsville, Towson, Pimlico, and Pikesville. The *Baltimore & Lehigh Narrow-Gauge R. R.* reaches Long Green, Bel-Air, and York, Pa.

Railroad Stations.—The *Union Station*, in Charles St., commonly called the *Charles St. Station*, is used almost exclusively for through travel on the *Pennsylvania R. R.* and its branches, the *Northern Central* and the *Philadelphia, Wilmington & Baltimore R. Rs.* The *Pennsylvania R. R.* and the *Western Maryland R. R.* have also inner stations at Calvert and Hillen Sts., devoted to local traffic. The *Western Maryland* and *Baltimore & Potomac R. Rs.* run into the Charles St. station. The station of the *Baltimore & Ohio R. R.*, and its *Philadelphia* branch, is in Camden St., near Howard ; of the *Baltimore & Lehigh R. R.*, cor. North Ave. and the bridge over Jones's Falls ; of the *Steelton* and *Catonsville* (steam) *R. Rs.*, at Calvert and Charles St. Stations.

Theatres and Amusements.—The *Lyceum*, in N. Charles St., is the best theatre. *Ford's Grand Opera-House*, in Fayette St. near Eutaw, is much frequented. The *Holliday St. Theatre*, opposite the City Hall, is a favorite resort. The *Academy of Music* (now *Harris's Academy*) is devoted to cheaper performances. Next door to it has been built the *Howard Auditorium* (formerly Oratorio Hall), a popular play-house, with wax-works. The *Front St. Theatre* is in Front St. near Gay. At the *Concordia Opera-House*, cor. Eutaw and German Sts., German opera and drama are occasionally given. Many lectures and public gatherings are held in *Levering Hall* of the Johns Hopkins University, cor. Eutaw and Little Ross Sts. Concerts and lectures are given at the *Peabody Institute* : *Lehman's Hall*, Howard St.; and *Young Men's Christian Association*. The *race-course* of the Maryland Jockey Club is at Pimlico, 2 m. from the N. W. boundary of the city, on the Western Maryland R. R.

Reading-Rooms.—At the *Peabody Institute*, cor. Charles and Monument Sts. (free from 9 A. M. to 10 P. M.) ; the *Mercantile Library*, in Charles near Saratoga Sts. (open from 10 A. M. to 10 P. M.) ; the *Maryland Institute*, cor. Baltimore and Harrison Sts. ; the *Baltimore Library* of the Maryland Historical Society, cor. Saratoga and St. Paul Sts., with a reading-room for members and introduced visitors ; the *Young Men's Christian Association*, cor. Charles and Saratoga Sts. The reading-room of the *Johns Hopkins University Library*, Little Ross St., may be visited by request ; and the *McCoy Library*, bequeathed to the University, is in Lanvale St. near Eutaw Pl. (open from 2 P. M. to 5 P. M. *The Enoch Pratt Free Library*, on Mulberry St. near Cathedral, endowed with $1,000,000, is open to the public free.

Art Collections.—A Department of Art, in connection with the Peabody Institute, is in process of organization. Good pictures are usually on exhibition (free) at the sales-galleries of *Myers & Hedian*, 46 N. Charles St. The private gallery of Mr. W. T. Walters, Mount Vernon Place, is one of the richest in America (open on Wednesdays, February, March, and April; 50 cts. admission, for the benefit of the poor.

Clubs.—These include the *Maryland Club*, cor. Charles and Eager Sts., which has a superb marble building that cost $240,000 ; the *Baltimore Club*, in Charles, between Read and Eager Sts.; the *University Club*, cor. Charles and Madison Sts.; and the *Athenæum Club*, cor. Charles and Franklin Sts.

Post-Office.—The Post-Office, a massive building, is on the square bounded by Calvert, Fayette, North, and Lexington Sts. Open from 8 A. M. to 11 P. M. Sundays, 8.30 to 10 A. M.

BALTIMORE, the chief city of Maryland, is picturesquely situated on the N. branch of the Patapsco River, 14 miles from its entrance into Chesapeake Bay, and about 200 miles from the Atlantic Ocean. The harbor is capacious, consisting of an inner basin for small vessels, and an outer harbor accessible to the largest ships. The entrance to the harbor is defended by Fort McHenry, which was unsuccessfully bombarded by the British fleet in the War of 1812. **Baltimore St.,** running E. and W., is one of the main business thoroughfares, and in it and Lexington, Howard, N. Eutaw, and N. Charles Sts., are located the principal retail stores. The principal wholesale houses are in Sharp, Hanover, and Howard Sts. *North Charles St.* is the most attractive and fashionable promenade; *Eutaw Place* is also popular. On Eutaw Place, 150 ft. wide, are many handsome dwellings, churches, and the *Altamont* hotel. The favorite drives are through Eutaw Place and Druid Hill Park to the Pimlico race-course, out Charles St. to Lake Roland (6 m.), returning by Mt. Washington, Green Spring Ave., and Druid Hill Park, and on the old York Road to Govanstown (4 m.), returning by Charles Ave.

The present site of Baltimore was chosen in 1729, and its name was given it (in 1745) in honor of Lord Baltimore, the proprietary of Maryland. In 1780 it became a port of entry. In 1782 the first pavements were laid in Baltimore St., and in the same year the first regular communication with Philadelphia was established through a line of stage-coaches. The charter of the city dates from 1797. The population, which at that time was 26,000, had increased by 1850 to nearly 200,000 ; in 1860 it was 212,418 ; in 1870, 267,354 ; in 1880, 332,190 ; and in 1890, 484,439. The commerce of the city is very active ; and through her two great arteries of traffic (the Baltimore & Ohio and the Northern Central Railroads) she is successfully competing for the trade of the North and Northwest. Large shipments of grain are made to Europe, and tobacco, cotton, petroleum, flour, canned goods, lumber, coal, cattle, and lard, are also exported. There are also iron-works, nail-factories, locomotive-works, brass and bell foundries, cotton-factories, and other industrial establishments (5,258 in 1890). In the production of artificial fertilizers Baltimore stands first in the United States. The canning of oysters, fruits, and vegetables, is estimated to reach the annual value of $10,000,000 ; and 500,000 hides are annually made into leather and sent to New England.

From the number of its monuments, Baltimore is often called "the Monumental City," and its chief glory in this line is the * **Washington Monument,** standing 100 feet above tide-water, in the heart of the city, at the intersection of Mt. Vernon and Washington Places. The base of the monument is 60 ft. square and 35 ft. high, supporting a Doric shaft 130 ft. in height, which is surmounted by a colossal statue of Washington, 16 ft. high. The total height is thus 181 ft. from the ground and 281 ft. above the river. It is built of brick, with an outer casing of white marble, and cost $200,000. From the balcony of the monument a magnificent * view of the city, harbor, and surrounding country is obtained (access by a circular staircase within; fee, 15c.). The

* **Battle Monument** stands in Monument Square, in Calvert St., between Fayette and Lexington Sts., and was erected in 1815 to the memory of those who fell defending the city from the British in September, 1814. The square sub-base on which the monument rests is 20 ft. high, with an Egyptian door at each front, on which are appropriate inscriptions, and representations (in *basso-rilievo*) of some of the incidents of the battle. The column rises 18 ft. above the base, is encircled by bands on which are inscribed the names of those who fell, and is surmounted by a female figure in marble, emblematic of the city of Baltimore. The *Wildey Monument* (in Broadway near Baltimore St.) is a plain marble pediment and shaft surmounted by a group representing Charity protecting orphans; it is dedicated to Thomas Wildey, founder of the order of Odd-Fellows in the United States. The *Wells and McComas Monument* (cor. Gay and Monument Sts.) commemorates two boys who shot General Ross, the British commander, Sept. 12, 1814. The *Poe Tombstone* stands in the churchyard of the Westminster Presbyterian Church, cor. Greene and Fayette Sts.

Facing the Washington Monument on the S. is the stately white-marble building of the * **Peabody Institute,** founded and endowed by George Peabody, the eminent London banker, and designed for the use of scholars and as a general library of reference, the books not to be taken from the room. It contains a free library of over 110,000 volumes, two lecture-halls, and a conservatory of music; and a Department of Art, to include art-collections and a school of art, is in process of organization. Also fronting the monument (cor. Charles and Monument Sts.) is the costly * **Mount Vernon Church** (Methodist), built of green serpentine, with outside facings of buff Ohio and red Connecticut sandstone, and 18 polished columns of Aberdeen granite. This is the most aristocratic residence-quarter of Baltimore, and surrounding the Place and on the adjacent streets are some of the finest private houses in the city. One block off (at the cor. of Park and Madison Sts.) is the * **First Presbyterian Church.** Its spire is 268 ft. high, with side towers 78 and 128 ft. high, and the interior is richly decorated.

The * **City Hall,** completed in 1875, is one of the finest municipal buildings in America. It fills the entire square inclosed by Holliday, Lexington, North, and Fayette Sts., is 225 by 140 ft., and cost $2,271,-135. It is of marble, in the composite style, 4 stories high, with French roof and an iron dome 260 ft. high. A balcony 250 ft. above the street affords a magnificent view of the city (visitors may ascend on Mondays from 10 A. M. to 3 P. M.). Near by (cor. Fayette and North Sts.) is the *U. S. Court-House,* a massive granite structure. The old * **Exchange,** in Gay St. between 2d and Lombard, is a large structure, with a façade of 240 ft. and colonnades of 6 Ionic columns on the E. and W. sides, and the whole surmounted by an immense dome. The **Post-Office** building, a handsome granite structure, plain but effective, is on the square bounded by Calvert, Fayette, North, and Lexington Sts. The *Stock Exchange,* on German St., is an elegant building. The *Corn and Flour Exchange,* cor. Holliday and 2d Sts., is a solid and handsome building; and the *Rialto Building,* cor. 2d and Holliday Sts.,

4

is a fine specimen of Renaissance architecture. The **Masonic Temple,** in Charles St. near Saratoga, is a stately stone edifice, completed in 1870 at a cost of $400,000. The *Odd-Fellows' Hall,* cor. Cathedral and Saratoga Sts., is a handsome building, containing a large library. The building of the *Y. M. C. A.,* cor. Charles and Saratoga Sts., is one of the finest in the city, and contains a library, reading-room, gymnasium, etc. Among business structures the offices of the *American* (S. W. cor. Baltimore and South Sts.) and the *Sun* (S. E. cor. same streets) are noteworthy. The *Merchants' Shot-Tower* (cor. Front and Fayette Sts.) is one of the landmarks of the city; it is 216 ft. high and 40 to 20 ft. in diameter, and contains 1,100,000 bricks.

Two of the finest churches in the city have already been mentioned. The most celebrated is the * **Cathedral,** cor. Mulberry and Cathedral Sts. It is of granite, in the form of a cross, 190 ft. long, 177 broad at the arms of the cross, and 127 high from the floor to the top of the cross which surmounts the dome. At the W. end rise 2 tall towers, crowned with Saracenic cupolas resembling the minarets of a Mohammedan mosque. It contains one of the largest organs in America, and 2 excellent paintings: " The Descent from the Cross," presented by Louis XVI, and "St. Louis burying his Officers and Soldiers slain before Tunis," the gift of Charles X of France. The Roman Catholic churches of *St. Alphonsus* (cor. Saratoga and Park Sts.), of *St. Vincent de Paul* (in N. Front St.), and of *St. Ignatius Loyola* (cor. Calvert and Read Sts.), are rich in architecture and decorations. *Grace Church* (Episcopal), cor. Monument and Park Sts.), is a fine specimen of Gothic architecture, in red sandstone. Close by, at the cor. of Read and Cathedral Sts., is *Emanuel Church* (Episcopal), also Gothic. *Christ Church* (Episcopal), is a beautiful marble structure, cor. of St. Paul and Chase Sts. *St. Paul's* (Episcopal), cor. Charles and Saratoga Sts., is the old parish church, the first Episcopal church in the city, and one of the oldest of any denomination; it is a fine Romanesque structure. Other fine Episcopal churches are * *St. Peter's,* of marble, cor. Druid Hill Ave. and Lanvale St., and *St. Luke's,* near Franklin Square. The **Unitarian Church,** cor. N. Charles and Franklin Sts., is an imposing structure, with a colonnade in front composed of 4 Tuscan columns and 2 pilasters which form the arcades. From the portico the entrance is by 5 bronze doors. The *Eutaw Place Baptist Church,* cor. Eutaw and Dolphin Sts., is noted for its beautifully proportioned marble spire, 186 ft. high. The *Brown Memorial Church* (Presbyterian), cor. Park and Townsend Sts., is a spacious marble edifice in the Gothic style; and the *Westminster,* cor. Green and Fayette Sts., is noteworthy for containing the grave and monument of Edgar Allan Poe. The *Associate Reformed Church* is a handsome Romanesque building of gray stone with a circular auditorium. It is on the cor. Maryland Ave. and Preston St. The *First Methodist Episcopal Church,* cor. St. Paul and 3d Sts., is a fine stone structure, rendered imposing by its lofty square tower or campanile in the Venetian style. Associated with it are the striking series of buildings belonging to the *Woman's College.* The whole group forms one of the most notable architectural features of the city. The

Hebrew *Synagogue*, in Lloyd near Baltimore St., is large and handsome. The new *Synagogue*, just completed, is an elaborate stone structure near the upper end of Madison Ave.

The * **Athenæum Building,** cor. Saratoga and St. Paul Sts., contains the *Library of the Medical and Chirurgical Faculty* (17,000 volumes), and the collections of the * *Maryland Historical Society*, comprising a library of 28,000 volumes, numerous historical relics, and some fine pictures and statuary (admission free). The *Mercantile Library* (36,000 volumes; open from 10 A. M. to 10 P. M.) is in the building in Charles, near Saratoga St. The **Maryland Institute,** designed for the promotion of the mechanic arts, is a vast brick structure, cor. Baltimore St. and Market Square. The first floor is used for the *Center Market;* the second floor, 260 ft. long, for classes in designing, painting, book-keeping, etc. It also contains a library (21,000 volumes), lecture-rooms, a school of design, etc. The **Johns Hopkins University,** endowed with $3,500,000, has extensive buildings on Howard, Eutaw, Little Ross, and Monument Sts., to which it has added very complete *Chemical, Physical, Geological, and Biological Laboratories* and an *Astronomical Observatory.* The principal object of the University is the higher education of post-graduates, in which it has been very successful. It also has a college course. The *Johns Hopkins Hospital* was opened in May, 1889, with large buildings, which cost nearly $2,000,000, in N. Broadway, and has all the modern improvements in hospital arrangement. The treatment is on the latest and best system, so as to enable it to take the same rank in medical science as the Johns Hopkins University does in letters. The *University of Maryland,* founded in 1807 with a law and medicine faculty, has a building modeled after the Pantheon in Rome, on the cor. Lombard and Greene Sts. The *State Normal School,* cor. Carrollton Ave. and Townsend St., is one of the finest buildings in the city. The *City College,* in N. Howard St., is a graceful edifice in the Collegiate-Gothic style. The *Enoch Pratt Free Library,* in Mulberry St., between Cathedral St. and Park Ave., is a building of white marble, opened January 4, 1885, and contains 105,000 volumes. It is in the Romanesque style, with a tower 98 ft. high rising from the middle of the façade. It has room for 200,000 books. There are also 4 branch libraries.

The markets of Baltimore are famous, and worth visiting. The *Lexington Market* is most convenient. It should be seen on Saturday night.

Prominent charitable institutions are the *Maryland Institution for the Instruction of the Blind,* a large marble building in North Ave. near Charles St., and the *Episcopal Church Home,* for the relief of the afflicted and destitute, in Broadway near Baltimore St. The * **State Insane Asylum** is a massive pile of granite buildings near Catonsville (6 miles from the city). The *Sheppard Asylum for the Insane,* founded by Moses Sheppard, a wealthy Quaker, occupies a commanding site near Towsontown, 7 miles from the city. The * **Bay View Asylum** (Almshouse) is a vast brick building, situated on a commanding eminence near the outskirts of the city on the Philadelphia road. To these we may add the *Baltimore Orphan Asylum,* on Stricker St., founded in 1801, for the maintenance of orphans under 9 years of age; the *Henry Watson's Children's Aid Society*, 72 Calvert St., receiving children between the ages of 8 and 15.

By the Western Maryland R. R. the visitor can reach many notable charitable institutions a few miles from the city: the *Mt. Hope Retreat for the Insane*, on the Reistertorin road, under the charge of the Sisters of Charity: it has 600 patients, extensive grounds, and fine buildings; the *McDonogh School for Poor Boys*, founded by John McDonogh, of New Orleans; and the pretty cottage buildings of the *Thomas Wilson Sanitarium for Sick Children*, to which daily excursion-trains are run in summer. Some of the largest Catholic educational institutions for girls are the *Mt. de Sales*, on the Frederick road; the *Notre Dame*, in Charles St. Ave.; and the *St. Agnes*, at Mt. Washington. In the northern section is the *Home for Incurables* and the *United States Marine Hospital*. The *Samuel Ready Orphan Asylum* is on the eastern limits. There are some fifty other prominent orphan asylums, hospitals, and homes for the aged and infirm in the city.

The water-works of Baltimore are extensive; among the reservoirs formed to supply the city we may mention *Lake Roland* (about 8 miles from the city, on the Northern Central R. R.), extending over 116 acres. The drive to this lake is through a romantic country, the valley of Jones's Falls being picturesque along its entire length. *Loch Raven*, a large lake, 5 miles in length, formed by damming the Gunpowder River, is the largest source of supply. The water is conducted by a tunnel blasted through the solid rock for 7 miles to *Lake Montebello*, about 2 miles from the city.

*** Druid Hill Park** (reached by many street-cars) is a beautiful pleasure-ground of 680 acres, situated in the northern suburbs of the city. The architectural decorations of the park are few; its charms lying chiefly in its rural beauty, its secluded walks, drives, and bridle-paths. The surface is undulating and well wooded, the trees being among the oldest and finest in any public park in America. Several of the eminences overlook the surrounding country, and from the *tower at the head of Druid Hill Lake there is a superb view of the river and harbor. *Patterson Park*, at the E. end of Baltimore St., embraces 200 acres, pleasantly laid out, and commands extensive views in every direction. The earthworks thrown up during the attack of 1814 still remain. The principal cemeteries are **Greenmont Cemetery,** in the N. part of the city (York road and Charles St. cars); **Loudon Park Cemetery,** 2 miles from the city *via* Catonsville horse or steam cars; **Loraine Cemetery,** on the Franklin and Windsor roads, by Powhatan R. R., 3 miles. All contain many monuments, and are picturesquely laid out.

*** Federal Hill** (reached by the South Baltimore horse-cars) is a commanding eminence on the S. side of the inner basin, and affords fine views of the city, river, and bay. It has been purchased by the city for a park, and contains a U. S. Signal Station. **Fort McHenry,** at the entrance of the harbor, is worth a visit; it is situated at the end of Whitestone Point, 3 miles from the City Hall, and is reached by S. Baltimore horse-cars and also by ferry from foot of Broadway. The sentinels will usually admit strangers. The **Railroad Tunnels,** by which all the railroads on the N. side of the city are connected with tide-water at Canton, are among the wonders of Baltimore. The Baltimore and Potomac Tunnel is, next to the Hoosac Tunnel, the longest in America (6,969 ft.) and the Union Tunnel is 3,410 ft. long. They were completed in 1873, at a cost of $4,500,000. The Belt Line Tunnel, now in course of construction by the Baltimore & Ohio R. R.. will be even more extensive. It is being made under Howard St. for its entire length, and will thence extend N. of Huntington Ave. and along the N. side of the city. This will give an exit for this railroad to Philadelphia without crossing the ferry, as at present, and will at the same time give a rapid transit system to the city.

Among the Bayside resorts most patronized are *Tolchester Beach,* 25 miles from Baltimore, in Kent County, and *Bay Ridge,* below Annapolis, about 32 miles distant. Both have hotels and bathing-houses, and steamers run to them twice a day.

Itineraries.

The following series of excursions has been prepared so as to enable the visitor whose time is limited to see as much of the city as possible in the least amount of time. Each excursion is planned to occupy a single day, but the visitor can readily spend more time as special features crowd upon his attention.

1. Get general view of the city and harbor from top of Washington Monument; note the famous Baryé bronzes (gift of Mr. Walters) in Mount Vernon Place, as well as the bronze statues of Chief-Justice Taney and George Peabody; visit the Peabody Institute Library and Art Gallery, and (if it be Wednesday in February, March, or April) the private galleries of Mr. W. T. Walters, No. 5 Mount Vernon Place W. Visit the buildings of the Johns Hopkins University, three blocks W. on Howard St. Guide may be obtained at the Registrar's office, who will show visitors the chemical, biological, and physical laboratories, library, gymnasium, and other features.

2. Take the cable street-car line and ride through the business portion of the city from Patterson to Druid Hill Parks. Visit both of these parks. Examine the wharves and shipping along the harbor and basin. Visit City Hall (fine view), Post Office, Battle Monument, and the Equitable Building.

3. Drive out Eutaw Place to Park and back by way of Mount Royal Ave. and Cathedral or Charles Sts. Thence by Eager St. to Broadway and visit the Johns Hopkins Hospital, which has the finest hospital buildings in the world. Also see the Wildey Monument.

4. By street-cars or carriage to Federal Hill—the place where General Benjamin F. Butler held the city during the war—and thence to Fort McHenry. The well-known Chesapeake potteries and the Columbian Iron Works, with Government war-vessels in course of construction, are near the latter. The vast plant of the Maryland Steel Company, with Bessemer converters, rolling-mills, and ship-yards, may be reached at Sparrows's Point by either steamer or rail.

5. If the weather is fine, a boat may be taken for an excursion on the harbor, or a trip to Tolchester Beach, Annapolis, or Bay Ridge.

6. A walk about the city. Through Baltimore St. examine the Baltimore & Ohio Building, cor. Calvert St. Thence through Charles St. to Mulberry; visit the Enoch Pratt Free Library. Continue walk out Charles St. to North Ave. and return by Calvert or St. Paul St., to see handsome residences and fine iron bridges over Jones Falls and the railroads.

7. Visit the beautiful country-seats N. of the city. Drive out Hartford road and visit Johns Hopkins estate "Clifton," now the property of the Johns Hopkins University. Drive through the Garrett estate to the York road or Charles St. Ave. Thence to Lake Roland and return by Roland Ave. or Green Spring Ave. and Druid Hill Park. The larger water-supply at Loch Raven, with its very picturesque surroundings, may be reached by the Baltimore & Lehigh narrow-gauge railroad from North Ave.

7. Baltimore to Washington.

THE traveler has a choice of two routes in going from Baltimore to Washington; the Washington branch of the *Baltimore & Ohio R. R.* and the *Baltimore & Potomac R. R.* The distance by the former is 40 miles, and time 45 minutes; by the latter, 43 miles, and time 1 to 1½ hr.: fare, $1.20. The country traversed is flat, with few picturesque and no very striking features. On leaving the Baltimore depot, the trains of the Baltimore & Potomac line pass through the great tunnels beneath the city, mentioned above; and just before entering Washington through another tunnel 1,500 ft.-long. By the Baltimore & Ohio line the splendid * *Washington Viaduct* is crossed (9 miles out). The first view of the Capitol in approaching Washington is very fine and should not be lost.

New York to Washington.—The regular Express trains run through in 6 to 8 hours. The *Limited Express* train of the Pennsylvania R. R., composed exclusively of palace cars, runs through in 5 hours and 5 minutes. The *Royal Blue Line Express* of the Baltimore & Ohio R. R., composed exclusively of palace cars, including dining-car, runs through in 5 hours. This is the fastest train in the United States.

8. Washington City.

Hotels.—The best on the American plan are the *Arlington*, in Vermont Ave., between H and I Sts.; *Arno*, 16th St. N. W. (between H and I Sts.); the *Ebbitt*, cor. F and 14th Sts., is a favorite with army and navy officers; the *Riggs*, cor. 15th and G Sts.; *Willard's*, cor. Pennsylvania Ave. and 14th St.; and *Wormley's*, 15th and H Sts. Other good hotels on the American plan are the *American House*, cor. Pennsylvania Ave. and 7th St.; the *Cochran*, 14th and K Sts.; the *Metropolitan*, in Pennsylvania Ave. near 6th St.; the *National*, cor. Pennsylvania Ave. and 6th St.; and the *Randall*, Pennsylvania Ave. and 15th St. The best on the European plan are *Chamberlain's*, cor. 15th and I Sts.; the *Normandie*, cor. 15th and I Sts.; the *Shoreham*, cor. 15th and H Sts.; the *St. James*, cor. Pennsylvania Ave. and 6th St.; and *Welcker's*, 15th St. near H. The *Hamilton* (cor. 14th and K Sts.) is a select family hotel.

Restaurants.—*Welcker's* (in 15th St. near H), and *Wormley's* (cor. 15th and

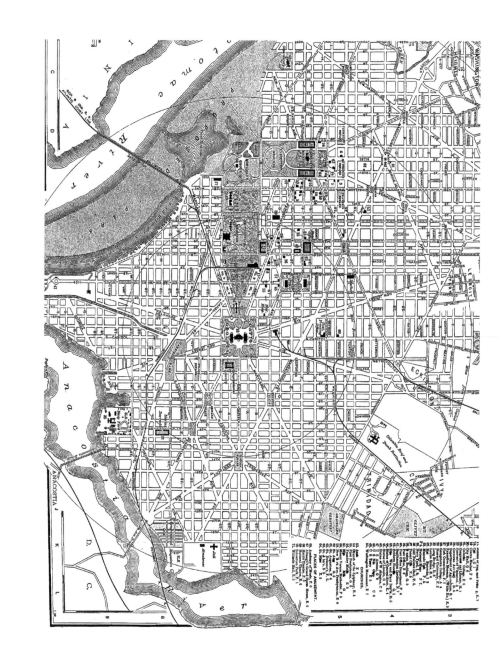

H), are excellent ; other good restaurants are *Moylan's*, Pennsylvania Ave., adjoining Willard's, and the *Losekam*, F St., bet. 13th and 14th Sts. ; *Harvey's* (cor. Pennsylvania Ave. and 11th St.) is noted for its oysters. Good lunch-rooms are the *Holly-Tree*, 518 9th St. ; and *Evans's*, in F St. near 9th St. The hotels on the European plan have restaurants attached. In the basement of the Capitol, under each House, is an excellent restaurant, open during the sessions of Congress. Visitors who prefer to obtain transient board or rooms can find good accommodations in all parts of the city.

Modes of Conveyance.—*Street-cars* (fare, 5c., 6 tickets for 25c.) afford easy access to all points. There are electric cars running to Tennallytown, Chevy Chase, and to Eckington. There are herdic lines in Pennsylvania Ave. and 15th St., and in F and G Sts. *Cabs* for one or two passengers, 75c. an hour ; three or four passengers, $1 an hour in the city limits ; by the trip, 25c. for each passenger for 1 mile. For excursions beyond the city limits it is better to hire carriages at the livery-stables or hotels. A *steamboat* for Mount Vernon leaves the 7th St. wharf daily at 10 A. M. *Ferry-boats* run to Alexandria hourly during the day from 7th St. wharf (fare, 15c.; round trip, 25c.).

Railroad Stations.—The station of the *Baltimore & Potomac R. R.* is a spacious and highly ornate building, cor. B and 6th Sts. The *Richmond & Danville, Chesapeake & Ohio*, and the *Richmond, Fredericksburg & Potomac* trains leave from the Baltimore and Potomac station. That of the *Baltimore & Ohio R. R.* is at the cor. New Jersey Ave. and C St.

Churches.—Those most visited by strangers are the *St. Aloysius* (Roman Catholic), cor. N. Capitol and I Sts., noted for its rich interior and fine choral music ; *St. Matthew's*, E. of Lafayette Square, usually attended by Catholic members of the Diplomatic Corps ; *St. Dominic's* (Roman Catholic), an imposing granite structure, cor. 6th and F Sts. ; *St. Augustine* (Roman Catholic), in 15th, between L and M Sts., noted for its music ; *St. John's* (Episcopal), fronting Lafayette Square on the N., a famous old church, attended by Presidents Madison and Monroe ; the *Church of the Epiphany* (Episcopal), in G St., between 13th and 14th ; the *Ascension* (Episcopal), of light stone, the finest church in the city, cor. Massachusetts Ave. and 12th St., N. W.; the *Metropolitan Methodist*, a splendid brown-stone edifice, cor. 4½ and C Sts.; the *Mount Vernon Methodist*, cor. 9th and K Sts.; the *Foundry Methodist*, in F St. near 14th ; the *First Presbyterian*, in 4½ St. near C St.; the *Church of the Covenant* (Presbyterian), cor. 18th St. and Connecticut Ave.; the *N. Y. Avenue Presbyterian*, in N. Y. Ave. near 14th St.; *Calvary* (Baptist), cor H. and 8th Sts.; *Garfield Memorial* (Christian), Vermont Ave. near N St.; *All Souls* (Unitarian), cor. 14th and L Sts.; *Church of Our Father* (Universalist), cor. 13th and L Sts.; and *First Congregational*, cor. G and 10th Sts.

Theatres and Amusements.—The *New National Theatre* is in E. St., near 14th St. *Albaugh's Grand Opera-House*, in 15th St., near Pennsylvania Ave., was opened in Nov., 1885. *Harris's Bijou Theatre* (formerly Ford's Opera-House) is in 9th St., near Pennsylvania Ave. *Lyceum Theatre* is at the cor. Pennsylvania Ave. and 11th Sts., N. W. *Academy of Music*, cor. D and 9th Sts., is used for entertainments. In *Masonic Hall* (cor. F and 9th Sts.) public parties and balls are often given. *Odd-Fellows' Hall*, in 7th St., between D and E Sts., and *Willard's Hall*, in F St. near 14th, are also used for lectures and concerts. The *Panorama of the Battle of Manassas*, cor. 15th St. and Ohio Ave., was opened Oct. 5, 1885, and is a remarkably realistic representation. *Schuetzen Park* is a German resort in 7th St., beyond the Howard University.

Reading-Rooms.—At all the leading hotels are reading-rooms well supplied with newspapers. The *Library of Congress*, in the Capitol, is open to visitors from 9 A. M. to 4 P. M. The *Army Medical Library*, cor. 7th and B Sts., is the finest medical library in the United States. The reading-rooms of the *Young Men's Christian Association* (N. Y. Ave., between 14th and 15th Sts.) are open (free) from 9 A. M. to 10 P. M. The *Patent-Office Library* is rich in scientific and mechanical works. At the offices of the Washington correspondents of leading American newspapers files of newspapers are usually accessible to the visitor.

Art Collections.—The *Corcoran Gallery of Art* (cor. Pennsylvania Ave. and 17th St.) has one of the richest collections in America (see p. 63). Admission free on Tuesdays, Thursdays, and Saturdays, also on Friday evening, from 7.30 to 10 P. M. ; on other days, 25c.

Post-Office.—The *City Post-Office* is in G St. between 6th and 7th Sts. Open from 6 A. M. to 11 P. M.; on Sundays from 8 to 10 A. M. and 6 to 7 P. M.

WASHINGTON CITY, the political capital of the United States, is situated on the N. bank of the Potomac River at its confluence with the Eastern Branch. Its site is an admirable one, consisting of an extensive undulating plain surrounded by rolling hills and diversified by irregular elevations which furnish advantageous positions for the various public buildings. The plan of the city is unique ("the city of Philadelphia griddled across the city of Versailles"), and is on a scale which shows that it was expected that a vast metropolis would grow up there. It covers an area 4½ miles long by 2½ broad, embracing nearly 9½ square miles. A very small portion of this, however, is as yet built upon. **Pennsylvania Avenue,** in that part of its course between the Capitol and the White House (1⅛ mile) is the busiest and most fashionable street in the city; it is 160 ft. wide, and on it or near it are many of the leading hotels, theatres, stores, etc. F St., between 7th and 15th Sts., is now the leading street for shops. **Seventh St.,** which intersects Pennsylvania Ave. about midway between the Capitol and the Treasury, is the next most important thoroughfare, and contains many handsome stores. *Massachusetts Ave.* extends entirely across the city (4½ m.), parallel with Pennsylvania Ave., and on portions of its course is lined with fine residences. *Maryland Ave.* leads S. W. from the Capitol to the Long Bridge, and N. E. to the Toll-gate. *Vermont* and *Connecticut Aves.* contain many handsome residences. *Fourteenth St.* is one of the most important of the cross-streets. Ninth St., from Pennsylvania Ave. to F St., is entirely a business street. The favorite drives are to the Zoölogical Park; to the Cabin John Bridge; to the Soldiers' Home; to the Heights of Georgetown; to the Little Falls of the Potomac (3 miles above Georgetown); to the Great Falls of the Potomac (17 miles from Washington); and across the river to Arlington, Alexandria, and the heights along the Virginia shore.

The best time to visit Washington and to see its most characteristic aspect is during the sessions of Congress. These begin on the first Monday in December, and last until March 4 in the odd-numbered years, and until June, July, or even October in the even-numbered years. During this period the galleries of the Senate and House of Representatives are open to visitors. The sessions of both Houses begin at noon and usually close before sunset, but sometimes they are prolonged far into the night. A flag displayed over the N. wing of the Capitol indicates that the Senate is in session; over the S. wing that the House is in session. When the sittings are prolonged into the night, the great lantern over the dome is illuminated. The best times for seeing the natural beauties of Washington are May, or early June, and October.

The site of Washington City, if not chosen by Washington himself, seems to have been selected through his agency, and it was he who laid the corner-stone of the Capitol. This was on Sept. 18, 1793, seven years before the seat of government was removed thither from Philadelphia. Under Washington's direction the city was planned and laid out by Andrew Ellicott. It appears to have been Washington's desire that it should be called the "Federal City," but the name of "the city of Washington" was conferred upon it on Sept. 9, 1791. The city was incorporated May 3, 1802. Its population in 1860 was 60,000 ; in 1870, 109,189 ; in 1880, 147,307 ; and by census of 1890, 230,392. This is increased during the sessions of Congress by a floating population amounting to many thousands. The commerce and manufactures of Washington are unimportant.

The Public Buildings [1] are the chief attraction of Washington, and the ** **Capitol** is not only the finest of these, but is probably the most magnificent public edifice in the world. It crowns the summit of Capitol Hill (90 ft. high), and consists of a main building 352 ft. long and 121 ft. deep, and two wings or extensions, each 238 by 140 ft. Its whole length is 751 ft. 4 in., and the area covered rather more than 3½ acres. The material of the central building is a light-yellow free-stone (painted white), but the extensions are pure white marble. The surrounding grounds, which are beautifully cultivated, and embellished with fountains and statuary, embrace about 50 acres, and are known as East and West Grounds. The main front is toward the E., and is adorned with three grand porticoes of Corinthian columns. On the steps of the central portico are groups of statuary by Persico and Greenough ; and on the esplanade in front of it is * Greenough's colossal statue of Washington. Colossal marble statues of Peace and War are on the r. and l. of the entrance; and over the doorway is a bass-relief of Fame and Peace crowning Washington with laurel. The W. front projects 83 ft., and is embellished with a recessed portico of 10 columns. This front, though not so imposing architecturally as the eastern, commands a fine view of the central and western portions of the city and of all the principal public buildings. The *Bronze Door*, which forms the entrance to the Rotunda from the E. portico, is worth attention. It was designed by Randolph Rogers, cast by Von Müller, at Munich, is 17 ft. high and 9 ft. wide, weighs 20,000 lbs., and cost $28,000. The work is in *alto-relievo*, and commemorates the history of Columbus and the discovery of America. There are also bronze doors at the entrance to the Senate wing, designed by Crawford, and completed (after his death) by Rinehart, of Baltimore. The *Rotunda* is 96 ft. in diameter and 180 ft. high. In the panels surrounding it are 8 large pictures, illustrating scenes in American history, painted for the Government by native artists; and over the 4 doors or entrances are *alti-rilievi* in stone. At a height of 107 ft. from the floor, there is painted, in imitation of *alti-rilievi*, a series of illustrations of American history, on a space 9 ft. high encircling the spacious wall. The floor is of freestone, supported by arches of brick, resting upon two concentric peristyles of Doric columns in the crypt below. The * *Dome* rises over the Rotunda in the center of the Capitol, and is the most imposing feature of the vast pile. The interior measures 96 ft. in diameter, and 220 ft. from the floor to the ceiling. Externally it is 135½ ft. in diameter, and rises 241 ft. above the roof of the main building, 307½ ft. above the base-line of the building, and 377 ft. above low tide. Visitors should not fail to make the ascent of the Dome. A spiral stairway, between the outer and inner shells

[1] All public buildings, including the Capitol and the several Departments, are open to the public every day (except Sundays), the Capitol from 9 A. M. to 5 P. M., but the Departments from 9 A. M. to 2 P. M. only. No fees are asked or expected for showing them, except to the regular guides.

(diverging to the l. from the corridor outside the N. door of the Rotunda)
affords easy access, and gives a favorable opportunity for inspecting,
from different points of view, the fresco-painting on the canopy overhead. This is the work of Brumidi ; it covers 6,000 ft. of space and cost
$40,000. All the figures (63 in number) are of colossal proportions, so
as to appear life-size when seen from the floor beneath. From the balustrade at the base of the canopy is obtained a magnificent* view of the
city and the surrounding country. From the gallery immediately underneath the fresco-gallery another spiral stairway leads up to the lantern
(17 ft. in diameter and 52 ft. high). This is surmounted by the tholus,
or ball, and this, in turn, by Crawford's fine bronze statue of Liberty,
19¼ ft. high.—Leaving the Rotunda by the S. doorway, the visitor finds
himself in the *Old Hall of Representatives* (now used as a "National
Statuary Hall"). This room, the noblest in the Capitol, is semicircular
in form, 96 ft. long and 57 ft. high to the apex of the ceiling. The 24
columns which support the entablature are of variegated green *breccia*,
or pudding-stone, from the Potomac Valley ; and the ceiling is painted in
panel, in imitation of that of the Pantheon at Rome. Light is admitted
through a cupola in the center of the ceiling. Over the S. door is a statue
of Liberty, by Causici, and an eagle by Valaperti. Over the N. door is a
statue, by Franzoni, representing History standing in a winged car, the
wheel of which, by an ingenious device, forms the dial of a clock. In 1864
the room was set apart as a National Statuary Hall, each State being
requested to send statues of two of its most eminent men. A number of
States have responded, and the Hall is slowly filling up, containing already a numerous array of statues and busts, of which those of Hamilton, Jefferson, Winthrop, General Greene, Livingston, Fulton, Samuel
Adams, Roger Williams, Ethan Allen, Vinnie Ream's statue of Lincoln,
and a plaster cast of Houdon's Washington, are most noteworthy.—The
corridor to the S. leads to the present *Hall of Representatives*, the finest
legislative chamber in the world, 139 ft. long, 93 ft. wide, and 36 ft. high.
The ceiling is of iron-work, with 45 stained glass panels on which are
painted the arms of the States. To the left of the marble desk of the
Speaker is a full-length portrait of Lafayette, and to the r. a full-length
portrait of Washington, copied from Stuart's, by Vanderlyn. Two
landscapes by Bierstadt, "The Discovery of the Hudson," and "Settlement of California," and a fresco by Brumidi, of Washington parting with
his officers, fill panels on the S. wall. The Strangers' Gallery (reached
by two grand marble stairways) extends entirely round the hall ; the
space not specially appropriated for the use of the diplomatic corps and
the reporters for the press is open to visitors. The *Speaker's Desk*, of
white marble, is very fine ; and the Lobby, or Retiring-Room, in the rear
of the desk, is a superb chamber. From the S. lobby of the Hall two
stairways descend to the basement, where are located the Refectory and
committee-rooms. The room of the *Committee of Agriculture* will repay
a visit; the walls and ceiling are painted in fresco by Brumidi.—The
Senate-Chamber, reached by the corridor leading N. from the Rotunda,
is somewhat smaller than the Hall of Representatives, being 113¼ ft.
long, 80¼ ft. wide, and 36 ft. high. It is very tastefully fitted up. The

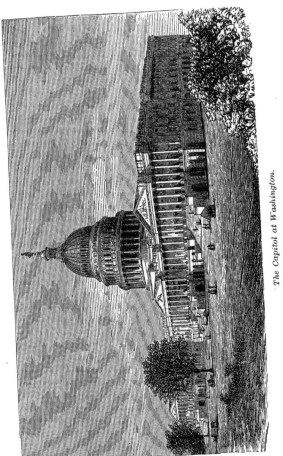

The Capitol at Washington.

visitors' galleries are reached by *marble stairways, which are among
the most striking architectural features of the Capitol.   The President's
and Vice-President's Rooms, the Senators' Retiring-Room, the Reception-
Room, and the Senate Post-Office, are beautiful chambers.   In the Vice-
President's Room hangs Rembrandt Peale's great portrait of Washington.   The *Marble Room* is particularly chaste and rich in its decorations; senators alone have the privilege of allowing visitors to enter it.
In the basement of the Senate Extension are committee-rooms, richly
frescoed and furnished, and the corridors are exquisitely painted.—The
Supreme Court-Room (formerly the Senate-Chamber) is reached by the
corridor leading N. from the Rotunda.   It is a semicircular apartment,
75 ft. long and 45 ft. high, decorated with rich Ionic columns of Potomac marble, and with busts of the former Chief-Justices.   Visitors are
admitted during the sessions of the Court (October to May, 12 to 4 P. M.).
Underneath the room is the apartment formerly occupied by the Court
and now devoted to the Law Library (100,000 volumes).—The *Library
of Congress* is reached by the corridor from the W. door of the Rotunda.
It occupies the entire W. projection of the Capitol; the main room is 91
ft. long and 34 ft. wide, ceiled with iron, and fitted up with fire-proof
cases.   The Library was founded in 1800; was burned by the British
in 1814; was again partially burned in 1851; and went into its present
rooms in 1853.   The collection, which is the largest in the United States,
now numbers more than 650,000 volumes, exclusive of pamphlets, and
is increasing at the rate of 10,000 or 15,000 volumes a year.   A separate building to cost $6,000,000 is being built on the square facing the
E. side of the Capitol.   It is in the style of the Italian renaissance, designed by J. L. Smithmeyer.   Accommodations for 8,000,000 books
have been arranged for.   All copyright books are, by law, required to
be deposited in this library, and the representation of American publi-
cations is by far the most complete in the country.   Books may be read
in the library by visitors, but not taken away (open from 9 A. M. to 4
P. M.).—The lighting, heating, and ventilating apparatus of the House
and Senate-Chambers are worthy of notice.   The total cost of the Capitol was $13,000,000.

The site of the Capitol was located in 1791: the corner-stone was laid in
Sept., 1793, by Washington. The wings were burned by the British in 1814. The
building was finished in 1827. Mr. Walter's design for its extension was commenced in 1851, and finished in 1865, with the completion of the Dome.

From the W. entrance of the Capitol Grounds, Pennsylvania Ave.
leads in 1¼ mile to the *U. S. Treasury* (cor. 15th St.), a magnificent building in the Ionic style, 468 ft. long and 264 ft. wide, 3 stories
high above the basement, erected at a cost of $6,000,000.   The E.
front has an Ionic colonnade 342 ft. long, modeled after that of the Temple of Minerva at Athens.   This front is of Virginia freestone; the rest
of the building is of Dix Island granite.   The W. front has side porticoes, and a grand central entrance with 8 monolithic columns of enormous size.   The N. and S. fronts are alike and are adorned with stately porticoes.   The building contains about 200 rooms, of which the
finest is the *Cash-Room*, extending through 2 stories, and lined through-

out with rich marbles. The *Gold-Room*, in which there are many millions of dollars in gold coin, may be seen by permit from the Treasurer. The *Bureau of Engraving and Printing*, cor. 14th and B Sts., S. W., near the Washington Monument, was completed in 1880, and is a branch of the Treasury Department. All the notes, bonds, etc., of the Government are printed here, and it is a place of much interest to visitors. The building is open to visitors from 9 A. M. to 2 P. M.

Just W. of the Treasury is the * **Executive Mansion** (usually called the " White House "). It is of freestone, painted white, 170 ft. long and 86 ft. deep, two stories high, with a portico on the N. side (main entrance) supported by 8 Ionic columns, and a semicircular colonnade on the S. side of 6 Ionic columns. The corner-stone was laid in 1792 ; the building was first occupied by President Adams in 1800 ; burned by the British in 1814 ; and restored and reoccupied in 1818. The grounds lie between 15th and 17th Sts., and extend to the Potomac, comprising about 75 acres, of which 20 are inclosed as the President's private grounds, are handsomely laid out, and contain a fountain and extensive conservatories. The *East Room* (open daily from 10 A. M. to 3 P. M.) is the grand parlor of the President ; it is a fine chamber 80 ft. long, 40 wide, and 20 high, richly decorated and furnished. The Blue, Red, and Green Rooms are on the same floor, and are elegant in their appointments. The *Executive Office* and the *Cabinet-Room* are on the 2d floor, as are also the private apartments of the family. N. of the White House is **Lafayette Square,** the finest public park in the city, laid out in winding paths and filled with trees and shrubbery. In the center is Clark Mills's * bronze equestrian statue of Gen. Jackson, remarkable for its delicate balancing, which was accomplished by making the flanks and tail of the horse of solid metal. On the S. E. cor. is the *Lafayette Monument*, with statues of Lafayette, Rochambeau, d'Estaing, De Grasse, and Duportail. It was erected in 1890, and was executed by the French sculptors Antoine Falguiere and Antonin Mercie.

Just W. of the White House (fronting on Executive Ave. cor. 17th St. and Pennsylvania Ave.) is the vast and ornate building of the * **State, War, and Navy Departments,** of granite, in the Roman Doric style, 567 ft. long and 342 ft. wide, 4 stories high, with lofty Mansard-roof. It has 4 façades, those on the N. and S. and those on the E. and W. respectively being counterparts. The State Department occupies the S. portion of the building ; and the Hall of the Secretary of State, the Embassadors' Saloon, and the Library (30,000 volumes) are splendid rooms. (Open to visitors from 10 A. M. to 2 P. M.) The War and Navy Departments occupy the N. and E. wings respectively of the building fronting the Executive Mansion. The War Department is open from 9 A. M. to 2 P. M., and the Navy Department from 9 A. M. to 2 P. M.

The office of the *Department of the Interior*, better known as the * **Patent-Office,** is a grand Doric building of marble, freestone, and granite, occupying 2 blocks in the central portion of the city (between 7th and 9th and F and G Sts.), 453 ft. long and 331 ft. wide, including porticoes, and 75 ft. high. The F St. portico (main entrance) is reached by broad granite steps, and consists of 16 Doric columns of immense size, upholding a classic pediment. The interior contains many noble rooms. The *Model-Room* (open from 9 A. M. to 4 P. M.) occupies the entire upper floor of the edifice, forming 4 large halls or chambers unequaled for extent and beauty on the continent. The total length of this floor is 1,350 ft., or rather more than a quarter of a mile; and it is filled with cases containing immense numbers of models representing every department of mechanical art. The frescoes on the ceiling of the S. Hall are much admired. On the second or main floor are the offices of the Secretary of the Interior, of the Indian Bureau, of the General Land-Office, and of the Commissioner of Patents.

The **Bureau of Education,** cor. 8th and G Sts., opposite the Interior Department, contains a library of 18,000 bound volumes, including all important pedagogical works and 100,000 pamphlets. The special function of this bureau is to increase the enlightened directive power of the people with regard to their schools, by means of annual and special reports, the materials for which are collected by extensive correspondence with the officials in charge of the State, city, and county public-school systems, etc.

The * **Post-Office Department,** on F St. opposite the Patent-Office, is an imposing edifice of white marble in the Italian or modified Corinthian style, 300 ft. long, 204 ft. wide, and 3 stories high, erected at a cost of $1,700,000. In the center of the 8th St. front is a bit of sculpture representing the railroad and the telegraph. The Postmaster-General's Office is in the 2d story on the S. side. The *Dead-Letter Office* (2d story N. side) contains some curious objects.

The * **Pension Building,** located in the square bounded by F, G, 4th and 5th Sts., occupies 80,000 square ft., and borders on Judiciary Square. It is 400 by 200 ft., and the walls are 75 ft. high. The architecture is Renaissance. Among the most notable features of the exterior decoration are the terra-cotta cornices with medallions and ornaments, and the band of sculpture in terra-cotta on the level of the second floor, 3 ft. in height, by 1,200 feet in length, representing an army in campaign, assisted by sailors and boats of the navy.

The * **Department of Agriculture** (open from 9 A. M. to 2 P. M.) occupies a spacious brick and brown-stone building in the Renaissance style, situated on the Mall at the foot of 13th St. It contains a library, a museum, an herbarium (with 25,000 varieties of plants), and extensive greenhouses. The grounds are tastefully laid out, and contain a great variety of trees and plants. The *Flower-Gardens* (in front of the main building) are adorned with statuary, and when in bloom present a memorable sight.—A short distance E. on the Mall is the * **Smithsonian Institution,** a beautiful red sandstone building in the Romanesque style, 447 by 150 ft., with 9 towers ranging from 75 to 150 ft. in height (reached from Penn. Ave. *via* 7th St.). This noble institution was founded by James Smithson, an Englishman, "for the increase and diffusion of knowledge among men." The building was commenced in 1847 and completed soon after. It contains a museum of natural history with numerous specimens, and ethnological collections, with many curiosities. S. E. of the main building is the *National*

Museum, which contains the Centennial exhibits of, and donations of foreign Governments to, the United States, the Washington relics, the Grant swords, etc., and the collections of minerals, etc. The grounds attached to the Institution (52½ acres) are beautifully laid out. Visitors are admitted from 9 A. M. to 4.30 P. M.—Also on the Mall, E. of the Smithsonian and just W. of the Capitol grounds, are the * **Botanical Gardens,** which consist chiefly of a series of conservatories filled with rare and curious plants, flowers, and fruits (free to visitors from 9 A. M. to 6 P. M.). N. of the conservatory stands the *Bartholdi Fountain,* so much admired at the Centennial Exhibition. The **Army Medical Museum and Library,** also on the Mall, immediately adjoining the National Museum, occupies the building at the N. W. of B and 7th Sts. S. W. (open from 9 A. M. to 3 P. M.). This museum was founded and a large portion of the medical and surgical specimens collected during the war of the rebellion. Since the close of the war the officers in charge have continued to collect specimens from the medical officers of the army at military posts, and they have received a number from physicians engaged in private practice. Opposite the Army Medical Museum, across 7th St., is the **Commission of Fish and Fisheries,** with its biological laboratory, extensive aquaria for the study and display of salt and fresh water fishes, and also one of the principal shad-hatching stations.

The * **U. S. Naval Observatory** (lat. 38° 53′ 38.8″, lon. 77° 3′ 1·8″ W. from Greenwich) occupies a commanding site on the bank of the Potomac at the foot of 24th St., with handsome grounds embracing 19 acres. It was founded in 1842, and is now one of the foremost institutions of the kind in the world. It possesses many fine instruments (including a 26-inch equatorial telescope), and a good library of astronomical works. Visitors are admitted at all hours, and are allowed to inspect the telescope and other instruments when they are not in use. It will be removed to a new observatory overlooking Rock Creek, N. of Georgetown, during the present year.

The *Weather Bureau* is on the cor. M and 24th Sts. N. W. It has a library of 12,000 volumes, and the instrument and indication rooms are of interest to visitors. The *U. S. Arsenal* is located amid pleasant grounds on Greenleaf's Point, at the confluence of the Potomac and the Eastern Branch (reached by 4½ St.). The buildings contain vast stores of arms and ammunition. The * **Navy-Yard** is situated on the Eastern Branch, about 1¼ mile S. E. of the Capitol (reached by Pennsylvania Ave. cars). It has an area of 27 acres, inclosed by a substantial brick wall, within which, besides officers' quarters, are vast foundries and shops, 2 ship-houses, and an armory. The *Naval Museum* (open from 9 A. M. to 4 P. M.) contains an interesting collection of arms, ammunitions, and relics. Other interesting features are the Experimental Battery and the fleet. Two blocks N. of the Navy-Yard are the *Marine Barracks,* the headquarters of the U. S. Marine Corps; and near by is the *Marine Hospital,* for sick and disabled sailors.

Noteworthy buildings not belonging to the Government are the *Baltimore Sun* building, on F St.; the *Court-House* (on 4½ St. near Lou-

isiana Ave.); the *Masonic Temple* (cor. F and 9th Sts.); *Odd-Fellows'
Hall* (in 7th St. between D and E Sts.); the spacious *Washington Market*,
fronting Pennsylvania Ave. between 7th and 9th Sts.; and the
churches and hotels already enumerated under their respective heads.
Columbian University has its main building cor. H and 15th Sts. N. W.
It was incorporated as a college in 1821 and as a university in 1873.
It has collegiate, law, and medical departments. On the cor. H St. and
Madison Place is the *Cosmos Club*, where "Dolly" Madison held court
and dispensed her gracious hospitality after her husband's death, and
across Lafayette Square on the cor. 17th St. is the *Metropolitan Club*,
with its fine home. The large hotels *The Shoreham* and *Normandie* are
in 15th St., while in the vicinity are the *Arlington*, *Arno*, and other
hotels.

The *Corcoran Art-Gallery* is a brick and brown-stone building
in the Renaissance style near the White House (cor. Pennsylvania
Ave. and 17th St.). It was founded by the late William W. Corcoran,
the banker, who deeded it to the people, and presented it with his
private art collection and an endowment fund of $900,000. It contains
nearly 200 paintings, most of them masterpieces; a fine collection
of casts, and among the marble statuary "The Greek Slave," by
Powers, and "The Dying Napoleon," by Vela; many rich bronzes, notably
those by Barye; and other interesting art specimens. Another
noble institution, founded and liberally endowed by Mr. Corcoran, is the
Louise Home (between 15th and 16th Sts. on Massachusetts Ave.), a
handsome building erected at a cost of $200,000, designed to furnish a
home to impoverished elderly ladies of education and good family.

The **Washington Monument,** which is the loftiest in the world,
except the Eiffel Tower, stands on the Mall near 14th St., and was dedicated
with appropriate ceremonies on Washington's birthday, 1885. Its
design contemplated, besides a spacious "Temple," or base, a shaft 600
ft. high; but after $230,000 had been expended in building it to a
height of 174 ft., funds gave out and the work was suspended. In 1876
Congress made an appropriation for the completion of the monument
on a new plan to be chosen by experts. Rapid progress was then made
on it. It is 555 ft. high. The top is reached by an interior stairway
and also by an elevator, which runs every half-hour from 9 A. M. to 5
P. M. on week-days. Clark Mills's colossal equestrian *Statue of Washington*
stands in Washington Circle, at the intersection of Pennsylvania
and New Hampshire Aves. H. K. Brown's colossal equestrian *Statue
of General Scott* stands at the intersection of Massachusetts and Rhode
Island Aves. with 16th St. The same artist's bronze equestrian statue
of *General Nathanael Greene* stands at the intersection of Maryland and
Massachusetts Aves. with 5th St. N. E. Ball's colossal bronze *Statue
of Lincoln* stands in Lincoln Park (in the E. part of the city); it was
erected by contributions of colored people. Another statue of Lincoln,
by Lot Flannery, stands in Judiciary Square, on 4½ St. On Pennsylvania
Ave. near 9th St. is a bronze statue of *General Rawlins*, by J.
Bailey, and at 10th St. is a marble statue of *Benjamin Franklin*. Vinnie
Ream's statue of *Admiral Farragut* stands in Farragut Square

(cor. Connecticut Ave. and I St.), and J. Q. A. Ward's fine equestrian statue of *General Thomas* stands in a circle at the intersection of 14th St. with Massachusetts and Vermont Aves., and in Luther Place near Thomas Circle is a bronze statue of *Martin Luther.* It is a replica of the celebrated figure of the Luther Memorial Group, by Reitschel, at Worms, Germany. At the Maryland Ave. entrance to Capitol Park is the bronze statue of *President Garfield*, by J. Q. A. Ward, erected by his comrades of the Army of the Cumberland in 1887. The pedestal was the gift of Congress, and has bronze recumbent figures of the student, warrior, and statesman. In Dupont Circle is a heroic bronze statue of *Admiral Samuel F. Dupont*, by Launt Thompson, which was erected in 1884 at a cost of $14,000. A bronze equestrian statue of *General McPherson* stands at the intersection of Vermont Ave. and 15th St. The *Naval Monument*, erected to the memory of the officers and seamen and marines who fell in the civil war, stands in the middle of Pennsylvania Ave., near the W. entrance to the Capitol Grounds. In the Capitol Grounds is a fine statue of *Chief-Justice Marshall*, by Story.

The * **Soldiers' Home** (for disabled soldiers of the regular army) occupies an elevated plateau 3 miles N. of the Capitol (reached by 7th St. cable-cars or by a charming drive, also by electric road). It consists of several spacious marble buildings in the Norman style, surrounded by a beautiful park of 500 acres. It has been the custom of several Presidents to occupy one of the smaller buildings of the Home as a summer retreat, and here President Lincoln passed some of the last hours of his eventful life. N. of the Home is a *National Cemetery*, in which 5,424 soldiers are buried. At the terminus of the electric road is the **Catholic University of America,** which, founded in 1889, is intended as an institution for post-graduate work. At present only a divinity faculty has been established, but in time faculties for philosophy, law, medicine, social and political science, and the natural sciences will be added. A building of Tennessee blue granite, ornamented with white stone handsomely carved, has already been erected. Its chapel, said to be "one of the most beautiful in America," deserves a visit, and its fine stained windows, with designs of sacred interest, were made in Munich. On the 7th St. road just beyond the city limits is the **Howard University,** founded in 1864 for the education of youth "without regard to sex or color," but patronized almost exclusively by negroes (700 students). The University building is a brick structure, painted white, situated on elevated ground, and surmounted by a tower, from which there is a fine view of the city and its environs. The **Columbia Institution for the Deaf and Dumb and National Deaf-Mute College** is in the park called Kendall Green, just beyond the northeastern boundary of the city (reached by the New York Ave. cars). This college is the only one in the world for deaf-mutes. Within the grounds is the beautiful bronze group of Gallaudet teaching a deaf child, by Daniel C. French. The **Government Asylum for the Insane** (of the Army, Navy, and District of Columbia) occupies a lofty eminence on the S. bank of the Anacostia (reached by crossing the

Navy-Yard bridge and ascending the heights beyond Uniontown). The building is in the Collegiate-Gothic style, 711 ft. long, and is surrounded by a park of 419 acres, from which there are noble views. (Admittance on Wednesday from 2 to 6 P. M.) In the *Congressional Cemetery (1 mile E. of the Capitol near the Eastern Branch) are the graves of Congressmen who have died during their term of service. Its situation is high, and it contains some noteworthy monuments. *Glenwood* is a pleasant rural cemetery about a mile N. of the Capitol. The celebrated **Long Bridge** crosses the Potomac into Virginia from foot of 14th St. It is a shabby structure about a mile long.

Georgetown, now called *West Washington*, is distant but 2 miles from the Capitol, and divided only by Rock Creek from Washington City, with which it is connected by 4 bridges and 2 lines of horsecars. The town is beautifully situated on a range of hills which command a view unsurpassed in the Potomac Valley. It is the port of entry of the District, and a line of steamships plies between it and New York. One of the chief points of interest is *Georgetown University*, at the W. end of the town. This is an old institution of learning (founded in 1789, and incorporated as a university in 1815), and the most famous belonging to the Roman Catholic Church in the United States. It is under the control of the Jesuits. The buildings are spacious, and contain a library of 35,000 volumes, among which are some extremely rare and curious books, some beautifully illuminated missals, and some rare old MSS.; an astronomical observatory, and a museum of natural history. In the rear of the college is a picturesque rural serpentine walk, commanding fine views. The *Convent of the Visitation* (in Fayette St. near the College) was founded in 1799, and is the oldest house of the order in America. It consists of several fine buildings in a park of 40 acres. Visitors admitted between 11 A. M. and 2 P. M. The *Aqueduct*, by which the waters of the Chesapeake & Ohio Canal are carried across the Potomac, will repay inspection. It is 1,446 ft. long and 36 ft. high, with 9 granite piers, and cost $2,000,000. There is a carriage-way above the water-course. *Oak Hill Cemetery,** on the N. E. slopes of the Heights, though containing but 30 acres, is one of the most beautiful in the country. It contains an elegant Gothic chapel with stained-glass windows and completely overgrown with ivy, the massive marble mausoleum of W. W. Corcoran, and several notable monuments. Many eminent men are buried here, among them Secretary Edwin M. Stanton and Chief-Justice Salmon P. Chase.

The grounds of the **Zoölogical Park** lie on both sides of Rock Creek, just N. of Woodley Lane in the suburbs. They comprise 166 acres, which were purchased by act of Congress in 1889, at a cost of nearly $200,000. The park is inclosed, and several structures suitable for the use of the animals have been erected. A number of North American animals have been placed here, and an excellent nucleus started for a national zoölogical garden. A number of the larger Rocky Mountain animals have been captured in the Yellowstone National Park and are now in the collection. The *Rock Creek Park*, comprising a tract of land of 2,000 acres, extending along both sides of Rock Creek,

5

adjoins the Zoölogical Park. It was purchased by Congress in 1890, and will be beautified with drives, walks, and similar features of land-scape gardening.

Arlington House, once the residence of George Washington Parke Custis, the last survivor but one of the Washington family, and later of Gen. Robert E. Lee, occupies a commanding position on the Virginia side of the Potomac, nearly opposite Georgetown (reached from Georgetown *via* Aqueduct Bridge). It stands more than 200 ft. above tide-water, and the view from the portico is among the best this part of the river affords. The lower rooms of the mansion are open to the public, but contain nothing of interest, the collection of pictures and relics having been removed. In the office of the Superintendent a register is kept for visitors, and a record of all who are buried in the *National Cemeteries* now located on the place. In front of the mansion is the grave of General Philip H. Sheridan. The graves of the white soldiers are W. of the house; those of the colored troops and refugees about ¼ mile N. There are about 15,000 in all.

Alexandria is situated on the S. side of the Potomac 7 m. below Washington (reached by railroad, or by ferry-boats hourly from 7th St. wharf). It is a quaint old town, dating from 1748, and is intimately associated with the life and name of Washington. In *Christ Church* (cor. Washington and Cameron Sts.) the pew in which he sat (No. 59) is an object of much interest. Pew No. 46 was occupied by Robert E. Lee when he resided at Arlington before the civil war. The Museum, Court-House, Odd-Fellows' Hall, and Theological Seminary are among the prominent buildings. On the outskirts of the city is a *National Cemetery*, in which nearly 4,000 soldiers are buried.

* Mount Vernon is 15 m. below Washington, on the Virginia side of the Potomac, and is reached by steamers which leave the 7th St. wharf daily at 10 A. M. (fare for the round trip, including admission to the grounds, $1.00). The sail down the river is delightful, and affords excellent views of the country around Washington. Mount Vernon, then known as the "Hunting Creek estate," was bequeathed by Augustine Washington, who died in 1743, to Lawrence Washington. The latter named it after Admiral Vernon, under whom he had served in the Spanish wars. George Washington inherited the estate in 1752. The central part of the mansion, which is of wood, was built by Lawrence, and the wings by George Washington. It contains many interesting historical relics, among which are the key of the Bastile, presented by Lafayette, portions of the military and personal furniture of Washington, portraits, and Rembrandt Peale's painting of "Washington before Yorktown." The *Tomb of Washington* stands in a retired situation near the mansion. It is a plain but solid brick structure, with an iron gate, through the bars of which may be seen the marble sarcophagi containing the remains of George and Martha Washington. The Mount Vernon domain (including the mansion and 6 acres), which had remained since the death of Washington in the possession of his descendants, was purchased in 1856 for the sum of $200,000, raised by subscription, under the auspices of the "Ladies' Mount Vernon Association," aided by the efforts of Edward Everett. It is, therefore, and will continue to be, the property of the nation.

Itineraries.

The following series of excursions has been prepared so as to enable the visitor whose time is limited to see as much of the city as possible in the least amount of time. Each excursion is planned to occupy a single day, but the visitor can readily spend more time as special features crowd upon his attention. Travelers by the Pennsylvania or Baltimore & Ohio R. Rs. can readily obtain the local guide-books of Washington furnished by these lines.

1. Take cars on Pennsylvania Ave. to Navy-Yard; same cars back to Capitol. During the sessions the Supreme Court and both Houses of Congress open at noon. Visit the Botanical Gardens at the foot of the Capitol Grounds, seeing the statue of Garfield and the Peace Monument on the way.

2. Visit the Washington Monument (foot of 13th St.), Bureau of Engraving and Printing, Agricultural Department, Smithsonian Institution, U. S. National Museum, and Fish Commission; these are all within walking distance of each other. The National Museum contains a lunch-room.

3. Visit the Treasury Department, State, War, and Navy Departments, the White House, and Corcoran Art-Gallery. A drive to the Zoölogical Park by way of Connecticut Ave. would pass the Church of the Covenant, the English Legation, and the statues of Admirals Farragut and Dupont.

4. Take F St. cars to Patent-Office, Dead-Letter Office, Bureau of Education. Walk two squares E. on F St. to Pension-Office. These departments close at 2 P. M., and the Soldiers' Home could be visited in the afternoon. It is reached by the electric and cable cars.

5. Take Pennsylvania Ave. cars to 7th St.; transfer to 7th St. cable-cars to wharf, where take steamer at 10 A. M. for Mount Vernon. On return trip stop at Alexandria; thence back to the city by either ferry-boat or train.

6. Visit Arlington, best reached by carriage. The Weather Bureau, Naval Observatory, and Fort Myer can be visited on the way.

9. The Hudson River.

The trip up the Hudson may be made either by railroad or steamboat, the latter affording the better opportunity for viewing the scenery; but the boats run only during the summer and autumn months. The **day boats** leave the pier at the foot of Vestry St. at 8.45 A. M., and from W. 22d St. 15 minutes later, reaching Albany at 6 P. M. The **night boats** start from the foot of Canal St. at 6 P. M., reaching Albany at 6 o'clock the next morning. The trip may be made to equal advantage by the Troy boats, which start from the foot of Christopher St. at the same hours. The steamer "*Mary Powell*" leaves the pier foot of Desbrosses St. daily at 3.15 P. M., and runs to Rondout and Kingston. A delightful excursion may be made by taking the morning boat to West Point, Cornwall, and Newburg, and returning on the afternoon boat (fare for the round trip, $1). The **New York Central & Hudson River R. R.** runs along the E. bank of the river all the way to Albany (143 miles), and, though the view from the cars is restricted for the most part to the western side of the river, the journey is nevertheless a most attractive one. The time to Albany is 2¼ to 4 hours. The **West Shore,** running to Albany on the west side of the river and to Buffalo, is also a popular line of travel. Time to Albany, 3 to 4 hours. The *Northern R. R. of New Jersey* also follows the west shore, terminating at Nyack.

THE trip up the Hudson River (especially by steamer) will afford the traveler advantageous views of some of the most picturesque scenery in America. The Hudson has been compared to the Rhine, and what it lacks in crumbling ruin and castle-crowned steep it more than makes up by its greater variety and superior breadth. George William Curtis says of it: "The Danube has in part glimpses of such grandeur, the Elbe has sometimes such delicately-penciled effects, but no European river is so lordly in its bearing, none flows in such state to the sea."

The first few miles of the steamer's course afford fine views of the harbor and city, of the Jersey shore, and the northern suburbs, including *Fort Washington.* Before the city is fairly left behind, the **Palisades** loom up on the left—a series of grand precipices rising in many places to the height of 300 ft., and stretching in unbroken line along the river-bank for more than 15 miles. The rock is trap, columnar in formation, and the summit is thickly wooded. In striking contrast, the right bank presents a continuous succession of beautiful villas standing in the midst of picturesque and exquisitely kept grounds, with a frequent sprinkling of villages and hamlets. Above Washington Heights (on the E. side) *Spuyten Duyvil Creek* enters the Hudson. *Mount St. Vincent* (10 miles beyond) is the seat of the Convent of St. Vincent, under the charge of the "Ladies of the Sacred Heart." The buildings present a striking appearance from the river, and among them is the castellated structure known as "Fonthill," formerly the residence of Edwin Forrest, the tragedian. **Yonkers** (17 miles on the E. side; *Getty House, Mansion House*) has a population of 32,033, and is beautifully situated on villa-crowned slopes at the mouth of the Neperan or Saw-Mill River. It is an ancient settlement, and was the home of the once famous Phillipse family, of which was Mary Phillipse, Washington's first love. The Manor House, a spacious stone edifice, built in 1682, is now the *City Hall;* and near by is Locust Hill, where the American troops were encamped in 1781. *Piermont* (22 miles on the W. side) is at the end of the Palisades; it takes its name from a mile-long pier which runs out from the shore to deep water; and 3 miles S. W. is the old town of *Tappan,* interesting as one of Washington's headquarters during the Revolution, and as the place where the unfortunate Major André was imprisoned and executed. The house occupied by Washington is still shown, and near by is the spot where André was executed (Oct. 2, 1780). At Piermont begins the widening out of the river into the broad and beautiful *Tappan Zee,* which is nearly 10 miles long and 4 miles wide at the widest part. On the E. bank, 26 miles from New York, is the village of *Irvington,* named in honor of Washington Irving, whose cottage of **Sunnyside** ("Wolfert's Roost") is close by, upon the margin of the river, but hidden from the traveler's view by the dense growth of the surrounding trees and shrubbery. The cottage is a quaint and picturesque structure, and the E. front is embowered in ivy, the earlier slips of which were given to Irving by Sir Walter Scott, at Abbotsford,

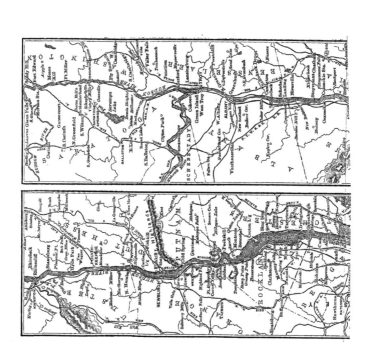

and planted by Irving himself. In the vicinity of Irvington are many fine residences, the most conspicuous of which is *Lyndehurst*, the old *Paulding Manor*, situated on a high promontory, and now the property of Jay Gould. Just above is *Tarrytown, which has many attractions, historic as well as scenic. It was at a spot now marked by an inscription in the village that André was arrested, and Tarrytown witnessed many fights between guerrillas during the Revolution. It takes its chief interest, however, from its association with Irving's life and writings. Here is the church which he attended, and of which he was warden at the time of his death (Christ Church); here he is buried (in the graveyard of the old Dutch Church, the oldest religious edifice in the State), and near by are the scenes of some of his happiest fancies, including the immortal Sleepy Hollow and the bridge rendered classic by the legend of Ichabod Crane. Opposite Tarrytown, at the foot and on the side of a beautifully wooded range of hills, is the pretty town of **Nyack,** a popular suburban place, with many handsome villa residences, and the terminus of the Northern R. R. of N. J., a branch of the Erie R. R.

Sing Sing (32 miles) is on the E. bank, occupying an elevated slope, and makes a fine appearance from the river. A State Prison is located here, and its vast stone buildings are conspicuous objects from the steamer (the railway passes beneath them). Many fine villas crown the heights above and around the village, looking down upon the Hudson, which at this point attains its greatest breadth. Four miles above, on the E. side, is *Croton Point*, a prominent headland projecting into the river. Here the Croton River enters the Hudson, and 6 miles up this stream is Croton Lake, which supplies the metropolis with water. The lake is formed by a dam 250 ft. long, 40 ft. high, and 70 ft. thick; and the water is conveyed to New York by the famous **Croton Aqueduct,** which is over 40 miles long, with 16 tunnels and 24 bridges. Another aqueduct of much greater capacity was completed in 1890. The lake may be reached by carriages from Sing Sing, or from Croton or Croton Falls station on the Harlem R. R., or by the New York & Northern R. R. Above Croton Point is *Haverstraw Bay*, another lakelike widening of the river, and as the boat enters it the Highlands begin to loom up in the distance. At the head of Haverstraw Bay are *Stony Point* (on the W.), a rocky peninsula on which are a lighthouse and the ruins of a famous Revolutionary fort, and *Verplanck's Point* (on the E.), notable as the spot where Henry Hudson's ship, the "Half Moon," first came to anchor after leaving Yonkers. Here also are remains of a small Revolutionary fort. Above, on the E. bank, is **Peekskill** (43 miles from New York), one of the prettiest towns on the Hudson, at the mouth of the Peek's Kill or Annsville Creek, and the site of the State camp of the National Guard; opposite which (reached by ferry) is *Caldwell's Landing*, memorable for the costly but futile search after the treasure which the famous pirate, Captain Kidd, was supposed to have secreted at the bottom of the river here. At this point the river makes a sudden turn toward the W., which is called "The Race."

We have now reached the *Highlands, and for the next 16 miles the scenery is striking. On the left is *Dunderberg* (Thunder Mountain,

1,120 ft. high), and at its base a broad deep stream which, a short distance above its mouth, descends to the river in a beautiful cascade. On the right is *Anthony's Nose* (904 ft. high), a rocky promontory whose base is penetrated by a railway-tunnel 200 ft. long. Lying in the river near this point is the picturesque *Iona Island*, a favorite picnic resort. Just above (on the right) is *Sugar-loaf Mountain* (765 ft. high), and near by, reaching far out into the river, is a sandy bluff on which Fort Independence once stood. At the foot of Sugar-loaf is *Beverly House*, where Benedict Arnold was breakfasting when news came to him of André's arrest, and whence he fled to the British vessel (the Vulture) anchored in the stream below. Passing swiftly on, the *Buttermilk Falls* soon come into view (on the left), descending over inclined ledges a distance of 100 ft. On the lofty bluff above is the spacious and handsome *Cranston's Hotel*, one of the favorite summer haunts of pleasure-seeking New-Yorkers. There is a special landing for passengers who wish to reach the hotel, and on the opposite river-bank is *Garrison's*, another popular summer resort, with fine hotels and picturesque surroundings.

* **West Point** (*West Point Hotel*), just above Cranston's (51 miles from New York), is one of the most attractive places on the river, and should be visited. It is the seat of the U. S. Military Academy, the buildings for which occupy a broad terrace, 175 ft. above the river, reached from the landing by a steep and costly road cut out of the solid cliff-side. The most noteworthy of the buildings are the Cadets' Barracks, the Academic Building, the Mess Hall, and the Library (26,000 vols.), in which is the Observatory. The Chapel and the Museum of Ordnance and Trophies are interesting. The buildings front the spacious Parade-Ground, smooth as a lawn and level as a floor; and the grounds are tastefully laid out, containing several fine monuments, and commanding a variety of pleasing views. The Cemetery is reached by a winding road; and from the crumbling walls of * *Fort Putnam* (on Mt. Independence, 600 ft. above the river) a view is obtained which will abundantly repay the labor of reaching it. The best time to visit West Point is during June, July, and August; the scenery being then at its best, and the military exercises of the Academy offering additional attractions. The "Commencement," or graduating exercises, occurs about the 3d week in June, and about June 20th the cadets go into camp for the summer.

Above West Point, on the same side, is *Cro' Nest*, one of the loftiest of the Highland group (1,400 ft.), and still above is * *Storm King* (1,529 ft. high), where an immense cantilever bridge is being erected across the river. Between Storm King and Cro' Nest lies the lovely vale of *Tempe ;* and opposite is the pretty little village of *Cold Spring*, behind which rises the massive granite crown of *Mount Taurus* ("Bull Hill"). Beyond, still on the E. side, the Highlands are continued in the jagged precipices of *Breakneck* and *Beacon Hill*, respectively 1,220 and 1,624 ft. in height. These mountains are among the most commanding features of the river scenery, and from the summit of the latter New York City may be seen. **Cornwall Landing** (*Grand View House, Mountain House*), a picturesque village on the W. bank, is a much-frequented summer resort on the river, and contains a number of hotels,

adjacent to which are fine drives.   Here the Highlands come to an end,
and the steamer enters the broad expanse of Newburg Bay, on the W,
shore of which is **Newburg,** a handsomely built city of 23,087 inhabitants.   This city is the northern terminus of a branch of the Erie R. R.
Newburg was the theatre of many interesting events during the Revolution, and * Washington's Headquarters, an old gray stone mansion S.
of the city, is still preserved as a museum of historical relics (admission
free).   Opposite Newburg is *Fishkill Landing*, a small but pretty village, with which it is connected by a steam ferry; and 15 miles above,
on the E. bank (75 miles from New York), is *·**Poughkeepsie,** the
largest city between New.York and Albany, built on an elevated plain,
nearly 200 ft. above the river, and backed by high hills.   There are
several fine churches, numerous handsome residences, and no less than
eight important educational institutions, including *Vassar College*, one
of the leading female colleges of the world.   The buildings of Vassar
College occupy a commanding site 2 miles E. of the city, the main building (500 ft. long) being modeled after the Tuileries.   N. of the city, on
an eminence overlooking the river, are the vast and stately buildings of
the Hudson River State Hospital for the Insane.   A cantilever bridge
crosses the Hudson at this point, over which the trains of the Central
New England & Western R. R. pass, connecting the coal regions of
Pennsylvania with New England.   Across the river from Poughkeepsie
(ferry) is *New Paltz Landing*, from which stages run 14 miles to * **Lake
Mohonk** (*Mountain House*), a delightful summer resort situated near
the summit of Sky Top, one of the loftiest of the Shawangunk Mountains, 1,243 ft. above the river.   The lake and its vicinity are extremely
picturesque, and the views from neighboring summits surprisingly fine.
(Lake Mohonk is also reached from New York *via* Wallkill Valley
branch of the Erie R. R. to New Paltz, and thence by stage, in 6
miles).

Five miles above Poughkeepsie, on the E. bank, is *Hyde Park*,
on a terrace ½ mile back from the river, containing several summer
boarding-houses, with fine country-seats in the vicinity.   On the W.
side, *Rondout* and **Kingston,** 2 miles above, now form one city, with a
large trade.   Here the *West Shore* road connects with the *Ulster & Delaware R. R.* into the heart of the mountain-region.   At *Rhinecliff*, opposite Kingston, is the Beekman House, nearly 200 years old, and an excellent specimen of an old Dutch homestead; and here a branch of the
Central New England and Western R. R. meets the Hudson.   A short
distance above is *Rokeby*, the estate of William B. Astor, and still
farther on, above Barrytown, are *Montgomery Place*, the residence of Edward Livingston, and *Annandale*, the villa of John Bard.   At **Catskill
Landing** (111 miles), a favorite point of departure for the mountains
(*see* Route 10), the *West Shore R. R.* connects with the *Catskill Mountain R. R.*   Four miles above (on the E. side) is the flourishing city of
Hudson, which is finely situated upon a bold promontory, at the head
of ship-navigation on the river.   From Prospect Hill (500 ft. high) there
is an incomparably fine view of the Catskills.   Five miles from Hudson,
in the Claverack Valley, are the **Columbia Springs,** a quiet rural

resort much frequented by invalids and others. From Hudson to Albany the scenery offers nothing calling for special notice.

Albany.

Hotels, etc.—*Delavan House* (in Broadway, near the R. R. depot), *Kenmore* (N. Pearl St.), and *Stanwix Hall* are first-class hotels. *American, Globe,* and the *Mansion House* are less expensive. Prices are from $2.50 to $4 a day. *Reading-rooms* at the State Library, in the Capitol; at the Young Men's Association, in Washington Ave.; and at the Y. M. C. A., in N. Pearl St. *Electric-cars* to different parts of the city and to Troy. There are three iron *bridges* to Greenbush.

Albany, the capital of New York State, is finely situated on three hills on the W. bank of the Hudson, six miles below the head of tide-water. It was founded by the Dutch as a trading-post in 1614 and called Fort Orange, and, next to Jamestown in Virginia, was the earliest European settlement in the original 13 States. Its present name was given it in 1664, in honor of the Duke of York and Albany (afterward James II). It was chartered in 1686, and made the State capital in 1798, since which time its population has increased from 5,349 (in 1800) to 94,923 in 1890. Albany has a large commerce from its position at the head of navigation on the river, as the *entrepot* of the great Erie Canal from the W. and the Champlain Canal from the N., and as the center to which several important railways converge.

Broadway is the important wholesale business thoroughfare near the river, while the chief retail stores are on Pearl St. *State St.* leads by a steep ascent from the *Federal Building* to Capitol Square, in which are the public buildings, and then extends for a mile beyond. The *** New Capitol,** to the W. of the site of the old Capitol, was begun in 1871, and is now occupied by the Legislature of the State. It is of Maine granite, in the renaissance style, and stands on elevated ground; its tower is to be 320 ft. high, and will command a fine view. The structure is 300 ft. N. and S. by 400 ft. E. and W., and with the porticoes, when completed, will cover more than three acres, and the walls are 108 ft. in height. The *** State Library,** in the W. side of the Capitol, contains 156,493 volumes, and an interesting collection of curiosities and historical relics. The University of the State of New York has its offices in the Capitol. The *** City Hall,** in Eagle St., foot of Washington Ave., is built of rose granite. The *City Building,* in S. Pearl St., contains various offices of the city government. The U. S. Government Building, cor. State St. and Broadway, is a handsome edifice. The State Armory is in Washington Ave., near Lark St. The *Young Men's Association,* with a library of 12,000 volumes and a read-ing-room, occupies *Harmanus Bleecker Hall,* in Washington Ave., the Association using the "Bleecker bequest" of $120,000 to build the hall. On the opposite corner is the fine office-building of the Delaware & Hudson R. R. The ** State Geological and Agricultural Hall* contains the natural history collection of the New York State Museum and those of the New York State Agricultural Society. The *Medical College,* cor. Eagle and Jay Sts., is a prosperous institution, with an extensive museum; and in State St. is the Law School of the Albany University. Of

the more than 50 churches in the city the ***Cathedral of the Immaculate Conception** (in Eagle St.) and the *Church of St. Joseph*
(cor. Ten Broeck and 2d Sts.) are the most noteworthy. The Cathedral seats 4,000 persons, and its stained-glass windows are among the
richest in America. A fine altar and other additions have been made.
All Saints Episcopal Cathedral (Bishop Doane) is situated at Elk and
Swan Sts. *St. Peter's* (Episcopal), cor. Lodge and State Sts., is a handsome Gothic structure, and owns a service of communion plate presented by Queen Anne to the Onondaga Indians. The *Second Reformed
Church*, in Madison Ave., the *First Presbyterian*, on the cor. of Willett
and State Sts., and *Calvary Baptist Church*, cor. State and High Sts.,
are recent and handsome edifices. The *Fort Orange* and *Albany Clubs*
have fine houses. **Dudley Observatory,** founded and endowed by
Mrs. Blandina Dudley, stands on Observatory Hill, near the N. limits of
the city. During 1892 the erection of a new observatory on an elevation above Lake Ave., near its junction with the New Scotland plank-
road, was begun. The *Penitentiary*, 1 mile W. of the city, is a model
prison, and near it stand the *Poor Houses. Greenbush*, with its famous
Van Rensselaer House, *East Albany*, and *Bath-on-the-Hudson* are populous suburbs on the opposite side of the river.

The *State Normal School* occupies a building in Willet St. fronting
Washington Park. The *Female Academy*, and the *Albany Academy*, for
boys, are leading schools. In this building Prof. Joseph Henry, with a
mile of wire, constructed the first electro-magnetic telegraph. The *High
School* (Academy Square) is at the head of the public-school system.

An interesting relic of the early days of the city is the old *Schuyler
House*, in Schuyler St. near S. Pearl, which was burned down in 1759,
and immediately rebuilt, portions of the original walls remaining. It
was the residence of Colonel Peter Schuyler, the first Mayor of Albany.
Washington Park, 81 acres, in the W. part of the city is a beautiful
pleasure-ground. It contains a bronze statue of Robert Burns, presented
by the Scottish citizens, and a bust of Dr. James H. Armsby. The sum
of $20,000 was bequeathed by Henry L. King for the erection of a
fountain.

*Troy (reached from Albany by railroad, steamboats, and electric-
cars) lies on the E. bank of the Hudson, 6 miles above Albany, and at the
head of river navigation. Its population in 1890 was 60,956, and it has
a large commerce, with manufactures of iron, steel, cars, and especially
extensive manufactures of collars, cuffs, and shirts. *River St.*, running
parallel with the river, is the chief thoroughfare; and near 1st, 2d, 3d,
and 5th Sts., are the finest churches and private residences. The *Savings-
Bank*, in State St., is an elegant edifice, costing $450,000; the new
City Hall is a fine building; and there are many handsome business
structures. A Soldiers' Monument, 90 ft. high, is on Washington
Square. Besides bronze tablets representing "The Cavalry," "The
Artillery," "The Monitor and Merrimac," and "The Infantry," it is
surmounted by a heroic bronze female figure, entitled "Calling to
Arms," by James E. Kelly. The *Athenæum* is a beautiful freestone
edifice, in which is the *Young Men's Christian Association Library.*

The buildings of *St. Joseph's Theological Seminary* (on Mt. Ida, E. of the city) are noble specimens of Byzantine architecture. The *Rensselaer Polytechnic Institute*, on 8th St., is one of the leading schools in the United States for instruction in civil engineering. There are several other important educational institutions. In West Troy is the great *Watervliet Arsenal*, with 40 buildings in a park of 105 acres. It is here that the heavy rifled cannon of the most advanced type are made.

10. The Catskill Mountains.

Kingston may be reached by the following routes : New York Central and Hudson River R. R. from New York or Albany to Rhinecliff, crossing the river by ferry ; Day Line Steamers from New York or Albany every week-day, to Rhinebeck, crossing the river by ferry ; Albany Day Line and steamer Mary Powell from New York ; and West Shore R. R. from New York or Albany. *Catskill* may be reached by the following routes : New York Central and Hudson River R. R. from New York or Albany to Catskill Station, crossing the river by ferry ; Day Line Steamers from New York or Albany every week-day ; Catskill Night Line, foot of Jay St., New York, leaving at 6 P. M. every week-day; and West Shore R. R. from New York or Albany. (1) The Catskill Mountain R. R. starts from *Catskill*, and carries passengers as far as Palenville, the terminus, where the Kaaterskill Clove debouches on the Hudson River Valley, a distance of sixteen miles. The intermediate stations between Catskill and Palenville are South Cairo, Cairo proper, Mountain House Station, and Otis Junction. At Otis Junction passengers take elevator-car to landing, which is near the Catskill Mountain House, and but a short walk from Hotel Kaaterskill. (2) From *Kingston* the mountains may be reached *via* Ulster & Delaware R. R. to Phœnicia, where connection is made with the Stony Clove & Catskill Mountain R. R. The New Grand Hotel is reached by this route direct from Kingston to Grand Hotel Station. On changing cars for the Stony Clove R. R., the tourist may proceed to Hunter, on Schoharie Creek, 14 miles distant. The Kaaterskill R. R. connects with the Stony Clove & Catskill Mountain R. R. at Kaaterskill Junction, near Hunter, and runs to Kaaterskill Station, and has its eastern terminus 1 mile from the Catskill Mountain House. It is also the station for Hotel Kaaterskill. This is the favorite route of entering the mountains from the west.

THE CATSKILLS, or Kaatskills, follow the course of the Hudson for 20 or 30 miles, lying W. of it, and separated from it by a valley 10 or 12 miles wide. Their chief interest lies in the beauty and variety of their scenery. In a field of very limited area, easy of access and soon explored, they present a multitude of picturesque objects which have long made them a favorite resort of artists and of all who find pleasure in the wild haunts of the mountains. Indian tradition singled them out as the favorite dwelling-place of spirits, and they, with the exception of the Hudson Highlands, are the only faëry ground that American literature has ventured to appropriate.

The village of **Catskill** (a station on the West Shore R. R.) occupies an elevated and attractive site at the mouth of Catskill Creek, and has some 4,920 inhabitants. The scenery in the neighborhood, especially along the banks of the creek, is very pleasing. The *Prospect Park Hotel*, near the landing, is spacious and handsome, located on a high plateau and surrounded by extensive grounds. In the village are several excellent hotels ; and about a mile to the W. is the *Grant House*, situated on a commanding elevation, with a noble outlook to the mountains.

(1) From **Catskill** the Catskill Mountain R. R. runs 6 trains a day to _Palenville._ The road follows Catskill Creek, and at South Cairo finds an opening in the foot-hills. Then, after skirting the base of Cairo Round-Top Mountain, the train stops at Mountain-House Station, whence the visitor was formerly driven up over a road where every turn brings into view new scenes of grandeur. The Otis Elevating Railway, taken at Otis Junction, now conveys the tourist to the summit, a distance of 1¼ mile, in ten minutes. The old and well-known * **Mountain House** is a spacious edifice, perched upon one of the terraces of Pine Orchard Mountain, at an elevation of 2,250 ft. above the river. From the broad rock platform in front of the hotel, a view of surpassing beauty may be obtained. Directly in front, the mountain falls almost perpendicularly to the plain; to the right, the broad Hudson winds through its noble valley; in the dim distance Albany may be descried with a glass; and on the horizon the Hudson Highlands, the Berkshire Hills, and the Green Mountains unite their chains, forming a continuous line of misty blue. The views from the Mountain House are said to embrace an area of 12,000 square miles, including portions of Vermont, Massachusetts, Connecticut, and New Jersey. At Sunset Rock, near the summit of South Mountain, 3,000 ft. above the sea, is the _Hotel Kaaterskill,_ the largest in the mountains, and commanding a magnificent extent of view. The _North Mountain_ is easily ascended from the Mountain House; the best view is from Table Rock, ¾ of a mile N. of the hotel. On the N. side of this rock a fine echo may be heard. Another favorite excursion is to the top of _South Mountain,_ which commands a fine view of the Catskill Pass and some distant peaks of New Jersey. * **High Peak** (6 miles W. of the Mountain House) is one of the loftiest of the Catskill summits, and should certainly be climbed in order to see the region fairly. The ascent is toilsome, but ladies often accomplish it, and the view from the summit (3,804 ft. high) well repays the labor of reaching it.

The _Two Lakes_ (North and South) are back of the Mountain House, on the road to the famous * **Kaaterskill Falls,** which are two miles distant. At the head of the falls is the _Laurel House,_ an excellent hotel, commanding fine views of the falls, of the country about, and of Round Top and High Peak in the immediate neighborhood. The falls are formed by the outlet of the lakes plunging into a deep hollow where the mountain divides like the cleft foot of a deer. The descent of the first cascade is 180 ft., that of the second 80, and below these there is another fall (the Bastion) of 40 ft. Below the falls the sides of the gorge rise in a succession of walls of rock to the height of 300 ft. or more. To see the falls to the best advantage, the visitor should descend the winding stairs leading from the terrace of the hotel (fee 25c.) and spend an hour or two in exploring the gorge and glen below. As the supply of water is limited, the stream has been dammed at the verge of the cliff and is only turned on at intervals for the benefit of visitors. Below the falls, the Kaaterskill has a devious and winding course of 8 miles to the Catskill, which it enters near the village. Some ruggedly picturesque scenery may be enjoyed by descending the glen to the road in the Clove, about a mile from the falls. *Sunset Rock*, commanding noble views, is reached from the Laurel House by a walk of a mile and a half through the forest.

A favorite excursion from either the Mountain House or the Laurel House is to **Haines's Falls,** a spot much frequented by artists. At the Haines House, near by, one pays the usual fee (25 c.) for viewing the scene. The fall has two leaps, the first of 150 and the second of 80 ft., with a third one below of 60 ft., and others still, so that in less than a quarter of a mile the stream falls 475 ft. From Haines's Falls a rugged and picturesque ravine, called the *Kaaterskill Clove*, traversed by a tolerable road, leads down to the plain below. In this ravine are the High Rocks, and 200 or 300 yards below are the beautiful **Fawn Leap Falls** (fee 25c.). At the mouth of the Clove is **Palenville** *Winchelsea House*), where there are many boarding-houses, and where artists most do congregate. Six miles from Palenville is the *Plattekill Clove*, a deep and rugged gorge, in which are the *Black Chasm Falls*, 300 ft. high. Another pleasant ride is along the ridge 5 or 6 miles to the entrance of the *Stony Clove*, and thence through the wilderness of this fine pass, within whose depths ice remains throughout the year. At the head of the Clove is *Roggen's Mountain Hotel*, a favorite resort for sportsmen. On the road from the Mountain House is the pretty little hamlet of *Tannersville* (*Fabian House* and the *Roggen's Mountain House*); and 4 miles W., beyond the entrance to Stony Clove, is *Hunter* (Breeze Lawn House), nestling in a narrow glen, with Hunter Mountain (4,082 ft. high) towering above it.

A branch of the Catskill Mountain R. R. runs to the mountain village of *Cairo* (10 miles), which is connected by stages with *South Durham* (16 m.), *Windham* (26 m.), *Prattsville* (36 m.), and in all of which are numerous summer boarding-houses. Near it are the celebrated Pratts' Rocks, on which are cut busts of the Pratts, who founded the town.

(2) From **Kingston** the Ulster & Delaware R. R. carries you along the banks of the beautiful Esopus, through many enchanting scenes, and up a grade of remarkable steepness. At W. Hurley (9 miles from Kingston) a stage conveys passengers a distance of nine miles to *Overlook Mountain House*, commanding extensive views over the Hudson Highlands and Valley. *Echo Lake* and *Plaatekill Falls* are respectively a mile and a half and three miles distant. At *Shokan* (18 miles) the line turns N. to *Phœnicia* (27 miles), where is the junction with the *Stony Clove & Catskill R. R.*, and the *Tremper House* is excellent. *Pine Hill*, 12 miles farther (*Guigou House*), commands a view of the Shandaken Valley; and next we reach *Grand Hotel Station*, the highest point of the railroad. Here is the *New Grand Hotel*, with its glorious prospect over the head-waters of the Delaware. The mountains in the distance are *Slide Mountain*, the highest of the Catskill group (4,200 ft.), and *Panther*, *Table*, and *Balsam Mountains*. From *Phœnicia* the Stony Clove and Catskill Mountain R. R. runs to Hunter, and from Kaaterskill Junction the Kaaterskill R. R. runs past Tannersville to *Kaaterskill Station* (p. 74).

11. New York to Boston via New Haven, Hartford, and Springfield.

The "Springfield Route," as this line is familiarly called, is composed of the New York, New Haven & Hartford R. R. to Springfield and thence *via* the Boston & Albany R. R. to Boston. Solid drawing-room car express trains leave either city at 9 and 11 A. M. and 4 (limited express) and 11 P. M. Distance, 234 miles ; fare, $5.

LEAVING the Grand Central Station, the train runs on the track of the New York & Harlem R. R. as far as Woodlawn Cemetery (12 miles), and then takes the New York, New Haven & Hartford R. R. which runs through several pretty suburban towns. *New Rochelle* (17 miles) ; *Larchmont* (18 miles) is a pleasant resort and invites many visitors during the summer months ; *Mamaroneck* (21 miles) and **Rye** (24 miles) are especially attractive, owing to their proximity to Long Island Sound, which affords excellent salt-water bathing, boating, etc. The American Steam Yacht Club has erected a fine club-house here. *Greenwich* (28 miles), the first station in Connecticut, is a picturesque old town, pleasantly situated on hill-slopes commanding fine views of Long Island Sound. It is noted for its great number of beautiful summer villas. The *Belle Haven Mansion* and the *Lenox House* are favorite houses, open only in summer. A short distance S. E. is *Indian Harbor*, where is the *Indian Harbor House*, an elegant and spacious summer hotel. **Stamford** (33 miles) is a favorite resort of New York merchants, many of whom have embellished its heights with handsome mansions and villas. The town is embowered in trees, and there are several fine churches and public buildings. *Shippan Point,* 2 miles S. of Stamford, is frequented in summer by many hundreds, who crowd the spacious *Ocean House* and numerous smaller places of entertainment. *S. Norwalk* (42 miles) is near the beautiful village of **Norwalk** (reached by horse-cars), which is also much resorted to in summer. Its harbor is a picturesque bay, which affords oysters in great abundance and of excellent quality. The hotels are the *City* and the *Mahackemo* in S. Norwalk, and the *Norwalk* in Norwalk. About 1½ mile S. of S. Norwalk, on Gregory's Point, is the *Dorlon House*, a delightful summer hotel on the Sound. **Westport** (45 miles) is the next station of importance. **Fairfield** (51 miles), still another popular summer resort, has the finest beach on the Sound, and supports two large hotels, the *Manor House* and the *New Merwin House*. The adjacent scenery is very attractive. *****Bridgeport** (*Sterling* and *Wilson ;* 56 miles) is a flourishing city of 48,866 inhabitants, situated on an arm of Long Island Sound, and noted for the extent and variety of its manufactures, chief among which are sewing-machines, leather, carriages, arms, cutlery, and locks. It is the southern terminus of the Housatonic and Naugatuck Divs., and has 15 churches, 8 banks, and daily newspapers. The city is handsomely built, and Golden Hill is crowned with fine villas. The Fairfield County Court-House is a handsome structure. *Black Rock* on the Sound is a much-frequented resort during the summer. Passing now the pretty villages of *Stratford* (59 miles) and *Milford* (64 miles), the train approaches New Haven (73 miles) across extensive salt-meadows.

New Haven.

Hotels, etc.—The *New Haven House*, cor. College and Chapel Sts., is the leading hotel. The *Elliott*, cor. Chapel and Olive Sts.; the *Tontine*, cor. Church and Court Sts.; and the *Tremont*, cor. Orange and Court Sts., are good. There are *Reading-rooms* at the Free Public Library of the Y. M. C. A. and at the Young Men's Institute in Chapel St. *Street-cars* traverse all parts of the city, and run to the suburbs. *Carriages* are allowed to charge 50c. for one passenger one course, unless an advance agreement is made, and each additional passenger 25c. *Steamboats* run to New York thrice daily (fare, $1).

New Haven, the largest city of Connecticut, is situated at the head of New Haven Bay, 4 miles from Long Island Sound, on a broad plain surrounded by rolling hills. It was settled in 1638, was incorporated as a city in 1784, and from 1701 to 1875 was one of the capitals of the State. It is the center of 5 branch railroads, and has a large coasting-trade. Its manufactures are very extensive, including machinery, hardware, locks, clocks, fire-arms, carriages, pianos, jewelry, India-rubber goods, etc., and involve a capital of over $16,000,000. The population in 1890 was 81,298. The city has numerous churches, many public schools, several banks, 5 daily newspapers, and numerous charitable institutions. New Haven is the center of an extensive railroad system. In addition to the main line, there are the Northampton to Shelburne Falls, Air-Line to Willimantic, Shore-Line to New London, and Berkshire Divs. to Pittsfield—all divisions of the New York, New Haven & Hartford R. R.

Chapel St., the principal thoroughfare, extends in a W. N. W. direction from end to end of the city. *State* and *Church* are also important business streets, and * **Hillhouse Ave.** is lined with handsome residences. *Temple St.* has perhaps the finest elms, and its four rows of trees form an arch over the roadway. Both *Prospect St.* and *Whitney Ave.* are fine streets with beautiful residences, and the latter forms a drive to Whitneyville Lake. The many magnificent elms with which its principal streets are planted has caused New Haven to be called the "City of Elms." The * **Public Green,** in the center of the city, is a fine lawn shaded by noble elms, and contains the *Center Church*, the *United Church*, and *Trinity Church*. Back of Center Church are the grave and monument of the regicide, John Dixwell. On the E. side of the Green is the * **City Hall,** a very handsome building, containing the municipal offices. The *Custom-House* (which contains the *Post-Office*) is a Portland brown-stone edifice in Church St. near Chapel St. The other principal public buildings are the *Court-House*, in Church St., the *Second Regiment Armory*, the *New Haven Hospital*, the *Orphan Asylum*, and the *County Prison*. The last two are in the W. part of the city. The *Union Depot* is a large brick building, fronting the harbor.

Across College St. from the Green are the grounds of * **Yale University,** one of the oldest and most important educational institutions in America. It was founded in 1701, established at New Haven in 1717, and in 1892-'93 had 184 instructors and 1,969 students. Besides its academic and scientific undergraduate departments, the University has law, medical, theological, and fine arts courses of graduate instruction. The

Yale College Views.

grounds include 9 acres, and contain many buildings. The most noteworthy buildings include * *Osborn Hall* (on the S. E. cor. of the square);
Library Buildings, with 200,000 volumes; the elaborate * *Art Building*
(at the S. W. cor. of the square), containing a fine collection of paintings, statuary, and casts; the *Alumni Hall* (on the N. W. cor.), used for
the annual examinations and graduates' meetings; and the dormitories,
Farnam Hall, *Durfee Hall*, *Laurance Hall*, and *Welch Hall*, with the
* *Battell Chapel* (on the N. E. cor.).   In Elm St., close by, are the two
handsome buildings of the *Divinity School*, with the dainty little *Marquand Chapel* and the Bacon Memorial Library between; near by in Elm
St. is the *Gymnasium*.   At the head of College St. is *Sheffield Hall*, and
in Prospect St. is *North Sheffield Hall*, where the engineering and physics
have their departments, containing the laboratories and collections of
the Sheffield Scientific School.   The Sheffield residence in Hillhouse Ave.
has been acquired by the University, and contains the biological department.   Beyond in Prospect St. is the Yale Observatory, containing a
6-inch heliometer and an 8-inch equatorial instrument.   In the * *Peabody Museum*, of which only the S. wing has been built, cor. Elm and
High Sts., are the collections of the University in geology, mineralogy,
and zoölogy, including the famous paleontological collection of Prof.
Othniel C. Marsh.   The *Sloane Physical Laboratory* is in Library St.;
the *Kent Chemical Laboratory*, cor. High and Library Sts.   * *Dwight
Hall* is an elegant edifice that was built for the use of the Y. M. C. A.
of the college.   The Athletic Grounds, about 1¾ mile from the college
campus on the Derby turnpike, are very large.   The halls of the several
secret societies are scattered through the city, and are of unique architecture.   There are numerous industries in New Haven, and one of the
most interesting features is the Winchester Firearms Factory.

   The *Old Burying-Ground* (on Grove St., near High) contains many
interesting and venerable monuments, and the *Evergreen Cemetery* (on
the bank of West River) is tastefully adorned.   *Sachem's Wood* (the
Hillhouse residence), at the head of Hillhouse Ave., is a pleasant spot,
and Lake Whitney is a favorite resort for boating.   One of the most
popular drives is down the E. side of the harbor to *Fort Hale*, an old
ruin dating from 1814, whence there is a fine view.   This is now included in the park system, and is connected by a drive with the other
parks.   Rising above the plain near the city are the lofty promontories
known as East and West Rocks.   *East Rock* (reached by cars) is 360
feet high, and commands a wide and beautiful view.   Three hundred
and fifty acres of it have been laid out by the city as a park, and it is
one of the finest in the country.   On the summit of the Rock is the
Soldiers' Monument, erected at a cost of $50,000, the lookout from the
top of it being 480 ft. above Mill River, which winds about the foot of
the Rock.   *West Rock* is 400 ft. high.   On the top there is a group of
bowlders called the "Judges' Cave," because Goffe and Whalley, two of
the judges of King Charles I of England, were secreted here for a while
in 1661.   Near the base of the rock on the N. are Wintergreen Fall
and Wintergreen Lake.   On the road thither is the *Springside Almshouse*, which cost $200,000.   *Savin Rock*, a bathing-place, with summer

hotels, on Long Island Sound, 4 miles S. W. of the city, is a popular resort (reached by cars from the Green).

The first important station beyond New Haven is *Wallingford* (86 miles), an important manufacturing town. *Meriden* (92 miles), *Berlin* (99 miles), and *Newington* (105 miles) are the other principal stations before reaching *Hartford* (110 miles).

Hartford.

Hotels, etc.—The *Allyn House*, the *City Hotel*, the *Hotel Capitol*, the *Hotel Heublein*, and the *United States Hotel* are the leading houses. The street-car system runs to West Hartford, East Hartford, Parkville, Wethersfield, and through the principal streets of the city. *Carriages* charge 50c. for one or two passengers to any point within the city limits. *Steamboats* run to New York daily. The *Post-Office* is in City Hall Sq. Railroads centering at Hartford are the *New York, New Haven & Hartford R. R.*, and its *Valley Div.*, running to Saybrook Junction, *New York & New England R. R.*, and the *Central New England & Western R. R.*

Hartford, the capital of Connecticut, is situated at the head of sloop navigation on the Connecticut River, 50 miles from Long Island Sound. It had a population in 1890 of 53,230, and, besides an immense manufacturing business, is one of the great centers of fire and life insurance, the assets of the various companies being $162,000,000. Its manufactures include iron and brass ware, bicycles, steam-engines, machinery, tools, sewing-machines, fire-arms, silver-plated ware, stone-ware, woolens, tobacco, etc. The city is regularly laid out, and comprises an area of about 10 square miles, intersected by Park River, which is spanned by numerous bridges. *Main St.*, running N. and S., is the leading thoroughfare. *State* and *Asylum Sts.* are active business streets. In the outskirts are many tasteful villas, and the city as a whole is remarkably well built.

The *Union Depot* is a fine structure, designed by Henry H. Richardson. S. of the depot, in the bend of Park River, is the beautiful * **Bushnell Park** (46 acres). At its entrance is a Memorial Arch, erected in 1885 to the memory of the soldiers and sailors who fell during the civil war. Also in the park is a bronze statue of General Israel Putnam, 8 ft. high, by J. Q. A. Ward, costing $14,000, and one to Dr. Horace Wells, a discoverer of anæsthesia, that cost $10,000. In the park is the * **State-House,** of marble, 300 ft. long by 200 ft. wide and 250 ft. high to the top of the dome, completed in 1878 at a cost, including the site, of $3,000,000. Besides chambers for the two Houses of the Legislature, it contains rooms for various State Departments and officials, the Supreme Court, and the State Library, one of the largest law-libraries in the country. In the Senate-chamber hangs the famous portrait of Washington by Gilbert Stuart, purchased from the artist, in 1800, by the State. In the office of the Secretary of State hangs the charter given to the colony by Charles II of England. Near the park is the *High-School*, a very handsome building, which cost $250,000. In Asylum St., near the depot, is the *Asylum for the Deaf and Dumb ;* it was founded in 1817, and was the first institution of the kind in America. The *Hartford Theological Seminary* occupies fine buildings just W. of the High-School. The *Retreat for the Insane* stands on elevated ground in the S. W. part of the city. The *Hartford Hospital*, in Hudson St. near the Retreat, is a handsome stone edifice. The *Hartford Orphan Asylum*, a fine building in the modern English style, stands just W. of the Capitol, and the *Old People's Home* in Jefferson St., is of interest. State-House Square, in the center of the city, is the site of the *Old State-House*, erected in 1794, which building, in the Grecian style, is now occupied as the City Hall of Hartford. In the same square is the *Post-Office Building*, also occupied by the U. S. Court, the Custom-House, and Internal Revenue Collector's Office. The *Cheney Building* (cor. Main and Temple Sts.), the *Phœnix Bank Building*, the *Hartford Fire Ins. Building*, the *Connecticut Mutual Life Ins. Co.* (opposite State-House Square) and the *Etna Life* (in Main St.) are very striking. The *Courant Building*, occupied by the *Hartford Courant*, the oldest newspaper in the United States, established in 1764, is in State St., opposite the Post-Office. The *State Arsenal* is in North Main St., and is a fine edifice. The *Opera-House* is at 395 Main St.. The *Wadsworth Athenæum*, in Main St., contains the Watkinson Free Library of Reference, the Hartford Library, and the Library of the Connecticut Historical Society, in all about 100,000 volumes. In the same building can be seen the rich collection of the Connecticut Historical Society, from 9.30 A. M. to 4.30 P. M., daily, except Sunday; also a good collection of paintings and statuary, from 9 A. M. to 4 P. M., Monday and Tuesday free. The *Hartford Club* occupies an old colonial mansion at 33 Prospect St.

Among the 36 churches are the * **Church of the Good Shepherd** (Episcopal), erected by Mrs. Samuel Colt as a memorial of her husband and children; *Christ Church* (Episcopal), cor. Main and Church Sts.; *Trinity Church* (Episcopal), the *Park Congregational*, Pearl St. *Congregational*, and *Asylum Hill Congregational Church*, First (*Center*) *Congregational Church*, the *Second* (*South*) *Congregational Church*, the *South Baptist*, and the Roman Catholic *Cathedral* in Farmington Ave. The buildings of * **Trinity College** stand on Rocky Hill, about a mile S. of the Capitol. When completed, they will form a quadrangle 1,050 ft. long and 376 ft. wide, inclosing 3 courts containing an area of 4 acres. The architecture is Early English, the design of William Burges, of London. The grounds (80 acres) are handsomely adorned. There are 20 instructors and 124 students.

Colt's Patent Fire-arms Manufactory is located on the banks of Connecticut River, in the S. E. portion of the city. The grounds extend from the river to Wethersfield Ave., upon which stands the Colt mansion ("Armsmear"), surrounded by immense greenhouses, graperies, etc. The Pratt & Whitney Co., whose tools are famous the world over, have their plant in Hartford. "Mark Twain" has a handsome residence in Farmington Ave. Close by are the houses of Charles Dudley Warner, Mrs. Harriet Beecher Stowe, and Mrs. Isabella Beecher Hooker. The *Ancient Burying-Ground*, containing the ashes of the first settlers, is in the rear of Center Church, in Main St. *Cedar Hill Cemetery* should be visited to see the Colt, the Beach, the Morgan, and other monuments,

6

and the fine prospect over the surrounding country. The *Old North* and *Spring Grove Cemeteries* are worth visiting. The favorite drives in the vicinity of Hartford are to *Tumble-Down Brook*, 8 m. W., on the Albany road; to *Talcott Mountain*, 9 m. W.; to the *Reservoirs*, on the Farmington road; to *Prospect Hill;* and to *Wethersfield* (4 m. S.), the most ancient town on the river. *Charter-Oak Trotting Park*, one of the famous race-courses in the country, is near Hartford. *East Hartford* (reached by a long bridge) contains some quaint old houses, and the long street shaded by elms for miles makes a very enjoyable drive.

Between Hartford and Boston the only places requiring mention are *Springfield* (136 miles) and *Worcester* (190 miles), both in Massachusetts. **Springfield** (*Massasoit House, Haynes' Hotel, Hotel Warwick, Cooley's Hotel*) is one of the prettiest among the smaller American cities, and is noted for the great variety of its industries. The New York, New Haven & Hartford R. R. has its northern terminus here, connecting at its magnificent station, which was one of the last works of H. H. Richardson, with the Boston & Albany R. R. going east and west. The Connecticut River R. R. runs north from Springfield, making connection at South Vernon with the Central Vermont system for Canada and the White Mountains. The population according to the census of 1890 was 44,179. It is situated on the Connecticut River, 26 miles N. of Hartford, is well built, and its wide streets are shaded with elms and maples. It rises from the Connecticut River in terraces, on which the many residences are located, overlooking the wide valley, while in the distance are the elevations of Mt. Holyoke and Mt. Tom, making its situation one of considerable beauty. The principal point of interest is the * *United States Armory*, located in spacious grounds on Armory Hill (reached by State St.), and commanding fine views. This establishment employs 700 hands, and 175,000 stand of arms are kept constantly in stock. During the civil war the works were run night and day, and over 800,000 guns were made, at a cost of $12,000,000. It is the largest establishment for the manufacture of small-arms by the Government in the United States. The *City Hall* contains a public hall seating 2,700 persons. The * *Court-House* is a massive granite structure costing $200,000, designed by H. H. Richardson; and the building of the *City Library* (containing 80,000 volumes and a museum of natural history) is very handsome. Court Square contains a *Soldiers' Monument*, given to the city by Gurdon Bell; also a statue of *Miles Morgan*, an early settler, erected by Junius S. Morgan. There are also several fine churches, of which the most noteworthy are the * *Church of the Unity* (State St.), the *Memorial Church, South Church, North Church* (Congregational), *Church of the Sacred Heart*, the *Cathedral of St. Michael* (Roman Catholic), and *Christ* (Episcopal) *Church*, a fine brownstone edifice; its Parish House contains a fine painting in glass of Mary Magdalen at the Tomb, by John La Farge. There are two *Cemeteries* in the city, both beautiful by the diversity of the surface and their numerous shade-trees. In *Peabody Cemetery* Dr. J. G. Holland, Samuel Bowles, and others, are buried. *Hampden Park* has fine race-tracks. The *Springfield Republican* is published here, and under the editorship of Samuel Bowles became one of the famous newspapers of the country. *Forest Park*, to the S. of the city, is the gift of O. H. Greenleaf, E. H. Barry (who donated his estate for this purpose), and other citizens. It contains 400 acres, and includes extensive ponds, in which are planted the Egyptian lotus and other rare aquatic plants. *Stearns Park* contains a statue of The Puritan, by Augustus St. Gaudens, which is a memorial to Deacon Samuel Chapin, one of the first settlers of Springfield and the ancestor of all the Chapins in the United States. At *Hampden Park* the annual game of football between Yale and Harvard Universities is played, and it is also the resort of frequent bicycle and other races.

Worcester (*Bay State, Colonnade, Waldo,* and *Lincoln*) is a large manufacturing center, the second city in Massachusetts in wealth and population, which celebrated in 1884 its two hundredth anniversary. The railroads which lead to and through the city are the Boston & Albany, Worcester Div. of New York, New Haven & Hartford R. R., Boston & Maine, New York & New England, and Fitchburg, giving large facilities for business and travel. An electric railway runs to Spencer, a distance of 12 miles, and another line is contemplated to go to Clinton, 17 miles. Its population in 1890 was 84,655, and its principal manufactures are of boots and shoes, machinery and tools, a great variety of metal and wood products, stone-ware, cars, carpets, etc. The principal staple is iron and steel wire, which in two establishments alone gives employment to over 4,000 workmen. The city is, in the main, regularly laid out with pleasant streets, *Main St.* being the leading thoroughfare. The *Union Passenger Station*, designed by H. H. Richardson, is one of the largest in New England. Near the center of the city is the Common, on which are a **Soldiers' Monument*, designed by Randolph Rogers, and a monument to Colonel Timothy Bigelow, a Revolutionary officer. Among the public buildings are two county *Court-Houses*, adjacent to each other on Lincoln Square, the *City Hall*, the *High-School*, the *Young Men's Christian Association*, and *Mechanics' Hall* (seating over 2,000). Near the Court-Houses is the building of the ** American Antiquarian Society*, containing a library of 85,000 volumes and a cabinet of antiquities. It has constructed an additional building in Salisbury St., on ground given by Stephen Salisbury. This is the resort of students from all portions of the country, and is specially rich in books and pamphlets bearing on the history of America. The *Free Public Library* (in Elm St.) has 70,000 volumes and a reading-room (open from 9 A. M. to 9 P. M.). The *Worcester Natural History Society* (in Foster St.) has interesting collections. Worcester is justly proud of its educational institutions, among which are *Clark University*, a post-graduate institution, and perhaps the only one of its kind in the United States, the *College of the Holy Cross* (Roman Catholic), the *Worcester Polytechnic Institute*, the *State Normal School*, the *Worcester Academy*, the *Highland Military Academy*, and the *Oread Institute* for young ladies. All these have fine buildings. The ** State Lunatic Asylum* is in the eastern part of the city, erected at a cost of $1,350,000. An elaborate park system has been laid out surrounding the city. On the W. is *Lincoln Park*, with interesting aquatic, horticultural, and arboricultural features, while on the N. W. is *Salisbury Park*, given to the city by Stephen Salisbury. Also on the N. is *North Park*, and on the E. side is *Normal Hill*. The city is in the midst of a region full of charming resorts for summer tourists. Among them may be specially mentioned *Lake Quinsigamond*, connected with Worcester by an electric railway, and *Mount Wachusett*

Beyond Worcester the train passes for 25 miles through a thickly settled region, with numerous small towns, and stops at *S. Framingham*, a thriving manufacturing village, and center of an important system of railways. Four miles beyond, near the foot of Cochituate Lake, whence Boston draws its water-supply, is the large village of *Natick*, celebrated for its shoe-manufactures. Next come the wealthy suburban towns of *Wellesley, Newton, Brighton*, and *Brookline*, and the train enters Boston over the Back Bay lands.

12. New York to Boston via Providence.

The "Shore Line Route," composed of the New York, New Haven & Hartford R. R. to Providence, and thence *via* the Old Colony R. R. to Boston, leaves New York from Grand Central Station, and Boston at Park Square Station. Express trains leave either city at 10 A. M., 1 P. M., 5 P. M. (limited express), and 12 midnight; the two latter trains running daily. Drawing-room cars are attached to day train, and sleeping-cars to night trains. Distance, 232 miles; fare, $5.

As far as *New Haven* (74 miles) this route is identical with Route 11. Beyond New Haven the road runs close along the shore of the Sound, passing several popular summer resorts. *Branford* (82 miles) has within its limits Branford Point, a favorite watering-place, on and near which are several large summer hotels. *Guilford* (90 miles) is a pretty town, built round a finely shaded public square, and noted as the birthplace of Fitz-Greene Halleck, the poet, who died there Nov. 17, 1867. *Guilford Point*, S. of the village, has a number of hotels and is a popular summer resort. *Saybrook Junction* (105 miles) is an old and quaintly rural village, whence the Valley Div. runs S. to the venerable town of *Old Saybrook*, and to the shore. Shortly beyond Saybrook the train crosses the Connecticut River, and, passing several small villages, of which *East Lyme* is a place of some resort, soon reaches **New London** (124 miles), a city of 13,757 inhabitants, pleasantly situated on the W. bank of the river Thames, and possessing one of the finest harbors in the United States. Above the city, on the E. side of the river, is the *U. S. Navy-Yard*, and on the W. side of the Thames, below the city, are *Fort Trumbull* and Fort Griswold. New London contains numerous silk and woolen mills, and manufactures machinery and hardware largely. It was formerly a seat of seal and whale fishing. A *City Hall* of polished freestone, a granite *Custom-House*, several fine churches and a great number of costly residences, are among the architectural features of the city. *Cedar Grove Cemetery* is pleasantly situated, and the ancient burial-ground of the town is of special interest to the antiquarian. The *Crocker House*, in the city, is first-class; and 2 miles S. at the mouth of the Thames is the famous

Pequot House, accommodating 500 guests, a fashionable summer resort along the Sound shore. Across the river from New London is *Groton,* where is a tall granite monument commemorating the cruel Fort Griswold massacre (Sept. 6, 1781). **Stonington** (135 miles) is the last station in Connecticut, and is much frequented in summer. It is a quiet, sleepy town, with quaint houses surrounded by beautiful grounds, and with notably good facilities for fishing, bathing, and boating. The *Hoxie House* is a well-patronized hotel, and there are several smaller ones. The "Stonington Line" of steamers plies daily to and from New York. Steamers also run to *Block Island,* on which are summer hotels; and several times daily to *Watch Hill Point,* which, after Newport and Narragansett Pier, is the most popular summer resort in Rhode Island. The Point is also reached by steamer from New London and from Westerly on the Providence Div. of the New York, New Haven & Hartford R. R. It is the extreme S. W. tip of Rhode Island, has a superb beach, and is surrounded by attractive scenery. The leading hotels are the *Atlantic,* the *Larkin,* the *Ocean,* the *Plimpton,* and the *Watch-Hill.*

From Stonington to Providence the distance is 50 miles, and there are a number of prosperous little towns on the way, none of which require special mention. At *Kingston* (158 miles from New York) a branch line diverges to *Narragansett Pier,* next to Newport the chief summer resort in Rhode Island, situated at the mouth of Narragansett Bay, and possessing one of the finest beaches on the Atlantic coast. Its *Casino,* by McKim, Mead & White, is a noteworthy building. Fishing and boating are excellent, and there are fine drives and views. The leading hotels are the *Atwood,* the *Atlantic,* the *Continental,* the *Delavan, Green's Inn, Hotel Berwick, Hotel Gladstone,* the *Massasoit,* the *Mathewson,* the *McSparron,* the *Metatoxet,* the *Ocean,* the *Revere,* the *Rockingham.* The *Tower Hill House* is 1½ mile from the Pier on Narragansett Heights, and 150 ft. above the bay. At *Wickford Junction* (165 miles from New York) connection is made with the *Newport & Wickford R. R. and Steamboat Line.*

Providence.

Hotels, etc.—The *Narragansett,* on Broad St., and the *Dorrance,* on Westminster St., are the best. The *City* and the *Central* are good. *Street-cars* run to all parts of the city and to the adjoining towns. Direct communication over College Hill is made by a cable-road, the first operated in New England. It starts at Market Square and extends to Red Bridge. Electric-cars run from Dorrance St., cor. Broad, to Pawtuxet. *Steamboats* daily to New York, twice daily to Newport, tri-weekly to Block Island in summer, and hourly to many shore resorts.

Providence, the second city of New England in wealth and population, and the chief city and one of the capitals of Rhode Island, is picturesquely situated on the northern arm of Narragansett Bay, known as Providence River. The river extends to the center of the city, making water communication to all points of easy access. The city extends easterly to the Seekonk River, on which the *Narragansett Boat-Club* has a fine boat-house. Providence was founded in 1636 by Roger Will-

iams, who had been banished from Massachusetts on account of his
religious opinions. It was incorporated in 1832, with a population of
17,000, and in 1890 had a population of 132,146. Providence is noted
for its fine colonial mansions, among which are the Dorr residence, in
Benefit St.; that of Henry G. Russell, in Brown St.; that of the late
John Carter Brown, in Benefit St.; that of Edward Carrington, in William St.; and that of Mrs. William Gammell, in Power St. The Rhode
Island Wheelmen's Club occupies the upper portion of a colonial building in S. Main St., the lower portion of which contains the offices of
the Providence National Bank, which celebrated its centennial anniversary in 1891. The interior of this building is well worth a visit. The
manufactures of Providence are very extensive, including "prints," cotton and woolen goods, iron, jewelry, etc. Among these are some which
are famous throughout the country—the *Gorham Manufacturing Co.'s
Works*, the *American Screw Co.*, the *Brown & Sharpe Manufacturing
Co.*, the *Providence Steam-Engine Co.*, the *Corliss Steam-Engine Works*,
the *Armington & Sims Engine Co.*, the *New England Butt Co.*, the
Rhode Island Locomotive Works, the *Nicholson File Co.*, etc. Providence is the center of more than 60 woolen and 100 cotton mills, and
has 40 banks.

The surface of the city is very irregular, and the sides and summits
of the hills are covered with dwelling-houses, surrounded by ornamental
gardens. *Westminster St.* is the main business thoroughfare, and extending from it to Weybosset St. is the *Arcade*, the largest of the kind
in the United States, 225 ft. long, 80 ft. wide, and 3 stories high.
Near by is the massive granite building of the *Custom-House* and *Post-
Office*. The *State-House* is a plain brick building at the corner of N.
Main and S. Court Sts. The *Union Station* is a large and handsome
brick building in the heart of the city, fronting on Exchange Place. The
improving of the terminal facilities of the railroads has led to the construction of an additional station. At the head of Exchange Place
stands the *City Hall,* one of the finest municipal buildings in New
England, erected at a cost of over $1,000,000. Directly in front of it
is the *Soldiers' and Sailors' Monument,* erected by the State
in memory of its citizens (1,741 in number) who fell in the civil war.
It was designed by Randolph Rogers, cost $60,000, and consists of a
base of blue granite with 5 bronze statues; while at the foot of Exchange Place is a fine equestrian statue in bronze of General Ambrose
E. Burnside. The **County Court-House** is an imposing edifice, cor.
College and Benefit Sts. The *Opera-House* and the *Butler Exchange* (in
Westminster St.) are fine structures. The *Rhode Island Hospital Trust
Co.* and the *Burrill Building* are large commercial structures in Westminster St. Of the churches the most noteworthy are * *St. Stephen's*
(Episcopal), with rich stained-glass windows; *Grace* (Episcopal), with
an exceedingly graceful spire; the quaint old *First Baptist*, belonging
to the oldest Baptist Society in America, founded in 1639; the *Church
of the Messiah*, near Olneyville Square, is a memorial to Arthur Gammell, erected by his mother; the *Roger Williams Baptist*, the *Union
Congregational*, the *First Universalist*, and the Roman Catholic churches

of *St. Joseph* and *St. Mary*.   The Cathedral of *St. Peter and St. Paul* is in High St.

On the heights (Prospect St.) in the E. section of the city are the spacious grounds and substantial buildings of *** Brown University,** an old and important institution of learning, founded in 1764. Its library numbers over 68,000 volumes, and is housed in a handsome fireproof building, which has room for 100,000 volumes more. The *Sayles Memorial Hall* is a beautiful building, erected by Hon. W. F. Sayles, in memory of his son, who died while a student of the University. *Slater Hall*, a stately building, erected by H. N. Slater, is used as a dormitory, while *Wilson Hall* is another fine building, founded by George F. Wilson. The Museum of Natural History is rich in specimens, and the Art Collection includes some good portraits. The grounds comprise 16 acres, and are shaded with grand old elms. The *Lyman Gymnasium*, recently erected, is a most complete structure of its kind. The *Friends' School*, an institution for both sexes, is on a hill overlooking Seekonk River, and from its cupola can be seen nearly every prominent place in the State. It was founded in 1819, by Moses Brown, and is under the care of the Friends' Yearly Meeting. The *Rhode Island Historical Society* has a fine brick and granite building, recently much enlarged, opposite the University grounds, in which are a valuable library and some interesting historical relics. The *Providence Public Library* (Snow and Moulton Sts.) has 45,000 volumes and many pamphlets, including a special collection of 8,000 pamphlets, reports, etc., on slavery and the rebellion, from the library of the late C. Fiske Harris. Providence has long been noted for its valuable private libraries. Chief among them is that of the late John Carter Brown, famous for its "Americana." The *** Athenæum** (cor. College and Benefit Sts.) contains a reading-room, a library of 50,000 volumes, and some valuable paintings, among which are portraits by Allston and Sir Joshua Reynolds, and Malbone's masterpiece ("The Hours"). The *Hope Club* has an elegant home in the aristocratic quarter of the city; and the *Providence Art-Club* has a charming club-house, with picture-gallery, reading-rooms, etc. The *Young Men's Christian Association* has an elegant brown-stone building that cost over $200,000. The *Butler Hospital for the Insane* is on the W. bank of Seekonk River; the *Dexter Asylum for the Poor* is situated on elevated land in the N. E. part of the city. The *Rhode Island Hospital* has stately buildings in the S. part of the city. The *Home for Aged Women* is in the S. E. part, and the *Home for Aged Men* in the S. W. part of the city. The *State Farm*, in Cranston, comprises 500 acres, and contains the State Prison, Workhouse, House of Correction, Almshouse, State Hospital for the Insane, and Reform School.

There are several public squares and small parks. *Roger Williams Park* is near the W. shore of Narragansett Bay, in the S. part of the city; it was devised to the city in 1871 by Betsey Williams, a descendant of Roger Williams, and to whose memory a large bronze statue has been erected. *Blackstone Park* is one of the new parks. *Prospect Terrace*, on Congdon St., commands an unrivaled view of the city. *** Swan**

Point Cemetery, tastefully laid out and ornamented, is on the W. bank of Seekonk River, near the Butler Hospital for the Insane.

At Cranston, 4 m. W. of Providence, is the famous *Narragansett Trotting Park*, now owned by the R. I. Society for the Encouragement of Domestic Industry, whose State Fair is annually held there. A favorite drive is to *Hunt's Mills* (3½ miles), where there is a beautiful brook with a picturesque little cascade. *Pawtuxet*, 5 miles from the city on the W. shore of the Bay, has a fine beach and excellent bathing. The Conant Thread Co. has its large plant here. In summertime, steamers leave Providence every fifteen minutes for the various resorts on the Bay, and 4 times daily for Newport. *Warwick Neck* (12 m.) is worthy of note for its summer residences. *Seaconnet Point* is also a summer resort, with cottages. Its reddish, cliff-like rocks are frequently chosen as subjects by artists. Near by are the club-houses of the West Island Club. * **Rocky Point,** midway between Providence and Newport, has an observatory, with an extensive view, on the summit of a hill near by. It is famous for its clam-bakes, sharing the honor with *Silver Spring*, higher up, on the E. shore of the Bay. *Squantum*, near Silver Spring, is owned by the Squantum Club, has been fitted up at an expense of $60,000, and is noted for the private clam-bakes of the Club. The Powham Club has a building here.

Between Providence and Boston the distance is 44 miles, and *en route* are half a dozen uninteresting manufacturing towns, chief among which is *Pawtucket*, now a city, 5 miles from Providence. Here are made immense quantites of calico, thread, tacks, rope, braid, etc., and there is a fine water-power. As the train nears Boston the suburban villages of *Hyde Park* and *Roxbury* are passed, and the train stops at the Park Square Station on Columbus Ave. near the Common.

13. New York to Boston via " Air-Line R. R."

This route is composed of the New York & New Haven R. R. to New Haven; the Air Line Division of same road from New Haven to Willimantic; and the New York & New England R. R. from Willimantic to Boston. Total distance, 213 miles; time, 6 hrs.; fare, $5. It is the shortest route between New York and Boston. Limited express trains leave New York or Boston at 3 P. M., due at either city at 8.40 P. M., with complete drawing-room car service on each train.

As far as New Haven this route is identical with Route 11. *Wallingford* (12 miles) is described on p. 80. **Middletown** (*Kilbourn, McDonough ;* 24 miles from New Haven, 98 from New York) is one of the most beautiful cities in Connecticut, with a population in 1890 of 9,013. It lies on the W. bank of the Connecticut River, and is well built. *Main St.* is the leading business thoroughfare, and *High St.* is lined with fine residences. Upon an eminence overlooking the city (reached by High St.) stand the buildings of the *Wesleyan University* (Methodist), the most striking of which are the Memorial Chapel, Rich Hall, and Judd Hall. In Rich Hall is the library (25,000 volumes), and in Judd Hall some rich natural history cabinets. The * view from the tower of the old chapel is extremely fine, and another scarcely inferior may be obtained from *Indian Hill Cemetery*, which contains some handsome monuments. The *Berkeley Divinity School* (Episcopal) is located in Main St.; its chapel is an exquisite specimen of Gothic architecture. The extensive buildings of the *State Hospital for the Insane* stand on a high hill S. E. of the city, and command a wide-extended view. *Willimantic* (128 miles) is a prosperous manufacturing village and railroad center, where are produced large quantities of thread, silk, cotton goods, etc. (population, 8,648). *Putnam* (151 miles) is another thriving manufacturing town, at the crossing of the Norwich Division. Daily stages run from Putnam to **Woodstock** (*Elmwood Hall*), one of the most beautiful villages in New England, delightfully situated amid wonderfully picturesque scenery. "Its like," says Mr. Beecher, "I do not know anywhere. It is a miniature Mt. Holyoke, and its prospect the Connecticut Valley in miniature." About a mile from the village is Woodstock Lake, skirted by primeval woods and abounding in fish. At *E. Thompson* (160 miles) a branch line diverges to the busy town of *Southbridge;* and then follow in rapid succession the stations of *Blackstone* (177 miles), *Wadsworth*, *Franklin* (186 miles), and *Walpole* (194 miles). Passing then through the suburban towns of *Dedham*, *Hyde Park*, and *Dorchester*, the train stops at the Boston depot (foot of Summer St.).

14. New York to Boston, via New York & Northern and New York & New England R. Rs.

This route is by Manhattan Elevated R. R. to Harlem River, N. Y. city; thence *via* New York City & Northern R. R. to Brewster; and thence *via* New York & New England R. R. to Boston. Fare to Boston, $5; time, 10 hours; distance, 245 miles. Principal stations, South Yonkers, 8 miles; Pocantico Hills, 23 miles; Mahopac, 44 miles; Carmel, 49 miles; *Brewster*, 54 miles; Danbury, 65 miles; Waterbury, 95 miles; Bristol, 110 miles; New Britain, 119 miles; Hartford, 128 miles; Manchester, 136 miles; Willimantic, 159 miles. Thence to Boston, see Route 13.

THIS route passes through populous and thrifty districts of New York, Connecticut, and Massachusetts. Though somewhat longer than rival routes, and not used for through night-travel, it is highly attractive on account of the highly cultivated country through which it passes. Beyond *High Bridge* (see p. 21), the first station reached is *South Yonkers*, described in Route 9. Passing through a number of small stations the train reaches Tarrytown (see p. 69), where it has three stopping-places. Nine miles beyond the track skirts *Croton Lake*, which is described on p. 69. Small stations intervene till the train reaches *Mahopac*, where passengers for the beautiful watering-place of *Lake Mahopac* alight. *Carmel*, 5 miles beyond, is the county-town of Putnam Co., N. Y., and besides the Court-House, has several banks, newspaper-offices, seminaries, and two excellent hotels. Here is *Lake Gleneida*, which makes the place attractive to summer visitors. A run of 5 miles brings the train to *Brewster* (54 miles from New York), where it takes the track of the New York & New England R. R., which runs due W.

to Fishkill Landing on the Hudson River. *Mill Plain* is a small station
on the border-line of New York and Connecticut.

Five miles E. is the borough of **Danbury** (*Turner House*), one of
the county-towns of Fairfield Co., Conn., with a population of 16,552.
The place is historically noted as having been burned by the British in
1777. Here is the crossing of the Danbury Div. of the New York,
New Haven & Hartford R. R. It is noted for hat-manufacturing, in
which upward of $2,000,000 is invested. It has 11 churches, 4 banks,
3 newspapers, and several excellent hotels. It is also largely patronized
as a summer resort on account of the beauty of the country, a characteristic, indeed, of the route of the New York & New England R. R. in
its whole- course through the State. At Hawleyville, 6 miles beyond,
connection is made here with the Shepaug, Litchfield & Northern R. R.,
which runs to *Litchfield* (see Route 38).

The train, passing a number of stations, reaches **Waterbury**
(*Cooley, Franklin,* and *Scoville ;* 95 miles). This city had a population
in 1890 of 28,646. Connection is made here with the Naugatuck
Div. of the New York, New Haven & Hartford R. R. It contains 5
banks, several famous schools, 5 newspapers, the *Bronson Library* of
40,000 volumes (free), a handsome City Hall, and 11 churches, of which
St. John's Episcopal Church is one of the most beautiful in the State,
the spire being 200 ft. high. In the center of the town is a fine public park, whence the streets radiate. The many fine private residences
embowered in shrubbery attract the eye of the stranger. The town is
the great center for the manufacture of watches (the *Waterbury Co.*
turning out 1,200 watches a day), also for brass and German-silver
works.

Passing four unimportant stations, the train reaches *Plainville*, a
small manufacturing town, where connection is made with the Northampton Div. of the New York, New Haven & Hartford R. R. Five
miles E. is **New Britain** (119 miles), with a population in 1890 of
19,007 (*Hotel Ruswin*), where connection is made with the Berlin branch
of the New York, New Haven & Hartford R. R. The building of the
Connecticut State Normal School is a notable structure. The town has
7 churches, 2 banks, a public library, and many manufactories of hardware, jewelry, locks, etc. The *Russell & Erwin Co.'s* shops occupy 5
acres. In the center of the city is a fine public park of 72 acres. Ten
miles beyond, Hartford (see p. 80) is reached. Eight miles farther brings
the traveler to *Manchester* (136 miles), which has a population of 8,222.
There are important manufactures of paper, cotton, woolen, and silk
goods here. In *South Manchester* is Cheney Bros.'s celebrated factory of
American silks. Nineteen miles E. the traveler reaches *Willimantic*,
159 miles. Thence the route is the same as in Route 13.

15. Steamboat Routes to Boston.

a. Via "Fall River Line."

THIS route is by steamer to Fall River, Mass., and thence by the Old Colony R. R. (time 10 to 13½ hrs.). The steamers Puritan, Pilgrim, Plymouth, and Providence of the Fall River Line, are among the finest in American waters, and there are few trips more enjoyable than that part of the present journey which is made on them. Their route in leaving New York (from Pier 28, foot of Murray St., at 4.30 P. M. in winter, 5 P. M. in spring and fall, and 6 P. M. in summer, annex boat from Brooklyn half an hour earlier) affords an excellent view of the harbor and city, of Brooklyn and the Long Island shore, of the islands in the East River (see p. 22), of the famous Hell-Gate, and of the tranquil waters of Long Island Sound. The greater part of the voyage is on the Sound, and when *Point Judith* is passed the steamer's destination is close at hand. The boats then proceed to Fall River without stopping, where passengers take the express train to Boston. A daily service is also established between Newport and New York during the summer by the steamers leaving New York at 5 P. M. and Newport at 9.15 P. M.

Newport.

From New York, Newport is reached *via* Route 12 from Grand Central Depot to *Wickford*, and thence by Newport & Wickford R. R. and Steamboat Line, fare, $4.30 (a through drawing-room car express train runs during the summer months between New York and Newport) ; or by Fall River boat every night (fare, $2). From Boston, *via* Old Colony R. R. (distance, 68 miles ; fare, $1.70). From Providence, by steamer twice daily in summer, once in winter.

Hotels.—The *Ocean House,* on Bellevue Ave., is the largest and most fashionable, and is generally open from June 15th to October 1st. The *Aquidneck House*, at the cor. of Pelham and Corne Sts., is cozy and quiet. The *Perry House,* opposite Washington Square, at the head of the Long Wharf, is much patronized by business men. *Brayton's* and the *Clifton House* are good second-class hotels, open all the year round. The range of prices at these hotels is from $2.50 upward a day. The private-cottage system largely prevails at Newport, and hotel-life is quite subordinate to it. Furnished cottages cost $500 to $5,000 for the season. Board in private houses is $10 to $20 a week.

Methods of Transportation.—Electric-cars start from the Post-Office every twenty minutes to One-Mile Corner, or Middletown ; to Morton Park (three quarters of a mile), every twenty minutes ; and to Gaston's Beach (1 mile), every twenty minutes in summer, and every half-hour in winter.

Newport, one of the most fashionable and frequented of all the American summer resorts, is situated on the W. shore of Rhode Island and on Narragansett Bay, 5 miles from the ocean. It is a port of entry, and has a fine harbor, the approach to which from the sea is charming. During the season it is the rendezvous of the New York Yacht Club, and frequent races take place. Newport was settled in 1639, incorporated in 1700, and as late as 1769 exceeded New York in the extent of its commerce ; but it suffered greatly during the Revolution, and never recovered its commercial importance. The old town lies near the water; but a new city of charming villas and sumptuous mansions has sprung up along the terraces which overlook the sea.

Of the places of interest within the city proper, the first is * **Touro**

Park, between Pelham and Mill Sts. Here is the * *Old Stone Mill* (sometimes called the "Round Tower"), whose origin was once the theme of discussion, and which is still asserted by some antiquaries to have been built by the Norsemen 500 years before the arrival of Columbus. The weight of evidence appears to favor the theory that it was erected by Governor Benedict Arnold, who died in 1678. Near the Old Mill is J. Q. A. Ward's fine bronze statue of Commodore M. C. Perry, who was a native of Newport. In Equality Park, in Broadway, stands the *Soldiers' and Sailors' Monument,* in bronze, by W. Clark Noble. The **State-House** (for Newport is one of the capitals of Rhode Island) is a venerable building (dating from 1739), fronting on Washington Square in the center of the town. In its Senate-chamber is one of Stuart's celebrated portraits of Washington. The *Perry Mansion,* occupied by Commodore Perry after his victory on Lake Erie, fronts on this square; and before it stands the bronze statue of Commodore O. H. Perry, by William G. Turner; also the *City Hall* and the *Perry House.* Other objects of historical interest are the *Jewish Cemetery,* in Touro St., and the *Synagogue,* erected in 1762, when there were many wealthy Jews in Newport, and still kept in order by a bequest of $15,000 left for that purpose by Abraham Touro. * **Trinity Church** (Episcopal), in Church St., is a venerable edifice, built in 1725, possessing a special interest from the fact that Bishop (then Dean) Berkeley often preached in it (1729 to 1731). The *First Baptist Church,* in Spring St., dates from 1638, and is said to be the oldest church in Rhode Island. The *Central Baptist Church,* built in 1735 by the Second Congregational Church, and purchased by the *Central Baptist Society* in 1847, stands on Clarke St., and adjoining it is the *Armory* of the Newport Artillery Company, organized in 1741. In Pelham St., opposite the old Stone Mill, is the *Channing Memorial Church,* built 1880–'81, which contains some fine stained-glass windows. The *Vernon House,* cor. Clarke and Mary Sts., was the headquarters of Rochambeau in 1780. The *Hazard Memorial School* is a large building under the charge of Sisters of the Roman Catholic Church, and is a conspicuous edifice. The * **Redwood Library** (Bellevue Ave.) is a substantial building in the Doric style; it contains 37,000 volumes, and some choice paintings and statuary. The *Historical Society,* in the Seventh-Day Baptist meeting-house, in Touro St. just above the Jewish synagogue, has a fine collection (admission free). Adjoining this in the rear is the museum and lecture-room of the *Newport Natural History Society.* The *People's Library* (free) is in Thames St., and contains 30,000 volumes. The *Opera-House,* on Washington Square, is a handsome edifice. On Bellevue Ave., near the Ocean House, is the *New Casino,* a commodious and picturesque building, which comprises a fashionable lounge, a club-house, a theatre, restaurant, and a tennis-ground. Balls and musical and dramatic entertainments are frequent, and concerts are given twice daily in summer. The concerts and theatrical entertainments are usually open to the public on payment of a small admission charge, but the club privileges of the house can only be obtained by introduction.

The surf-bathing at Newport is unsurpassed. There are four fine beaches, of which * **First** or **Gaston's Beach** is the one principally used. It is ½ mile from the Ocean House, and stages and electric-cars run regularly to and fro. The pavilion is a handsome and commodious structure. furnished with a restaurant, hot and cold baths, and numerous bathing-houses. **Sachuest Beach** (Second) is about a mile E. of the First, and is used only by the more adventurous, the breakers being very heavy. At the W. end of this beach is * **Purgatory,** a dark chasm 160 ft. long, 50 ft. deep, and from 8 to 14 ft. across.   *Third Beach* is a long, secluded strip of sand, and beyond it are the picturesque * **Hanging Rocks,** within whose shadow Bishop Berkeley is said to have written his "Minute Philosopher."   *Bailey's Beach*, at the foot of Bellevue Ave., is also used in calm weather. The *Forty Steps*, leading from the bluff to the rocks beneath, are at the foot of Narragansett Ave.

The famous **Cliffwalk** extends along the Atlantic Ocean from Easton's Beach to Bailey's Beach, a distance of nearly 3 miles. It passes through the grounds of those whose estates extend to the water; and by a clause in the old deeds in which fishermen's rights were granted, a highway must be kept open forever. The Walk is in full view of many of the handsome villas, including those of Mrs. William Gammell, Robert Goelet, Ogden Goelet, Louis L. Lorrillard (formerly Miss Catherine Wolfe's), Cornelius Vanderbilt, Fairman Rogers, William Astor, William K. Vanderbilt, W. W. Astor, Ogden Mills, F. W. Vanderbilt, and others. *Morton Park*, a pretty tract of land at the southerly end of the electric-car route, in Coggeshall Ave., was presented to Newport by Levi P. Morton, formerly Vice-President of the United States. The Park is not large, but the city is constantly making expenditures for its improvement. Adjoining it are the *Polo Grounds*, where matches are played during the summer months.

The grand drive of Newport is * **Bellevue Avenue,** 2 miles long, and, during the fashionable hours, thronged with costly equipages. Bellevue Ave. extends from the Jewish Cemetery to Bailey's Beach, and Ocean Ave. is then reached, where the "ten-mile drive" continues from Bailey's Beach along the south shore, by Grave's Point, Bateman's, Castle Hill, Fort Adams, Brenton's Cove, along the crest of Halidon Hill with its superb views of the harbor, the islands, and the bay. Another drive is by the *West Road*, from Broadway to Bristol Ferry, a distance of 9 miles, by Lawton's Valley, the coal-mines, and Portsmouth Grove, with full view of the bay. The East Road extends from Broadway to Stone Bridge, a distance of 12 miles. Another drive is along Paradise Road, from Second Beach, by Hanging Rocks to Indian Ave., and then along Indian Ave., continuing N. along East Shore. The * **Spouting Rock** (reached by Bellevue Ave.) is a popular resort of excursion-parties. It is a deep cavern, running back from the sea into the rocky cliffs, and is quiet enough in ordinary weather; but after a S. E. storm the waves rush madly in and dash through an opening in the roof, sometimes to the height of 50 ft. The view from the cliffs above is considered one of the finest that Newport affords. Another favorite excursion is to the **Glen,** a quiet and sequestered retreat, where an old mill stands near a pond. It is 7 miles out, on the Stone Bridge road. The *Pirate's Cave,* 4½ miles from the city, on the road to Brenton's Point, and *Miantonomo Hill,* 1¼ mile, are often visited. *Lily Pond,* the largest sheet of spring-water on the island, is easily reached from Spouting Rock. **Fort Adams,** near Brenton's Cove, 3¼ miles from the city, is one of the largest and strongest fortresses in the United States, mounting 460 guns. Three times a week occur what are called the "fort days," when the band discourses its best music. On *Coasters' Harbor Island* is the building formerly used as the county poor-home, but now occupied by the U. S. Government as a training station for naval apprentices. The Naval War College, built during 1892, is on the same island. *Canonicut Island,* opposite Fort Adams, is rapidly becoming very popular as a summer resort, and many cottages have been built there within recent years. It is connected by ferry with Newport, and at *Jamestown,* as the landing-place is called, are *Bay View Hotel, Bay Voyage House, Canonicut Park Hotel, Gardner House, Prospect House,* and *The Thorndyke.* *Brenton's Cove* is approached by a causeway leading to Fort Adams, and affords a charming view of the city. *Goat Island,* opposite the city wharves, is the headquarters of the torpedo division of the U. S. Naval Service. *Lime Rock,* famous as the home of Ida Lewis, lies in the harbor S. of Goat Island. A popular excursion is by Providence steamer to *Rocky Point* (see p. 88). Daily steamboat excursions may be also made to *Block Island* and *Narragansett Pier.*

Beyond Newport the steamer from New York plows the lovely waters of Narragansett Bay, and soon stops at **Fall River** (*Mellen House, Wilbur House*), one of the large manufacturing cities of Massachusetts, with a population in 1890 of 74,398. Cotton-cloth is the great article of manufacture, and more spindles are said to be in operation than in any other American city. Fall River is well built, many of the edifices being of granite, and the vast factories are worth inspecting. North and South Main St. is the principal thoroughfare. Here passengers take the cars of the Old Colony R. R. and are conveyed to Boston (49 miles) in about 1¼ hour. The route is through a well-cultivated and populous farming country. Many towns and villages cluster along the line, of which the principal are **Taunton** (*City Hotel*), another prosperous manufacturing city, with 25,448 inhabitants; and thence by way of Stoughton and Canton to Park Square Station. There is another route from Fall River to Boston *via* Bridgewater. It is the same as the one here described, except that Taunton is not passed.

b. Via "Stonington Line."

Next to the Fall River Line this is the most popular of the steamboat routes to Boston (time, 13 hours). The Maine and New Hampshire leave daily (except Sundays) from Pier 36, North River, one block above Canal St., at 5 P. M. The route is nearly the same as that of the Fall River boats; but the distance traveled by steamer is shorter, and the occasionally stormy ocean-passage around Point Judith is avoided. At **Stonington** (see p. 85) passengers are transferred to the cars of the Providence Div. of the New York, New Haven & Hartford R. R., and the remainder of the route is identical with Route 12.

c. Via "Norwich Line."

The boats of this line run from Pier 40, North River, daily (Sundays excepted) at 5 P. M. (time, 13½ hours), by Long Island Sound to New London and Norwich, Conn., which is reached in the early morning. **New London** has already been described on p. 84. Here the cars of the Central Vermont R. R. are taken, and in 13 miles we reach **Norwich** (*Wauregan House*), a beautiful city, with a population of 16,156, lying between the Yantic and Shetucket Rivers, which there unite and form the Thames. The city is laid out in broad avenues, bordered with fine trees, and the churches, public buildings, and private houses are very attractive. Washington St. and Broadway are lined with handsome dwellings, surrounded by shade-trees and ornamental gardens. The **Court-House** is in Union Square. Main St. is the leading business thoroughfare. The *Free Academy* is an imposing building near the Parade, or Williams Park (reached by Broadway). Near by is the *Park Congregational Church*, and in Washington St. is the ivy-clad *Christ Church* (Episcopal). *St. Patrick's Cathedral*, in Broadway, is a fine specimen of church-architecture. The *Yantic Cemetery* and the old burying-ground contain some interesting monuments; and in the ancient Indian burying-ground in Sachem St. a granite obelisk marks the grave of Uncas. Near Greenville is the battle-field, where a granite block marks the site of Miantonomo's capture; and a drive of 5 miles toward New London leads to *Mohegan*, where a remnant of the aborigines still live. The once famous *Falls in the Yantic* have been sacrificed to the need of water-power for factories. The capital invested in manufacturing is large, and the principal products are worsted, printing-presses, cotton-goods, fire-arms, paper, locks, stoves, and various articles of iron and steel. At Norwich the line of the New York & New England R. R. is reached, and the route thence is identical with Route 13.

16. Boston.

Hotels.—The *Brunswick* (cor. Boylston and Clarendon Sts.), the *Copley Square Hotel* (Huntington Ave.), the *Thorndike* (Boylston and Church Sts.), and the *Vendome* (cor. Commonwealth Ave. and Dartmouth St.), are among the finest in the country. Other houses on the American plan are the *Abbotsford*, 186 Commonwealth Ave.; the *American House*, in Hanover St., centrally located; the *Langham*, 1697 Washington St., cor. Worcester, at the S. End; the *Quincy House*, Brattle Square; the *Revere*, on Bowdoin Square; the *Tremont*, cor. Tremont and Beacon Sts.; and the *United States*, in Beach St. cor. Lincoln. The rates charged at these hotels vary, according to the location and reputation of the house, from $3 to $5 per day. Of the hotels on the European plan, the *Adams House* (Washington St., between Boylston and West Sts.), the *Parker House* (School St., opposite the City Hall), the *Victoria* (Dartmouth and Newbury Sts., in the Back Bay district), and *Young's Hotel* (Court Ave.), are the best. Rooms at these houses are from $1 to $5 per day. Among the less expensive hotels on the European plan are the *Boston Tavern*, Washington St. near Bromfield; and the *Crawford House*, cor. Court and Brattle Sts. Rooms at these hotels are from $1 to $3 a day; meals *à la carte* in restaurants attached or elsewhere.

Restaurants.—The restaurants of the *Adams House, Parker House*, and *Young's Hotel*, are famous. *Copeland's* (128 Tremont St.) and *Weber's* (in Temple Place) are much frequented by ladies. At *Ober's* (4 Winter Place) will be found the French *cuisine*, and in Van Rensselaer Place, off Tremont St., are two French restaurants, where a *table-d'hôte* is served for 60c. *Dooling's* (Temple Place), the *Le Déjeuner* (132 Tremont St.), the *Moulton* (24 Summer St.), the *Oak Grove Farm* (413 Washington St.), *Park's* (Bosworth St.), the *Quincy Café* (Brattle St.), the *St. Nicholas* (10 Province St.), the *Thorndike*, and *Vercelli's* (200 Boylston St.), an Italian restaurant, are all good. There is an excellent *café* in the Women's Educational and Industrial Union, 264 Boylston St. Good restaurants are attached to all the railway-stations.

Modes of Conveyance.—The *street-car* system of Boston is very complete, all the long-distance routes running electrical cars. *Carriages* are in waiting at the depots and at stands in various parts of the city. The fares are regulated by law, and are as follows : For one passenger a course in city proper, 50c. ; from points S. of Dover St., or W. of Berkeley St. to points N. of State, Court, and Cambridge Sts., $1 ; each additional passenger, 50c. Complaints of overcharges should be made to the Superintendent of Hacks, City Hall. There are 2 *ferries* to East Boston—North Ferry, from Battery St. to Border St.; and South Ferry, from Eastern Ave. to Lewis St. (fare, 2c.). The Winnisimmet Ferry connects the city with Chelsea (fare, 5c.). The Herdics and the cabs of the Boston Cab Co. have largely taken the place of hacks ; fares, 25, 35, and 50 cts. a course, or $1 an hour.

Railroad Stations.—The *Boston & Maine R. R. (Southern Division) Station*, Causeway St. near Lowell St., is of brick, trimmed with Nova Scotia freestone, 700 ft. long and 205 ft. wide. Just beside it, in Causeway St., stands the station of the *Boston & Maine R. R. (Eastern Division)* ; and a few paces from the latter is the station of the *Fitchburg R. R.* The *Boston & Albany Station* is in Kneeland St., between Lincoln and Utica Sts.; the station of the *Boston & Maine (Western Division)* is in Haymarket Sq., at the end of Washington St.; that of the *Providence Branch of the Old Colony R. R.* (Shore Line route) is in Columbus Ave. near the Common, known as the Park Square Station ; that of the *Old Colony R. R.* is at the cor. of Kneeland and South Sts.; that of the *Boston, Revere Beach & Lynn* (narrow gauge) in Atlantic Ave.; and that of the *New York & New England R. R.* in Atlantic Ave., at the foot of Summer St.

Theatres and Amusements.—The *Tremont Theatre*, Tremont St. near Mason, is the most attractive in the city ; the *Boston Theatre*, in Washington St. near West St., is the largest in New England. The *Globe Theatre*, in Washington St. near Essex, is devoted to star performances, and the *Park Theatre* (opposite) is largely devoted to farcical comedy. The *Columbia Theatre*, in Washington St. near the Boston & Albany R. R. bridge, is a beautiful playhouse (constructed in 1891), and is very popular. The *Bowdoin Square Theatre*, in Bowdoin Square, furnishes first-class entertainments. The *Palace Theatre*, in Court St. near Scollay Square, is devoted to variety performances at low prices. The *Boston Museum*, Tremont St. near School St., contains the oldest theatre in the city. In the Museum are pictures, casts, wax-figures, and curiosities of all sorts (admission, 35c.). The *Howard Athenæum*, Howard St. near Court St., gives variety-shows. The *Bijou*, devoted mainly to vaudeville and comic opera, is a few doors from the Boston Theatre. The *Hollis St. Theatre* stands on the old church site, between Washington and Tremont Sts., and is devoted to comedy and society drama. *Music Hall*, Hamilton Place and Winter St., is one of the finest in the country, and here the *Symphony Concerts* are given throughout the winter season ; *Bumstead Hall*, in the same building, entrance on Winter St., is dedicated to piano and minor concerts. *Mechanics' Hall*, in Huntington Ave., has an immense stage and auditorium, suitable for great gatherings ; opera, grand concert, the annual Mechanics' Fair, horse, dog, and poultry shows, and intercollegiate athletic sports are held here. Lectures and concerts are given at Music Hall ; at *Tremont Temple*, in Tremont St., opposite the Tremont House ; at *Association Hall*, cor. Boylston and Berkeley Sts. ; at *Horticultural Hall*, 100 Tremont St. ; at *Chickering Hall*, in Tremont St. near West ; and at the *Hawthorne Rooms*, in Park St. At Horticultural Hall are also held the flower-shows for which Boston is famous. *Horse-races* occur at Beacon Park and Mystic Park, in the suburbs of the city.

Reading-Rooms.—In the leading hotels are reading-rooms (supplied with newspapers) for the use of guests. The Public Library building, in Copley Square, is the finest in the United States, and was constructed at a cost of about $3,000,000, and was opened in 1893. It is free to all, and contains a larger num-

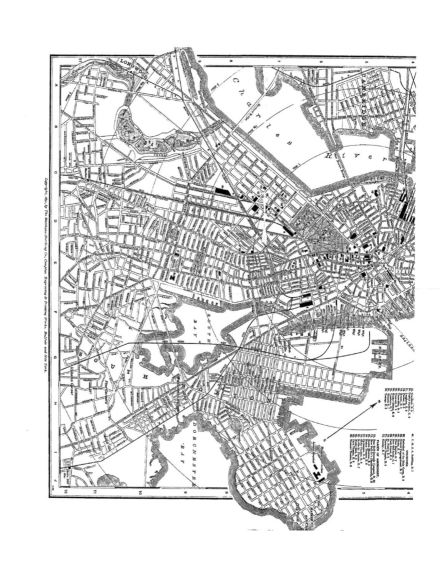

ber of books than any other American library, except the Congressional at Washington. The *Athenæum*, in Beacon St. near Bowdoin, has excellent reading-rooms, but introduction by a member is necessary. Free reading-rooms may be found at the *Young Men's Christian Union*, 48 Boylston St., and the *Young Men's Christian Association*, cor. Boylston and Berkeley Sts. The free reading-rooms of the *Women's Union* are at 264 Boylston St.

Art Collections.—The *Museum of Fine Arts*, St. James Ave. and Dartmouth St., contains an extensive collection of pictures, statuary, casts, and antiquities (admission, 25c.; free on Saturday and Sunday afternoons). Exhibitions are held at the *Boston Art Club* (opposite Art Museum), the *St. Botolph Club Art Gallery*, Newbury St. (by tickets from members for special exhibitions). Good pictures, engravings, etc., may be seen (free) at the sales galleries of *Williams & Everett*, 190 Boylston St.; at *Doll & Richards*, 2 Park St.; and at *J. Eastman Chase's*, 7 Hamilton Place.

Clubs.—The *Temple Club*, 35 West St., is the oldest in the city. The *Somerset Club* has a fine house in Beacon St. The *Suffolk Club* is at 4½ Beacon St. The *Union Club* owns a house in Park St., containing a valuable library. The *Central Club*, the *St. Botolph Club*, at 2 Newbury St., and the *Tavern Club* (artistic and literary), 4 Boylston Place; and the *Puritan Club*, at cor. of Mt. Vernon and Joy Sts. The *New England Woman's Club* has quarters in Park St. The *Algonquin Club* is in Commonwealth Ave., north side, between Exeter and Fairfield Sts., and the *Art Club*, Dartmouth, cor. of Newbury St. The *Athletic Club*, Exeter St. cor. Blagden, has an immense and admirably furnished building, with a fashionable membership.

Post-Office.—The Post-Office is in Devonshire St., between Milk and Water Sts. It is open for the delivery of letters from 7.30 A. M. to 7.30 P. M. The Back Bay P. O., Pierce Building, Huntington Ave., is most convenient for visitors up-town, and Station A, Washington St. near Brookline, for visitors to the South End.

BOSTON, the capital of Massachusetts, and chief city of New England, is situated at the W. extremity of Massachusetts Bay, in latitude 42° N. and longitude 71° W. The city embraces Boston proper, East Boston, South Boston, Roxbury, Dorchester, Charlestown, Brighton, and West Roxbury, containing in all about 22,000 acres. Boston proper, or old Boston, occupies a peninsula of some 700 acres, very uneven in surface, and originally presenting three hills, Beacon, Copp's, and Fort, the former of which is about 130 ft. above the sea. The Indian name of this peninsula was Shawmut, meaning "Sweet Waters." It was called by the earlier settlers Trimountain or Tremont. *East Boston* occupies the W. portion of Noddle's or Maverick's Island. Here is the deepest water of the harbor, and here the ocean-steamers chiefly lie. *South Boston* extends about 2 m. along the harbor, an arm of which separates it from Boston proper. At South Boston are the large docks and warehouses of the N. Y. & N. E. R. R. Near the center are Dorchester Heights, which afford a fine view of the city, bay, and surrounding country. The city is connected with Charlestown by the Charles River bridge, 1,503 ft. long, and the Warren bridge, 1,300 ft. long; and with Cambridge by the West Boston bridge, which crosses Charles River from Cambridge St., Boston, and is 2,756 ft. long, with a causeway 3,432 ft. long. Craigie's bridge, 2,796 ft. long, extends from Leverett St. to E. Cambridge; from this bridge another, 1,820 ft. long, extends to Prison Point, Charlestown. Harvard Bridge, 2,166 ft. long, extends from West Chester Park to Old Cambridge, and from this a fine view of both cities may be obtained. South Boston is reached by the Federal St. bridge, about 500 ft. long, and the South Boston bridge, 1,550

7

ft. long, also by the Broadway bridge. A causeway, built across Back
Bay on a substantial dam 1½ m. long, extends from the foot of Beacon
St. to Sewall's Point in Brookline.   The harbor is a spacious indentation of Massachusetts Bay, embracing about 75 square miles, including several arms.   There are more than 50 islands or islets in the harbor, and it offers many picturesque views.

In the older portions of the city the streets are irregular, and
generally narrow, though somewhat has been done toward widening
and straightening them since the fire.   Those in the section built
on the made land of Back Bay are wide, well paved, regularly laid
out, and present a handsome appearance. *Washington, Tremont,* and
Winter Sts. are the principal thoroughfares for general retail stores.
Park and *Boylston Sts.* have the newest shops and best display. *State
St.* is the financial center, and contains the headquarters of the leading
bankers and brokers. *High St.* and adjacent streets are the largest boot
and shoe markets in the world; and in Franklin, Chauncey, Summer, and
the neighboring streets are the great wholesale dry-goods establishments.
Commonwealth Ave., in the heart of the Back Bay or fashionable

quarter, is one of its finest streets. It is 240 ft. wide, and through the center run rows of trees and several statues, noticeable among which are those of Leif Eriksen the Norseman, by Anne Whitney, and William Lloyd Garrison, the former of which looks out on the most striking feature of the great chain of parks, the Back Bay Fens, a botanic garden of much beauty. Boulevard driveways lead to the other parks, that about Chestnut Hill Reservoir, that about Jamaica Pond, Bussey Park, the Arnold Arboretum, and Franklin Park. Marine Park, at City Point, with its pleasure-pier 3,000 ft. long, and the Charles River Embankment, are also noticeable features of the park system. The beauty of its surroundings is such that there are pleasant drives out of Boston in almost any direction. The most popular drive is to Brookline or around Chestnut Hill Reservoir (5 miles), while that through Franklin Park and beyond, in Blue Hill Ave., shows more natural beauty.

The first white inhabitant of Boston was the Rev. John Blackstone, supposed to have been an Episcopal clergyman, and to have arrived in 1623. Here he lived alone until 1630, when John Winthrop (afterward the first Governor of Massachusetts) came across the river from Charlestown, where he had dwelt with some fellow-immigrants for a short time. About 1635 Mr. Blackstone sold his claim to the now populous peninsula for £30, and removed to Rhode Island. The first church was built in 1632; the first wharf in 1673. Four years later a postmaster was appointed, and in 1704 (April 24th) the first newspaper, called the "Boston News Letter," was published. The "Boston Massacre" occurred March 5, 1770, when 3 persons were killed and 8 wounded by the fire of the soldiery. On Dec. 16, 1773, the tea was destroyed in the harbor, and Boston bore a conspicuous part in the opening scenes of the Revolution. The city was incorporated in 1822, with a population of 45,000, which had increased to 136,881 in 1850, to 177,840 in 1860, and 250,526 in 1870. By the annexation of the suburbs of Brighton, Charlestown, and W. Roxbury, the population had increased in 1880 to 362,839, and in 1890 was 448,477. On the 9th of November, 1872, one of the most terrible conflagrations ever known in the United States swept away the principal business portion of Boston. The district burned over extended from Sumner and Bedford Sts. on the S. to near State St. on the N., and from Washington St. E. nearly to the harbor. About 775 of the finest buildings in the city were destroyed, causing a loss of $70,000,000.

Perhaps the most interesting and attractive spot in Boston is the *Common, a park of 48 acres in the heart of the city, surrounded by an iron fence, laid out in sloping lawns and winding walks, and shaded by magnificent trees. The Common is considered to date from 1634, and by the city charter it is made public property forever. A pond and fountain, on the site of the ancient and historic *Frog Pond*, occupy a central point in the grounds. On Flagstaff Hill, overlooking the Pond, is the *Soldiers' and Sailors' Monument*, 90 ft. high, with 4 statues of heroic size at the base, surmounted by a colossal figure of America standing on a hemisphere, and guarded by 4 eagles with outspread wings. Near the monument stood the famous *Old Elm*, which antedated the birth of the city, and was finally blown down in the gale of Feb. 15, 1876. Facing Tremont St., on the Mall, is the monument to Crispus Attucks and others killed in the Boston massacre. Near Park St. is the beautiful *Brewer Fountain*, of bronze, cast in Paris, and the *Cogswell Fountain* is opposite West St.—The *Public Garden, separated from the Common by Charles St., was dedicated to the public in 1859, but the chief improvements have been made in the last twelve years. It comprises 22 acres, beautifully laid out, and contains Thomas Ball's noble equestrian statue of Washington, W. W. Story's bronze statue of Edward Everett, a statue of Charles Sumner, by Thomas Ball, one representing "Venus rising from the Sea," and the beautiful monument in honor of the discovery of ether as an anæsthetic. There is also a monument to Colonel Cass, one of the heroes of the late war. In the center is a serpentine pond covering 4 acres and crossed by a handsome bridge.

N. of the Common is Beacon Hill, on the summit of which stands the *State-House, an imposing edifice 173 ft. long and 61 ft. deep, with a stately colonnade in front, and surmounted by a gilded dome. On the terrace in front are statues of Daniel Webster and Horace Mann. On the entrance floor (Doric Hall) are Ball's statue of Governor Andrew, busts of Samuel Adams, Lincoln, and Sumner, and a collection of battle-flags. In the Rotunda, opening off Doric Hall, are Chantrey's statue of Washington, copies of the tombstones of the Washington family in Brington church-yard, England, and many historical relics. The * view from the dome (open when the General Court is not in session) is very fine, including the city, the harbor and ocean beyond, and a vast extent of country. The new part now building in the rear will be even more striking than the other. In Beacon St., near the State-House, is the *Boston Athenæum, an imposing freestone edifice in the Palladian style, containing a library of more than 150,000 volumes, a reading-room, and some choice pieces of sculpture. The Athenæum was incorporated in 1807, and is richly endowed. The *American Academy of Arts and Sciences* has its rooms and library (15,000 volumes) in the Athenæum building. Near the Athenæum is *Pemberton Square,* the site of an old Indian burying-ground, between which and Somerset St. is the new *County Court-House ;* and on the farther slope of Beacon Hill is *Louisburg Sq.,* containing statues of Columbus and Aristides. In the Park Square is the bronze group "Emancipation," presented by Moses Kimball. In Somerset St., near Beacon St., are the offices of the *Boston University,* founded in 1869 by Isaac Rich, who bequeathed it $2,000,000. The *Historic and Genealogical Society's* building (Somerset St.) contains 18,000 volumes, and a collection of antiquities.

Opposite the Common, at the cor. of Tremont and Boylston Sts., is the imposing **Masonic Temple.** Opposite the Temple is the Gothic building of the *Young Men's Christian Union.* From this point Boylston St. leads W. past the Public Garden to the aristocratic Back Bay. Beyond the Garden, in Berkeley St., between Boylston and Newbury Sts., is the fine building of the *Society of Natural History,* with a library of 20,000 volumes, and valuable cabinets (open to the public from 9 A. M. to 5 P. M.; free on Wednesdays and Saturdays; on other days, 25c. admission). Near by is the *Institute of Technology,* and at the cor. of Clarendon St. is the *Hotel Brunswick.* The fine new building of the Young Men's Christian Association is at the cor. of Boylston and Berkeley Sts. Close by is the *Second Church* (Unitarian), with a rich interior; nearly opposite which (cor. Clarendon St. and Huntington Ave.) is *Trinity Church (Episcopal),

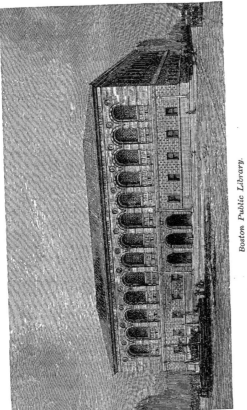

Boston Public Library.

with chapel, one of the largest, finest, and most splendidly decorated churches in America, finished in 1877 at a cost of $750,000. One block beyond (cor. Boylston and Dartmouth Sts.) is the new *** Old South Church**—church, chapel, and parsonage. The interior is extremely ornate, and the tower is 248 ft. high. A block S., in Copley Square (cor. Dartmouth St. and Huntington Ave.), is the *** Museum of Fine Arts,** a substantial red-brick building, elaborately adorned with terra-cotta bas-reliefs, copings, and moldings. In the lower halls are statuary, casts, and Egyptian antiquities; and.in the upper galleries a library and one of the richest collections of paintings and engravings in the country. (Admission free on Saturdays and Sundays after 12 noon; other days, 25c.; closed till noon on Mondays.) The building of the Boston Art Club stands near the Old South Church, while the building of the new **Public Library,** on the square, was completed and occupied in 1893. It is open to the public every day in the week, and, next to the Library of Congress, the largest in America. It contains 500,000 volumes, 275,000 pamphlets, and the valuable Tosti collection of engravings. In Newbury St., cor. of Exeter, is the *Normal Art School.* In Huntington Ave. near Dartmouth St. is the Massachusetts Charitable Mechanics' Association. The ** First Baptist Church,* cor. Clarendon St. and Commonwealth Ave., is a massive stone edifice in the form of a Greek cross, with a campanile 176 ft. high, surrounded near the top with a frieze containing colossal statues in high relief, after designs by Bartholdi. The *First Church* (Unitarian), cor. Berkeley and Marlborough Sts., has a richly decorated interior. Close by, at the cor. of Berkeley and Newbury Sts., is the elegant little *Central Congregational Church,* cruciform in shape, with a stone spire 240 feet high; near which, in Newbury St., is the ornate *Emmanuel Church* (Episcopal). The *Arlington St. Church* (Unitarian), fronting the Public Garden, with a fine chime of bells; the *Church of the Advent,* at the cor. of Mt. Vernon and Brimmer Sts.; and the *Hollis St. Church,* Newbury St., cor. Exeter, are noteworthy buildings. The depot of the Providence Division, Old Colony R. R., in Columbus Ave. near the Common, is worthy of attention. The English High and Boston Latin Schools face on Montgomery St., and the *Girls' High and Latin School* is on West Newton St.

In Dock Square, in the heart of the business quarter, stands *** Faneuil Hall,** the most interesting building in the United States next to Independence Hall, Philadelphia. This famous edifice, the "cradle of liberty," was erected in 1742 and presented to the town by Peter Faneuil, a Huguenot merchant. Destroyed by fire in 1761, it was rebuilt in 1768, and enlarged to its present dimensions in 1805. The basement is a market with shops. The public hall is on the second floor, and adorned with a full-length portrait of the founder, and with portraits of Washington, Samuel Adams, J. Q. Adams, Webster, Everett, Lincoln, John A. Andrew, Henry Wilson, and Charles Sumner. It is now used for great political gatherings and public demonstrations. Just E. of Faneuil Hall is **Quincy Market,** a vast granite building, 530 by 50 ft., and 2 stories high; and near by (in State St.) is the massive and stately * **U. S. Custom-House,** of granite, in the form of a Greek cross, with handsome porticoes on either front, erected from 1837 to 1849 at a cost of $1,076,000. The *Merchants' Exchange,* in which is the *Stock Exchange,* is on the site of the old *Merchants' Exchange,* 55 State St. It is a structure of large size and massive architecture. Close by, in India St., is the *Chamber of Commerce* building. At the head of State St., in Washington St., is the *Old State-House,* built in 1748, and often mentioned in Revolutionary annals, but now remodeled inside and outside. The lower floors are given over to business uses, but the upper floors are devoted to an historical museum ; open daily, admission free. Just above, in Court Square, is the old *County Court-House,* a fine building of Quincy granite. In rear of the Court-House, fronting on School St., is the * **City Hall,** one of the most imposing edifices in the city. It is of white Concord granite, in the Italian Renaissance style, and surmounted by a Louvre dome 109 ft. high. The interior is striking, and on the lawn in front are Greenough's bronze statue of Franklin, and Ball's bronze statue of Quincy.

Opposite the City Hall is the *Parker House,* and just above (at the cor. of School and Tremont Sts.) is the venerable. *King's Chapel,* built in 1754 by the Episcopalians on the site of the first church of that sect in Boston. Adjoining it is the first burying-ground established in Boston, containing the graves of Isaac Johnson, "the father of Boston," Governor Winthrop, John Cotton, and other distinguished men. On Tremont St. to the right of School St. is a granite building in which are the rooms of the *Massachusetts Historical Society,* with a library of 28,000 volumes, and many valuable MSS., coins, maps, charts, portraits, and historical relics. Close by is the *Boston Museum* (admission 35c.), containing pictures, casts, wax-figures, and curiosities from all parts of the world. Turning down Tremont St. to the left from School St., we pass *Tremont Temple* (used for lectures, readings, etc.), and soon reach * **Horticultural Hall,** an ornate white granite structure, in which frequent floral shows are held ; also fairs, concerts, and lectures. Just beyond is **Music Hall,** one of the finest in the country, and seating nearly 3,000 persons (entrances on Hamilton Place and Winter St.). Opposite is the famous *Park Street Church* (Congregational), founded in 1809 ; adjoining which is the *Old Granary Burying-Ground,* in which are buried Peter Faneuil, Samuel Adams, John Hancock, and other distinguished men. Near Temple Place is the church of *St. Paul's* (Episcopal), of gray granite in the Ionic style, with a classic portico of 6 columns. Still beyond is the Masonic Temple, already described (see page 100) ; and farther still (cor. Berkeley St.) is **Odd-Fellows' Hall,** a white granite building of chaste and pleasing design. At the corner of Concord St. is the *Methodist Church,* a quaint structure, with 2 spires. In Harrison Ave. near Concord St. is the **City Hospital,** a spacious granite edifice standing in 7 acres of grounds; and near it is the Roman Catholic *Home for Orphans,* the *Church of the Immaculate Conception* (famed for its music and its fine interior), and *Boston College,* a Jesuit institution with many pupils. Fronting on Franklin Square, in this vicinity, is the New England Conservatory of Music, with rooms for more than 500 women students. Also in Harrison Ave. is the *Church of St. James* (Roman Catholic), in the purest form of the classical basilica, with richly adorned interior. At the corner of Washington and Malden Sts. is the *Cathedral of the Holy Cross (Roman Catholic), the largest and finest church edifice in New England. It is in the mediæval Gothic style, 364 ft. long and 170 ft. broad, with stained-glass memorial windows of artistic beauty, and a magnificent high altar of marble and onyx.

Returning to the business quarter, we find at the cor. of Washington and Milk Sts. the *Old South Church, an historic relic of much interest. It was built in 1729, and was used as a place of meeting by the heroes of '76, and subsequently by the British as a place for cavalry-drill. It barely escaped the flames in the great fire of 1872, and immediately afterward was leased to the Government for a post-office. It is used for lectures, and contains an historical collection, open daily, admission 25c. In the square bounded by Milk, Water, and Devonshire Sts. and Post-Office Square, is the *Post-Office and Sub-Treasury, of granite, highly ornate in style, and is said to be the finest building in New England. The upper stories are used by the *U. S. Sub-Treasury ;* the Cash-room is very richly adorned.

The new buildings and "blocks" erected in the burned district since 1872 comprise some of the finest commercial structures in America. Among them are the *Brewer Building,* on Devonshire, Franklin, and Federal Sts. ; the *Franklin Building,* cor. Franklin and Federal Sts. ; the *Rialto Building,* cor. Devonshire and Milk Sts.; the *Simmons Building,* cor. Congress and Water Sts. ; the *Cathedral Building,* in Winthrop Square; and especially those of the *N. Y. Mutual Life Ins. Co.* (cor. Milk and Pearl Sts.), the *Equitable Life Assurance Co.* (cor. Milk and Federal Sts., opposite the Post-Office), the *New England Mutual Life Ins. Co.* (cor. Milk and Congress Sts.), the *Fiske Building,* State St., the *Ames Building* (cor. Washington and Court Sts.), and the *Sears Building* (cor. Washington and Court Sts.).

On **Copp's Hill,** in the N. E. part of the city, is the old *North Burying-Ground,* the second established in Boston, and still sacredly preserved. Here lie three fathers of the Puritan Church, Drs. Increase, Cotton, and Samuel Mather. *Christ Church,* in Salem St. near Copp's Hill, is the oldest in the city, having been erected in 1722. In the tower is a fine chime of bells, and here was the signal-light for Paul Revere.

Of the charitable institutions of Boston, the *Perkins Institution for the Blind* is famous. It was founded in 1831 by Dr. Samuel G. Howe, and occupies spacious buildings on Mt. Washington, in S. Boston. Near by on the hill is *Carney Hospital,* managed by the Sisters of Charity. The *Massachusetts General Hospital* is a vast granite structure on Charles River, between Allen and Bridge Sts. The *City Hospital* has been described previously. The *Marine Hospital* (for invalid seamen) occupies a commanding site in Chelsea, and is a spacious and stately building. The *Soldiers' Home* is located on the top of Powder-Horn Hill, Chelsea. The *U. S. Naval Hospital* is near by. The *House of Industry* and the

Almshouse are on Deer Island, in the harbor; and the *House of Correction* and *Lunatic Asylum* in S. Boston.

Itineraries.

The following series of excursions has been prepared so as to enable the visitor whose time is limited to see as much of the city as possible in the least amount of time. Each excursion is planned to occupy a single day, but the visitor can readily spend more time, as special features crowd upon his attention.

1. Visit the State-House dome, where an excellent view of the city may be obtained; Doric Hall (State-House); the Common; Public Garden; a short walk westward on Boylston St. brings one to Trinity Church, the new Old South Church, and the Public Library; then take electric-car[1] at Copley Square and ride to Chestnut Hill Reservoir, passing Commonwealth Ave., and along Beacon St., past the finest residences; stroll around Chestnut Hill Reservoir, and return by electric-car, stopping at the corner of West Chester Park and Beacon Sts. to obtain the splendid view of the Charles River from the Harvard bridge, and to the Museum of Fine Arts, in Copley Square (open daily from 12 M. to 5 P. M.; admission 25c.; free on Saturdays and Sundays).

2. Visit the Old South Church, cor. Milk and Washington Sts.; the Old State-House, cor. State and Washington Sts.; and Faneuil Hall—in all of which are many interesting pictures and relics; the Merchants' Exchange, in State St., where the Stock Exchange is located; the Chamber of Commerce Building, in India St.; then take East Boston car, on Devonshire St. (which crosses State St.), and connect with ferry for East Boston, thus passing across the upper harbor, and, returning by the same route, stop at Copp's Hill, the home and burial-place of many Revolutionary heroes, and visit Christ Church, from the tower of which Paul Revere was signaled to warn the people of the march of the British to Concord and Lexington, April 18, 1775.

3. Take Bunker Hill cars on Washington or Tremont Sts., and ride to Bunker Hill Monument; visit the Monument; a short walk down the hill brings one to the Navy-Yard, which is filled with interesting and instructive objects, especially the Museum; return to the city proper by cars which pass the Navy-Yard, and visit the Boston Museum and King's Chapel, both on Tremont St., and the new Court-House, in Pemberton Sq., close at hand.

4. Take steamer at Rowe's Wharf, Atlantic Ave., reached by cars from Washington St., and go to Pemberton Landing, Hull, thence by train along Nantasket Beach, at the mouth of the harbor, to Nantasket, the popular seaside resort of Boston; take barge to Cohasset, passing along the famous Jerusalem road, lined with the villas of wealthy Boston merchants, and in sight of the celebrated Minot's Ledge Lighthouse; return to Boston by the same route. Boats leave, during the summer season, almost hourly, and trains and barges connect with every boat. The fare by boat and train from Boston to Nantasket is

[1] Visitors should bear in mind that the electric-cars stop only at points designated by a white band on the electric poles.

25 cents. Charts of the harbor are usually distributed, free, on the boats.

5. Take cars on Washington St. to Franklin Park, the largest tract in the Boston park system; park wagons may be chartered for drives all over the park for 25 cents a passenger; carriages may then be hired at moderate rates for a drive to Forest Hills Cemetery, one of the most beautiful burial-places in the country; thence to the Arnold Arboretum, in West Roxbury, another tract in the chain of parks; thence to and around Jamaica Pond, and through Brookline, a town adjoining Boston, which is filled with the costliest residences and most beautiful drives; thence through Massachusetts Ave., and the Back Bay Park, along the Charles River embankment and Commonwealth Ave.

6. Take street-cars running from Park Square or Bowdoin Square to Harvard Square, Cambridge; visit the homes of Longfellow and Lowell (the former being on Brattle St., which leads from Harvard Square, and the latter a few minutes' walk beyond, on Elmwood Ave., off Brattle St.), the many beautiful buildings of Harvard University, including Memorial Hall, the Museum, Austin Law School, and Hemenway Gymnasium; then take street-cars to Boston.

7. Take train on Boston & Maine R. R., Eastern Division (Causeway St.), and ride to Marblehead; thence by carriage along the beautiful north shore, through the towns of Beverly and Manchester-by-the-Sea, where are many elegant summer residences; returning to Marblehead, take a stroll about the quaint old town, and go by ferry to the Neck —a famous summer resort; return by Boston & Maine R. R. to Boston.

The environs of Boston are remarkably attractive. On almost all sides lie picturesque and venerable towns, and the country between, even when not strictly beautiful, is never flat and tame. Charlestown, Brighton, Jamaica Plain, and W. Roxbury were annexed in 1875, and now form part of the city. Roxbury and Dorchester had been previously annexed. In all of them are the fine villa residences of Boston merchants, and other features of interest which make them worth a visit. At *Charlestown*, on the N. (reached by street-cars from Scollay Square), is the famous *** Bunker Hill Monument,** occupying the site of the old redoubt at Breed's Hill, and commemorative of the eventful battle fought on the spot, June 17, 1775. It is a plain but massive obelisk of Quincy granite, 30 ft. square at the base, and 221 ft. high. From the observatory at the top, reached by a spiral flight of 295 steps, is obtained a magnificent view, including the entire vicinity of Boston. The monument was dedicated June 17, 1843, in the presence of President Tyler and his Cabinet, on which occasion Daniel Webster delivered an oration which is considered his finest oratorical effort. In the house near the monument is a fine statue of General Warren, who was killed on the Hill; and a stone marks the spot where he fell. The *U. S. Navy-Yard* is also located in Charlestown. It comprises about 100 acres, and contains, among other objects, the longest rope-walk in the country, and an immense dry-dock. In *Chelsea* (connected with Boston by ferry, and with Charlestown by a bridge over the Mystic River) are Woodlawn Cemetery, Marine Hospital, Soldiers'-Home, and Naval Hospital, which have already been described; also the prominent manufacturing plants of the Low Art-Tile Co., the Forbes Lithograph Manufacturing Co., and the pottery where the famous Chelsea *faience* was made. *Nantasket Beach*, an hour's sail by steamers from Rowe's Wharf, 340 Atlantic Ave., is a justly celebrated summer resort. *Brighton*, a station on the Boston & Albany R. R., 5 miles W. of the city proper, is famous for its cattle-market. *Point Shirley*, 5 miles from Boston, affords a pleasant drive. The most direct route is by way of the E. Boston ferry.

***Brookline** is a beautiful town on the Boston & Albany R. R. (reached also by the Mill-Dam from Boston). In it is the *Brookline Reservoir*, with a capacity of 120,000,000 gallons. About 1 m. distant, on the boundaries of Brookline, Brighton, and Newton, is the great *Chestnut Hill Reservoir*, with a capacity of 800,000,000 gallons. From Boston to and around this point is a favorite drive. *Lexington* and *Concord* are reached by the Arlington branch of the Boston & Maine R. R. from the Lowell depot. *Concord* may also be reached by the Fitchburg R. R.

***Cambridge,** one of the most renowned of the American academic cities, lies about 3 miles W. of Boston (street-cars from Bowdoin Square and Park Square), and has a population of 70,028. It covers an extensive area, generally level, and is laid out in broad streets and avenues, lined with elms and other shade-trees. Its greatest attraction is **Harvard University,** the oldest and one of the most richly endowed institution of learning in America. (Harvard College, the original institution and still its legal name, eventually grew by the accretion of supplemental departments, so that now the term *Harvard University* is applied to the whole institution, the original name, Harvard College, being retained by the academic department only.) It was established by an act of the Legislature of Massachusetts Bay in 1636, but nothing was done till John Harvard, dying in Charlestown in 1638, left half his property to endow the proposed college. It embraces, besides its collegiate department, law, medical, dental, mining, scientific, art, and theological schools. In 1892–'93 there were 294 instructors and 2,966 students in the University, of whom 1,985 were in the college proper. The University lands in various parts of Cambridge comprise 60 acres. The college yard contains about 15 acres, tastefully laid out and adorned by stately elms, and on the W. and N. side are beautiful gates of wrought iron and pressed brick recently given to the University. Here, forming a large quadrangle, are clustered 21 buildings of brick or stone, from 2 to 5 stories high. The most notable of these are *Matthews's Hall*, a large and ornate structure used as a dormitory; *Massachusetts Hall*, an ancient building, dating from 1720; *Holden Chapel ; Harvard, University, Gray's, Sever,* and *Boylston Halls ; Appleton Chapel ; Thayer Hall*, and *Dane Hall*, for the law school. * *Gore Hall*, within the quadrangle, contains the College Library (300,000 volumes and about the same number of pam-

Views
of
Harvard University.

MASSACHUSETTS HALL

CAMBRIDGE: HEMENWAY GYMNASIUM.

BUSSEY AGRICULTURAL INSTITUTION: JAMAICA PLAIN.

HARVARDS MONUMENT.

SEVER HALL: CAMBRIDGE.

CAMBRIDGE: GORE HALL.

MEMORIAL HALL & SANDERS THEATRE.

CAMBRIDGE: AGASSIZ MUSEUM.

phlets, the principal of an aggregate of 40 collections of books, constituting together the University Library, which now contains nearly 400,-
000 bound volumes, and the largest collection of maps in America).
N. of the quadrangle is * *Memorial Hall*, erected by the alumni and
friends of the University in commemoration of the students and graduates who lost their lives during the civil war. It is a handsome edifice
of brick and Nova Scotia stone, 310 ft. long by 115 wide, with a tower
200 ft. high. It is one of the finest college buildings in the world, and
cost $575,000. It contains, besides Memorial Hall proper, a theatre
and a spacious dining-hall. The old President's house in the college
yard, known as Wadsworth House, was used by Washington for a day
or two before he transferred his headquarters to the Craigie House.
Near the college yard are the *Gymnasium*, the * *University Museums*
(open every day from 9 A. M. to 5 P. M., and on Sunday afternoons),
which consist of the Museum of Comparative Zoölogy (known commonly
as the Agassiz Museum), and those of ethnology, mineralogy, botany,
physical geography, and American archæology (usually styled the Peabody Museum); and about ¼ mile N. W. are the *Botanical Garden* (containing a valuable herbarium) and the *Observatory*. The Women's College, known as the *Harvard Annex*, is not a part of Harvard University,
but its instructors are all professors of the University. Its home is in
the Fay House, fronting the Washington Elm.

W. of the college yard is *The Common*, on which stands a granite
monument erected by the city in honor of her soldiers who fell in the
civil war. Near the Common is *Christ Church* (Episcopal), a venerable edifice. In this church Washington worshiped. It is near the
first burying-ground where the first and later President of Harvard College, Washington Allston, Richard H. Dana, and other noted Cambridge
men, are buried. In the vicinity of the Common are also the *First
Unitarian* Church, and the * *Shepard Memorial Church* (Congregational), erected in honor of Thomas Shepard, who was pastor at Cambridge from 1635 to 1649. These churches are both descendants of
the original church established in Cambridge, one representing the congregation and the other the church-members of a body which remained
one till the Unitarian controversy divided them. In front of the latter
is the famous * *Washington Elm*, beneath which Washington assumed
the command of the American army in 1775, and which is thought to be
300 years old. The principal statues in Cambridge are those of John
Harvard (ideal), near Memorial Hall; one of Josiah Quincy, a former
president of the college, in Sander's Theatre; and one of John Bridge,
a Puritan settler, on the Common. Many structures erected before the
Revolution are still standing, among them the house used by Washington for his headquarters, known as the Craigie House, and recently inhabited by the poet Longfellow; and *Elmwood*, the home of James
Russell Lowell. Longfellow Park, opposite his old home, still keeps
open the poet's favorite view. On Main St. stands the * *City Hall*, the
gift of a former Cambridge resident. The same donor also endowed
the *Ringe School of Manual Training* and the *Cambridge Public Library*, standing on opposite corners of Broadway and Irving Sts. The

library contains manuscripts and memorials of authors who have lived in Cambridge.

The cemetery of *Mount Auburn is in Cambridge (1 mile from Harvard Square, 4 miles from Boston), and is one of the oldest and most beautiful in America. It contains 125 acres, and is embellished by landscape and horticultural art and many beautiful and costly monuments. The gateway was designed after an Egyptian model; and near it is the *Chapel,* an ornamented Gothic edifice, containing statues of Winthrop, Otis, John Adams, and Judge Story. Central, Maple, Chapel, Spruce, and other leading avenues afford a circuit of the entire grounds, with a view of the principal monuments. The *Tower,* 60 ft. high, in the rear of the grounds, is 187 ft. above Charles River, and commands a fine view. It is reached by Central, Walnut, and Mountain Avenues. Numerous lakes, ponds, and fountains in various parts of the cemetery add to its beauty.

17. Boston to Portland via "Eastern Shore."

By the phrase " Eastern Shore " is meant that part of the New England coast lying between Boston and Portland. The *Boston & Maine R. R.* traverses the entire distance. Distance from Boston to Portland, 108 miles. There is also a line of steamers from Boston to Portland daily from India Wharf, except Sundays, at 7 P. M., and from Portland to Boston daily, except Sundays, at 7 P. M.

LEAVING Boston by the Eastern Division, *Chelsea* (5 miles) is speedily reached, and affords a convenient point from which to visit the Chelsea or **Revere Beach,** a favorite summer resort of the less well-to-do classes of Boston, who throng it on Sundays and holidays. It is reached from the city by street-cars and also by the Boston, Revere Beach & Lynn R. R., and affords a delightful promenade and drive as well as excellent sea-bathing. **Lynn** (11 miles from Boston) is a flourishing city of 55,727 inhabitants, situated on the shore of Massachusetts Bay, and surrounded by pleasing scenery. It contains some handsome churches and school-houses, numerous fine villas of Boston merchants, a costly Soldiers' Monument, and a very fine City Hall. *High Rock* is a commanding eminence in the center of the city, which affords a wide-extended view. *Pine Grove Cemetery* is a beautiful rural burying-ground. Four miles from Lynn (reached by stage) is **Nahant,** a bold promontory connected with the mainland by narrow ridges of sand and stone thrown up by the ocean, above which the highest point rises 150 ft. A large and splendid hotel was built in 1824, and numerous summer residents filled the place with their cottages, making it the most fashionable watering-place in New England; but the hotel was burned down in 1861, and since then the tide of pleasure-seekers has gone in other directions, especially toward Swampscott and Marblehead. Several small hotels (*Hotel Nahant, Hood Cottage*) still remain, however, and the villas give it a gay aspect in summer. The beach of Nahant (1½ mile long) is hard and smooth, shelving gently, and with a splendid surf. There are many natural wonders and curiosities in the vicinity; and on the N. side the *Garden of Maolis* (entrance, 25c.) offers a picturesque retreat where good fish or clam dinners may be enjoyed. Nahant is reached from Boston by steamer as well as by railway.[1]

A mile beyond Lynn the train reaches **Swampscott,** which is to Boston what Long Branch was to New York, the favorite summer resort of its wealthiest citizens. The leading hotels are the *Blaney, Hotel Preston*, the *Lincoln House*, and the *Ocean House.* The shore is lined with tasteful villas, and wealth has fairly turned poverty out of the place. There are 3 beaches of varying length, and picturesque headlands jut out into the sea. The bathing is excellent, with no under-tow, and the water is thought to be warmer than at Nahant or Rye Beach. The permanent residents are chiefly engaged in the cod and haddock fishing, and supply the market with fresh fish. The Swamp-scott Branch of the Boston & Maine R. R. diverges here to Marblehead, passing *Phillips Beach, Beach Bluff*, and *Clifton*, all popular resorts.

Salem (*Essex House*; 16 miles from Boston) is a venerable town of 30,801 population, and the site of the first permanent settlement in the old Massachusetts colony. Many interesting historical associations clus-ter around Salem, and every period in her annals has been illustrated by some important event or illustrious name. The year 1692 is remarkable as the date of the witchcraft delusion at Salem village, now a part of Danvers, for which several persons were tried and executed. In the Court-House are deposited the documents that relate to these curious trials. The house is still standing, built in 1634 (Roger Williams's house, at the corner of Essex and North Sts.), in which some of the preliminary examinations were made. The place of execution is in the western part of the city, an eminence overlooking the city, harbor, and surrounding shores, and known as *Gallows Hill.* A pleasant drive of 5 or 6 miles will enable the visitor to examine the several places of inter-est mentioned in Charles W. Upham's work on the subject. Many in-teresting memorials of Nathaniel Hawthorne are to be seen in Salem. His *birthplace* is at 21 Union St.; the Turner House, at 32 Turner St., has been identified as "The House of the Seven Gables"; the Custom-House, where "The Scarlet Letter" was begun; and his desk is in the *Old First Meeting-House*, built in 1634, which is adjoining the Essex Institute. *Plummer Hall* is a handsome building in Essex St., contain-ing the library of the Salem Athenæum (22,000 volumes), and that of the Essex Institute (75,000 volumes, and a large collection of newspa-pers, pamphlets, manuscripts, and various historical relics). In rear of Plummer Hall is preserved the frame of the original first meeting-house, the oldest church edifice in New England, dating from 1634. *East India Marine Hall* contains the fine natural history collections of the Essex Institute; while the ethnological museum of the East India Marine Society is in the new East Hall (open to the public every week-day from 9 A. M. to 5 P. M.). The visitor to Salem should not fail to take the horse-cars to *Peabody* (2 miles distant), to visit the *Peabody In-stitute*, in which are deposited many interesting works of art, and the

---

[1] Fuller particulars of these beaches, as well as of other popular resorts men-tioned in this Guide, may be found in "Appletons' Illustrated Hand-book of American Summer Resorts."

various memorials of the founder, George Peabody, among which may be mentioned the portrait of Queen Victoria, the Congress Medal, etc. A short distance in one direction from the Institute building is the house in which Mr. Peabody was born, and about the same distance in an opposite direction, in *Harmony Grove Cemetery,* is his grave.

Four miles from Salem (reached by Marblehead Branch of the Boston & Maine R. R.) is the quaint and interesting old town of **Marblehead,** built on a rugged, rocky promontory, which juts far out into the sea and forms an excellent harbor. This spot was one of the first settled in New England, the town of Marblehead having been incorporated by the Puritan colony just 15 years after the landing of the Pilgrims at Plymouth. So bleak and bare are the Marblehead rocks, that Whitefield asked in wonder, "Where do they bury their dead?" There are many queer houses still standing which were built and occupied before the Revolution; the most noteworthy being the *Bank Building,* which is supposed to have been built in 1768 for Colonel Jeremiah Lee, and which is a fine specimen of the palatial homes of the nabobs of the last century. A hundred years ago Marblehead was, next to Boston, the most populous town in Massachusetts, and had a large maritime commerce. Now its character has almost wholly changed from the olden time, for it has become a center of the shoe-manufacture. The *Old Fort* is a plain, hoary-looking edifice, standing on the rugged slope of the promontory looking toward the sea. *Marblehead Neck* (*Follet and Nomepashemet*), easily reached by boats across the harbor, or by a ride of 2 miles along shore, is a favorite resort; many campers-out flock hither to pitch their canvas tents and spend a few weeks of the heated term, and the place is growing in popularity. *Lowell Island,* 2 miles distant, is reached by boat from Marblehead, and is used as a sanitarium. It is noted for the purity of its air and the beauty of its views, and has a large summer hotel on it.

Two miles beyond Salem on the main line is *Beverly,* an ancient town (population 10,821) now busy with shoe-factories. The Beverly beaches are very pleasant; and from here to *Manchester* (*Masconomo*) the strip of coast is lined with beautiful residences standing amid ornamental gardens. *Magnolia* (*Hesperus*) and other attractive places are along the Gloucester Branch of the Eastern Div. of the Boston & Maine R. R., running as it does to the extreme point of Cape Ann. Next comes *Ipswich* (28 miles), and 9 miles beyond this point is **Newburyport** (*Hancock* and *Wolf Tavern*), an old, historic town, built on an abrupt declivity of the Merrimac River, 3 miles from the ocean, having a population of 13,947. Like Salem and Marblehead, it is one of those antique coast towns which have to a large extent lost their maritime importance, while preserving the relics and mementos of a former commercial prosperity. The principal industries now are the manufactures of boots and shoes, cotton-goods, and silver-ware. The *Marine Museum* (in State St.) contains a number of these mementos. The *Public Library,* founded by Josiah Little, was further endowed by George Peabody, and contains 27,000 volumes. There are several fine churches, and many quaint houses of the olden time, and the visitor should not fail to see J. Q. A. Ward's bronze statue of Washington. Steamers leave every hour in summer, from the city wharf, for **Salisbury Beach,** which is one of the best on the coast. It extends about 6 miles, from the Merrimac to the Hampton River. Twenty years ago there was nothing there but the lonely breaker; but now a good hotel (the *Seaside House*) and many summer cottages have peopled it. The shore descends very gradually, and on its long slope the people of the surrounding country have had an annual reunion every September for more than 100 years. At the mouth of the Merrimac is *Plum Island,* extending 9 miles in length and 1 in width; a favorite resort is the *Plum Island Hotel,* just remodeled. From *Hampton* (10 miles beyond Newburyport) stages run 3 miles to **Hampton Beach,** a muchfrequented resort, with numerous summer cottages (*Boar's Head Hotel*) that are generally thronged in summer. The bathing and fishing at Hampton Beach are capital, the scenery charming, and the drives in the vicinity pleasant. **Rye Beach,** the most fashionable of the New Hampshire beaches, is reached by stage in 3 miles from N. Hampton, or by a delightful drive of 7 miles from Portsmouth. The bathing is excellent, the surf being particularly fine and without any undertow. The hotels are the *Farragut and Atlantic Hotel* and the *Sea-View Hotel;* there are also a number of boarding-houses and a colony of cottages.

Portsmouth (56 miles from Boston), having a population of 9,-827, the only seaport of New Hampshire, is situated 3 miles from the mouth of the Piscataqua River, and, excepting the narrow strip connecting it with the mainland, is entirely surrounded by water. The harbor is deep and safe, and in it are many islets, some accessible by bridges. Here is the point of departure for the well-known summer-resort, the Isles of Shoals. Portsmouth is a singularly venerable and tranquil-looking place, with beautifully shaded streets, ancient buildings, large gardens, and home-like residences. Among the objects of special interest are the old church of St. John, the Athenæum, Governor Wentworth's mansion (at Little Harbor), and the tomb of Sir William Pepperell, which is near the Navy-Yard. The *Kittery United States Navy-Yard* is located upon Continental Island (reached by ferry from foot of Daniel St.), and contains, besides the usual ship-houses, shops, etc., a very fine balance dry-dock. The hotels of Portsmouth are the *Kearsarge*, the *Langdon*, and the *Rockingham*. The *Wentworth*, a large summer hotel, is on New Castle Island, about 2 miles from the city.

The Isles of Shoals.

These are a group of eight rugged islands, lying about 9 miles off the coast, and reached in summer from Portsmouth by steamer Oceanic, making several trips daily, and connecting with all trains on the Concord & Montreal R. R. and Boston & Maine R. R. The isles are small in extent, the largest, Appledore, containing only 250 acres. As the steamboat approaches, they separate into distinct elevations of rock, all having a bleak and barren aspect, with little vegetation, and with jagged reefs running far out in all directions among the waves. *Appledore* rises in the shape of a hog's back, and is the least irregu-

lar in appearance. Its ledges rise some 75 ft. above the sea, and
it is divided by a narrow, picturesque little valley, wherein are situated the *Appledore House* and its cottages, the only buildings on the
island. Just by Appledore is *Smutty Nose* or *Haley's Island*, low, flat,
and insidious, on whose sullen reefs many a stalwart vessel has been
dashed to destruction. About ¼ mile beyond is *Star Island*, once the
site of the little village of Gosport, now occupied by the *Oceanic House*,
with its cottages. On the W., toward the mainland, is *Londoner's Isl-
and*, jagged and shapeless, with a diminutive beach ; while 2 miles away
is the most forbidding and dangerous of all these islands, *Duck Island*,
many of whose ledges are hidden beneath the water at high tide, and at
low tide are often seen covered with the big white sea-gulls, which shun
the inhabited isles. *White Island*, the most picturesque of the group,
is about a mile S. W. from Star Island, and has a powerful revolving
light on it which is visible for 15 miles around.

Nine miles N. E. of Portsmouth is the hamlet of *York*, near which
is **York Beach** (*Garrison House, Marshall House*), a popular place
of resort in summer. *Cape Neddick* runs out into the sea at the N.
end of the beach, and a short distance inland is *Mt. Agamenticus*, from
the summit of which there are fine views of the White Mountains, of
the ocean, and of the harbors of Portsmouth and Portland. *Bald-Head
Cliff* is a remarkable rocky promontory, 5 miles N. of York Beach, of
peculiar conformation and commanding noble views. Beyond, stretching away to Wells, is the long *Ogunquit Beach*.

Beyond Portsmouth the train crosses the Piscataqua River into
Maine and soon reaches Wells, whence stages run 6 miles to **Wells
Beach** (*Bay-View House* and *Maxwell House*). The beach is 6 miles
long, is covered with snipe and curlew, and is a great rendezvous for
sportsmen. In the woods are partridges and woodcock, and a large
trout-stream crosses the beach. From Kennebunk (90 miles) the branch
runs into the Port (4½ miles), passing the stations of *Parsons, Kennebunk Beach*, and *Grove Station*. *Kennebunkport* is a pleasant old village, on the river, only a mile above its mouth, where old ship-yards
used to be. On the neighboring roads you may ride to Monsam River
Falls ; or up to the Shelter Community at Alfred ; or to the ancient
trees of the now desolate camp-grounds ; or to tranquil old Kennebunk ;
or along the rocky shore of Lake Arundel to Cape Porpoise, on one side,
or down the sands to Lord's Point and Hart Beach, on the other ; or to
the cities of Biddeford and Saco, on the N. **Biddeford** (37 miles
from Portsmouth) is a thriving city of 14,443 inhabitants, opposite
Saco, near the mouth of the Saco River, which here falls 55 ft. and
furnishes a fine water-power to both places. The **Saco Pool,** which
is in Biddeford, though usually spoken of in connection with Saco, is a
deep basin scooped out of the solid rock, about ¼ mile from the sea,
with which it is connected by a narrow passage. It is emptied and
filled with each changing tide, and is reached from Biddeford by a
steamer which runs twice daily in summer.

Portland.

Hotels, etc.—The *Falmouth House*, in Middle St., and the *Preble House*, in Monument Square, are the best. The *City, St. Julian, United States*, and *West End*, are good. Rates are from $2 to $3 a day. *Street-cars* run from the Union Station through the main streets and to the suburbs. *Reading-rooms* at the Public Library and at the Y. M. C. A., in Congress St.

Portland, the commercial metropolis of Maine, is picturesquely situated on a high peninsula at the S. W. extremity of Casco Bay, and is one of the most beautiful cities in the country. It was settled in 1632, and has had a steady growth; but on the night of July 4, 1866, a great fire swept away half the business portion, destroying $10,000,000 worth of property. The entire district destroyed by the fire has since been rebuilt. The streets are profusely embellished with trees. The population in 1860 was 26,341, in 1880, 33,810, and in 1890 had increased to 36,425. For a city of its size, Portland has exceptionally fine public buildings. The * **City Hall** is one of the largest and finest municipal structures in the country. Its front, of olive-colored freestone, elaborately dressed, is 150 ft. long, its depth is 221 ft., and it is surmounted with a graceful dome 160 ft. high. The **Post-Office** is a beautiful building of white Vermont marble, in the mediæval Italian style, with a portico supported by Corinthian columns. The **Custom-House,** erected at a cost of $485,000, is a handsome granite structure, with elaborate ornamentation within. *Kotzschmar Hall* is a public hall with a seating capacity of 500, and is used for public entertainments. The *Society of Natural History* has a good building, with a fine collection of birds, fishes, reptiles, shells, and minerals. The *Public Library*, incorporated in 1867, contains 35,000 volumes, and, with the *Maine Historical Society*, occupies a handsome building, the gift of James P. Baxter, costing $100,000. The *Marine Hospital*, erected in 1855 at a cost of $80,000, is an imposing edifice.

No visitor to Portland should fail to ascend the * **Observatory** on Munjoy's Hill, in order to enjoy the famous view from the top. Near the Observatory is the *Eastern Promenade*, whence there is a pleasing outlook over the city and harbor. Congress St. leads thence to the *Western Promenade* on Bramhall's Hill. Each of these promenades is 150 ft. wide, and planted with rows of trees. From Bramhall's Hill is a noble prospect, as on a clear day may be seen Mounts Washington, Kearsarge, and others of the White Mountain range. *Lincoln Park*, in the center of the city, contains about 2½ acres. *Evergreen Cemetery*, containing 55 acres, is about 2½ miles distant (reached by electric-cars from Monument Square). *Deering Park* (the City Park) contains 50 acres, and is easy of access by street-car or on foot. There are many pleasant drives in the vicinity; we may mention only those to Cape Elizabeth (*Cape Cottage*), around Deering's Woods, along Falmouth Foreside; and the two Promenades: the surrounding scenery is most enchanting. *Diamond Island* is a favorite spot for picnics, and is noted for its groves of noble trees; and *Peak's Island* is embowered in foliage, and contains several small summer hotels. *Cushing's Island* is reached by ferry from the city, and contains the *Ottawa House*.

8

Among the notable and interesting churches of the city are the Episcopal Cathedral of St. Luke, the great Catholic Cathedral of the Immaculate Conception, with a spire 16 ft. higher than Bunker Hill Monument; the old First Parish Church (Unitarian), with heavy walls of granite and a quaint clock-tower; the Second Parish Church, of stone; and the First Baptist Church. The house in which Longfellow was born still stands at the cor. Fore and Hancock Sts., once a fashionable quarter, but now the dingiest part of the town, amid docks and elevators and railways. It is occupied by several Irish families. Up in the busy residence quarter, in Congress St., stands the ancestral Wadsworth mansion, Longfellow's abiding-place when he visited Portland in later years. Next door is the Preble House, erected by an Italian architect in 1806 for the home of Commodore Preble, of Tripoli fame. In the Monument Square, which is the center of the city, is the imposing Soldiers' Monument. At the crossing of Congress and State Sts. is a bronze statue of Henry W. Longfellow, by Franklin Simmons, which cost $12,000 and was erected by public subscription.

Portland is the point of departure for steamers, giving frequent communication with Harpswell and the islands in Casco Bay; with Bath, Boothby, Mt. Desert, and Machias, and all other points along the Maine coast and Penobscot River. The boats of the International Steamship Company run to Eastport and St. John, N. B., making connection for Grand Menan Island, the city of Fredericton, Halifax, N. S., Charlottetown, P. E. I., and all parts of the Maritime Provinces. The Portland Steam Packet Company runs a daily line between Portland and Boston.

18. Boston to Portland via Boston & Maine R. R.

THE distance is 116 miles, and the fare $2.50, through tickets; $3, with privilege of stopping over at way-station. Twelve miles from Boston is *Reading*, a manufacturing village. **Andover** (23 miles) (*Elm House, Mansion House*) is an academic town, settled in 1643. The *Theological Seminary* (Congregational) is the leading institution of that sect in America, and the Phillips Academy is of wide reputation and still older date (founded in 1778). The Albert Female Seminary and the Punchard High-School rank high among schools of the kind. The scenery about Andover is very pleasing, and the society of the place is refined and cultivated. **Lawrence** (*Franklin House*) is 26 miles from Boston, and is one of the largest manufacturing cities of Massachusetts, with a population of 44,654. Its prosperity dates from 1845, when a dam was thrown across the Merrimac River (on both sides of which the city is built), giving a fall of water of 28 ft., and furnishing power for the numerous mills and factories located here. The leading manufactures are cotton-cloth, woolens, shawls, paper, flour, and files. The vast mills are separated from the city by the canal which distributes the water-power. This canal runs parallel with the river for a mile, at a distance of about 400 ft. from it, and another canal has been cut on the S. side. The *Common* is a tasteful little park of 17¼ acres, surrounded by handsome buildings, including several churches and the city and county buildings. The finest church in the city is *St. Mary's* (Roman Catholic), a beautiful stone structure in the Gothic style. There are a number of good public libraries, several of which pertain to the mills. On Prospect Hill is a neat park, with attractive views. *South Lawrence* is a busy manufacturing suburb across the river.

Beyond Lawrence the train skirts the bank of the Merrimac for 7 miles to **Haverhill** (*Hotel Webster*), another lively manufacturing city, beautifully situated on hills which slope gently down to the river. Shoe-making is the leading industry, and in this Haverhill ranks next to Lynn. The city is well built and contains 27,412 inhabitants. The Public Library is a very handsome building, with 20,000 volumes. The City Hall (in Main St.) is also handsome, and N. of it is a fine white-marble Soldiers' Monument, erected in 1869. A mile N. E. of the city is *Lake Kenoza*, a pretty little sheet, named and celebrated by the poet Whittier, who was born in Haverhill in 1807. Beyond Haverhill the train enters New Hampshire and soon reaches *Exeter* (51 miles), a pretty, elm-shaded village, seat of another Phillips Academy and of the Robinson Female Seminary. There are some important factories and machine-shops here, neat county buildings, and many tasteful residences. Seventeen miles beyond (with several intervening villages) is the busy little city of **Dover** (*American House*), the oldest place in New Hampshire, settled in 1623. It now has 12,790 inhabitants, 11 churches, and extensive manufactories, chiefly of cotton-cloth and boots and shoes. The Cocheco Mills and Print Works are among the largest of the kind in the country. From Dover a branch of the Boston & Maine R. R., runs in 28 miles to Alton Bay on Lake Winnepesaukee (see Route 22).

The stations next after Dover (Rollinsford, Salmon Falls, North Berwick, and Wells) are small. About a mile from Wells is *Wells Beach*, already described (see p. 112). From *Kennebunk* (90 miles) stages run to **Cape Arundel** (*Ocean Bluffs Hotel*), a bold promontory which is much resorted to in summer on account of its excellent bathing, boating, and fishing. Nine miles beyond Kennebunk the train crosses the Saco River between the twin manufacturing cities of **Saco** and **Biddeford** (see p. 112). Four miles E. of Saco (reached by trains on the Boston & Maine R. R.) is **Old Orchard Beach,** the finest in New England, and, after Swampscott and Rye, the most frequented and fashionable. It is nearly 10 miles long, is hard and smooth as a floor, shelves gently to the water, and affords unsurpassed surf-bathing. The hotels are the *Aldine, Everett, Fiske, Lawrence, Ocean, Old Orchard*, and *Sea-Shore*. The fishing in the vicinity is excellent, and sufficient game is always to be found to tempt the sportsman. Eight miles beyond Saco is Scarborough station, whence stages run 3 miles E. to **Scarborough Beach** (*Atlantic House, Kirkwood House*, and large boarding-houses). There are good bathing on the beach and fishing and hunting near by; and the place has a large summer patronage. **Cape Elizabeth,** on the S. side of Portland Harbor, may be considered a part of Portland, from which it is reached by a pleasant drive. It is a delightful summer resort, with excellent bathing and fishing. The hotels are the *Cape Cottage* and the *Ocean House*.

19. Portland to the White Mountains.

a. Via Maine Central R. R.

THE White Mountain Div. of the Maine Central R. R. runs through the famous Crawford Notch, the heart of the White Mountain region, and·offers some of the finest scenery to be found in America. Observation-cars, open on all sides, are run on the mountain section of the road. In the close cars, seats on the right are most desirable. Between Portland and the mountains there are many points of interest. *Cumberland Mills*, the station name for the thriving city of Westbrook, is noted for its paper-mill—the largest in the world—where most of the paper for the popular magazines is made. **Sebago Lake** (17 miles from Portland) is a beautiful sheet, 12 miles long by 9 miles wide, with very deep, cool, and clear waters. A number of islands dot its surface, and its shores are diversified and pleasing, with half a dozen towns nestling here and there. At its N. W. end it connects by the Songo River with *Long Lake*, a river-like body of water nearly 14 miles long and only 2 miles wide. The distance between the two lakes is but 2½ miles "as the crow flies," but the Songo River makes 27 turns and thus secures for itself the length of 6 miles. A steamer during the summer season makes daily the round trip between *Sebago Lake Station* and *Harrison*, at the northern end of Long Lake (34 miles). A lock in the Songo River raises the steamers and other craft plying upon these waters from the level of the lower to the upper lake. The trip to Harrison and return, including landings at *Naples, Bridgton*, and *North Bridgton*, is made in about 8 hours, and affords a very agreeable excursion. From Bridgton, stages run 1 mile west to **Bridgton Center** (*Bridgton House*), a prettily situated village, which is becoming popular as a summer resort. In the vicinity are numerous small ponds, and the summit of *Pleasant Mt.* (8 miles distant) affords a beautiful view. From Harrison daily stages run to *South Paris*, on the Grand Trunk R. R. (14 miles; fare, $1).

Beyond Lake Sebago the train follows the valley of the Saco River, amid pleasing scenery. *Cornish* (32 miles) is at the confluence of the Saco and Ossipee Rivers, and 3 miles beyond a fine view is obtained of the Great Falls of the Saco, where the river descends 72 ft. in successive pitches. **Fryeburg** (50 miles) is a pretty village on the river, surrounded by attractive scenery, and much resorted to in summer, when it is the headquarters of the Chautauquan movement in northern New England. From this place there are stages to various points in the White Mountains. Near the village are several eminences from which fine panoramic views of the distant White Mountains may be had. Beyond Fryeburg the State of New Hampshire is entered. At *North Conway* (60 miles), where connection can be made with the Boston & Maine R. R. (Northern Div.), beautiful views are had from the cars, and at *Intervale Station*, 2¼ miles beyond, the road enters and follows for some distance the charming Conway Intervale. From *Glen* (65 miles), stages run in two miles to Jackson and in 14 miles to the *Glen House* (see p.

122). At this point, on clear days, look to the right for a fine view of the summit of Mt. Washington, with its gleaming white buildings. *Bartlett* (71 miles) is a thriving lumbering village, situated amid picturesque scenery in a smiling intervale walled in by lofty hills. At *Bemis* (77 miles) the famous ride through the Notch may be said to commence. Near here the steeper grade of the road begins, and the remainder of the journey to the summit at the Crawford House is only accomplished by a continuous ascending grade of 116 ft. to the mile. The road rises above the valley at some points to an elevation of over 300 ft., affording much finer views than any to be obtained from the highway. Three miles from Bemis is the *Frankenstein Trestle*, a graceful iron structure spanning a gorge 500 ft. wide, and commanding a lovely valley-view to the S., and to the N. a noble view of Mt. Washington and the summits of the Presidential Peaks. Crossing the Trestle, the road soon curves to the left toward Mt. Willey, revealing the summit of that mountain in the W., and the shoulder of Mt. Webster in the E., these two eminences forming the walls of the Notch proper. Another and deeper gorge between Mts. Willey and Willard is crossed on an iron trestle, just beyond which a most lovely *view is obtained of the secluded Willey Valley. The train now follows the contour of Mt. Willard, soon enters the narrowest part of the Notch, dashes through the Great Cut and then through the Gate of the Notch, emerging upon the plateau of the *Crawford House* (see p. 126). Four miles from Crawford's the train reaches the *Fabyan House* (91 miles), where connection is made with the Mt. Washington R. R. for the summit of Mt. Washington. *Fabyans* is the center of White Mountain travel, and tourists after fingering here can proceed to *Zealand Junction* (92 miles), a mile farther along the line of the Maine Central, where connection is made with the Profile Franconia Notch R. R., for the *Profile House, Maplewood,* and *Bethlehem,* all attractive resorts in the Franconia Range. On the way to Zealand an admirable view of the Ammonoosuc Falls is obtainable from the car windows on the left. At *Twin Mountain* (94 miles) there is the Twin Mountain House. The next station is *Quebec Junction,* where a branch of the Maine Central R. R. leads N. to Jefferson, Lancaster, Colebrook, Quebec, and other points, while the main line continues on through *Whitefield* (104 miles), a mountain village, where beautiful views of the west side of the Presidential Range may be obtained, thence on to *Lunenburg* (111 miles), just over the Connecticut River in Vermont, where connection is made with the St. Johnsbury & Lake Champlain R. R. for St. Johnsbury, Burlington, Newport (Vt.), and Montreal. *Jefferson* (105 miles ; see p. 126), the next station N. of Quebec Station, is one of the most picturesque and beautiful of the White Mountain resorts. Stages are at the station on the arrival of trains to convey passengers to *Jefferson Hill House, The Waumbek,* and other first-class hotels. Seven miles farther N. is *Lancaster* (112 miles), occupying a position on this side of the mountains similar to North Conway on the east slope. Lancaster is a thriving village of 3,373 inhabitants. The *Lancaster House* is the chief hotel.

b. Via Grand Trunk R. R.

This route runs near the bases of the principal White Mountain peaks, and, like the preceding route, affords a succession of grand and impressive views. The stations are: Falmouth, 5 miles; Yarmouth, 11; Pownal, 18; New Gloucester, 22; Danville Junction, 27; Mechanics' Falls, 36; S. Paris, 47; Norway, 49; Bethel, 70; Gorham, 91. All these are small villages or hamlets, most of which require no further mention at our hands. **Bethel** is a lovely village in the Androscoggin Valley, with mineral springs, numerous summer boarding-houses, and fine views of the mountains. From Bethel to Gorham the views are wonderfully varied and striking, including Mts. Moriah, Washington, Madison, Adams, and Jefferson, and other towering peaks. From **Gorham** (described on p. 124) the whole White Mountain region is easily accessible.

20. Boston to the White Mountains.

a. Via Eastern Division Boston & Maine R. R.

THIS is the shortest and quickest route from Boston to the White Mountains, the distance to North Conway (138 miles) being made in less than 6 hrs. (fare $4). As far as **Portsmouth** (57 miles), it is identical with Route 17. At *Conway Junction* (10 miles beyond Portsmouth) the Mountain Division diverges from the main line and passes in 12 miles to **Rochester**, a large manufacturing village of 7,396 inhabitants, where four railroads meet. *Milton* (87 miles) and *Union* (93 miles) are small hamlets, frequented in summer. From *Wolfboro Junction* (97 miles) a branch railroad runs in 12 miles to **Wolfboro** on Lake Winnepesaukee (see Route 22). Stations: *Wakefield*, 99 miles; *Ossipee*, 111; *Center Ossipee*, 115; and *West Ossipee*, 121. The scenery now becomes more pleasing, and Chocorua looms up on the left. **Conway** (*Conway House*, *McMillan Hotel*, and *Sunset Pavilion*) stands at the vestibule of the mountain-region, and commands noble views. It is more quiet than North Conway, and as all the objects of interest near the latter can be as well visited from Conway, many prefer it to its more frequented neighbor. *Echo Lake*, the *Cathedral*, and *Diana's Bath* (described in connection with N. Conway) are as near to Conway; and excursions may be made to other points of interest—to *Chocorua Lake* (8 miles), to *Champney's Falls*, *Conway Center*, and *Chatham*. At **North Conway** (see p. 121) connection is made with trains of the Maine Central R. R. without change of cars, for the Crawford and Fabyan Houses, Jefferson, and Lancaster, passing through the White Mountain Notch, and for Glen House.

b. Via Boston & Maine R. R. and connections.

As far as **Concord,** N. H. (75 miles), this route is described in Route 27. Leaving Concord, a number of small stations are passed in succession. Just beyond *Tilton* (90 miles from Boston) a fine view of the Sandwich Range is had on the left, and from this point the scenery is very attractive. **Laconia** is a busy manufacturing town on Lake Winnesquam, from which the summit of *Mt. Belknap* (see Route 22) may be reached in 8½ miles. **Weirs** (105 miles) is described in Route 22. Here connection is made with the Lake Winnepesaukee steamers ; and N. Conway may be reached by crossing to Wolfboro and taking the preceding route. Stations: *Meredith,* 109 miles ; *Ashland,* 117 ; and *Bridgewater,* 120. **Plymouth** (*Pemigewasset House*) lies in the lovely Pemigewasset Valley, on the outskirts of the White Mountain region, and amid charming scenery. There are several natural curiosities within excursion-distance, and 4 miles N. E. (ascended by carriage-road) is ** **Mt. Prospect** (2,963 ft. high), affording what is said to be the finest view south of the mountains. Several small stations are now passed, and then (145 miles from Boston) comes **Warren** (*Tip Top House*), a much-visited highland village, 9 miles from ** **Mt. Moosilaukee,** the highest peak in New Hampshire outside of the White and Franconia groups, and commanding magnificent views. From the *Prospect House* may be seen the valley of the Connecticut, the White and Franconia Mountains, Lake Winnepesaukee, nearly the whole of Vermont and New Hampshire, and several Canadian peaks. Stations: *E. Haverhill,* 151 miles ; *Haverhill* and *Newberry,* 156 ; *Woodsville,* 166 ; and **Wells River** (169 miles), the junction of the Boston & Maine (Passumpsic Div.) and the Montpelier & Wells River R. R. with the present route. At **Wing Road** (192 miles) the present route diverges to the *Twin Mountain House* (201 miles) and the *Fabyan House* (206 miles). (See Route 21.) At *Bethlehem Junction* (196 miles) the Profile and Franconia Notch R. R. connects, and runs to the Profile House. At the Fabyan House the train connects with the railway up Mt. Washington, and with the Maine Central R. R.

c. *Via Boston & Maine R. R.*

This route takes Lake Winnepesaukee *en route.* As far as **Dover** (68 miles) it is the same as Route 18. The line from Dover traverses the Cocheco Valley, passing *Rochester* (see p. 118), *Farmington* (86 miles), and *New Durham* (92 miles), and stops at **Alton Bay** (96 miles). From Alton Bay the traveler may go by steamer to *Weirs,* and thence by Route *b ;* or to *Wolfboro,* and thence by Route *a* to *North Conway ;* or to *Center Harbor,* and thence by stage through Sandwich to West Ossipee and thence by Route *a* to North Conway. (See Route 22.)

21. The White Mountains.

The routes from Boston to the White Mountains are described in Route 20. From Portland, in Route 30. From New York the White Mountains may be best and most quickly reached *via* Route 30 to Wells River, and thence to the Twin Mountain, White Mountain, and Fabyan Houses, as in Route 20, *b* (drawing-room cars through) ; or (2.) *Via* New Haven or New London to Norwich, Worcester, and Nashua, to Concord, and thence as in Route 20, *b* (a solid through drawing-room car express train runs between New York and Fabyans by way of Connecticut River Line during the summer months). (3.) *Via* steamer to New London, thence *via* Route 31 to Brattleboro and Wells River, and from the latter point as in Route 20, *b.* (4.) *Via* Albany, Rutland, and Bellows Falls, to Wells River (Routes 37 and 29), and thence as in preceding route. From Montreal or Quebec *via* Grand Trunk R. R. to Gorham. Also from Montreal *via* Canadian Pacific, Boston & Maine, and Maine Central R. Rs. to Fabyan.

THE WHITE MOUNTAINS (the "Switzerland of America") rise from a plateau in Grafton and Coòs Counties, New Hampshire, about 45 miles long by 30 broad, and 1,600 ft. above the sea. Some 20 peaks of various elevations rise from the plateau, which is traversed by several deep, narrow valleys. The peaks cluster in two groups, of which the eastern is known locally as the White Mountains, and the western as the Franconia Group. They are separated by a table-land varying from 10 to 20 miles in breadth. The principal summits of the eastern group are Mounts Washington (6,293 ft. high), Adams (5,819), Jefferson (5,736), Clay (5,554), Monroe (5,396), Madison (5,381), Franklin (4,923), Pleasant (4,781), Clinton (4,331), and Webster (3,930). The principal summits of the Franconia Group are Mounts Lafayette (5,269 ft.), Moosilaukee (4,810), Liberty (4,472), Cherry Mountain (3,600), and Pleasant (2,018). Near the S. border of the plateau rise Whiteface Mountain (4,057 ft.), Chocorua Peak (3,508), Mount Ossipee (2,956), and Red Hill (2,038); and in the S. E., Mount Kearsarge (3,270). With the exception of the Black Mountains of North Carolina, several of these peaks are the highest elevations in the United States E. of the Rocky Mountains. Multitudes of little streams force their way down steep glens from springs far up the mountain-sides, and flow through narrow valleys among the hills. The courses of these rivulets furnish irregular but certain pathways for the rough roads that have been cut beside them, and by which the traveler gains access to these wild mountain-retreats.

The aboriginal name of the White Mountains was *Agiochook* or *Agiocochook*, signifying "Mountain of the Snowy Forehead and Home of the Great Spirit." The first white man to visit them, according to Belknap, the State historian, was Walter Neal, in 1632. The Notch was discovered in 1771, the first inn was erected in 1803, a bridle-path to the summit of Mount Washington was cut in 1819, and the first hotel was opened in 1852. Since this latter date the popularity of the mountains has steadily increased, and each summer they are largely visited by tourists. As to the time to visit them, Starr King recommends the early summer. "From the middle of June to the middle of July, foliage is more fresh; the cloud-scenery is nobler; the meadow-grass has a more golden color; the streams are usually more full and musical; and there is a larger proportion of the 'long light' of the afternoon, which kindles the atmosphere into the richest loveliness. The mass of visitors to the White Mountains go during the dog-days, and leave when the finer September weather sets in, with its prelude touches of the October splendor. In August there are fewer clear skies; there is more fog; the meadows are appareled in more sober green; the highest rocky crests may be wrapped in mists for days in succession; and a traveler has fewer chances of making acquaintance with a bracing mountain-breeze. The latter half of June is the blossom-season of beauty in the mountain-districts; the first half of October is the time of its full-hued fruitage."

In describing the mountains, we shall begin at North Conway, the S. E. portal, and proceed by the usual routes to different points, describ-

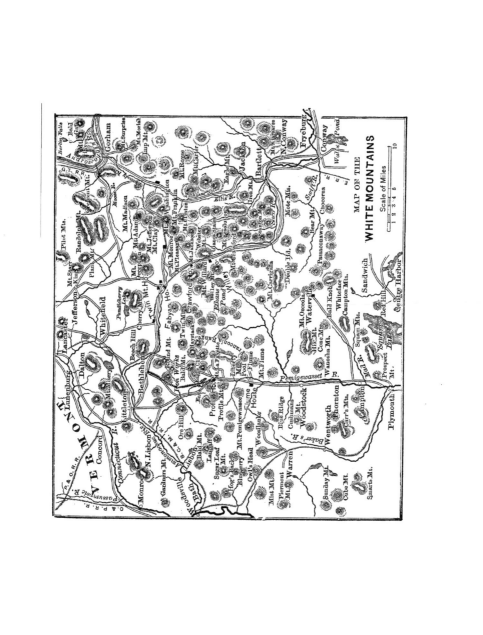

MAP OF THE
WHITE MOUNTAINS

Scale of Miles

ing in connection with each the various features of interest in its neighborhood. Of course, the tourist can arrange his routes differently, and still find the description equally serviceable.

North Conway.

This is one of the prettiest towns in the entire mountain-region, and is a favorite rendezvous for artists and tourists who wish to be within easy excursion-distance of the mountains, while avoiding the excitement and expense of the larger hotels. It is beautifully situated on a terrace overlooking the intervales of the Saco, and is surrounded on all sides by mountains. On the E. the rugged Rattlesnake Ridge walls it in, Kearsarge or Pequawket rising in lonely dignity a little to the N.; on the W. are the Moat Mountains, with the peak of Chocorua in the distance; and on the N. and N. W. almost the whole line of the White Mountains proper, crowned in the center with the dome of Mount Washington, closes in the view. The leading hotel of the village is the *Kearsarge House*, a large and well-kept house with accommodations for 300 guests. The *Intervale House* 1½ mile N. of the village, is large and excellent, and commands a unique and beautiful view of the mountains up a long valley-vista. Other hotels are the *Eastman House*, the *McMillan House*, the *North Conway House*, the *Randall House*, and the *Sunset Pavilion*. The rates at all these houses are from $3 to $4 per day. There are also many boarding-houses in the village ($7 to $14 a week).

There are several points of interest in the immediate neighborhood of North Conway. On the opposite side of the river, and about 3 miles distant, are *The Ledges*, a series of tall cliffs from 100 to 900 ft. high, which extend along the mountain-side for 4 or 5 miles. A figure of a horse (called the "White Horse") can be fancied on the side of one of these cliffs, and is visible from the village. *Echo Lake is a beautiful sheet of water lying at the very base of the cliffs (Moat Mountain), and is celebrated for the distinctness of the echo which it throws back. Above Echo Lake (reached by climbing the cliff) is the *Cathedral (3 miles from N. Conway), a cavity in the granite. The wall gradually inclines outward, forming a magnificent Gothic arch 40 ft. long, 20 wide, and 60 high, with noble forest-trees constituting the outer wall. A little north of the Cathedral, approached by a pleasant woodland path, is *Diana's Bath*, a crystal pool, 10 ft. in diameter and more than 10 ft. deep, overhung with trees, and having a beautiful cascade just below. Favorite drives are *Around the Square* (5 miles) and the *Thorn Hill Drive*, which ascends one of the spurs of Thorn Mountain and affords a fine view. *Mt. Kearsarge (or Pequawket), 3 miles from the village, is 3,270 ft. high, and is easily ascended. Parties of 2 or 3 persons are carried from North Conway to the foot of the mountain for 50c. each. A fair bridle-road leads to the summit. As this is the highest peak S. of the mountains in this direction, the view from its summit is extremely fine, embracing the whole White Mountain range, and an especially good view of Mt. Washington. The sharp peaks of Chocorua, with the Moat Mountain in the foreground, can also be seen with great distinctness; the course of the Saco River can be

traced almost from its source, as it winds among the intervales, and
finally bends away into Maine; and in the broad level expanse toward
the S. E. the eye is caught by Sebago Lake, Lovewell's Pond, and numerous smaller bodies of water. There is a small building on the summit,
at which those wishing to see the gorgeous sunset and sunrise views
can remain overnight.

North Conway to Gorham.

Many picturesque stage-routes, which used to form the only modes of communication from point to point, still afford enjoyable rides. The road from
North Conway to the Glen House (20 miles), and thence to Gorham, traverses
some of the most striking scenes.

For a few miles after leaving North Conway the road passes up the
valley of the Saco, amid delightful scenery, with Mt. Kearsarge looming
up grandly on the right and presenting an endless variety of forms. At
Bartlett the old stage-road to the Notch diverges to the W., while the
one we are pursuing runs nearly due N. At the crossing of the Ellis
River, the former site of the *Goodrich Falls* is seen. These falls were
among the heaviest and finest in the White Mountains, but were spoiled
in 1875 by the erection of a mill. In seasons of high water they are
still imposing. For the next mile the road is bordered by heavily wooded hills, between which occasional glimpses are had of the summits of
the Washington range, and the little hamlet of **Jackson** is reached.
There is a church here (Baptist), several hotels (*Arden House, Glen
Ellis Hotel, Gray's Inn, Jackson Falls House, Thorn Mountain House,*
and *Wentworth Hall*), and some half a dozen houses. From the portico
of the first-mentioned hotel there is a noble view of the surrounding
mountains, with *Iron Mountain* (2,736 feet) on the right, and the bold
peak of *Tin Mountain* (1,650 feet) on the left. Within three minutes'
walk of the hotel are *Jackson Falls*, a romantic cascade on the Wild-Cat
Brook. . In this vicinity is some of the best trout-fishing to be found
among the mountains, and the place is much frequented by sportsmen
and artists. On leaving Jackson, there is an impressive view of the
dark gorges, which open miles away toward Mt. Washington, and then
the road ascends through the desolate *Pinkham Notch*, filled with an almost unbroken forest. About 7 miles beyond Jackson, a path to the
right leads to the Glen Ellis Falls, which are quite near the road, and a
little farther on is the entrance to the Crystal Cascade. Here the Peabody River is crossed twice in quick succession, and a further ride of 3
miles brings us to our destination.

The **Glen House,** built in the spring of 1885, fronts the Peabody
River and the Washington range, to which it is nearer than any other
hotel in the mountains, five of the highest peaks being in full view
from the portico. Directly in front are the outworks and huge shoulder
of Mt. Washington itself. Next comes Mt. Clay (5,554 ft.), rising over
the huge "Gulf of Mexico"; then Jefferson (5,736 ft.); then Adams
(5,819 ft.); and, finally, Madison (5,381 ft.). From the hotel, parties on
Mt. Washington may readily be seen with the aid of a glass. (Stages
connect with *Gorham* (8 miles), on Grand Trunk R. R., and *Glen* Sta-

tion. (14 miles), on Maine Central R. R., and twice a day with the Mt. Washington R. R.   Fare, $6.)

In the vicinity are many points of interest.   The *Garnet Pools*, about ½ mile distant, near the Gorham road, are a series of basins in the Peabody River, some of them 15 and 20 ft. deep, worn in the granite rock by the action of the water.   ***Thompson's Falls** are a series of picturesque cascades in an affluent of the Peabody River, 2 miles from the hotel, on the road to North Conway.   The view of Mt. Washington and Tuckerman's Ravine, from the upper fall, is the finest that is obtained from any point.   *Emerald Pool*, noted for its quiet sylvan beauty, is a short distance from the road just before reaching Thompson's Falls.   On the North Conway road a path through the woods leads to the **Crystal Cascade,** " an inverted liquid plume," 80 ft. high, situated near the mouth of Tuckerman's Ravine, The best view of the cascade is not from the foot, but from a high moss-covered bank opposite.   A mile beyond (4 miles from the hotel), a plank-walk to the left leads to the ***Glen Ellis Fall,** where the Ellis River slides 20 ft. over the cliff at a sharp angle and then plunges 60 ft. into a dark-green pool below.   This is one of the loveliest cascades in the entire region.   ***Tuckerman's Ravine** is a tremendous chasm in the S. side of Mt. Washington, whose frowning walls, 1,000 ft. high, are plainly visible from the hotel.   It is filled, hundreds of feet deep, by the winter snows, through which a brook steals as summer draws near, gradually widening its channel until it flows through a grand snow-cave, which was found, by actual measurement one season, to be 84 ft. wide on the inside, 40 ft. high, and 180 ft. long.   The snow forming the arch was 20 ft. thick.   The engineers of the carriage-road up Mt. Washington dined in that snow-arch July 16, 1854.   After rain the cliffs back of the ravine present an appearance which has gained for them the name of the "Fall of a Thousand Streams."   The ravine is reached from the Glen by Thompson's Path, which diverges from the carriage-road about 2 miles up the mountain (distance 4½ miles), or by a rugged and difficult path by the brook-side from Crystal Cascade.   The more common way of visiting it, however, is to descend into it from the summit of Mt. Washington. The distance from the summit to the bottom of the ravine is about a mile.

The **Carriage-road up Mt. Washington** from the Glen was, until the completion of the steam-railway, the easiest and most popular way of reaching the summit, and is still preferred by many.   The road was begun in 1855 and completed in 1861, and is a noble piece of engineering.   The average grade is 12 ft. in 100, and the steepest, which is 2½ miles from the base, is 16 in 100 for a short distance only. The tolls are, for each person on foot, 32c.; on horseback or in carriages, 80c.   The fare for a seat in one of the regular mountain-carriages, which leave the Glen House morning and afternoon, is $5 for the round trip, $3 either way.   The time required for the ascent is about 3 hours, and for the descent 1½ hour.

The Glen House to Gorham.

Stages run from the Glen House to Gorham to connect with all trains on Grand Trunk R. R., which is 8 miles N. E. (fare, $1.50). The ride is down the valley of the Peabody, with fine mountain-views. About 2¼ miles from the Glen is a bridge over the Peabody River, by crossing which and proceeding to a point near a farmhouse, ¼ mile from the road, the traveler may see **The Imp,** a peak of Mt. Moriah, so named from the marked resemblance which the summit bears to a grotesque human countenance.

Gorham (*Alpine House, Eagle House, Randolph Hill House*), the N. E. gateway to the mountain-region, is a thriving village, situated in a broad and beautiful valley at the confluence of the Androscoggin and Peabody Rivers, 800 ft. above the sea. It is a station on the Grand Trunk R. R., whose repair-shops are located here. The scenery in the vicinity of the village is remarkably striking, both in the views of the mountain-ranges and isolated mountains, and of rivers and waterfalls. The range of Mts. Moriah, Carter, and The Imp, in particular, is seen to great advantage. Mt. Carter is one of the highest and Mt. Moriah the most graceful of the larger New Hampshire hills; the best view of them is from the Alpine House. The noble chain of hills to the N. W. of Gorham is known as the Pilot Range; while on the E. and S. E. the valley is walled in by the stalwart and brawny Androscoggin Hills. ***Mt. Hayes,** the highest of these latter (2,500 ft.), is directly N. E. of the village, and may be ascended by a path leading directly to the summit in two hours. "The picture from the summit can not be sufficiently praised. The view of Adams and Jefferson, sweeping from the uplands of Randolph, will never be forgotten. And Mt. Washington shows no such height, or grandeur, when seen from any other point." ***Mt. Surprise,** a spur of Mt. Moriah, fronts Mt. Hayes on the opposite side of the valley, and a bridle-path leads from the village to its summit (2½ miles). This bridle-path was formerly feasible for horses, but it has been allowed to get so much out of repair as to be no longer safe even for pedestrians without the aid of a guide. Mt. Surprise is 1,200 ft. high, and its summit affords a grand view of the "Presidential group" (Adams, Washington, Jefferson, and Madison). "There is no other eminence where one can get so near to these monarchs, and receive such an impression of their sublimity, the vigor of their outlines, their awful solitude, and the extent of the wilderness which they bear upon their slopes." The highest summits of the range rise directly against the eye, with no intervening ridge or obstacle. **Mt. Moriah,** 4,065 ft. high, can now only be ascended on foot, though there was once a good bridle-path. The ascent is tedious but not otherwise difficult, and the view from the summit is very striking.

Randolph Hill, 600 ft. higher than the valley, is reached by a pleasant carriage-drive of 5 miles from Gorham, and from its summit a superb view is obtained of the whole northerly wall of the Mt. Washington range. From the foot of Randolph Hill a path has been "blazed" through the forest to the summit of *Mt. Madison,* which may be ascended with guides. The summit is 5,381 ft. high, and the outlook which it affords is only inferior to that from the peak of Mt. Washington. It is possible for a strong pedestrian (with guides) to start from Gorham early in the morning, and, ascending Mt. Madison, pass over its summit, around or over the sharp pyramid of Adams, over Jefferson, between the humps of Mt. Clay, and reach the hotels on the summit of Mt. Washington before sunset. This route would lie among and over the largest mountains of the White Mountain range, and would afford a continuous succession of unrivaled views. From the *Lead-Mine Bridge,* 4 miles E. of the village, a pleasing view is obtained of the Androscoggin, dotted with islands in the foreground, with the mountains in the distance. It should be visited between 5 and 7 o'clock P. M., in order to see the sun set behind the mountains. An extremely attractive drive of 6 miles along the W. bank of the river leads to the * **Berlin Falls,** where the whole volume of the Androscoggin pours over a granite ledge, descending nearly 200 ft. in the course of a mile. The best views of the cataract are obtained from a jutting rock near the lower end, and from the bridge above which spans the narrowest part of the stream.

Stages leave Gorham, on the arrival of the trains, for the Glen House (8 miles; fare, $1.50). The ascent of Mt. Washington may be made from Gorham in a day *via* the Glen House.

Gorham to the Notch.

Since the completion of the carriage-road on the E. side, and of the railroad on the W. side, nearly all the travel through the mountains passes over Mt. Washington, and comparatively few tourists go by the old stage-routes. These, however, have not lost their charm, and whoever can spare the time should certainly make the trip from Gorham to the Notch, *via* the "Cherry Mountain Road," now to be described. The scenery along its entire length is grander than is afforded by any other route among the mountains. The distance is 32 miles. There is no regular stage, but mountain-wagons can be hired at Gorham on reasonable terms. The beauties of the road begin almost before the village is left behind. It takes in the glorious outlook from Randolph Hill, of which we have already spoken; it commands every slope and summit of the Mt. Washington range from the N.; and for 12 miles of the way they are all in view at once, with no intervening hills to break the impression of their majesty. The mountain-forms are much grander on the northerly than on the southern side, and the road we are traversing commands the finest views obtainable in this direction. "From the village of *Jefferson* (Starr King House), through which this Cherry Mountain road runs, not only is every one of the great White Mountain group visible, but also the Franconia Mountains, the side of the Willey Mountain, in the Notch, the line of the nearer Green Mountains beyond the Connecticut—in fact, a panorama of hills to the northwest and north almost as fine as the prospect in that direction from the summit of Mt. Washington." The finest point of view is * **Jefferson Hill**

(17 miles from Gorham), which is becoming one of the most frequented resorts in the White Mountain region. Here are the *Jefferson Hill House,* the *Mountain House,* the *Plaisted House,* the *Starr King,* the *Waumbek House,* and numerous boarding-houses. The rates at the hotels are $9 to $18 per week; at the boarding-houses, $7 to $12. The view of the mountains, above described, is incomparably fine from the Waumbek House; and from the piazza, with a glass, people on the summit of Mt. Washington can be distinctly seen, and the trains moving up and down the steep side. The remainder of the road to the Crawford House (16 miles) is scarcely inferior in scenic grandeur to that already described, and the entrance of the Notch is extremely fine. The White Mountain House and the Fabyan House are passed *en route.* There is a shorter route from Gorham to the Notch than the preceding, but it is much less attractive, and in engaging the carriage care should be taken to stipulate for Jefferson Hill.

The Crawford House and Vicinity.

The **Crawford House** is a large and popular summer-hotel, situated on a little plateau 1,899 ft. above the sea, and facing the Notch. It bears the name of the earliest hosts of these mountain-gorges, and is near the site of the old Notch House, one of the first taverns opened in the White Mountain region. The Crawford House and adjacent hotels are now connected with the outside world by two lines of railway, and passengers can run through from Boston to the very doors (see Routes 19 and 20). The station of the Maine Central R. R. stands a few rods from the front of the hotel. Within a stone's-throw of the hotel and of each other there are two springs, one of which discharges its waters into the Saco, while the other empties into a tributary of the Ammonoosuc, and reaches the sea through the Connecticut. In front of the house, near the gate of the Notch, is a tiny lake, which forms the head-waters of the Saco.

A favorite excursion from the hotel is the ascent of * **Mt. Willard,** which is easily made by a road 2 miles long, either in carriages or on foot. The summit is 2,570 ft. high, and commands a wonderful view of the tremendous gulf of the Notch, and of the mountain-peaks far and near. Speaking of the view of the Notch from this point, Bayard Taylor says, "As a simple picture of a mountain-pass, seen from above, it can not be surpassed in Switzerland." Near the summit of the mountain, on the S. side, is the *Devil's Den,* a dark, cold cave, about 20 ft. deep, 15 high, and 20 wide, only accessible by means of ropes. *Gibbs's Falls* are a series of romantic cascades, reached by a walk of half an hour from the hotel, along the aqueduct by which it is supplied with water, and then along the brook-side. The falls are about ¼ of a mile from where the aqueduct issues from the brook. *Beecher's Falls* (named after Rev. Henry Ward Beecher, who is said to have taken an involuntary bath in one of the basins) are situated on the slope of Mt. Lincoln, and may be reached by an easy path through the woods to the right of the hotel. They consist of a series of beautiful cascades extending for ⅓ mile along a mountain-brook. From some shelving rocks at the head of the uppermost fall, called the "Flume Cascade," there is a fine view of the summit of Mt. Washington.

The favorite excursion from the Crawford House is through the famous *** Notch,** which is seen to the best advantage as approached from this direction, the giant masses of Webster, Willard, and Willey being directly in front. The Notch is a tremendous gorge or rift in the mountains, which rise on either side to the height of 2,000 ft., and which, in one spot, called the "Gateway," are only 22 ft. apart. The Saco River runs through it, and also the Maine Central R. R., which along the slopes of Mount Willey is 300 ft. above the stage-road. The *Elephant's Head* is a rocky bluff on the E. side of the Notch, about ½ mile from the hotel, the supposed resemblance of which to an elephant's head, as seen from the hotel piazza, gives it its name. Just within the Gate, a view is obtained of the *Old Maid of the Mountain*, a great stone face on a spur of Mt. Webster. An overhanging rock on the same side of the road is called the *Devil's Pulpit*, and on the face of this the profile of *The Infant* is visible to imaginative minds. Directly opposite the Devil's Pulpit is another profile called the *Young Man of the Mountain ;* and far up the slopes of Mt. Willard is the black mouth of the Devil's Den, already mentioned. *The Flume* is a portion of a little mountain-stream, to the left of the road about ¾ mile from the hotel. A little farther down the Notch is the *** Silver Cascade,** the finest waterfall on the W. side of the mountains. The stream, the upper part of which is visible from the piazza of the hotel, descends 800 ft. in the course of a mile, 400 of which are nearly perpendicular. The best view is from the bridge, near which the current rushes through a narrow flume, like that already described. Passing down the Notch between Mts. Willey and Webster, we come to the *Willey House* (3 miles from the hotel), where the whole Willey family, 9 in number, were crushed by an avalanche from which they were trying to escape, August 28, 1826. A rock 30 ft. high split the avalanche and saved the house from which they fled to their death. The house is occupied, and a small entrance-fee is charged, but there is nothing inside to interest. Three miles beyond the Willey House, on Avalanche Brook, a small mountain-stream emptying into the Saco, is the *** Sylvan Glade Cataract,** regarded by many as the finest waterfall in the mountains. It is 2 miles from the road, in a steep ravine, whose cliffs, crowned with a dense forest of spruce, are singularly grand. The cascade leaps first over 4 rocky ledges, each about 6 ft. high, and then glides at an angle of 45° down a solid bed of granite 150 ft. into the pool below. It is about 75 ft. wide at the base, and 50 at the summit. A mile above the cataract, there are several other falls, the finest of which is called the *Sparkling Cascade*. This is the limit of the ordinary excursions, but it is quite worth while to engage a vehicle and drive farther along the old stage-route to North Conway. As we proceed down the Saco we pass through a dense forest and come in succession to the *Giant's Stairs*, 5,500 ft. high ; *Mt. Resolution*, 3,436 ; and *Mt. Crawford*, 3,130. Next, the *Mt. Crawford House*, once the most popular of the mountain inns, is passed ; and ½ mile beyond we cross *Nancy's Brook and Bridge*, so named after a young woman who perished here from exposure when in pursuit of a faithless lover. Near by is the grave of Abel Crawford, "patriarch of the mountains."

The * **Bridle-Path up Mt. Washington** from the Crawford House commands finer views than any other route, leading over the summits of Mts. Clinton, Pleasant, Franklin, and Monroe. It can only be traversed on foot, but the path is plain and safe, *except in case of a fog*, when great caution should be exercised, as several fatal accidents have occurred.

The Fabyan and Twin Mountain Houses.

The **Fabyan House** is an old and popular hotel in the mountains, with all the appointments and conveniences of a first-class city hotel. It stands on the Giant's Grave, a lofty, grave-shaped mound, and commands a noble view of the whole White Mountain range. It is at the junction of the Maine Central and Concord & Montreal R. Rs., White Mt. division, and is also the nearest of the large hotels to the lower terminus of the Mt. Washington Railway. It is likewise a convenient point for excursions to Mt. Willard, the Notch, and the Willey House, the intervening 5 miles being over a good road with fine views. Between the railroads (about half a mile from the Fabyan House) is the *Mount Pleasant House*, attractively situated, and charging moderate prices. About a mile beyond Fabyan's is the *White Mountain House*, long known as an old and favorite hostelry, pleasantly located in the midst of an open tract of country. In the rear is a fine view of the Presidential peaks, and in front, beyond the Ammonoosuc, rises the lofty range which connects the Great Notch with Franconia. There are pleasant rambles in the neighborhood, and varied views from the adjacent hills. The once famous *Lower Ammonoosuc Falls* have been spoiled by the erection of a saw-mill above. The * *Upper Ammonoosuc Falls* (3½ miles from the Fabyan House on the road to Mt. Washington) are very fine. The **Twin Mountain House** is a large and highly popular hotel, 5 miles W. of Fabyan's, pleasantly situated on heights above the Ammonoosuc River. The Concord & Montreal R. R. has a station here, and it is a convenient point from which to visit the various places of interest.

Bethlehem (the *Bellevue, Highland House, Prospect House, Sinclair House, and Strawberry Hill House*) is a popular summer resort on the Profile & Franconia Notch R. R., readily reached from Bethlehem Junction, on the Concord & Montreal R. R., or from Zealand Junction, on the Maine Central R. R. It is the highest village E. of the Rocky Mountains, and is the meeting-place of the American Hay-Fever Association. The **Maplewood House** (1 mile from Bethlehem) is reached by Narrow Gauge R. R. from Bethlehem Junction or Zealand Junction, the station being Maplewood Station. The house accommodates 500 guests, and offers fine views. Several excursions may be made from Bethlehem, the best being to the summit of *Mt. Agassiz* (1¾ mile distant). The Profile & Franconia Notch R. R. runs from Zealand Junction, on the Maine Central R. R., to the Profile House (see page 130).

Mount Washington.

The ascent of Mt. Washington by the Carriage-Road from the Glen House is described on p. 123.　The ascent by the Crawford House Bridle-Path on p. 126.

The *Mt. Washington Railway* connects with the Concord & Montreal R. R., near *Marshfield* (Ammonoosuc Station).　The distance from Marshfield to the summit is about 3 miles.　The fare for the ascent or descent is $3 ; round trip, $6 ; trunks are not charged extra.　The Mt. Washington Railway was begun in 1866, and opened in 1869, and was the model for that up the Rigi.　The grade is enormous, being 3,596 ft. in 3 miles, and in places 1 foot in 3.　The track is of three rails bolted to a trestle-work of heavy timber.　The third or center rail is like a wrought-iron ladder with rounds 4 inches apart.　Into this fits a cog-wheel which fairly pulls the train up the mountain.　The seats for the passengers are so swung as to be horizontal, whatever may be the inclination of the track.　The safety of the train is secured by independent, self-acting brakes.　The time occupied by the ascent is 1½ hour, but the slow progress is forgotten in the splendid panorama of the gradually widening views.

The summit (6,293 ft. high) is an acre of comparatively level ground, on which stand the *Mount Washington Summit Hotel* ($1.50 for each meal, and the same for a night's lodging), the old Tip-Top House (which is no longer in use), the engine-house of the railway, and the U. S. Signal-Service observatory.　At this station, formerly occupied in winter, observers have recorded a temperature of 59° below zero, while the wind blew with a velocity of 190 miles an hour. Visitors to Mount Washington should always go well clad.　The range of the thermometer, even in midsummer, is from 30° to 50°.　It frequently falls as low as 25°, and sometimes to 20°, or 12° below freezing. The tourist should spend one night on the summit, in order to see the wonderful sunrise and sunset views.　The enjoyableness of the trip is greatly increased by going up the mountain one way and down the other (up by railway and down by stage, or *vice versa*).

The view from Mt. Washington is incomparably grand.　In the W., through the blue haze, are seen in the distance the ranges of the Green Mountains ; the remarkable outlines of the summits of Camel's Hump and Mount Mansfield being easily distinguished when the atmosphere is clear.　To the N. W., under your feet, are the clearings and settlement of Jefferson, and the waters of Cherry Pond ; and, farther distant, the village of Lancaster, with the waters of Israel's River. The Connecticut is barely visible ; and often its appearance for miles is counterfeited by the fog arising from its surface.　To the N. and N. E., only a few miles distant, rise boldly the great northeastern peaks of the White Mountain range—Jefferson, Adams, and Madison —with their ragged tops of loose, dark rocks.　A little farther to the E. are seen the numerous and distant summits of the mountains of Maine. On the S. E., close at hand, are the dark and crowded ridges of the mountains of Jackson ; and beyond, the conical summit of Kearsarge, standing by itself on the outskirts of the mountains ; and, farther over,

9

the low country of Maine and Sebago Lake, near Portland. Still farther, the ocean is distinctly visible from sunrise to about 10 o'clock, weather permitting. Farther to the S. are the intervales of the Saco, and the settlements of Bartlett and Conway, the sister ponds of Lovewell, in Fryeburg; and, still farther, the remarkable three-toothed summit of Chocurua, the peak to the right being much the largest, and sharply pyramidal. Almost exactly S. are the shining waters of the beautiful Winnepesaukee, seen with the greatest distinctness on a favorable day. To the S. W., near at hand, are the peaks of the southwestern range of the White Mountains: Monroe, with its two little Alpine ponds sleeping under its rocky and pointed summits; the flat surface of Franklin, and the rounded top of Pleasant, with their ridges and spurs. Beyond these, the Willey Mountain, with its high, ridged summit; and, beyond that, several parallel ranges of high, wooded mountains. Farther W., and over all, is seen the lofty, bare summit of Mt. Lafayette, in Franconia. There used to be an observatory on the summit, and there is a magnificent view in all directions. There are also special points whence fine outlooks are obtained.

The Franconia Mountains.

These mountains, though in popular estimation inferior in interest to the eastern cluster, are really not so, except it be in the wonders of the mountain ascents; and even in this the panorama from the summit of *Lafayette* is scarcely less extensive or less imposing than the scene from the crown of Mount Washington, while the exquisite little lakes, and the singular natural eccentricities in the Franconia group, have no counterpart in the other. They lie W. S. W. of the White Mountains, from which they are separated by the Field, Willey, and Twin Mountain ranges, and consist of sharp and lofty peaks, covered almost to their summits with dense forests. The name is usually applied to all the mountains around the Notch, but belongs, more properly, to the majestic range on the E. side. The **Franconia Notch** is a fine pass between the Franconia and Pemigewasset ranges, 5 miles long and ½ mile wide, walled in by precipitous cliffs, filled with forests, and traversed by the crystal waters of the upper Pemigewasset River. Harriet Martineau said of it, "The Franconia Defile is the noblest mountain-pass I saw in the United States."

The **Profile House** is the headquarters of the Franconia range, and is the largest summer hotel in the White Mountain region. It is situated in a narrow glen between two lakes, near the N. end of the Franconia Notch, 1,974 ft. above the sea.

From New York the best route is *via* Route 30 to Wells River, and thence by Route 20 *b*. Drawing-room cars run without change. The Profile & Franconia Notch R. R. connects at Bethlehem Junction with the Concord & Montreal R. R., and runs in about 10 miles to the Profile House (six trains a day). The Profile House may also be reached from Plymouth *via* the Pemigewasset Valley Branch of the Concord & Montreal R. R. to North Woodstock, and thence by a stage ride of 10 miles (stage fare, $2).

Of the many objects of interest in the neighborhood of the Profile House, one of the most charming is * **Echo Lake,** a diminutive but very deep and beautiful sheet of water about ¼ mile N. of the hotel, entirely inclosed by high mountains. From the center of this fairy-water, a voice, in ordinary tone, will be echoed distinctly several times, and the report of a gun breaks upon the rocks like the roar of artillery. The Indian superstition was, that these echoes were the voice of the Great Spirit, speaking in gentleness or in anger. The best time to visit the lake is toward evening, when the flush of sunset is on the mountains. *Eagle Cliff* is a magnificently bold and rocky promontory almost overhanging the hotel on the N. Directly opposite Eagle Cliff, and forming the southern side of the Notch, is **Profile Mountain** (or Mt. Cannon), 2,000 ft. above the road, and 4,107 ft. above the sea. Away up on its crown is a group of mighty rocks which, as seen from the hotel, bears an exact resemblance to a mounted cannon. The mountain is ascended by a difficult footpath in about 2 hours, and the view from the summit is surpassingly fine, including the surrounding peaks, the towering heights of Washington and his peers, the softly swelling hills sloping away to the S., and the lovely valley of the Pemigewasset. It is upon this mountain, also, that we find the * **Profile,** or "Old Man of the Mountain." This strange freak of Nature, so admirably counterfeiting the human face, is 90 ft. long from the chin to the top of the forehead, and is 1,200 ft. above the road, though far below the summit of the mountain. It is formed of three distinct masses of rock, one forming the forehead, another the nose and upper lip, and a third the chin. The rocks are brought into the proper relation to form the profile at one point only, namely, upon the road through the Notch, ¼ mile S. of the hotel (indicated by a sign-board). The face is boldly and clearly relieved against the sky, and, except in a suspicion of weakness about the mouth, has the air of a stern, strong character, well able to bear, as he has done unflinchingly for centuries, the scorching suns of summer and the tempest-blasts of winter. Passing down the road a little way, the Old Man is transformed into "a toothless old woman in a mob-cap," and soon after melts into thin air, and is seen no more. Hawthorne has found in this scene the theme of one of the pleasantest of his "Twice-Told Tales," that of "The Great Stone Face." At the base of the mountain, immediately under the ever-watchful eye of the Old Man, is the exquisite little **Profile Lake,** sometimes called the "Old Man's Washbowl," where boats can be procured. It is full of the finest trout, and near by is the *Trout House,* where several hundred of this beautiful fish are kept for breeding purposes. From the shore of this lake the best view of Eagle Cliff is had. There is a carriage-road from the Profile House to the summit of *Bald Mountain,* 1¾ mile distant, whence a noble view is obtained without undergoing the fatigue consequent upon the ascent of the more lofty peaks. * **Mt. Lafayette** is the monarch of the Franconia kingdom, towering skyward to the height of 5,269 ft. Its lofty pyramidal peaks are the chief objects in all views for many miles around. The summit is reached from the Profile House in 3¾ miles by a good bridle-path. On the summit stand the walls of an old house, erected as a shelter for visitors. From this point is obtained "a view more beautiful, in some respects, though it may be less grand and majestic, than that from Mt. Washington." The Green

Mountains are plainly seen, as well as the entire White Mountain range; the peak of Katahdin cleaves the air to the N. E., and to the S. the Pemigewasset Valley shows its contour for a distance of 40 miles.

Walker's Falls, reached by following for ¼ mile a rivulet which crosses the road 2⅓ miles S. of the hotel, is one of the most picturesque of the mountain cascades, though the volume of water is not very great, nor the height of the fall at all remarkable. Half a mile farther up the stream is a larger fall. A mile farther S. is **The Basin,** a granite bowl, 60 ft. in circumference and 15 ft. deep, filled with cold, pellucid water. It lies near the road-side, where the Pemigewasset has worn deep and curious cavities in the solid rock. The water, as it flows from the Basin, falls into most charming cascades; and at the outlet, the lower edge of the rocks has been worn into a remarkable likeness of the human leg and foot, called the " Old Man's Leg." Across the brook below the basin, is thrown a bridge of logs, which enables the visitor to reach a path leading ½ mile to a succession of the most exquisitely lovely *Cascades* in this whole region. These cascades should be followed to the point where they end in a waterfall (*Tunnel Falls*) 30 ft. high. About 1½ mile beyond the Basin (5 miles from the Profile House) is the **Flume House,** beautifully situated at the head of the valley, with Mt. Liberty in front and Mt. Pemigewasset behind, and good guides can be obtained there. The views northward toward the Notch, and southward toward the Pemigewasset Valley, are surpassingly fine. Opposite the hotel a path through the forest leads ⅘ of a mile to ***The Pool,** a wonderful excavation in the solid rock, smooth as though hewn by human hands. It is about 150 ft. wide and 40 ft. deep, the water entering by a cascade, and escaping through the rocks at its lower extremity; from the top of the rocks above to the surface of the pool the distance is nearly 150 ft.

*The Flume, one of the most famous of all the Franconia wonders, is ¾ of a mile from the hotel, and is reached by a carriage-road leading to the part of the *Cascade* below the Flume. The cascade is a continuous succession of gentle rapids, 600 ft. long, and at its upper end is the entrance to the flume itself, which is a rugged ravine 700 ft. long, with precipitous, rocky walls 60 ft. high, and not more than 20 ft. apart. Through this grand fissure comes the little brook which we have just seen; and a plank-walk leads along its bed to the upper end of the ravine, where the walls approach within 10 ft. of each other. At this point, about half-way up, a huge granite bowlder, several tons in weight, used to hang suspended between the cliffs, where it had been caught in its descent from the mountain above. But the storms of the summer of 1883 swelled the little brook into a furious torrent, that dislodged from its resting-place this wonder of the Flume, and precipitated it into the gorge, a thousand feet below. The **Georgianna Falls** (or *Harvard Falls,* as they are sometimes called) are of greater magnitude than any yet discovered in these mountains. They plunge over the precipice in two leaps of 80 ft. each, and are reached by a path from a small farmhouse about a mile S. of the Flume House, on the Plymouth road (guide at the farmhouse).

22. Lake Winnepesaukee.

Lake Winnepesaukee is reached from New York, Boston, or Portland, by any of the routes described in Routes 19, 20, or 30, and at the beginning of Route 21. The best approach for those who wish to make the tour of the lake on the way to the White Mts. is *via* Boston & Maine (main line) and Dover & Winnepiseogee branch to Alton Bay, whence the steamer Mt. Washington runs to Wolfboro, Weirs, and Center Harbor. The Lady of the Lake runs from Wolfboro to Weirs and Center Harbor twice a day.

LAKE WINNEPESAUKEE, the largest and most beautiful sheet of water in New Hampshire, lies in the two counties of Belknap and Carroll, and is a sort of portal to the White Mountain region from the S. It is very irregular in form, its extreme length from N. E. to S. W. being about 25 miles, and its width varying from 1 to 10 miles. Its waters are wonderfully pure and translucent, numerous islands are dotted over its surface, and lofty hills and mountains close it in on all sides. Its name is of Indian origin, and means "The beautiful water in a high place," or, as some maintain, "The Smile of the Great Spirit."

Alton Bay (*Winnepesaukee House*) is the most southern point of the lake, and is situated at the head of a narrow estuary, which appears more like a river than a lake. There are several points of interest in the vicinity. From *Sheep Mountain*, 2 miles N., there is a fine view of the lake; also from *Prospect Hill* and *Mt. Major*. *Lougee Pond*, 7 miles S. W., is noted for its tame fish; and *Merry-Meeting Lake*, 7 miles E., is a beautiful sheet of water. The pleasantest excursion, however, is to the summit of ** **Mt. Belknap,** 10 miles distant. The fare for a party in a regular mountain-wagon is $1.50 each, and the excursion occupies an entire day. The view from the summit is very fine. The distance from Alton Bay to Wolfboro is 11 miles, and to Center Harbor 30 miles; so that the sail includes nearly the entire length of the lake. **Wolfboro** (*Bellevue Hotel, Hotel Glendon,* and *Kings' Wood House*) is situated on two beautiful slopes of land rising from the lake. It is the most important point on the lake, and has 3,020 inhabitants. A branch of the Eastern Div. Boston & Maine R. R. connects Wolfboro with North Conway; and steamers run to Weirs, Center Harbor, and Alton Bay. A highly popular excursion from Wolfboro is to **Copple Crown Mountain,** 2,100 ft. high, and 6¼ miles distant. Carriages from the hotel run to within a mile of the summit, from which point horses may be obtained, or the ascent may be easily made on foot. The carriage-fare is $1.50 for each person of a party. The view from the summit is very fine. The lake can be seen for nearly its entire length. To the S. is a vast level expanse, dotted with lakes and villages and patches of woodland; Belknap and Gunstock, with the mountains of the Merrimac Valley, stretch away toward the W., with the Ossipee and Sandwich ranges closing in the head of the lake; and almost due N. Chocorua looms up in massive grandeur, with the distant peak of Mt. Washington above its shoulder. The ocean is visible to the S. E. on a clear day. A mile N. of Copple Crown (6 miles from Wolfboro) is a smaller mountain called *Tumble-Down Dick*, which is more easily ascended, and affords a scarcely inferior view.

The sail from Wolfboro to Center Harbor affords a constant succession of striking views. First Ossipee and Chocorua attract the attention as they loom up against the northern horizon; and then, about midway of the lake, the dim but majestic peak of Mt. Washington is seen 40 miles away. **Center Harbor** (20 miles from Wolfboro and 10 from Weirs) is a very small village, but, being a highly popular summer resort, has commodious hotels so located as to command charming views of the lake and vicinity (*Moulton Hotel, Senter House*). There are also smaller hotels and boarding-houses. Stages leave Center Harbor daily for Sandwich, Center Sandwich, and West Ossipee. Steamers run to Weirs, Wolfboro, and Alton Bay. The drives in the vicinity of the village are very attractive, but the chief objects of interest are Red Hill and Squam Lake. *** Red Hill** is a remarkably beautiful eminence, 2,038 ft. high, about 6 miles N. W. of the lake. Carriages run to the foot of the hill, where horses are always in readiness to convey passengers to the summit by a bridle-path 1½ mile long. In order to obtain the best views, the ascent should be made in the forenoon, or in the afternoon between 3 and 5 o'clock. At the latter hour the view of the lake and its islands is charming. The view as a whole is one of the finest in New England, and has been compared to that from the summit of Mt. Holyoke. *** Squam Lake,** lying W. from Red Hill and 2 miles N. W. of Center Harbor, is another lovely sheet of water. It is about 6 miles long and 3 miles wide at its widest part, and, like Winnepesaukee, is studded with a succession of romantic islands. This lake abounds in fish, and is noted for the limpid purity of its water. The drive around Squam Lake from Center Harbor (21 miles) affords a delightful day's excursion.

From Center Harbor a steamer runs several times daily to **Weirs** (10 miles), the principal point on the W. side of the lake. This short sail is delightful, and from a point about 3 miles below Center Harbor is obtained the finest view on the lake. Weirs is simply a station on the Concord & Montreal R. R., where the trains connect with the steamboats. Near Weirs is *Endicott Rock*, supposed to have been set up as a monument or boundary by the surveyors sent out in 1652 by Governor Endicott of Massachusetts.

23. Portland to Mt. Desert.

a. All-Rail Route via Maine Central.

The all-rail route to Mt. Desert is *via* Maine Central (main line) to Bangor, thence by Mt. Desert Branch to Mt. Desert Ferry station ; distance, 179 miles ; fare, $5.00 limited, or $8.50 round-trip.

THE towns from Portland to Bangor are described in Route 24. From the latter-named city the Mt. Desert Branch crosses the Penobscot River into the town of *Brewer*. Thence it runs through the small towns of *Holden* and *Dedham* to **Ellsworth,** a place of 4,804 inhabitants, the county-seat of Hancock County, and a port of entry. The place contains a court-house, custom-house, several banks, six churches, and three newspaper offices. It is an important center of lumbering and ship-building. Thence passing through the village of Hancock, the road reaches its terminus at Mt. Desert Ferry, where steamboats connect with Bar Harbor, Mount Desert, a distance of 8 miles over smooth water.

b. Via Maine Central R. R., and Steamer from Rockland.   Fare, $4.

The distance from Portland to Rockland by the Maine Central R. R. is 89 miles. On the line, 29 miles from Portland, is **Brunswick,** a thriving town at the head of tide-water on the Androscoggin River, noted as the seat of *Bowdoin College* (incorporated in 1794). The college buildings, situated amid a beautiful grove of pine-trees near the station, are worthy of a visit, and the gallery of paintings is famous. A few miles beyond Brunswick is **Bath,** a busy little city of 8,723 inhabitants, situated on the Kennebec River, 12 miles from the sea. Ship-building is the leading industry, and one of the fine steel cruisers forming part of the " new navy " was built here. There are several fine churches and other buildings. At Bath the cars are carried by ferry across the Kennebec River, and passing on to the rails of the Knox & Lincoln Branch of the Maine Central R. R. reach Rockland in 49 miles, passing *en route* the small towns of *Wiscasset* (on the Sheepscot River), *Newcastle, Damariscotta, Warren*, and *Thomaston*. The latter contains the Maine State Prison. **Rockland** is a city of 8,174 inhabitants, situated on Owl's-Head Bay, an inlet of Penobscot Bay. The town is well built, and the adjacent scenery is remarkably picturesque. Rockland contains 8 churches, a public library, and a custom-house that cost $175,000. The chief articles of export are lime and granite. Ship-building, also, is a leading industry. Near by is the *Bay Point Hotel*, on the Rockland Breakwater. At Rockland, passengers for Mt. Desert take the steamer, which pursues a devious course across Penobscot Bay and through intricate channels to the island, stopping by the way at **Castine,** a pretty and wealthy village, situated on a narrow peninsula projecting into the bay, and much resorted to in summer for its coolness, seclusion, and boating and fishing facilities. This line has the advantage for those who like only a few hours' travel on the boat.

c. Via Steamer from Portland.   Fare, $4.

A popular way of reaching Mt. Desert is by steamers twice a week (Tuesday and Friday) from Portland, landing at Bar Harbor. The steamer leaves Portland on the arrival of the train, leaving Boston at 7 P. M. (Fare from Boston, $5.50.)

Mount Desert.

Mount Desert lies in Frenchman's Bay, just off the coast of Maine, about 110 miles E. of Portland, and 40 miles S. E. of Bangor. The island is 14 miles long and 8 miles wide at the widest part, and has an area of 100 square miles. At its northern end it approaches so nearly to the mainland that a bridge affords permanent connection between the two; and nearly midway it is pierced by an inlet known as Somes's Sound, which is 7 miles long. "The island," says Robert Carter, in his "Summer Cruise," "is a mass of mountains crowded together, and seemingly rising from the water. As you draw near, they resolve themselves into 13 distinct peaks, the highest of which is about 2,000 ft. above the ocean. Certainly only in the tropics can the scene be excelled—only in the gorgeous islands of the Indian and Pacific Oceans. On the coast of America it has no rival, except, perhaps, at the Bay of Rio Janeiro." The mountains are mainly upon the southern half of the island, and lie in 7 ridges, running nearly N. and S. The highest peak is *Green Mountain ;* and the next, which is separated from Green Mountain by a deep, narrow gorge, is called *Newport.* The western sides of the range slope gradually upward to the summits, but on the east they confront the ocean with a series of stupendous cliffs. High up among the mountains are many beautiful lakes, the largest of which is several miles in length. These lakes, and the streams that flow into them, abound in trout. There are several harbors on the islands, the best known of which are Southwest, Northeast, and Bar Harbor.

Bar Harbor (*Belmont, Grand Central, Hotel des Isles, Louisburg, Lynam, Malvern, Rodick, St. Sauveur, West End Hotel*) is on the E. shore of the island, opposite the Porcupine Islands, and derives its name from a sandy bar which connects Mt. Desert with the largest of the Porcupine group. The village is known locally as "East Eden," and is the favorite stopping-place for travelers. The scenery in the neighborhood is pleasing, and many excursion-points are near. The first excursion should be to the summit of *Green Mountain,* and by the Green Mountain R. R. the expedition is made very comfortably. There is a fine hotel on the mountain-top, and a good road leads from the village (in 4 miles) to the hotel, and enables vehicles to ascend the entire distance. Pleasure-parties frequently prefer to ascend on foot, and it is customary to remain overnight at the hotel in order to view the sunrise from this altitude (1,527 ft.). The view from the summit is very fine, embracing the whole of Mt. Desert, Frenchman's Bay with its many islands, the boundless ocean on the one hand, and a vast stretch of the Maine coast on the other. **Eagle Lake,** so named by Church, the artist, is visible at intervals during the entire ascent of the mountain ; and, half-way up, a short *détour* from the road will bring the tourist to its pebbly shore. *Mt. Newport* is ascended from the Schooner Head road, and *Kebo,* which may be reached in half an hour from the hotels at Bar Harbor, affords a pleasing prospect. The several points along the coast to which the visitor's attention is directed are The Ovens, which lie 6 or 7 miles up the bay, and Schooner Head, Great Head, and Otter Creek Cliffs, lying on the seaward shore of the island. *The Ovens* may be reached by boat or by a pleasant drive of 7 miles through the woods. They are a series of cavities worn in the cliffs by the action of the tides, some of which are large enough to contain 30 or 40 people at a time. They can only be visited at low tide, and are then a favorite picnic-ground for summer residents in Bar Harbor. The *Via Mala* is a curious archway in one of the projecting cliffs. *Schooner Head,* so named from the fancy that a mass of white rock on its sea-face has the appearance of a small schooner, is on the seaward side of the island, 4 miles S. of

Bar Harbor. The *Spouting Horn* is a wide chasm in the cliff, which extends down to the water and opens to the sea through a small archway below high-water mark. At high tide, and especially in stormy weather, the waves rush through this archway and send a spout of water far above the summit of the cliff. *Great Head, 2 miles S. of Schooner Head, is the highest headland between Cape Cod and New Brunswick. It is a bold, projecting mass, whose base has been deeply gashed by the waves. Still farther S. are the Otter Creek Cliffs, situated near a small stream known as Otter Creek. The most interesting feature of these cliffs is *Thunder Cave* (reached from the road by a superb forest-walk). The cave is a long, low gallery in the cliff-side, into which the waves rush with impetuous force, and, dashing themselves against the hollow cavity within, produce a sound closely resembling thunder. In fair weather the sound is apparent only when near, but in great storms it may be heard distinctly at the distance of 7 miles. About 9 miles S. W. of Bar Harbor is *Jordan's Pond, a beautiful lake 2 miles long and ½ mile wide, surrounded by picturesque mountain scenery and abounding in fish. *Cromwell's Cove* is 1½ mile S. of the village. The Pulpit, the Indian's Foot, and the Assyrian (a rock figure in one of the cliff-sides) are in this vicinity.

Somes's Sound, which divides the lower portion of the island into two distinct sections, possesses many attractions for those who admire bold headlands. It is usual to ascend the Sound in boats from Southwest Harbor; but explorers sometimes drive to Somesville, a neat little village at the head of the Sound (8 miles from Bar Harbor, and 6 miles from Southwest Harbor), and there take boats for a sail down stream. The Sound cuts through the center of the mountain-range at right angles between Dog Mountain and Mt. Mansfield, and has striking views on either hand. *Eagle Cliff* is one of the cliffs of Dog Mountain, and rises perpendicularly to a height of nearly 1,000 ft. *Fernalds' Point*, on the W. shore of the Sound, is the site of the ancient Jesuit settlement of St. Sauveur, and near by is Father Biard's Spring. The Sound affords excellent fishing and boating, though it is necessary to guard against the sudden gusts which at times rush down from the mountains.

Southwest Harbor (*Claremont House*, *Freeman House*, and *Island House*) is less picturesque in its surroundings than the eastern and northern shores of the island, but there are several points of interest in the vicinity. Chief of these is the Sea-Wall (3 miles S. W.), a *cheval-de-frise* of shattered rock skirting the shore for the distance of a mile, and against which the sea beats with tireless impetuosity. Beach Mountain (affording a noble view), Dog Mountain, Flying Mountain, Mt. Mansell, and Sargent's Mountain may all be ascended from Southwest Harbor. *Long Lake* is 2½ miles N. W.; *Denning's Lake*, about 3 miles N.; and *Seal Cove*, 5 miles W. These are all in the neighborhood of fine scenery, and the lakes abound in fish.

24. Portland to Moosehead Lake.

The regular route *via* Newport, Dexter, and Greenville is described below. Round-trip tickets from Boston, $15.

THE distance from Portland to Bangor *via* Maine Central R. R. is 136 miles; to Augusta, 62 miles. The first important station after leaving Portland is *Brunswick*, 29 miles (see Route 23 *b*). Beyond Brunswick the train crosses the Androscoggin and passes in 27 miles to *Gardiner* (*Evans House, Young's Hotel*), a leading center of the lumber industry, with a population of some 5,491, at the mouth of the Cobbosseecontee River. It has abundant water-power, and paper, flouring, saw, and planing mills. Four miles beyond Gardiner, on the banks of the Kennebec, is *Hallowell* (*Hallowell House*), a town of 3,181 inhabitants, with extensive quarries in the neighborhood, from which large quantities of granite are exported by river and rail. It contains also a cotton-factory, an oil-cloth factory, three national banks, and a public library. Two miles above, at the head of navigation on the Kennebec, is **Augusta** (*Augusta House, Cony House*), the capital of the State of Maine. It is a beautifully situated and well-built city of 10,527 inhabitants, owing much of its loveliness to a great abundance of shade-trees and shrubbery. Among the noteworthy buildings are the * *State-House*, built of white granite, and one of the finest public edifices in New England; the *Court-House* of Kennebec County; the *State Insane Asylum*, a handsome granite structure on the heights E. of the river; and the *Kennebec Arsenal*, with well-arranged grounds and neat buildings. This place is noted as the home of James G. Blaine. The great dam across the Kennebec, ½ mile above the city, is 584 ft. long, and furnishes immense water-power.—Beyond Augusta several small stations are passed, and then come *Waterville* (81 miles from Portland), a beautiful town of 7,107 inhabitants near the Ticonic Falls of the Kennebec, seat of Colby University (Baptist); *Burnham*, whence the Belfast Div. runs in 33 miles to **Belfast,** a prosperous maritime city of 5,294 inhabitants, on Penobscot Bay; and *Newport* (108 miles), whence the Dexter Branch runs 30 miles N. to Dexter, Dover, and Foxcraft, the junction point with the Bangor & Aroostook R. R. running to Greenville (47 miles), the only stations *en route* requiring mention being *South Sebec*, whence stages run in 6 miles to **Sebec Lake,** a beautiful sheet of water 12 miles long and abounding in fish; and Milo, where a branch railroad runs to Katahdin Iron-Works (19 miles). At the latter-named point there are good river and lake fishing, an excellent hotel, and mineral springs. The railroad passes through a thinly-settled and picturesque country and affords many fine mountain-views. **Bangor** (*Bangor House, Bangor Exchange, Penobscot Exchange*, and *Windsor Hotel*), the third city of Maine, and one of the great lumber-marts of the world, is situated at the head of navigation on the Penobscot River, 60 miles from the ocean, and contains 19,103 inhabitants. It is solidly and handsomely built, and very wealthy for its size. Besides the lumber industry, for which all the vast forest country above is laid under contribution, ship-building is carried on, as also a large business in roofing-slates, ice, hay, potatoes, moccasins, steam boilers and machinery, etc. The granite *Custom-House* and *Post-Office* is a handsome structure. The opera-house is one of the finest in New England. The *Bangor Theological Seminary*, situated in the higher part of the city, and several of the churches, are noteworthy edifices. *Norombega Hall*, with seats for 2,000 persons, is on the Kenduskeag Bridge; the lower story is used as a market. A dam thrown across the Penobscot River furnishes water and power in abundant supply for the city. From Bangor the route continues *via* Maine Central R. R. to Mattawamkeag and Vanceboro, connecting at Mattawamkeag with the Canadian Pacific Short Line, running westward past Lake Megantic, famous for its sports with rod and gun, continuing on to Montreal.

Steamers run daily in summer between Bangor and Boston. There is also railway and steamboat connection with Bar Harbor and intermediate points.

Moosehead Lake.

Moosehead Lake, the largest in Maine, lies among the northern hills on the verge of the great Maine forest, and may be reached from Montreal by the Canadian Pacific Ry. It is 38 miles long, and at one point is 14 miles wide, though near the center there is a pass which is not more than a mile across. It is 1,023 ft. above the sea, into which it empties by way of the Kennebec River. Its waters are deep, and furnish ample occupation to the angler in their stores of trout and other fish. The best time for visiting Moosehead Lake, or any portion of the Maine woods, is from the 15th of May to the 15th of June (*before* "fly-time"), and from August 10th to October 10th (*after* "fly-time").

Greenville is a small hamlet on the S. shore, with several small but good hotels, and the only permanent settlement on the lake. Small steamers leave Greenville daily for Mt. Kineo and the other end of the lake, the passage to which affords a panoramic succession of fine scenery. On the W. side **Mt. Kineo** (1,958 ft.) overhangs the water with a precipitous front over 800 ft. high. On a long peninsula, jutting out from its base into the lake, the popular *Mt. Kineo House*, a large and fine hotel, is situated, and close by are the best fishing-grounds on the lake. The mountain is easily ascended (with a guide) from the hotel, and its summit reveals a picture of forest beauty well worth the climbing to see. The lake is visible from end to end, and to the northeast Katahdin stands out in massive grandeur against the horizon. About 18 miles N. of Mt. Kineo the landing place at the end of the lake is reached, whence a portage, 2 miles long, leads across to the Penobscot River. This river may be descended in canoes in 7 to 10 days to Oldtown, and for those who enjoy roughing it the journey will prove a genuine "experience." "Birches," as the boats are called, and guides may be procured either at Greenville or at the Mt. Kineo House. By this approach *Mt. Katahdin* (5,385 ft. high) is seen in much finer outline than from the E., and may be ascended from the river with the canoe-guides.

25. Portland to the Rangeley Lakes.

THE route is *via* the Maine Central R. R. to Farmington, thence by Sandy River R. R. to Phillips (18 miles), and thence by the Phillips & Rangeley R. R. to Rangeley (29 miles). As far as Brunswick the route is the same as Route 23 *b.* Beyond Brunswick, a number of small villages are passed, lying amid a rich farming and grazing region. At *Livermore Falls* the Androscoggin River is reached. From *Wilton* stages run daily to *Weld,* a small village on the shore of a mountain-surrounded lake. **Farmington** is a frontier town of 3,207 inhabitants. (Another route from Portland to Farmington is *via* Lewiston, and is 10 miles shorter than the preceding.)

The Sandy River R. R. (2 ft. gauge) connects with Maine Central and traverses between Farmington and Phillips one of the most beautiful sections of the State, passing as it does the entire distance along the banks of the Sandy River through what is called the "Garden of Maine," on account of its splendid farms. **Phillips** (*Barden House*) is about half-way, and is an attractive resort, being near some trout-streams, and with *Mt. Blue,* 5 miles, and *Saddleback Mountain,* 8 miles distant, both of which command fine views. The latter is 4,000 ft. high, and from its summit may be seen the whole Rangeley region, the White Mountains, the valley of the Upper Kennebec, and portions of Canada. Phillips is now connected with Rangeley by means of the narrow gauge Phillips & Rangeley R. R., which runs through a beautiful country, commanding magnificent views of mountain scenery.

The Rangeley Lakes.

This remote and romantic series of lakes lies in the N. W. corner of Maine, within the borders of its great forest-region, and in what is perhaps the most picturesque portion of the State. It consists of several distinct lakes connected by narrows and streams, extending from the Oquossoc or Rangeley Lake (1,511 ft. above the sea) to Lake Umbagog (1,256 ft. above the sea), forming one continuous water-way for a distance of nearly 50 miles ; embracing 80 square miles of water-surface, and abounding in blue-back trout and other game-fish. Each lake has its individual name, but the chain is known collectively as "The Rangeley Lakes"; and there is probably no equally accessible portion of the country which offers such attractions to sportsmen, and especially to trout-fishers. It is claimed that there are two distinct species of trout in these waters, one of which is found nowhere else, and produces specimens weighing as much as 10 pounds, while the smaller kind is caught with an ease and in quantities which can be equaled in no other known locality.

At the head of Rangeley Lake, 37 miles from Farmington, is **Rangeley** (*Rangeley Lake Hotel*), which draws many summer visitors. From this point connection is made twice daily by two little steamers, with the *Mountain-View House,* at the foot of the lake, and with the *Outlet,* whence a short and easy "carry" leads to Indian Rock, the headquarters of the Oquossoc Angling Association. *Indian Rock* is a famous old Indian camping-ground, and is the favorite resort of sportsmen, being the most central point of the region, and within half a mile of the lakes Mooselucmaguntic and Cupsuptic. Lake Oquossoc, or Rangeley, is 7 miles long and 2 miles wide at the widest part; and is surrounded by forest-clad hills. Lake Mooselucmaguntic is the largest of the series, and is 10 miles long and 2 to 4 wide. At *Haines' Landing* is the *Mooselucmaguntic House*. Four small steamers ply on the lakes, forming an almost continuous and connecting line from the head of Rangeley to the foot of Umbagog. Traveling in this remote wilderness is difficult, and guides should be procured by those who leave the more frequented localities. **Upton** is a small town of 232 inhabitants at the foot of Lake Umbagog. A small steamer runs thence in 13 miles to *Errol Dam* (*Umbagog House*), a lumbermen's village in New Hampshire, at the head-waters of the Androscoggin.

Another route from Portland to the Rangeley Lakes is *via* Grand Trunk R. R. to *Bethel*, and thence by stage to *Cambridge*, at the foot of Lake Umbagog (distance, 26 miles ; fare, $2.50) ; or to *Andover*, and thence to the S. arm of Richardson Lake, connecting there with steamer running in connection with others for the Upper Lakes. The railroad journey is described in Route 19 *b*. The stage-route traverses a wild but picturesque region.

26. Portland to Montreal and Quebec.

Via Grand Trunk R. R. Distance to Montreal, 297 miles. To Quebec, 317 miles.

THE Grand Trunk R. R. is an important thoroughfare, and connects the maritime city of Portland with the St. Lawrence and the Great Lakes of the interior. Its route traverses a fertile and productive country, for the most part under fine cultivation, the streams in its vicinity affording to the manufacturer water-power of the greatest value, and to the tourist a variety of picturesque and romantic scenery. As far as **Gorham** (91 miles), the entrance from this direction of the White Mountain region, it has already been described in Route 19 *b*. Beyond Gorham it follows the line of the Androscoggin and the Upper Ammonoosuc to *Groveton Junction* (122 miles), and thence passes into the valley of the Connecticut, reaching the banks of that river at *North Stratford* (134 miles), the last station in New Hampshire. From N. Stratford connection is made with the Quebec Div. of the Maine Central R. R., which passes to *Colebrook* (*Monadnock House*), near which is *Upper Mt. Monadnock*, and from which it is easy to reach the * **Dixville Notch,** 10 miles S. E. This remarkable pass is much narrower than either of the great Notches in the White Mountains, and no portion of the White Mts. surpasses it in sublimity or in a certain desolate and wild grandeur. It is 1½ mile long, and about half-way through is a lofty projecting pinnacle called *Table Rock*, 600 ft. high, from which one can look into Maine, Vermont, and Canada. The *Dix House* is a summer hotel at the mouth of the Notch.

Beyond N. Stratford the route enters Vermont and passes in 15 miles to **Island Pond,** where the railway company has erected handsome buildings, and where the border custom-house is located. Eleven miles beyond, the train passes *Norton Mills* (160 miles), and enters the Dominion of Canada. At *Lennoxville* (193 miles), connection is made with the Boston & Maine R. R., and 3 miles beyond is **Sherbrooke,** the most important station between Portland and Montreal, only 16 miles from Lake Memphremagog (see Route 30). At *Richmond* (221 miles) the Quebec Branch diverges, while the main line runs almost due west in 76 miles to Montreal (297 miles). *St. Hyacinthe* (262 miles) is a quaint old French-Canadian city on the Yamaska River, with a fine cathedral and famous Jesuit college. At *St. Lambert*, the train crosses the St. Lawrence on the magnificent Victoria Bridge and enters **Montreal** (see Route 60).

The *Quebec Branch* runs N. E. from Richmond to Quebec in 96 miles, traversing a thinly populated but picturesque region, and stopping at a number of small stations, of which the principal are *Danville* (12 miles from Richmond) and *Arthabaska* (32 miles). From the latter a branch road runs in 35 miles to *Doucet's Landing*, on the St. Lawrence, which is connected with *Three Rivers* on the N. bank of the river by means of a ferry. The train stops at *Point Levi*, opposite Quebec, and passengers cross the St. Lawrence in ferry-boats. **Quebec** (see Route 60).

27. Boston to Montreal via Lowell and Concord.

Via the Boston & Maine, Concord & Montreal, and Central Vermont Railroads. Distances : to Lowell, 26 miles ; to Concord, 75 ; to Montreal, 334.

THIS route traverses the most populous portion of three States, passing very many cities, towns, and villages, of which only the most important can be even mentioned. **Lowell** (*American, Merrimac, Washington*) is 28 miles from Boston, and is the third city of Massachusetts in point of population (77,696), and one of the most noted manufacturing cities in the Union. It is situated on the Merrimac, at the mouth of the Concord, and the source of its prosperity is the Pawtucket Falls in the Merrimac, which have a descent of 30 ft., and furnish water-power to the extent of about 10,000 horse-power. The city is regularly laid out and well built, Belvidere, the E. part, being the handsomest portion. The principal public buildings are the *Court-House*, the *City-Hall*, and several fine churches and school-houses. The vast mills are among the most noteworthy structures. There are several tastefully ornamented public squares ; and in one of them (on Merrimac St.) is a monument erected to the memory of Ladd and Whitney, who fell in the attack upon the 6th Massachusetts in Baltimore, April 19, 1861. Near this monument is * Rauch's fine bronze statue of "Victory," erected as a memorial of the Lowell men who fell in the civil war. Beyond Lowell the line follows the Merrimac to Concord, entering New Hampshire just beyond *Tyngsboro* (38 miles), and soon after reaching **Nashua** (*Hotel Dexter, Latin Hotel, Tremont House*), a pretty manufacturing city of 19,311 inhabitants, at the confluence of the Merrimac and Nashua Rivers. Here the cars pass on to the tracks of the Concord & Montreal R. R., and in 17 miles reach **Manchester** (*Elva House, Hotel Belmont, Manchester House*, and *Windsor Hotel*), the largest city of New Hampshire, with a population in 1890 of 44,126, and extensive manufactures, chiefly of prints. The water-power is furnished by a canal around the Amoskeag Falls of the Merrimac, and on the canal are located the immense factories. In the city are a number of neat public squares, several fine churches, and a public library with 20,000 volumes.

Nine miles beyond Manchester is *Hooksett*, the site of several cotton-factories and extensive brick-yards. Here the Merrimac is crossed on a bridge 550 ft. long. W. of the town is *Pinnacle Mountain*, the summit of which commands broad views. Nine miles from Hooksett is **Concord** (*Eagle Hotel*), the capital of New Hampshire, handsomely built on the sloping W. bank of the Merrimac River, with streets regularly laid out and shaded with an abundance of trees. The city is celebrated for its carriage-manufactories and for the superior quality of the granite quarried in the vicinity, some of the finest structures in the country being built of it. Main St. and State St. are the leading streets. The * *State Capitol* is a fine building of Concord granite, situated in a square bounded by Main, State, Park, and Capitol Sts. The *City Hall and Court-House* is a brick structure on Main St., N. of the Capitol. The *State Prison* is a granite building on Main St., and the *Asylum for the Insane* has handsome buildings in the W. part of the city. The population of Concord in 1890 was 17,004.

At Concord the train takes the Boston & Maine R. R. (Concord Div.), and passes in 69 miles to *White River Junction*, with numerous small stations *en route*. Near *Franklin* (19 miles) Daniel Webster was born in 1782. From *Potter Place* (31 miles) stages run in 4 miles to **Mt. Kearsarge,** from the summit of which (2,943 ft. above the sea) there is a noble view. (This must not be confounded with the White Mountain peak of the same name, described on page 121.) About half-way from base to summit is the Winslow House, a commodious summer hotel. At *W. Lebanon* (67 miles) the train crosses the Connecticut, on a bridge which commands fine views of the river, and enters **White River Junction,** the converging point of 4 important railroads. There is a good restaurant in the depot, and trains usually stop long enough for a meal to be eaten.

At White River Junction the Central Vermont R. R. is taken, and the train passes on into Vermont, following the White River for 25 miles, and crossing it several times. Sharon Station (13 miles from the Junction) is opposite the village of *Sharon*, where Joseph Smith, the founder of Mormonism, was born in 1805. The scenery now becomes more bold and rugged, the hills increase in height, and beyond *W. Randolph* (32 miles) the highest peaks of the **Green Mountains** come into view, on the left. At *Roxbury* (46 miles) the road leaves the White River, and, crossing the summit of the pass (1,000 feet above the sea), reaches the source of Dog River and descends to *Northfield* (53 miles), where is located the Norwich University, a military college. Ten miles beyond Northfield is *Montpelier Junction*, whence a short branch road runs to **Montpelier** (*Pavilion Hotel*), the capital of

Vermont, beautifully situated on the Winooski River, in a narrow valley surrounded by hills. The village is compactly built, and has a population of about 4,160. The * *State Capitol* is a fine edifice of light-colored granite, in the form of a cross, the main building being 72 ft. long, and each of the wings 52 ft. The main building is 113 ft. deep, and 124 ft. high to the top of the dome, which is surmounted by a graceful statue of Ceres. The entrance is approached from a Common by granite steps in terraces. In the portico is a marble statue of Ethan Allen, by the Vermont sculptor, Larkin G. Mead ; and in the building are historical and geological cabinets, a State Library with 15,000 volumes, and the flags carried by the Vermont volunteers during the civil war. *Mt. Hunga* is 7 miles from Montpelier, and from it may be had a very fine view. A carriage-road has been constructed to within ½ mile of the summit. The picturesque *Benjamin's Falls* are within a mile of Montpelier.

From *Waterbury* (7 miles beyond Montpelier) stages run in 10 miles to **Stowe** (*Mt. Mansfield House*), a much-frequented summer resort, delightfully situated on a plain surrounded by noble mountain scenery. Favorite excursions from Stowe are to *Moss-Glen Falls* (3 miles), *Gold Brook* (3 miles), *Bingham's Falls* (5 miles), *Morrisville Falls* (8 miles), and *Smuggler's Notch* (8 miles). The latter is a wild and picturesque pass between Mts. Mansfield and Stirling. But the great excursion is to the top of * **Mt. Mansfield,** the loftiest peak of the Green Mountains (4,389 ft. high). Its summit, as seen from Stowe, is likened to the upturned face of a giant, showing the Forehead, Nose, and Chin in three separate peaks. The *Nose* has a projection of 400 ft., and the *Chin* all the decision of character indicated by a forward thrust of 800 ft. A good carriage-road carries the tourist from Stowe to the *Summit House* at the base of the Nose, whence a steep and rugged path leads to the top, the view from which is little if at all inferior to that from Mt. Washington. The Chin is 400 ft. higher than the Nose, and may be ascended from the Summit House by a path 2 miles long. The view is in all respects similar to that from the Nose. One night, at least, should be spent at the Summit House in order to enjoy the glorious sunrise and sunset views.

From *Ridley's Station* (5 miles beyond Waterbury) carriages run in 6 miles to **Camel's Hump,** the second highest of the Green Mountain peaks (4,077 ft. high). A carriage-road extends about half-way to the summit, and the remainder of the ascent may be made either on horseback or on foot. The view closely resembles that from Mt. Mansfield, but this noble peak itself now forms one of the most striking features of the landscape. The beautiful *Bolton Falls* are near Ridley's Station.

Beyond this point, the route traverses the picturesque valley of the Winooski, and at *Williston* (91 miles) emerges into a more open country. On the right of the cars are now visible the summits of the Green Mountains ; on the left, beyond Lake Champlain, those of the Adirondacks. At *Essex Junction* (94 miles) a branch road runs in 8 miles to *Burlington* (see p. 149). (see p. 149). The main line continues N. with the Green Mts. constantly in view on the right, and Lake Champlain frequently in sight on the left. **St. Albans** is 121 miles from White River Junction, and 265 miles from Boston. It is built upon an elevated plateau 3 miles from Lake Champlain, and is one of the prettiest villages in the country. "St. Albans," says Mr. Beecher, "is a place in the midst of greater variety of scenic beauty than any other I remember in America." The public square of 4 acres in the center of the village is an ornamental ground, surrounded by the principal buildings. The extensive shops of the Central Vermont R. R. are located at St. Albans, and the village is noted as the market-place of the great butter and cheese business of Franklin Co. Magnificent views are obtained from *Aldis Hill* ($\frac{1}{2}$ mile N. E. of the village) and from *Bellevue Hill* (2 miles S. W.). Ten miles N. E. of St. Albans (on the Missisquoi Valley Div.) are the **Missisquoi Springs** and **Sheldon Springs,** both supposed to cure cancer. There are nearly 20 medicinal springs in Sheldon. The hotels are *Congress Hall* and the *Portland House.* Population, 1,365.

Beyond St. Albans the route reaches *Swanton Junction,* whence a branch line diverges to *Rouse's Point* and *Ogdensburg,* passing the *Alburgh Springs,* whose waters are a specific for cutaneous diseases. On the main line to Montreal, 12 miles N. of St. Albans, are *Highgate Springs,* another valued mineral water; and 3 miles beyond, the train crosses the boundary and enters the Dominion of Canada, passing 6 or 8 small stations. At *St. Johns* (42 miles from St. Albans) the Grand Trunk R. R. is taken, and the train passes in 27 miles to **Montreal** (see Route 60).

28. Boston to Montreal via Nashua and Concord and Plymouth, N. H.

Distances : To Nashua, 39 miles ; to Concord, 75 miles ; to Montreal, 342 miles. This route is, perhaps, the pleasantest that can be taken to reach Montreal and the St. Lawrence. Frequent trains leave Boston, the express leaving 7.15 P. M., arriving at Montreal 7.55 A. M., and the morning train leaving Boston 9 A. M., arriving at Montreal 8.35 P. M.

ON leaving *Nashua,* which has been previously described (page 142), the train runs north through Manchester to Concord, and thence due north to Lake Winnepesaukee (page 133), a beautiful sheet of water touched by the railway on its western shores. The views obtained from the cars of this lake and the islands with which it is dotted are one of the attractions which make this route so popular. *Plymouth, N. H.,* is the next important city reached, from which there are branch lines to the summer resorts in the White Mountains, and from this point to Wells River the line runs through a picturesque country, with the White Mountains close on the right of the railway. Sharply defined in the foreground is Lafayette, and to the south the Profile range ; to the east and north, Cherry Mountain and the Lancaster range; while between Lafayette and Cherry tower the grander summits of famed Mount Washington and the Presidential range. The route continues to *St. Johnsbury,* in Vermont, a flourishing town, in which the most noted manufactory is that of Fairbanks's scales (see p. 153). Proceeding north the

10

road follows the Passumpsic River for some 'distance, presenting many
favorable views of the picturesque Green Mountains, until *Newport*, at
the head of Lake Memphremagog, is reached.   Thousands of tourists
from all over the United States and Canada visit Lake Memphremagog.
Lovely islands dot its surface; rugged hills frown down upon it and are
mirrored in its limpid depths, and high above all tower the two famous
promontories of Elephantis and Owl's Head.   The dense forest looks
dark and almost forbidding, but those shadowy woods have re-echoed
the merry laugh and jests of many happy voices whose owners have
gathered at the common point in quest of that closely pursued object—
pleasure.   From Magog a steamer makes a daily trip round the lake,
touching at many points, including Georgeville, the Revere House, near
Elephantis; the Mountain House, at the foot of Owl's Head, and Newport.   This excursion by steamer forms a most enjoyable side-trip, for
the tourist can stop at Magog, make the circuit of the lake and see its
many beauties, and from Newport go by Boston & Maine R. R. to
either Boston, the White Mountains, or to Montreal.   From the steamer
one has a fine view of the lake, its picturesque surroundings and
islands, the numerous handsome summer residences upon the shores,
and all points of interest, including the mountains.   From Newport the
line turns westwardly to *Richford*, and thence by *Sutton Junction* in
the Province of Quebec, and by Farnham to St. John's and Montreal,
crossing the St. Lawrence River over the Canadian Pacific cantilever
bridge at Lachine.

29. Boston to Montreal via Rutland and Burlington.

Via Fitchburg R. R., and Central Vermont R. R.   Distance to Fitchburg, 50
miles ; to Bellows Falls, 114 miles ; to Rutland, 166 miles ; to Burlington, 234
miles ; to Montreal, 329 miles.

LEAVING Boston by the Fitchburg R. R. (depot on Causeway St.
near the Warren Bridge), the train passes Charlestown, Somerville, and
Cambridge (described in connection with Boston), and in 10 miles reaches
Waltham, a flourishing manufacturing town of 18,707 inhabitants on
the Charles River, noted as the site of the Waltham Watch Company's
Works, which are the most extensive in the world.   The first cotton-
mill in the United States was erected at Waltham in 1814.   Near the
village is Prospect Hill (480 ft. high), affording broad views.   Ten miles
farther (20 miles from Boston) is **Concord** (*Thoreau House*), a handsome manufacturing village of 4,427 inhabitants, on both sides of the
Concord River.   Here, on April 19, 1775, the same day as the battle of
Lexington, blood was shed, and the great drama of the Revolution begun.
A granite obelisk, 25 ft. in height, marks the spot.   **Lexington** is 11
miles from Concord by a branch road.   On the village green stands a
monument, erected by the State, to the memory of the 8 men who were
killed in the battle.   *Ayer Junction* (36 miles) was formerly called Groton
Junction, and is a thriving village and railroad center.   **Fitchburg**
(*American House, Fitchburg*) is a busy manufacturing city of 22,037 inhabitants, built along the Nashua River, which affords a fine water-power.

Its principal manufactures are machinery and agricultural implements, paper, chairs, and cotton goods. A bronze monument, in memory of her soldiers who fell in the civil war, has been erected by the city, from designs by Millmore. Rollstone Hill and Pearl Hill, near the city, afford fine views.

At *Westminster* the train passes northward, and several small stations are passed in quick succession, of which the principal is *Winchendon* (18 miles). Just beyond Winchendon the State line is crossed, and the train enters New Hampshire, stopping at *Fitzwilliam* (Cheshire House, Fitzwilliam Hotel), a hilly town, watered by several streams and ponds well stocked with fish. Five miles beyond is *Troy*, whence stages run in 5 miles to * **Monadnock Mountain,** in the town of Jaffrey. It is 3,169 ft. high, and from its summit 40 lakes and a large number of villages are in view, while the scenery immediately around is grand and beautiful. A large summer hotel has been erected half-way up the mountain. **Keene** (42 miles; *Cheshire House, Eagle House*) has thriving manufactures of leather, boots and shoes, furniture, organs, etc., and is said to be one of the handsomest villages in New England. It is built on a flat E. of the Ashuelot River, and has broad and pleasantly shaded streets; population, 7,446. *Walpole* (60 miles) is a pretty village near the base of Mt. Kilburn, much resorted to in summer on account of its scenic attractions. From the summit of Derry Hill an extensive and pleasing view may be had. Four miles beyond Walpole the train crosses the river into Vermont, and stops at **Bellows Falls** (*Commercial House*), a well-known railroad center and summer resort; population, 3,092. The Falls are a series of rapids in the Connecticut, extending about a mile along the base of a high and precipitous hill, known as *Mt. Kilburn*, which skirts the river on the New Hampshire side. At the bridge which crosses the river at this place the visitor can stand directly over the boiling flood; viewed from whence, the whole scene is very effective. In the immediate neighborhood are the *Abendquis Springs,* highly tonic and possessing medicinal properties. *Fall Mountain Hotel* is located near the springs at the base of Mt. Kilburn, and is a pleasant resort for invalids. There is a good path from the hotel to the *Table Rock* on the top of the mountain, from which an extended view of the valley of the Connecticut is had.

From Bellows Falls the route is *via* the Rutland Div. of the Central Vermont R. R., which passes through the marble district, through the Green Mountains, and near the shore of Lake Champlain, affording fine views along nearly the whole line. At *Bartonsville* (9 miles from Bellows Falls) the ascent of the mountains begins, and between this and *Chester* (13 miles) is a deep ravine spanned by a bridge. At *Healdville* (33 miles) the grades become heavy, and in a mile the train reaches *Summit*, the highest point on the line. In the 18 miles between Summit and Rutland there is a descent of 1,000 ft. **Rutland** (*Bardwell House, Bates House*) is a prosperous town of 11,760 inhabitants, at the junction of the present route with the Bennington & Rutland and Delaware & Hudson Systems, 166 miles from Boston, 230 from New York (*via* Harlem Division of N. Y. C. & H. R. R. R. and Benning-

ton & Rutland Railway), and 68 from Burlington. The town is picturesquely situated, contains some fine public and commercial buildings, including the State Workhouse and the extensive Howe Scale Works, has
numerous quarries and marble-works in its vicinity now organized into
one corporation, the largest in the world of its kind, and which dictates
the price of marble throughout the United States, and is a center from
which several pleasant excursions may be made. The road to *Killington Peak* (7 Miles E.) is unattractive, and the ascent arduous, but the
view from its summit, which is 4,221 ft. high, is extremely fine. *Mt.
Ida*, too, is near by, and beyond Killington Peak, as seen from Rutland,
are *Mt. Pico* and *Castleton Ridge*, shutting out the view of Lake Champlain. Another pleasant excursion from Rutland is to the **Clarendon
Springs,** 6 miles distant. These mineral springs are a highly popular
resort, and the hotel can accommodate 250 guests.

Sutherland Falls (6 miles N. of Rutland) is the site of large marble-
works, and 3 miles beyond is *Pittsford*, noted for its beds of iron-
ore and extensive marble-quarries. Seventeen miles from Rutland is
Brandon, a manufacturing village of 3,810 inhabitants, with marble-
quarries, vast deposits of excellent bog iron-ore, and several factories
where mineral paint is made from mines in the vicinity. It is pleasantly situated, near fine scenery, and draws many summer visitors.
From *Salisbury* (10 miles beyond Brandon) stages run in 5 miles to
Lake Dunmore, a lovely mountain-lake, nestling at the foot of
the loftiest range of the Green Mountains, and almost surrounded by
bold hills, seen here in verdant slopes and there in rocky bluff and precipitous cliff. It is about 4 miles long and 1½ mile wide at the widest
part, and its clear and limpid waters afford excellent bathing, boating,
and fishing. On the W. shore is a summer hotel with cottages. The
drives in the vicinity are exceptionally pleasant. Six miles beyond Salisbury is the picturesque and handsomely built village of **Middlebury,**
situated on Otter Creek at some fine falls in that stream, and surrounded
on all sides by most attractive mountain scenery. It has a population
of about 2,793, and is distinguished as the seat of *Middlebury College*,
founded in 1800. The college has 3 large stone buildings in the midst
of extensive grounds, with a library of 14,000 volumes and a small
natural-history collection. The favorite excursions from Middlebury
are to *Belden's Falls* (2¼ miles), to *Lake Dunmore* (8 miles), to *Bristol*
(11 miles), and to *Snake Mountain* (10 miles), from the summit of
which there is a remarkably fine view of the Green Mountain range
from Mt. Mansfield to Rutland, of the clustering Adirondack peaks, of
the northern part of Lake George, and of Lake Champlain, from Ticonderoga to the great bay above Burlington. On the summit are a small
hotel and a wooden tower 80 ft. high. The famous *Bread-Loaf Inn*
at Ripton is reached by stage from Middlebury in 8¼ miles.

Fourteen miles beyond Middlebury (47 from Rutland) is **Vergennes,** the oldest city in Vermont (incorporated in 1783), and one of
the smallest in the Union, with a population in 1890 of 1,773. It is at
the head of navigation on Otter Creek, 8 miles from Lake Champlain,
and near the *Falls*, which have a descent of 37 ft. Commodore

McDonough's fleet, which won the naval battle of Lake Champlain (Sept. 11, 1814), was fitted out at Vergennes.

Burlington (*Van Ness and American House, Burlington Hotel*), the largest city of Vermont, is situated upon the E. shore of Lake Champlain, on ground which rises from the water to a height of 367 ft. The first permanent settlement at Burlington was made in 1783, and it was formerly one of the great lumber-marts of the country. In 1865 the township was divided into the city of Burlington and the town of South Burlington. The city grew rapidly, for an Eastern city, and had in 1890 14,590 inhabitants. It has several of the largest mills in the country for planing and dressing lumber, and extensive manufactories of articles of wood, as of doors, packing-boxes, spools, etc., as also of cotton and marble. The city is regularly laid out and handsomely built, and many of the residences and churches are noticeable for their beauty. The *Cathedral of St. Mary* (Roman Catholic) is a large and striking structure; and *St. Paul's Church* (Episcopal) is a fine old stone building, in the Gothic style, with windows of stained glass. The *Court-House* and the *Custom-House and Post-Office* are handsome buildings, on the public square in the center of the city. The *City Hall* and the *Fletcher Library* (containing 18,000 volumes) are also on the square. On Church St. is a spacious and handsome *Opera-House*, completed in 1879. The depot of the Vermont Central R. R., near the wharf, is an extensive building. Other buildings of interest are the *Lake View Retreat* (a private insane asylum), and the *Providence Orphan Asylum* (Roman Catholic). The *University of Vermont*, whose buildings crown the summit of the hill back of the city, was incorporated in 1791, organized in 1800, and is open to both sexes. The corner-stone was laid by Lafayette in 1825. A statue of Lafayette by J. Q. A. Ward is on the college green. In 1865 the State Agricultural College was united with it. It has a library of 28,000 volumes, and a museum containing upward of 50,000 specimens in natural history. The *Billings Library*, given to the college at a cost of $150,000, contains the collection (28,-950 volumes) belonging to the late Geo. P. Marsh, which cost $25,000, and is the best collection of books in the northern languages in the world. The view from the dome of the university building is superb, and has been pronounced the finest lake-view in America, and is perhaps equaled by the * view obtained from the top of the costly *Mary Fletcher Hospital*, a little to the N. E. The 10 miles width of the lake makes an admirable foreground for the towering Adirondack peaks on the W., while to the E. the chain of the Green Mountains lifts against the sky, and N. and S. lies a great expanse of lake. Near the university is the *Green-Mount Cemetery*, where Ethan Allen lies, under a granite shaft 42 ft. high, surmounted by a marble statue of the old hero. *Lake View Cemetery*, in the N. W. part of the city, directly on the shore of the lake, is one of the finest in the State.

From Burlington to *Essex Junction*, the distance is 8 miles, and the train passes *en route* the picturesque falls of the Winooski River. From Essex Junction the route is the same as in Route 27. The distance from Burlington to Montreal is 95 miles.

30. New York to Montreal and Quebec by the Connecticut Valley.

Via New York, New Haven & Hartford R. R., Connecticut River R. R., Central Vermont R. R., and Boston & Maine R. R. Distances : New York to Springfield, 136 miles ; to Bellows Falls, 220 miles ; to White River Junction, 260 miles ; to Montreal, 617 miles (*via* Sherbrooke), 561 miles (*via* Montpelier and St. Albans); to Quebec, 637 miles. Trains leaving New York at 9 A. M. and 4 P. M. make connection at Springfield for all northern points by way of the Connecticut River Line.

As far as **Springfield, Mass.,** this route has already been described in Route 11. Leaving Springfield, the train passes over level meadow-lands along the Connecticut River, and in 4 miles reaches *Chicopee,* a handsome town of 14,050 inhabitants, noteworthy as the site of the Ames Manufacturing Co., which produces so many fine arms and bronzes. Here were cast the bronze doors of the Senate wing of the Capitol at Washington (see p. 57), and Ball's equestrian statue of Washington, in the Public Garden at Boston. Four miles beyond is **Holyoke** (*Hotel Hamilton* and *Windsor Hotel*), which possesses the greatest water-power in New England, being the site of the great dam of the Holyoke Water-Power Co. The river has a fall here of 60 ft. in ¾ of a mile, and is dammed by an immense structure 1,000 ft. in length and 30 ft. in height, built of wood spiked to the rock of the river-bed and covered with plates of boiler-iron. This dam throws the water into a canal which distributes it to the various factories. Holyoke had 35,637 inhabitants in 1890, is well built, and boasts of one of the finest City Halls in Massachusetts, and of a handsome soldiers' monument. Beyond Holyoke the scenery grows more picturesque, the hills on either side beginning to assume the name and aspect of mountains ; and just beyond Smith's Ferry the train passes between Mt. Tom (on the left) and Mt. Holyoke (on the right), and stops at **Northampton** (*Mansion House* and *The Norwood*), which is said to be the most beautiful village in America. The village is built on a rising grade about a mile W. of the river. Its streets are laid out with picturesque irregularity, and abound in shade-trees of venerable age and noble size ; among them is *Edwards's Elm,* so named after the celebrated Jonathan Edwards. Near the center of the village is the *Smith College for Women,* founded by Miss Sophia Smith, of Hatfield, and endowed with a fund of about $500,000. A large art-gallery, filled with choice paintings and statuary, the gift of a wealthy citizen, a large scientific building, a gymnasium, and a music-hall are connected with the college. The free *Public Library* (with 20,000 volumes) is lodged in *Memorial Hall,* erected in memory of the men of Northampton who fell during the civil war. On an eminence W. of the village is *Round Hill,* where the historian George Bancroft and Joseph G. Cogswell once had a boys' school. On the same hill is the *Clarke Institution for Deaf-Mutes* (endowed with $300,000), and near by are the buildings of the *State Lunatic Asylum.* A *City Hospital* has just been erected and endowed with $50,000. The *Smith Charities* annually disburse the interest of

$1,250,000 for benevolent purposes. The vicinity of Northampton is the most beautiful portion of the Connecticut Valley, and attractive drives lead in all directions. This city is famous as an educational center. Within a radius of 8 miles are located Williston Seminary for Boys, at Easthampton; Amherst and the Massachusetts Agricultural Colleges at Amherst; Smith Academy, at Hatfield; and Mt. Holyoke Seminary, at South Hadley.

On the opposite side of the river from Northampton, 3 miles distant, is * **Mt. Holyoke,** "the gem of Massachusetts mountains." It can be reached by private conveyance, crossing the river at Hockanum Ferry, either by ascending the carriage-road to the top or to the foot of the inclined railway (600 ft. long) by which passengers are carried up the steepest part of the mountain. On the summit, 1,120 ft. above the sea, stand the *Prospect House* and an observatory. The view from the Prospect House has been often pronounced by tourists the finest in America. *Leeds* is a suburb of Northampton, with silk and other manufactures.

The view from *Prospect House* embraces 10 mountains in 4 States, and about 40 villages. "On the W., and a little elevated above the general level, the eye turns with delight to the populous village of Northampton, exhibiting in its public edifices and private dwellings an unusual degree of elegance. A little more to the right the quiet and substantial villages of Hadley and Hatfield ; and still farther E., and more distant, Amherst, with its colleges, observatory, cabinet, and academy, on a commanding eminence, form pleasant resting-places for the eye. Facing the S. W., the observer has before him, on the opposite side of the river, the ridge called Mt. Tom, rising 200 ft. higher than Holyoke, and dividing the valley of the Connecticut longitudinally. The western branch of this valley is bounded on the W. by the Hoosac range of mountains, which, as seen from Holyoke, rises ridge above ridge for more than 20 miles, checkered with cultivated fields and forests, and not unfrequently enlivened by villages and church-spires. In the N. W., Graylock may be seen peering above the Hoosic ; and, still farther N., several of the Green Mountains, in Vermont, shoot up beyond the region of the clouds in imposing grandeur. A little to the S. of W., the beautiful outline of Mt. Everett is often visible. Nearer at hand, and in the valley of the Connecticut, the insulated Sugar-Loaf and Mt. Toby present their fantastic outlines, while far in the N. E. ascends in dim and misty grandeur the cloud-capped Monadnock."

Mt. Tom is about 5 miles S. of Northampton, on the same side of the river. It is 200 ft. higher than Mt. Holyoke, but is comparatively seldom visited on account of the difficulty of the ascent. *Mt. Nonotuck,* the northern peak of the Mt. Tom range, is easily reached from the Mt. Tom Station. On its summit is a well-kept hotel, and the view is nearly if not quite equal to that from Mt. Holyoke.

About a mile beyond Northampton the train passes in sight of **Hadley,** a venerable and interesting old village, lying in the Great Bend of the Connecticut, which here makes a *détour* of 7 miles in order to accomplish a mile of direct distance. Here the regicides, Goffe and Whalley, were long concealed. Hadley is connected with Northampton by a bridge across the river. Here we take our last view of the river until South Vernon is reached, 33 miles distant. From *S. Deerfield* (11 miles from Northampton) a carriage-road leads to the Mountain House on the summit of *Sugar-Loaf Mountain,* a conical peak of red sandstone rising almost perpendicularly 500 ft. above the plain, and commanding broad and pleasing views. This peak is said to have been the head-quarters of King Philip during the Indian wars, and the valley which it overlooks was the scene of some of the bloodiest incidents of those cruel wars. On the battle-field of Bloody Brook, where Captain La-throp with 80 youths, "the flower of Essex County," were drawn into an ambuscade and slain, a monument has been erected. The train passes in sight of the monument, and in 5 miles reaches *Deerfield,* a pretty village near the foot of *Deerfield Mountain,* which is 700 ft. high, and commands a much-admired view. Stages run S. E. in 2 miles to *Sunderland,* whence a carriage-road leads to the summit of *Mt. Toby,* from which another beautiful view may be had. A tower 63 ft. high, containing rooms for a night's lodging, stands on the crest.

Nineteen miles above Northampton is the beautiful village of **Greenfield** (*American House, Mansion House*), with elm-shaded streets and garden-surrounded villas. The hill-ranges in the neighborhood open fine pictures of the valleys and windings of the great river; and the vicinity abounds in delightful drives. This is one of the most popular summer resorts in the valley. Directly E., on the Connecticut, is Turner's Falls, the site of an immense water-power, second only to that of Holyoke; and frequent excursions are made to the Coleraine, Shelburne, and Leyden Gorges. Just beyond *Bernardston* (7 miles from Greenfield) the river again comes in sight, and soon after the train crosses the boundary-line and enters the State of Vermont. From *South Vernon* (14 miles from Greenfield) the summit of Mt. Monadnock (see p. 147) may be seen 30 miles E. through the valley of the Ashuelot. Here the Connecticut River R. R. ends, and the Connecticut River, Vermont Valley & Sullivan County R. R. is taken to **Brattleboro,** 43 miles from Northampton (*Brooks House*), a handsomely-built village on the W. side of the Connecticut at the mouth of Whetstone Creek. The situation is very fine, and the scenery and drives in the vicinity are romantic and pleasing. The Vermont Asylum for the Insane is located here, and numerous factories, including the Estey Cottage-Organ Works, the largest in the world. In the cemetery is a costly monument to James Fisk, Jr., and from Cemetery Hill there is a fine view of the Connecti-cut Valley and of Wantastiquet and Mine Mts. on the opposite side of the river. Across the river (reached by bridge) is the pretty town of *Hinsdale,* New Hampshire.

Twenty-four miles above Brattleboro (several small stations *en route*) is **Bellows Falls,** which has been described elsewhere (see p. 147). From here the train passes N. by *Charlestown, Claremont,* and *Windsor.* The latter is a pretty highland village, with considerable manufactures and trade, and surrounded by attractive scenery. At the Windsor House guides and horses may be procured for the ascent of **Ascut-ney Mountain,** in 5 miles by a good road. Ascutney (or "Three Brothers") is an isolated peak, 3,163 ft. high, and the view from its summit is the finest and most extensive of any in Eastern Vermont. At *White River Junction* (40 miles from Bellows Falls, 154 from Boston, and 260 from New York) the regular Montreal through route diverges from the present route and proceeds *via* Montpelier and St. Albans (see Route 27). Such Montreal passengers as prefer it can continue on present route to either Newport or Sherbrooke, *via* Boston & Maine R. R.

Just beyond the Junction the train crosses White River and passes in 4 miles to *Norwich*, whence stages run in ¾ mile to *Hanover*, *N. H.*, the seat of **Dartmouth College,** one of the most famous institutions of learning in America. It was founded in 1769; and Daniel Webster, Rufus Choate, and Chief-Justice Chase were among its alumni. · The college buildings are grouped around a square of 12 acres in the center of the plain on which the village stands. The most notable are Dartmouth Hall, Reed Hall, Culver Hall, Wilson Hall (containing the library of more than 75,000 volumes), and the new Gymnasium. The college includes, besides the literary department, a medical school, the Thayer School of Civil Engineering, and the New Hampshire College of Agriculture and the Mechanic Arts. The New Hampshire Agricultural Experiment Station is also connected with the college. Still running N. along the boundary between Vermont and New Hampshire, the train passes several small stations, and in 30 miles reaches **Newbury,** one of the prettiest towns in the upper Connecticut Valley. It is built on a terrace 100 ft. above the river, and is much visited on account of its celebrated Sulphur Spring and its beautiful scenery. The great *Ox-Bow* of the Connecticut and *Mt. Pulaski* are both in this township. The latter is easily ascended from the village, and affords a noble view. **Wells River** (4 miles above Newbury) is at the junction of the Maine Central R. R. to the White Mts. (see Route 20 *b*).

The scenery now becomes more rugged and impressive, and fine views are had from the car-windows on either side. Numerous small villages are passed, and then, at the head of the Connecticut Valley, comes **St. Johnsbury** (*St. Johnsbury House, Avenue House*), the most important and attractive town in this portion of Vermont. Many of the dwellings are elegant, there are several fine churches, and the Court-House is a handsome structure. In front of the latter is a Soldiers' Monument designed by Larkin G. Mead. The Athenæum contains 10,000 volumes and an art-gallery. The *Fairbanks Museum of Natural Science*, given to this town by Colonel Franklin Fairbanks, is a fine building, with an endowment ample, with judicious usage, to provide for the maintenance of the museum for all time. Its cost was $250,000. Here the Connecticut Valley ends, and the train passes on through a picturesque hill-country toward Lake Memphremagog. From *W. Burke* (16 miles from St. Johnsbury) carriages run in 6 miles to **Willoughby Lake,** a lovely sheet of water 7 miles long and from ¼ to 2 miles wide, lying between Mt. Annanance (2,638 ft. high) and Mt. Hor (1,500 ft. high), and teeming with muscalonge and trout. A good hotel stands on the lake-shore, whence a bridle-path leads to the summit of Mt. Annanance, the view from which is extremely fine. Other small stations are passed, and the train speedily reaches **Newport** (*Memphremagog House*), at the head of Lake Memphremagog. The village is built on Pickerel Point, and contains 1,673 inhabitants. *Prospect Hill*, just S., affords a fine view of the lake and surrounding elevations ; and *Jay Peak* (12 miles W., 4,018 ft. high) commands a view which includes the Green and White Mts., Lake Champlain, and the Adirondacks. Pleasant excursions are to *Clyde River Falls* (2 miles), *Bear Mountain* (7 miles), and *Bolton Springs* (15 miles). The latter are in Canada.

Lake Memphremagog.

Lake Memphremagog is a beautiful sheet of water, 30 miles long and 2 to 4 miles wide, lying partly in Vermont and partly in Canada. Its shores are rock-bound and indented with beautiful bays, between which jut out bold, wooded headlands, backed by mountain-ranges. Muscalonge, a fish peculiar to these waters, and trout are taken here in perfection. A steamer leaves the pier at Newport every morning and afternoon for Magog, at the other end of the lake. At Magog it connects with the Central Vermont and Canadian Pacific R. Rs., connecting with the main lines from Montreal to the N. W. and from Quebec to the N. E. Tourists can thus proceed northward without returning to Newport. In ascending the Lake, *Indian Point*, the *Twin Sisters*, and *Province Island* are passed within a few miles of Newport. E. of Province Island, and near the shore, is *Tea-Table Island*, a charming rural picnic resort; and on the W. shore the boundary-line between Vermont and Canada strikes the lake. About half-way down the lake, on the W. side, is the *Mountain House*, nestling in a lovely nook at the foot of * **Owl's Head** (3,270 ft. high). A footpath leads from the hotel to the summit, which can be reached in 1 to 2 hours. The view in clear weather is very extensive, including the entire length of Memphremagog, the White Mountains, Lake Champlain, Willoughby Lake and Mountain, the St. Lawrence River, and the white pinnacles of Montreal. At and near the Mountain House are the best fishing-grounds on the lake; and *Fitch's Bay* and *Whetstone Island*, *Magoon Point*, *Round* and *Minnow Islands*, are in the vicinity, affording pleasant picnic and excursion points for visitors sojourning there. *Skinner's Island* and *Cave*, said to have been the haunt of Uriah Skinner, " the bold smuggler of Magog," during the War of 1812, are also near by. *Balance Rock*, on the S. shore of Long Island, is frequently visited. The E. shore of the lake, in this vicinity, is much improved and adorned with some handsome summer villas. About a mile N. of the hotel, on the W. side, is a series of precipitous cliffs 700 ft. high, and the water beneath is of unfathomed depth. *Mt. Elephantis* (or Sugar-Loaf) is seen to advantage from Allen's Landing; its outline is supposed to resemble that of an elephant's head and back. *Concert Pond*, W. of Mt. Elephantis, abounds in brook-trout, and attracts numerous visitors. *Georgetown*, 20 miles from Newport and 12 from Magog, has a hotel and several stores, and is a favorite summer resort with the Canadians. **Magog** is a small hamlet at the N. end of the lake, where the Memphremagog discharges its waters through the Magog River into the St. Francis. There is excellent trout-fishing in the vicinity of Magog; and from the summit of *Mt. Orford*, 5 miles W., and reached by carriage-road, an exceedingly striking outlook is obtained over the somber and far-stretching Canadian forests.

From Newport the Canadian Pacific R. R. runs N. W. to Montreal in 65 miles, passing *Richford*, *W. Farnham*, and *St. Johns*. This road forms a part of the **Boston & Montreal Air Line**, which follows Route 27 to White River Junction, the present route thence to Newport, where the Canadian Pacific R. R. is taken. The total distance from Boston to Montreal by this route is 314 miles.

Soon after leaving Newport an arm of the lake is crossed, and the train speedily passes the frontier and enters Canada, traversing for many miles the Eastern Townships, "as beautiful a tract of country perhaps as any on the continent, both with regard to mountain and lake scenery, beautiful rivers, and fertile valleys." *Massawippi* (20 miles from Newport) is near the lovely and fish-teeming *Lake Massawippi*, and beyond this the train follows the Massawippi River for 16 miles, reaching **Sherbrooke** (40 miles from Newport), an important station on the Grand Trunk R. R. The route from Sherbrooke to Quebec and Montreal is described in Route 26.

31. New London to Brattleboro.

Via New London Northern Div. of the Central Vt. R. R.   Distance, 121 miles.

THIS route crosses the two States of Connecticut and Massachusetts, and forms part of a popular through route from New York to the north. **New London** is described in Route 12. It is connected with New York by the Shore Line Route and by the Norwich Line of steamers. The present route runs N., following the Thames River for 13 miles to **Norwich,** which has already been described (see p. 95). A seat on the right-hand side of the car on this portion of the route will afford some pleasing views. *Willimantic* (30 miles from New London) is a busy manufacturing town at the crossing of the New York & Boston Air Line route (see Route 13). Beyond Willimantic the train follows the Willimantic River, and passing several small stations reaches **Stafford,** celebrated for its mineral springs, one of which is regarded as one of the best chalybeate springs in the United States. The Indians estimated the curative properties of these springs very highly, and the whites have used them for more than a hundred years. The springs and a large hotel (the *Stafford Springs House*) are close by the depot on the W. side of the track. The village is 2 miles distant.

· Ten miles beyond Stafford the train crosses the State line and enters Massachusetts. *Monson* (61 miles from New London) is the first station in Massachusetts, and is near some excellent granite-quarries. *Palmer* (65 miles) is at the junction of the Boston & Albany R. R., and has fine water-power which is extensively used for manufactories. Stations: *Barrett's Junction* at the crossing of the Athol Branch of the Boston & Albany R. R., *Belchertown* (76 miles), and *Amherst* (85 miles). Just beyond Belchertown a fine view of the Connecticut Valley and Mt. Holyoke appears on the W. of the road. **Amherst** (*Amherst House*) is charmingly situated, and is noted for its college, its picturesque surroundings, and its refined and cultivated society. It is irregularly built upon a hill, commanding extensive views, and has a population of 4,512. Its leading interest is paper-manufacturing. *Grace Church* (Episcopal) and the *First Congregational* are fine edifices.

Amherst College was founded in 1821, and is one of the leading educational institutions of New England. Its buildings occupy an eminence on the S. side of the village, and command a prospect of exceeding beauty. The college collections in zoölogy, botany, geology, mineralogy, etc., are among the richest in the country, and are accessible to visitors. The cabinet of minerals collected by Charles U. Shepard is of great value, and is said to be surpassed only by those of the British Museum and the Imperial Cabinet at Vienna; and the collection of 20,000 specimens of fossil tracks of birds, beasts, and reptiles is without a rival. The Memorial Chapel is a fine building, and so are Walker and Williston Halls. The *Massachusetts Agricultural College* has extensive and handsome buildings about a mile N. of the village green, and possesses, besides other objects of interest, the Durfee Plant-House, containing many rare and beautiful plants. Founded in 1866, this institution has become the most successful agricultural school in the country. Amherst is within excursion distance (7 miles) of Northampton and Mt. Holyoke (see p. 151).

Beyond Amherst the scenery is very pleasing, and may be enjoyed from the left-hand side of the cars. From *Leverett* (90 miles) there is an impressive view of Mt. Toby (see p. 151). *Miller's Falls* (100 miles) is at the crossing of the Fitchburg R. R. (Route 25). *Northfield* (109 miles) is an attractive village, and the last station in Massachusetts. Just beyond it, the train crosses the Connecticut River, affording fine views from the bridge, and passes to *South Vernon* (111 miles). From this point the route is the same as that described in Route 30, and passes on to the White Mts., the Green Mts., and Canada.

32. Boston to Plymouth.

Via Old Colony R. R., Kneeland St. Station.   Distance, 38 miles.

FROM Boston the route passes *Braintree* to *South Braintree* (11 miles), and thence to *North Abington* (18 miles). The next station, *Abington* (20 miles), is noted for its shoe-factories. A short branch line runs thence in 7 miles to the ancient town of *Bridgewater*. Beyond Abington the road traverses the great forest and lake region of the Old Colony, skirts the W. shore of Plymouth Harbor, and stops at Plymouth.

Plymouth (*Samoset House*) is a flourishing manufacturing village of 7,314 inhabitants on Cape Cod Bay. Its interest is chiefly historical, and it will be forever famous as the landing-place of the Pilgrim Fathers (Dec. 22, 1620), and as the site of the first settlement made in New England. ***Plymouth Rock,** on which the Pilgrims first landed, is in Water St., and is covered by a handsome granite canopy, in the attic of which are inclosed the bones of several men who died during the first year of the settlement. A portion of the rock has been placed in front of Pilgrim Hall, and surrounded by an iron fence. ***Pilgrim Hall** is in Court St., and contains a large hall, the public library, portraits and busts, and many interesting relics of the Mayflower pilgrims and other early settlers of Massachusetts. Near the Hall are the County Court-House and House of Correction, both fine buildings. · The

Town Green is at the end of Main St. *Leyden St.*, the oldest street in New England, runs E. from Town Square to the water. The **＊Burying Hill,** where some of the Pilgrims were interred, is a place of much interest. It contains some ancient and venerable tombs, and commands a wide view. *Cole's Hill*, W. of the canopied rock, is noted as the spot where nearly half the Mayflower pilgrims were buried the first winter; but no trace of their graves remains. The *National Monument to the Pilgrims*, the corner-stone of which was laid Aug. 1, 1859, stands on a high hill near the Samoset House. It consists of a granite pedestal 40 ft. high, surrounded by statues 20 ft. high; and surmounted by a colossal granite statue of "Faith," 40 ft. high. The environs of Plymouth are very attractive, and in the township are about 200 ponds, one of the largest, *Billington Sea*, being stocked with fish. It is about 2 miles from the village.

In returning, after leaving Plymouth, the first station of importance is *Kingston* (4 miles, and 42 miles from Boston), where the South Shore branch turns to the north along the line of the coast. Kingston is connected with the harbor by Jones River, and is a pleasant summer residence, with fine drives in all directions. Six miles farther brings us to *Duxbury*, a finely situated town on bay and harbor, with a succession of beaches. It has interesting historical associations, and this section of the old colony was allotted to John Alden, the youngest of the Pilgrims. An old house of his, built 225 years ago, is still standing. *Marshfield* (34 miles from Boston) was the home of Daniel Webster, and the old homestead is a great attraction for visitors. The celebrated "Brant Rock," a noted gunning station, is in the immediate vicinity. Then *Marshfield Center* and *East Marshfield* are passed, and the quaint old shore town of *Scituate* is reached. The shores rise into a succession of four sand-cliffs, and form landmarks for mariners as well as objects of interest in scenery. There is fine bay-fishing, with boating and gunning for sea-fowl in the vicinity. Five miles beyond is *Cohasset* (22 miles from Boston), the summer home of Robson, Crane, and other actors, and where Lawrence Barrett spent his vacations. Near by is the noted Minot Ledge, upon which is built one of the famous lighthouses of the world. Passing westward *Old Colony House* is reached, where a branch runs N. along the *Nantasket Beach* to *Pemberton*, a distance of 4 miles. The beach is a great natural breakwater that completely protects the lower Boston Harbor. Nantasket is also connected with Boston by steamboat, and is much frequented during summer by pleasure-seekers from the metropolis. Resuming the main line we pass in quick succession *Hingham*, originally settled in 1635, and where the oldest occupied church in the United States is, then *Weymouth* (12 miles from Boston), where, in 1623, occurred the attack of Miles Standish upon the assembled Indian chiefs; and *Braintree*, an old historic town, where Thayer Academy is situated. *Quincy*, a beautiful town of 16,723 inhabitants, noteworthy as the home of the Adams and Quincy families, is reached, and eight miles farther, after crossing Neponset River, we again reach Boston. The return route is some seven miles longer than the line followed in going to Plymouth.

33. Boston to Cape Cod.

Via Old Colony R. R. main line, Cape Cod Division. Distance, Boston to Provincetown, 120 miles. Fare, $3.

As far as *S. Braintree,* this is the same as described in Route 31. The next station is *Holbrook* (15 miles), a small manufacturing village, and 5 miles farther is **Brockton,** a prosperous town of 27,294 inhabitants, with extensive factories of shoes, furniture, carriages, etc. **Bridgewater** is 27 miles from Boston, and is the site of extensive iron-foundries, rolling-mills, machine-shops, and brick-yards. The Bridgewater Iron-Works are among the largest on the continent. At *Middleboro* (36 miles) the Cape Cod Div. of the Old Colony R. R. begins. Stations: *S. Middleboro* (42 miles), *Tremont* (45 miles), *Wareham* (49 miles), and *Buzzard's Bay Station* (54 miles). From the latter, a branch road diverges to Wood's Holl, whence steamers run to Martha's Vineyard (see Route 34). The present route continues on past the small stations of *Bourne* (58 miles), *Bournedale* (59 miles), and *Sandwich* (62 miles). At Sandwich the Cape begins, and extends E. about 35 miles, with a width rarely exceeding 8 miles, and then bends N., and gradually N. W., extending about 30 miles farther. The curve still continues around to the W., S., and E., inclosing the fine landlocked harbor of Provincetown. This latter portion does not average half the width of the former, and is greatly indented by bays both on the outer and inner sides.

"The ride throughout the Cape," says Samuel Adams Drake, "affords the most impressive example of the tenacity with which a population clings to locality that has ever come under my observation. To one accustomed to the fertile shores of Narragansett Bay or the valley of the Connecticut, the region between Sandwich, where you enter upon the Cape, and Orleans, where you reach the bend of the forearm, is bad enough, though no desert. Beyond this is simply a wilderness of sand. The surface of the country about Brewster and Orleans is rolling prairie, barren, yet thinly covered with an appearance of soil. Stone-walls divide the fields, but from here down the Cape you will seldom see a stone of any size in g ing 30 miles. . . . Eastham, Wellfleet, and Truro grow more and more forbidding, as you approach the *Ultima Thule,* or land's end. . . . . It was something to conceive, and more to execute, such a tramp as Thoreau's (from Orleans to Provincetown on the ocean side of the Cape), for no one ought to attempt it who can not rise superior to his surroundings, and shake off the gloom the weird and widespread desolateness of the landscape inspires. I would as lief have marched with Napoleon from Acre, by Mt. Carmel, through the moving sands of Tentoura."

Seven miles beyond Sandwich is *W. Barnstable,* whence stages run in 6 miles to *Cotuit Port,* on the S. shore, a favorite resort of sportsmen. *Yarmouth* (75 miles) is near a camp-meeting ground, and is the junction of a branch road which runs in 4 miles to **Hyannis** on the S. shore, which is becoming a popular summer resort. Beyond Yarmouth are the small stations of *S. Yarmouth* (80 miles), *Harwich* (84 miles, with branch to *Chatham*), *Brewster* (89 miles), *Eastham* (97 miles), and *Wellfleet* (109 miles). Near Truro (114 miles) is one of the most fatal beaches on the New England coast; and on Clay Pounds, on the outer shore of Truro, is the famous *Highland Light,* 200 ft. above high-water mark, and provided with Fresnel lenses. Six miles beyond Truro is

Provincetown, a thriving fishing village, with a magnificent land-locked harbor, which is frequently crowded with shipping seeking a haven of refuge. Near here are the principal cod and mackerel fisheries on the coast, and nearly all the inhabitants are in one way or another connected with the sea-going business. From the summit of High Pole Hill there is a fine view, with the Atlantic Ocean on one side and Massachusetts Bay on the other. *Race Point* is the outermost land of the Cape, and has a revolving light 150 ft. above high water. It is reached from Provincetown by a walk of 3 miles across the sand-dunes. "Standing here," says Mr. Drake, "I felt as if I had not lived in vain. I was as near Europe as my legs would carry me, at the extreme of this withered arm with a town in the hollow of its hand. For centuries the storms have beaten upon this narrow strip of sand, behind which the commerce of a State lies intrenched. The assault is unflagging, the defense obstinate. Fresh columns are always forming outside for the attack, and the roll of ocean is forever beating the charge. Yet the Cape stands fast, and will not budge."

34. Boston to Martha's Vineyard and Nantucket.

Via Old Colony R. R. to Wood's Holl (78 miles), and thence by steamer.

FROM Boston to *Buzzard's Bay Station* (54 miles) this route is the same as Route 33. Beyond, the train runs along the shore of Buzzard's Bay, passing the small stations of *Monument Beach, Pocasset,* and *Falmouth.* A mile S. E. of the latter is the popular summer resort of **Falmouth Heights** (*Tower's Hotel*), a line of high and picturesque bluffs fronting on Vineyard Sound, with a good beach and other attractions. At *Wood's Holl* (72 miles from Boston), where the Marine Biological Laboratory was founded in 1888, connection is made with the steamer for Martha's Vineyard, 7 miles distant.

Martha's Vineyard.

From New York Martha's Vineyard is reached *via* Fall River steamers to Fall River (Route 15), thence *via Fall River Branch* of Old Colony R. R. to *New Bedford,* and thence by a steamboat sail of 30 miles to Cottage City. Total distance, 225 miles ; fare, $4. Another route is by steamer to Fall River, as before, thence by Old Colony R. R. to Wood's Holl, where the steamer from New Bedford calls.

Hotels.—The *Narragansett, Naumkeags, Pawnee, Sea View,* and *Wesley* are the leading hotels. Prices, from $2 to $4 a day. Smaller hotels, in the village of Oak Bluffs, are the *Island,* the *Central,* and the *Wesley.*

Martha's Vineyard is an island 20 miles long and 6 miles in average width, lying off the S. coast of Massachusetts, and separated from the mainland by Vineyard Sound. Its surface is generally level, though there are elevations rising to the height of 150 ft. above the sea. The soil is generally light, and a great part of the surface is covered with low forests. The inhabitants, of whom there were 4,369 in 1890, are chiefly engaged in navigation and fishing. Martha's Vineyard was discovered by Bartholomew Gosnold in 1602, was settled by Thomas Mayhew in 1642, and suffered much from the British during the Revolutionary War. Of late years it has become noted for its annual camp-meetings and as a summer resort.

Near the Sea-View Hotel is the great Methodist * **Camp-Meeting Ground,** where 20,000 to 30,000 people are gathered every August. The grounds are regularly and tastefully laid out, and comprise a tabernacle capable of seating 5,000 persons. E. of the camp-ground, on bluffs 30 ft. high, overlooking the sea, the village of **Oak Bluffs** was laid out in 1868, and has become a popular summer resort; besides the hotels named above, it contains numerous cottages of summer residents. A narrow-gauge railway from Cottage City connects Oak Bluffs with Edgartown and Katama, and the *Sea-View Boulevard*, an admirable drive along the coast, runs to the same places. **Edgartown** (*Ocean-View Hotel* and *Sea-Side House*) is a neat village, 6 miles E. of Oak Bluffs, containing several churches, a town-hall, the county buildings, and the Martha's Vineyard National Bank. Its harbor is well sheltered, and at the entrance is a lighthouse showing a fixed light 50 ft. above the sea, erected on a pier 1,000 ft. long. Beyond Edgartown the railway and the boulevard extend to **Katama Bay,** noted for its clam-bakes and for its attractive scenery. The *Mattakesett Lodge* here is one of the best hotels on the island. A short distance W. of Oak-Bluffs is the East Chop Light, whence a fine view is obtained of **Vineyard Haven,** one of the most celebrated harbors on the coast. * **Gayhead,** the westerly end of Martha's Vineyard (20 miles from Oak Bluffs), is a spot well worth the attention of the visitor. It was pronounced by Edward Hitchcock one of the most remarkable geological formations in America.

Nantucket.

Nantucket is about 30 miles from Martha's Vineyard, whence it is reached by steamer twice a day during the summer and once in winter from Wood's Holl. The island is of an irregular triangular form, about 16 miles long from E. to W., and for the most part from 3 to 4 miles wide. It has a level surface in the S., and in the N. is slightly hilly. The soil is light, and, with the exception of some low pines and the shade-trees in the town, the island is treeless. Farming and fishing are the chief occupations of the inhabitants (of whom there were 3,268 in 1890), the surrounding waters abounding in fish of various kinds. The climate in summer is remarkably cool, and the island is fast becoming a favorite summer resort.

The town of **Nantucket** (*Bay-View, Nantucket, Ocean House, Springfield Hotel, Surfside House,* and *Veranda Hotel*), on the N. of the island, was at one time the chief whaling-port of the world, and increased rapidly in size and prosperity until 1846, when it was visited by a severe conflagration that destroyed nearly a million dollars' worth of property. After this the whale-fishery, and with it the prosperity of the town, rapidly declined; and until the stream of summer visitors began to flow in, it had a distinct air of decrepitude and decay. It is picturesquely situated, and presents an appearance from the water which is hardly confirmed on closer scrunity. The streets are cleanly, and, having trees and flower-gardens, are often pretty and cheerful. The roofs of many houses are surmounted by a railed platform, a reminder of old whaling-times. The town contains 9 churches, a town-hall, a national bank, a savings-bank, 5 public halls, a custom-house, and several good public schools. Among the schools is the celebrated Coffin School, founded and endowed by Sir Isaac Coffin, Bart. In the *Athenæum* is a public library of 5,000 volumes, and some interesting relics of whales and whaling. Three excursions must be made from the town before one can say that he has "seen Nantucket." One is to a cliff at the North Shore whence a wide view is had ; and another to the beaches of the South Shore, where the waves roll in grandly after a storm. The third is **Siasconset** (pronounced Sconset), a quaint fishing hamlet on the S. E. shore of the island, 11 miles from Nantucket, with which it is connected by railroad. It contains *Atlantic House* and *Ocean-View Hotel*, and a number of cottages for summer residents. On *Sankoty Head*, 1 mile N. of Siasconset, there is a lighthouse, and from the eminence on which it stands the Atlantic is visible on all sides of the island.

35. Boston to Hoosac Tunnel and Troy.

Via Fitchburg R. R. Distance, to Hoosac Tunnel, 135 miles ; to North Adams, 143 miles ; to Troy, 191 miles.

As far as and including **Fitchburg** (50 miles), this route has been described in Route 29. (There is another division of the Fitchburg R. R., passing through Framingham, S. Marlboro, and Clinton, by which the distance from Boston to Fitchburg is only 37 miles.) Soon after leaving Fitchburg the train reaches *Wachusett*, whence stages run in 6 miles to **Mt. Wachusett,** from the summit of which (2,018 ft. high) there is one of the grandest views to be obtained in all New England. It is said that 300 villages and portions of 6 States are included in it. Stations, *Ashburnham* (60 miles), *Gardner* (65 miles), *Baldwinville* (71 miles), and *Athol* (82 miles). *Miller's Falls* (97 miles), where the Connecticut and Deerfield Rivers are crossed, and the beautiful village of *Greenfield* (105 miles) have been described in Route 30. Beyond Greenfield the route follows the Deerfield River, passing amid extremely picturesque scenery, the most striking feature of which is the narrow and romantic * Deerfield Gorge, traversed just before reaching the village of *Shelburne Falls* (119 miles). At Shelburne Falls the Deerfield River makes a descent of 150 ft. in a few hundred yards, roaring through a narrow channel. The scenery beyond is very charming, and at *Charlemont* (128 miles) the Hoosac Mountains are in full view. Passing now for 8 miles through a savage, rugged, and desolate region, the train stops for a moment at Hoosac Tunnel Station, and then plunges into the profound darkness of the tunnel. The * **Hoosac Tunnel** is the longest in the United States, and is one of the most wonderful achievements of modern engineering. It pierces the solid micaceous slate of the Hoosac Mountain, is 4¾ miles long, was nearly 20 years in constructing (1855 to 1874), and cost the State of Massachusetts about $16,000,000. The cut-stone façade of the entrance is worthy of notice.

11

Just beyond the W. end of the tunnel is the town of *N. Adams,* and the route continues through Williamstown (see p. 169), Pownal, Petersburg, Hoosic Falls, Johnsonville, and Lansingburg to **Troy** (in 48 miles). This forms a through route between Boston and the West, to obtain which the vast expenditures for the Hoosac Tunnel were incurred.

36. Boston to Albany and the West.

Via Boston & Albany R. R.   Distance to Albany, 202 miles.

THIS is the most popular passenger route from Boston to the West, and traverses some of the most attractive portions of New England. The Boston & Albany R. R. was among the first constructed in America, being completed to Worcester in 1835, to Springfield in 1839, and to Albany in 1842. Its completion was the occasion of festivities in both Boston and Albany, the memory of which has not yet died out. As far as **Springfield** (99 miles) the route is the same as Route 11 taken in reverse. Immediately after leaving Springfield the train crosses the Connecticut River on a long bridge, and follows the Agawam River to **Westfield,** a beautiful town of 9,805 inhabitants, encircled with hills. In the center of the village is a neat public square, surrounded by churches and other buildings, and adorned with a soldiers' monument. The State Normal School here has a wide reputation. The New York, New Haven & Hartford R. R. crosses here. Beyond Westfield the route leads up Westfield River, amid picturesque scenery, which rapidly becomes mountainous. At *Chester* (27 miles from Springfield) the grades become very heavy, and the train enters the Berkshire hills, which are described in Route 38. The scenery along all this portion of the route is extremely fine, and at Summit the track is 1,211 ft. above the sea. *Becket, Washington,* and *Hinsdale* are high-perched mountain towns; and *Dalton* (146 miles from Boston) is a manufacturing village on the W. side of the range of hills that has just been crossed. Five miles beyond Dalton, in the heart of the Berkshire hills, is **Pittsfield** (described in Route 38). Three miles beyond Pittsfield is *Shaker Village,* one of the settlements of the curious sect of Shakers; and a short distance N. of the village is the mountain where, according to tradition, the Shakers hunted Satan through a long summer night, and finally killed and buried him. Eleven miles beyond Pittsfield the road crosses the State line and enters New York State, running by *Chatham,* where connection is made with the Harlem Div. of the New York Central and Hudson River R. R., *Kinderhook,* and *Schodack* to *Greenbush,* whence the train crosses a fine bridge and enters **Albany** (see p. 72).

37. Albany to Rutland.

a. Via Delaware & Hudson R. R.   Distance, 100 miles.

FROM Albany to *Whitehall* (78 miles) this route is described in Route 44. Beyond Whitehall the line runs N. E. to *Fairhaven* (86 miles), where there are extensive slate-quarries, and *Hydeville* (88 miles), a pretty village at the foot of *Lake Bomoseen,* a beautiful sheet of water, 8 miles long and 1 to 1¼ wide, famed for its boating and fishing. Three miles farther is **Castleton** (*Lake Bomoseen House*), a neatly-built village, situated on a plain near the Castleton River, and surrounded by pleasing scenery. The township in which it is located is noted for its slate-stone, which is extensively quarried, and from which is made an imitation of marble "so perfect that it challenges the closest scrutiny." There are 5 churches in the village and a State Normal School. *W. Rutland* (97 miles) is noted for its vast marble-works, and stages run thence in 4 miles to *Clarendon Springs* (see p. 148). **Rutland** (101 miles) has already been described (see p. 147).

b. Via Delaware and Hudson R. R. (*Rutland & Washington Division*).
Distance, 98 miles.

As far as *Eagle Bridge* (23 miles) this route follows the Fitchburg R. R., passing *Lansingburg,* a thriving manufacturing village on the Hudson River, and *Schaghticoke* (13 miles), a manufacturing village on the Hoosic River, which furnishes a fine water-power. *Salem* (41 miles) is a pretty village on White Creek. From this station the road makes a *détour* into Vermont, and runs near the boundary for some miles until at *Granville* (60 miles) it again enters New York, finally leaving the State near Poultney (68 miles). **Poultney** is a beautiful village, noted for its coolness in summer, and then much resorted to. It lies amid varied and picturesque scenery, and the walks and drives in the vicinity are very attractive. Among its many pleasant excursions are those to the *Gorge*, the *Bowl*, *Carter's Falls*, *Lake Bomoseen* (see p. 162), and *Lake St. Catherine* (or Austin). The latter is 3 miles from Poultney, is about 6 miles long, and has a summer hotel at its lower end. Daily stages run in eight miles to **Middletown Springs** (*Montvert Hotel*), one of the most famous mineral springs in Vermont. The waters are impregnated with iron, and are an excellent tonic. Two miles beyond Poultney the present route connects with the preceding one at Castleton, and proceeds to Rutland in 12 miles.

38. The Housatonic Valley and the Berkshire Hills.

The point of departure for the trip up the Housatonic is **Bridgeport**, Conn., which is reached from New York *via* Route 11 (fare, $1.15) or by steamboat daily from Pier 35 East River (fare, 50c.). From Bridgeport to Pittsfield the distance is 111 miles, and the fare $3.00. But the through fare from New York to Pittsfield is only $3.40.

THE Housatonic River rises in Berkshire Co., Massachusetts, and flowing S. enters the State of Connecticut, where, after winding through Litchfield Co., and forming the boundary between New Haven and Fairfield Counties, it meets the tide-water at Derby, about 14 miles from Long Island Sound. The sources of the stream are more than 1,200 ft. above the level of the sea, and in its course of 150 miles it offers some exquisitely beautiful scenery. The railway runs along its bank for about 75 miles. "Of all the railroads near New York," says Mr. Beecher in his "Star Papers," "none can compare, for beauty of scenery, with that from Newtown up to Pittsfield, but especially from New Milford to Lenox."

Bridgeport, our point of departure, has already been described (see p. 77). For some miles after leaving Bridgeport the route traverses a level and thinly-settled country, destitute of picturesque features ; but at *Newtown* (19 miles) the hills begin to show mountainous symptoms, and the traveler obtains glimpses of forest-clad hills and lovely intervales. **New Milford** (35 miles) is a large and beautiful village, with broad, well-shaded streets, and surrounded by delightful scenery. It has some popularity as a summer resort, and is also the site of several manufactories. From New Milford to the terminus of the road, the scenery is ever changing and of rare beauty. **Kent** (48 miles, *Elmore House*) is a quiet little village, with the river running through it, situated in the midst of the charming Kent Plains. Hatch and Swift Lakes or Ponds are visible from the cars ; and on a lofty plateau, W. of Kent, are the *Spectacle Ponds*, a pair of twin lakelets, of oval shape, fringed by dense woods and connected by a narrow strait. From the lofty hill just above them the view is grand. *Cornwall Bridge* (57 miles) is a small manufacturing village surrounded by exquisite scenery. Daily stages run thence to *Litchfield*, said to be the most beautiful village in Connecticut, and to *Sharon*. From *W. Cornwall* (61 miles) stages run to *Goshen*, a pretty highland town, celebrated for its butter and cheese. **Falls Village** (67 miles) is at the Great Falls of the Housatonic, which are the largest and finest in the State, descending 60 ft. over a ledge of limestone. About 2 miles N. W. of the village is *Mt. Prospect* (reached by carriage-road), from the summit of which there is a fine view over the valley and the outlying villages. At the foot of this hill is a deep fissure in the rocks, known as the *Wolf's Den*.

In his "Star Papers," Mr. Beecher writes lovingly of all this region, and we quote a paragraph which may prove useful to the tourist : "If one has not the leisure for detailed exploration, and can spend but a week, let him begin, say, at *Sharon* (reached by stage from Cornwall Bridge on the Berkshire Div. of the New York, New Haven & Hartford R. R.) or at *Salisbury* (reached from *Canaan* by the Central New England & Western R. R., or by stage from Falls Village). On either side to the E. and to the W. ever-varying mountain-forms frame the horizon. There is a constant succession of hills swelling into mountains, and of mountains sinking into hills. The hues of green in the trees, in grasses, and in various harvests, are endlessly contrasted. There are no forests so beautiful as those made up of both evergreens and deciduous trees. At Salisbury, you come under the shadow of the Taconic range. Here you may well spend a week, for the sake of the rides and the objects of curiosity. Four miles to the E. are the Falls of the Housatonic, very beautiful and worthy of much longer study than they usually get. . . . On the W. of Salisbury you ascend *Mt. Riga* to *Bald Peak*, thence to *Brace Mountain*, thence to the *Dome*, thence to that grand ravine and its wild water, *Bash-Bish*, a ride in all of about 18 miles, and wholly along the mountain-bowl. On the E. side of this range, and about 4 miles from Salisbury, is *Sage's Ravine*, which is the antithesis of *Bash-Bish*. Sage's Ravine, not without grandeur, has its principal attraction in its beauty ; Bash-Bish, far from destitute of beauty, is yet most remarkable for grandeur. I would willingly make the journey once a month to see either of them. Just beyond Sage's Ravine, very beautiful falls may be seen just after heavy rains, which have been named *Norton's Falls*. Besides these and other mountain scenery, there are the *Twin Lakes* on the N. of Salisbury, and the two lakes on the S., around which the rides are extremely beautiful."

Just beyond *Canaan* (73 miles), a pretty village at the intersection of the Berkshire Div. of the New York, New Haven & Hartford and Central New England & Western R. Rs., the train crosses the boundary-line of Massachusetts and enters the renowned

Berkshire Hills,

"a region not surpassed in picturesque loveliness, throughout its whole longitude of 50 miles and its average latitude of 20 miles, by any equal area in New England, and perhaps not in all this Western world." From *Sheffield* (138 miles), a quiet town at the base of the Taconic Mountains, the ascent of *Mt. Washington* is easily made and affords a far-viewing prospect. This mountain was once a part of the great Livingston Manor, and its summit overlooks the rich and lordly domain once included in that now forgotten name. Six miles above Sheffield is **Great Barrington** (*Berkshire House, Collins House, Commercial House*), of which Mr. Beecher says that it "is one of those places which one never enters without wishing never to leave it. It rests beneath the branches of great numbers of the stateliest elms. It is a place to be desired as a summer residence." The Congregational and Episcopal Churches, and the High School, are handsome buildings, and there are several fine villas in the outskirts. The Congregational Church, the gift of Mrs. Mark Hopkins, is one of the finest country churches in New England, and was erected at a cost of upward of $150,000. *South Egremont*, 4 miles S. E. of Great Barrington, is reached by daily stage from Great Barrington. The *Mt. Everett House* here is an excellent summer hotel, situated just under the lofty crest of Mt. Everett, whose summit may be scaled by way of "its vast, uncultivated slope, to a height of 2,000 ft." From the summit the view is exceedingly fine, taking in half the whole stretch of the Housatonic Valley, the Catskills, and the Hudson. The trout-fishing in the vicinity of S. Egremont is exceptionally good. The *Berkshire Soda Springs* are about 3 miles S. E. of Great Barrington, amid wild and romantic scenery. "Next to the north of Great Barrington," says Mr. Beecher, "is **Stockbridge**, famed for its meadow-elms, for the picturesque beauty adjacent, for the quiet beauty of a village which sleeps along a level plain just under the rim of hills. If you wish to be filled and satisfied with the serenest delight, ride to the summit of this encircling hill-ridge, in a summer's afternoon, while the sun is but an hour high. The Housatonic winds in great circuits all through the valley, carrying willows and alders with it wherever it goes. The horizon on every side is piled and terraced with mountains. Abrupt and isolated mountains bolt up here and there over the whole stretch of plain, covered with evergreens." The distance by railway from Great Barrington to Stockbridge is 8 miles, but it is only 6¼ miles by the highway, and this latter should be chosen, if the tourist have time. The entire ride is through the most delightful scenery, and about half-way is *Monument Mountain*, one of the special attractions of the vicinity. The view from the summit is very fine, resembling that from Mt. Everett. Stockbridge contains many handsome villa residences. The *Stockbridge House* is an excellent hotel, open only during the summer, and *Edwards Hall*, the house in which Jonathan Edwards wrote his treatise on "The Freedom of the Will," is also open as a summer hotel. Both are situated on the main street of the village, and near by are an elegant Italian fountain, a fine soldiers' monument, and a memorial monument to Jonathan Edwards. Among the most interesting features of Stockbridge are the old burying-ground of the Mohegan Indians, and the fine antique mansion built by Judge Theodore Sedgwick and afterward occupied by his famous daughter Catharine. There is a handsome stone library building containing 5,000 volumes, and the Hon. David Dudley Field has presented the town with a bell-tower of stone, containing a silvery chime of bells and a clock. The Episcopal stone Church was the gift of Mrs. Charles Butler, of New York, a native of Stockbridge, and is a fine building, and the Congregational Church is also a noteworthy structure. The *Stockbridge Casino* is an attractive building. On the heights above the village formerly stood an old Mission-House, erected early in the last century, but the site has now been given to the city as a park, by Cyrus W. Field. The view from these heights is one of the loveliest imaginable. The drives in the vicinity of Stockbridge are extremely picturesque, and there are several points of interest besides Monument Mountain, already mentioned. The drive from Stockbridge to Lenox (6 miles), passing by the famous "bowl," in the extreme northerly part of Stockbridge, is considered one of the finest drives in the Berkshires. About 3 miles N. is *Lake Mahkeenac* (formerly called "Stockbridge Bowl"), a capacious basin of crystal-clear water, on whose margin Hawthorne once lived for a year and a half. About 1¼ mile from the village is the wonderful *Ice-Glen*, piercing the northern spur of Bear Mountain. "In its long and awesome corridors and crypts, formed by massive and gloomy rocks, and huge but prostrate trees, the explorer may sometimes find masses of ice in the heart and heat of midsummer."

Six miles N. of Stockbridge is the flourishing town of **Lee** (*Morgan House, Norton*), which owes its prosperity to its extensive paper-mills and woolen fabrics. It is also celebrated for its marble, which is among the best in the world. Large quantities of it were used in constructing the newer portions of the Capitol at Washington, and the public buildings at Philadelphia. The village contains several fine churches and private residences, and there are many attractive drives in the vicinity. That down the valley of the Hopbrook and up the mountain to *Monterey* is said not to be excelled in beauty in any part of Europe. *Tyringham*, 5 miles from Lee, is a lovely town lying snugly in a valley, on each side of which rise the hills. *Fernside*, a popular summer resort, was for many years the home of the Tyringham Shakers, who owned 4,000 acres in one body, and they were among the first societies of that sect to be organized. From what was their "Holy Ground" on the hill, one of the finest views in Berkshire is obtained. *Otis,* 13 miles E. from Lee, has a number of lovely drives in the Farmington River Valley, and 3 large lakes and reservoirs. It is quite a summer resort, as is *Sandsfield,* 8 miles S. of Otis. The views from these towns are grand. Four miles beyond Lee we come to **Lenox** (*Curtis House*), a favorite resort of Bostonians and New-Yorkers. It "is known for the singular purity and exhilarating effects of its air, and for the beauty of its mountain scenery." Beecher's "Star Papers," from which the above is a quotation, were written in a house which stood near the site now occupied by General Rathbone's mansion. Fanny Kemble Butler, who long resided here, said of the graveyard at Lenox: "I will not rise to trouble any one if they will let me sleep there. I will only ask to be permitted, once in a while, to raise my head and look out upon this glorious scene." Lenox having ceased to be the shire town of the county, the former court-house, a handsome building, now accommodates a free library and reading-room, public hall, town offices, etc. The *Club-House* is attractive by its broad piazzas and excellent tennis-grounds. Trinity Episcopal Church, of which at the laying of its corner-stone, in 1886, President Arthur took part, is the fashionable church during the season. A substantial drinking fountain, made of Italian and Tennessee marble, was erected in Main St. in 1885, to the memory of Miss Emma Stebbins, the sculptor. The residences of Charles Lanier, William D. Sloane, Morris K. Jesup, John H. Parsons, F. A. Schermerhorn, and General John F. Rathbone, are attractive specimens of architecture. There are numerous pleasant excursions from Lenox, a popular one being to the summit of *Bald Head* (carriages all the way), which commands a very fine view of the village, and of the valley to the south, including Monument Mountain. Other excursions are to the *Ledge, Mattoon Hill,* and *Perry's Peak,* an isolated summit 6 miles from the town, over 2,000 ft. high, and overlooking a vast range of country from the Catskills to the Green Mountains.

Six miles above Lenox is situated **Pittsfield** (*Maplewood,* open only in summers, *American House, Berkshire House,* and *Burbank Hotel*), a flourishing city of 17,281 inhabitants, the capital of the Berkshire region. It is beautifully situated on a lofty plateau, with the Taconics on the W. and the Hoosacs on the E.; and contains many handsome public and commercial buildings and private dwellings. The *Court-House* is a costly white-marble edifice, and the *Roman Catholic Church* one of the finest in Western Massachusetts. There are several handsome churches, including the spacious and costly *Methodist Church* and the famous *First Church,* of which "Fighting Parson Allen" of Revolutionary fame was once pastor, and later the well-known John Todd filled its pulpit. Other churches are the *St. Stephen's P. E. Church, Unity, Second Advent,* and *South (Cong.) Church.* The Rev. Jonathan L. Jenkins, of the First Church, and Rev. Wilberforce Newton, of St. Stephen's Church, are well known as leaders in the movement tending to the unification of churches. The *Maplewood Hotel,* half a mile from the depot, stands in the midst of spacious grounds, and the

Miss Salisbury School is a celebrated private school. In the park, near the center of the town, is a *Soldiers' Monument*, which was dedicated with imposing ceremonies on September 24, 1872. The * *Berkshire Athenæum* is a unique building of bluestone, freestone, and red granite, and contains a library of 20,000 volumes, a museum, cabinets, and reading-rooms, all free. The Berkshire Historical Society and the Wednesday Morning Club meet within its walls. Here also is the headquarters of the Agassiz Association, which has enrolled 20,000 members. There is, too, a fine *Academy of Music*. The drives in the vicinity are very fine, especially those to Lenox (6 miles), to Williamstown (20 miles), described below, and to Lebanon Springs (9 miles). On the mountain-road thither is *Lake Onota*, a romantic sheet of water, about 2 miles W. of Pittsfield, and a favorite excursion. Other drives which the stranger should not miss are those to *Wacbnah Falls*, in Windsor (10 miles), and to *Potter Mountain* (6 miles). *Ashley Pond*, from which the water-supply of the town is drawn, lies S. E. on the crest of the Washington Hills; and near by is *Roaring Brook*, a wild mountain-torrent that dashes down the side of the mountain through a rugged cleft known as *Tories' Cave*. N. of Onota, on the slopes of the Taconics, are the romantic *Lulu Cascade; Balance Rock*, a huge and nicely-poised bowlder; and, on the plateau of a giant crest above, a lovely mountain lakelet called *Berry Pond*. About 3 miles N. of Pittsfield is *Pontoosuc Lake* (reached by electric-cars from the city), a beautiful lake, with boats, steamers, etc., and about 2½ miles S. is the *South Mountain*, from the summit of which there is a fine view. Six miles W. is *Richmond*, a pleasant resort; and here *Perry's Peak*, 2,077 ft. high, commands a fine view. *Dalton*, 4 miles E. of Pittsfield, is a busy paper-manufacturing town, and just on the Pittsfield line is the mill where the Government bank-note paper is manufactured. *Lanesboro*, a cozy town, has a number of delightful drives, and is noted as the birthplace of Josh Billings the humorist, and other eminent men, Governor George N. Briggs, Judge Shaw, of the Massachusetts Supreme Court, etc. It has Episcopal, Methodist, Baptist, and Congregationalist churches. It has extensive iron-ore mines, and marble in abundance. Within a few years it has come into prominence as a summer resort. **Lebanon Springs** (mentioned previously) are among the most famous and frequented in the country, and the waters are regarded as remedial for rheumatism, liver-complaint, and cutaneous affections. Two miles from the Springs is **Shaker Village,** founded over a century ago by the disciples of Ann Lee, and now the headquarters of the United Society of Believers in Christ's Second Appearing—better known as "Millennial Church." On Sundays, in the summer, their singular form of worship may be witnessed.

At Pittsfield the Pittsfield Div. of the railroad comes to an end, and the region N. of it (known as "Northern Berkshire") is penetrated by the Pittsfield & North Adams Branch of the Boston & Albany R. R., which we follow to Williamstown. If the tourist have time he can make his trip much more enjoyable by hiring a suitable conveyance and taking the highways instead of the railway. The road from Pittsfield to Williamstown through Lanesboro and New Ashford (20 miles) presents a continuous panorama of beautiful scenery, and other drives are scarcely less attractive.

On the railroad, the first noteworthy station above Pittsfield is *Cheshire* (10 miles), famous for butter, cheese, and lumber. For many years the inhabitants were almost unanimously Democratic in politics; and on the invitation of the eccentric Elder Leland, to show their appreciation of President Jefferson, they made him a present, on January 1, 1802, of an ,enormous cheese weighing 1,235 pounds. From this point to N. Adams the road follows the valley of the Hoosac River, with the lofty Saddleback Range on the W. for the greater part of the way. *Adams* (5 miles from Cheshire) is one of the best points from which to visit *Greylock Mountain,* which rises majestically to the height of 3,535 ft., and is the highest elevation in Massachusetts. The ascent is easily made by a carriage-road, and the view is surpassingly grand, taking in all the Berkshire Hills. On the S. are the valleys of the Hoosac and Housatonic; to the E. are the Green Mountains on the N., and the Catskills on the W., and Mts. Monadnock, Tom, and Holyoke. Adams is a thrifty manufacturing town, with extensive gingham-mills, and a factory for "xylonite," employing 500 hands. Five miles above Adams is **North Adams** (*Mansion House, Richmond House, Wilson House*), a busy manufacturing village, having shoe interests. Next to this the principal industries are the extensive print-works and gingham-mills. The town is the terminus of the Pittsfield & North Adams Div. of the Boston & Albany R. R. on the S. About a mile E. of the village is the *Natural Bridge*, a roof of marble, through and under which Hudson's Brook has excavated a tunnel 15 ft. wide and 150 long. In the ravine of this brook there are several picturesque points; but next in interest to the bridge itself is a columnar group of rocks, which at its overhanging crest assumes the aspect of gigantic features, and is called *Profile Rock*. The *Cascade* is in a romantic glen 1¼ mile from the hotel, and is 30 ft. high. About 2 miles S. is the W. entrance to the **Hoosac Tunnel** (see p. 161). The old stage-road across the Hoosacs from N. Adams to the E. end of the tunnel (8 miles) affords an interesting mountain-drive. Hawthorne says of it: "I have never driven through such romantic scenery, where there were such variety and boldness of mountain-shapes as this; and, though it was a sunny day, the mountains diversified the view with sunshine and shadow, glory and gloom."

Five miles W. of N. Adams is the academic **Williamstown** (*The Greylock* and *Taconic Inn*), beautifully situated in a mountain-inclosed valley, and noted as the site of *Williams College*, founded in 1793, and a highly prosperous institution. The college buildings are the only architectural features of the town, and embrace 15 structures, of which the finest is Hopkins Memorial Hall. The library of 20,000 volumes is in Lawrence-Hall; and the residence of President Carter is opposite West College, on the main street. Morgan Hall, a beautiful 3-story building of cut stone, built at a cost of $100,000, is the gift of the Hon. Edwin D. Morgan; Clark Hall, given by Edward Clark, and the observatory E. of the College Hall, presented by David Dudley Field, are fine structures. Three new laboratories are being built at a cost of $150,000. The late President Garfield was a graduate. Near by is *Mills's Park*, an inclosure of 10 acres, in which a marble shaft, surmounted by a globe, marks the spot where Samuel J. Mills and his associate students met by a haystack in 1807 to consecrate themselves to the work of foreign missions. There is a bronze soldiers' monument on a granite pedestal in the main street. Among the many attractive resorts in the vicinity of Williamstown are *Flora's Glen*, the *Cascades*, and *Snow Hole*, a gorge in the mountain where the snow never entirely melts. *Sand Springs*, 1½ mile N. of the village, is a resort where the waters of the spring are thought to be efficacious in cutaneous diseases, and bathing-houses are provided. *Mount Hopkins* (2,800 ft. high) is a short distance S. of Williamstown, and is often ascended for its broad and striking view. The ascent of Greylock is often made from this side. *The Hopper* is a stupendous gorge between Greylock, Prospect, and Bald Mountains, through which flows the picturesque Hopper Brook.

39. New York to Vermont via Harlem Division of the New York Central & Hudson River R. R.

THE Harlem division of the N. Y. Central R. R., in connection with the Lebanon Springs and Bennington & Rutland R. Rs., forms a short and popular route to Vermont and the North. It skirts the eastern portions of all those counties lying upon the Hudson River and traversed by the Hudson River R. R. (see Route 9). The stations and towns along the line are, for the most part, inconsiderable places, many of them having grown up with the road. The country traversed is varied and picturesque in surface, much of it being rich agricultural land; but it does not compare with the river route in scenic attractions.

Leaving the Grand Central Depot, the train passes through long tunnels under the city, and at 134th St. (4 miles) crosses the Harlem River. *Fordham* (9 miles) is the seat of St. John's College, a noted Roman Catholic institution; and 1½ mile W. is *Jerome Park*, once the finest race-course in America. *Williams Bridge* (10½ miles) is on the Bronx River. One mile beyond is *Woodlawn Cemetery*, one of the most beautiful of the many cemeteries in the neighborhood of New York. *White Plains* (22 miles) was the scene of the eventful Revolutionary battle of Oct. 28, 1776. Stations, *Chappaqua* (33 miles), *Bedford* (39 miles), and *Katonah* (42 miles). From *Golden's Bridge* (44 miles) a branch road runs in 7 miles to **Lake Mahopac,** a highly popular summer resort. The lake is 1,000 ft. above the sea, is 9 miles in circumference, with very irregular shores, and is the center of a group of 22 lakes, lying within a circle of 12 miles' radius, and amid pleasing scenery. The boating on the lakes is excellent, and the fishing good. The drives are fine, and there are many pleasant excursions. *Thompson's Hotel* is first class, and there are many boarding-houses.

Beyond Golden's Bridge, the train passes the small stations of *Brewster's* (52 miles), *Paterson* (60 miles), *Pawling* (64), and *Dover*

Plains (76). Beyond the latter, the scenery becomes mountainous and fine. From *Amenia* (84 miles), stages run in 4 miles to *Sharon* (see p. 164); and at *Boston Corners* (99 miles) the Berkshire Hills come in sight on the right. *Copake* (104 miles) is only 2 miles from the Bash-Bish Fall (see p. 164). At *Chatham* (127 miles) connection is made with the Boston & Albany R. R. (see Route 36), by which the distance to Albany is 24 miles, and with the Lebanon Springs R. R., which runs N. in 58 miles to Bennington, Vt., where connection is made with the Bennington & Rutland R. R., which runs in 55 miles to Rutland. Nineteen miles beyond Chatham the train reaches **Lebanon Springs** (see p. 168). At *Petersburg* (166 miles) the Fitchburg R. R. is intersected, and shortly beyond the train enters the State of Vermont, and soon reaches **Bennington** (*Putnam House, Stark House*), one of the prettiest towns in the State. It is situated in a picturesque mountain-inclosed valley, 800 ft. above the sea, is solidly and handsomely built, and contains 6,391 inhabitants. Manufacturing is extensively carried on, the chief products being cotton goods and knit underclothing. *Bennington Center*, one mile distant, is the Revolutionary village, and was the site of the old Catamount Tavern, which was burned in 1871. *Hoosac*, N. Y., the adjoining township, was the scene of the battle of Bennington (Aug. 16, 1777), in which a detachment of the British forces under Col. Baum was utterly defeated by the Green Mountain Boys, led by the intrepid Col. Stark. About 2 miles from Bennington, by footpath (4½ by carriage-road), is * *Mt. Anthony*, on whose summit is a tower from which a broad and beautiful view may be obtained. Among numerous and pleasant drives in the vicinity are those to *Petersburg*, to *Prospect Mt.*, and to *Big Pond*.

Between Bennington and Rutland a mountainous region, affording much pleasing scenery, is traversed. There are several pretty towns *en route*, of which the only one requiring mention is **Manchester** (*Equinox House, Mansion House*), a beautiful village nestling in a valley between the Green and Equinox ranges. Many visitors are attracted thither in summer by its pure and invigorating air, fine scenery, trout-fishing, and driving. A noticeable feature of the village is its white-marble pavements, there being numerous marble-quarries in the vicinity. *Mt. Æolus* is 5 miles from the village, and to the S. E. is *Stratton Mountain*. Near the latter is *Stratton Gap*, a beautiful glen, which furnished the subject of one of A. B. Durand's best paintings. * **Mt. Equinox** (3,706 ft. above the sea) is ascended by a road from the village, and is noted for its glorious views, the following points being visible in clear weather: Lakes George and Champlain, Kearsarge and the Franconia Mountains in New Hampshire, Greylock in Massachusetts, Killington Peak in Vermont, and the Catskill Mountains and Saratoga village in New York. *Skinner Hollow* is a deep gulf on the S. side of the mountain, containing a cave in which the snow never entirely melts, a stream which finds an outlet through a cavern, and a marble-quarry. **Rutland** (240 miles) is described on page 147.

40. New York to Buffalo and Niagara Falls.

Via New York Central & Hudson River R. R. (drawing-room and sleeping-cars, and vestibuled trains). To Buffalo, 440 miles; fare, $9.25. To Niagara Falls, 448 miles; fare, $9.25.

FROM New York City to Albany (143 miles) this route has already been described in Route 9. The railway runs close along the E. bank of the Hudson River, affording good views of the river itself and of the opposite bank, and the continuous view of the river and its scenery from the cars makes the journey a most attractive one. Seats on the left-hand side of the cars should be obtained going N.; on the right-hand side going S. The railroad crosses the river to Albany on a fine bridge, whence it traverses from E. to W. the entire length of New York State, passing through the rich midland counties. It has two termini at the E. end, the main line at Albany and the branch at Troy, the branches meeting after 17 miles at Schenectady. It then continues in one line to Syracuse (148 miles from Albany), where it divides and is a double route to Rochester, whence the Niagara Falls branch diverges to the Falls and the main line passes on to Buffalo, where connection is made with the Lake Shore & Michigan Southern and the Michigan Central R. Rs. The quadruple tracks of the road are laid with steel rails, and drawing-room cars and sleeping-cars are attached to the through trains, which leave New York for Chicago. The great Erie Canal traverses the State from Albany to Buffalo nearly on the same line with the railroad, and often in sight from the cars.

Leaving Albany (which is described on p. 72), the train passes *W. Albany*, with its extensive machine-shops and cattle-yards, and in 17 miles reaches **Schenectady** (*Edison House, Carley House*), a city of 19,902 inhabitants, situated on the right bank of the Mohawk River, on a spot which once formed the council-grounds of the Mohawks. It is one of the oldest towns in the State, a trading-post having been established here by the Dutch in 1620, and is the seat of *Union College*, founded in 1795, and of important wool-manufactures, iron-works, and railroad-shops. Here the Saratoga & Champlain Div. of the Delaware & Hudson R. R. diverges and leads to Saratoga Springs and Lakes George and Champlain (see Routes 44 and 46). Leaving Schenectady, the train crosses the Mohawk River and the Erie Canal on a bridge nearly 1,000 ft. long, and traverses a rich farming country to *Amsterdam* (33 miles) and *Fonda* (43 miles). From Fonda a railway runs in 26 miles to *Northville*, where connection is made with daily stages which run in 29 miles to *Lake Pleasant* in the Adirondack region (Route 47). From *Palatine Bridge* (55 miles) carriages run in 8 miles to *Sharon Springs* (Route 52). *Fort Plain* (58 miles) is a flourishing village 2 miles from old Fort Plain of Revolutionary memory; and *St. Johnsville* (64 miles) is a prosperous manufacturing town on the banks of the Mohawk, with fine scenery in the vicinity. Ten miles beyond is **Little Falls** (*Girvan House*), which is remarkable for a bold passage of the river and canal through a wild and most picturesque defile. The river falls 45 ft. in half a mile, and affords a water-power which is extensively used in manufactures. Twelve miles S. W. of Little Falls is *Richfield Springs* (Route 43). Stations, *Herkimer* (81 miles), the starting-point of the Adirondack & St. Lawrence R. R., *Ilion* (83 miles), and then (95 miles from Albany) comes the large and handsome city of **Utica** (*Bagg's, Butterfield House, St. James*), situated on the S. bank of the Mohawk, on the site of old Fort Schuyler (built in 1756). The city has 44,007 inhabitants, extensive and varied manufactures, and is the center of an important railway and canal system. Genesee St. is the leading thoroughfare; on it are the handsome *City Hall* and many fine commercial buildings, churches, and private residences. The *State Lunatic Asylum* is a spacious building on a farm W. of the city (reached by horse-cars). At Utica direct connection is made for *Richfield Springs* by the Delaware, Lackawanna & Western R. R. (Route 43). Utica is the chief terminus of the Rome, Watertown & Ogdensburg R. R., which is the gateway to the Adirondack region from the W., and which is also a prominent route to the Thousand Islands and Montreal.

An easy and popular excursion from Utica is *via* Rome, Watertown & Ogdensburg R. R. to Trenton Falls (distance, 17 miles). **** Trenton Falls** (*Kanyahoora Hotel, Moore's Hotel*) are situated on the W. Canada (or Kanata) Creek, a tributary of the Mohawk. The descent of the stream, 312 ft. in a distance of 2 miles, is by a series of half a dozen cataracts, which have worn for themselves out of the lime-stone hills a bed which at some points is 200 ft. below the level of the surrounding country. The ravine is very narrow, with precipitous walls, and the path along the bottom, which was hewn out at consider-able cost and is kept in admirable order, is passable only at low water. During high water the path along the cliff must be followed, and affords some striking views of the profound chasm below and of the torrent which in time of flood rages along with the force and tumult of a Niagara. It is difficult to say whether the falls are most impressive in times of high or low water, but those who can should see them under both conditions. The usual way of visiting them is by a stairway which descends the precipice a few rods from Moore's Hotel (fee 25c.). From the platform at the foot a pathway, difficult in places but entirely safe, leads up the ravine past * *Sherman's Fall* (33 ft. high), * *High Falls* (40 ft. high and extremely beautiful), *Mill-Dam Fall* (14 ft. high), and the *Alhambra, a great natural hall or amphitheatre which "has been the despair of artists and descriptive writers." At *Rocky Heart* most vis-itors turn back, but the adventurous may pass on to *Prospect Fall* (20 ft. high), at the head of the chasm. An easier way of reaching Pros-pect Fall is by a walk or drive of 3 miles along the cliff from the hotel.

From Utica the Rome, Watertown & Ogdensburg R. R. runs N. to *Sackett's Harbor* (104 miles from Utica) on Lake Ontario, connecting at Watertown with branch to *Clayton* (108 miles) and *Ogdensburg* (134 miles) on the St. Law-rence River. From Clayton steamers run in connection with the trains to **Alexandria Bay** (see Route 60). *Boonville* (35 miles from Utica) is a con-venient entrance to the **John Brown Tract,** which forms the S. portion of the great Adirondack wilderness. Guides and outfit may be obtained at Boonville. The *Fulton Lakes* (Route 47) are 26 miles N. E. of Boonville, and there are many other lakes in the vicinity abounding in fish.

Beyond Utica the train passes in 14 miles to **Rome** (*Arlington House* and *Stanwix Hall*), a thriving city of 14,99J inhabitants at another junction of this route with the Rome, Watertown & Ogdensburg R. R. and of the Erie and Black River Canals. Large railroad-shops and rolling-mills are located here, there is excellent water-power, and Rome is one of the best lumber-markets in the State. There are a few fine buildings and many handsome residences.

The *Rome, Watertown & Ogdensburg R. R.* runs N. W. from Rome to *Watertown* (73 miles), *Cape Vincent* (96 miles), and *Ogdensburg*, 141 miles. From Cape Vincent a steamer runs twice daily in summer to *Alexandria Bay* (see Route 60). and there is a steam-ferry to *Kingston*, Can. This is a through route from New York to Kingston, Alexandria Bay, and the Thousand Islands. Fare from New York to Cape Vincent, $8.20.

Leaving Rome the train passes *Verona* (118 miles), with a mineral spring, and *Oneida* (122 miles), which is about 6 miles from **Oneida Lake,** a beautiful sheet of water 19 miles long and 6 miles wide, abounding in fish, and surrounded by a highly cultivated country. **Chittenango** (133 miles, *White Sulphur Springs Hotel*) lies at the entrance of the deep and narrow valley through which the waters of Cazenovia Lake are discharged into Oneida Lake, and is noted for its iron and sulphur springs, which are much frequented by invalids. Fifteen miles beyond Chittenango is **Syracuse** (*Globe Hotel, Hotel Mowry,* and *Vanderbilt House*), with a population of 88,143, and important manufactures and trade. It is pleasantly situated at the S. end of *Onondaga Lake* (which is 6 miles long and about 1½ wide), and to the S. lies the reservation of the few remaining Onondaga Indians. The *Government Building, Court-House, Clerk's Office, Onondaga Co. Savings-Bank* and *Syracuse Savings-Bank, High School, Home for Old Ladies, State Asylum for Idiots,* and the *Orphan Asylum,* are all fine. The *Penitentiary* is on a hill in the N. E. part of the city. The *Syracuse University* (Methodist) has three fine buildings and an observatory on a hill to the E. The celebrated *Von Ranke Historical Library* is located here. *St. Mary's* (Catholic) *Cathedral, St. Paul's* (Episcopal) *Cathedral,* and the *May Memorial Church* (Unitarian) are also fine. The famous *Salt Springs* are on the shore of the lake. Through Syracuse also pass the West Shore, the Delaware, Lackawanna & Western, and the Rome, Watertown & Ogdensburg (Phœnix Line to Oswego). Between Syracuse and Rochester the New York Central R. R. has two lines: the "Old Road" *via* Canandaigua, 104 miles long, and the "Direct Route," 81 miles long.

Syracuse to Rochester via "Direct Route."

This route runs parallel with the Erie Canal nearly all the way through a level country, with numerous small towns along the line, but none that require mention. *Lyons* (193 miles) is the largest, and is a pretty village of 6,228 inhabitants, capital of Wayne Co., which produces a great quantity of dried fruit. In a hill-side near *Palmyra* (207 miles) Joseph Smith claimed to have found the golden plates of the Mormon Bible. *Rochester* (229 miles) is described beyond.

Syracuse to Rochester via " Old Road."

The distance by this route (Auburn Branch) is 104 miles, and is traversed only by local trains.   Leaving Syracuse the train passes several minor stations, and in 18 miles reaches *Skaneateles Junction,* where connection is made by rail (5 miles) with *Skaneateles,* a thriving village at the foot of * **Skaneateles Lake,** a charming water 16 miles long and 1 to 1½ wide, 860 ft. above the sea, and surrounded by hills rising 1,200 ft. above the surface.   Boating and fishing are excellent, and the lake is much visited in summer, when a steamer plies. between Skaneateles and the village of *Glen Haven* at the S..end. - About 10 miles S. E. of Skaneateles is the picturesque and romantic *Otisco Lake,* 4 miles long and embosomed amid lofty hills.   Nine miles beyond Skaneateles Junction is **Auburn** (*Osborne House, Owasco Inn*), a city of 25,858 inhabitants, situated near Owasco Lake, which finds its outlet through the town.   Genesee St. is the principal thoroughfare, and nearly all the streets are pleasantly shaded.   On Genesee St. are the *County Court-House,* and the churches of *St. Peter* (Episcopal), *St. Mary's* (Roman Catholic), and the *First Presbyterian.*   The *Theological Seminary* (Presbyterian) has substantial stone buildings in the N. E. part of the city.   Near the station is the vast and massive *State Prison,* covering 18 acres of ground, which are inclosed by a stone wall 30 ft. high. Auburn was long the home of William H. Seward, and his grave is in the cemetery on Fort Hill (reached by Fort St.).   **Owasco Lake** is 3 miles S. of Auburn, and is a favorite summer resort.   It is 11 miles long, about a mile wide, and surrounded in part by bold hills.   A steamer plies during the summer between *Owasco Village* and *Moravia.*

At *Cayuga* (11 miles beyond Auburn) the train crosses Cayuga Lake by a bridge nearly a mile long, affording a fine view from the cars to the left.   From this point the Lehigh Valley R. R. runs S. in 38 miles to Ithaca, and steamboats also ply upon the lake.   At Aurora, 12 miles S. of Cayuga, is *Wells College,* where Mrs. Cleveland, the wife of the President, was educated.   At the S. end of Cayuga Lake is **Ithaca** (*Clinton House, Ithaca Hotel*), the seat of Cornell University, one of the most beautiful cities in the State, and surrounded by charming and picturesque scenery.   The buildings of *Cornell University,* on the hills E. of the village, 400 ft. above the lake, are worth a visit.   This institution was founded in 1865, and in 1892–'93 had a faculty of 145 instructors and an attendance of over 1,665 students.   Its libraries aggregate 130,000 volumes.   In the immediate vicinity of the village there are said to be no less than 15 cascades and waterfalls, varying from 30 to 160 ft. in height, 5 of them being over 100 ft.   The beautiful **Ithaca Fall,** 150 ft. broad and 160 ft. high, is about a mile distant in *Ithaca Gorge,* which is said to contain more waterfalls within the space of a mile than any other place in America.   The celebrated * **Taghkanic Falls** are 10 miles from Ithaca, and may be reached by a pleasant drive along the shore of the lake, by the lake-steamers, or by the Lehigh Valley R. R.   Near the Falls is the *Cascade House.*   Taghkanic Creek flows through a comparatively level country until about 1½ mile-

from the lake it encounters a rocky ledge lying directly across its
course. But the stream has succeeded in excavating for itself a channel
from 100 to 400 ft. in depth and 400 across at its lower extremity.
Through this chasm the waters hurry on to the precipice, where they
fall perpendicularly ·215 ft. into a rocky basin, forming a cataract more
than 50 ft. higher than Niagara. At the· bottom ·of the Fall the walls
of the ravine are nearly perpendicular and 400 ft.· high, but paths and
stairways assist the passage. (Ithaca is also reached by the Delaware,
Lackawanna and Western R. R.· and the Lehigh Valley R. R., Routes
43 and 45.)

Five miles beyond Cayuga is the manufacturing village of *Seneca
Falls*, pleasantly situated at the falls. of the Seneca River; and 10
miles farther is the academic city of **Geneva** (*Franklin House, The
Kirkwood*), beautifully situated at the foot of Seneca Lake, and noted
for its educational institutions, of which *Hobart College* (Episcopal) is
the most important. ***Seneca Lake,** one of the largest and most
beautiful in New York State, is 35 miles long and 1 to 4 miles wide, is
very deep, and never freezes over. Steamboats run several times daily
in summer from Geneva to *Watkins*, at the S. end of the lake, stopping *en route* at *Ovid* and *Dresden*. Near Watkins is the famous
Watkins Glen (see Route 59).

Twelve miles beyond Geneva are the **Clifton Springs** (*Clifton·
Springs Sanitarium*), one of the most frequented resorts on the line of
the New York Central R. R. The waters are sulphurous in character,
and are considered efficacious in bilious and cutaneous disorders. Ten
miles farther is *Canandaigua* (*Canandaigua Hotel*), a pretty village·of
8,229 inhabitants at the N. end of Canandaigua Lake, 28 miles from
Rochester. **Canandaigua Lake** is 16 miles long, narrow and deep,
is bordered by numerous vineyards, abounds in fish, and is much
visited in summer. Steamers run down the lake to *Seneca Point* and
Woodville. Canandaigua is the northern terminus of the 'Northern
Central R. R. of the Pennsylvania System (see Route 59). Between
Canandaigua and Rochester there are no important stations.

Rochester.

Hotels, etc.—The *Powers* (W. Main St.), the *Livingston* (Exchange St.),
the *New Osburn* (S. St. Paul St.), and the *Whitcomb House* (cor. E. Main and
Clinton Sts.), are first-class houses. *Brackett House, Congress Hall*, and *Waverly House*, are less expensive houses. *Street-cars* on the principal streets
and to the suburbs; *stages* to adjacent towns. *Post-Office* in North Fitzhugh St.

Rochester is on both sides of the Genesee River, 7 miles from its
mouth in Lake Ontario. Soon after it enters. the city the river makes
a rapid descent, and there is a perpendicular fall of 96 ft. near the
center, and one of 25 ft. and one of 84 ft. on the N. To the water-
power thus afforded the prosperity of the city is attributable, and
it contains several large flour-mills. In addition to the coal and iron
trade, other important industries are clothing, boots and shoes, engines, agricultural implements, trees, and garden ·and flower seeds.
The immense *nurseries in which these latter are produced are well

worth a visit. Rochester was first settled in 1810, was incorporated as
a city in 1834, and in 1890 had a population of 133,896. The streets are
nearly all laid out at right-angles, many of them are well paved with
stone, and most of them are bordered with shade-trees. *Main St.*, *State
St.*, *Lake Ave.*, and *East* and *West Aves.*, are the principal thoroughfares.
At the cor. of W. Main and State Sts. are the ***Powers Buildings,**
a tubular block of stores, built of stone, glass, and iron, 7 stories high.
In the upper halls is a fine collection of paintings, and on the top is a
tower (open to visitors) from which may be obtained a fine view of the
city and its surroundings. Opposite is the *Wilder Building*, 12 stories. On State State is the *Ellwanger and Barry Building*, 8 stories.
Near the Powers Buildings is the *Arcade.* On W. Main St. is the *County Court-House*, in which is the *Law Library* of 20,000 volumes. Back
of the Court-House is the **City Hall,** a building of gray sandstone,
138 by 80 ft., with a tower 175 ft. high. In the same vicinity is the
Free Academy, a large brick building with sandstone trimmings. Next
to it is *Rochester Savings-Bank*, cor. W. Main and Fitzhugh Sts. Other
noticeable buildings are the *Lyceum Theatre* and the *Genesee Valley
Club.* Warner's Building, in the Gothic style, 7 stories high, is a fine
structure. The finest church edifices are the *First Baptist*, in Fitzhugh St.; the *First Presbyterian*, in Spring St.; *Second Presbyterian*,
in Fitzhugh St.; *St. Patrick's Cathedral* (Roman Catholic), in Frank
St.; the *Asbury M. E. Church;* and the *Third Presbyterian Church.*
The *Young Men's Christian Association Building*, just erected, cost
$150,000. The ***University of Rochester** was founded by the
Baptists in 1850, and now has 15 instructors. It is situated in the E.
part of the city (in University Ave.), where it has 24½ acres, and occupies three massive buildings. The library contains 26,000 volumes,
and the geological cabinets, collected by Henry A. Ward, are among
the finest in the country. The library and cabinets are in a handsome fire-proof building. There is also a Baptist *Theological Seminary*, founded in 1850. Its library numbers more than 25,000 volumes, including 4,600 which constituted the library of Neander, the
German church historian. The *City Hospital* (West Ave.) has a fine
building with accommodations for 150 patients. *St. Mary's Hospital*
(in West Ave.) is an imposing edifice of stone, with accommodations
for 500 patients. The *State Industrial School* is an extensive brick
building surrounded by grounds 42 acres in extent, about one mile N.
from the center of the city. The Western New York *Institution for
Deaf-Mutes* is in N. St. Paul St., with 150 pupils. Other points of
interest are ***Mount Hope Cemetery,** picturesquely situated in the
S. part of the city (reached by horse-cars). From Mt. Hope may be
seen the Hemlock Water Reservoir (with a large fountain), supplying
the city with water from Hemlock Lake, 30 miles S.; and the cut-stone
Aqueduct, 848 ft. long with a channel 45 ft. wide, by which the Erie
Canal is carried across the Genesee River. A boulevard, which is a
continuation of Lake Ave., 300 ft. wide and 6 miles long, extends to
Lake Ontario, making a noble drive.

The ***Genesee Falls** are seen to the best advantage from the E.
12

side of the stream. The railroad-cars pass about 100 yards S. of the most southerly fall, so that passengers in crossing lose the view. To view the scene properly, the visitor should cross the bridge over the Genesee above the mill, and place himself immediately in front of the fall. The first fall is 80 rods below the Aqueduct, and is 96 ft. high. From Table Rock, in the center of it, Sam Patch made his last and fatal leap. The river below the first cataract is broad and deep, and is spanned by two fine iron bridges, with occasional rapids to the second fall, where it again descends perpendicularly 25 ft. A short distance below is the third fall, which is 84 ft. high. Below the third falls, navigation is good to Charlotte and Ontario Beach, about 5 miles, making a delightful trip in summer. These places are connected with Rochester by means of an electric railway. *Irondoquoit Bay* and the other beaches are much visited in summer, and daily steamers cross the lake to Toronto (70 miles).

The distance from Rochester to Buffalo is 69 miles. Of the five small towns *en route* the only one requiring mention is **Batavia,** a pretty village of 7,221 inhabitants, noted as the site of the *State Institution for the Blind*, one of the finest structures of its kind in the country.

Buffalo.

Hotels, etc.—The best are the *Iroquois* (absolutely fire-proof), Main, Eagle, and Washington Sts.; the *Niagara*, on Front and Porter Aves.; the *Tifft House*, the *Genesee*, and the *Hotel Broezel*. *Horse* and *electric cars* run through the principal streets and to the suburbs. *Steamboats* run to the principal ports on the Great Lakes (see Route 104). *Reading-rooms* at the Buffalo Public and Grosvenor Libraries, in Lafayette Square, and at the Y. M. C. A., Mohawk, Pearl, and Genesee Sts. *Post-Office* at the cor. of Washington and Seneca Sts.

Buffalo, the third city in size in the State of New York, is at the mouth of Buffalo River and head of Niagara River, at the E. end of Lake Erie, and possesses the largest and finest harbor on the lake. It is the terminus of the Erie Canal, the New York Central R. R., the Erie R. R., the West Shore R. R., the Delaware, Lackawanna & Western R. R., the Lehigh Valley R. R., and sixteen other railroads, connecting it with all parts of the country. The city has a water-front of several miles upon the lake and rivers. Its commerce is very large, as its position at the foot of the great chain of lakes makes it the entrepot for a large part of the traffic between the East and the great Northwest. The lake navigation of the city is one of the most important elements of business. The manufactures of starch, soap, and lumber, of iron, tin, brass, and copper ware are large. Malting and brewing are carried on, and the cattle, lumber, and coal and iron interests have developed rapidly. Natural gas is introduced. Buffalo was first settled in 1801; it became a military post during the War of 1812, and was burned by a force of Indians and British in 1814; and it was incorporated as a city in 1832. Since the completion of the Erie Canal in 1825 its growth has been very rapid, and its population in 1890 was 255,664.

Buffalo, in the main, is handsomely built. Its streets are broad and

Grain Elevators at Buffalo.

straight, and for the most part laid out regularly. *Main, Niagara, Seneca, Broadway, North, Linwood,* and *Delaware Aves.* are the principal thoroughfares. The streets in most portions are bordered with shade-trees. Shade-trees adorn the public squares, named respectively Niagara, Lafayette, Franklin, Johnson's, Prospect, and the Terrace. The three former are in the busiest section of the city. Lafayette contains the Soldiers' Monument, built at a cost of $50,000, and bank and insurance buildings. A portion of the river front is a bold bluff, called *The Front,* affording fine views of the city, river, lake, Canada shore, and the hilly country to the S. E. On this bluff stood Fort Porter, and several companies of U. S. infantry are stationed here in barracks. The prominent public buildings are: the *Custom-House* and *Post-Office,* a large but plain freestone edifice, at the cor. of Washington and Seneca Sts.; the *Board of Trade Building,* in Seneca St.; the *State Arsenal,* in Broadway; the *Erie County Penitentiary,* a capacious building of brick and stone; and the *General Hospital,* in High St. The *City and County Hall,* a splendid and spacious granite edifice fronting on Franklin St., was completed in 1880 at a cost of nearly $1,400,000. Several of the bank buildings are imposing edifices, especially those of the Erie County, the Buffalo, and the Iron Banks, and the Western Savings-Bank. St. Paul's Cathedral (Episcopal), in Pearl St., which was destroyed by an explosion of gas, May 10, 1888, only the walls and tower, containing a fine chime of bells, being left standing, has been restored; *St. Joseph's Cathedral (Roman Catholic) is in Franklin St., of blue-stone trimmed with white-stone, in the Gothic style, with a chime of 42 bells; the *Delaware Ave. Methodist Episcopal,* in Delaware St.; *Trinity (P. E.), in Delaware Ave.; the *First Presbyterian Church,* in the Circle; the *Congregational Church,* cor. Elmwood and Bryant Sts.; *Calvary* (Presbyterian), in Delaware St.; and the *St. Louis* (French R. C.), in Main St. The leading educational institutions are the *Medical College,* of the University of Buffalo, in Main St., and the *Niagara Medical College; Canisius College,* a Jesuit institution, occupying a handsome building of stone and brick in Washington St. near Tupper; the *Buffalo Law School,* organized as a department of Niagara University in 1887; *St. Joseph's College,* on the terrace in the rear of St. Joseph's Cathedral, conducted by the Christian Brothers; *St. Mary's Academy,* on the same square, in Franklin and Church Sts.; the *Buffalo Female Academy* (Protestant), in Delaware Ave.; the *Heathcote School,* in Pearl St., and *St. Margaret's School,* both of the Episcopal Church; and the *State Normal School,* in Jersey St., a large and imposing building. The *Buffalo Library,* in Lafayette Square, is a massive structure, containing a circulating library of 70,000 volumes; and in the same building is the Buffalo Historical Society, with a library of 10,000 volumes and cabinets, the Buffalo Fine Arts Academy and School of Art, and the Society of Natural Sciences, with a very valuable collection of minerals, a good botanical and conchological cabinet, and a complete set of Henry A. Ward's fossil casts. *Grosvenor Library,* also in Lafayette Square, is a public library for reference, founded by a bequest of Seth Grosvenor, of Buffalo. It contains about

33,000 volumes, chiefly important books not easy of access elsewhere.
The **Music Hall,** in Main St.; the *Star Theatre*, in Mohawk and
Pearl Sts.; the *Academy of Music*, in Main St.; *Corinne's Lyceum*, in
Washington St.; and the *Court St. Theatre*, are the principal places of
amusement. The *Church Charity Foundation* (Episcopal), in Rhode
Island St. near Niagara, embraces a home for aged and destitute women,
and an orphan ward. The *Ingleside Home*, with a building in Michigan
St., is designed for the reclamation of fallen women, and has been very
successful since its organization in 1869. The *Buffalo Orphan Asylum*
(Protestant) has commodious buildings in Virginia St.; and the *St. Vincent Female Orphan Asylum*, cor. Main and Riley Sts., and the *St. Joseph's
Boys' Orphan Asylum*, at Limestone Hill, are large Roman Catholic
institutions. The **State Insane Asylum,** in Forest Ave., standing in grounds of 203 acres, adjoins the Buffalo Park. Visitors are
allowed to enter every Thursday. The *Fitch Institute* is at the cor. of
Michigan and Swan Sts. The buildings of the *Women's Educational
and Industrial Union* and the *Women's Christian Association* are in
Niagara Sq.

A superb public **Park**, or system of parks, has been designed and
laid out by Frederick Law Olmsted, the architect of Central Park in
New York City. The land embraces about 512 acres, and is divided
into three plots, situated in the western (river front), northern, and eastern parts of the city, with broad boulevards, forming a continuous drive
of over 10 miles. The **Forest Lawn Cemetery,** bounded on two
sides by the Park, contains some fine monuments, among them that of
President Fillmore, the Indian chief Red Jacket, and of Gen. Alfred J.
Myers, founder of the United States Signal Service Corps. Near the entrance is the elaborate *Crematory*, erected in 1883. At Black Rock, the
northern part of Buffalo (reached by Niagara St.), the magnificent **International Bridge,** completed in 1873 at a cost of $1,500,000,
crosses the Niagara River to the Canadian villages of Fort Erie and
Victoria. *Niagara Falls* (see p. 181) are 22 miles from Buffalo *via*
N. Y. Central R. R., 23 miles *via* Erie R. R., and 26 miles *via* Canada
Division of the Michigan Central R. R.

No visitor should leave Buffalo without having seen the great canal-
basins, the parks, the grain-elevators, and some of the iron-works. The
Buffalo River may be seen to good advantage by going to the foot of
Main St., where excursion boats run to the resorts on the lake and river.
The large Union Passenger Station of the N. Y. Central and Erie R. Rs.,
and the immense coal-chutes of the various roads, are also worth a visit.

Rochester to Niagara Falls.

At Rochester the Niagara Falls branch of the N. Y. Central R. R. diverges from the main line and runs to the Falls in 77 miles. A very large
portion of the through travel and traffic between the East and the West
passes over this line by way of the Niagara Cantilever Bridge and the
Michigan Central R. R. to Detroit. Rochester is described on p. 176.
Leaving Rochester the train runs through a rich agricultural region,
passing two or three small stations to *Brockport* (17 miles from Roch-

ester), a pretty village of 3,742 inhabitants on the Erie Canal, containing the fine building of the State Normal School. *Albion* (30 miles) is another attractive village, capital of Orleans County, with a handsome Court-House and a costly Soldiers' Monument. *Medina* (41 miles) is noted for its quarries of dark-red sandstone, known as "Medina sandstone"; and 16 miles beyond is **Lockport** (*American, Grand,* and *Niagara*), a city of 16,038 inhabitants, famous for its limestone-quarries and its manufacture of flour. It is situated at the point where the Erie Canal descends by ten double locks from the level of Lake Erie to the Genesee level. These locks may be seen from the cars. By means of them an immense water-power is obtained, which is utilized by the factories and flour-mills. Nineteen miles beyond Lockport is *Suspension Bridge* (418 miles from New York), which has been regarded as one of the triumphs of modern engineering. Over this the trains of the Grand Trunk R. R. cross the Niagara River within full view of the Falls and of the Whirlpool. It is 821 ft. long from tower to tower, is 245 ft. above the water, and was finished in 1855, at a cost of $500,000. A carriage and foot way is suspended 28 ft. below the railway-tracks. A still more remarkable triumph of engineering, however, is the *Niagara Cantilever Bridge,* built by the Michigan Central R. R. and completed in November, 1883, over which all trains of this road pass. It is located about 300 ft. above the Railroad Suspension Bridge, just over the head of the Whirlpool Rapids and in full front of the cataract. It is constructed entirely of steel, and is the first bridge of the kind built in the world. Its essential principle is that of a trussed beam supported at its center on a steel tower, the landward end being securely anchored. The cantilever arms meet each other at the center of the bridge. The bridge is designed to bear a running load of a ton per foot. The total length of the bridge is 895 ft.; length of fixed span, 125 ft.; height of abutments, 50 ft.; height of clear span above the river, 245 ft.; length of clear span across the river, 500 ft.; height of steel towers, ·130 ft.; length of cantilevers, 375 and 395 ft.; total weight resting on columns, 1,600 tons. There is a double track over the bridge. The engineers were Messrs. C. C. Schneider and Edmund Hayes.

Niagara Falls.

Hotels, etc.—On the American side are the *International Hotel* (European plan) and the *Cataract House* (both close to the Falls, and near the rapids), *Kattenbach's* (fronting the State Park and rapids), and the *Prospect House* (cor. of Union and 2d Sts.), transferred from the Canadian shore on the formation of Queen Victoria Park. The *Clifton House*, on the Canada side, is a high class hotel.

Carriages at reasonable rates may be secured either on the train or at the carriage-stands. Park or Reservation carriages may also be secured at a small charge. The legal tariff is $2 per hour, but special terms can be made. Besides the price agreed upon for the carriage, the tourist will have to pay all tolls. There are scarcely any points of interest connected with the Falls which are not within walking distance; especially if a day or two can be devoted to the American side and the same length of time to the Canadian side. Moreover, in the number and variety of the attractions seen, the pedestrian will be apt to enjoy an advantage over the carriage-traveler. *Street-cars* run to Suspension Bridge.

The Falls of Niagara are situated on the Niagara River, about 22 miles from Lake Erie and 14 miles from Lake Ontario. This river is the channel by which all the waters of the four great upper lakes flow toward the Gulf of St. Lawrence, and has a total descent of 333 ft., leaving Lake Ontario still 231 ft. above the sea. From the N. E. extremity of Lake Erie the Niagara flows in a N. direction with a swift current for the first 2 miles, and then more gently with a widening current, which di-

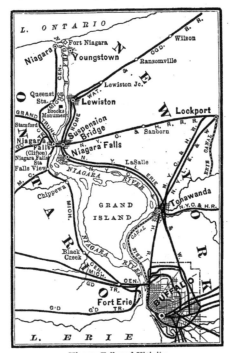

Niagara Falls and Vicinity.

vides as a portion passes on each side of Grand Island. As these unite below the island, the stream spreads out to 2 or 3 miles in width, and appears like a quiet lake studded with small, low islands. About 16 miles from Lake Erie the current becomes narrow and begins to descend with great velocity. This is the commencement of the Rapids, which continue for about a mile, the waters accomplishing in this distance a fall of 52 ft. The Rapids terminate below in a great cataract, the descent of which is 164 ft. on the American side and 158 ft. on the Canadian. At this point the river, making a curve from W. to N., spreads out to an extreme width of 4,750 ft., embracing Goat and the Three Sister Islands. Goat Island, which extends down to the brink of the cataract, occupies about one fourth of this space, leaving the river on the American side about 1,100 ft. wide, and on the Canadian side about double this width. The line along the verge of the Canadian Fall is much longer than the breadth of this portion of the river, by reason of its horseshoe form, the curve extending up the central part of the current. The waters sweeping down the Rapids form a grand curve as they fall clear of the rocky wall into the deep pool at the base. In the profound chasm below the fall, the current, contracted in width to less than 1,000 ft., is tossed tumultuously about, and forms great whirlpools and eddies as it is borne along its rapidly descending bed. Dangerous as it appears, the river is here crossed by small row-boats, which are reached from the banks above by an inclined railroad, and the Maid of the Mist, a small steamer, makes frequent trips to the edge of the Falls. For 7 miles below the Falls the narrow gorge continues, varying in width from 200 to 400 yards. The river then emerges at Lewiston into a lower district, having descended 104 ft. from the foot of the cataract.

The gorge through which the Niagara River flows below the Falls bears evidence of having been excavated by the river itself. Within comparatively recent years changes have taken place by the falling down of masses of rock, the effect of which has been to cause a slight recession of the cataract, and extend the gorge to the same degree upward toward Lake Erie. Thus in 1818 great fragments descended at the Horseshoe Fall, and since 1855 several others, which have materially changed the aspect of the Falls. Table Rock, once a striking feature, has wholly disappeared. Sir Charles Lyell estimates the rate of recession to be about a foot a year, but the rate is not uniform. For several successive years there will be no apparent change; and then, the soft underlying strata having been gradually worn away, great masses of the upper harder ones fall down, causing a very noticeable change in a very brief time. At the present site of the Falls a layer of hard limestone rock, of the formation known as the Niagara limestone, covers the surface of the country, and forms the edge of the cataract to the depth of between 80 and 90 ft. Professor James Hall, State geologist of New York, points out that, after a further recession of about 2 miles, this limestone layer, with the soft layers under it, will have been swept away, and the Fall will become almost stationary on the lower sandstone formation, with a height of only 80 feet. As, however, it will take rather more than 10,000 years to excavate this 2 miles, the tourists of our day need feel no alarm lest the stupendous torrent dwindle beneath their gaze! In regard to the volume of water which passes over the Falls, Sir Charles Lyell estimates it at 90,000,000,000 cubic ft. an hour.

*Goat Island is the point usually visited first. It is reached by a bridge 360 ft. long, the approach to which is about midway in the State reservation. The bridge itself is an object of interest, from its apparently dangerous position. It is, however, perfectly safe, and is crossed constantly by heavily laden carriages. The view of the * **Rapids** from the bridge is one of the most impressive features of the Niagara scenery. The river descends 52 feet in a distance of three quarters of a mile by this inextricable turmoil of waters. Below the bridge, a short distance from the verge of the American Fall, is *Chapin's Island*, so named in memory of a workman who fell into the stream while at work on the bridge. He lodged on this islet and was rescued by a Mr. Robinson, who gallantly went to his relief in a skiff. About midway of the stream the road crosses *Bath Island*. From the island end of the bridge three paths diverge, that to the right being the one usually followed.

A short walk brings us to the foot-bridge leading to **Luna Island,** a huge rock-mass of some three quarters of an acre, lying between the Center Fall and the American Fall. The exquisite lunar rainbows seen at this point, when the moon is full, have given it the name it bears. The width of the **American Fall** from Luna Island is over 1,100 ft., and the precipice over which it plunges is 167 ft. high. Just beyond Luna Island a spiral stairway (called "Biddle's Stairs," after Nicholas Biddle, of United States Bank fame, by whose order they were built) leads to the foot of the cliff. From the foot of the stairs, which are secured to the rocks by strong iron fastenings, there are two diverging paths. That to the right leads to the * **Cave of the Winds,** a spacious recess back of the Center Fall. Guides and water-proof suits for visiting the Cave may be obtained at the stairs (fee, $1.00), and the excursion is well worth making. You can pass safely into the recess behind the water, to a platform beyond. Magical rainbow pictures are found at this spot; sometimes bows of entire circles, and two or three at once, are seen. A plank-walk has been carried out to a cluster of rocks near the foot of the fall, and from it one of the best * views of the American Fall may be obtained. The up-river way, along the base of the cliff toward the Horseshoe Fall, is difficult, and much obstructed by fallen rocks. It was from a point near Biddle's Stairs that the renowned jumper, Sam Patch, made two successful leaps into the waters below (in 1829), saying to the throng of spectators, as he went off, that "one thing might be done as well as another." Reascending the stairs, a few minutes' walk along the summit of the cliff brings us to a bridge leading to the islet on which stood the famous Terrapin Tower, which, having become dangerous, was blown up with gunpowder in 1873. The view of the * **Horseshoe Fall** from this point is surpassingly grand. The mighty cataract has a contour of 3,010 ft., with a perpendicular plunge of 158 ft., and it is estimated that 15,000,000 cubic feet of water pass over the ledges every minute. One of the condemned lake-ships (the Detroit) was sent over this Fall in 1829; and though she drew 18 ft. of water, she did not touch the rocks in passing over the brink of the precipice, showing that the water is at least 20 ft. deep *above* the ledge. Lately the rocks have broken away, altering the appearance of

Niagara Falls from Prospect Park.

the Horseshoe Falls. Also, the building of the water-works pump-house has changed the formation of the water, to carry it away from this point; so tourists now can not get behind the main sheet of water.

At the other end of Goat Island (reached by a road from the Horseshoe Fall), a series of graceful bridges leads to the * **Three Sisters,** as three small islets lying in the Rapids are called. The islands are rugged masses of rock, covered with a profuse and tangled vegetation, and afford fine views of the Rapids at their widest and wildest part. On Goat Island, near the Three Sisters, is the *Hermit's Bathing-Place,* so called after Francis Abbott, "the Hermit of Niagara," who used to bathe here, and who was finally drowned while doing so. At the foot of Grand Island, near the Canada shore, is *Navy Island,* which was the scene of some interesting incidents in the Canadian Rebellion of 1837–'38, known as the Mackenzie War. *Chippewa,* which held at that period some 5,000 British troops, is upon the Canadian shore, nearly opposite. It was near *Schlosser Landing,* about 2 miles above the Falls, on the American side, that during the war the American steamer Caroline, which had been perverted to the use of the insurgents, was set on fire and sent over the Falls, by the order of Sir Allan McNab, a Canadian officer. Above Navy Island is *Grand Island* (17,000 acres), somewhat noted as the spot on which, in 1820, Major Mordecai M. Noah founded "Ararat, a city of refuge for the Jews," in the vain hope of assembling there all the Hebrew population of the world.

The State of New York purchased, in 1885, the property bordering the Falls, and laid out **Niagara Park,** to be controlled by a State Commission, empowered to remove all obstructions to the view, and to improve the grounds. No charge is made for admission to Niagara Park. A "vertical railway," running on a steep incline, leads from the Park to the base of the cliff; and from its foot the river may be crossed in the steamboat the *Maid of the Mist* (trip, 50c.). The passage across the river is perfectly safe, and is worth making for the very fine * view of the Falls obtained in mid-stream. A winding road along the cliff-side leads from the landing on the Canadian side to the top of the bluff near the bridge. By climbing over the rocks at the base of the cliff on the American side (turn to the left after descending the railway), the tourist may penetrate to a point within the spray of the American Fall, and get what is perhaps, on the whole, the finest view of it to be had.

The usual way of crossing to the Canadian side is over the * **Suspension Bridge,** which arches the river about ⅓ of a mile below the Falls, and is one of the curiosities of the locality (fee for pedestrians, 25c.). It was finished in 1869, at a cost of $175,000; is 1,190 ft. from cliff to cliff, 1,268 ft. from tower to tower, and 190 ft. above the river; and it was widened in 1888, all the wooden parts of the structure being replaced by iron. It was carried away by the gale of January 10, 1889, but has been rebuilt. The tower on the American side is 100 ft. high, and that on the Canadian side 105 ft. Here is situated the *Clifton House,* from which a fine view of the Falls is obtained.

A road to the left from the bridge terminus leads along the cliff, affording good views of the American and Center Falls. A short distance above the terrace near the Falls is the spot still called *Table Rock, though the immense overhanging platform originally known by that name has long since fallen over the precipice. From this point a general view of the Falls is obtained, and that of the Horseshoe Fall is incomparably grand. The concussion of the falling waters with those in the depths below produces a spray that veils the cataract two thirds up its height. Above this impenetrable foam, to the height of 50 ft. above the Fall, a cloud of lighter spray rises, which, with the prevailing southeast wind blows the spray over the buildings and surroundings. The appropriateness of the name Niagara ("Thunder of Waters") is very evident here. Continuing on through Queen Victoria Park we come to *Dufferin Islands*, from which one of the finest views of the rapids above the Falls is to be had.

Guides and water-proof suits for the passage under the Horseshoe Falls may be procured at Table Rock (fee, 50 c.). This passage (which no nervous person should attempt) is described as follows in "Picturesque America" : "The wooden stairways are narrow and steep, but perfectly safe ; and a couple of minutes brings us to the bottom. Here we are in a spray-land indeed ; for we have hardly begun to traverse the pathway of broken bits of shale when, with a mischievous sweep, the wind sends a baby cataract in our direction, and fairly inundates us. The mysterious gloom, with the thundering noises of the falling waters, impresses every one ; but, as the pathway is broad, and the walking easy, newcomers are apt to think there is nothing in it. The tall, stalwart negro, who acts as guide, listens with amusement to such comments, and confidently awaits a change in the tone of the scoffers. More and more arched do the rocks become as we proceed. The top part is of hard limestone, and the lower of shale, which has been so battered away by the fury of the waters that there is an arched passage behind the entire Horseshoe Fall, which could easily be traversed if the currents of air would let us pass. But, as we proceed, we begin to notice that it blows a trifle, and from every one of the 32 points of the compass. At first, however, we get them separately. A gust at a time inundates us with spray ; but the farther we march the more unruly is the Prince of Air. First, like single spies, come his winds ; but soon they advance like skirmishers ; and, at last, where a thin column of water falls across the path, they oppose a solid phalanx to our efforts. It is a point of honor to see who can go farthest through these corridors of Æolus. It is on record that a man, with an herculean effort, once burst through the column of water, but was immediately thrown to the ground, and only rejoined his comrades by crawling face downward, and digging his hands into the loose shale of the pathway. Professor John Tyndall has gone as far as mortal man, and he describes the buffeting of the air as indescribable, the effect being like actual blows with the fist."

Termination Rock is a short distance beyond Table Rock at the verge of the Fall. The spray here is blinding and the roar of waters deafening.

Below the Falls are several points of interest, which are best visited on the American side. The first of these is the *Suspension Bridge*, which spans the gorge 2 miles below the Falls, and supports railway-tracks, a roadway, and footways. The bridge is 245 ft. above the water, and supported by towers on each bank, the centers of which are 821 ft. apart. It was built in 1855 by John A. Roebling, and cost $500,000. The fee for crossing the bridge is 25c. for pedestrians, which confers the right to return free on the same day. From one side of this bridge a fine distant view of the Falls is had, and from the other a bird's-eye view of the seething and tumultuous * *Whirlpool Rapids.* Three hundred feet above may be seen the Michigan Central R. R. Cantilever Bridge (see p. 181). By descending the elevator which leads from the top to the base of the cliff near the site of the old Monteagle House, a nearer view is obtained of these wonderful Rapids, in which the waters rush along with such velocity that the middle of the current is 30 ft. higher than the sides. Three miles below the Falls is the * **Whirlpool,** occasioned by a sharp bend in the river which is here contracted to a width of 220 ft.

Since the acquisition of the land near the Falls on the American side by the State of New York, the labors of the commission appointed to lay out the ground have caused many changes in the old landmarks, and still more are contemplated. Unsightly buildings have been removed, roads and pathways laid out, and trees planted where they add to the beauty of the scene. The great tunnel in course of construction, by means of which power will be obtained from the waterfall, is on the American side.

In the vicinity of Niagara is *Lewiston* (7 miles N.), at the head of navigation on Lake Ontario; and directly opposite (on the Canadian side) is Queenston. *Queenston* is well worth a visit, and affords a pleasant drive from the Falls. It is historically as well as pictorially interesting. Here General Brock and his aide-de-camp fell, October 11, 1812. * **Brock's Monument,** which crowns the heights above the village, is 185 ft. high, surmounted by a dome of 9 ft., which is reached by a spiral flight of 250 steps inside. The remains of Brock and his comrade lie in stone sarcophagi beneath, having been removed thither from Fort George. This is the second monument erected on the spot, the first having been destroyed by the scoundrel Lett, in 1840. At *Drummondville*, 1½ mile W. of the Falls, is a tower which overlooks the battle-field of Lundy's Lane. At the mouth of the Niagara River, on the Canadian side, is *Niagara on the Lake*, a favorite summer resort (*Queen's Royal Hotel*).

41. New York to Buffalo and Niagara Falls.

Via Erie Railway. Distances : To Middletown, 64 miles ; to Port Jervis, 88 : to Susquehanna, 193 ; to Binghamton, 215 ; to Elmira, 274 ; to Hornells-ville, 332 ; to Buffalo, 423 ; to Niagara Falls, 442. The time to Buffalo or Niagara Falls is about 14 hours, and the fare $9.25.

THE Erie Railway is one of the greatest triumphs of engineering skill in this or any other country, and affords some of the grandest and most varied scenery to be found E. of the Rocky Mountains. Prior to its construction, portions of the line were considered impassable to any other than a winged creature, yet mountains were scaled or pierced and river-cañons passed by blasting a path from the face of stupendous precipices ; gorges of fearful depth were spanned by bridges swung into the air ; and broad, deep valleys crossed by massive viaducts. The road was begun in 1836 and completed in 1851, and has cost to date upward of $60,000,000. Palace drawing-room and sleeping cars are attached to all the through trains.

The terminal station in Jersey City is reached by ferry from foot of Chambers St. and W. 23d St. Leaving Jersey City the train traverses a series of salt marshes, and in 17 miles reaches **Paterson** (*Hamilton House, Franklin*), a busy manufacturing city of 78,347 inhabitants, situated on the right bank of the Passaic River immediately below the falls. It was founded in 1791 by Alexander Hamilton, in the cotton interest, and its cotton-factories are now very extensive. Its most important interest, however, is silk-manufacturing, for which it has 30 factories, employing about 8,000 persons, and turning out a product of $4,000,000. It has also extensive manufacturing interests in velvet, woolen, linen, locomotives, heavy machinery. The total product of Paterson manufactures is upward of $20,000,000, employing a capital of about $10,000,000. Next to Newark it is the largest manufacturing city of New Jersey. The *Passaic Falls* have a perpendicular descent of 50 ft., and the scenery in the vicinity is very picturesque. A small and rugged park surrounds them; and on a hill in the vicinity are a costly Soldiers' Monument and a belvedere tower whence there is a fine view. Beyond Paterson the route traverses a fertile but uninteresting country, and just this side of *Suffern* (32 miles) crosses the boundary-line and enters New York State. From Suffern a branch line runs in 18 miles to Piermont on the Hudson River (see p. 68). Here the beautiful Ramapo Valley begins, and the scenery becomes increasingly picturesque and impressive. *Ramapo* (34 miles) is near Torne Mountain (1,189 ft.), from the summit of which there is a wide-extended view. During the campaign of 1777 Washington often ascended this mountain to watch the movements of the British army and fleet around New York. Beyond *Sloatsburg* (36 miles), on the right, are seen the ruins of the Augusta Iron-Works, where was forged the chain that was stretched across the Hudson to check the advance of the British ships. The next station is *Tuxedo*, well known for its park, which contains the cottages of many of the fashionable set of New York city. At *Turner's* (48 miles) a branch road diverges to Newburg on the Hudson (see p. 71). From *Monroe* (50 miles) and also from *Greycourt* (54 miles) stages run in 8 to 10 miles to * **Greenwood Lake** (*Brandon, Traphagen,* and *Windermere*), a highly popular summer resort, which is also reached from New York *via* New York & Greenwood Lake R. R. (48 miles). This "miniature Lake George" is a beautiful, river-like body of water, 10 miles long and 1 mile wide, nearly inclosed by mountains, and offering some extremely picturesque scenery. Its waters are clear and deep, and abound in fish. A small steamer plies on the lake, making daily trips. In the vicinity are the smaller but scarcely less charming Lakes Macopin, Sterling, and Wawayanda. **Turner's** is the most attractive station on this portion of the line, and is near some lovely little lakes. The view from the hill N. of the station is superb, the Hudson River, with Fishkill and Newburg, being in sight.

From Greycourt another branch line runs to Newburg in 18 miles, and the Lehigh & Hudson R. R. diverges to Warwick. On the main line, 6 miles beyond Greycourt, is the pretty little village of *Goshen*, one of the capitals of Orange County, and celebrated for its milk, butter, and cheese.   Here the Wallkill Valley R. R. diverges and runs in 43 miles to Kingston and Rondout on the Hudson (see p. 71), passing *New Paltz*, whence stages run in 6 miles to *Lake Mohonk* (see p. 71).   Seven miles beyond Goshen, at the crossing of the New York, Ontario & Western R. R., is the busy manufacturing village of **Middletown** (see p. 194).   Four miles beyond, at *Howell's* (71 miles), the most picturesque section of the line begins, and fine views are had all the way to Port Jervis.   On approaching *Otisville* (76 miles), the eye is attracted by the bold flanks of the Shawangunk Mountain, the passage of which great barrier (once deemed insurmountable) is a fine achievement of engineering skill.   A mile beyond Otisville, after traversing an ascending grade of 40 ft. to the mile, the road runs through a rock-cutting 50 ft. deep and 2,500 ft. long. This passed, the summit of the ascent is reached, and thence we go down the mountain's side many sloping miles to the valley beneath, through the midst of grand and picturesque scenery.   Onward the way increases in interest, until it opens in a glimpse, away over the valley, of the mountain-spur known as the *Cuddeback ;* and at its base the glittering water is seen, now for the first time, of the Delaware & Hudson Canal.   Eight miles beyond Otisville we are imprisoned in a deep cut for nearly a mile, and, on emerging from it, there lie spread before us (on the right) the rich and lovely valley and waters of the *Neversink*.   Beyond sweeps a chain of blue hills, and at their feet, terraced high, gleam the roofs and spires of the town of *Port Jervis* (88 miles) ; while to the S. the eye rests upon the waters of the Delaware, along the banks of which the line runs for the next 90 miles.

Port Jervis (*Delaware House, Fowler House*) is situated at the confluence of the Delaware and Neversink Rivers, and contains 9,327 inhabitants.   Extensive railroad-shops are located here, and it is the terminus of the E. division of the Erie road.   The scenery in the vicinity is delightful, and the village itself is a very pretty one. Riding, driving, hunting, and fishing may be enjoyed to any extent, and many summer visitors are attracted to it.   *Point Peter* is ascended from the village, and affords a pleasing outlook over the Delaware and Neversink Valleys.   Six miles distant are the *Falls of the Sawkill*, where a mountain-brook is precipitated 80 ft. over two perpendicular ledges of slate-rock into a wild and romantic gorge.   **Milford,** a lovely mountain-surrounded town, six miles below Port Jervis, with which it is connected by stages, has become a favorite summer-resort, with several good hotels.   Shortly beyond are the beautiful falls of the *Raymondskill*, and there are fine trout-streams in the neighborhood.

Three miles beyond Port Jervis the train crosses the Delaware into the State of Pennsylvania, which it traverses for 26 miles to Delaware Bridge, where it again enters New York.   Near *Shohola* (107 miles) some of the greatest obstacles of the entire route were encountered, and for several miles the roadway was hewed out of the solid cliff-side at a cost of $100,000 a mile.   *Lackawaxen* (111 miles) is a pretty village at the confluence of the Lackawaxen Creek and Delaware River.   Here the Delaware is spanned by an iron suspension-bridge supporting the aqueduct by which the D. & H. Canal crosses the river. At *Narrows-burg* (122 miles) the river is compressed by two points of rock into a channel 100 ft. deep. Beyond Narrowsburg for some miles the scenery is uninteresting and the stations unimportant. Near *Callicoon* (136 miles) is the romantic and trout-teeming Callicoon Brook; and *Hancock* (164 miles) is attractively situated. At *Deposit* (177 miles) the train leaves the valley of the Delaware and begins the ascent of the high mountain-ridge which separates it from the lovely valley of the Susquehanna. As the train descends into the latter valley there opens suddenly ōn the right a * picture of rare and bewitching beauty. This first glimpse of the Susquehanna is esteemed one of the finest points of the varied scenery of the Erie route. A short distance below, the train crosses the great * *Starucca Viaduct,* 1,200 ft. long and 110 ft. high, constructed at a cost of $320,000, and spanning the Starucca Valley with 18 arches. From *Susquehanna* (193 miles) the viaduct itself is a most effective feature of the valley views. **Susquehanna** (*Starucca Hotel,* at the station) contains the vast repair-shops of the company, and is one of the stopping-places for meals.

For a few miles beyond Susquehanna the route still lies amid mount ain-ridges, but these are soon left behind, and the train enters upon a beautiful hilly and rolling country, thickly dotted with villages and towns. At *Great Bend* (201 miles) the Delaware, Lackawanna & Western R. R. comes in from the Pennsylvania coal-fields. *Kirkwood* (206 miles) claims rivalry with Sharon, Vt., as the birthplace of Joseph Smith, the Mormon prophet. Nine miles beyond Kirkwood is **Binghamton** (*Bennett, Crandell, Exchange,* and *Lewis*), an important railroad and manufacturing center, pleasantly situated on a wide plain in an angle formed by the confluence of the Susquehanna and Chenango Rivers. It contains 35,005 inhabitants, and is a leading seat of the coal and iron industry. Six railways converge here, and, besides large manu-facturing interests, there is an extensive trade with the adjacent country. The *Court-House* (on Court St.) is a handsome building, the *Bank Building* (cor. Court and Chenango Sts.) is another, and there are several fine churches. The *New York State Military Store-House* is a fine structure. The *Asylum for the Chronic Insane* is a vast stone structure on a commanding eminence a mile from the city (reached by street-cars). Other interesting institutions are the *Susquehanna Orphan Asylum* and *St. Mary Orphan Asylum.* This place is one of the largest cigar-manufacturing cities in the United States, and it is also noted for its leather and boot and shoe interests. On the far-view-ing *Mt. Prospect* is a popular water-cure hotel.

Twenty-two miles beyond Binghamton is **Owego** (*Ahwaga House, Central House*), a prosperous town of 9,008 inhabitants, situated on the Susquehanna at the mouth of Owego Creek, and surrounded by pleasing scenery. The Lehigh Valley R. R. connects here, and the Cayuga Division of the Del., Lack. & Western R. R. runs N. E. 35 miles to Ithaca (see Route 40). *Evergreen Cemetery* is on the N. side of the Susquehanna River on a hill 200 ft. high, which commands fine views. On Owego Creek, a short distance from the village, is *Glenmary*, once the home of N. P. Willis and the place where he wrote his charming "Letters from under a Bridge." Beyond Owego, passing several small stations, of which *Waverly* (256 miles) is the principal, the train runs in 37 miles to **Elmira** (*Delevan, Rathbun,* and *Wyckoff*), the largest city of the Southern Tier, with 29,708 inhabitants, and extensive manufactures, among which are the car-shops of the Erie R. R., the *Elmira Iron and Steel Rolling-Mills,* and the engine-works of B. W. Payne & Sons. *Water St.* is the business thoroughfare. The *Court-House* is a handsome edifice, and the *Elmira Female College* has a building in the N. portion of the city. E. of the city is the *Elmira Water-Cure,* which is well patronized. The *State Reformatory* and the *Southern Tier Orphans' Home* are also located here. The Northern Central R. R. (Route 59) intersects the present route at Elmira. The Lehigh Valley R. R. also comes in from the coal-regions of Pennsylvania; and the Elmira, Cortland & Northern R. R. runs in 50 miles to Ithaca (see Route 40). The manufactures are largely on the increase, and among them the La France Fire-Engine Works are noted. **Corning** (291 miles) is a busy manufacturing village of 8,550 inhabitants, on the Chemung River. The Rochester Div. of the Erie R. R. diverges here from the main line toward the N. W., and the Tioga Div. of the Erie R. R. enters from the S. W. The Delaware, Lackawanna & Western R. R. intersects the city from the W. and from New York.

The Rochester Division of the Erie R. R. runs N. W. to Rochester in 95 miles and to Attica (111 miles), where a junction is made with the Buffalo Div. described below. The distance from New York to Buffalo by this route is 433 miles, being 10 miles longer than the route *via* Hornellsville. There are many small towns between Corning and Rochester, of which the most important is *Bath,* a thriving manufacturing village, surrounded by a rich and populous agricultural country. *Avon* (76 miles from Corning and 19 from Rochester) is noted as the site of the much-frequented **Avon Springs** (*Avon Springs Hotel, Livingston, Sanitarium*). The springs, 3 in number, are about a mile S. W. of the station. and the Lower Spring discharges 54 gallons a minute. The waters are saline-sulphurous, are taken both internally and. in the form of baths, and are considered remedial in rheumatism, indigestion, and cutaneous diseases. **Rochester** (386 miles from New York) is described in Route 40.

Beyond Corning the main line runs for 2 miles parallel with the Rochester Division, then passes six small stations, and in 41 miles reaches **Hornellsville,** a place of 10,996 inhabitants, with extensive repair-shops, engine-houses, etc. Here the Buffalo and Western Divisions diverge; the former running N. W. to Buffalo and Niagara Falls (described below), and the latter running almost due W. to Dunkirk on Lake Erie. The section between Hornellsville and Dunkirk is the least attractive of the Erie line, the country being comparatively unsettled, and no important towns having grown up within it. Soon after leaving Hornellsville the train enters the valley of the Canisteo River, on the banks of which are the hamlets of Almon and Alfred. At *Tip-Top Summit* the road reaches its highest point (1,760 ft. above tide-water), and the descent is begun into the valley of the Genesee. The country *en route* is peculiarly wild and lonely, desolate and somber forest tracts alternating with the stations and little villages along the line. At *Cuba Summit* the train crosses the Alleghany water-shed, 1,680 ft. above the sea. and just beyond are many brooks and glens of rugged beauty. Passing *Olean* (396 miles) and *Carrollton* (408 miles) the route enters the Reservation of the Seneca Indians (embracing 42 square miles) and follows the wild banks of the Alleghany River, flow-ing amid hills as wild and desolate as itself. At *Salamanca* (414 miles) the New York, Lake Erie & Western R. R. connects with the Erie and forms the route taken by the "through trains" to the West (see Route 68). Beyond Salamanca the Erie traverses for 47 miles a dreary and monotonous forest region, and reaches its terminus at **Dunkirk,** a place of 9,416 inhabitants on Lake Erie, 460 miles from New York. Dunkirk has a safe and commodious harbor, protected by a breakwater, considerable trade, and some manufactures. Connection is made here with the Lake Shore & Michigan Southern R. R., which runs from Buf-falo W. to Cleveland and Chicago (see Route 67).

The Buffalo Division.

Leaving Hornellsville the train passes a number of small stations, and in 30 miles (362 miles from New York) reaches **Portage** (*Ingham Hotel*), the most attractive point on the entire Erie line. Here are the * **Portage Falls,** 3 in number, and each of sufficient beauty to re-pay the tourist for the journey from New York. They are formed by the descent of the Genesee River from the plateau on which it has flowed tranquilly for many miles to the lake-level. The Upper or *Horseshoe Falls* are just below the R. R. bridge and have a vertical de-scent of 70 ft. Half a mile below is the * *Middle Fall,* where the river plunges 110 ft. into a chasm formed by perpendicular ledges of rock. The action of the water has worn a cave or hollow in the W. bank, which is called the *Devil's Oven.* In times of high water this cavern is submerged; but when the river is low it will hold 100 people. For 2 miles below the Middle Fall the river rushes through a deep and narrow gorge, and at the * *Lower Falls* roars down a wonderful series of cas-cades and rapids, descending 150 ft. in ½ mile. The railroad crosses the river on an iron * bridge, 818 ft. long and 234 ft. high. The Upper Falls are visible from the bridge, but no idea of their grandeur can be formed until they are seen from below. Also visible from the bridge is the long *Aqueduct* by which the Genesee Canal crosses the river.

Six miles beyond Portage is *Gainesville* (368 miles), whence a rail-way runs in 7 miles to the lovely **Silver Lake,** where the sea-serpent was said to have been seen in 1855. *Warsaw* (375 miles) is a pleasant village, at the entrance of the romantic O-at-ka Valley, and surrounded by rich pastoral scenery. It is much visited in summer. There are numerous salt wells and mines in this vicinity. At *Attica* (392 miles) the Rochester Div. (see above) joins the main line, which then passes on to **Buffalo** (423 miles) and **Niagara Falls** (442 miles). Both these places are described in Route 40.

42. New York to Buffalo and Niagara Falls.

Via West Shore R. R.

Distances : Albany, 141 miles ; Utica, 232 miles ; Syracuse, 278 miles ; Palmyra, 338 miles ; Rochester, 359 miles ; Elba, 387 miles ; Buffalo, 428 miles ; Suspension Bridge, 450 miles.

THE West Shore R. R. traverses the W. shore of the Hudson, and west from Albany runs nearly parallel with the N. Y. Central R. R., touching many of the same cities.

The W. shore of the Hudson has already been described in Route 9, but a brief account of the route pursued by the West Shore R. R. to Albany will be of interest. Leaving the R. R. depot at Jersey City or *Weehawken*, the trains reach **Hackensack** (8 miles), the first station of importance, passing in rear of the Palisades. This is the county town of Bergen County, and contains a population of 6,004. It is also on the line of the New York, Susquehanna & Western and New Jersey & New York Railroads. It is a considerable manufacturing place, and has 12 churches and several banks and newspaper-offices. *Tappan* (19 miles) is a small village about 1¼ mile W. of the Hudson River, where the unfortunate Major André was tried and executed as a spy, in October, 1780. *Nyack* (24 miles) is described on page 69. At **Haverstraw** (33 miles), a town of 5,170 inhabitants, situated on Haverstraw Bay, and well known for its extensive brick-works, and at the foot of the Ramapo Hills, the road strikes the river-shore. The scenery is exceedingly picturesque here, showing the Highlands in the distance. Passing *Caldwell's, Iona Landing,* and *Fort Montgomery,* * **West Point** is reached (47 miles). This beautiful spot, noted as the seat of the U. S. Military Academy, and as a charming summer resort, is described on page 70. *Cornwall Landing,* 5 miles above, is one of the most popular summer-places on the river. * **Newburg,** on the W. shore of Newburg Bay (see page 71), is a prosperous city, round which many interesting Revolutionary memories cluster (23,087 population). Here the West Shore R. R. connects with a branch of the Erie R. R. and with the N. Y. & New England R. R. (by ferry to Fishkill Landing, on the E. side of the river). *Marlboro, Highland,* and *West Park,* are small stations intervening before we reach **Esopus** (80 miles), a thriving village of 3,448 population. Kingston, 8 miles beyond, is described on page 71. Here the railway connects with the Ulster & Delaware and Wallkill Valley Railroads, giving easy access to several popular resorts in the Catskill Mountains. **Catskill,** 22 miles N. (4,920 population), is the point of departure for the mountains *via* the Catskill Mts. R. R., which connects here (see page 71). The branch to Albany (see page 72) diverges from *Coeyman's Junction,* 18 miles N., reaching that city in a run of 13 miles.

After leaving the Junction (128 miles N. of New York), the first station of importance is the city of * **Schenectady** (see Route 40), the West Shore station being in what is known as South Schenectady (152 miles). At *Rotterdam Junction,* 7 miles beyond, connection is made

13

with the Fitchburg R. R., giving a through line from Boston and other eastern points to the West. The road now begins to skirt the S. bank of the Mohawk River. There are a number of small stations, *Patter-sonville, Fort Hunter, Auriesville, Fultonville,* and *Downing,* before we reach *Canajoharie,* a thriving town of 2,089 inhabitants (190 miles). From *Fort Plain,* 3 miles beyond, to **Utica,** the railroad passes through the same towns with the New York Central & Hudson River R. R. (see Route 40). Small towns intervene, among which may be mentioned *Oneida Castle,* where connection is made with the New York, Ontario & Western R. R., and *Canastota,* a thrifty manufacturing town of 2,774 inhabitants, through which also pass the Elmira, Cortland & Northern R. R., the New York Central & Hudson River R. R., and the Erie Canal. *Chittenango* is described in Route 40, and there are no towns of moment till we reach the city of **Syracuse** (278 miles) (see Route 40). Here the railroad makes connection with its Chittenango branch, the Delaware, Lackawanna & Western, and the Rome, Water-town & Ogdensburg Railroads. *Weedsport* (299 miles) has 1,580 population, and is a place of some manufacturing interest. It is on the Erie Canal, and the Auburn Div. of Lehigh Valley R. R. passes through it. At *Montezuma* (307 miles) the navigable outlet of Cayuga Lake flows into the Erie Canal. *Clyde, Lyons, Newark, Palmyra, Macedon, Fairport,* and *Pittsford,* are the most important towns on the road between Weedsport and Rochester, all being thriving manu-facturing places. At Genesee Junction, 7 miles before we reach Roches-ter, the West Shore R. R. makes connection with the Western New York & Pennsylvania R. R. **Rochester** (359 miles) is described in Route 40. Between this city and Buffalo (428 miles), described in Route 40, there are no places of much importance on the present route. From Buffalo the West Shore R. R. runs over the track of the Sus-pension Bridge Division of the Erie R. R. to *Suspension Bridge* (see Route 40), where it makes through connection with the Great Western Division of the Grand Trunk R. R. for points West.

The *New York, Ontario & Western R. R.* runs to Cornwall, and thence passes in a N. W. direction to Oswego on Lake Ontario, running through cars from New York to the Thousand Isles, nearly bisecting the State. The route is through a less densely settled portion of the State, but the country is very beautiful and picturesque, the sport with rod and gun excellent along the whole line of the road from Cornwall, and the many small towns which dot the route are rapidly attracting summer visitors who wish to unite economy with pleasure and comfort. After leaving Cornwall (52 miles), a number of small stations intervene before we reach *Campbell Hall,* whence the Wallkill Valley R. R. runs to Kingston on the Hudson. The first station of importance is **Mid-dletown** (78 miles), a flourishing town of 11,977 population. It is at the junc-tion with the Erie R. R., and the New York, Susquehanna & Western R. R., has several iron and woolen manufactures, and is the seat of the *Homœopathic State Insane Asylum.* At *Sidney* (200 miles) is the junction with the Susque-hanna Division of the Delaware & Hudson R. R., and the New Berlin Branch. **Norwich,** a town of 5,212 population (225 miles), is at the junction with the Delaware, Lackawanna & Western R. R., and contains a number of blast-fur-naces, tool-works, machine-shops, foundries, breweries, tanneries, etc. It has a handsome stone court-house, eight churches, and several banks and news-paper offices. At *Randallsville* (244 miles) is the crossing with the Rome & Clinton Div. of the Delaware & Hudson R. R. *Oneida Castle,* 23 miles farther on, is at the crossing with the West Shore R. R., and at *Oneida* (267 miles), we reach the junction with the N. Y. Central R. R.   At *Central Square* (298 miles) and *Fulton* (313 miles), connections are made with the Rome, Watertown & Ogdensburg R. R.   In a ride of 12 miles farther we reach the beautiful city of **Oswego** on the lake (322 miles, see Route 55).

43. New York to Buffalo.

Via Delaware, Lackawanna & Western R. R.

Distances : Paterson, 15 miles ; Manunka Chunk, 77 miles ; Water-Gap, 88 miles ; Scranton, 144 miles ; Binghamton, 207 miles ; Elmira, 263 miles ; Bath, 298 miles ; Mt. Morris, 346 miles ; Rochester and Pittsburg Junction, 363 miles ; Buffalo, 409 miles.   Time, 12¼ hours.

PASSENGERS take the ferry-boat from the foot of Barclay St. or Christopher St. to the station in Hoboken.   Leaving Hoboken, the train passes through the Bergen Tunnel to *Passaic*, at the head of tide-water and navigation on the Passaic River, a few miles below the Great Falls, where the river cleaves through the mountains.   The population of Passaic is 13,028, and the city contains many important manufacturing establishments, dye and print works, woolen and worsted works, bleacheries, planing-mills, and foundries.   **Paterson** (*Franklin, Hamilton*) is a busy manufacturing town, described in Route 41 (page 188).

At *Washington* is the junction of the Morris & Essex Div. with the main line, the former passing on to *Easton* (see Route 45), the latter leading in 11 miles to **Manunka Chunk,** just before reaching which the train pierces through the Manunka Chunk Mt. by the Voss Gap Tunnel, 1,000 feet long.   Here the Belvidere Div. of the Pennsylvania R. R. comes in.   Eleven miles farther on, through a rugged and romantic country, the road leads to the

Delaware Water-Gap.

Hotels, etc.—The *Kittatinny House,* standing on the mountain-side above the railway station, is an old and favorite resort.   The *Water-Gap House* is a spacious hotel on the summit of Sunset Hill.   The *Mountain House* and the *Glenwood* are smaller.   Prices at these hotels are from $2 to $3 a day, $10 to $16 a week.   *Fare* from New York to the Water-Gap, $2.55 ; excursion, $3.80. It can be reached from Philadelphia by Route 55, and from New York *via* the New York, Susquehanna & Western R. R.   Distance, 98 miles.

The Delaware Water-Gap is where the Delaware River, after a journey of about 200 miles through a wild, rugged, and romantic country, forces its way through the Kittatinny or Blue Mountains.   The Gap is about 2 miles long, and is a narrow gorge between walls of rock some 1,600 ft. in height, and so near to each other at the S. E. entrance as hardly to leave room for the river and the railroad.   The valley N. of the Blue Ridge and above the Gap bore the Indian name of Minnisink, or "Whence the waters are gone."   "Here a vast lake once probably extended; and whether the great body of water wore its way through the mountain by a fall like Niagara, or burst through a gorge, it is certain that the Minnisink country bears the mark of aqueous action in its diluvial soil, and in its rounded hills, built of pebbles and bowlders."

. Of the two grand mountains which flank the mighty chasm of the

Gap, the one on the Pennsylvania (W.) side is named *Minsi*, in memory of the Indians; that on the New Jersey (E.) side bears the name of *Tammany*, an ancient Delaware chief, who was canonized during the last century, and proclaimed the patron saint of America. Mt. Minsi is soft in outline, and densely wooded, but Tammany exhibits vast, frowning masses of naked rock. Successive ledges, or geological terraces, mark the face of Minsi, and upon the lowest of these, 200 ft. above the river, stands the old Kittatinny House. The stream that issues beneath the hotel and falls in a cascade into the river has come down the mountain-side through a dark and picturesque ravine. Far up the ascent it takes its rise in the *Hunter's Spring*, a cool and sequestered spot, reached by a path from the hotel. Under the name of *Caldeno Creek* it continues its downward course by cascade and waterfall to the river. Along the face of Minsi, about 500 ft. above the river, runs a grand horizontal plateau of red shale, extending for several miles along the mountain, and known as the *Table Rock*. Extensive views are obtained from this point, and the Caldeno flows over the ledge at an angle of 45 degrees in a charming succession of miniature falls and rapids. The rocky strata beneath are densely covered with moss, which gives the spot its name of *Moss Cataract*. Below the cataract, in a secluded, deeply-shaded glen, is the placid rock-basin known as *Diana's Bath;* and at a still lower range the stream dashes at *Caldeno Falls* over a rugged, rocky precipice. All these points are reached from the Kittatinny House by a path marked in *white* lines on rocks and trees. The summit of *Mt. Minsi* is reached from the hotel by a path 3 miles long, marked by *red* lines. The ascent is easy, and the view from the summit the finest to be obtained in this region. Paths diverging from the main path to the summit lead to various points of interest. A short distance from the hotel a path marked with *blue* lines, and turning off to the left, leads to the *Lover's Leap*, whence the best view of the Gap is obtained. Half a mile farther, a *white*-lined path to the right leads to *Hunter's Spring*, already mentioned; and still beyond a *yellow*-lined path (to the left) leads to *Prospect Rock* (2 miles from the hotel), whence another noble view is obtained. *Mt. Tammany* may be ascended from the hotels by a rugged path 2½ miles long, but it should not be undertaken except by a vigorous climber. The view from the summit is fine, but does not differ materially from that from the summit of Mt. Minsi. On the apex of the lofty peak is a picturesque mountain-lake, of which popular superstition declares that it has no bottom.

The best near view of the Gap is obtained by descending the river in a boat to *Mather's Spring*, on the New Jersey shore (1½ mile from the hotel). The *Indian Ladder Bluff*, at the foot of Mt. Tammany, the *Cold Air Cave*, *Benner's Spring*, and the *Point of Rocks*, are favorite excursion-points along the river. A few miles above the Gap the Delaware is joined by the Bushkill Creek, upon which is one of the most beautiful waterfalls of the district—the *Bushkill Falls*. On a small affluent of the same stream are the *Buttermilk Falls* and the picturesque *Marshall Falls*. All these falls are within 7 miles of the hotels. There is a pleasant drive from the Gap up the *Cherry Valley*.

The Water-Gap is traversed on a narrow shelf between the river and mountain, and as the train emerges at the N. end it crosses Broadhead Creek, and passes through a cut in Rock Difficult, so called from the difficulty encountered in making a passage through its flinty mass. *Stroudsburg* is the first station beyond the Gap, and is a pleasant summer resort. At *Spragueville* the ascent of the Pocono Mt., the E. slope of the Alleghanies, begins, the grade for 25 miles being at the rate of 65 ft. to the mile. Just beyond *Oakland* the Pocono Tunnel is traversed near the top of the mountain, a point from which the view, extending more than 30 miles, is most impressive. At *Tobyhanna* the descent of the W. slope of the mountains begins. Just beyond *Moscow* the valley of Roaring Brook is entered, and the train descends by steep grades into the Lackawanna Valley and soon reaches **Scranton** (see Route 52). Beyond Scranton the train runs N. to *Great Bend*, a small village on the Erie R. R. and the Susquehanna River. Fourteen miles farther on is **Binghamton,** described in Route 41, where the Syracuse and Oswego Div. branches off. (Route 55.)

Richfield Springs.

At Binghamton, the *Utica Division* runs to Chenango Forks and to **Utica,** on the N. Y. Central R. R. From *Richfield Junction* (13 miles from Utica) a branch line runs in 21 miles to **Richfield Springs,** a popular summer resort in Otsego County, near the head of Schuyler's Lake. Trains run direct from New York to the Springs during the summer season. The village is neat, but the hotels constitute Richfield. The leading hotels are the *Spring House* and the *Earlington,* which face each other on opposite sides of the main street. Smaller houses are the *Canadarago, Davenport, National, Tuller House,* and others. There are 17 mineral springs near the village, the most important being that within the grounds of the *Spring House.* The waters are considered especially efficacious in diseases of the skin. There are delightful walks and drives in the vicinity of the Springs, and fine boating and fishing on **Schuyler's Lake,** which is 1 mile from the village. This lake is 2½ miles long, and is inclosed by gentle hills which combine with it in many attractive landscapes. The *Schuyler Lake House* is celebrated for its fish and game dinners. Stages run several times daily to *Otsego Lake,* connecting with the steamers for Cooperstown; also connecting at Springfield Center with stages to *Cherry Valley* and *Sharon Springs.*

Twenty miles beyond Binghamton the train reaches **Owego,** whence the Cayuga Div. runs to *Ithaca* (34 miles) (see Route 40, page 175). **Elmira** is the next place of importance, and is described in Route 40. At **Bath,** connections are formed with the Bath & Hammondsport R. R.; at *Mt. Morris* with the Western, New York & Pennsylvania R. R. for Portage and intermediate points; and at the *Rochester & Pittsburg Junction* with the line running N. to Le Roy and Rochester and S. to Wyoming and Warsaw. For **Buffalo,** see Route 40, page 178.

44. New York to Montreal via Saratoga Springs and Lake Champlain.

Via the N. Y. Central & Hudson River or West Shore R. R., or during summer by the steamboat (Route 9) to Albany or Troy, and thence *via* Delaware & Hudson R. R.   Distances : To Albany, 143 miles ; to Saratoga Springs, 181 ; to Whitehall, 219 ; to Rouse's Point, 341 ; to Montreal, 383.   This is the shortest and most direct route between New York and Montreal, and the through trains make the journey in about 12 hours.   Drawing-room and sleeping cars are attached to the through trains.

As far as Albany or Troy this route has already been described in Route 9.   At Albany (143 miles) the cars take the track of the Delaware & Hudson R. R. and run N. past the Rural Cemetery to *W. Troy* (149 miles); *Cohoes* (152 miles), a busy manufacturing city at the crossing of the Mohawk River; *Waterford* (154 miles), a large manufacturing village on the Hudson; and *Albany Junction* (155 miles), where the Albany Div. joins the main line from Troy, 6 miles distant. *Round Lake* (168 miles) is a celebrated Methodist camp-meeting ground; and 6 miles beyond is **Ballston Spa** (*Sans-Souci Hotel, Ballston Spa House*), a fashionable and frequented resort, noted for its mineral springs, whose fame, however, has been overshadowed by the more popular Saratoga waters.   The village of 3,527 inhabitants is situated upon the Kayaderosseras Creek, and contains several factories.   Seven miles beyond (181 from New York) the train stops at

Saratoga Springs.

Hotels, etc.—The hotels of Saratoga are among the largest in the world. The *Grand Union Hotel*, in Broadway, is a vast building, and with its grounds occupies a large village block.   The grounds are beautifully shaded by large elms.   In these grounds are held the famous garden parties, when the whole interior is converted into a fairy spectacle by the use of profuse decorations.   In the handsome ball-room of this hotel is Adolph Yvon's celebrated painting of "The Genius of America."   The *United States Hotel*, a few hundred feet farther north, on the same side of Broadway, is a brick edifice, surrounded by wide piazzas and beautiful promenades, and is richly furnished.   *Congress Hall*, on the opposite side of Broadway, is also a large brick building. The beautiful ball-room of this hotel is connected with it by a light and airy bridge spanning Spring St.   The *Windsor Hotel*, on the brow of the hill overlooking Congress Spring Park, has one of the most desirable locations at Saratoga.   It is a quiet and aristocratic house.   The *Clarendon*, opposite, is a large wooden structure with ample accommodations.   The *Kensington* is a new hotel in Union Ave., one block from Congress Spring Park.   These hotels are the most noted, and charge from $3 to $5 a day.   Besides these hotels there are many smaller ones : The *Adelphi*, the *American*, the *Columbian*, the *Continental*, the *Everett*, the *Heustis House*, the *Holden House*, the *Kenmore*, the *Vermont House*, and the *Waverly*.   The above hotels are opened only during the summer season,   The *Worden Hotel*, cor. Broadway and Division St., is open all the year around, and is an excellent hotel.   *Dr. Strong's Remedial Institute*, in Circular St., is open during the entire year, and is near the principal springs and hotels.   Besides those enumerated, there are more than 50 smaller hotels and many boarding-houses, at any of which good board can be obtained for from $5 to $25 a week.   From the railroad station to most of the hotels is only a short walk, and the principal hotels have stages conveying passengers to and from the station free of charge.

Saratoga Springs is situated upon the last mountain of the eastern spur of the Adirondacks, to the E. of which stretches a level plateau. It has an altitude of about 400 ft. above the sea-level, and the air is dry and healthful. The natural advantages of the village have been supplemented by a liberal public policy on the part of the corporation, and its system of public instruction and public water-supply, the sewage system, the fire and police departments, and all the varied appliances for health, comfort, and safety demanded by modern living, are not surpassed by any city in the United States.

Along the southern and eastern edge of the mountain, upon which the village is partially situated, is a ravine, where, at frequent intervals, gush forth mineral springs of varied character, whose medicinal properties have made them famous as far as civilization extends. Saratoga is yearly visited by tourists, and those in search of health and pleasure, from all parts of the world. Its resident population is 11,975, according to the census of 1890; but during the season, which lasts from June 1st to October 1st, there are often not less than 30,000 strangers in the village. The principal avenues are shaded by magnificent elms and other native trees. There is no more brilliant spectacle than Saratoga during the height of the season. All the principal hotels have large orchestras that play morning and evening, and balls and hops are of nightly occurrence.

The medicinal properties of the High Rock Spring were known to the Iroquois Indians at the period of Jacques Cartier's visit to the St. Lawrence in 1535. In 1767 Sir William Johnson was carried thither on a litter by the Mohawks, and he is believed to have been the first white man to visit the spring. The first log cabin was erected in 1773, and the first framed house in 1784 by Gen. Philip J. Schuyler, who in the same year cut a road through the forest to the High Rock from Schuylerville. Hotels began to be erected about 1815, and since then the fame of the Springs has spread so widely that, in addition to the hosts of visitors, immense quantities of the waters are bottled and sent to all parts of the United States and Europe. The name Saratoga (Indian, *Saraghoga*) signifies " the place of the herrings," which formerly passed up the Hudson into Saratoga Lake.

The principal street of the village is *Broadway*, in which the large hotels are situated. The *North Broadway* section is that part of the village where the most costly and beautiful cottages are situated, many of them of considerable architectural pretensions. In this section is **Woodlawn Park,** the celebrated country seat of Henry Hilton, of New York city. It is the largest private park in America, having more than twice the area of Central Park in New York city. There are about 30 miles of hard, smooth roads winding through its picturesque confines, and many small lakes and sparkling rivulets adorn the landscape. While this park is maintained at private expense, the owner generously throws it open to the public, who may enjoy its miles of fine park roads and beautiful vistas of forest and lawn. It is one of the favorite drives around Saratoga.

The most popular drive is to **Saratoga Lake,** 3½ miles to the E. of the village. It is reached by way of Union Ave., a wide boulevard lined with rows of elms; and in the afternoon, the favorite hour for driving, it presents a very gay scene with hundreds of handsome pri-

vate equipages. The other principal streets are Circular St., Lake Ave.,
Caroline St., Philadelphia St., Spring St., and South St. **Congress
Spring Park** is a low ridge in the shape of a horseshoe encircling
the lower ground on which the Congress and Columbian Springs are
situated. It is near the principal hotels, is shaded by noble trees, and
laid out in smooth walks, and is the favorite ramble. A brass band
plays several times daily, and tennis tournaments, fireworks, concerts,
and other entertainments are given here during the season. A small
entrance fee is charged. There are two small camps of Indians and
Canadian half-breeds in Saratoga during the summer, where basket-
work, moccasins, and other goods of Indian making can be bought, and
where many forms of amusement are provided for children. In one of
these camps is a circular railway, a favorite resort for exercise. There
are in all at Saratoga upward of 40 springs, with such varied constituents that one is astonished that such differences could exist in waters
that come to the surface of the earth within such a comparatively small
area. They range in quality from strong cathartic and diuretic to refreshing table waters, sparkling with natural carbon dioxide gas. Of
these springs the *Congress* is the most celebrated. It is situated in
Congress Spring Park, and was discovered in 1792. It is a cathartic
water. Near it and in the same inclosure is the *Columbian Spring*,
quite different in character from the Congress, and a valuable tonic.
The *Hathorn Spring*, about 300 ft. farther N. of Congress Spring on
Spring St., is one of the strongest cathartic springs at Saratoga. Its
waters contain nearly 100 grains of mineral constituents and about
50 cubic inches of carbon dioxide gas to every pint. A few feet N. of
the Hathorn Spring is the *Patterson Spring*, a mild cathartic water
agreeable to the taste. The *Hamilton Spring*, cor. Spring and Putnam
Sts., and the *Washington Spring*, in the grounds of the Clarendon
Hotel, are both similar in their properties to the Columbian Spring. In
a small park between Caroline St. and Lake Ave. are located the *Pavilion and United States Springs*, agreeable to drink and of tonic properties. Next N. on the opposite side of Lake Ave. is the *Imperial
Spring*, considered a desirable table water. A stone's throw to the E.
is the *Royal Spring*, a fine table water, coming from a depth of more
than 600 ft. Following Spring Ave. to the N. the *Seltzer Spring* is
reached, similar in quality to the German seltzer-water; opposite to
which is the Saratoga *Magnetic Spring*. In connection with this spring
is a large and commodious bath-house. Immediately N. of the Seltzer
is the *High Rock Spring*, the oldest known of the Saratoga Springs. It
was discovered in 1767 by Sir William Johnson, the first white man
who visited Saratoga. It bubbles up through an aperture in a conical
rock 4 ft. high, formed by deposits of the mineral substances of the
water. The *Star Spring*, a saline cathartic water, is situated a few
hundred ft. N. ; and just beyond it is the well-known *Empire Spring*,
a saline water agreeable to the taste, and highly prized for its medicinal
qualities. N. of the Empire is the *Saratoga "A" Spring*, and then
comes the *Red Spring* and the *Elixir Spring*. The last two are owned
by the same company, and in connection with them is an extensive

bath-house.   Along the same valley and about $\frac{1}{4}$ mile farther N. is the celebrated *Excelsior Spring*, situated in large and beautiful grounds, in which also is the *Union Spring*.   At this point, within an area of a few acres, over 10 mineral springs have been discovered, the waters of which are not all utilized; the whole property being owned by the proprietor of the Excelsior Spring, which is deemed the most valuable.   Its waters are saline and mildly cathartic.   About $\frac{1}{2}$ mile from the Excelsior Spring are the *Eureka Mineral Spring* and the *White Sulphur Spring*.   About 2 miles to the S. of the village is another remarkable group of springs, among which the most celebrated are the *Geyser*, *Carlsbad*, *Champion*, and *Lafayette*, all cathartic waters, and the *Kissingen* and *Vichy*, the two later named for the celebrated springs of Europe whose qualities they very closely resemble.   Both of these waters are delicious beverages, and are used as table waters very extensively.   These springs are reached by a beautiful drive down Ballston Ave. and also by electric railway.   They are surrounded with picturesque parks and lakes.   Most of this group of springs are what is known as spouters, the water being thrown many feet into the air from the tubing by the force of the carbon dioxide gas that continually escapes from them.

The *Putnam Spring*, on the N. side of Philadelphia St., is now the site of the **Saratoga Baths,** recently erected by a wealthy Saratogian. It is one of the finest bath-houses in the country, costing $130,000, and in it are given all species of baths—Turkish, Russian, Roman, etc. It also has a large swimming pool.   The waters of the Putnam Spring are utilized, and mineral baths are given to those who desire them. The establishment is fitted entirely with porcelain tubs.

Among the other buildings of interest at Saratoga, aside from the splendid hotels and fine private residences, are the *Academy*, and the fine *Armory* of the Saratoga Citizens' Corps, both situated in Lake Ave.   The *Pompeiian House*, adjoining the Windsor Hotel in South Broadway, is a reproduction of the celebrated House of Panza, destroyed at Pompeii in A. D. 79.   The reproduction is supposed to be historically correct, and it is visited by scholars and students from all parts of the country.   It is justly regarded as one of the principal objects of interest at Saratoga.   The *Temple Grove Seminary* for young ladies, in Circular St., is a well-known educational institution.   All the public-school buildings are fine brick structures and a credit to the village.   The *Town Hall*, cor. Broadway and Lake Ave., contains most of the public offices of the town and village.   A fine *Convention Hall*, in Broadway, adjoining Congress Spring Park, capable of accommodating 5,000 people, was built during 1892.   It contains a suite of rooms fitted up for the exclusive use of the State Court of Appeals, which meets in June of each year.   Churches of almost every denomination are found in the village, and several newspapers are published.

The **Saratoga Race-Course,** in Union Ave., is one of the best-known courses in the country, and the Southern and Western horse-owners find their animals greatly benefited by a season at Saratoga.

The races are conducted from about the middle of July to the middle of August, 3 or 4 days each week, and are well patronized by residents and visitors. Upward of $100,000 has recently been expended upon this property in various improvements. Beyond the race-course and immediately adjoining it are the handsome grounds of *Yaddo*, the summer residence of Spencer Trask, of New York city. There are several miles of heavily shaded and picturesque drives through the grounds, along the shores of a succession of beautiful little lakes, which are open to the public.

To **Saratoga Lake,** a beautiful body of water, 8 miles long by 2½ wide, 3½ miles W. from Saratoga, is the favorite afternoon drive. At this lake is situated the famous *Moon's Lake House*, and the lake has been the scene of many interesting regattas. An electric railway, having its terminus on the corner of Broadway and West Congress St., extends to Saratoga Lake, the Race-Course, and to "The Geysers," where are located the Kissingen, Vichy, Geyser, Carlsbad, and other springs. Around the lake are several road-houses, the most famous being *Moon's, Crum's*, and *Riley's*. The two latter vie with each other in the excellence with which they prepare the famous Saratoga bass and potatoes. Game of all kinds can also be procured at these road-houses.

Twelve miles E. of Saratoga is the *Saratoga Battle-Ground*, where two important battles of the Revolution were fought, and the scene of the surrender of Gen. John Burgoyne. At Schuylersville a magnificent monument has been erected by the State and national Governments to commemorate these historical occurrences, and it is well worth a visit.

Mt. McGregor, 6 miles N. of the village, and lying about 1,000 ft. higher, now famous as the place at which Gen. Grant died, is reached in about half an hour by the Mt. McGregor R. R., a narrow-gauge road which ascends to the summit of the mountain by many devious curves. On the mountain is a large hotel (*Hotel Balmoral*) in the midst of a park of 1,000 or more acres of land. Two picturesque lakes are situated in the immediate vicinity of the hotel, upon the mountain-top, well stocked with fish and equipped with good boats. Mt. McGregor is a desirable resort for those troubled with hay-fever. From the brow of the mountain, in front of the hotel, a most magnificent panorama of the upper valley of the Hudson is obtained. Trains leave Saratoga for Mt. McGregor several times daily.

Leaving the spacious depot at Saratoga, the train runs N. E. through an uninteresting country, and in 16 miles reaches **Fort Edward** (*Eldridge, St. James*), whence a branch runs to Glens Falls and Caldwell on Lake George (see Route 46). Beyond Fort Edward there is no important station until Whitehall (219 miles from New York) is reached. **Whitehall** is a lumbering village of 4,434 inhabitants, situated at the head or S. end of Lake Champlain. It lies in a rude, rocky ravine at the foot of Skene's Mt., and was a point of much importance during the French and Indian Wars and the Revolution, but contains nothing now to detain the traveler. From Whitehall one route to Montreal passes N. E. *via* Castleton and Rutland, Vt. (see Routes 29 and

37). The present route runs almost due N. along the W. shore of Lake Champlain. The lake and the principal points of interest on its shores are described in Route 46. Here we shall only mention the special features which make the railway journey enjoyable.

Soon after leaving Whitehall the fine scenery begins (seats should be obtained on the right-hand side of the cars). The R. R. track runs close along the margin of the lake at the foot of steep bluffs, with fine views across the water of the Vermont shore. At *Fort Ticonderoga* (see Route 46) a branch line diverges and runs in three miles to Baldwin on Lake George (Route 46). At *Addison Junction*, 2 miles from Ticonderoga, connection is made with a railroad which connects with the Central Vermont R. R. at Leicester Junction. From *Westport* stages run to Elizabethtown and other places in the Adirondacks. From Ticonderoga to *Port Kent* (Route 46), a distance of about 55 miles, the scenery is beautiful, the train running now on high terraces, now through deep rock-cuttings, now at the base of towering cliffs, and affording exquisite lake-views. Port Kent is one of the entrances to the Adirondack region, a branch railroad runs from here to Keeseport (6 miles), stopping on the way at Ausable Chasm (3 miles) (Route 47, III). Between Port Kent and *Plattsburg* (309 miles from New York) the scenery is less impressive, but fine views are had of the distant mountains. Three miles S. of Plattsburg is Bluff Point, the station for *Hotel Champlain*, the largest hotel on the lake. Plattsburg is described in Route 46. Beyond Plattsburg the route leaves the lake and traverses a comparatively flat and uninteresting country. At *Rouse's Point* (334 miles) the train takes the track of the Grand Trunk R. R. and passes in 50 miles to Montreal (see Route 27).

45. New York to Buffalo and Niagara Falls.

Via Lehigh Valley R. R.

Distances : Easton, 75 miles ; Bethlehem, 87 miles ; Allentown, 92 miles ; Mauch Chunk, 121 miles ; Wilkesbarre, 175 miles ; Towanda, 257 miles ; Ithaca, 311 miles ; Geneva, 351 miles ; Buffalo, 454 miles ; Niagara Falls, 465 miles.

PASSENGERS leave New York by ferry-boat from foot of Liberty St. for *Jersey City*, thence by Lehigh Valley train through *Bound Brook* (32 miles) and *Flemington Junction*, where connection is made for *Flemington* (52 miles) by branch of this road. Continuing on main line, *Lansdown* is reached, where another branch connects for *Clinton* (60 miles). Passing through *Pattenburg*, immediately beyond where the Musconetcong Tunnel, nearly a mile long, pierces the mountain, the Musconetcong Valley is reached. After passing several small towns the passenger arrives at *Phillipsburg* (75 miles, 8,644 inhabitants), an iron manufacturing town on the Delaware River opposite Easton, Pa., with which it is connected by three bridges. At Phillipsburg connection is made with the Belvidere Division of the Pennsylvania R. R. (See Route 54.) **Easton** (75 miles ; *Franklin House*, *Paxinosa Inn*, and *United States Hotel*) is situated on some steep hills, at the confluence of the Delaware and Lehigh

Rivers and Bushkill Creek. It is a well-built and wealthy town, with a population of 14,481, and extensive iron-works, mills, distilleries, etc. The Court-House, the County Prison, and the Opera-House are handsome buildings, and there are several fine churches. To the E. on College Hill is * *Lafayette College*, a richly endowed institution, with 28 instructors and 297 students, an extensive library, and fine scientific collections. Pardee Hall is a handsome building, and from its tower there is a noble view. The curious *Durham Cave* is near Easton, and *Mt. Jefferson* is an abrupt peak in the center of the town. From Easton the Lehigh Valley R. R. runs along the Lehigh River amid pleasing scenery, and in 12 miles reaches **Bethlehem** and **South Bethlehem** (87 miles; *Eagle, Wyandotte*, and *Sun*), with a combined population of 17,064 inhabitants: Bethlehem is noted as the chief seat in the United States of the Moravians, or United Brethren, who settled here under Count Zinzendorf in 1741. The old Moravian buildings for the most part still remain, and the principal ones, which are built of stone and stand in Church St., near Main, are in a good state of preservation and are still occupied. The *Moravian Church* is a spacious stone structure capable of seating 2,000 persons. Near the church is the *Moravian Boys' School*, and there is also a *Moravian Female Seminary* of high repute, founded in 1749, and still flourishing. The *Sun Hotel* was opened as an inn in 1760, and, though greatly enlarged in 1851, still retains its ancient and massive walls. On a spur of the Lehigh Mts. above the town is the * *Lehigh University*, founded in 1865 and liberally endowed by the Hon. Asa Packer. From the park around the buildings there is a view of 20 miles. At this point also is located the extensive plant of the Bethlehem Iron Co., employing more than 4,000 hands. This industry has in the past few years been brought into prominence on account of the contracts received from the United States Government for the manufacture of guns, heavy armor-plate, etc. Five miles beyond Bethlehem the train reaches **Allentown** (*American, Eagle*, and *Hotel Allen*), a thriving city of 25,228 inhabitants, built upon an eminence on the west bank of the Lehigh River. It is regularly laid out and well built, with electric-car service on the streets. It is one of the most enterprising cities in eastern Pennsylvania, and has within its boundaries extensive blast-works, furnaces, rolling-mills, silk-mills, furniture-factories, barb-wire, thread-works, etc. The County Court-House and County Prison are handsome edifices, and several of the school-buildings are noteworthy. *Muhlenberg College* (Lutheran) stands amid ample grounds in the S. E. part of the city. * *Mammoth Rock*, 1,000 ft. high and commanding broad views, is near the city, as are also several mineral springs. At this point connection is made with the Philadelphia & Reading R. R. for Reading, Lancaster, Columbia, Lebanon, and Harrisburg. Leaving Allentown the Lehigh Valley R. R. runs by a number of huge blast-furnaces, and in three miles reaches *Catasaqua* (95 miles), a thriving village, population 3,704, with vast iron-works, furnaces, and car-shops. *Hokendauqua* (96 miles) and *Coplay* (97 miles) are also the site of immense iron-works, and at the latter point are also large cement-works. *Slatington* (108 miles) is in the midst of the most extensive slate deposits ever discovered. The slate on the Capitol at Washington, D. C., ½ inch in thickness, came from this place. The village is charmingly situated, and is a pleasant summer resort. Two miles beyond Slatington is the **Lehigh Water-Gap,** a picturesque gorge in which the Lehigh River flows through the Blue Mountains. Steep, forest-clad cliffs rise from the water on either side, and there is barely room in the narrow pass for the river, railroad, highway, and canal. The scenery in this vicinity is remarkably wild and impressive. Four miles beyond Lehigh Gap we reach *Lizzard Creek Junction,* from which a branch of the Lehigh Valley R. R. extends through *Orwigsburg* and *Schuylkill Haven* to *Pottsville* (149 miles). (See Route 56.) Continuing on the main line from Lizard Creek Junction we reach *Lehighton* (117 miles), which is situated on the Lehigh River at the mouth of Mahoning Creek. The old Moravian Cemetery is on the hill, from which may be had a fine view of the Mahoning Valley) and at the foot of which 12 settlers were murdered by the Indians in 1775. At *Weissport,* on the opposite side of the river, formerly stood Fort Allen, erected by Benjamin Franklin in 1756 as a frontier defense. Its site is now occupied by the Fort Allen Hotel. At *Packerton* (119 miles) are the vast scales, 123 feet long, with a capacity of 103 tons, which weigh loaded coal trains while in motion. At this point are also located the car-shops of the Lehigh Valley R. R. Co. Just beyond the train crosses the Lehigh River on an iron bridge, runs along the base of Bear Mountain, and stops at Mauch Chunk.

Mauch Chunk.

Mauch Chunk (*American, Broadway,* and the *Mansion*) is noted for being situated in the midst of some of the wildest and most picturesque scenery in America, the town lying in a narrow gorge between and among high mountains, its foot resting on the Lehigh River and its body lying along the hillsides. The town is but one street wide, and the valley is so narrow that the dwelling-houses usually have their gardens and outhouses perched above the roof, and there is barely room for the railroad, street, river, and canal, which pass through the gorge side by side. The chief architectural features of the village are * *St. Mark's Church* (Episcopal), a fine edifice of cream-colored stone with stained-glass windows and massive tower, and the fine railroad building erected by the Packer estate. *Prospect Rock* is a projecting bluff near the Mansion House, from which a pleasant view may be had; but the view from *Flag-staff Peak,* just above, is much finer, and the ascent is easily made.

Mauch Chunk lies in the very heart of the Pennsylvania coal-region, and the coal-traffic sends many trains through the town every day and a constant procession of canal-boats. The coal-mines which supply this traffic are situated in the Wyoming, Hazelton, Beaver Meadow, Mahanoy, and Lehigh regions. The coal from Panther-Creek Valley used to be brought this distance by the celebrated "Switch-Back" Gravity Road (the *Mauch Chunk & Summit Hill R. R.*). The "Switch-Back" is now used only as a pleasure road. It is run by gravity. The cars are drawn to the top of Mt. Pisgah by a powerful engine on the summit, whence they descend 6 miles, by gravity, to the foot of Mt. Jefferson, where they are again taken up by means of a plane, which ascends 462 ft. in a length of 2,070 ft., and then run on to Summit Hill. From that point the cars return, all the way, by the "back-track," or gravity road, to Mauch Chunk, landing the passengers but a short distance from the spot where they began the ascent over Mt. Pisgah. Several passenger-trains daily run between the station at the foot of Mt. Pisgah and the mines; and the excursion is both novel and enjoyable. The time required for the circuit is about three hours; fare, round trip, 75c. An omnibus, connecting with the trains, runs from the Mansion House to the foot of the inclined plane (fare, 25c.). The first plane is 2,322 ft. long, and leads to the summit of **Mt. Pisgah** (850 ft. above the river), from which a noble view is obtained. Mt. Jefferson is the highest point on the road, which descends thence on a slight grade to **Summit Hill,** on which is a mining village of 2,816 inhabitants, with a church, several hotels, and other evidences of civilization. Summit Hill is a good deal resorted to in summer. Beyond Summit Hill the center of the coal-region is reached. Visitors desirous of enjoying the experience of being "down in the mines" can do so by lying over here for a few hours. The return to Mauch Chunk is by a descending grade of 96 ft. to the mile, and the entire 9 miles is traversed in about 25 minutes.

Two miles beyond Mauch Chunk, and on the main line of the Lehigh Valley R. R., is * *Glen Onoko* (123 miles; *Wahneta Hotel*), a wild and beautiful ravine on the side of Broad Mt. It is 900 yards long and from 40 to 80 ft. wide, and presents a continuous succession of cascades, rapids, and pools, which afford a fine spectacle in seasons of high water. From the upper end of the Glen a path leads to the *Rock Cabin* and to *Packer's Point*, whence there is an extensive view. *Glen Onoko*, in point of natural scenery and picturesque beauty, surpasses anything of its kind in the country, not excepting the far-famed Watkins Glen, and is visited annually by thousands of tourists. At *Penn Haven Junction*, 6 miles above Glen Onoko, three branch roads of this line diverge to the coal regions, where are situated many large towns, prominent among which are Mahanoy City, Shenandoah, Ashland, Shamokin, Audenried, Hazleton, and Freeland. Leaving Penn Haven Junction on the main line, and still following the Lehigh River, the next place of importance reached is *White Haven* (145 miles), population 1,634, an important lumbering village. Here the ascent of the mountains begins with heavy grades, passing *Glen Summit* (156 miles) where Glen Summit Hotel is, which stands at the apex of the great divide, from whose summit the waters flow westward into the Susquehanna and eastward into the Lehigh Rivers. It is situated at an altitude of 2,000 ft. above the sea, overlooking mountain peaks and ranges, valleys and basins; is altogether a peerless mountain summer resort. A number of handsome cottages now dot the mountain-sides, and more are in course of construction. In point of architecture no two are alike, and all are models of beauty. The walks and drives, which are broad and well kept, lead in every direction, and are traversed daily by the sojourners.   An especially fine drive is that to *Bear Creek* (8 miles), another beautiful little summer resort, which is also reached by railroad (the Bear Creek Branch) from Bear Creek Junction, a few miles north of White Haven.

  . The summit of the mountain is reached at *Fairview* (159 miles), and the descent to the Wyoming Valley commences.   *Newport* (166 miles) stands high and affords a magnificent view of the Wyoming Valley, the Susquehanna River being visible for more than 20 miles from its entry through Lackawannock Gap, near Pittston, to its exit through Nanticoke Gap, near Shickshinny.   Nine miles beyond *Newport*, picturesquely situated on the Susquehanna River, in the center of the Wyoming Valley, is **Wilkesbarre** (175 miles; *Wyoming Valley Hotel, Luzerne House, The Exchange,* and *Bristol House*), a prosperous city of 37,718 inhabitants, with broad, well-shaded streets, and handsome public and private buildings.   The Court-house, County Prison, Opera-House, and Y. M. C. A. Building are all fine structures.   There is also a very good public library and several costly churches.   Many fine villa residences front upon the esplanade along the river.   Back of the city and about 2 miles distant is *Prospect Rock*, which is 750 feet high, and affords the best view of the entire Wyoming Valley.   A small steamer runs on the Susquehanna from Wilkesbarre to Nanticoke (7 miles), and affords fine views of the lower portion of the valley, which, however, are best seen by a drive along the river road.   A bridge across the river connects Wilkesbarre with **Kingston,** 4 miles above which, near the hamlet of Troy, is the site of *Forty Fort,* where the unfortunate battle of Wyoming was fought.   Near by is the **Wyoming Monument,** a massive granite obelisk, 63 feet high, with appropriate inscriptions.   About 3 miles above Forty Fort is *Queen Esther's Rock*, so called from the half-breed Indian woman (Queen of the Senecas), who there avenged her son's death by tomahawking 14 American soldiers with her own hand.

  The Valley of Wyoming is about 20 miles long and 3 miles wide, being formed by two parallel ranges of mountains, averaging on the west about 800 and on the east 1,000 ft. in height.   It is traversed by the Susquehanna River, which enters its upper end through a bold mountain-pass known as the Lackawannock Gap, and passes out of its lower end through another opening in the same mountain called Nanticoke Gap.   The river is in most places about 200 yards wide, and from 4 to 20 ft. deep; it moves with a very gentle current, except at the rapids or when swollen with rain or melted snows.   Near the center of the valley it has a rapid, called Wyoming Falls, and another at the lower gap, called the Nanticoke Falls. - Several tributary streams fall into it on each side, after traversing rocky passes, and forming beautiful cascades as they descend to the plain.   Describing this valley, the elder Silliman says : "Its form is that of a very long oval or ellipse.   It is bounded by grand mount-ain-barriers, and watered by a noble river and its tributaries.   The first glance of a stranger entering it at either end, or crossing the mountain-ridges which divide it (like the Happy Valley of Abyssinia) from the rest of the world, fills him with peculiar pleasure, produced by a fine landscape, containing richness, beauty, and grandeur. . . . Few landscapes that I have beheld can vie with the Valley of Wyoming."   The Massacre of Wyoming, which has given the valley a melancholy prominence in history, and which forms the theme of Campbell's "Gertrude of Wyoming," occurred on July 3, 1778.   The settlers, who had pre-viously been at variance on account of being interested in charters from differ-ent authorities, had, at the outbreak of the Revolution, united in an effort to

form a home-guard for self-protection.   Two of the companies thus formed
were ordered to join General Washington, and a third, imperfectly organized
and equipped, was unequal to the terrible need that soon arose.   A body of 400
British and 700 Indians, chiefly Senecas, under Colonel John Butler, entered the
valley June 30, 1778 ; and the inhabitants, having taken refuge in Fort Forty
(so called from the number in one of the bands of settlers), gave battle on the
3d of July and lost.   Then followed the terrible massacre, which, though it was
exaggerated at the time, has had few parallels in American history.   Neither
age nor sex was spared, and but few of the ill-fated people escaped by fleeing
over the mountains to Stroudsburg.   The village of Wilkesbarre was burnt, and
its inhabitants either killed, taken prisoners, or scattered in the surrounding forests.   Upward of 300 persons are estimated to have perished on that fatal day.

At Wilkesbarre connection is made with the Delaware & Hudson
R. R. for Scranton from all Lehigh Valley trains.   At Wilkesbarre
also the Harvey's Lake Branch diverges from the main line at a very
heavy grade through a natural glen, noted for its picturesque scenery,
at the present terminus of which is Harvey's Lake, the largest body of
fresh water at that altitude east of the Rocky Mountains.   *Harvey's
Lake* is a summer resort, fast coming into popular favor and extensively patronized by picnic parties.   It is in the very heart of the lumbering regions of eastern Pennsylvania ; much lumber is cut, and there
are numerous saw-mills in continual operation, with capacity of many
thousands of feet daily.

Nine miles beyond Wilkesbarre is **Pittston** (184 miles), population 10,302, at the head of the valley on the Susquehanna, just below
the mouth of the Lackawanna Creek.   W. of the town are the Lackawanna Mountains, filled with rich coal-mines, which here find an outlet.   A prominent object of interest in the vicinity is *Campbell's Ledge*,
from which a charming view of the valley is obtained.   One mile beyond Pittston, at *L. & B. Junction*, trains of the Bloomsburg Division
of the Delaware, Lackawanna & Western R. R. connect for Scranton.
(See Route 52.)

On the portion of the main line of the Lehigh Valley R. R. beyond
L. & B. Junction the scenery continues varied and pleasing and at
times impressive.   *Tunkhannock* (207 miles) is the capital of Wyoming
County, and is picturesquely situated on the Susquehanna at the mouth
of Tunkhannock Creek.   From Triangle Hill, near by, there is a broad
view.   Connection is here made with the Montrose R. R. for Montrose.
The Susquehanna River between Pittston and Towanda is noted as being the finest bass-fishing stream in the United States.

Still following the Susquehanna River amid changing forest and
hill scenery, the train passes a number of small stations and reaches
Towanda (257 miles), a busy manufacturing town of 4,169 inhabitants, situated on the river at the mouth of Towanda Creek.   It is much
visited in summer, and has a lucrative trade in farm and dairy produce
with the surrounding region.   At this point the State Line & Sullivan
Branch extends for a distance of 34 miles into the lumbering regions
of Sullivan County.   Fifteen miles beyond Towanda is **Athens** (272
miles), a flourishing village at the confluence of the Susquehanna and
Chemung Rivers.   It occupies the site of the important Indian village
of *Diahoga*, which was the gathering-place of the Tory-Indian forces

that perpetrated the massacre of Wyoming. Near by is *Spanish Hill*, on which ancient Spanish coins are said to have been found. Crossing the Chemung River at Athens, the train reaches *Sayre* (274 miles), the headquarters of the Northern Division of the Lehigh Valley R. R. Twe miles beyond is *Waverly* (276 miles), where the train enters the Stato of New York. Reaching **Elmira** (294 miles), connection is made with the Erie Railway for all points E. and W. (For Elmira, see Route 41.)

At Sayre another branch line diverges, running through *Owego* (293 miles), *Freeville*, and many smaller towns to *Auburn* (360 miles) (see page 175), connecting there with the New York Central & Hudson River R. R. This branch, continuing from Auburn, extends to *Weedsport* (369 miles), connecting there with the West Shore R. R., thence to *Sterling* (388 miles), where connection is made with the Rome, Watertown & Ogdensburg R. R., and ending at *Fair Haven* (390 miles), on Lake Ontario.

Leaving Sayre on the main line, the road traverses rich farming districts, passing through *Van Etten, Spencer, West Danby, Newfield*, and reaches *Ithaca* (311 miles). (See page 175.) Leaving the city, the line continues along the shore of Cayuga Lake, reveling in scenes of enchanting loveliness as it gradually surrounds the hills on its way to Seneca Lake, passing *Taghanic Falls*, one of the most noted, possibly, of all the falls in this part of the State. Its interesting features are the deep ravine, its extraordinary height, and sharply defined outlines. The view of the lake and the surrounding country at this point is truly grand. The water breaks over a clear-cut table rock, falling vertically 215 feet. Twelve miles beyond Taghanic Falls, *Sheldrake* is reached, where Cayuga Lake Hotel stands, a well-managed summer resort. Cayuga Lake is about 40 miles long, varying from 1 to 4 miles wide, and is one of the most remarkable inland bodies of water in the world. It is 441 feet above tide-water, and 196 feet above Lake Ontario. Continuing on the main line, we reach Geneva (351 miles), at the foot of Seneca Lake. (See Route 40.) From this point the line of the Lehigh Valley R. R. continues, passing through *Clifton Springs* (363 miles), till *Rochester Junction* is reached, where, by branch line of the Lehigh Valley R. R., connection is made for **Rochester,** 14 miles distant (400 miles from New York). (See Route 40.) Continuing on the main line from Rochester Junction, *Batavia* (417 miles) is reached, where the train is divided—one portion running to **Buffalo** (454 miles), and the other portion to **Niagara Falls** (465 miles) and **Suspension Bridge.**

46. Lake George and Lake Champlain.

THE direct approach to Lake George is by Route 44 to Fort Edward, whence a branch road runs in 15 miles to Caldwell, at the head of the lake, passing through **Glens Falls** (*American, Rockwell*), a village of 9,509 inhabitants, situated on the Hudson River, at a fine cataract 50 ft. high, at which many travelers might feel an interest in stopping before they reach Lake George. The falls are very fine, and the spot is of peculiar interest as the scene of some of the most thrill-

14

ing incidents of Cooper's romance, "The Last of the Mohicans," which all lovers of American literature will remember. About 2 miles from the lake, in a dark glen, the road passes in sight of the *Williams Monument,* a plain marble shaft erected on the spot where Col. Ephraim Williams, founder of Williams College, fell in a battle with the French and English, Sept. 8, 1755. Near by is the storied *Bloody Pond,* into which the bodies of the slain were cast after the battle, tingeing its waters for many years (according to the legend) with a sanguine hue. The approach to the lake is very impressive, fine but fleeting glimpses being caught of its gleaming waters and blue hills. Finally, as the train reaches **Caldwell** (*Fort William Henry Hotel, Lake House, The Crosbyside, Carpenter's*), the whole glorious scene bursts upon the view. Caldwell is a small village at the S. end of Lake George, much visited in summer, and chiefly noted for its hotels. The *Fort William Henry Hotel* stands on the site of the old Fort William Henry, remnants of which are still visible, and from its spacious piazzas an unrivaled view of the lake is obtained. About ½ mile to the S. E. are the picturesque ruins of Fort George, and the outlook embraces French and Prospect Mountains, and Rattlesnake Hill, all of which may be ascended from the village. Many persons spend the season at Caldwell, making excursions to the various points of interest on the lake. The fishing is excellent, and pleasure-boats may be obtained in any numbers.

Lake George is a picturesque sheet of water in Warren and Washington Counties, N. Y., 33 miles long from N. E. to S. W., and from ¾ of a mile to 4 miles wide. It is the most famous and most frequented of American lakes, and is remarkable alike for the pellucid clearness of its water, its multitude of little islands, popularly supposed to correspond in number with the days of the year, and the beautiful scenery of its banks. The lake is bordered on either side by high hills, which here recede from the undulating shore, there lift their wooded crests in the distance, and again hang rugged cliffs over the water, or project bold promontories into its placid depths. It empties to the N. into Lake Champlain, from which it is separated by a narrow ridge only 4 miles wide; and, except in its widest part, seems more like a river than a lake. The Indian name of Lake George was "Andiatarocte," which meant "the tail of the lake," while "Horicon" (meaning "silvery waters") is a fanciful title given by James Fenimore Cooper, who objected to its present name. When the French discovered it, early in the seventeenth century, they named it "Le Lac du St. Sacrement" (Lake of the Holy Sacrament), but its English conquerors called it after King George II, then on the throne.

Lake George fills a conspicuous and romantic place in American history. For more than a century it was a channel of communication between Canada and the settlements on the Hudson. In the French and Indian War it was repeatedly occupied by large armies, and was the scene of several battles. In an engagement near the S. end of the lake, September 8, 1755, between the French and the English, Colonel Williams, of Massachusetts, the founder of Williams College, was killed, Baron Dieskau, the French commander, severely wounded, and the French totally defeated (see above). In 1757 Fort William Henry, at the same end of the lake, was besieged by the French General Montcalm, at the head of 8,000 men. The garrison capitulated after a gallant defense, and were barbarously massacred by the Indian allies of the French. In July, 1758, the army of General Abercrombie, about 15,000 strong, passed up the lake in 1,000 boats, and made an unsuccessful attack on Ticonderoga. A year later (July, 1759) General Amherst, with an almost equal force, also traversed the lake, and took Ticonderoga and Crown Point. The head of Lake George was the depot for the stores of the army of General Burgoyne before he began his march to Saratoga.

A steamer leaves daily from Caldwell for Baldwin at the N. end of the lake and returns in the afternoon (fare either way $2, which entitles the passenger to return free the same day). Leaving the pier in front of Fort William Henry Hotel, the steamer touches at the docks of the Lake House, and then crosses to *Crosbyside*, opposite Caldwell, and the site of the spacious and popular *Crosbyside Hotel*. About a mile N. of Crosbyside is *St. Mary's on the Lake*, the summer retreat of the Paulist Fathers. The nearest island to Caldwell, about 1 mile distant, is *Tea Island*, so called from a "tea-house" once erected there for the entertainment of visitors, but of which only the stone walls now remain. This island is covered with noble trees, and bordered with picturesque rocks, and is a favorite resort for picnic and boating parties. A mile and a half beyond is **Diamond Island,** so named on account of the beautiful quartz-crystals found on it in abundance. Here, in 1777, was a military depot of Burgoyne's army, and here a severe skirmish occurred in that year between the garrison and a detachment of New England militia, in which the latter were signally worsted. Next beyond are the two diminutive islets known as the *Two Sisters*, and along the E. shore is *Long Island*, which appears from the boat to be part of the main shore. Just above is *Ferris's Bay*, where Montcalm moored his boats and landed his troops in 1757. **Dome Island,** a richly wooded island, is about 10 miles from Caldwell, near the center of the widest part of the lake. Putnam's troops took shelter here, while he went to apprise General Webb of the movements of the enemy at the mouth of *Northwest Bay*, which here runs in to the W. A little W. of Dome Island is the "Hermitage," or * *Recluse Island*, on which a neat villa has been erected among the trees, and a graceful bridge thrown over to a little dot of an island close at hand, named *Sloop Island*, from its fancied resemblance to a sloop, when seen from a certain point of view. *Pilot Mountain* is a precipitous peak on the E. shore, at the foot of which are the *Trout Pavilion* and the *Kattskill House*, favorite resorts for anglers and sportsmen. Near these hotels are the best fishing-places on the lake, and the wooded mountains in the rear afford good hunting. From this point, the steamer runs nearly due N. to Bolton, passing between Dome and Recluse Islands, already mentioned. **Bolton** is a village on the W. shore, the largest on the lake after Caldwell, and has fine hotels (the *Bolton, Lake View,* and *Mohican House*). A bridge connects Bolton with *Green Island*, on which is the *Sagamore*, a charmingly situated hotel. Back of the village is *Prospect Mountain*. *Ganouskie Bay* extends for 5 miles above Bolton, and is closed in on the E. by *Tongue Mountain*, which comes in literally like a tongue of the lake, into the center of which it seems to protrude, with the bay on one side, and the main passage of the waters on the other. On the right or E. shore, nearly opposite the Tongue, is the bold semicircular palisade called *Shelving Rock.* Passing this picturesque feature of the landscape, and afterward the point of Tongue Mountain, we come to *Fourteen-Mile Island,* at the entrance of the "Narrows," where there is a large hotel. On the mainland, about 1 mile S. of Fourteen-Mile Island, is *Shelving-Rock Fall,* situated on a small stream, which empties into Shelving-Rock Bay. It is a very picturesque cascade, and is much resorted to by picnic-parties. At * **The Narrows** the shores of the lake approach each other, the space between being crowded with islands. This is the most picturesque and striking portion of the lake scenery, and enthusiastic visitors have declared it to be unsurpassed for beauty by any of the famous lakes of Switzerland or Scotland. On the E. is *Black Mountain,* the highest of the peaks that line the lake-shore. It is well wooded at its base, though frequent fires have swept over its surface, while the summit of the mountain stands out rocky and bare. Its height is 2,878 ft., and the view from the summit is very extensive. The ascent is easy from either Black Mountain Point or Hulett's Landing. Beyond Black Mountain the steamer passes *Sugar-Loaf Mountain,* on the E.; *Bosom Bay,* with the little village of Hulett's Landing; and *Deer's-Leap Mountain,* on the W., said to be so named from the tragical fate of a buck, which, being hotly pursued by a hunter and his dogs, leaped over the precipitous side of the mountain facing the lake, and was impaled on a sharp-pointed tree below.

Emerging from the Narrows on the N., we approach a long, projecting slip of fertile land known as *Sabbath-Day Point.* This spot is memorable as the scene of a fight in 1756 between the colonists and a party of French and Indians. The former, sorely pressed, and unable to escape across the lake, made a bold defense and defeated the enemy, killing very many of their men. In 1776 Sabbath-Day Point was again the scene of a battle between some American militia and a party of Indians and Tories, when the latter were repulsed and some 40 of their number were killed and wounded. This part of Lake George is even more charming in its views, both up and down the lake, than it is in its numerous historical reminiscences. On a calm, sunny day the romantic passage of the Narrows, as seen to the S., is wonderfully fine; while in the other direction is the broad bay or widening of the lake, entered as the boat passes Sabbath-Day Point. On the W. side of this widening of the lake (which is here 4 miles across) is the picturesque little village of **Hague,** from which is seen the ridge called the "Three Brothers." It has several good hotels, and near it are some excellent bass-fishing grounds and two trout-streams. Below Hague the lake contracts again to a narrow pass between the precipitous *Anthony's Nose,* on the E., and *Rogers's Slide,* on the W. Rogers's Slide is a rugged and steep promontory, about 400 ft. high, down which the Indians, to their great bewilderment, supposed the bold ranger, Major Rogers, to have slid, when they pursued him to the brink of the cliff. A short distance N. of it, on a bold promontory, is the *Rogers Rock Hotel,* one of the largest and best on the lake. Stages run from the hotel to Addison Junction, where connection is made with the White Mountain trains (Route 44). Beyond Rogers's Slide the lake is narrow, the shores low and uninteresting, and soon the voyage terminates at *Baldwin* (33 miles from Caldwell). E. of the landing is the low-lying *Prisoners' Island*, where, during the French War, those taken captive by the English were confined; and to the N. is *Lord Howe's Point*, where the English army under Lord Howe, consisting of 16,000 men, landed previous to the attack on Ticonderoga.

From the steamboat-landing at Baldwin, a branch of the Delaware & Hudson R. R. (see Route 44) runs to Fort Ticonderoga, on Lake Champlain, 5 miles distant. At Ticonderoga village, about midway between the two lakes, the stream which discharges Lake George into Lake Champlain tumbles down a rocky descent in a highly picturesque fall.

Lake Champlain.

Lake Champlain lies between New York and Vermont, and extends from Whitehall in the former State to St. John's in Canada. It is 126 miles long, and varies in breadth from 40 rods to 12½ miles. Its outline is very irregular, the shores being indented by numerous bays, and there are upward of 50 islands and islets. Its depth varies from 54 to 399 ft., and vessels of 800 or 1,000 tons navigate its whole extent. The principal rivers entering the lake are Wood Creek, at its head; the outlet of Lake George, the Ausable, Saranac, and Chazy, from New York; and Otter, Winooski, Lamoille, and Missisquoi, from Vermont. The outlet of the lake is the Sorel or Richelieu River, sometimes called the St. John's, which empties into the St. Lawrence, and, with the Chambly Canal, affords a passage for vessels to the ocean. On the south it communicates, by means of the Champlain Canal, with the Hudson River at Troy. Navigation is usually closed by ice about the end of November, and opens early in April. The waters of the lake abound with bass, pickerel, muscalonge, and other varieties of fish. This lake, filling a valley inclosed by high mountains, is celebrated for its magnificent scenery, which embraces the Green Mountains of Vermont on the E. and the Adirondack Mountains of New York on the W. Several pleasant villages and watering-places, with one or two important cities, are situated on its shores, and it has always been one of the most attractive features of the Northern Tour.

A writer in "Picturesque America" institutes the following comparison between the sister lakes : "On Lake George the mountains come down to the edge of the waters, which lie embowered in an amphitheatre of cliffs and hills ; but on Lake Champlain there are mountain-ranges stretching in parallel lines far away to the right and left, leaving between them and the lake wide areas of charming champaign country, smiling with fields and orchards and nestling farmhouses. There are on Lake Champlain noble panoramas ; one is charmed with the shut-in sylvan beauties of Lake George ; but the wide expanses of Lake Champlain are, while different in character, as essentially beautiful. It is in every way a noble lake. Ontario is too large—a very sea ; Lake George is perhaps too petty and confined ; but Lake Champlain is not so large as to lose, for the voyager upon its waters, views of either shore, nor so small as to contract and limit the prospect." The name of the lake is derived from that of Samuel de Champlain, the French Governor of Canada, who discovered it on the 4th of July, 1609.

Whitehall, at the head or S. end of the lake, has already been described in Route 44. The Lake Champlain steamers used to start from

Whitehall, but, since the completion of the railway along the W. shore (described in Route 44), they come no higher than Fort Ticonderoga (24 miles below). The narrowness of this upper portion of the lake gives it much more the appearance of a river than of a lake. For the first 20 miles the average width does not exceed ½ mile, and at one point it is not more than 40 rods across. Fort Ticonderoga is the point where the lake widens and becomes a lake in fact as well as in name.

Fort Ticonderoga (*Fort Ticonderoga Hotel*) is a station on the lake at the foot of Mt. Defiance, whence a branch railroad runs in 5 miles to Baldwin on Lake George (as described above), and whence the Lake Champlain steamers run daily in summer to Plattsburg. Ticonderoga village is 2 miles from the landing, and the ruins of the famous old * **Fort Ticonderoga** are on a high hill about a mile to the N. The view from the crumbling ramparts is extremely fine; and a still finer one may be obtained from the top of * *Mt. Defiance*, which is easily ascended from the village. *Mt. Independence* lies in Vermont opposite Ticonderoga, about a mile distant; remains of military works are still visible there. *Mt. Hope*, an elevation about a mile W. of Ticonderoga, was occupied by Burgoyne previous to the recapture of the fort in 1777.

Fort Ticonderoga was first built by the French in 1756, and was called by them "Carillon." We have already mentioned Abercrombie's attempt to capture it in 1758, and Lord Amherst's more successful campaign in the following year. (See "Lake George.") The French, being unable to hold the fort, dismantled and abandoned it at the approach of the English forces; and soon afterward Crown Point was also abandoned. The English enlarged and greatly strengthened the two fortifications, expending thereon $10,000,000, which was at that time an immense sum for such a purpose. The fort and field-works of Ticonderoga embraced an area of several miles. After the cession of Canada to the English, in 1763, the fort was allowed to fall into partial decay; and at the outbreak of the Revolution it was one of the first strongholds captured by the Americans. Colonel Ethan Allen, of Vermont, at the head of the Green Mountain Boys, surprised the unsuspecting garrison, penetrated to the very bedside of the commandant, and, waking him, demanded the surrender of the fort. "In whose name and to whom?" exclaimed the surprised officer. "In the name of the great Jehovah and the Continental Congress!" thundered the intrepid Allen, and the fort was immediately surrendered. Afterward, however, in the campaign of 1777, Burgoyne easily reduced it by placing a battery of artillery on the summit of *Mount Defiance*, on the south side of the Lake George outlet and 750 feet above the lake, from which shot could be thrown into the midst of the American works. After the surrender of Burgoyne, the fort was dismantled, and from that time was suffered to fall into ruin and decay.

Leaving the landing at Fort Ticonderoga the steamer runs N. to *Shoreham*, on the Vermont shore, and thence crosses the lake to the village of *Crown Point*, with fine mountain-views all the way. Six miles below, on the W. side, is the rugged promontory of * **Crown Point,** which was the site of Fort St. Frederic, erected by the French in 1731, and of a much stronger work subsequently erected by the English, the massive ruins of which are still plainly visible. The history of this fort is strikingly similar to that of Fort Ticonderoga, the fate of either fortress generally determining that of the other. In 1759 the English took possession of the whole region; in 1775 Crown Point was taken by Ethan Allen at the time he captured Ticonderoga;

and in 1777 Burgoyne retook it and made it his chief depot of supplies in the advance to Saratoga. A lighthouse now stands on the peak of the promontory, but otherwise all is desolation. Fine views are obtained from the bastions of the old fort. Beyond Crown Point, on the Vermont shore, is *Chimney Point.* Between them the lake is very narrow, but opens out above into the broad Bulwagga Bay, on the W. shore of which is the pretty village of *Port Henry* (20 miles from Fort Ticonderoga), with extensive iron-works and ore-beds. Just beyond Port Henry the scenery is exceedingly fine. To the E. the Green Mountains with their lofty peaks, Mt. Mansfield and Camel's Hump, rise against the distant horizon; and on the W. "the Adirondack Hills mingle their blue tops with the clouds." Eleven miles below Port Henry, on Northwest Bay, is **Westport** (*Westport Inn*), an entrance to the Adirondack region by stages to Elizabethtown. Ten miles below Westport, on the same side, is the small village of *Essex*, and between them the steamer passes ***Split Rock,** where a portion of the mountain, ¼ acre in extent and 30 ft. high, is isolated by a remarkable fissure and converted into an island. Leaving Essex, the steamer passes out into the broadest reach of the lake, bears over toward the Vermont shore, passes the islets called the Four Brothers and Rock Dunder, and soon reaches the beautiful city of **Burlington** (described in Route 29). The view of the city as approached from the lake is remarkably pleasing. Leaving Burlington, the steamer runs across the lake 10 miles to **Port Kent,** where tourists take the Keeseville, Ausable Chasm & Lake Champlain R. R. in visiting the famous **Ausable Chasm.** From this point, whether on land or water, the views in every direction are striking and beautiful. The most interesting feature of the town is the old stone mansion of Elkanah Watson, on a hill near the lake. Port Kent is one of the entrances to the Adirondack region.

Three miles beyond Port Kent, the Ausable River comes in on the W., and 5 miles farther the steamer enters the narrow channel between the mainland and *Valcour Island*, which was the scene of the desperate naval battle between Arnold and Carleton, in 1776. The large *Hotel Champlain* is seen on the bluff overlooking the lake, 3 miles south of Plattsburg. All steamers make landing at the hotel, which is a convenient place for travelers to stop at in making the Northern tour. Beyond, the steamer enters Cumberland Bay, and stops at **Plattsburg,** a prosperous and beautiful village of 7,010 inhabitants on the W. shore of the lake, at the mouth of the Saranac River, 168 miles from Albany. A branch railroad runs from Plattsburg to Ausable (20 miles), and this is a favorite entrance to the Adirondacks. The Chateaugay R. R. runs up the Saranac Valley 17 miles to *Dannemora*, the site of Clinton Prison; and from this point S. W. by Mts. Johnson and Lyon to the trout-teeming *Chazy Lake* (4 miles long by 1½ wide, and 1,400 ft. high), near *Bradley Pond*, also famous for its abundance of trout, and to *Lyon Mt. Station.* The terminus of the road is at Saranac Lake, 39 miles farther into the wilderness. Stages run from Lyon Mt. Station to "Ralph's" and "Merriles's," on **Chateaugay Lake.** This station is the site of a valuable iron-mine, and is nearly in the center of the great Chateaugay wilderness, 4 miles from the upper Chateaugay Lake (5 miles long and 2 miles wide), whence a navigable stream leads in 4 miles to the *Lower Lake*, 2½ miles long. *Loon Lake* (54 miles) lies three miles from the station of that name, and **Saranac Lake** (73 miles), is one of the most delightful resorts of the region. This is now the best route for the northern Adirondack region.

Cumberland Bay, on which Plattsburg is situated, was the scene of the victory of Macdonough and Macomb over the British naval and land forces, under Commodore Downie and Sir George Provost, familiarly known as the *Battle of Plattsburg*. It occurred on Sept. 11, 1814, and resulted in the capture of the British fleet and the defeat of their army with the loss of 2,500 men.

The Lake Champlain steamers run to North Hero. Daily steamers run across the lake in 25 miles to **St. Albans** (Route 27), passing among beautiful islands.

47. The Adirondacks.

THIS remarkable tract, which, thirty years ago, was known even by name only to a few hunters, trappers, and lumbermen, lies in the northern part of New York State, between Lakes George and Champlain on the E., and the St. Lawrence on the N. W. It extends on the N. to Canada and on the S. nearly to the Mohawk River. The mountains rise from an elevated plateau, which extends over this portion of the country for 150 miles in latitude and 100 in longitude, and is itself nearly 2,000 ft. above the level of the sea. Five ranges of mountains, running nearly parallel, traverse this plateau from southwest to northeast, where they terminate on the shores of Lake Champlain. The most westerly, which bears the name of the Clinton Range, though it is also sometimes called the Adirondack Range, begins at Little Falls and terminates at Trembleau Point, on Lake Champlain. It contains the highest peaks of the entire region, the loftiest being Mt. Marcy (or Tahawus), 5,334 ft. high, while Mts. Seward (4,334), McIntyre (5,112), McMartin, White-face (4,871), Dix Peak (4,916), Colden (4,772), Santanoni (4,644), Snowy Mountain, and Pharaoh are among other of the prominent peaks. Though no one of these peaks attains to the height of the loftiest summits of the White Mountains of New Hampshire, or the Black Mountains of North Carolina, their general elevation surpasses that of any range east of the Rocky Mountains. The entire number of mountains in the Adirondack region is supposed to exceed 500, of which only a few have received separate names. They are all wild and savage, and covered with the "forest primeval," except the stony summits of the highest, which rise above all vegetation but that of mosses, grasses, and dwarf Alpine plants.

In the valleys between the mountains lie many beautiful lakes and ponds, more than 1,000 in number. The general level of these lakes is about 1,500 ft. above the sea, but Lake Perkins, the highest of them, has nearly three times that elevation. Some of them are 20 miles in length, while others cover only a few acres. The largest of these lakes are Long Lake, the Saranacs, Tupper, the Fulton Lakes, and Lakes Col-

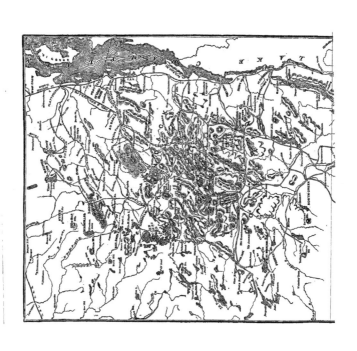

den, Henderson, Sanford, Blue Mountain, Raquette, Forked, Newcomb, and Pleasant. "Steep, densely wooded mountains," says a writer in "Picturesque America," describing these lakes, "rise from their margins; beautiful bays indent their borders, and leafy points jut out; spring brooks trickle in; while the shallows are fringed with water-grasses and flowering plants, and covered sometimes with acres of white and yellow water-lilies. The lakes are all lovely and romantic in everything except their names, and the scenery they offer, in combination with the towering mountains and the old and savage forest, is not surpassed on earth. In natural features it greatly resembles Switzerland and the Scottish Highlands, as they must have been before those regions were settled and cultivated." This labyrinth of lakes is connected by a very intricate system of rivers, rivulets, and brooks. The Saranac and the Ausable run in nearly parallel lines toward the N. E., discharging their waters into Lake Champlain. They define upon the map the position of the valleys, which have the same general arrangement throughout the whole chain, and to some extent the position of the ranges of mountains also. In the other direction, the Boreas, the Hudson, and the Cedar Rivers, which all unite below in the Hudson, define the extension of the valleys of the Ausable and its branches on the southern declivity of the great plateau; and farther W. the chain of lakes, including Long Lake, Raquette Lake, and the Fulton Lakes, lie in the same line with the valley of the Saranac, and mark its extension from the central elevation of the plateau toward the S. W. The largest and most beautiful river of the Adirondack region—its great highway and artery—is the Raquette, which rises in Raquette Lake, in the W. part of Hamilton County, and, after a devious course of 120 miles, flows into the St. Lawrence.

The mountains of the entire region are covered with forests, groves of birch, beech, maple, and ash, succeeding to the evergreens, among which the most common are the hemlock, spruce, fir, and cedar, with the valuable white pine intermixed with and overtopping the rest. In the lower lands along the streams a denser growth of the evergreens is more common, forming almost impenetrable swamps of cedar, tamarack or hackmatack, and hemlock. In these woods and mountain solitudes are found the panther, the great black bear, the wolf, the wild-cat, the lynx, and the wolverine. The moose is said to be extinct, but deer are abundant; and so, also, are the fisher, sable, otter, mink, muskrat, fox, badger, woodchuck, rabbit, and several varieties of the squirrel. There are scarcely any snakes, and none large or venomous. Among the birds are the grand black war-eagle, several kinds of hawk, owl, loon, and duck; the crane, heron, raven, crow partridge, and numerous smaller birds. The salmon-trout and the speckled trout swarm in the lakes, and the latter also in the brooks and rivers. The lake-trout are caught sometimes of 20 pounds and more in weight; but the speckled trout are seldom large.

There are several routes by which the Adirondack region may be reached. No. I is *via* the Adirondack R. R. from Saratoga to North Creek. This takes the traveler from S. W. to N. E., into the southern

Adirondacks. No. II is *via* Delaware & Hudson R. R. from Albany to Plattsburg, where connection is made with the Chateaugay R. R. to Saranac, or a branch to Ausable. No. III is *via* Port Kent, reached by steamer from Burlington (see Route 29); and No. IV is *via* Westport on Lake Champlain, reached by Delaware & Hudson R. R. to Fort Ticonderoga, and thence by steamer. From Boston the Vermont Central R. R. will take the traveler to Burlington, whence steamers run to Plattsburg and Westport.

During the summer of 1892 the Adirondack & St. Lawrence R. R. was opened to the public. It extends over a distance of 212 miles, beginning at *Herkimer*, where it makes connection with the New York Central & Hudson River and West Shore R. Rs., then N. to *Trenton Falls* (24 miles), with stage connection for Moore's Hotel. **Remsen** is a junction with the Rome, Watertown & Ogdensburg R. R., and just beyond the Black River is crossed. Four miles beyond is *Honnedaga*, the station for the Adirondack League Club, whose club-house is on Honnedaga Lake, 19 miles distant. Another house belonging to this club is situated on Moose Lake, 5 miles from Old Forge. At *Fulton Chain*, 58 miles from Herkimer, there is stage connection for Old Forge House and steamboat landing, a mile and three-quarters distant, where water communication may be had by way of steamer to the head of Fourth Lake, and by small boat to the head of Fifth Lake; and a short and easy route is thus opened to the Raquette, Forked, Blue Mountain, and Long Lakes by the Fulton Chain. Parties taking small boats at the head of Fourth Lake may go through to Raquette Lake by making three short portages, and from Raquette the other lakes may be easily reached by steamer or small boats. *Lake Lila* is the station for Dr. W. Seward Webb's private preserve, bearing the Indian name of Ne-ha-sa-ne Park. *Bog Lake, Horseshoe Pond, Childwold*, with stage for Childwold Park House and Cottages, are passed, and *Tupper Lake Junction* (113 miles) is reached at the junction of the Northern Adirondack R. R., beyond which is *Saranac Inn, Paul Smith's, Rainbow Lake*, and *Loon Lake*. Here the hotel is on the shore of the lake, and from its observatory a magnificent prospect of the Green Mountains of Vermont, and of many of the lofty peaks of the Adirondacks to the south, may be had. The Chateaugay R. R. connects at Loon Lake, whence the route is N. to *Malone*, where it connects with the Vermont Central R. R. for Montreal by way of Valleyfield and Coteau Junction. Through trains, with full sleeping, drawing-room, and dining car service, run daily from New York to Montreal over this line. With the railroads penetrating the Adirondacks in all directions, the older methods of travel by stage-coaches will gradually disappear; but as there are still many who prefer this means of visiting this great park, we retain the description of the older routes.

Outfit and Guides.—Nearly all traveling in the Adirondacks is done by means of boats of small size and slight build, rowed by a single guide, and made so light that the craft can be lifted from the water and carried on the guide's shoulders from lake to lake or from stream to stream. Competent guides, steady and intelligent men, can be hired at all the hotels for $2 to $3 a day, who will provide boats, tents, and everything requisite for a trip. Each traveler should have a guide and a boat to himself, and the cost of their main-tenance in the woods is not more than $1 a week for each man of the party. The fare is chiefly trout and venison, of which an abundance is easily procured. A good-sized valise will hold all the clothes that one person needs for a two months' trip in the woods, besides those he wears. The following list comprises the essentials of an outfit for a man: A complete undersuit of woolen or flannel, with a "change"; stout trousers, vest, and coat; a felt hat; two pairs of stockings; a pair of common winter-boots and camp-shoes; a rubber blanket or coat; a pair of pliable buckskin gloves, with cha-mois-skin gauntlets tied or buttoned at the elbow: a hunting-knife, belt, and a pint tin cup; a pair of warm blankets, towel, soap, etc. A lady's outfit should comprise: A short walking-dress, with Turkish drawers fastened with a band tightly at the ankle; a flannel change throughout; thick balmoral boots, with rubbers; a pair of camp-shoes, warm and loose-fitting; a soft felt hat, rather broad in the brim; a water-proof or rubber coat and cap; a pair of buckskin gloves with armlets of chamois-skin or thick drilling, sewed on at the wrist of the glove and buttoned near the elbow so tightly as to prevent the entrance of flies; and a net of fine Swiss mull as a protection against mosquitoes, gnats, etc.

I. Saratoga Springs to Schroon, Blue Mountain, Long, Raquette, and Fulton Lakes, and Indian Pass.

The Adirondack R. R. runs northward from Saratoga Springs to North Creek (57 miles; fare, $2.50). It is a most picturesque route, running straight up the lovely Kayaderosseras Valley, from Saratoga. It crosses the Sacandaga River by a bridge 450 ft. long and 96 ft. high, and passes near *Corinth Falls,* where the Hudson, with a width of only 50 ft., makes a leap of 60 ft. over the precipice. · At *Hadley* (22 miles) passengers leave the train for **Lake Luzerne,** which, with the village of the same name, lies just across the Hudson. Lake Luzerne is a small but exceedingly picturesque body of water, and is a popular summer resort, and a favorite excursion from Saratoga Springs. There are sev-eral hotels in the village, of which the *Wayside House* and *Rockwell's* are the best; and the fishing, hunting, and boating are excellent. From *Potash Hill,* near the lake, an admirable view is obtained, and Lake George is only 10 miles distant (reached by a good road). From *Thur-man* (35 miles from Saratoga) stages run in 9 miles to Caldwell, at the head of Lake George, by way of Warrensburg (see Route 46); and from *Riverside* (49 miles) stages run in 6 miles to **Pottersville,** which is only a mile from Schroon Lake (fare, $1).

*** Schroon Lake** is 10 miles long and about 2¼ wide, and is sur-rounded by lovely scenery. A boat plies on its waters, connecting the landing at Pottersville with *Schroon Lake Village,* the principal sum-mer resort in this vicinity. The village lies on the W. shore of the lake, and, besides numerous boarding-houses, has several summer-hotels (the *Lake, Leland, Ondawa,* and *Taylor*). The boating and fishing on the lake are unsurpassed, and excursions may be made (with guides) to the summit of *Mt. Pharaoh,* to the top of *Mt. Severn,* and to the beau-tiful **Paradox Lake** (*Pyramid Lake House*), which lies 4 miles above the N. end of Schroon Lake. At the foot of Mt. Pharaoh is *Pharaoh Lake,* famous for the abundance of its trout.

Daily stages run from N. Creek *via* Indian Lake to *** Blue Mount-ain Lake,** one of the loveliest of the Adirondack chain, 3 miles in diameter, irregular in shape, and 2,000 ft. above the sea. It is surrounded by dense forests, and in the lake and adjacent ponds. are abundance of trout. The hotels are the elegant *Prospect House*, the *Blue Mt. Lake House*, and the *Blue Mt. House*. **Blue Mountain** (3,824 ft. high) overlooks the lake, and is ascended by a well-defined trail from the Long Lake road. The view from its summit is extremely fine.

Stages also run from N. Creek to Minerva, Tahawus, Newcomb, and Long Lake Villages. *Tahawus* (Lower Iron Works) is a decayed hamlet in the very heart of the Adirondack mountain-system. At this point the road to Long Lake turns directly W. to the village of **New-comb** (*Half-way Hotel, Wayside House*), which is a good point at which to procure guides, boats, and camp equipage. Near Newcomb are *Lakes Harris, Delia,* and *Catlin,* and 12 miles W. is *Long Lake Village* on Long Lake. **Long Lake** is the longest, though not the widest, of the Adirondack lakes, and for 13 miles almost resembles a river. The scenery on the lake itself is varied and exquisite, and from it a noble view can be had of *Mt. Seward,* 4,334 ft. high. On the E. of the lake (3½ miles from the head) is *Long Lake Village* (*Austin, Long Lake Hotel,* and the *Sagamore,* about ½ a mile above), the center of supplies and the starting-point for routes in all directions. Many guides live here, and the vicinity is much frequented by sportsmen. From *Long Lake* a series of ponds and carries leads to **Little Tupper Lake,** a lovely and sequestered sheet of water, 6 miles long, dotted with islands and girt by rugged, precipitous shores. This can also be reached by the Adirondack & St. Lawrence R. R. It affords excellent sport. On *Sand Point,* near the foot of the lake, is the *Pine Grove House.* Blue Mountain Lake is 8 miles S. from Long Lake Village *via* South Pond. *Grampus Lake* is reached from Long Lake by ascending the Big Brook, and the Handsome and Mohegan Ponds may be visited from Grampus. At the S. W. end of Long Lake is *Owl's Head Mountain* (2,825 ft. high). At its base lie Clear Pond and Owl's Head Pond.

From the S. end of Long Lake a route of 10 miles with two carries leads through *Forked Lake* (*Forked Lake House*), a sheet of water 5 miles long, to * **Raquette Lake,** the last of the chain of lakes we have been following. Shortly after leaving Long Lake the picturesque *Buttermilk Falls* are passed, and the entire journey is through the midst of fine forest scenery with occasional mountain-views. Raquette Lake is 12 miles long and 5 miles wide at the widest part, and its surface is sprinkled with the most beautiful little islets. Dense forests close in on every side, and as it is comparatively unfrequented it makes rich returns to the sportsman. A short distance from its N. end are the ruins of *Cary's Hotel,* which is connected by the Carthage road with Long Lake Village (14 miles). *Hathorn's Forest Cottages* are at South Beach, and *Blanchard's Wigwams* is at the head of Marryatt's Bay. *Bennett's,* nearer the Blue Mountain, is small, but well suited for invalids. There is a multitude of lakelets and ponds in the vicinity of Raquette Lake, but only two or three require special mention. *Shallow Lake* (reached by an inlet from Marryatt's Bay) and *Queer Lake* (2 miles S.) are noted for trout. *Beach's Lake,* a fine sheet of water, 3½ miles long, lies 4 miles N. W., and is reached by a long and tedious "carry." A series of ponds and carries leads from Beach's Lake to Little Tupper Lake (already described), but the route is long (15 miles) and difficult. *Salmon Lake* is N. of Beach's Lake, and may be reached by a carry of 2 miles. A pull of 4 miles up the Brown Tract Inlet, from the S. W. point of Raquette Lake, and a carry of 1½ mile lead to the upper or eighth of the chain of **Fulton Lakes,** which extend southwestward into the "John Brown Tract" (see p. 173). From the *Eighth Lake* a short portage leads to the *Seventh,* from which the *Sixth* can be reached by boat. There is a portage between *Sixth* and *Fifth,* and also one between *Fifth* and *Fourth.* * **Fourth Lake** is the largest of the chain; it is studded with islands and surrounded by rugged and precipitous shores. Hemlock grows down to the edge of the water; and in the undisturbed repose of the waters the fringes of foliage are clearly reflected. In the center of the lake is a beautiful rocky islet known as *Elba.* There is a passage for boats into *Third Lake,* close by which Bald Mountain frowns down; and the passage continues open to *Second Lake. Second* is hardly distinguishable from *First Lake,* there being a mere sand-bar between them. This section is seldom visited save by hunters and fishermen, to whom it yields rich returns. It is now more easily reached from *Fulton Chain* on the Adirondack & St. Lawrence R. R.

The route from the Raquette to the **Saranac** region is from the N. end of Long Lake, over Johnson's Carry to *Mother Johnson's Raquette Falls House.* A few rods from the house are the *Falls,* around which boats are hauled (1¼ mile, $1.50 a load). Seven miles farther (13 from Long Lake) the route leaves the river by Stony Creek to *Stony Creek Pond,* then by Indian Carry (1 mile, 75 cents a load) to *Corey's,* at the S. end of the Upper Saranac Lake. At the upper end of the lake railroad connection is readily had with either the Adirondack & St. Lawrence or the Chateaugay R. Rs.

From Tahawus, a picturesque road leads 11 miles N. to the hamlet of *Adirondack* (or Upper Iron Works), once a thriving mining town, but now in ruins, save the *Club House,* used as the headquarters of a hunting and fishing club. It lies in the midst of singularly wild and impressive scenery. A mile S. of the village is *Lake Sanford,* skirted by the road from Tahawus, and 5 miles long. On the N. are Henderson Mountain and Lake, and beyond these (2½ miles from the village) are the lovely *Preston Ponds,* which afford as good trout-fishing as is to be had in the entire region. *Mount Seward,* one of the loftiest of the group (4,334 ft.), lies 8 miles to the N. W., and *Mount Marcy* on the N. E., the monarch of the region (Sub-Route III), may be ascended by an easy path 12 miles long. The trail to the summit leads past the exquisite *Lake Colden,* and near **Avalanche Lake,** which is one of the highest of the Adirondack lakes (2,846 ft. above the sea). The greatest attraction, however, and perhaps the finest sight in the Adirondacks, is the * **Indian Pass,** a stupendous gorge between Mts. Wallface and McIntyre, in the wildest part of that lonely and savage region which the Indians rightly named "Conyacraga," or the Dismal Wilderness, the larger part of which has never yet been visited by white men, and which still remains the secure haunt of the wolf, the panther, the great black bear, and the rarer lynx and wolverine. The springs which form the source of the Hudson are found at an elevation of about 2,900 feet above the sea, in rocky recesses, in whose cold depths the ice of winter never melts entirely away. Here, in the center of the pass, rise also the springs of the Ausable, which flows into Lake Champlain, and whose waters reach the Atlantic through the mouth of the St. Lawrence, several hundred miles from the mouth of the Hudson; and yet, so close are the springs of the two rivers, that "the wild-cat, lapping the waters of the one, may bathe his hind-feet in the other; and a rock rolling from the precipice above could scatter spray from both in the same concussion." In freshets, the waters of the two springs actually mingle. The main stream of the Ausable, however, flows from the N. E. portal of the pass; and the main stream of the Hudson from the S. W. The latter is locally known as the Adirondack River, and after leaving the pass flows into Lakes Henderson and Sanford. The Indian Pass is reached from Adirondack by an easy and well-marked trail; and after traversing it the visitor may descend in 10 miles by a path blazed on the trees to *N. Elba.* A long and arduous trail leads from Adirondack to Keene Valley by way of the Ausable Ponds.

II. From Plattsburg to Saranac, Placid, and Tupper Lakes.

From Plattsburg (reached by Delaware & Hudson R. R.) the Chateaugay R. R. runs directly to *Saranac Lake,* passing *Loon Lake* (where stages for *Loon Lake House* meet all trains), *Bloomingdale* (66 miles from Plattsburg), whence stages run to *Crystal Spring, Big Clear Pond,* and to *Paul Smith's* (5 miles). From *Saranac Lake* (73 miles from Plattsburg) stages run (9 miles) to *Lake Placid.*

From *Saranac Lake* station of the Chateaugay R. R. a stage-ride of 1 mile takes the visitor to the *Hotel Ampersand* or to **Saranac Lake House,** a comfortable hotel on the *Lower Saranac Lake,* which is 7 miles long and 2 miles wide, studded with islands, 52 in number. This is one of the best places to procure guides, boats, and camp-equipage. The Saranac River connects the lake with *Round Lake,* 3 miles W. Round Lake is about 2 miles in diameter, is a beautiful sheet of water, dotted with islets, and is famous for its storms. It is in turn connected with the *Upper Saranac Lake* by another stretch of the Saranac River. A small steamer plies on the Lower Saranac, running from Saranac Lake House to the rapids below Round Lake, whence row-boats carry the passengers over to the Upper Saranac Lake. A fine view can be obtained of Round Lake and the surrounding mountains, and a guide will conduct the traveler to the summit of *Ampersand Mountain,* whence the view is superb. At the foot of the mountain, on the S., lies the sequestered *Ampersand Pond,* where Agassiz, Lowell, and Holmes used, a few years ago, to pitch their "Philosophers' Camp." A short "carry" of ¼ mile conducts from Round Lake to the *****Upper Saranac Lake,** the largest and one of the most beautiful of the Adirondack lakes. It is 8 miles long and from

1 to 3 miles wide, and its surface is studded with little islands. At its head is **Saranac Inn,** reached by stage (14 miles) from Saranac Lake station, situated in the vicinity of good hunting-ground and excellent fishing-waters. The **Ampersand Hotel** on Saranac Lake has been built as a winter resort, and the *Adirondack Cottages* (1 mile from the lake) are specially designed for patients suffering from pulmonary troubles. A small steamer plies on the lake, making the circuit from the Saranac Inn twice daily, and touching at the Sweeney Carry, and Corey's. The "Route of the Nine Carries" conducts from the Upper Saranac to *St. Regis Lake*, on which is situated * **Paul Smith's,** the best known hotel in the Wilderness. It is a very popular resort as an outfitting-point, for which it offers many advantages. It has a telegraph-office, and is connected by stage-road with *Bloomingdale*, on the Chateaugay R. R.; with Rogers, at the terminus of the Ausable branch of the Delaware & Hudson R. R.; Saranac Lake station (14 miles); with Paul Smith's station, on the Northern Adirondack R. R.; and with Paul Smith's station, on the Adirondack & St. Lawrence R. R. (4 miles). St. Regis Lake is one of the most picturesque of the group, and is surrounded by numerous small ponds. A short distance N. E. lies the *Rainbow Lake*, a favorite resort with fishermen.

A second route is by a branch of the Delaware & Hudson R. R., which runs from Plattsburg S. W. to *Rogers*, on the Ausable River (20 miles, fare $1). The ride to Saranac Lake (37 miles), if this old route is taken, is amid picturesque scenery, **Whiteface Mountain,** the great outpost of the Adirondacks, being in sight for a considerable portion of the way. At the village of *Ausable Forks*, 3 miles from the railway terminus, the traveler can (by hiring a special conveyance) turn off into a road which leads through the famous * **Wilmington Pass,** and can regain the main road about 2 miles before it reaches Saranac Lake. The mountain is easily ascended from *Wilmington*, where guides may always be had. A carriage-road leads nearly to the summit, and the remainder of the ascent is on horseback. There is a house near the summit, where a comfortable night's lodging may be obtained. The * view from Whiteface (4,871) ft. high, is indescribably grand, only surpassed by that from the top of Mount Marcy. The "Pass" is 6 miles beyond Wilmington. It is a profound chasm cloven boldly through the flank of Whiteface, scarcely wide enough for the road and the river, and 2 miles long. Through the Pass flows the Ausable River, with a succession of rapids and cataracts, and on either side rises a majestic mountain-wall, so high that the crowded row of pines along its broken and wavy crest is diminished to a fringe. At the foot of Whiteface, on the S. W. side (reached by a road which branches off to the right just beyond the Notch), lies * **Lake Placid** (reached as above by stage from *Saranac Lake* station, and also by daily coach from Westport, see p. 215), one of the loveliest lakes of the Wilderness, and the central point of interest, as well as the geographical center of the Adirondack Mountain region, 5 miles long and about 2 miles wide. There are several large hotels here (*Castello Rustico, Lake Placid House, Stevens House*, and *Whiteface Inn*), and it is a favorite summer resort. One of the best views of Whiteface is obtained from Lake Placid, and near its southern shore is one of the curiosities of the region, *Paradox Pond,* whose outlet in high water flows back on the lake. Also near by is *Mirror Lake (Mirror Lake House).* Two miles S., on the road to Elizabethtown, is the hamlet of *North Elba (Mountain View House),* close to which (on the S.) are the house and farm "of John Brown, of Ossawattomie," who lies buried close by. The best route to Lake Placid is as above, by Chateaugay R. R. to Saranac, and thence by stage.

The Chateaugay R. R., after leaving Plattsburg, runs W. to *Chazy Lake* (28 miles), whence it connects with a steamer running to the northern extremity of the lake, landing the tourist at the Chazy Lake House. Six miles beyond is *Lyon Mountain* Stages may be taken here for Ralph's and Merrill's, and steamer for Indian Point House. *Loon Lake* (54 miles) is the station for the *Loon Lake House,* which is reached by stage; also reached by the Adirondack & St. Lawrence R. R. *Round Pond* (58 miles) is a convenient point for visiting the famous fishing-grounds at Jones, Round, and Mud Ponds in the immediate vicinity. Three miles farther along the route is *Rainbow,* where stages meet the train for the fish-abounding Rainbow Lake (*Rainbow House*), 3 miles long and about 1 mile wide. *Bloomingdale* (66 miles) is the next important station, and stages for *Paul Smith's Hotel, Crystal Spring House,* and *Big Clear Pond House* are taken here. The terminus of the road is *Saranac Lake* (see p. 222).

The "Round Trip" from Paul Smith's comprises a circuit of about 45 miles, including the St. Régis and Saranac Lakes, and the principal adjacent points of interest, and affording every variety of locomotion known to the Wilderness, without enough of any to become wearisome. The route is as follows : By boat across Lower St. Regis Lake, Spitfire Pond, and Upper St. Regis Lake, with connecting streams (4 miles) ; on horse or foot over the St. Germain Carry, 1¼ mile (horse $1.50) ; boat across Big Clear Pond (2 miles) ; carriage to Saranac Inn at head of Upper Saranac Lake, 3 miles (boat $1.50, passengers 50c. each) ; on small steamer through Saranac Lake, 8 miles (fare $1) ; short carry to Round Lake (50c.) and thence by river and Round Lake (3 miles) ; by steamer on Saranac River and Lower Saranac Lake to Saranac Lake (9 miles) ; carriage from Saranac Lake to Paul Smith's (14 miles). From Saranac Lake the "Round Trip" may be made in reverse order.

At Saranac Lake three great Wilderness routes diverge: one N. to Paul Smith's, already described in the "Round Trip"; another W., to the Tupper Lake region; and a third S., to Long and Raquette Lakes. The route to the Tupper Lakes is from Saranac Lake or the Prospect House to the S. W. end of the Upper Saranac Lake, whence *Sweeney's Carry* leads across in 3 miles to the Raquette River (boats hauled across for $1.50). From the W. terminus of the carry a small steamer runs down the river to McClure's on *Tupper Lake,* which is so named from the guide or hunter who first discovered it. This may now be also reached by the Northern Adirondack or the Adirondack & St. Lawrence R. Rs. from Moira. It is a lovely sheet of water, 7 miles long by 1 to 3 miles wide, surrounded by primeval forests, and containing many picturesque rocky islands, covered with evergreens. At its head the wild and little-explored Bog River flows into the lake over a romantic cascade, which forms one of the great attractions of the

Adirondacks, being a famous place for trout. A steamer leaves Tupper Lake daily for Sweeney's Carry, connecting with the steamboat on Saranac Lake. From the N. end of the lake the Raquette River may be descended to the pretty *Piercefield Falls* (9 miles), or to *Big Wolf Pond* (10 miles). The latter is reached by turning off from the river into Raquette Pond 2 miles below Lake Tupper, and then ascending Wolf Brook to *Little Wolf Pond*, whence a carry of ¼ mile leads to Big Wolf Pond. From the S. end of Tupper Lake an easy route leads 9 miles S. through the lovely *Round Pond* to **Little Tupper Lake** (see Sub-Route I). From Little Tupper a series of small ponds and carries leads E. to Long Lake (16 miles).

From the Moodys a route leads over a 3-mile carry to *Horseshoe ·Pond*, and thence with an occasional short carry 12 miles farther to the dismal and deer-abounding **Mud Lake**, 4 miles in circumference. This lake is covered in their season with lily-pads and. margined with rank wild grass, which attracts deer in greater numbers than any other spot in the Wilderness. It is also said to have been a favorite feeding-ground of the moose before they were exterminated. Seven miles N. of Mud Lake is **Cranberry Lake,** one of the largest of the Adirondack series, being 15 miles long and 1 to 5 miles wide. It discharges to the N. through the Oswegatchie River, and is usually visited from the W. via *Gouverneur*, a station on the Rome, Watertown & Ogdensburg R. R., 108 miles from Rome (see Route 40). .The distance from Gouverneur to the lake is 36 miles. *Silver Lake* (reached by diverging to the N. W. from the Mud Lake route), *Pleasant Lake*, and numerous other ponds and lakelets of great beauty lie scattered in every direction over this remote and desolate region.

The usual route from the Saranac region to Long Lake and Raquette Lake is *via Corey's*, at the S. end of the Upper Saranac Lake (2 miles from Sweeney's Carry), whence the *Indian Carry* leads across in 1 mile (75c. a load) to the *Stony Creek Ponds* (sometimes called Spectacle Ponds). The *Hiawatha Hotel* stands at the S. terminus of the carry close by the first of the Stony Creek Ponds, which are three in number and discharge by Stony Creek into the Raquette. The river is entered at a point 20 miles from Tupper Lake and about 13 miles from Long Lake. The route is up-stream, and in 7 miles *Mother Johnson's* tavern is reached. A few rods above the house are the *Raquette Falls*, about 12 ft. high, around which the boats are hauled (1¼ mile, $1.50 a load). About 4 miles above Johnson's Carry, the mouth of Cold River, coming down from Mt. Seward, is passed; and a mile beyond the boat enters *****Long Lake** (see Sub-Route I), with Buck Mt. on the right and the Blueberry Mts. on the left.

III. Port Kent to Ausable Chasm and the Saranac Lakes.

Port Kent is on the W. shore of Lake Champlain, 12 miles above Plattsburg and nearly opposite Burlington, with which it is connected by steamer (see Route 46). Here the Keeseville, Ausable Chasm & Lake Champlain R. R. connects with trains and steamers and conveys passengers to the *Lake-View House* (3 miles) and to *Keeseville* (5 miles). The Lake-View House stands on a far-viewing eminence just above the hamlet of Birmingham, and is the most convenient point from which to visit the Chasm. Near Keeseville, the Ausable

15

River flows over the Alice Falls, and then descends a line of swirling rapids to the *Birmingham Falls, where it plunges over a precipice 70 ft. high into a semicircular basin of great beauty. A few rods farther down are the Horseshoe Falls, near which the gorge is entered from above by a stairway of 166 steps leading down a cleft in the rock (fee 50c.). Below this the stream grows narrower and deeper, and rushes through *Ausable Chasm, where at the narrowest point a wedged bowlder cramps the channel to the width of 6 or 8 ft. Still lower down the walls stand about 50 ft. apart and are more than 100 ft. high, descending to the water's edge in a sheer perpendicular line. The chasm is nearly 2 miles long, and from the main stream branches run at right angles through fissures, some of which offer very striking and beautiful effects. Stairways, walks, and galleries enable the visitor to reach the principal points of interest; and with the aid of boats constructed especially for the purpose the entire Chasm can be traversed. The entrance fee of 50c. entitles the traveler to visit all points reached by the galleries and walks, including the boat-ride from Table Rock to the Pool; for the boat-ride from the Pool down the rapids to the basin at the end of the gorge an additional fee of 50c. is charged.

Stages run from the Lake-View House to Keeseville, Ausable Forks, St. Regis, and Saranac Lakes.

IV. Westport to Elizabethtown and Keene Valley.

Westport is on the shore of Lake Champlain, a few miles S. of Port Kent, and is reached by railway (Route 44) or by steamer from Fort Ticonderaga. It is described in Route 46. From Westport stages run 8 miles W. to **Elizabethtown,** a favorite summer resort, lying within the borders of the mountain-region amid singularly picturesque and impressive scenery. The *Mansion House*, the *Valley House*, and the *Windsor* are good hotels, and there are several boarding-houses. The village stands on a plateau, closed in on all sides by lofty hills and mountain-peaks, most of which may be ascended without difficulty. The *Hurricane Peak* (3,763 ft.) lies 5 miles W., and may be ascended with guides. The view from its summit is one of the most pleasing that the Adirondacks afford. The *Giant of the Valley* (4,530 ft.) is also sometimes ascended from this place, though the route is long and difficult; and a singularly lovely view may be obtained from *Cobble Hill* (1,963 ft.), a dome-like elevation about a mile S. W. of the town. *Raven Hill* is a lofty peak to the E., from the summit of which Lake Champlain and the Green Mountains are combined in a noble view. A delightful drive from Elizabethtown is down Pleasant Valley to the romantic cascades of the Boquet River; or 8 miles S. W. to the *Split-Rock Falls*, where a mountain-brook descends 100 ft. through a rugged and resounding gorge. There is good fishing in the vicinity of the village, in the Boquet River, in Black and Long Ponds, and in the trout-abounding New Pond.

Elizabethtown is the center from which several important stage-routes diverge. The State Road through the mountains here intersects the Great Northern Highway which runs S. to Schroon Lake (32 miles) and N. to Keeseville (21 miles). Stages connect with all trains at Westport, 8 miles distant (fare, $1). Daily stages run to the head of Keene Valley (16 miles, fare $1.50). Daily stages also run to Saranac Lake (distance, 35 miles; fare, $3), *via* Keene, N. Elba, and Lake Placid. Stages run to Schroon Lake *via* Root's on Mondays, Wednesdays, and Fridays, returning alternate days (fare, $2.50).

About 10 miles W. of Elizabethtown is the beautiful *** Keene Valley,** nestling between two lofty mountain-ranges, and watered by the S. branch of the Ausable River. From the village of *Keene* or *Keene Center* at the N. end to Beede's at the S. end the valley is nearly 8 miles long; and at different points in it are *St. Hubert's Inn*, the *Tahawus House*, *Maple-Grove Cottage*, *Crawford's*, the *Beede House*, and several others. St. Hubert's Inn, at the S. end, affords an excellent starting-point for several interesting excursions. Close by are ** Roaring-Brook Falls*, where a brawling mountain-brook dashes over a cliff 500 ft. high in a succession of cascades. Four miles away is the romantic *Hunter's Pass;* and nearer at hand is the lovely *Chapel Pond*, nestling at the base of Giant of the Valley, Camel's Hump, and Bald Peak, which almost close it in. About 5 miles from St. Hubert's Inn, reached by an excellent macadamized road through the primitive forest, are the lonely and sequestered *** Ausable Ponds,** which are among the loveliest of the smaller Wilderness lakes. They are separated from each other by an easy "carry" a mile long. Near the Lower Pond are the beautiful *Rainbow Falls;* and it is only 7 miles from the Upper Pond to the summit of *** Mount Marcy,** the monarch of the Adirondack group. Guides and an outfit may be obtained at the hotels in Keene Valley, and the ascent, which with the return requires 2 days, well repays the labor. The trail itself is wonderfully picturesque, and the view from the summit (5,337 ft. high) embraces the entire Adirondack region, together with Lake Champlain and the Green Mountains of Vermont.

Other routes to the summit of Mt. Marcy are from Adirondack by a trail 12 miles long; from Root's Inn by a road, bridle-path, and trail in 20 miles; and from *Scott's* by way of the Indian Pass in 23 miles. All these are described in Sub-Route V.

The stages from Elizabethtown to Saranac cross Keene Valley at its N. end, traverse the picturesque pass between Pitch-Off and Long Pond Mountains to *Cascade Lakes* (the *Cascade House*), 4 miles from Keene. Seven miles farther (22 miles from Elizabethtown) is the *Mountain-View House*, commanding fine views and a convenient center for excursions. About 2 miles beyond are John Brown's farm and grave, near which is the hamlet of *N. Elba*, and in 2 miles more *Lake Placid* is reached. Lake Placid and the route thence to Saranac (13 miles) are described in Sub-Route II.

V. Schroon Lake to Elizabethtown and Keeseville.

From Pottersville (see p. 219) the Great Northern Highway runs almost due N. to Elizabethtown (32 miles) and Keeseville (53 miles). Stages run on this route three times a week (Tuesdays, Thursdays, and Saturdays, returning alternate days), leading for the first 10 miles along the shores of Schroon Lake, and then up the valley of Schroon

River, passing for the greater part of the entire distance amid picturesque and striking scenery. Ten miles from Schroon Lake the stage stops at **Root's Inn,** one of the favorite resorts of sportsmen. Roads from Ticonderoga (23 miles distant) and Crown Point (18 miles distant) intersect at this point; and several of the Adirondack attractions are within easy excursion-distance. *Mt. Marcy* (Sub-Route III) is visited by a wagon-road leading in 10 miles to Mud Pond, whence a forest bridle-path leads in 9 miles to the foot of the mountain, and a well-defined trail to the summit. Beyond Root's the road traverses the beautiful Schroon Valley to its head, climbs the mountain-pass, and descends into Pleasant Valley, passing the Split-Rock Falls. *Elizabethtown* (32 miles from Schroon) is described in Sub-Route IV. Beyond Elizabethtown the road traverses the picturesque gorge of Poke-o'-Moonshine, and passes in 22 miles to *Keeseville* (Sub-Route III). At Elizabethtown and Keeseville connection is made with the usual routes into the Adirondack lake-region.

VI. Skeleton Tours or "Round Trips."

The following tour can be made in ten days or two weeks, and will embrace the most striking "sights" of the Wilderness: From Crown Point, on Lake Champlain, to *Root's Inn* (see above), 18 miles; thence to *Tahawus*, 20 miles; thence to *Long Lake,* 20 miles. From Tahawus to *Adirondack*, 11 miles. From Adirondack to the summit of Mt. Marcy (with guides); also to the *Indian Pass*, the most majestic natural wonder, next to Niagara Falls, in the State. From the Indian Pass to *N. Elba*, on the Elizabethtown road (10 miles through the woods). From N. Elba to Saranac Lake. From Saranac Lake to Keeseville and the famous *Ausable Chasm.*

The following is a very popular "round trip" (all tramping), which embraces the "Heart of the Adirondacks": Start from Beede's at the head of Keene Valley, and go by forest-path 4 miles to Lower Ausable Pond; boat through pond 2 miles; trail 1 mile to the Upper Ausable Pond; boat through Upper Pond 3 miles; trail 7 miles to summit of Mt. Marcy through Panther Gorge and past Cathedral Rocks; trail 7 miles to Lakes Colden and Avalanche, past Lake Perkins or "Tear of the Clouds" (the highest of the Adirondack waters, 4,312 ft. above the sea) and Opalescent Flume; from Lake Colden 5 miles past Calamity Pond to Adirondack; thence up the Hudson 5 miles to the Indian Pass; thence down the Ausable to N. Elba and John Brown's grave, 10 miles; thence to Lake Placid, 5 miles; through Lake Placid, 5 miles; to summit of Whiteface Mt., 3½ miles; thence to Wilmington, and out by any route that may be selected. This trip takes in some of the wildest and most characteristic scenery of the Adirondacks, including the wonderful Indian Pass, the hardly less wonderful Panther Gorge on the E. of Mt. Marcy, and that great geological curiosity, the mammoth trap-dike of Lake Avalanche. There are good "camps" at Upper Ausable Pond, Panther Gorge, Lake Perkins, near the summit of Mt. Marcy, and at Lake Colden.

VII. Lake Pleasant.

Lake Pleasant is in Hamilton County, New York, on the borders of the Adirondack region, and is reached from Amsterdam on the New York Central R. R. by a stage or carriage ride of 50 miles. There are numerous lakes in the vicinity besides Pleasant, the chief of which are Round and Piseco; and the Saranac region is connected with Lake Pleasant by intermediate waters and portages. Deer and other game are abundant in the forests, and fine trout may be taken in all the brooks and lakes. *Sageville* is a thriving little village, situated on elevated ground between Lakes Pleasant and Round, and the *Lake Pleasant House* there is a favorite resort. The *Sturgis House* is at the outlet of the lake. **Piseco Lake** is larger than Lake Pleasant, and lies about 8 miles W. *Raquette Lake* (see p. 220) is 30 miles distant by boat on Jessup's River and Indian and Blue Mt. Lakes. Guides and camp equipage may be obtained at the hotels.

48. Long Island.

LONG ISLAND, part of the State of New York, is 120 miles in extreme length from E. to W., with an average width of 14 miles, and an area of 1,682 square miles. It is bounded on the N. by Long Island Sound, which separates it from Connecticut, and on the S. by the Atlantic Ocean, while East River separates it from New York City. The northern half of the island is agreeably diversified with hills, but the surface is, for the most part, strikingly level. The coast is indented with numerous bays and inlets; and delicious fresh-water ponds, fed by springs, are everywhere found on terraces of varying elevation. These little lakes, and the varied coast-views, give Long Island picturesque features which, if not very imposing, are certainly of a most attractive and pleasing character, heightened by the rural beauty of the numerous quiet little towns and charming summer villas. Along the southern shore of the island, which is a network of shallow, landlocked waters extending 70 miles, fine shooting and fishing may be had. Hotel and boarding-house accommodation is ample. The leading city of Long Island (*Brooklyn*) and the two resorts most frequented by visitors from New York (*Coney Island* and *Rockaway Beach*) have already been described in Routes 1 and 2.

a. Long Island R. R. Main Line and Branches.

The entire railway system of Long Island is under one management, but there are three divisions so distinct that it will be convenient to follow them here. The Main Line of the Long Island R. R. extends along the central line of the island, branching at its E. end, as the island itself does. There are three depots at the W. end: one in Long Island City reached from New York by ferries from James's Slip, foot of New Chambers St., E. R., and from the foot of E. 34th St.; and two in Brooklyn (at the cor. of Flatbush and Atlantic Aves., and in Bushwick Ave. cor. of Montrose).

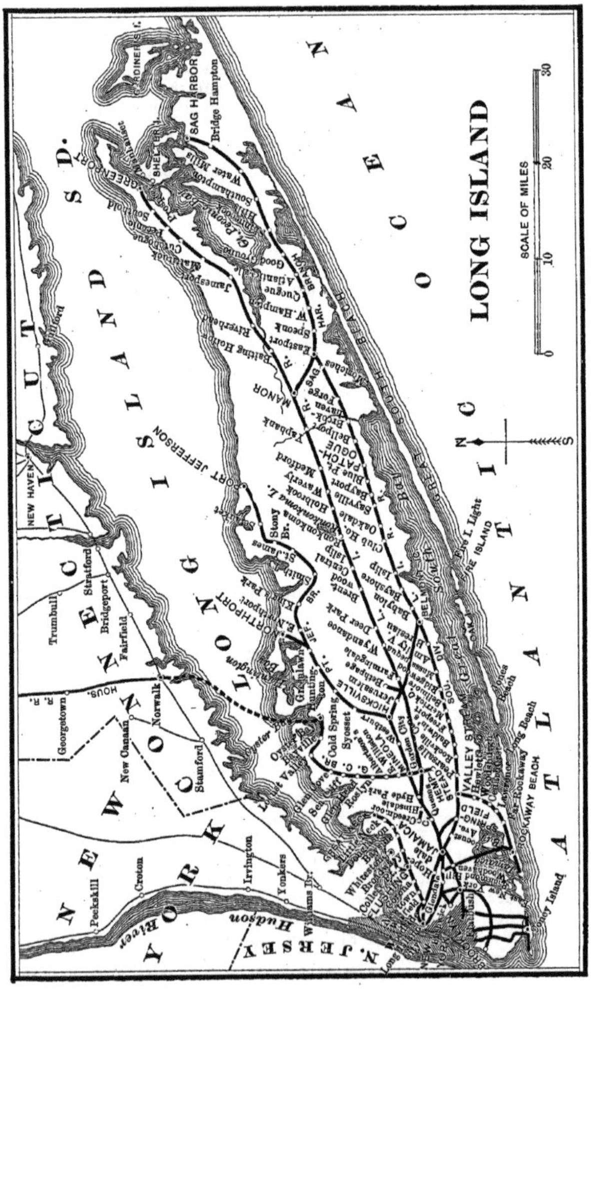
LONG ISLAND
SCALE OF MILES
0
10
20
30
ATLANTIC OCEAN
LONG ISLAND SD.
NEW YORK
CONNECTICUT
N. JERSEY
Hudson River
GREAT SOUTH BEACH
SAG HARBOR
Bridge Hampton
Southampton
Riverhead
Patchogue
Babylon
Jamaica
Flushing
Coney Island
Rockaway Beach
Long Beach
Fire I. Light
Port Jefferson
Northport
Huntington
Stamford
Norwalk
Bridgeport
Stratford
New Haven
Yonkers
Irvington
Croton
Peekskill

Leaving Long Island City, the train passes several small suburban villages and in 10 miles reaches **Jamaica** (*Jamaica Hotel*), an interesting old town on Jamaica Bay, settled in 1656, and containing now about 5,361 inhabitants. The South Shore Division passes through the town, and it is connected with East New York and Brooklyn by the Atlantic Div., which runs frequent local trains. The hamlets of *Hollis* and *Queens* (13 and 14 miles) are popular places of residence; and, just beyond, the road branches, one branch going to *Garden City* (19 miles), the residence-city built by the late A. T. Stewart as a model for suburban homes, while the main line passes on to *Mineola* (19 miles), where the *Oyster Bay* Branch crosses. There is a fine Episcopal cathedral and collegiate institute at Garden City, and the model houses are well worthy of notice. Just S. of Garden City is the ancient village of *Hempstead* (21 miles), situated on the borders of the wide-spreading Hempstead Plains.

The Oyster Bay Branch runs N. by Garden City and Mineola to **Roslyn** (23 miles from New York), a pretty village at the head of Hempstead Bay. Near Roslyn is *Cedarmere*, for many years the country residence of the late William Cullen Bryant. It is a spot of great though quiet picturesque beauty, overlooking the Bay and the Connecticut shore across the Sound. Bryant's grave is in the adjacent cemetery. Near Roslyn are many lovely lakelets, and a short distance S. E. is Harbor Hill, the highest land on Long Island, from the summit of which (319 ft. high) there is a pleasing view. Four miles beyond Roslyn is **Sea Cliff**, on a picturesque bluff overlooking the Sound, and noted for its camp-meetings. The Sea Cliff House accommodates 400 guests at reasonable prices ; there are over 100 cottages, and the camp-meeting tabernacle seats 4,000. Five miles beyond Roslyn is **Glen Cove** (*Allen House*), which is a highly popular resort in summer. Glen Cove, Sea Cliff, and Roslyn are also reached from New York by steamer from Peck Slip (Pier 24), East River (Sub-Route *c*). Six miles beyond Glen Cove is **Oyster Bay** (also called *Syosset Bay*), a deep inlet from the Sound, which is numerously visited in summer. This place was the landing point of the through trains which connect Boston, Philadelphia, Baltimore, and Washington without charge of cars. The *Bay View Hotel* is a small summer hotel, and there are many farmhouses where board may be had at moderate rates.

Beyond Mineola the Main Line runs to *Westbury* (22 miles), and *Hicksville* (25 miles), which is named for Elias Hicks, the Quaker schismatic, who lived and preached in this region from 1771 to 1830, riding 10,000 miles on his missionary journeys, and preaching over 1,000 times. The Port Jefferson Branch diverges here and runs N. E. in 34 miles to Port Jefferson.

The Port Jefferson Branch runs N. E. from Hicksville, and in 4 miles reaches *Syosset*. From *Cold Spring* (3 miles beyond Syosset), stages run to **Cold Spring Harbor** (*Laurelton House, Forest Lawn House*), which attracts many summer visitors. Three miles beyond Cold Spring is *Huntington*, a village of 3,028 inhabitants, pleasantly situated on Huntington Harbor : and 6 miles farther is **Northport** (*Northport House*), situated on another deep inlet from the Sound. Near Northport is the famous Beacon farm, comprising 1,000 acres on the borders of the Sound. Stations : St. Johnland, Smithtown, Stony Brook, and Setauket. **Port Jefferson** (34 miles from Hicksville) is the terminus of the road, and is a village of 2,026 inhabitants.

Beyond Hicksville the Main Line runs nearly due E. to *Farmingdale* (30 miles), *Brentwood* (41 miles), *Central Islip* (43 miles), and *Ronkonkoma* (48 miles). One mile N. of Lakeland is **Ronkonkoma Lake,** a peculiar sheet of water about 3 miles in circumference, situated nearly in the center of the island. The perch-fishing in the lake is good, the boating excellent, and near the shore are several hotels and boarding-houses. Ten miles beyond Ronkonkoma is *Yaphank*. At *Manor* (65 miles) the road forks, the Montauk Division running S. of the Great Peconic Bay to Sag Harbor, while the Main Line runs N. E. in 29 miles to Greenport. Between Manor and Greenport are the pleasant villages of *Riverhead* (73 miles), *Mattituck* (83 miles), *Cutchogue* (85 miles), and *Southold* (90 miles), each of which attracts many summer visitors. **Greenport** (*Booth House, Wyandank Hotel*) is a lively village near the E. end of the island, with a snug harbor and a large fishing-fleet. It affords excellent still-water bathing, boating, sailing, and fishing; and in their season wild ducks are abundant. Greenport may also be reached, as a general thing, by steamers from New York, New London, and Hartford. Daily stages run in 9 miles to *Orient Point*, where there is a large summer hotel, and *Orient Village*, where there are several summer boarding-houses. Ferry-boats connecting with every train run from Greenport to **Shelter Island,** on which are two spacious summer hotels (*Manhasset House, Prospect House*). The island is about 14. miles long by 4 wide, and has a gently diversified surface, with fresh-water lakelets and picturesque bays. It is also the site of a Methodist camp-meeting ground, and is being rapidly improved. Eight miles E. lies *Gardiner's Island* (3,300 acres), on which the pirate Kidd buried vast treasures, part of which was recovered in 1699 by the Earl of Bellamont, colonial Governor of Massachusetts.

The *Montauk Division* diverges at Manor and runs by *Eastport* (70 miles), *Speonk* (73 miles), and *W. Hampton* (75 miles), to **Quogue** (78 miles), which is another popular summer resort with several large boarding-houses. It is situated on Shinnecock Bay, and the bathing and fishing are good. Stations, *Southampton* (90 miles) and *Bridgehampton* (95 miles), and then, 100 miles from New York, the train stops at **Sag Harbor,** a prosperous village with several excellent hotels, situated at the head of the picturesque Gardiner's Bay. It was once a leading whaling-station, but its maritime importance has long since ceased, though its coasting-trade is still large. A steamer runs from Pier 25, East River, New York, to Sag Harbor, Greenport, and Orient, 3 times a week (Tuesday, Thursday, and Saturday), and another tri-weekly line connects Sag Harbor with New London and Hartford, Conn. Daily stages from Bridgehampton run in 6 miles to **East Hampton,** "the quietest of all quiet towns," with quaint old houses, and a street of noble elms, which were planted at the instigation of Dr. Lyman Beecher, who was pastor here from 1798 to 1810. The surf-bathing at Napeague Beach (1 mile from the village) is excellent. A short distance N. E. is the sequestered village of *Amagansett*. About 15 miles E. (reached by private conveyance) is *Montauk Point*, the eastern extremity of Long Island. On it is a lighthouse with a powerful revolving light.

b. *Montauk Division.*

This division of the Long Island R. R. has the same terminal stations in Brooklyn and Long Island City as the main line (see Sub-Route *c*).

Jamaica (10 miles) has already been described. From *Valley Stream* (16 miles) a branch road runs S. W. to Rockaway Beach, passing the beach-villages of *Woodsburgh* (19 miles) and *Far Rockaway* (21 miles), with their spacious summer hotels and cottages. **Rockaway Beach** is described on p. 24. Near Pearsall's the Long Beach Branch diverges, and runs to **Long Beach,** a favorite point for summer excursions, with a vast hotel, bathing, music, etc. On the main line, 20 miles from Valley Stream, is **Babylon** (2,768 population), which is much resorted to for its fishing. Here there are several comfortable hotels and many summer boarding-houses. From Babylon a small steamer runs 8 miles across the Great South Bay to **Fire Island** (*Surf House*), which offers the attractions of surf and still-water bathing, boating and sailing, superb fishing, and cool ocean-breezes, and draws many summer visitors. The beach is admirable, and occupies the W. end of a remarkable sand ridge which is only a few rods wide but runs for 40 miles along the coast to Quogue Neck, inclosing a series of broad bays and estuaries. Six miles beyond Babylon is **Islip** (*Lake, Pavilion*), a pretty village on the Great South Bay, containing many tasteful villas. It is much visited in summer, and besides the hotels there are a number of board-ing-houses. From **Patchogue** (54 miles from New York; *Eagle, Laurel, Ocean Avenue*), a prosperous village about a mile from the Great South Bay, on the line is *Bellport*, a much-visited village on Bell-port Bay. Beyond this station is the pretty village of **Center Moriches** (*Beachview, Ocean, Riverside*), which is extensively visited in summer. Both fishing and hunting are excellent, and surf-bathing may be enjoyed by sailing across the Bay to the outer beach, where is a sum-mer hotel. There are many summer boarding-houses in Moriches and also in E. Moriches, which lies across the Tenillo River.

c. *The North Shore.*

The North Side Div. of the Long Island R. R. begins at Long Island City, and its depot is reached from New York by ferries from foot of New Chambers St., E. R., James Slip and E. 34th St., and from Brook-lyn by horse-cars. Leaving Long Island City (or Hunter's Point) the train passes the pretty suburban towns of *Woodside* (3 miles), *Winfield* (4 miles), and *Newtown* (5 miles), and in 8 miles reaches **Flushing,** a beautiful village of 19,803 inhabitants at the head of Flushing Bay, near the entrance of Long Island Sound. Many business men from New York make their homes in Flushing, and the village is noted for its wealth and culture, for its umbrageous streets and finely kept gardens, and for its educational institutions. In the N. part of the village is a neat monument to the soldiers of the country who died in the civil war. The extensive nurseries of the Parsons & Sons Co. here are worth a visit, and the drives in the vicinity are very attractive. At Flushing the road divides, one branch running by *Bayside, Douglaston*, and *Little Neck* to *Great Neck* (6 miles beyond Flushing), while the other diverges to *College Point* and *Whitestone* (3 miles beyond Flushing). All these points and the adjacent localities are much visited in summer, and several of them may be reached from New York by steamer, leaving Pier 24, East River, at 4

P. M. for Whitestone, Great Neck, Sea Cliff, Mott's Dock, Sands Point, Glen Cove, Glenwood, and Roslyn, returning next morning. At Creedmoor, 5 miles beyond Flushing, is the famous *Creedmoor Rifle-Range*, the most perfectly appointed in America (reached by frequent trains from Hunter's Point). The range and grounds belong to the National Rifle Association, and contain 30 targets that can be shot at from 50 to 1,200 yards. There are two hotels near the range.

49. New York to Easton.

As far as Denville either of two routes can be taken: The *Boonton Branch* of the Delaware, Lackawanna & Western R. R. is that taken by the through trains from New York, but is much less interesting than the one described below. The only important stations passed are **Paterson** (Route 41) and **Boonton** (30 miles), a manufacturing town of 3,307 inhabitants on the Rockaway River and Morris·Canal, in the midst of a very mountainous region. By this route *Denville* is 34 miles from New York. The other is by Morristown on the same road. Distance, 85 miles; time, 3½ hours.

PASSENGERS take the ferry-boat from foot of Barclay St. or Christopher St. to the station in Hoboken. Leaving Hoboken, the train traverses the Bergen Tunnel, and passes in 8 miles to **Newark** (see p. 30) across wide marshes. Three miles beyond Newark is **Orange** (*Mansion House, Park House*), a beautiful city of 18,844 inhabitants, surrounded by lovely scenery, and a favorite suburban home of business men from New York. A short distance to the W. lies *Llewellyn Park*, a small inclosure laid out in the best style of landscape-gardening and containing fine villas and mansions which have the park in common. A little farther W. is the *Orange Mountain*, the crest of which is crowned by costly residences standing amid highly cultivated grounds. From various points of the mountain there are remarkably fine views, including the lake and mountain region of New Jersey and New York City and Harbor. The drives about Orange are extremely picturesque.

Two miles beyond Orange the train stops at *S. Orange*, and at *Milburn*, 3 miles farther, rounds the extremity of Orange Mt. and begins the steep ascent of Second Mt., on the crest of which is **Summit** (*Blackburn House, Park House*), a popular summer resort, noted for the extent and beauty of its views. At *Madison* (26 miles) the road first enters the borders of the mountain-region, which continually grows more picturesque as the train proceeds westward. The Drew Theological Seminary is located at Madison, and 4 miles beyond is **Morristown,** with 8,156 inhabitants (*Mansion House, U. S. Hotel*), the capital of Morris County, on the Whippany River, upon a plain surrounded by hills. It is noteworthy as having been, during the Revolution, the headquarters of the American army upon two occasions. In the rear of the Court-House the ruins of old Fort Nonsense may still be seen; and *Washington's Headquarters*, owned and preserved by the State, is ¼ mile E. of the village. In the public square is a Soldiers' Monument, and on Pigeon Mt. is the vast and massive *State Insane Asylum*, built of granite at a cost of $3,000,000. Beyond Morristown the train crosses Morris Plains and at *Denville* (37 miles) meets the Boonton Branch.

Five miles beyond Denville is the prosperous little manufacturing city of **Dover** (*Jolly House*), whence a branch road runs to the pleasant village of *Chester* in 13 miles. From *Drakesville* (47 miles from New York) a branch road runs in 4 miles to *****Lake Hopatcong,** loftily situated among the Brookland Mts., 725 ft. above the sea. The lake is about 9 miles long by 4 miles wide, is dotted with islands, affords excellent fishing, and is surrounded by varied and beautiful scenery. The name (Hopatcong) means "Stone over the Water," and was given it by the Indians on account of an artificial causeway of stone which once connected one of the islands with the shore, but which is now submerged. Two small steamers ply on the lake, and there are several summer hotels (*Hotel Breslin, Mount Arlington House, Nolan's Point Villa*). Near the former is *Southard's Peak,* from the summit of which the Delaware Water-Gap and the Bloomfield Mts. are both visible. Four miles beyond Drakesville is *Stanhope* (52 miles), whence stages run in 2½ miles to **Budd's Lake** (also called *Lake Senecawana*), a beautiful sheet of water nearly circular in form, 3½ miles in circumference, deep, clear, and abounding in fish, and surrounded by a picturesque country, with fine views and mountains in the distance. This attractive spot is much frequented by excursion parties during the summer season. Schooley's Mt. is 8 miles distant, and Lake Hopatcong is easily visited from Budd's Lake. From *Hackettstown* (61 miles) stages run in 3½ miles to *****Schooley's Mountain** (*Dorincourt* and *Heath House*), a favorite summer resort of New-Yorkers. It is not an isolated peak, but a ridge of considerable extent, Budd's Lake being upon one part of its summit. It is about 1,200 ft. high, and even amid the "August ardors" its air is cool, pure, and bracing. The drives in the vicinity are delightful, and the scenery picturesque and pleasing. Another route from New York to Schooley's Mt. (and also to Lake Hopatcong) is *via* High Bridge Branch of the Central R. R. of New Jersey.

Washington (71 miles) is the junction of the line *via* Morristown with the Main Line of the Delaware, Lackawanna & Western R. R., and passes on *via* Stewartsville and Phillipsburg to **Easton,** Pa. (85 miles from New York), which is described in Route 45. The main line from Washington leads in 11 miles to *Manunka Chunk,* just before reaching which the train passes through the Manunka Chunk Mt. by the Voss Gap Tunnel, 1,000 ft. long. At Manunka Chunk the Belvidere div. of the Pennsylvania R. R. comes in. The Lackawanna route, with the continuation of the route to the Delaware Water-Gap, Scranton, Binghamton, and Owego, is described in Route 43.

50. Philadelphia to Harrisburg and Pittsburg.

By the Pennsylvania R. R. Distances : to Downingtown, 32 miles ; to Lancaster, 69 ; to Middletown, 96 ; to Harrisburg, 105 ; to Huntingdon, 203 ; to Altoona, 237 ; to Pittsburg, 354. This (formerly the Pennsylvania Central R. R.) is now part of the Pennsylvania R. R. system, which includes upward of 2,366 miles of railway. The limited express is formed of vestibuled cars, composed of drawing-room, dining-room, smoking, and sleeping-cars. Through trains, with vestibuled drawing-room, dining, and sleeping cars, run without change from New York *via* Philadelphia to Chicago, Cincinnati, St. Louis, and Louisville. The time from Philadelphia to Pittsburg is about 9¼ hours.

LEAVING the station in Philadelphia (cor. Market and Broad Sts.), the train passes through a pleasant suburban region and enters one of the richest agricultural districts in America, which is traversed for nearly 100 miles. The size and solidity of the houses and barns, and the perfection of the cultivation, will be apt to remind the tourist rather of the best farming districts of England than of what he usually sees in the United States. *Paoli* (20 miles) was the scene of a battle fought Sept. 20, 1777, in which the British under Gen. Gray surprised and defeated the Americans under Gen. Wayne. The battle is commonly called the "Paoli massacre," because a large number of the Americans were killed after they had laid down their arms. A marble monument, erected in 1817, marks the site of the battle-field. Beyond Paoli the scenery grows more picturesque, and fine views are had of the beautiful Chester Valley. *Downingtown* (32 miles) is the terminus of the Chester Valley Branch of the Reading R. R., and is near the marble-quarries which supplied the marble from which Girard College (Philadelphia) was built. At *Coatesville* (38 miles) the W. branch of the Brandywine is crossed on a bridge 885 ft. long and 75 ft. high. *Parkesburg* (44 miles) and *Christiana* (48 miles) are busy manufacturing villages. *Gap* (51 miles) is so named because it lies in the gap through which the road passes from the Chester Valley to the Pequea Valley. The scenery in the vicinity is attractive. **Lancaster** (*City, Cooper, Lancaster,* and *Stevens*) is situated near the Conestoga Creek, which is crossed in entering the city. It was incorporated in 1818, and was the seat of the State government from 1799 to 1812. It is now a prosperous manufacturing city of 32,011 inhabitants, containing many fine buildings, public and private. The *Court-House* (on E. King St.) is an imposing edifice with a Corinthian portico; and the *County Prison* (also on E. King St.) is a handsome building in the Norman style. *Fulton Hall,* near the market-place, is a noteworthy structure, used for public assemblies. On James St. are the substantial buildings of *Franklin and Marshall College* (German Reformed), organized in 1853 by the union of Marshall with the old establishment of Franklin College, which was founded in 1787. It is also the seat of the *Theological Seminary* of the Reformed Church. The institutions together have about 25,000 volumes in the libraries, and 200 students. The oldest turnpike road in the United States terminates at Lancaster, to which it runs from Philadelphia. Besides its large cotton-mills and cork-works, Lancaster has a watch-factory, a comb-factory, and extensive manufactures of axes, carriages, railroad-iron, etc. The city is the metropolis of a rich agricultural county, and the center of a vast tobacco-trade. It contains upward of 100 large tobacco-warehouses and packing establishments.

The only stations between Lancaster and Harrisburg which require mention are *Middletown* (96 miles), on the Susquehanna River, at the mouth of Swatara Creek, with extensive tubular iron-works and machine-shops, and *Steelton,* the location of the great Pennsylvania Steel Works, a town of 9,250 inhabitants, 3 miles below Harrisburg. **Harrisburg** (*Bolton, Lochiel,* and *United States*), the capital of Pennsylvania, is on the E. bank of the Susquehanna River, which is here a mile wide and spanned by 2 bridges. Harrisburg was laid out by John Harris in 1785, was incorporated as a borough in 1791, became the State capital in 1812, received a city charter in 1860, and in 1890 had a population of 39,385. The city is handsomely built, and is surrounded by magnificent scenery. The * *State-House*, finely situated on an eminence near the center, is a handsome brick building 180 ft. long by 80 ft. wide, with a circular Ionic portico in front surmounted by a dome commanding a fine view. In the second story is the State Library of 66,000 volumes, with numerous portraits and cabinets of curiosities. On each side of the State-House is a smaller building of similar design devoted to Government uses, and in the grounds is a beautiful *Soldiers' Monument* in honor of those who fell in the Mexican War. The *State Arsenal* is a spacious building a short distance outside of the city limits, surrounded by a grove of trees about 5 acres in extent. The *Court-House*, in Market St., is a brick edifice surmounted by a dome, and the *State Lunatic Asylum* is a vast and imposing building 1¼ mile N. of the city. The other principal public buildings are the Executive Mansion, the market-houses, county prison, the United States *Post-Office*, an opera-house, and several churches. *Front St.*, overlooking the river, contains many of the finest residences. *Harris Park*, at the intersection of Front St. and Washington Ave., is the spot where John Harris, Indian trader, and father of the founder of the city, was bound to a tree by the Indians, about the year 1719, who were about to burn him to death when a rescuing party arrived. *Harrisburg Cemetery* (reached by State St.) occupies a commanding situation and affords fine views. At the intersection of Dauphin and State Sts. is an obelisk, 110 ft. high, to the soldiers of Dauphin County who fell in the late war.

About 5 miles above Harrisburg the railroad crosses the Susquehanna on a splendid bridge 3,670 ft. long; the view from the center of this bridge is one of the finest on the line. Near Cove Station, 11 miles from Harrisburg, the Cove Mt. and Peter's Mt. are seen, and from this point to within a short distance of Pittsburg the scenery is superb, and in places grand beyond description. *Duncannon* (120 miles) is at the entrance to the beautiful Juniata Valley, which is followed for about 100 miles to the base of the Allegheny Mts. The landscape of the Juniata is in the highest degree picturesque; the mountain background, as continuously seen across the river from the cars, being often strikingly bold and majestic. The passage of the river through the Great Tuscarora Mt., 1 mile W. of *Millerstown* (138 miles), is especially fine. Four miles beyond *Mifflin* (154 miles) the train enters the wild and romantic gorge known as the * **Long Narrows,** which is traversed by the railway, highway, river, and canal. *Mount Union* (191 miles) is at the entrance of the gap of Jack's Mt. Three miles beyond is the famous Sidling Hill, and still farther W. the Broad Top Mt. *Huntingdon* (203 miles) is a flourishing village on the Juniata, finely situated and surrounded by beautiful scenery.

The *Huntingdon & Broad Top R. R.* runs S. W. from Huntingdon to Mt. Dallas, connecting at that point with the Bedford Div. of the Pennsylvania R. R. *Bedford* (53 miles from Huntingdon) is a pretty village on the Raystown branch of the Juniata, whence stages run in 1¼ mile to the **Bedford Springs** (*Springs Hotel*). The springs are pleasantly situated in a picturesque mountain glen, and their great altitude and delightful summer climate, together with the beautiful mountain scenery of the neighborhood, have long made them a popular resort for pleasure-seekers as well as invalids. The waters are saline-chalybeate, and are considered beneficial in dyspepsia, diabetes, incipient consumption, and skin diseases.

At *Petersburg*, 7 miles W. of Huntingdon, the railroad parts company with the canal and follows the Little Juniata, which it again leaves at *Tyrone* (223 miles) to enter the Tuckahoe Valley, famous for its iron-ore. At the head of the Tuckahoe Valley and at the foot of the Alleghanies is **Altoona** (*Arlington, Central, Logan House*), a handsome city of 30,337 inhabitants, built up since 1850, when it was a primitive forest, by being selected as the site of the vast machine-shops of the Pennsylvania R. R. The trains usually stop here for refreshments, and many travelers arriving here in the evening remain over-night in order to cross the Alleghanies by daylight. Just beyond Altoona the ascent of the Alleghanies begins, and in the course of the next 11 miles some of the finest scenery and the greatest feats of engineering on the entire line are to be seen. Within this distance the road mounts to the tunnel at the summit by so steep a grade that, while in the ascent double power is required to move the train, the entire 11 miles of descent are run without steam, the speed of the train being regulated by the "brakes." At one point (the Horseshoe) there is a curve as short as the letter U, and that, too, where the grade is so steep that in looking across from side to side it seems that, were the tracks laid contiguous to each other, they would form a letter X. The road hugs the sides of the mountains, and from the windows next to the valley the traveler can look down on houses and trees dwarfed to toys, while men and animals appear like ants from the great elevation. Going W. the left-hand, and coming E. the right-hand, side of the cars is most favorable for enjoying the scenery. The summit of the mountain is pierced by a tunnel 3,612 ft. long, through which the train passes before commencing to descend the W. slope. The much-visited **Cresson Springs** are 2¼ miles beyond the tunnel, 3,000 ft. above the sea. There are 7 springs here, and the waters are highly esteemed, but the place is visited rather for the delicious coolness of its summer climate than for the curative virtues of its mineral waters. The thermometer rarely reaches 75° during the hottest part of the hottest days of summer; and the nights are so cool that blankets are requisite for comfortable sleep. The hotels (of which the *Mountain House* is the principal) and the cottages accommodate about 2,000 guests. The drives in the vicinity are very attractive; and the Pennsylvania R. R. runs special trains at small cost for the benefit of those who wish to view the magnificent scenery along the mountain division of the road.

In descending the mountains from Cresson the remains of another railroad are constantly seen, sometimes above and sometimes below the track followed by the trains. This was the old Portage R. R. by which, in the ante-locomotive days, loaded canal-boats were carried over the mountain in sections by inclined planes and joined together at the foot. The stream which is almost continuously in sight during the descent is the Conemaugh Creek, which is crossed by a stone viaduct near *Cone-maugh Station* (273 miles), the terminus of the mountain division of the road. *Johnstown* (276 miles) is a busy manufacturing borough at the confluence of the Conemaugh and Stony Creeks. The Cambria Iron-Works, seen to the right of the road, are among the most extensive in America. At *Blairsville Intersection* (300 miles) the road branches, the main line running to Pittsburg by *Latrobe* (313 miles) and *Greensburg* (323 miles); while the Western Division runs to Allegheny City by *Blairsville* (303 miles). The scenery along both routes is pleasing but not striking.

Pittsburg.

Hotels, etc.—The *Monongahela House*, cor. Water and Smithfield Sts.; the *Hotel Anderson*, cor. 6th St. and Penn Ave.; the *Hotel Duquesne* (on the European plan), in Smithfield St. near 5th Ave., are the principal hotels. Other good houses are the *Central*, cor. Smithfield St. and 3d Ave.; the *St. Charles*, cor. Wood St. and 3d Ave.; and the *Hotel Boyer*, Duquesne Way and 7th St.

Restaurants.—*Hotel Duquesne; Newell's*, 5th Ave., above Wood St.; *E. Reineman*, 505 Wood St.; and for ladies, *A. J. Hagan*, 607 Smithfield St.

Horse, cable, and *electric cars* run on the principal streets and to the suburbs. *Reading-rooms* at the Mercantile Library in Penn St. near 6th, and at the Y. M. C. A., cor. Penn Ave. and 7th St. *Post-Office* at the cor. of Smithfield St. and 4th Ave.

Pittsburg, the second city of Pennsylvania in population and importance, and one of the chief manufacturing cities in the United States, is situated at the confluence of the Alleghany and Monongahela Rivers, which here form the Ohio. The city occupies the delta between the two rivers, with several populous suburbs annexed in 1872 and 1874, and the population in 1890 was 238,617. The city was laid out in 1765 on the site of the old French Fort du Quesne, famous in colonial annals, and on its capture by the British the name was changed to Fort Pitt, in honor of William Pitt. The city charter was granted in 1816. The city is substantially and compactly built, and the main thoroughfares are brilliantly illuminated by arc and incandescent electric lights. Nine bridges span the Alleghany River and 5 the Monongahela, and several new bridges are projected. From its situation, Pittsburg enjoys excellent commercial facilities, and has become the center of an extensive commerce with the Western States; while its vicinity to the inexhaustible iron and coal mines of Pennsylvania has made it a great manufacturing center. The extent of its steel, glass, and iron manufactures has given it the appellation of the "Iron City," while the heavy pall of smoke that formerly overhung it, before the introduction of natural gas, caused it to be styled the "Smoky City." The stranger should not fail to visit its great manufacturing establishments, particularly those of iron, steel, electric supplies, and glass.

Liberty, Wood, Market, Smithfield, 5th Ave., Penn, and *6th Sts.* are the principal business streets, and they contain many handsome buildings. Among the public buildings are the *** Municipal Hall,** ·cor. Smithfield St. and Virgin Alley, costing $750,000, with a granite front and a massive tower; the *Custom-House* and *Post-Office*, a fine structure of stone, cor. Smithfield St. and 4th Ave.; the **United States Arsenal,** standing in ornamental grounds in the N. E. section of the city,

and the *Masonic Temple* on Fifth Ave. Of the 170 churches, the
most imposing is the Roman Catholic *Cathedral of St. Paul,* a large
edifice of brick, with 2 spires and a dome over the choir. ***Trinity
Church** (Episcopal) is a fine building in the English-Gothic style, in
6th Ave. near Smithfield St. *St. Peter's* (Episcopal), in Grant St., is also
a handsome structure. The *First Presbyterian,* near Trinity Church, is
a massive stone edifice with 2 towers ; and the *United Evangelical* (German) church, cor. Smith Ave. and Smithfield St., is a handsome building.
Other notable church edifices are the *First Baptist,* the *Third Presbyterian,* and also the *English Evangelical Church.* Many other fine stone
churches have been recently built. The *Duquesne Club* has a handsome
house in 6th Ave. near Smithfield St. The spacious building of the **Mercantile Library** is in Penn St. near 6th St. ; it contains 19,000 volumes and a reading-room. The *Pittsburg Art Association* meets at the
Pittsburg Club Theatre. The *Young Men's Christian Association* has
a fine building corner of Penn Ave. and 7th St. The upper rooms are
occupied by the *School of Design for Women.* There are in the city
two theatres, an Opera-House, an Academy of Music, and several public
halls. The *West Pennsylvania Exposition Society* buildings, with a main
building 540 by 150 feet, and a machinery hall 300 by 200 feet, were
first opened in 1889. The **Court-House,** finished in Romanesque
style, was erected at the cost of about $2,500,000. The main tower is
320 ft. high. The Jail is connected by a stone arched bridge thrown
over Ross St. It was one of Henry H. Richardson's last designs.

The *Pittsburg Female College* (Methodist) and the *Pennsylvania Female College* (Presbyterian) are both flourishing institutions. The *High
School* is a handsome building. Most of the public-school buildings are
large and substantial. Among the principal charitable institutions are
the *Western Pennsylvania Hospital,* a large building located on the side
of the hill fronting the Pennsylvania R. R., with a department for the
insane at Dixmont, on the Pittsburg, Fort Wayne & Chicago R. R. ; the
City General Hospital ; the *Homœopathic Hospital,* on 2d Ave. above
Smithfield St., an institution built at a cost of $225,000, containing
rooms for patients ; the *Mercy Hospital,* in Stephenson St. ; the *Episcopal Church Home ;* and the Roman Catholic *Orphan Asylum.* The
Convent of the Sisters of Mercy (Webster Ave. cor. of Chatham) is the
oldest house of the order in America.

Birmingham is a portion of the city lying across the Monongahela
from Pittsburg (reached by bridge). An inclined-plane R. R. (fare, 6c.)
leads to the summit of Mt. Oliver (250 ft. high) ; another inclined plane
leads to the summit of Mt. Washington (370 ft. high) ; also another to
Duquesne Heights, from all of which fine views may be had. Besides
these there are three other inclined railways leading to other summits of
the city. *Manchester,* now a part of Allegheny City, is 2 miles below
Pittsburg, on the Ohio. Here is located the *Riverside Penitentiary ;*
and the *Passionist Monastery of St. Paul* and the *Franciscan Convent*
are near by. *East Liberty,* a part of the city known as the East End, 5
miles from the Court-House, on the Pennsylvania R. R., is a thriving
suburb, containing fine residences, and affording a delightful drive of

Steel-Works near Pittsburg, Pa.

many miles over fine roads to the E. wards of the city. Two cable lines run to East Liberty, fare 5 c. The East End is the finest part of the city. Fifth Ave. contains many fine residences. The city has acquired about 500 acres of land for a public park, the greater part being the gift of Mrs. Mary E. Schenley. It is situated about 3 miles from the Court-House, being beautifully located. At *Braddock*, on the Pennsylvania R. R., 10 miles E. of the city, are the Bessemer Steel-Works, owned by Carnegie, Phipps & Co. A fine *Public Library*, the gift of Andrew Carnegie, is the principal building in the place. On the opposite side of the Monongahela River are the Homestead Steel-Works, owned by the same firm. This plant employs many thousands of hands, and is one of the best-equipped works in the country, being well worth a visit.

In the vicinity of Pittsburg are many important manufacturing works, the Pittsburg Plate-Glass Works being one of the largest and most important, having works at Creighton and Tarentum, 20 miles out on the West Penn. Div., and also having large works at Ford City, on the Allegheny Valley R. R.

Allegheny City (*Hotel Federal*) is situated on the W. bank of the Alleghany River, opposite Pittsburg, with which it is connected by 9 bridges, 2 of which are fine suspension-bridges. Its manufacturing interests are large. In 1890 the city had a population of 105,287. The *City Hall* is on the square at the crossing of Ohio and Federal Sts., and opposite it is a fine library building, erected by Andrew Carnegie, and called "The Carnegie Free Library." * **St. Peter's** (R. C.) is the finest church in the city. Other notable churches are *Trinity* (Evangelical Lutheran), *North Avenue Methodist Episcopal, Second United Presbyterian, Sandusky Street Baptist*, etc. The * **Western Penitentiary,** a large stone building on the banks of the Ohio, known also as the *Riverside Penitentiary*, is nearly completed. Visitors are admitted from 2 to 4½ P. M. every day except Saturdays and Sundays. The *Western Theological Seminary* (Presbyterian) was established here in 1827. It is situated on one of the finest streets in the city, fronting the Park, with fine dwellings for the professors on either side. The *Theological Seminary of the United Presbyterian Church*, established in 1826, and the *Allegheny Theological Institute*, organized in 1840 by the Synod of the Reformed Presbyterian Church, are also located here. The *Western University*, founded in 1819, formerly located in Pittsburg, has a valuable geological and natural-history collection, and 275 students. The *Allegheny Observatory*, situated on an elevated site N. of the city, is a department of this institution. Among the principal charitable institutions is the *Allegheny General Hospital*, located in Stockton Ave., opposite the parks; also the *Home for the Friendless*, a building situated on Washington St., just off the Public Park. The *Public Park* lies around the center of the city; it contains 100 acres, and is adorned with several tiny lakelets, numerous fountains, and a monument to Humboldt. On the lofty crest near the Alleghany, in the E. part of the city, stands the * **Soldiers' Monument,** erected to the memory of the 4,000 men of Allegheny County who lost their lives in the civil war. It

16

consists of a graceful column, surrounded at the base with statues of an
infantry-man, a cavalry-man, an artillerist, and a sailor, and surmounted
by a bronze female figure of colossal size. Here are also the *Hampton
Battery Monument*, and a monument to Alexander Humboldt erected by
the German residents. A fine view is obtained from this point. Electric-cars run to various parts of the city. Fine views of the city and
country may be had from the surrounding hills.

51. Philadelphia to Central New York, Buffalo, and Niagara Falls.

Via the Philadelphia & Reading R. R. to Bethlehem, and thence *via* Lehigh
Valley R. R. This route affords a great variety of scenery, and enables the
tourist to visit the large iron and other industrial works in the Lehigh Valley
and the most interesting portions of the coal regions in the State of Pennsylvania.

THE Philadelphia depot of the Philadelphia & Reading R. R. is at
the cor. of 12th and Market Sts. For 6 miles the road runs through
the northern suburbs of the city and then enters Montgomery County,
which it traverses for many miles, entering then the rich farming and
dairy region of Bucks County. *Gwynedd* (19 miles) is a Welsh village
of some importance. North of this is a short tunnel; and after passing *Sellersville* (32 miles) the railway runs through the Landis Hills,
sometimes called the Rock Hills, by a tunnel 2,200 feet long. This
divides the waters of the Schuylkill and Delaware Rivers. From the
summit, 1 mile W. of the station, a fine view of Limestone Valley and
Quakertown is obtained. *Hellertown* (52 miles) has extensive iron-works,
and in the vicinity are other extensive iron and zinc works. Near here
there are fine views of the hills skirting the Lehigh Valley. Four miles
beyond Hellertown is **Bethlehem.** Here the Lehigh Valley R. R. is
taken, the eastern terminus of which is New York city. (For a description of the entire route beyond Bethlehem see Route 45.)

52. Philadelphia to Albany, N. Y.

By the Bethlehem Branch of the Philadelphia & Reading R. R. to Bethlehem; thence by the Lehigh & Susquehanna Div. of the Central R. R. of New
Jersey to Scranton; thence by the Pennsylvania Div. of the Delaware & Hudson Co.'s R. R. to Nineveh; and thence by the Susquehanna Div. of the same
road to Albany. Distances: to Bethlehem, 56 miles; to Mauch Chunk, 88; to
Wilkesbarre, 142; to Scranton, 160; to Green Ridge, 163; to Carbondale, 177;
to Nineveh, 231; to Albany, 350; to Saratoga Springs, 388. This is a popular
route from Philadelphia to Saratoga Springs and Montreal, and the variety of
scenery which it offers makes it very attractive in summer.

As far as Bethlehem (56 miles) this route is over the Bethlehem
Branch of the Philadelphia & Reading R. R. From Bethlehem to
Scranton the Lehigh & Susquehanna Div. of the Central R. R. of New
Jersey runs on the side of the Lehigh River to **Scranton** (*Forest
House, Wyoming House*), a flourishing city of 75,215 inhabitants, occu-
pying the plateau at the confluence of Roaring Brook and the Lacka-
wanna River. It is handsomely laid out, with broad, straight streets,

and contains many fine residences and public buildings, but its general appearance is somber. Its importance is due to its situation in the most northern of the anthracite basins, and to its railroad facilities. The Delaware, Lackawanna & Western R. R. (Route 43) connects here, and there are several other important lines. The trade in mining supplies is extensive, and the shipments of coal are immense. Its iron-manufactures are very important, and there are vast blast-furnaces, rolling-mills, foundries, machine-shops, glass-works, silk-mills, etc. Lackawanna, Penn, Washington, and Wyoming Aves. are the principal business streets. In the suburb of Dunmore is the Forest Hill Cemetery, whence fine views are obtained. At *Green Ridge* (2 miles beyond Scranton, and which is also connected with the center of the city by an electric railway, with grades at places of 300 ft. to the mile) the train passes on to the track of the Pennsylvania Division of the Delaware & Hudson R. R., and, ascending the valley of the Lackawanna amid numerous collieries and mining-villages, in 15 miles reaches **Carbondale,** a city of 10,833 inhabitants, at the N. end of the anthracite-coal region, near several extremely rich coal-mines. The chief object of interest here is the *Gravity Railroad,* a series of inclined planes on which coal-trains are sent over the Moosic Mountains to and from *Honesdale* (16 miles), on the Delaware & Hudson Canal, with no impelling force but gravity, save at one point. Beyond Carbondale the road traverses a mountainous, rugged, and sparsely-settled region, crosses the Alleghanies at an elevation of 2,500 ft., and descends amid picturesque scenery to the valley of the Susquehanna. Near *Jefferson Junction* (35 miles from Carbondale) the Erie R. R. (Route 41) is crossed, and the Albany train passes on by several small stations to *Nineveh* (231 miles from Philadelphia and 119 from Albany). Here the Susquehanna Div. of the Delaware & Hudson R. R. from Binghamton is taken, and the train passes N. E. up the smiling valley of the Susquehanna River by a number of pretty villages and hamlets. From *Afton,* 5 miles beyond Nineveh, stages run to *Vallonia Springs,* a picturesque highland village, 700 ft. above the river, and surrounded by beautiful scenery. The waters are impregnated with sulphur, iron, and magnesia, and are beneficial in cutaneous diseases. At *Sidney* (247 miles) the New York, Ontario & Western R. R. is intersected.

One mile beyond *Colliers* (75 miles from Albany and 67 from Binghamton) the Cooperstown & Charlotte Valley R. R. diverges, and runs north in 16 miles to **Cooperstown** (*Ballard, Central, Hotel Fenimore*), a village of 2,657 inhabitants, at the south end of Otsego Lake. The beautiful situation of the village, high up in the hills, with a bracing atmosphere and delightful scenery, renders it a charming summer resort, and attracts many visitors. Cooperstown was the home of J. Fenimore Cooper, the novelist, and his pen has rendered the whole region classic. "The same points still exist which Leather-Stocking saw; there is the same beauty of verdure along the hills; and the sun still glints as brightly as then the ripples of the clear water." The site of the old Cooper mansion (burned in 1854) is still pointed out; and the *Tomb of Cooper* is near Christ Church, which

also contains beautiful memorial windows.   The *Cooper Monument* is
in Lakewood Cemetery, a mile from the village; it is of Italian
marble, 25 ft. high, and is surmounted by a statue of Leather-Stocking.
Two miles from the village, on the W. shore of the lake, is *Hannah's
Hill* (named after Cooper's daughter), whence a fine view is obtained.
On the E. shore (2 miles from the village) is *Mt. Vision,* which commands a very beautiful view of the lake and of the country adjacent.
* *Rum Hill* (7 miles distant) is said to command a prospect of over 60
miles.   *Leather-Stocking's Cave* is on the E. shore, 1¼ miles from the
village; and the *Leather-Stocking Falls* (or Panther's Leap) are on the
same side, at the head of a wild gorge.   The *Mohegan Glen* is on the
W. shore (3 miles from the village), and contains a series of small but
picturesque cascades.   There are many pleasant drives in the vicinity
of Cooperstown; and highways lead to *Cherry Valley* (13 miles), to
Richfield Springs (14 miles), and to *Sharon Springs* (20 miles).   **Otsego Lake** is about 9 miles long and 1 to 1½ wide, and is described by
Cooper as "a broad sheet of water, so placid and limpid that it resembles a bed of the pure mountain atmosphere compressed into a setting
of hills and woods.   Nothing is wanted but ruined castles and recollec-
tions, to raise it to the level of the Rhine."   The shores are bold and diversified, and the clear waters teem with fish.   Two small steamers ply on
the lake, affording a delightful excursion, and connecting at the upper
end with stages for Cherry Valley and Richfield Springs.   (See Route 43.)

Beyond Colliers the road passes a number of small villages, crosses
the watershed between the Susquehanna and the Mohawk, and descends
by gentle grades into the latter valley.   At *Cobleskill* (305 miles from
Philadelphia and 45 miles from Albany) a branch line diverges and
runs N. W. in 14 miles to **Sharon Springs,** in a valley 1,100 ft.
above the sea-level (*Pavilion Hotel, Sharon House,* and *Union Hotel*),
which are visited by more than 10,000 invalids and pleasure-seekers
annually.   The village is situated in a narrow valley surrounded by
high hills, and is chiefly noted for its mineral springs, of which there
are four : chalybeate, magnesia, white sulphur, and blue sulphur.
These, together with a spring of pure water, are near each other and
near a wooded bluff W. of the village, and flow into a small stream
below.   The waters are pure and clear, and though they flow for a quarter of a mile from their source with other currents, they yet preserve
their own distinct character.   They tumble over a ledge of perpendicular rocks, with a descent of 65 ft., in sufficient volume and force to turn
a mill.   The Magnesia and White Sulphur Springs closely resemble the
White Sulphur Springs of Virginia.   The waters are drunk to a considerable extent, especially the Magnesia ; but the specialty of the place is
its baths, for which there are spacious and admirably appointed bath-
houses (40c. a bath).   Besides the water-baths, mud-baths are administered (in which the patient is covered with mud saturated with sulphur
and heated to about 110°).   These baths are considered remedial for
rheumatism and kindred ailments.   Other baths, prepared by mixing the

magnesia-water with extract of pine from the Black Forest of Germany, are administered for pulmonary, neuralgic, and paralytic diseases. There are pleasant drives and rambles in the vicinity of the hotels, and from the summit of the hill over the village a beautiful view may be obtained, including the Mohawk Valley, the Adirondacks, and the Green Mountains of Vermont. Sharon Springs is connected by stage (9 miles) with Canajoharie, on the West Shore, and with Palatine Bridge, on the New York Central R. Rs. (see Route 40).

Cherry Valley, a pretty little village at the head of Cherry Valley Creek, is 9 miles from Sharon Springs by railway and 7 miles by road. It is a place of great interest as the scene of one of the most atrocious massacres that have ever disgraced any war. Here, in August, 1778, the Tories and Indians fell upon the unprotected settlers, and, without making any distinction of age or sex, either killed or took captive the entire population. A monument now marks the site of the old fort and the grave of the slaughtered settlers. The valley is a popular but not fashionable summer resort, and besides the hotels there are numerous houses at which board may be obtained at from $7 to $12 a week. In the village is a young ladies' academy, the first principal of which was the Rev. Solomon Spaulding, whose fanciful antiquarian novel, written solely for his own amusement, was made the basis of the "Book of Mormon." Near the center of the township is *Mt. Independence*, a rocky eminence rising 2,000 ft. above the sea. On a small creek near by (2 miles from the village) are the *Teka-harawa Falls*, a picturesque cascade 160 ft. high. In the vicinity of these falls (1½ mile from the village) are the *Cherry Valley White Sulphur Springs*, which are becoming a popular resort. In the village of Salt Springsville, near by, are a number of salt-springs; and there are also chalybeate and magnesia waters in the vicinity. Cherry Valley is famous for the coolness, salubrity, and tonic effect of its summer climate.

On the main line, 6 miles beyond Cobleskill and 39 miles from Albany, is **Howe's Cave** (*Pavilion Hotel*), the most remarkable cavern known, after the Luray Cavern of Virginia and great Mammoth Cave of Kentucky. It was discovered in 1842 by Lester Howe, who is said to have penetrated to a distance of 12 miles, but the farthest point usually visited is about 4 miles from the entrance. The entrance is near the hotel (fee, including guide, $1.50). A stairway descends from the entrance to the Reception Room, after which follow in succession Washington Hall, the Bridal Chamber, the Chapel, Harlequin Tunnel, Cataract Hall, Ghost Room or Haunted Castle, and Music Hall. Stygian Lake is crossed in a boat, and beyond are Plymouth Rock, Devil's Gateway, Museum, Geological Rooms, Uncle Tom's Cabin, Grant's Study, Pirate's Cave, Rocky Mts., Valley of Jehoshaphat, Winding Way, and Rotunda. As far as the lake the cave is lighted with gas, and beautiful stalactites and stalagmites are everywhere seen. There are other remarkable caves in this vicinity, the most noteworthy of which is *Ball's Cave*, 4 miles E. of Schoharie.

Three miles beyond Howe's Cave is *Central Bridge*, whence a branch line runs in 5 miles to the pretty hill-village of *Schoharie*, then 6 miles farther to *Middleburg ;* and 9 miles farther is *Quaker Street*, where through passengers for Saratoga and the north who wish to save the détour by Albany take a branch road which runs N. E. *via* Schenectady to **Saratoga Springs** in 37 miles. Between Quaker Street and Albany the train runs for a considerable portion of the way in sight of the far-viewing Helderberg Mountains, passing through *Altamont*, 17 miles from Albany on the slope of the mountains, where superb views are to be had (the *Kushaqua*); then, descending the picturesque valley of Norman's Kill, passes 5 small stations, it reaches Albany (350 miles), where connection is made with railroads leading in all directions.

53. Philadelphia to Erie.

By the Philadelphia & Erie Division of the Pennsylvania R. R. Distances: to Harrisburg, 108 miles ; to Sunbury, 159 ; to Williamsport, 199 ; to Lock Haven, 223 ; to Emporium, 297 ; to Corry, 409 ; to Erie, 446. Two through trains daily run on this line, making the journey in 24 hours, and this is a favorite route from Philadelphia to Western New York and the Oil Regions of Pennsylvania.

From Philadelphia to Harrisburg this route follows the Pennsylvania R. R. and has been described in Route 50. From Harrisburg to Sunbury (57 miles) it follows the Northern Central R. R., and this section is described in Route 59. *Sunbury* (159 miles) is pleasantly situated on the E. bank of the Susquehanna River, at the intersection of the Philadelphia & Erie and Northern Central Railways. The former road is taken here, and the train passes in 2 miles to the attractive village of *Northumberland*, built upon a point of land formed by the confluence of the N. and W. branches of the Susquehanna. The Bloomsburg Div. of the Delaware, Lackawanna & Western R. R. connects here, and by means of it a pleasant excursion can be made to the Wyoming Valley (Route 45). *Milton* (171 miles) is a thriving village at the junction of the present route with the Catawissa Div. of the Philadelphia & Reading R. R. (Route 56). About 10 miles beyond Milton the two railroads cross each other and run on nearly parallel lines to Williamsport. The scenery along this portion of the road is strikingly picturesque. **Williamsport** (*Hepburn House, Park Hotel*) is a city of 27,132 inhabitants, picturesquely situated on the W. Branch of the Susquehanna, surrounded by high hills and much fine scenery. The streets are wide and straight, lighted with gas, and traversed by street-cars. The business quarter is substantially built, and numerous handsome residences and gardens make the place attractive. The suburbs of Rocktown and Duboistown lie across the river under the Bald Eagle Mts., and are connected with the city by a graceful suspension bridge. The county buildings are handsome structures, and *Trinity Church* is a very fine edifice. The *Dickinson Seminary*, with spacious buildings in Academy St., is a noted educational institution. Williamsport owes its prosperity to the lumber business, of which it is a leading mart. The great *Susquehanna Boom* extends from 3 to 4 miles up the river, has a capacity of 300,000,000 ft. of lumber,

and in spring is filled with pine and hemlock logs. The annual shipments of lumber average 250,000,000 ft., and there are vast saw-mills,
planing-mills, machine-shops, etc.

Leaving Williamsport, the train crosses in succession the Lycoming
Creek and the W. Branch, and still following the river passes in 25 miles
to **Lock Haven** (*Fallon House*), a city of 7,358 inhabitants, also famous
as a lumber-mart. Immense numbers of logs are annually received in the
boom here, and furnish employment to extensive saw-mills. The charming scenery about Lock Haven, especially that of the adjacent Bald
Eagle Valley, attracts many summer visitors. Beyond Lock Haven the
road runs for 28 miles through wild scenery to **Renovo** (*Renovo Hotel*),
a creation of the railroad, which here has extensive construction-shops
and foundries. The· village lies in a beautiful, mountain-surrounded
valley, and the loveliness of the scenery combined with the excellent
trout-fishing in the adjacent streams has made it a popular summer
resort. A few miles beyond Renovo the railroad leaves the Susquehanna
and for the next 50 miles traverses what, until its construction, was an
unknown land even to its nearest neighbors—a favorite refuge of out-
lawed criminals. It is the section of country known as the *Great Horseshoe of the Alleghanies*, which encompassed and isolated it, and it is still
a desolate wilderness save where a few straggling settlements have
sprung up along the railway. *Cameron* (292 miles) is a small village
near some rich veins of bituminous coal. Five miles beyond is **Emporium,** a lumbering town of 2,147 population, on the Driftwater, a
tributary of the Susquehanna, built in a valley, the sides of which rise
abruptly to the height of 700 to 1,000 ft. Valuable salt-springs have
been discovered in the vicinity, and it is expected that the manufacture
of salt will prove profitable. At Emporium the Western New York &
Pennsylvania R. R. (Route 54) diverges. Twenty-one miles beyond Emporium is the flourishing village of **St. Mary's,** surrounded by numerous veins of the richest bituminous coal, and near deposits of iron-ore and
fire-clay, with abundance of timber at hand. There are 2 religious houses
here: St. Mary's Convent of Benedictine Nuns and St. Mary's· Priory, a
Benedictine monastery. The convent is the oldest of the order in the
United States and is called the "Mother House." *Wilcox* (343 miles) is
noted as the site of one of the largest tanneries in the world; and *Kane*
(352 miles) is where the road leaves the *Wild-cat Country*, or "unknown
land." It is situated on the Big· Level, a narrow plateau which forms
the boundary from N. to S. of the great coal and oil region of Northwestern Pennsylvania, and is the summit whence trains descend by heavy
grades to the level of Lake Erie. *Warren* (381 miles) is an attractive
town of 4,322 inhabitants, at the confluence of the Conewango and the
Alleghany River, at the head of navigation on the latter. It is the site
of extensive tanneries, has an abundance of light sandstone for building
purposes, and lies between the coal and iron and the oil regions of Pennsylvania. The Dunkirk, Alleghany Valley & Pittsburg R. R. connects
at *Irvineton* (396 miles), where also the Rochester Div. of the Western,
New York & Pennsylvania R. R. comes in from the Oil Regions (see
Route 57). **Corry** (409 miles) is at the junction of the Philadelphia &

Erie and the New York, Lake Erie & Western Railways. It came into existence as a result of the discovery of oil, and prior to June, 1861, its site was covered with forest. The first house was erected in August, 1861; the great Downer Oil Works were erected shortly afterward, and the place has now a population of 5,677, with 8 churches, 2 banks, several good hotels, and 2 daily papers. Beyond Corry are *Union* (420 miles) and *Waterford.* (428 miles), and the road traverses a pleasant farming country to its terminus at Erie (446 miles from Philadelphia).

Erie (*Ellsworth House, New Moore House,* and *Reed House*) is a city and port of entry on Lake Erie, with a population of 40,634, a flourishing commerce, and extensive manufactures. It stands upon a bluff commanding a fine view, and is laid out with broad streets crossing each other at right angles. *The Park* is a finely shaded inclosure in the center of the city, surrounded by handsome buildings, and intersected by State St., which is the principal business thoroughfare. In the Park are a *Soldiers' Monument*, with 2 bronze statues of heroic size, and 2 fountains; and near by is the *Court-House*, a building in the classic style, and the *City Hall*, in the West Park. The *Park Opera-House* is a handsome edifice, and the *Custom-House* a white-marble building near the water. The *Pennsylvania Soldiers' and Sailors' Home* is situated on *Garrison Hill. St. Vincent Hospital* (Catholic) and the *Hamot Hospital* and *Home for the Friendless* (Protestant) are flourishing institutions. The *Union Depot* is of brick, in the Romanesque style, 480 ft. long, 88 ft. wide, and 2 stories high, and is surmounted by a cupola 40 ft. high. The city has 28 churches, 16 public schools, a public library of 6,000 volumes, 3 daily and 11 weekly newspapers. It is the station and winter-quarters of the *U. S. Steamer Michigan*, the only U. S. naval vessel on the chain of Great Lakes. A Government building for the U. S. Court, Post-Office, etc., is being erected at an expense of $150,000. The *Erie Cemetery*, in Chestnut St., comprises 75 acres beautifully laid out with walks and drives, and adorned with trees, flowers, and shrubbery. The harbor is the best on the Lakes, being 4½ miles long, from 1¼ to 2 miles wide, and 9 to 25 ft. deep, and is inclosed by Presque Isle, lying in front of the city. The harbor is protected by 3 lighthouses, 2 at the entrance and 1 on Presque Isle. The Pennsylvania Railway has docks furnished with railroad tracks, so that the transfer of merchandise takes place directly between the vessels and the cars. The principal articles of shipment are lumber, coal, iron-ore, fish, and grain. The leading manufactures are of iron-ware, machinery, cars, leather, brass, furniture, organs, boots, shoes, etc. It was from Erie that Perry's fleet sailed on the occasion of his memorable victory, and thither he brought his prizes. Several of his ships were sunk in Misery Bay, and the hull of the Niagara is still visible in fair weather. At Erie the traveler can take the Lake Shore & Michigan Southern R. R., or the New York, Chicago & St. Louis R. R., and go E. to Buffalo (88 miles), or W. to Chicago (451 miles); or by the Erie & Pittsburg R. R. S. to Pittsburg (151 miles), and the Philadelphia & Erie R. R. of the Pennsylvania System S. E. to Philadelphia (441 miles); or, he can take one of the steamers of the Erie & Western Transit Co. for Buffalo, Cleveland, Detroit, and all Lake Superior ports.

54. Philadelphia to Buffalo.

By the Pennsylvania R. R. to Harrisburg ; thence by the Northern Central R. R. to Sunbury ; thence by the Philadelphia & Erie Div. of the Pennsylvania R. R. to Emporium ; and thence by the Western New York & Pennsylvania R. R. to Buffalo. Distances : to Harrisburg, 105 miles ; to Sunbury, 159 ; to Williamsport, 190 ; to Emporium, 297 ; to Buffalo, 418. This is the shortest route between Philadelphia and Western New York.

As far as **Emporium** (297 miles) this route is identical with the preceding one. At Emporium the Western New York & Pennsylvania R. R. is taken, and the train runs N. through a sparsely-settled forest-region to *Port Alleghany* (322 miles), a small village on the Alleghany River. Beyond Port Alleghany the river is followed amid rugged scenery to **Olean** (348 miles), where the Erie R. R. (Route 41) is crossed. Olean is an important shipping-station at the head of navigation on the Alleghany River. Twelve miles beyond Olean is *Ischua*, E. of which is the Oil Creek Reservation of the Seneca Indians. Near *Franklinville* (369 miles) is the pretty Lime Lake, which may be seen from the cars on the left; and during the remaining 49 miles the road traverses a pleasant agricultural district.

55. Philadelphia to Lake Ontario.

By the New York Div. of the Pennsylvania R. R. to Trenton ; thence by the Belvidere Div. to Manunka Chunk ; and thence by the Delaware, Lackawanna & Western R. R. to Oswego. Distances : to Trenton, 34 miles ; to Manunka Chunk, 102 ; to Delaware Water-Gap, 113 ; to Scranton, 169 ; to Binghamton, 232 ; to Syracuse, 311 ; to Oswego, 346 ; to Buffalo, 434 ; to Niagara Falls, 456. This is a direct route from Philadelphia to the Delaware Water-Gap, Schooley's Mt., Central and Western New York, Buffalo, and Niagara Falls. There is only one change of cars between Philadelphia and Oswego, at Manunka Chunk.

FROM Philadelphia to **Trenton** (34 miles) this route is described in Route 3 *a.* At Trenton the Belvidere Division is taken, and the train follows the N. bank of the Delaware River to Manunka Chunk amid varied and picturesque scenery. Four miles beyond Trenton the New Jersey Lunatic Asylum is passed, and 5 miles farther is *Washington's Crossing*, where General Washington made the celebrated passage of the Delaware, when he surprised and defeated the Hessians at Trenton (Dec. 26, 1776). *Lambertville* (50 miles) is a large manufacturing village of 4,142 inhabitants, with a fine water-power derived from a feeder of the Delaware & Raritan Canal. Beyond Lambertville the scenery is very pleasing, and 8 small stations are passed before reaching *Phillipsburg* (85 miles), where connections are made with the Delaware, Lackawanna & Western R. R.. For description of Phillipsburg, see Route 45. Fourteen miles beyond Phillipsburg is *Belvidere (American House)*, a pretty village situated on both sides of Pequest Creek, where it empties into the Delaware. It has a fine water-power, with considerable manufactures, and a population of 1,768. **Manunka Chunk** (102 miles) is the junction with the main line of the Delaware, Lackawanna & Western R. R., and passengers for the north here change cars. (The

route from New York to Manunka Chunk over this road is described in
Route 43.) *Delaware* (105 miles) is the last station in New Jersey, the
train crossing the Delaware into Pennsylvania on a long bridge. All
trains stop at Delaware for refreshments. Eight miles beyond Delaware
station is the celebrated **Delaware Water-Gap,** for a description
of which, see Route 43, p. 195. The first station beyond the Gap is
Stroudsburg (117 miles). At *Spragueville* (121 miles) the ascent of the
Pocono Mt. begins; and the road, after passing *Oakland* (129 miles),
Tobyhanna (143 miles), and *Moscow* (156 miles), reaches **Scranton** (see
Route 52). Beyond Scranton the train runs N. to *Great Bend* (212
miles) on the Erie R. R. and the Susquehanna River, and thence to
Binghamton (see Route 41).

Leaving Binghamton, the train follows the Chenango River for 10
or 12 miles, then ascends the Tioughnioga River, and then traverses a
rich farming region to Syracuse on the Syracuse and Binghamton Div.
There are numerous villages *en route*, but the only ones requiring mention are *Cortland* (275 miles), a pretty place of 8,590 inhabitants, seat of
a State Normal School; and *Homer* (278 miles), a prosperous village,
near the Little York Lakes. **Syracuse** (311 miles) has been already
sufficiently described in Route 40. Beyond Syracuse the road skirts the
W. shore of Onondaga Lake, and soon reaches **Oswego** (*Doolittle House,
Lake Shore Hotel*). Oswego is the largest and handsomest city on
Lake Ontario, with a population in 1890 of 21,842, and extensive commerce and manufactures. Immense quantities of grain and lumber are
received and shipped here, and, with the exception of Rochester, more
flour is made here than in any other city in the State. *Kingsford's Oswego Starch Factory* is reputed to be one of the largest in the world, and
there are important foundries, iron-works, knitting-factories, box-shook
manufactories, malt-houses, etc. The city is divided by the Oswego
River, which is spanned by 3 iron drawbridges. The streets are regularly laid out with a width of 100 ft., and contain many fine public and
commercial buildings and private residences. There are two public
parks, one on each side of the river, which, as well as the residence-
streets, are beautifully shaded. The principal public buildings are the
Custom-House and Post-Office, of Cleveland limestone, costing $120,000;
the *City Hall* and the *County Court-House*, of Onondaga limestone; the
State Armory, of brick, with stone and iron facings; and the *City Library*, costing $30,000 and containing 12,000 volumes. There are also
several handsome school-buildings, and 2 public halls. The *Deep Rock
Spring* (in First St., W.), discovered in 1865, has attained a wide celeb-
rity, and the spacious *Doolittle House* has been erected over it to accommodate invalids and others. The naturally good harbor of Oswego has
been artificially improved, and now has 3 miles of wharfage. It is defended by *Fort Ontario*, a strong work on the E. shore (open to visitors).
Oswego is the terminus of the New York, Ontario & Western R. R. and
of the Delaware, Lackawanna & Western R. R. It is also the headquarters of the Rome, Watertown & Ogdensburg R. R., on which road
it is located about midway between the *Thousand Isles* and *Niagara
Falls*.

56. Philadelphia to Reading, Pottsville, and Williamsport.

By the main line of the Philadelphia & Reading R. R. and the Catawissa Div. Distances : to Reading, 58 miles ; to Port Clinton, 78 ; to Pottsville, 93 ; to Tamaqua, 98 ; to Williamsport, 199. The Phila. & Reading R. R. connects the great anthracite coal-fields with tide-water. The road was finished in 1842 at a cost of over $16,000,000. It traverses the valley of the Schuylkill River a distance of 58 miles to Reading, and thence 35 miles to Pottsville. In 1886 the Pennsylvania R. R. was opened from Philadelphia to Reading and Pottsville.

THE passenger station in Philadelphia is at the cor. of 12th and Market Sts., and, leaving the city, the fine stone bridge over the Schuylkill is crossed in full view of Fairmount Park, Laurel Hill, and other objects of interest mentioned in our description of Philadelphia. The Schuylkill River is now followed, and in 17 miles the train reaches *Bridgeport*, opposite which is **Norristown** (*Montgomery House*), a handsomely built town of 19,791 inhabitants, with a fine marble Court-House, several handsome school-buildings, and important manufactures. The *Chester Valley Branch* runs in 22 miles from Bridgeport to Downingtown on the Pennsylvania R. R. (Route 50). *Valley Forge* (23 miles) is memorable as the headquarters of Gen. Washington and the American army during the dismal winter of 1777. The building occupied by Washington is still standing near the railroad, whence it can be seen. **Phœnixville** (28 miles) is a flourishing town of 8,514 inhabitants, noted for its rolling-mills and furnaces. The Phœnix Iron-Works are among the largest in America, and it was here that the iron dome of the Capitol at Washington was made. Just beyond Phœnixville the train traverses a tunnel 2,000 ft. long, and passes in 12 miles to *Pottstown*, a pretty tree-embowered village of 8,514 inhabitants, surrounded by charming scenery. The railroad passes through one of its streets and crosses the Manatawny Creek on a lattice bridge 1,071 ft. long. **Reading** (*American, Highland, Mansion House*) is the third city of Pennsylvania in manufactures and the fifth in population, which in 1890 was 58,661. It is pleasantly situated on an elevated and ascending plain, backed on the E. by Penn's Mt. and on the S. by the Neversink Mt., from both of which flow streams of pure water, abundantly supplying the city. The streets cross each other at right angles, and in the center of the city is *Penn Square*, on which are the chief hotels and stores. The *Court-House*, on N. 6th St., is a very handsome edifice with a fine portico supported by 6 columns of red sandstone. The *City Hall* is at the cor. of Franklin and S. 5th Sts., and near by is a public library with 6,500 volumes. The *County Prison* is a substantial structure in Penn St. ; and the *Grand Opera-House* and the *Academy of Music* are fine buildings. Of the 31 churches the most noteworthy are *Trinity* (English Lutheran), an antique building with a spire 210 ft. high, and *Christ* (Episcopal), an imposing Gothic edifice of red sandstone in N. 5th St., with a spire 202 ft. high. Reading is the seat of the *Diocesan* (*Episcopal*) *College*, the *Stewart Academy*, the *Reading Academy*, and numerous schools. The inhabitants of this district are chiefly of German origin, and a dialect of

German, known as Pennsylvania Dutch, still prevails to some extent.
The city is especially noted for its iron manufactures. The shops of
the Phila. & Reading R. R. employ 2,800 men. The principal places of
interest in the vicinity of Reading are the *Mineral Spring,* 1½ mile E. ;
the park, *Penn's Commons,* Mt. Penn ; the Mt. Penn Gravity R. R.,
Neversink R. R. (electric), Inclined Plane Gravity R. R., *Antietam Lake,*
and *White Spot,* on Penn's Mt., 1,000 ft. above the river, famed for its
view. Nine miles from Reading, on the *Lebanon Valley Branch,* is the
celebrated Water-Cure Home at *Wernersville ;* and 54 miles on the same
line is Harrisburg, where connection is made for Carlisle, Pa., and for
Gettysburg, the scene of the greatest battle of modern times.

Beyond Reading the road still follows the Schuylkill, and in 20 miles
reaches *Port Clinton* (78 miles), a pleasant place at the mouth of the
Little Schuylkill. Here the Little Schuylkill Branch of the Reading R. R.
connects. From Port Clinton the Pottsville trains pass on by *Auburn*
(83 miles) and *Schuylkill Haven* (89 miles) to **Pottsville,** the terminus
of the Phila. & Reading main line. Pottsville is situated upon the edge
of the great Schuylkill coal-basin, in the gap by which the river breaks
through Sharp's Mt. The annual yield of the Schuylkill coal-fields is
about 8,000,000 tons, and this enormous product is conveyed to market
by the Reading R. R. and the Schuylkill Canal. The city dates from
1825, and in 1890 had a population of 14,117. The chief public buildings are the Court-House, Jail, Town-Hall, Union Hall, and Opera-House.
The coal-traffic is the principal source of the city's prosperity, but there
are also extensive foundries, rolling-mills, and machine-shops. The
great collieries lie to the N. and N. E., and are reached by numerous
branch roads which converge upon Pottsville.

At Pottsville the through trains for Williamsport take the Schuylkill Valley Branch of the Philadelphia & Reading R. R., which traverses a wild and desolate region for twenty miles to **Tamaqua** (98
miles), a prosperous town of 6,054 inhabitants, attractively situated on
the Little Schuylkill, in the midst of a rich coal-region, from which
it draws a large trade. Beyond Tamaqua the train traverses for fifty
miles a rugged and mountainous region which is fairly gridironed with
the numerous intersecting branches of the great coal roads. The scenery
of this section of the route is varied and impressive, and the Catawissa
Valley, which is traversed for 30 miles, offers scenes of singular beauty.
Catawissa is 145 miles from Philadelphia, and is picturesquely situated at the confluence of the Catawissa Creek and the Susquehanna
River. Nine miles beyond is **Danville,** a flourishing manufacturing
town of 7,998 inhabitants. The Montour Iron-Works here make vast
quantities of railroad iron, and on a hill near by is a State Insane Asylum with extensive buildings. *Milton* (171 miles) is the junction of the
present route with the Philadelphia & Erie R. R., which is described in
Route 53.

57. Pittsburg to Titusville and Buffalo. The Pennsylvania Oil Regions.

By the Alleghany Valley R. R. and Pittsburg Div. of the Western New York & Pennsylvania R. R. Distances : to Red Bank, 64 miles ; to Oil City, 132 ; to Titusville, 150 ; to Corry, 177 ; to Chautauqua Lake, 207; to Buffalo, 269. Through trains from Pittsburg to Buffalo *via* Oil City and Brocton accomplish the distance in 12 hours.

PITTSBURG has been described, and the route thither from Philadelphia, in Route 50. Leaving the Union Depot, the train passes for several miles among smoke-discolored factories and iron-works, and then reaches the Alleghany River, whose banks are followed for more than 100 miles amid picturesque and varied scenery. *Kittaning* (46 miles) is a flourishing manufacturing borough of 3,095 inhabitants, in the midst of a rich coal-region, which is extensively worked. From *Red Bank* (64 miles) the Low Grade Division of the Alleghany Valley R. R. runs in 110 miles to Driftwood on the Phila. & Erie R. R. (Route 53), passing the remote forest-town of *Brookville*, which offers great attractions to the sportsman. *Brady* (69 miles), *Parker* (83 miles), *Emlenton* (89 miles), and *Kennerdell* (108 miles) are small stations. All along this section of the route the apparatus of oil-wells, some in operation and others deserted, may be seen from the cars. Sixteen miles beyond Kennerdell is **Franklin,** a city of 6,221 inhabitants, built on the site of the old French *Fort Venango*, at the confluence of French Creek and the Alleghany River. Several railroads connect here. Nine miles beyond Franklin is **Oil City** (*Arlington, Exchange,* and *National*), the center and headquarters of the Oil Region. It is situated on the Alleghany River at the mouth of Oil Creek, the city being built along a narrow shelf between the river and a high bluff which is crowned with residences. Oil City was founded in 1860, incorporated in 1871, and in 1890 had a population of 10,932. It is not particularly attractive to the eye, but it will afford the visitor in a few short rambles the best opportunity of witnessing the various operations of obtaining, refining, barreling, gauging, and shipping the precious petroleum. The wells in the vicinity yield 600 barrels daily, and about 2,000,000 barrels are annually sent thence to market. The great iron tanks for storing the oil are worth a visit.

From Oil City the Pittsburg Div. of the Western New York & Pennsylvania R. R. follows the Alleghany River to *Irvineton* (50 miles), where the Philadelphia & Erie Div. of the Pennsylvania R. R. is intersected (see Route 53). The most important points on this line are *Oleopolis* (9 miles), once a flourishing center of trade, *Tionesta* (20 miles), and *Tidioute* (35 miles), the latter being quite a manufacturing place. The scenery on this division of the road is highly picturesque.

From Oil City the train for Buffalo follows the valley of Oil Creek, famous as the scene of the earliest " operations in oil." The old derricks and tanks are still standing, mementos of a former activity, and an occasional pumping-well is seen, while most of the stations *en route* are decadent relics of a lost prosperity. Eighteen miles beyond Oil City the train reaches **Titusville** (*European, Mansion,* and *United States*), a city of 8,073 inhabitants, and the largest place in the Oil Regions. It is situated in a broad and beautiful valley, through which flows Oil Creek. The streets are broad and well paved; the business blocks are of brick and stone; and there are quite a number of fine residences. The place owes its rapid growth and prosperity mainly to the oil-wells in the vicinity, which are very productive; and here are the capacious refineries of the Standard Oil Company. Besides the oil-works there are extensive iron-works, foundries and machine-shops, and various other manufactories. The Union & Titusville Branch runs in 25 miles from Titusville to *Union City* on the Philadelphia & Erie R. R. (see Route 53); and the Dunkirk, Alleghany Valley & Pittsburg R. R. runs in 91 miles to *Dunkirk* (see Route 41).

Leaving Titusville, a picturesque ride of 28 miles past a number of small villages brings us to **Corry** (see Route 53). Six miles beyond Corry the train crosses the boundary-line between Pennsylvania and New York, and then in about 20 miles reaches *Mayville*, at the head of **Chautauqua Lake,** the highest navigated body of water E. of the Rocky Mountains (1,291 ft. above the sea), and one of the most frequented of summer resorts. At *Chautauqua* (3 miles from Mayville), the headquarters of the famous Chautauqua Literary and Scientific Circle, the National Sunday-school Assembly hold their annual session during July and August; and at *Point Chautauqua* (1½ mile from Mayville) the National Baptist Association have extensive grounds. Several steamers ply on the lake, and there are numerous hotels. From Mayville the ride to *Brocton* (89 miles from Oil City) is through a pleasant country. At Brocton the train takes the Lake Shore & Michigan Southern R. R. to **Buffalo** (see Route 40).

58. Harrisburg to the Cumberland Valley.

By the Cumberland Valley R. R., which runs S. W. from Harrisburg to Winchester. Distances : to Carlisle, 18 miles ; to Shippensburg, 41; to Chambersburg, 52 ; to Hagerstown, 74 ; to Martinsburg, 94 ; to Winchester, 116.

LEAVING the Harrisburg station of the Pennsylvania R. R., the train crosses the Susquehanna, and passes for several miles amid strikingly picturesque scenery. *Mechanicsburg* (8 miles) is a pretty town of 3,691 inhabitants, with several neat churches, a number of prosperous factories, and the favorably-known educational institution, the Irving Female College. A branch railroad extends south from this place to Dillsburg, passing through Williams Grove, where the Grangers' picnic is held annually during the last week in August. This is held to be the largest gathering of farmers in the country, as well as the largest and best display of agricultural implements. Beyond Mechanicsburg the scenery is very pleasing, with the Kittatinny or Blue Mountains on the right and South Mountain on the left; and in 8 miles the train reaches *Gettysburg Junction*, where connection is made with the Harrisburg & Gettysburg R. R. for the famous battle-field, passing by the way the *Mt. Holly Springs* (*Holly Inn*), a summer resort with picturesque scenery, pleasant walks and drives, and good fishing in the adjacent streams. A mile beyond the junction is **Carlisle,** a borough of 7,620 inhabitants, situated nearly in the center of the Cumberland Valley. The surrounding country is level, productive, and highly cultivated. The town is well built, with wide and well-shaded streets, and a public square on which front the county buildings and public edifices of a superior order. In the square is a handsome *Monument* erected to the memory of the soldiers of Cumberland County who fell in the civil war. *Dickinson College*, founded in 1783, and now under the care of the Methodists, is one of the oldest and most flourishing institutions in the State. It has plain buildings in Main St., W. of the public square, with valuable scientific collections and a library of 26,000 volumes. The *Government Training-School for Indians* is also in Carlisle. The *Carlisle Barracks* were built in 1777 by the Hessian prisoners captured at Trenton, and have accommodations for 2,000 men. Washington's headquarters were at Carlisle in 1794, at the time of the Whisky Rebellion; and the town was shelled by the Confederates on the night of July 1, 1863, during Lee's invasion of Pennsylvania. It was captured by the Southern troops, who at the same time occupied Mechanicsburg and advanced to within 4 miles of Harrisburg. At the base of Pisgah Mt., 14 miles N. of Carlisle, are the **Perry Warm Springs,** a quiet and inexpensive resort amid attractive scenery. The waters have a temperature of 70° to 72°, and when taken internally are aperient and diuretic. They are most esteemed as a bath, and employed in this way are beneficial in diseases of the skin. The Springs are also reached by stage in 12 miles from Duncannon on the Pennsylvania R. R. (see Route 50).

Eleven miles beyond Carlisle, on the railroad, is *Newville*, whence stages run to the *Doubling Gap Springs*, a quiet resort, and connection is also made for the *Cloverdale Lithia Springs*. The adjacent scenery of the Doubling Gap, where the Blue Mt. turns on itself and forms a gigantic *cul-de-sac*, is peculiarly picturesque and striking. *Shippensburg* (41 miles) is the market and shipping-point for the productive farming region of which it is the center, and has a population of 2,188. The Cumberland Valley Normal School stands on a far-viewing hill to the N. At *Mont Alto Junction* connection is made for *Waynesboro* and the resort *Mont Alto*. Eleven miles beyond Shippensburg is **Chambersburg** (*National, Washington*), a borough of 7,863 inhabitants pleasantly situated on the Conecocheague Creek. The surrounding country, which forms part of the great limestone valley at the S. E. base of the Blue Mts., is populous and highly cultivated. The town is well built, the houses being mostly of brick or stone; and there are manufactories of cotton, wool, flour, paper, and iron. The *Court-House* is a handsome edifice, and *Wilson College* (for women) is a flourishing institution. Chambersburg was captured and set on fire by the Confederates under Gen. Early, on July 30, 1864, during a raid into Pennsylvania. Two thirds of the town was destroyed, inflicting a loss of $2,000,000.

Ascending the valley from Chambersburg, the train soon reaches the pretty village of *Greencastle* (63 miles), and 5 miles beyond crosses the famous Mason and Dixon's Line and enters the State of Maryland. Six miles beyond the line is **Hagerstown,** capital of Washington County, with a population of 10,118. It is pleasantly situated on the W. bank of Antietam Creek, 22 miles above its entrance into the Potomac, at the intersection of the present route with the Western Maryland R. R. and the Washington Co. Branch of the Baltimore & Ohio R. R. Here also connection is made with the Shenandoah Valley R. R., on which road are the Caverns of Luray, the Grottoes, and the Natural Bridge. The city is regularly laid out and well built, with a handsome Court-House, erected at a cost of $77,000. It is surrounded by a rich agricultural region, from which it draws considerable trade, and there are prosperous foundries and factories. About 7 miles S. of Hagerstown is the *College of St. James* (Episcopal). Hagerstown was the scene of several severe conflicts during the civil war, being captured a number of times by the Confederates and as often retaken by the National forces. *Williamsport* (81 miles) is where Lee recrossed into Virginia after the battle of Gettysburg (see Route 59). Here the train crosses the Potomac on a long bridge, enters W. Virginia, and passes in 14 miles to **Martinsburg,** a town of 7,226 inhabitants, on the Tuscarora Creek, where it crosses the Baltimore & Ohio R. R. (see Route 69). There are here a commodious Court-House, a Town-Hall, a Market-House, spacious agricultural fair grounds, and machine-shops of the Baltimore & Ohio R. R. Twenty-two miles beyond, **Winchester** (116 miles from Harrisburg) is reached. It is the terminus of the road, and is a place of 5,196 inhabitants. Stage connection may be made to *Capon, Rock Enon, Orkney,* and other noted Virginia springs. Near here was the scene of several of Gen. P. H. Sheridan's important battles during the civil war, as well as of his famous ride, described so graphically in Thomas Buchanan Read's poem. Both National and Confederate cemeteries are located in Winchester.

59. Baltimore to Niagara Falls.

By the Northern Central R. R. (Pennsylvania R. R. System) through Maryland, Pennsylvania, and Western New York, intersecting all the great lines of E. and W. travel. At Harrisburg it crosses the Pennsylvania R. R. ; at Williamsport, the Philadelphia & Erie Div. of the Pennsylvania R. R. ; at Elmira, the Erie R. R. ; and terminates at Canandaigua on the N. Y. Central R. R. It is a favorite route of travel from the South to Niagara Falls and all the great Northern resorts, and runs drawing-room and sleeping cars on all the through trains. Distances : to Hanover Junction, 46 miles ; to York, 57 ; to Harrisburg, 85 ; to Sunbury, 138 ; to Williamsport, 173 ; to Ralston, 202 ; to Elmira, 256 ; to Watkins Glen, 278 ; to Canandaigua, 325 ; to Rochester, 354 ; to Buffalo, 422 ; to Niagara Falls, 431. The time from Baltimore to Canandaigua is 18 hours ; to Buffalo, 23 hours ; to Niagara Falls, 24 hours.

THE terminal station in Baltimore of the Northern Central R. R. is the *Union Depot*, in Charles St. The Maryland section of the road traverses a rich but monotonous farming region, with numerous small stations *en route,* but nothing to call for special notice. Just beyond *Freeland* (34 miles) the train enters Pennsylvania. From *Hanover Junction* (46 miles) a branch diverges in 30 miles to

Gettysburg.

From Philadelphia, Gettysburg is reached *via* Pennsylvania R. R. to Hanover, and thence by Western Maryland R. R. Total distance, 136 miles. From New York it is reached *via* Philadelphia, or by Route 48 to Harrisburg, thence by Cumberland Valley R. R. to Gettysburg Junction, and thence by Philadelphia & Reading R. R. Total distance from New York, 250 miles.

Hotels.—*Eagle Hotel, Globe, McClellan House, Springs Hotel.*

Gettysburg is a borough of 3,221 inhabitants, capital of Adams Co., and is pleasantly situated on a gently rolling and fertile plain, surrounded by hills, from which extensive and pleasing views are obtained. The *Court-House* and *Public Offices* are commodious brick structures, and the residences are generally neat and substantial. *Pennsylvania College,* founded in 1832, and the *Lutheran Theological Seminary,* founded in 1825, are among the institutions of the place. Both have large and beautiful buildings, and the former has a library of 10,000 volumes, and the latter a library of 11,000 volumes. One mile W. of the borough are the **Gettysburg Springs,** whose waters, denominated Katalysine, have acquired a wide reputation for their medicinal qualities. They are said to resemble the celebrated Vichy water, and are considered remedial in gout, rheumatism, dyspepsia, and affections of the kidneys. The *Springs Hotel* accommodates the patients who resort here during the summer for treatment.

The chief interest of Gettysburg is historic, and this it is that attracts tourists from all parts of the world. A great battle, perhaps the most important of the civil war, was fought here on the 1st, 2d, and 3d of July, 1863, between the National forces under General Meade and the Confederate army under General Lee. The battle is described below, and it is only necessary now to point out the principal objects of interest. *Cemetery Hill,* so named from having long been the site of the village cemetery, forms the central and most striking feature at Gettysburg. Here were the Union headquarters, and standing on its crest the visitor has the key to the position of the Union forces during those eventful three days of July. Flanking Cemetery Hill on the W., about a mile distant, is *Seminary Ridge,* on which were General Lee's headquarters and the bulk of the Confederate forces. Other spots usually visited are *Benner's Hill, Culp's Hill, Round Top,* and *Little Round Top ;* also *Willoughby Run,* where Buford's cavalry held A. P. Hill's column in check during two critical hours. The *****National Cemetery,** containing the remains of the Union soldiers who fell in the battle of Gettysburg, occupies about 17 acres on Cemetery Hill adjacent to the village cemetery, and was dedicated with imposing ceremonies, and an impressive address by President Lincoln, Nov. 19, 1863. A *****Soldiers' Monument,** dedicated July 4, 1868, occupies the crown of the hill, 60 ft. high, and is surmounted by a colossal marble statue of Liberty. At the base of the pedestal are 4 buttresses bearing colossal marble statues of War, History, Peace, and Plenty. Around the monument, in semicircular slopes, are arranged the graves of the dead, the space being divided by alleys and pathways into 22 sections: one for the regular army, one for the volunteers of each State represented in

17

the battle, and three for the unknown dead. The number of bodies interred here is 3,564, of which 994 have not been identified. Near the entrance to the cemetery is a bronze statue of Major-General John F. Reynolds, who was killed in the first day's fight. Opposite the cemetery, an observatory, 60 ft. high, has been erected, commanding a fine view. A monument to the Pennsylvania troops was erected in 1887.

The **Battle of Gettysburg** was fought July 1, 2, and 3, 1863, between the Union army under General Meade and the Confederate Army of Northern Virginia under General Lee. Having resolved upon an invasion of the North, the Confederates had early in June concentrated a force of nearly 100,000 men, including 15,000 cavalry, in the vicinity of Culpepper, Va. They moved down the valley of the Shenandoah, and on the 24th and 25th crossed the Potomac in two columns, which, uniting at Hagerstown, Md., pressed on toward Chambersburg, Pa. The Union army, having broken up its camp opposite Fredericksburg and moved N., crossed the river farther down on the 28th, on which day Hooker, having resigned the command, was succeeded by Meade. Lee's communications being threatened, he resolved to concentrate his whole force at Gettysburg, already (unknown to him) occupied by a part of the Union army under Reynolds. The first collision occurred on July 1, about 2 miles N. W. of Gettysburg, between the Confederate advance under A. P. Hill and a reconnoitering party of cavalry (afterward supported by infantry) sent out by Reynolds. The Union forces, at first superior, were soon outnumbered, and were driven back in confusion through Gettysburg, losing 5,000 prisoners and as many killed and wounded. The Confederate loss in killed and wounded was probably somewhat greater, in prisoners much less. Both sides hurried up their forces, and on the morning of the 2d the bulk of the two armies was in position, the Union on Cemetery Ridge S. of Gettysburg, and the Confederate on Seminary Ridge opposite (to the west), except Ewell's corps, which lay 2 miles distant at the foot of Culp's Hill on the Union right. The forces present or close at hand were about equal, each numbering from 70,000 to 80,000 infantry and artillery. Lee resolved to attack the Federal position. The main attack was made by Longstreet's corps on the Union left, where considerable ground was gained. On the right Ewell effected a lodgment within the Union intrenchments. The Union loss in this action was fully 10,000, half in Sickles's corps, which lost nearly half its numbers. Lee determined to continue the assault on the 3d. Early in the morning Meade took the offensive against Ewell, and forced him from the foothold which he had gained, but of this Lee was not informed. The Confederates spent the morning in preparation, and at 1 o'clock opened fire from 120 guns, which was immediately returned, though Meade, owing to the rugged nature of the ground, was able to use at once only 80 of his 200 guns. After two hours the Union fire was gradually suspended, and Lee, supposing that their batteries had been silenced and that the infantry must be demoralized, ordered the grand attack of the day, which was directed against the Union center. The attacking column numbered about 18,000, consisting of Pickett's division and Pettigrew's brigade. Though met by a terrible fire of artillery and musketry, it pressed on, Pettigrew reaching within 300 yards of Hancock's line, when he was driven back in disorder ; while Pickett's division charged through Webb's front line among the Federal batteries, where for a quarter of an hour there was a struggle with pistols and clubbed muskets. The Union troops hurried from all sides and drove the enemy back down the slope, not one in four escaping. Meade with his left then drove back Hood from the ridge he had won the preceding day. The Confederate loss this day was about 16,000 in killed, wounded, and prisoners ; the Union loss was about 3,000. Both armies remained inactive the next day, and during the night Lee began his retreat to the Potomac, which he reached on the 7th. Here he was compelled to halt by the swollen stream. On the 12th Meade came in front of the Confederate intrenchments, but an attack was postponed till the 14th, when Lee was found to be safe on the other side, having succeeded in crossing during the night. The Union loss at Gettysburg was 23,190, of whom 2,834 were killed, 13,713 wounded, and 6,643 missing. The Confederate loss has never been officially stated ; but by the best estimates it was about 36,000, of whom about 5,000 were killed, 23,000 wounded, and 8,000 unwounded prisoners. The entire number of prisoners was about 14,000.—APPLETONS' CONDENSED AMERICAN CYCLOPÆDIA.

On the main line, 11 miles beyond Hanover Junction, is the ancient city of **York,** situated on Codorus Creek, and containing 20,793 inhabitants. York was settled in 1741, incorporated in 1787, and the Continental Congress sat here from Sept. 30, 1777, to July, 1778. During the Confederate invasion of Pennsylvania in 1863 it was occupied by Early, who levied a contribution of $100,000 on the citizens, but left the place unharmed. The city is pleasantly situated in a rich agricultural region, and is regularly laid out, with streets crossing each other at right angles. At the intersection of Main and George Sts., the leading thoroughfares, is Center Square. The Court-House is a handsome edifice with granite front and Corinthian columns. York contains several large car-shops, some of the most extensive manufactories of agricultural implements in the country, a shoe and a match factory, and the Codorus paper-mills. The train traverses the streets of York for some distance, descends the rich Codorus Valley, and a few miles below Harrisburg reaches the Susquehanna River, which is followed as far as Williamsport amid extremely beautiful scenery. From *Bridgeport* (83 miles) a long bridge crosses the river to **Harrisburg,** the capital of Pennsylvania, which is described in Route 50. The scenery along the line from Harrisburg to Williamsport is very fine, but none of the stations possess any special attractions for the tourist. *Sunbury* (138 miles) is at the intersection of the present route with the Philadelphia & Erie R. R. (Route 53), and has already been described. **Williamsport** (178 miles) is the converging point of the present route, Route 53 and Route 56, and has been described.

Leaving Williamsport, the train ascends the narrow valley of Lycoming Creek, and traverses for many miles a picturesque and sparsely-settled region, dear to sportsmen. The station of *Trout Run* (191 miles) is near a fish-abounding stream; and 11 miles beyond is **Ralston,** a sequestered hamlet, 1,800 ft. above the sea, and surrounded by lofty hills covered with primeval forest. The scenery is extremely picturesque; many romantic cascades are found in the mountain-gorges, and near by are numerous trout-streams which afford excellent sport. The McIntire Coal-Mine is just N. of the village, and the gravity railroad up the mountain-side is a great curiosity. *Minnequa* (219 miles) is near the **Minnequa Springs** (*Minnequa House*), which have become popular as a summer resort. They are situated in a lovely mountain-surrounded valley, 1,500 ft. above the sea, with excellent trout-brooks in the vicinity, and abundance of game in the adjacent woods. The waters contain oxide of iron, are tonic in quality, and are said to be efficacious in dyspepsia, rheumatism, consumption, and cutaneous diseases. *Fassett* (247 miles) is the last station in Pennsylvania, and a short distance beyond the train crosses into New York and passes in 9 miles to **Elmira,** at the intersection of the Erie R. R. (see Route 41). Beyond Elmira the line traverses a quiet rural district, and passes in 22 miles to *Watkins* and the famous

Watkins Glen.

From New York, Watkins is reached by the Erie R. R. (Route 41), or by the Delaware & Lackawanna Route (Route 43), to Elmira (273 miles), and from Elmira by the Northern Central R. R. (22 miles). It can also be reached by the N. Y. Central R. R. to Geneva on Seneca Lake, and thence either by steamer on Seneca Lake, or by the Fall Brook R. R. From Philadelphia by the Philadelphia & Reading R. R. and connecting roads, or by Pennsylvania R. R. to Harrisburg, and thence by the Northern Central R. R. (300 miles). From Baltimore *via* the Northern Central R. R. (278 miles).

Hotels.—In the village are the *Jefferson House* and the *Fall Brook House*, open all the year. Near the entrance of the Glen is the *Glen Park Hotel*. Within the Glen itself is the *Glen Mountain House*. The rates at these houses range from $2 to $4 a day; reduction by the week. The *Lake Shore House*, near Seneca Lake, charges lower rates.

Watkins is a village of 2,604 inhabitants, at the head of Seneca Lake, and within the shadow of Glen Mt. *Franklin St.*, running parallel with the mountain-ridge, leads in ¼ mile from the station to the entrance of the ** Glen*, which is simply a vertical rift or gorge in a rocky bluff some 700 or 800 ft. in height, through which tumbles a roaring mountain-brook. The length of the Glen is about 3 miles, and the cliffs at the deepest part of the gorge have an altitude of nearly 300 ft. First entering a huge amphitheatre to which there is no apparent exit, the visitor follows the path to its W. end, where he finds that, instead of meeting, the walls of rock overlap each other, leaving a narrow passage, through and up which he passes by steep stairways, running diagonally along the face of the wall, braced strongly to it, and also propped firmly from beneath. This first section is called *Glen Alpha*, and at its upper end are the *Minnehaha Falls*, beyond which the path traverses the narrow gorge, called the *Labyrinth*, to the *Cavern Cascade*, and ascends the *Long Staircase*, which is flung at an angle of 90 degrees across the tremendous chasm. From the head of the Long Staircase a path ascends a succession of steep stairways to a shelf of mountain on the N. side of the ravine. On this shelf is perched the **Mountain House,** consisting of a cottage built in the style of a Swiss *chalet*, on one side of the gorge, while on the other side (connected by a graceful iron suspension-bridge) is the main building. The *chalet* is a favorite point for rest and refreshment, and is in all respects one of the most attractive features of the Glen scenery. Its balconies overhang the gorge, with trees jutting up above them from ledges in the rocks below; and the visitor looks down from his advantageous position into depths of the Glen that remain inaccessible. Close at hand is Captain Hope's *Glen Art Gallery* (admission, 25c.), containing upward of 100 paintings by himself, chiefly of the Glen scenery.

Leaving the Mountain House, the path descends gradually almost to the bed of the stream, through the gloomy *Glen Obscura*, and, passing the *Sylvan Rapids*, enters the ** Glen Cathedral*, an enormous amphitheatre, which is considered the most imposing feature of the wonderful gorge. It is 1,000 ft. long, with a floor as level as if paved with human hands, and walls rising to a height of nearly 300 ft. In the center is the lovely *Pool of the Nymphs*, and at the W. end (called the " Chancel ") the *Central Cascade* pours its waters into the gorge over a ledge 60 ft.

high. From the N. side of the Cathedral the *Grand Staircase* leads to the *Glen of the Pools*, so named from the number of its water-worn basins. Beyond the Glen of the Pools the *Giant's Gorge* is reached, at the upper end of which are the exquisite *** Rainbow Falls,** where three cascades drop from one rocky ledge to another, foaming and seething, while to one side a thin stream, falling from a great height, spreads itself out like a silver mist, and mingles its waters with those in the rock-bound channel far below. The path passes behind the fall and leads up another stairway to the *Shadow Gorge*, which is narrow, rugged, and somber, the pathway being hewn out of the cliff-side, and at the head of which the **Pluto Falls** plunge over the rocky parapet into a deep, black pool. Near the Pluto Falls there is a rustic stairway leading to *Glen Arcadia*, from the entrance of which there are beautiful views both up and down the gorge, and at the head of which are the *Arcadian Falls*, spanned by a bridge from which the retrospect down *Elfin Gorge* is remarkably fine. Next above is *Glen Facility*, near the head of which is the iron bridge of the Fall Brook Railway, 450 ft. long and 165 ft. above the stream; and then come *Glen Horicon, Glen Elysium*, with steep wooded banks 400 ft. high, and *Glen Omega*, with *Omega Falls*. The last two are little more than open forest-glades, and contain scarcely a hint of the wild scenes below.

Those who do not care to retrace their steps to the entrance, on their return (the descent of the stairs is even more trying than the ascent), can leave the Glen at the Mountain House by a path leading out to the open country and through the beautiful Glenwood Cemetery, from the heights of which is presented one of the finest landscape views in this part of the State.

Havana Glen, 3 miles S. of Watkins (reached by carriages), is 1¼ mile long, ascends 700 ft., and is preferred by some to the Watkins Glen. It is very picturesque, is more airy and open, and is quite easily traversed; and yet it is not wanting in those elements of gloom and vastness which are the peculiar characteristics of its better-known rival. The same system of stairways and ladders prevails as at Watkins, but these aids of progress are fewer and the paths broader. (Admission, 25c.) The *Montour Glen* and the *Excelsior Glen* are also in the vicinity of Havana and Watkins, and are very striking.

Beyond Watkins the line skirts the W. shore of Seneca Lake (see Route 40) for 12 miles, and soon reaches *Penn Yan* (301 miles), a pretty village of 4,254 inhabitants at the foot of **Lake Keuka** (formerly called *Crooked Lake*), which is a beautiful sheet of water, 18 miles long, 1½ mile wide at the widest part, 718 ft. above the sea, and 277 ft. above Seneca Lake, which is only 7 miles distant. At the foot or N. end it is divided by a promontory into two branches, one 5 and the other 8 miles long. The scenery along the shores is extremely picturesque, and the waters are clear and full of fish.

From Penn Yan several steamers run twice a day each to *Hammondsport*, a neat village at the head of the lake, whence the Bath and Ham-

mondsport R: R. runs 10 miles to *Bath* on the Erie R. R.   Hammondsport is the center of an extensive grape-growing and wine-making region, and the adjacent hill-slopes are clothed with vineyards.   In the
cellars of the Urbana and Pleasant Valley Wine Cos. are hundreds of
thousands of bottles of Catawba, Isabella, claret, and native cham-
pagnes.   *Grove Springs*, a well-known summer resort, is 5 miles N. of
Hammondsport.   Beyond Penn Yan the Northern Central R. R. passes
several obscure hamlets, and soon reaches its terminus at **Canandaigua** (Route 40), on the N. Y. Central R. R.   From this point the traveler
can go W. to Rochester, Buffalo, and Niagara Falls, or E. to Albany and
Troy.   The route in both directions is described in Route 40.

60. The St. Lawrence River.

THE trip down the St. Lawrence usually begins at **Kingston** (*City
Hotel, Hotel Frontenac*), a flourishing Canadian city of 19,264 inhabitants,
at the foot of Lake Ontario, on the line of the Grand Trunk R. R., 172
miles from Montreal, 343 from Quebec, 392 from Detroit, and 469 from
Portland.   It is reached from New York *via* Route 40 to Rome, and
thence *via* the Rome, Watertown & Ogdensburg R. R. to *Cape Vincent* (distance, 347 miles; fare, $8.80).   From Cape Vincent a steam-
ferry connects with Kingston.   The Royal Mail steamers of the Richelieu & Ontario Navigation Co. leave Kingston daily at 5 A. M., and
reach Montreal at 6.30 P. M.

The Thousand Islands.

Almost immediately after leaving Kingston or Cape Vincent the
steamer enters that portion of the St. Lawrence known as the *Lake of the
Thousand Islands*, from the groups of islands and islets amid which it
threads its tortuous way toward Ogdensburg.   According to the Treaty
of Ghent these islands are 1,692 in number, and they extend for 40
miles below Lake Ontario.   They are of every imaginable shape, size,
and appearance, some of them barely visible, others covering many
acres; some only a few yards long, others several miles in length; some
presenting little or nothing but bare masses of rock, while others are so
thickly wooded that nothing but the most gorgeous green foliage is to
be seen in summer, while in autumn the leaves present colors of different hues hardly imaginable.   You pass close to, and often near
enough, to cast a pebble from the deck of the steamer on many of
these circular little islands, whose trees, perpetually moistened by the
water, have a most luxuriant leaf, their branches overhanging the current.   The lighthouses which mark out the channel are a picturesque
feature, but they are drearily alike—fragile wooden structures about 20
ft. high, uniformly whitewashed.   Many summer visitors remain at *Gan-
anoque* (International Hotel), on the Canadian side of the river, and at
Clayton (*Frontenac, Hubbard,* and *Walton*), opposite (17 miles from
Cape Vincent); but the chief resort of the Thousand Islands is **Alexandria Bay** (*Thousand Island House* and *Crossmon House*), on the
New York side of the river, the hotels being the most conspicuous

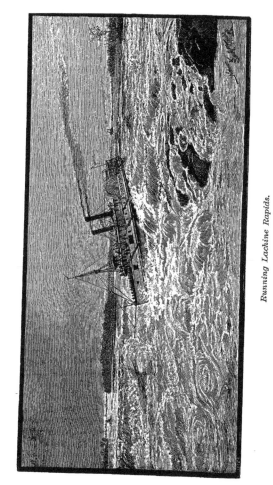

Running Lachine Rapids.

feature (12 miles from Clayton). On the islets near the bay are numerous elegant villas, among them one owned by George M. Pullman, of palace-car fame. The boating is excellent, and the fishing in the vicinity is very fine, including pickerel, muscalonge, black bass, and dory. There are also myriads of wild fowl in their season. About 8 miles S. E. of Alexandria Bay are the romantic *Lakes of Theresa* (Clear, Crystal, Mud, Butterfield, and Lake of the North), with good fishing, and shores and islands abounding in rare minerals. Frequent steamers ply between Cape Vincent, Clayton, and Alexandria Bay, on the arrival of trains at the two former, and a steamer leaves the Bay twice a day for a round trip, touching at *Thousand Island Park*, the largest of the islands, *Wells Island*, and other landings. *Morristown* is a post-village of New York, below Alexandria, and 14 miles from Ogdensburg. On the Canadian shore opposite (reached by ferry-boat) lies *Brockville*. At this point in the river the Lake of the Thousand Islands ends, and we come upon the open river, 2 miles wide.

Thirteen miles from Brockville, on the Canadian side, lies **Prescott,** and immediately opposite (connected by steam-ferry) the flourishing American city of **Ogdensburg** (*Seymour House*). The city has 11,662 population, is attractively situated and handsomely built, and is connected by railroad with a number of prominent points east, west, and south. It is the N. terminus of the Rome, Watertown & Ogdensburg R. R. (see Route 40), and is at the W. end of the Central Vermont R. R. A few miles below Ogdensburg the descent of the first rapids (*Gallopes Rapides*) is made, and immediately afterward of the *Rapide de Plat*. The descent of these rapids is made with full steam on, and there is scarcely anything to indicate that the steamer is not pursuing its usual placid course. Thirty miles below Ogdensburg is *Louisville*, whence stages run to **Massena Springs** (*Harrowgate Hotel, Hatfield House*), 7 miles distant. These springs are on the banks of the Raquette River, in New York State, and are five in number, the largest being named St. Regis, in honor of the tribe of Indians who first discovered its virtues. They are a popular resort in summer, their attractiveness being greatly enhanced by the beautiful scenery by which they are surrounded, and by their proximity to the Long Sault Rapids, about 5 miles distant. The springs are reached directly by Niagara Falls line of the Rome, Watertown & Ogdensburg R. R. from Ogdensburg, or by Central Vermont R. R. by stage from Brasher.

Dickinson's Landing is at the head of the famous * **Long Sault Rapids,** which are 9 miles in length, and through which a raft will drift in 40 minutes. Here the tourist experiences the celebrated sensation known as "shooting the Rapids." Until 1840 this passage was considered impossible; but by watching the course of rafts down the river a channel was discovered, and steamboats then attempted it for the first time, under the guidance of the Indian pilot *Teronhiahéré*. Some of the pilots are still Indians, and they exhibit great skill and courage in the performance of their dangerous duties. Yet no one need fear the undertaking, for there has never yet occurred a fatal accident in making this course. The *Cornwall Canal*, 11 miles long, enables vessels to go round the Rapids in ascending the river. *Cornwall* is a thriving town at the foot of the Rapids, opposite which is the large Indian village of *St. Regis.* Just below this place the St. Lawrence, now entirely in Canada, expands into **Lake St. Francis,** which is 25 miles long and about 5 miles wide, and is dotted with islets, especially at the lower end. *Coteau du Lac,* 30 miles below Cornwall, is at the head of the *Coteau Rapids,* which, 9 miles below, take the name of the *Cedars,* and, still farther on, of the *Cascades.* At the foot of the Cascades is *Beauharnois,* at the lower end of a canal 11¼ miles long, around the Rapids. The village is prettily situated on a bay, and is a favorite resort for picnics from Montreal. The expanse of the river from this point to the head of the Lachine Rapids is called **Lake St. Louis,** which is 12 miles long by 5 wide. One of the most noticeable features of this lake is *Nun's Island,* 5 miles below Beauharnois. It was formerly an Indian burying-ground, but is now the property of the Grey Nunnery at Montreal, and in a high state of cultivation. *Lachine* is at the head of the *** Lachine Rapids,** which, though the shortest, are the most turbulent and dangerous on the river. "In the descent of these we are wrought to a feverish degree of excitement, exceeding that produced in the descent of the Long Sault. It is an intense sensation, and, although in reality perfectly safe, is terrible to the faint-hearted, exhilarating to the brave. Opposite Lachine is the quaint Indian village of *Caughnawaga,* where dwell the descendants of the once powerful Iroquois Nation. The immense steel bridge spanning the St. Lawrence at this point is justly considered one of the engineering triumphs of the century. It was built by the Canadian Pacific Railway, is about a mile long, with two channel spans of 408 feet, and lofty enough to allow free passage to the largest steamers. From this bridge a fine view is obtained of the rapids, the villages upon either shore, the loftier structures of Montreal, and the distant mountains. As we reach calm water again, we can fairly distinguish in the growing night the prim form of the Victoria Bridge, and the spires, domes, and towers of Montreal, the commercial metropolis of British North America."

Montreal.

From New York, Montreal is reached by either the Connecticut Valley (Route 30), or most directly by the N. Y. Central R. R. and connections. From Philadelphia by Route 32. From Boston by Route 28 or Route 29. From Portland by Route 26. From Quebec by Grand Trunk R. R., Canadian Pacific R. R., or by steamer.

Hotels.—The leading hotels are the *Windsor* (in Dominion Sq.), the *St. Lawrence Hall* (in St. James St.), the *Balmoral* (in Notre-Dame St., West), and the *Richelieu Hotel* (on Jacques Cartier Sq.).

Modes of Conveyance.—*Street-cars* traverse the city in different directions, and afford easy access to principal points. *Carriages* wait at the stations and steamboat-landings, and at various stands in the city. Their charges are : One-horse carriage for 1 or 2 persons, 25c. a course within the city limits, or 75c. an hour ; for 3 or 4 persons, 40c. a course, $1 an hour. Two-horse carriage, for 1 or 2 persons, 40c. a course, 75c. an hour ; for 3 or 4 persons, 50c. a course, $1.25 an hour. *Stages* run to all the adjacent villages.

Montreal, the largest city and commercial metropolis of British North America, is situated on an island of the same name, at the confluence of the Ottawa and St. Lawrence Rivers, in lat. 45° 31′ N. and lon. 73° 35′ W. It derives its name from Mont Réal, or Mount Royal, which rises 700 ft. above the river, and closes the city in on that side. Including its suburbs, Montreal stretches along the river for 4 miles from S. E. to N. W., and for some distance extends from one to two miles inland. The houses are built of a grayish limestone from adjacent quarries, and with its tall spires and glittering roofs the city presents a picturesque panorama. The quays of Montreal are built of solid limestone, and, with the locks and wharves of the Lachine Canal, they present for about 2 miles a display of continuous masonry. One of the handsomest buildings among the many notable structures in the city is the stately stone station recently completed by the Canadian Pacific Railway Co., and known as the " Windsor St. Station." Situated at one corner of Dominion Sq., it overlooks the site of the ice-palaces and the frolicking-ground during Montreal's winter carnivals. The general offices of the company were moved here upon the completion of the station. The spacious and handsome brick depot erected in 1887 by the Grand Trunk Railway Co., in St. James St., is also an attractive building. The important business streets are St. James, McGill, Notre-Dame, St. Paul, St. Catherine, and Commissioners' Streets. The fashionable promenades are St. James, St. Catherine, and Sherbrooke Sts.

The first visit of a European to Montreal dates from 1535, when Jacques Cartier arrived, who named its mountain. In 1642 arrived the first installment of European settlers, and the original Indian name ("Hochelaga") gave place to the French one of "Ville Marie." This name, in due course of time, was replaced by the present one. By the capitulation of Montreal to Gen. Amherst, in 1760, all Canada was transferred from the French to the British crown. It was captured by the Americans under General Montgomery, in November, 1775, and held until the following summer. In 1779 Montreal contained about 7,000 inhabitants, and, with the annexation of several adjacent villages, it had increased in 1891 to 216,650. The commerce of Montreal is very large, as, though it is 500 miles from the sea, its advantageous position at the head of ship-navigation, and at the foot of the great chain of improved inland waters, has made it the chief shipping-port of the Dominion of Canada.

The **Victoria Square** is a neat public ground at the intersection of McGill and St. James Sts., containing a fountain and a bronze statue of Queen Victoria. Fronting on the square are a number of fine buildings, including the *Albert Buildings*. Of the numerous public buildings in the city one of the handsomest is the * **Bonsecours Market,** a spacious stone edifice in the Doric style, fronting on the river and on St. Paul St. The *Custom-House*, between Commissioners' St. and the river, is built on the site of the fort erected by M. de Maisonneuve in 1642, which, with the cabins clustered around it, formed the nucleus of "Ville Marie." The *Post-Office* is a beautiful cut-stone edifice in St. James St., near the Place d'Armes. The * **Court-House,** in Notre-Dame St., is a large and beautiful building in the Ionic style, 300 by 125 ft., erected at a cost of over $300,000. It contains a law library of 15,000 volumes. Behind it is the *Champ de Mars*, a fine military parade-ground. The * **City Hall** is a spacious and splendid edifice at the head of Jacques Cartier Square ; in it are the offices of the various civic and corporation functionaries. The handsome building of the **Bank of Mont-**

real, in St. James St., is a fine example of the Corinthian style. Fronting on Place d'Armes Square are several of the principal banks and the building of the New York Life Assurance Co. In St. James St., E. of Victoria Square, are the elegant **Molsons Bank,** the *Merchants' Bank,* the *Post-Office,* the principal Fire and Life Insurance Offices, and other notable structures. The huge *Victoria Skating-Rink,* in Drummond St., is used in summer for horticultural shows, concerts, etc. *Mechanics' Institute,* in St. James St.; is a plain structure in the Italian style, with an elaborately decorated lecture-room. The *Windsor Hotel* is one of the finest edifices of the kind in America.

Few American cities equal Montreal in the size and magnificence of its church edifices. The Roman Catholic Parish Church of *Notre-Dame,* fronting on the Place d'Armes, is, next to the Cathedral of Mexico, the largest on the continent, being 255 ft. long and 135 ft. wide, and capable of seating from 10,000 to 12,000 persons. It is of stone, in the Gothic style, and has six towers, one at each corner and one in the middle of each flank. The two on the main front are 220 ft. high, and in one of them is a fine chime of bells, the largest of which (the "Gros Bourdon") weighs 24,900 pounds. The view from the tower, which is in summer open to visitors (25c.), is extensive. Even this huge structure is surpassed in size by the **Cathedral of St. Peter,** now nearly finished, at the cor. of Dorchester and Cathedral Sts., after the plan of St. Peter's at Rome. It is 300 ft. long by 225 ft. wide at the transepts, and is surmounted by 5 domes, of which the largest is 250 ft. high, supported on 4 piers (each 36 ft. thick) and 32 Corinthian columns. The portico is surmounted by colossal statues of the Apostles, and affords entrance to a vestibule 200 ft. long and 30 ft. wide. The interior colonnades support lines of rounded arches, and there are 20 minor chapels. **Christ Church Cathedral** (Episcopal), in St. Catherine St., is the most perfect specimen of English-Gothic architecture in America. It is cruciform, built of rough Montreal stone with Caen-stone facings, and is surmounted by a spire 224 ft. high. *St. James's* (Roman Catholic), in St. Denis St., is a very elegant structure in the pointed Gothic style. *St. Patrick's Church* (Roman Catholic) occupies a commanding position at the W. end of Lagauchetière St. It has seats for 5,000 persons, and its handsome Gothic windows are filled with stained glass. The **Church of the Gesù** (Jesuit), in Bleury St., has the finest interior in the city. The nave (75 ft. high) is bordered by rich composite columns, and both walls and ceilings are beautifully painted and frescoed. The *Notre-Dame de Lourdes* is richly decorated in Romanesque style; the *Bonsecours,* originally built in 1673, rebuilt in 1771, near the great market, is the oldest and most picturesque church in the city. There are also chapels attached to all the nunneries. Besides Christ Church Cathedral, the principal Episcopal churches are *Trinity,* a fine stone edifice in the early English-Gothic style, in St. Denis St.; *St. George's,* in Dominion Square; *St. James the Apostle,* in St. Catherine St.; *St. Martin's,* in Upper St. Urbain St.; and *St. Stephen's,* in College St. **St. Andrew's Church** (Presbyterian), on Beacon Hall Hill, and *St. Paul's* (Presbyterian), in Dorchester St., are beautiful specimens of Gothic architecture. Near

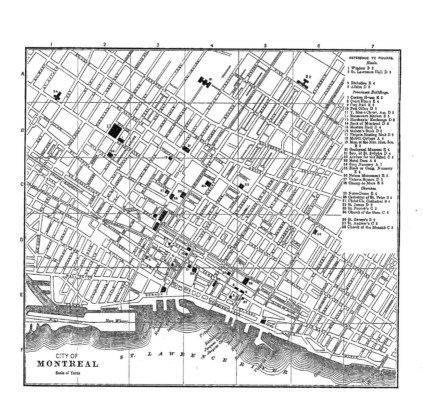

REFERENCE TO FIGURES.

Hotels.

1 Windsor B 2
2 St. Lawrence Hall D 3

3 Richelieu E 4
4 Albion D 3

Prominent Buildings.

7 Custom House E 2
8 Court House E 4
9 City Hall E 3
10 Post Office D 3
11 Y. Men's Ch'n, As. D 3
12 Bonsecours Market E 3
13 Merchants' Exchange D 3
14 Bank of Montreal D 4
15 Masonic Hall D 4
16 Molson's Bank D 3
17 Victoria Skating Rink B 2
18 McGill College A 4
19 Mus. of the Nat. Hist. Soc. B 3

20 Geological Museum E 4
21 Sem. of St. Sulpice D 4
22 Asylum for the Blind C 3
23 Hotel Dieu A 6
24 Grey Nunnery A 1
25 Black or Cong. Nunnery E 4

26 Nelson Monument E 5
27 Victoria Square D 3
28 Champ de Mars E 3

Churches.

29 Notre-Dame E 4
30 Cathedral of St. Peter B 2
31 Christ Ch. Cathedral B 4
32 St. James D 6
33 St. Patrick's C 2
34 Church of the Gesu C 4

36 St. George's D 3
37 St. Andrew's C 3
38 Church of the Messiah C 3

CITY OF
MONTREAL
Scale of Yards

ST. LAWRENCE RIVER

the former is the *Church of the Messiah* (Unitarian), a lofty and spacious building. The *Wesleyan Methodist*, in Dorchester St., is in the English-Gothic style; and to the same denomination belongs the still more stately edifice recently completed at the cor. of Alexander and St. Catherine Sts.

Chief among the educational institutions is the University of **McGill College,** which is beautifully situated at the base of Mount Royal, overlooking the city. The museum of this college is one of the finest in the country. The *Museum of the Natural History Society*, another valuable collection, is at the cor. of Cathcart and University Sts. (admission, 25c.). Some distance W. of McGill College, in Sherbrooke St., is the *Seminary of St. Sulpice*, for the education of Catholic priests. In the same street, at the foot of Laval Ave., is the immense establishment of the Christian Brothers. The Seminary of St. Sulpice (founded in 1657), adjoining the church of Notre-Dame, is 132 ft. long by 29 deep, and is surrounded by spacious gardens and court-yards. The *Asylum for the Blind*, in St. Catherine St. near St. George, has a fine chapel in the Romanesque style with richly frescoed interior. The ***Hôtel Dieu,** founded in 1644, is just outside the city limits (reached by Mance St.). This establishment is under the charge of the Sisters of St. Joseph. The *Montreal General Hospital* and the *Deaf and Dumb Asylums* (Protestant and Catholic) are noble charities. The *Fraser Institute*, a free public library and art-gallery (cor. of Dorchester St. and Union Ave.), contains a fine collection of French works presented by the Orleans and Bonaparte families.

The *** Grey Nunnery,** founded in 1862 for the care of aged and infirm persons and children, is a vast cruciform building in Dorchester St., West. The *Black* or *Congregational Nunnery*, in Notre-Dame St. near the Place d'Armes dates from 1659, and is devoted to the education of young girls. At Hochelaga (at the N. W. end of the Montreal horse-car line) is the great *convent of the Holy Names of Jesus and Mary*. The stranger desirous of visiting any of the nunneries should apply to the Lady Superior for admission.

Not as beautiful as the Canadian Pacific's rival structure, but fully as wonderful from an engineering point of view, is the *** Victoria Bridge,** which spans the St. Lawrence, connecting the city on the island with the mainland to the S. Its length is 9,194 ft., or nearly 2 miles. It rests, in this splendid transit, upon 24 piers and 2 abutments of solid masonry, the central span being 330 ft. long. The massive iron tube through which the railway-track is laid is 22 ft. high and 16 ft. wide. The total cost of the bridge was \$6,300,000. It was formally opened with great pomp and ceremony by the Prince of Wales, during his visit to America in the summer of 1860. The *Water-Works*, a mile or so above the city, are extremely interesting for their own sake, and for the delightful scenery in the vicinity. The former *Government House*, near Jacques Cartier Square, and the *Nelson Monument* near by, are objects of interest, though the monument is in a rather dilapidated condition. The *Mount Royal Cemetery* is 2 miles from the city, on the N. slope of the mountain, a broad avenue gradually ascending to this pleasant spot. The best views of Montreal and its neighborhood are obtained by taking the famous drive "*Around the Mountain*," 9 miles long. The * *Mt. Royal Park* should be taken *en route*. A Botanic Garden is in contemplation. No visitor to Montreal should fail to see the **Lachine Rapids** (see present route), taking the 7.55 A. M. train (from Bonaventure station) to Lachine, getting on the steamer there, and returning through the Rapids to Montreal, arriving at 9.30 A. M.

The Richelieu & Ontario Navigation Co. run daily steamers during the summer months to Quebec and the lower river ports. Distance to Quebec, 180 miles; fare, first class, $3. *Varennes* (15 miles below Montreal) lies between the St. Lawrence and Richelieu Rivers. It is connected with Montreal by a steamboat line, and is coming into notice on account of its mineral springs. **Sorel** (45 miles from Montreal) is situated at the confluence of the St. Lawrence and Richelieu Rivers, and is the first point at which the through steamers for Quebec make a landing. It is a small city, but there is good fishing in the vicinity, and in the autumn excellent snipe-shooting. Five miles below Sorel the river expands into **Lake St. Peter,** which is 25 miles long and 9 wide, and very shallow, except in the main channel, which is crooked and narrow, but which will permit the passage of the largest ships. This lake is noted for its storms, in which the immense lumber-rafts that may be constantly seen drifting down stream are sometimes wrecked. **Three Rivers** (*Du Fresne's Hotel*) is about half-way between Montreal and Quebec, and is the third city in size in the E. section of the Province. It is at the mouth of the St. Maurice River, which runs through a rich lumber-district, and brings to Three Rivers large quantities of logs and manufactured lumber. The city contains 8,334 inhabitants and several fine buildings. The **St. Leon Springs,** which are among the most famous in Canada, are reached by a stage-ride of 26 miles from Three Rivers (fare, $1.50); and the *Falls of the Shawanegan*, 30 miles up the St. Maurice River, may be visited by engaging canoes and guides for the purpose. The Falls have a sheer descent of 150 ft., and in magnitude are second only to Niagara. Below Three Rivers there is nothing worthy of notice until Quebec comes in sight, looming up majestically from the river.

Quebec.

From New York, Quebec is reached by Route 30 (637 miles; fare, $14.50), or by N. Y. Central and connections to Montreal; thence by Grand Trunk or Canadian Pacific Railway. From Boston by Route 27, connecting with Route 30 at White River Junction; or *via* Portland. From Portland by Route 26. The Grand Trunk Railway. has its terminus at Point Levi, and passengers cross to Quebec by ferry. The Quebec & Lake St. John Railway runs to Roberval (190 miles). The Quebec Central Railway runs to Beauce and Sherbrooke. The Intercolonial Railway runs to Halifax, N. S. (678 miles).

.**Hotels.**—The *Florence*, in St. John St.; the *St. Louis Hotel*, in St. Louis St.; *Mountain Hill House*, in Mountain St.; and *Blanchard's*, in the Lower Town. The Canadian Pacific Railway is building a large hotel.

Modes of Conveyance.—*Street-cars* (fare 5c.) traverse the streets along the river in the Lower Town and extend to the suburbs. A second line runs along St. John St. in the Upper Town. *Carriages* or *calèches* may be hired at the livery-stables, and on the cab-stands near the hotels and markets. The *calèche*, a two-wheeled one-horse apparatus, is the usual vehicle, and costs about 75c. an hour. *Ferries* connect the city with South Quebec, New Liverpool, and Point Levi, on the opposite side of the St. Lawrence, and run three times a day to the Isle of Orleans. An elevator runs from Champlain St. to Dufferin Terrace in the summer only.

Quebec, the oldest and after Montreal and Toronto the most important city in British America, is situated on the N. W. bank of the St. Lawrence River, at its confluence with the St. Charles, nearly 300 miles from the Gulf of St. Lawrence. The city is built on the N. extremity of an elevated tongue of land which forms the left bank of the St. Lawrence for several miles. Cape Diamond, so called from the numerous quartz-crystals formerly found there, is the loftiest part of the headland, 333 ft. above the stream, and is crowned with the vast fortifications of the *Citadel*. These occupy about 40 acres, and were once considered so impregnable that they obtained for Quebec the appellation of the "Gibraltar of America." From the citadel a line of wall runs W. toward the cliffs overhanging the valley of the St. Charles, and is thence continued around the brow of the promontory till it connects once more with Cape Diamond near the Governor's Garden. This circuit is nearly 3 miles in extent. The city is divided into the Upper and Lower Town, the ascent from the latter being by a very steep and winding street (Mountain St., or Côte de la Montagne). The Upper Town comprises the walled city with the two suburbs of St. Louis and St. John, between the walls and the Plains of Abraham. The Lower Town, where most of the leading wholesale houses are situated, is built around the base of the promontory. A very large part of the city within the walls, or the Upper Town proper, is taken up with the buildings and grounds of the great religious corporations. Over the remaining irregular surface, not covered by fortifications, are crowded the quaint mediæval streets and dwellings, built generally of stone, two or three stories high, and roofed, like the public buildings, with shining tin. The five original gates in the city-wall were removed some years ago, but three others of a more ornamental character have since been built. *Kent Gate*, named in honor of the Duke of Kent, father of Queen Victoria, is situated in St. Patrick St. *St. Louis Gate* is in St. Louis St., and *St. John's Gate* is in St. John St.

The site of Quebec was visited by Cartier in 1535, and the city was founded by Champlain in 1608. It was taken by the English in 1629, and restored to France by the treaty of 1632. In 1690 the neighboring English colonies made an unsuccessful maritime expedition against it ; and in 1711 the attempt was renewed, with no better success. In 1734 the city had, including its suburbs, 4,603 inhabitants. In 1759, during the Seven Years' War, the English, under General Wolfe, attacked the city and bombarded it. On Sept. 13th took place the first battle of the Plains of Abraham, in which both Wolfe and Montcalm, the French commander, fell, and England gained at one blow an American empire. The French, indeed, recaptured the city the next spring, but at the treaty of peace in 1763 Louis XV ceded the whole of New France to the English. In December, 1775, a small American force, under General Montgomery, attempted its capture, but failed, after losing 700 men and their commander. The population of the city at that time was only 5,000. In 1861 it was 59,990, and in 1871, 59,699, the decrease being attributed to the withdrawal of the British troops forming the garrison. In 1891 the population of Quebec was 63,090, according to the census. Quebec has a large maritime commerce, and is one of the greatest lumber and timber markets on the American Continent. The principal articles of manufacture are saw-mill products, boots and shoes, bakery products, confectionery, furniture, foundry products and machinery, worsted goods, cutlery, ropes and twines, tobaccos, leather, paper, etc.

The point to which the attention of the stranger in Quebec is first directed is * **Dufferin Terrace,** which lies along the edge of the cliff, towering 200 ft. above the river, and overlooking the Lower Town. Part of it occupies the site of the old Château of St. Louis, built by Champlain in 1620, and destroyed by fire in 1834. Dufferin Terrace, which was opened to the public in June, 1879, by the Marquis of Lorne and Princess Louise, is an unequaled promenade over ¼ mile long. The outlook from the Terrace is one of the finest in the world, and is of itself worth a trip to Quebec. The *Esplanade*, near the St. Louis Gate, is another attractive promenade; and the walk along the Ramparts, between the St. Louis Gate and St. John's Gate, affords prospects rivaled by few in America. The view from the **Grand Battery,** near the Laval University, is considered by many to be finer even than that from Dufferin Terrace; and that from the vast balcony of the University building is still more impressive. The *Place d'Armes*, or Parade-Ground, is a pretty little park adorned with a fine fountain, lying between Dufferin Terrace and the **Anglican Cathedral,** which is a plain, gray-stone edifice, surmounted by a tall spire, standing in St. Ann St. on the site to which tradition points as the spot where Champlain erected his first tent. Adjoining the Cathedral is the rectory and the pretty little *Chapel of All Saints*. Des Carrières St., running S. from the Place d'Armes, leads to the *Governor's Garden*, containing an obelisk 65 ft. high to the memory of Wolfe and Montcalm. Des Carrières St. also leads to the inner *glacis* of **The Citadel,** a powerful fortification, covering 40 acres of ground on the summit of Cape Diamond.

Market Square, on which has recently been erected a beautiful bronze fountain, is in the center of the Upper Town, surrounded by more or less striking buildings. On the E. side is the * **Basilica of Quebec** (formerly the Cathedral), a spacious cut-stone building, 216 ft. long and 180 ft. wide, and capable of seating 4,000 persons. The exterior of the edifice is very plain, but the interior is richly decorated, and contains several original paintings of great value by Vandyke, Caracci, Hallé, and others. In this Basilica lie the remains of Champlain, the founder and first governor of the city. Adjoining the Basilica on the N. are the quaint buildings of the *Seminary of Quebec*, founded in 1663 by M. de Laval, first bishop of Quebec. The Seminary Chapel was destroyed by fire in 1887, but has since been rebuilt. All the rare and priceless works of art were burned. The * **Laval University,** founded in 1852, occupies three very imposing buildings. They are of cut stone, 576 ft. long (the main building being 286 ft.), five stories high, and costing $240,000. The chemical laboratory is spacious, fire-proof, and provided with complete apparatus; the geological, mineralogical, and botanical collections are very valuable; the museum of zoölogy contains upward of 1,300 different birds and 7,000 insects; and the museum of the medical department is especially complete. The *Library* numbers more than 95,000 volumes, and the * *Picture*

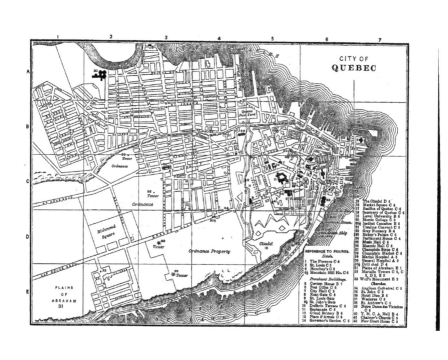

CITY OF
QUEBEC

REFERENCE TO FIGURES.

Hotels.
1 The Florence C 4
2 St. Louis C 5
4 Henchey's C 5
4a Mountain Hill Ho. C 6

Prominent Buildings.
5 Custom House B 7
6 Post Office C 6
7 City Hall C 5
8 Kent Gate C 5
9 St. John's Gate
9a St. John's Gate C 5
10 Dufferin Terrace C 6
11 Esplanade C 5
12 Grand Battery B 6
13 Place d'Armes C 6
14 Governor's Garden C 6

15 The Citadel D 5
16 Market Square C 6
17 Basilica of Quebec C 4
18 Seminary of Quebec C 6
19 Laval University B 6
20 Morrin College C 5
20a Institut Canadien B 5
21 Ursuline Convent C 5
22 Grey Nunnery B 4
24 Parliament House C 4
25 Bishop's Palace C 6
26 Music Hall C 5
28 Masonic Hall C 5
27 Champlain Stairs C 6
28 Champlain Market C 6
29 Marine Hospital A 2
30 General Hospital A 2
30a Drill shed D 4
31 Plains of Abraham E 1
32 Martello Towers C 5, C
    5, D 5, E 2
33 Wolf's Monument E 2

Churches.
34 Anglican Cathedral C 5
35 St. John C 5
36 Hotel Dieu B 6
37 Wesleyan C 5
38 St. Andrew's C 5
39 Notre Dame des Victoires
    C 6
40 Y. M. C. A. Hall B 5
41 Chalmer's Church C 5
42 New Court House C 5

PLAINS
OF
ABRAHAM
31

Richmond
Square

Ordnance Property

Citadel
15

Tower
Ordnance

Tower
Ordnance

Tower

Tower

Gallery (always open to the public) is one of the richest in Canada, and contains many important works of art. On the W. side of Market Square is the site of the old Jesuits' College buildings. *Morrin College* occupies the old stone prison at the cor. of St. Ann and Stanislas Sts. In this building are the library (14,500 volumes) and museum of the *Quebec Literary and Historical Society,* with its rich collection of MSS. relating to the early history of the country. The *High School* on the Cape has 200 students, an excellent library, and a small natural history collection.

In Donnaconna St., off Garden, is the * **Ursuline Convent,** a series of buildings surrounded by beautiful grounds. It was founded in 1639, and now has 40 nuns, who are devoted to teaching girls, and also to painting, needle-work, etc. The parlor and chapel are open to visitors, and in the latter are rare carvings on ivory and some fine paintings by Vandyke, Champagne, and others. The remains of the Marquis de Montcalm are buried here in an excavation made by the bursting of a shell within the precincts of the convent. His skull is preserved in the parlor of the chaplain. The *Grey Nunnery* is a spacious building on the *glacis* W. of the ramparts, and contains about 75 Sisters. The * *Chapel* adjoining the nunnery is a lofty and ornate Gothic edifice, with a rich interior. Near by (in St. John St. near St. Clair) is the Roman Catholic church of *St. John,* which replaces the large structure destroyed by fire some years ago. The * **Hôtel Dieu,** with its convent and hospital, stands in Palace St., near the rampart. It was founded in 1639 by the Duchess d'Aiguillon, and in 1875 comprised 45 Sisters of the Sacred Blood of Dieppe, who minister gratuitously to 10,000 patients yearly. In the Convent Chapel are some valuable paintings. The *Black Nunnery* is in the suburbs of St. Roch's. Application to the Lady Superiors will usually secure admittance to the nunneries.

The **Post-Office** is a handsome stone edifice at the cor. of Buade and Du Fort Sts. Near by is the *Cardinal's Palace,* a stately and handsome structure. Other noteworthy buildings in the Upper Town are the *City Hall,* the *Quebec Academy of Music,* the *Masonic Hall* and the *Garrison Clab,* in St. Louis St. On Grande Allée are situated the *Skating-Rink,* the * **Parliament and Departmental Buildings,** which were begun in 1878, and the *Armory* and *Exhibition Building,* a beautiful structure. The *Court-House,* finished in 1888, is on the cor. of St. Louis St. and Place d'Armes. Among noteworthy churches are the *Methodist Church* in St. Stanislas St., a fine specimen of the flamboyant Gothic style; *St. Matthew's* (Episcopal), in St. John St.; *St. Andrew's* (Presbyterian), at the intersection of St. Ann and St. Stanislas Sts., a spacious stone structure in the Gothic style; *Chalmers* (Presbyterian), in St. Ursule St.; *St. Patrick's* (Roman Catholic), in McMahon St., and *St. Sauveur* and *St. Roch's,* in the suburbs. The fine and spacious *Y. M. C. A. Hall* is in St. John St., just without St. John's Gate. There are a good library, lecture-room, and reading-rooms, etc., opposite to which is the Montcalm Market Square and Hall. The *Institut Canadien* is in Fabrique St., and in Ann St. is situated the *Woman's Christian Association.*

Just N. of Dufferin Terrace is the head of Mountain Hill. St., which descends to the Lower Town. To the right is a picturesque stairway, called the * *Champlain Steps*, or Côte de la Montagne, which leads down to the church of *Notre-Dame des Victoires*, erected in 1690 on the site of Champlain's residence. S. of the church is the *Champlain Market*, a spacious structure on the river-bank, near the landing of the river steamers. *St. Peter St.*, running N. between the cliff and the river, is the main business thoroughfare of this quarter, and contains the great commercial establishments, banking-houses, wholesale stores, etc. *St. Paul St.* stretches W. on the narrow strand between the cliff and the St. Charles, amid breweries and manufactories, till it meets, near the mouth of the St. Charles, *St. Joseph St.*, the main artery of the large suburb of St. Roch's. On the banks of the St. Charles are the principal shipyards, once so thriving; and the numerous coves of the St. Lawrence, from Champlain St. to Cape Rouge, are filled with acres of vast lumberrafts. On the opposite shore of the St. Lawrence are the populous towns of South Quebec, New Liverpool, and Point Levi, which present a scene of activity scarcely surpassed by the city itself. The * **Custom-House** is reached from St. Peter St. by Leadenhall St., and occupies the very apex of the point made by the confluence of the St. Lawrence and the St. Charles Rivers. It is an imposing Doric edifice, with a dome, and a façade of noble columns, approached by a long flight of steps. The **Marine and Emigrants' Hospital,** built on the model of the Temple of the Muses on the banks of the Ilissus, is near the St. Charles River, and ¼ mile farther up the river is the **General Hospital.** This institution was founded in 1693, and is under the charge of the nuns of St. Augustine. Overlooking St. Roch's suburbs is the *Jeffery Hale Hospital.* The *Finlay Asylum* is on Ste. Foye road.

The suburbs of St. Louis and St. John stretch S. and W. along the plateau of the Upper Town, and are constantly encroaching on the historic Plains of Abraham. They contain many handsome private residences, and several large conventual establishments and churches. The best approach to the *Plains of Abraham* is by Grande Allée, which commences at the St. Louis Gate and the *Martello Towers,* four circular stone structures erected in 1807–'12 to defend the approaches to the city. On the spot where Wolfe fell in the memorable battle of September 13, 1759, stands **Wolfe's Monument,** a modest column appropriately inscribed. A short distance to the left is the path by which his army scaled the cliffs on the night before the battle; it is somewhat shorn of its rugged character, but is still precipitous and forbidding. On the Plains, near the Ste. Foye road, stands the *Monument* commemorating the victory won by the Chevalier de Lévis over General Murray in 1760. It is a handsome iron column, surmounted by a bronze statue of Bellona (presented by Prince Napoleon), and was erected in 1854. About 3 miles out on the St. Louis road are **Mount Hermon Cemetery,** 32 acres in extent, beautifully laid out on irregular ground, sloping down to the precipices which overhang the St. Lawrence, and *St. Patrick's* (R. C.) *Cemetery.*

Within excursion distance of Quebec are several points of interest

Citadel at Quebec.

which the tourist should not fail to visit. The *Isle of Orleans* (reached by ferry) is a beautiful spot, and the drive around it a succession of noble views. There are also pleasant drives to *Spencer Wood*, the beautiful residence of the Lieutenant-Governor of the Province, and to *Château Bigot*, an antique ruin, standing in solitary loneliness at the foot of the Charlesbourg Mountain. *Lorette*, an ancient village of the Huron Indians, is reached by a 9-mile drive along the banks of the Little River road. The falls near the village are very picturesque, and Lake St. Charles, a famous fishing-place, is only a few miles off. The *** Falls of Montmorenci,** 8 miles below Quebec, are 250 ft. high and 50 ft. wide, and are wonderfully beautiful. The Falls may now be visited by rail, the Montmorency and Charlevoix Railway running twice a day èach way. A short distance above the Falls is the "Mansion House," in which the Duke of Kent passed the summer of 1791; and about 1 mile above are the curious *Natural Steps*, a succession of ledges cut by the river in the limestone rock, each about 1 ft. high, and as regularly arranged as if they were the work of human hands. The *** Falls of Chaudière** (10 miles) are reached *via* Point Levi, on the opposite side of the St. Lawrence. The rapid river plunges, in a sheet 350 ft. wide, over a precipice 150 ft. high, presenting very much the look of boiling water, whence its name, *Chaudière*, or caldron.

The regular tour of the St. Lawrence usually ends at Quebec, but the Lower River is well worth visiting by all lovers of fine scenery. The excursion may be made by the steamers of the Quebec Steamship Co., one of which leaves Quebec on alternate Tuesdays at 2 P. M., and runs to Pictou, Nova Scotia, stopping at intermediate ports. The distance to Gaspé, at the mouth of the river, is 443 miles; to Charlottetown, 784 miles; to Pictou, 829 miles. The steamers of this line make no stoppage between Quebec and Father Point (142 miles), but the intermediate points may be reached by railway or by local steamboat lines.

The steamers of the Richelieu & Ontario Navigation Co. leave Quebec at 7.30 A. M. on Tuesdays, Wednesdays, Fridays, and Saturdays for St. Paul's Bay, Les Eboulements, Murray Bay, Rivière du Loup (Cacouna), Tadousac, Ha ! Ha ! Bay, and Chicoutimi. At noon on Saturdays for Murray Bay and Rivière du Loup.

The Quarantine for Quebec is at *Grosse Isle*, 30 miles down; and 6 miles below is a group of islets, of which the chief, united by a belt of lowland, are *Crane Island* and *Goose Island*. They are the resort and breeding-place of numerous ducks, geese, and teal, to say nothing of smaller game. **Château Richer** is a thriving little village on the N. shore, much resorted to by sportsmen. Ducks, partridges, and snipe abound in the vicinity, and on the streams near by good trout-fishing may be had. A few miles below Château Richer is **Ste. Anne de Beaupré,** noted for its *Church of Ste. Anne*, in which miraculous cures are said to be effected by the relics of the saint, which are exhibited at morning mass. This church contains a variety of paintings, and is well worth a visit. The celebrated **Falls of Ste. Anne** are on the Ste. Anne River, 3 or 4 miles from the village. The lower fall is 130

18

ft. high, and below it the water rushes down through a rugged, somber, and picturesque ravine. The *Valley of St. Féréol*, the *Seven Falls*, and *Mt. Ste. Anne*, are other objects of interest in the neighborhood. Eight miles below Ste. Anne is **Cape Tourment,** a bold promontory, from the summit of which there is a superb view. A little beyond are the frowning peaks of *Cape Rouge* and *Cape Gribaune*. From Goose Island to the Saguenay River the St. Lawrence is about 20 miles wide. The water is salt, but clear and deep, and the spring-tides rise and fall 18 ft. The black seal, the white porpoise, and the black whale are sometimes seen. *St. Paul's Bay*, 55 miles from Quebec, is a popular resort, and claims to offer more attractions to the tourist, the poet, or the naturalist, than any other parish in the Province. It is surrounded by grand scenery. **Murray Bay** (82 miles below Quebec) is a popular watering-place, surrounded by wild scenery, and noted for the fine fishing in Murray River, and the Gravel and Petit Lakes. There are several hotels and large boarding-houses here, besides summer cottages, and a daily steamer from Quebec renders it easy of access. **Rivière du Loup** is a favorite summer resort on the S. shore of the river, 112 miles from Quebec. It is situated at the mouth of the Du Loup River, and commands a fine prospect of the St. Lawrence, which at this point is 20 miles wide. About a mile from the village is a * waterfall, where the Du Loup, after rushing for a while over a rocky bed, dashes in a sheet of foam over a precipice 80 ft. high. *Lake Temiscouata* is reached from Rivière du Loup by the Grand Portage Road, a distance of 36 miles. Only a few cabins dot the shores on this lovely lake, and it is just the place for the seeker after solitude and trout. **Cacouna**, 6 miles below Rivière du Loup, is the favorite summer resort of the Canadians, and is a very attractive village, combining picturesque scenery, good hotels, fine hunting and fishing, and admirable sea-bathing, for at this point the water of the St. Lawrence is almost as salt as that of the ocean. The *St. Lawrence Hall* is a large, first-class hotel, overlooking the river (with a capacity of 600 guests). The *Mansion House* is also very comfortable, and there are several large summer boarding-houses. The air of Cacouna is pure and bracing, and remarkably cool in summer; and there is much pleasing scenery in the vicinity of the village. Nearly opposite Cacouna is the mouth of the **Saguenay River** (see Route 61), which is one of the most striking points on the entire river. Just below (148 miles from Quebec) is **Trois Pistoles,** at the mouth of the river of the same name, famed for its fish. There are two hotels here, and several summer cottages, and the scenery in the vicinity is very pleasing. Thirty miles below Trois Pistoles are the island promontory and harbor of *Le Bic* (the Eagle's Beak), an ancient landing-place, still honored. Near it is *L'Islet au Massacre*, remembered as the scene of the bloody massacre of all but 5 out of 200 Micmac Indians by their Iroquois foes. **Rimouski** (180 miles from Quebec) has an extensive government wharf, and contains a splendid Cathedral, a number of handsome houses, and two good hotels. This is a place at which the tourist should stop, for the scenery of the valley of the Rimouski is extremely beautiful, and the trout-fishing unrivaled. Twenty miles below Rimouski is **Metis,** the site of the largest and longest of the government wharves. It is noted as a whale-fishing station. Some 50 miles farther down, we reach the *Point de Monts*, on the N. coast, and *Cape Chatte*, a few miles above Ste. Anne, the most northerly town on the S. coast of the St. Lawrence. Here are the last approaches of the two shores. Beyond the Point de Monts the N. shore makes a sharp turn to the northward, and in that direction we speedily have a sea-horizon. Rounding now the great shoulder of the Province of Quebec, we come, on the E. side, to **Cape Rosier,** passing meanwhile the S. W. half of the desert *Anticosti Island.* Here ends our present tour. Those who pursue the journey to Pictou and Halifax soon enter the region described in Route 65. (Also see for detailed description THE CANADIAN GUIDE-BOOK, Part I, Eastern Canada, published by D. Appleton & Co.)

61. The Saguenay River.

Steamers leave Quebec at 7.30 A. M. on Tuesdays, Wednesdays, Fridays, and Saturdays for Chicoutimi, at the head of navigation on the Saguenay; and on Wednesdays, Thursdays, and Saturdays for Ha ! Ha ! Bay. The trip from Quebec to the mouth of the Saguenay includes some of the most impressive portions of the St. Lawrence scenery, and is described in Route 59. It should be mentioned that the steamers usually make the trip *up* the Saguenay during the night, so that the best views of the river are obtained on the return-voyage the next day. Distances : Quebec to Tadousac, 134 miles ; Tadousac to Ste. Marguerite River, 15 miles ; to St. Louis Isle, 19 ; to Little Saguenay River, 27 ; to St. John's Bay, 32 ; to Eternity Bay, 41 ; to Trinity Bay, 48 ; to Cape Rouge, 56 ; to Cape East, 63 ; to Cape West, 65 ; to St. Alphonse, 72 ; to Chicoutimi, 100.

THE Saguenay is the largest tributary of the St. Lawrence, and unquestionably one of the most remarkable rivers in the world. Its headwater is Lake St. John, 40 miles long and nearly as wide, which, although 11 large rivers fall into it, has no other outlet than the Saguenay. The original name of this river was Chicoutimi, an Indian word signifying deep water ; and its present one is said to be a corruption of Saint Jean Nez. The course of the Saguenay—between lofty and precipitous cliffs, and in its upper part amid rushing cataracts—is about 140 miles from Lake St. John to the St. Lawrence, which it enters 120 miles below Quebec. Large vessels ascend as far as Chicoutimi, 98 miles from the mouth of the river. The Saguenay is a nearly straight river, with grand precipices on either side for almost its entire length, and a peculiarly stern, somber, savage, and impressive aspect. Says Bayard Taylor: "The Saguenay is not properly a river. It is a tremendous chasm, like that of the Jordan Valley and the Dead Sea, cleft for 60 miles through the heart of a mountain wilderness. . . . Everything is hard, naked, stern, silent. Dark-gray cliffs of granite gneiss rise from the pitch-black water ; firs of gloomy green are rooted in their crevices and fringe their summits ; loftier ranges of a dull indigo hue show themselves in the background ; and over all bends a pale, cold, northern sky."

Tadousac is a small village situated a short distance above the mouth of the Saguenay, 135 miles from Quebec. Apart from its attractions as a watering-place, it is interesting as the spot on which stood the first stone-and-mortar building ever erected by Europeans on the

Continent of America. The scenery here is wild and romantic in the extreme, and the adjacent waters abound in excellent salmon and trout. The *Tadousac Hotel* is a large, admirably kept, and comfortable house, and there are several summer cottages. Near the hotel are the ancient buildings of the Hudson Bay Co., and just E. is the quaint old Chapel of the Jesuit Mission, erected in 1746. The steamer stops long enough at Tadousac to afford the passengers ample opportunity for seeing the sights. Just above Tadousac is the pretty little cove of *L'Anse à l'Eau*, which is a fishing-station, and here begins one of the most somber and desolate stretches of the river. The banks on either hand consist of immense perpendicular cliffs, which are evidently prolonged far below the surface of the water. Now and then a massive promontory encroaches upon the channel, and at rarer intervals the river widens out into what are called bays, but would scarcely be called coves on any other stream. About 15 miles above Tadousac, after passing *Point Crêpe*, the *Ste. Marguerite River*, famous for its salmon-fisheries, comes in on the right; and 2 miles beyond the steamer skirts the shore of the desolate *St. Louis Isle*, in whose deep waters salmon-trout abound. About 30 miles above Tadousac is **St. John's Bay,** which is 3 miles long and 2 wide, and on the shore of which is one of the few small settlements that the Saguenay can boast. Nine miles above is * **Eternity Bay,** the most striking feature of the river-scenery. It is a narrow cove, flanked at the entrance by two precipices, each rising almost perpendicularly 1,600 ft. above the water. The steepest is * **Cape Trinity,** so called because of the three distinct peaks on its N. summit; and that on the other side is *Cape Eternity*. Speaking of these awful cliffs, Bayard Taylor says : " I doubt whether a sublimer picture of the wilderness is to be found on this continent." Farther on, * **Statue Point,** a grand bowlder, 1,000 ft. high, is noticeable for a cave half way up its face, utterly inaccessible from above or below, having an orifice probably 40 ft. in diameter. Still farther above is *Le Tableau*, a lofty plateau of dark-colored granite, 600 ft. high and 300 wide, smooth as though cut by the hand of art, and terminating suddenly in a single perpendicular rock 900 ft. high. A few miles beyond is the entrance to **Ha ! Ha ! Bay,** which runs 7 miles S. W. from the Saguenay, and is a mile wide. Ha ! Ha ! Bay was so named because of the delightful contrast which the first French voyagers there beheld after the awful solitude of the lower river. Its upper end is surrounded by undulating meadow-lands, and on its shores are the two small villages of *St. Alphonse* and *St. Alexis*. **Chicoutimi** (about 20 miles above Ha ! Ha ! Bay) is the head of navigation on the river, and is a place of considerable trade. It has a good hotel, a cathedral and convent, and a stone college of ambitious pretensions. The Chicoutimi River swarms with fish, and, just before it enters the Saguenay, plunges over a granite ledge 50 ft. high. Nine miles above Chicoutimi begin the *Rapids of the Saguenay*, said to be little inferior in grandeur to those of the Niagara, and a great deal longer. *Lake St. John* is 60 miles W. of Chicoutimi, and is reached by a good road, and from Quebec by the Quebec & Lake St. John's Railway direct.

Parliament Buildings, Ottawa.

62. Ottawa.

From Montreal, Ottawa is reached by steamer up the Ottawa River (101 miles), or by Grand Trunk Railway to Coteau, and thence *via* the Canada Atlantic Railway, or by the Canadian Pacific Railway. From Toronto by Grand Trunk Ry. and Canadian Pacific Railway *via* Brockville. From New York by Route 40 to Utica, and thence *via* Rome, Watertown & Ogdensburg R. R. to Ogdensburg, which is opposite Prescott, and thence by the Canadian Pacific Railway (438 miles), or by the Adirondack & St. Lawrence R. R. to Coteau. From Boston *via* Boston & Maine R. R. to Portland, and then *via* Grand Trunk Railway or *via* the Central Vermont R. R.

Hotels, etc.—The *Grand Union, Russell*, and *Windsor House* are in the central part of the city. Rates are from $2 to $4 a day. *Horse* and *electric cars* connect the city with the towns across the river (fare, 5c.).

OTTAWA, the capital of the Dominion of Canada, is situated on the S. bank of the Ottawa River, at the mouth of the Rideau. It is divided into an Upper and Lower Town by the Rideau Canal, which passes through it and connects it with Kingston, on Lake Ontario. The canal is crossed within the city limits by three bridges, one of stone, one of stone and iron, and one of wood, and has 8 massive locks. Bridges also connect Ottawa with the suburban towns of Hull and New Edinburgh, on the opposite sides of the Ottawa and Rideau Rivers respectively. The latter suburb is now a part of the city proper. The streets are wide and regular, the principal ones being *Sparks, Wellington, Elgin, Rideau*, and *Sussex*. The former is the popular promenade, and contains the leading retail-shops, etc. Ottawa was originally called Bytown, in honor of Colonel By, of the Royal Engineers, by whom it was laid out in 1827. It was incorporated as a city under its present name in 1854, and was selected by Queen Victoria as the seat of the Canadian Government in 1858. It has grown rapidly since the latter date, and had a population in 1891 of 44,154. The city is the entrepot of the lumber-trade of the Ottawa and its tributaries.

Ottawa is substantially built, containing many stone edifices, but the * **Government Buildings** are the chief feature of the city. They form three sides of a quadrangle on an eminence formerly known as Barrack Hill, 150 ft. above the river, and cost $4,000,000. The N. side of the quadrangle is formed by the *Parliament House*, which is 472 ft. long and 572 ft. deep from the front of the main tower to the rear of the Library, the body of the building being 40 ft. high and the central tower 180 ft. The Departmental Buildings run S. from this, forming the E. and W. sides of the quadrangle; the Eastern block is 318 ft. long by 253 ft. deep, and the Western 211 ft. long by 277 ft. deep. An additional block was subsequently completed on the S. side. They contain the Government bureaus, the Model-Room of the *Patent-Office* being in the South block. The buildings are in the Italian-Gothic style. The arches of the doors and windows are of red Potsdam sandstone, the external ornamental work of Ohio sandstone, and the columns and arches of the legislative chambers of marble. The roofs are covered with green and purple slates, and the pinnacles are ornamented with iron trellis-work. The legislative chambers are capacious and richly furnished, and have stained-glass windows. The *Senate Hall* is reached to the right from the main entrance (which is under the central tower). The vice-regal canopy and throne are at one end of this hall, and at the other are a marble statue and a portrait of Queen Victoria, together with full-length portraits of George III and Queen Charlotte by Sir Joshua Reynolds. The *Chamber of Commons* is reached to the left from the entrance, and contains some beautiful marble columns and arches. The *Library* is a handsome polygonal structure on the N. front of the Parliament House, containing at present over 100,000 volumes. The quadrangle is neatly laid out and planted with trees, and has a massive stone wall along its front. The *Geological Museum* is in Sussex St. **Rideau Hall,** the official residence of the Governor-General, is in New Edinburgh, across the Rideau River.

After the Government Buildings, the most important edifice in the city is the Roman Catholic **Cathedral of Notre-Dame,** which is a spacious stone structure, with double spires 200 ft. high. The interior is imposing, and contains a painting ("The Flight into Egypt") which is attributed to Murillo. Other handsome church edifices are *St. Andrew's* (Presbyterian) and *St. Patrick's* (Roman Catholic). The *Ottawa University* (Roman Catholic) has a large building in Wilbrod St., and the *Ladies' College* (Protestant), a very handsome one in Albert St. The Ontario Government has built commodious *Normal* and *Model Schools* on the S. of Cartier Square. The **Grey Nunnery** is an imposing stone structure at the cor. of Water and Sussex Sts. The Grey Nuns have a large school on Rideau St. The Nuns of the Congregation de Notre-Dame have a large boarding and day school just W. of Cartier Square. There are in the city four convents, two hospitals, three orphan asylums, and a Magdalen asylum. The 8 massive locks of the *Rideau Canal*, within the corporation limits, are worth a visit.

The scenery in the vicinity of Ottawa is picturesque and grand. At the W. extremity are the **Chaudière Falls,** where the Ottawa River plunges over a ragged ledge 40 ft. high and 200 ft. wide. In the great *Chaudière* (or caldron) the sounding-line has not found bottom at 300 ft. Immediately below the falls is a suspension-bridge, from which a superb view is obtained. One mile above the city are the *Little Chaudière Falls*, 13 ft. high, and 2 miles above are the rapids known as *St. Remoux*. The *Des Chênes Rapids*, 8 miles above Ottawa, have a fall of 9 ft. The **Rideau Falls,** two in number, are N. E. of the city on the Rideau River, and are very attractive, though eclipsed by the grandeur of the Chaudière.

The **Ottawa River,** the chief tributary of the St. Lawrence, and the largest stream, with the exception of the Saskatchewan and the Mackenzie Rivers, wholly within the Dominion, is navigable both above and below Ottawa. A morning boat runs down the river to Montreal, making the distance in about 10 hours. At *Grenville*, on this route, the traveler takes the cars around the Long Sault and Carillon Rapids to *Carillon* (12 miles), and at *Lachine* the famous **Lachine Rapids* are run by the steamer. Above Ottawa the river was formerly navigated for 188 miles by steamers, but the portages were numerous and the route by no means continuous, and has been discontinued. The Canadian Pacific Railway now absorbs the river traffic, as it passes through the valley of the Ottawa in its route through to the Great Northwest of Canada.

63. Manitoba.

How to reach.—(1) By Canadian Pacific Railway and its connections from Quebec, Montreal, Ottawa, or Toronto. (2) Or, in summer, by Lake steamers from Buffalo, Detroit, Sarnia, or Southampton, to Duluth, thence by Northern Pacific, and St. Paul, Minneapolis & Manitoba, and Canadian Pacific Railways to Winnipeg ; or, by steamer to Port Arthur, Lake Superior (Route 105), and thence by Canadian Pacific Railway ; or (3), by Route 87 or 88 to St. Paul, and thence by the Northern Pacific & Manitoba, and St. Paul, Minneapolis & Manitoba R. Rs., either through the " Park region," passing Barnesville and Fargo, to Gretna, where connection is made with the Canadian Pacific Railway to Winnipeg ; or, "By the shores of Minnetonka," through Morris, Barnesville, and Crookston, to St. Vincent, Emerson, and Pembina, whence the Canadian Pacific and Northern Pacific Railways run to Winnipeg. (For detailed description see THE CANADIAN GUIDE-BOOK, Part II, Western Canada, published by D. Appleton & Co.

MANITOBA, a province of the Dominion of Canada, lies just N. of Minnesota and Dakota, and is in the form of a parallelogram, 280 miles long E. and W. by 268 miles in breadth ; area, 73,956 square miles ; population 154,506, of whom many are of Indian origin. Since 1870 a considerable immigration from the Eastern Provinces and from Europe has set in, and the whites are rapidly becoming the dominant element in the population. The Indians of Manitoba are all settled on reservations and have taken to agricultural pursuits.

The general surface of Manitoba is a level prairie, about 800 ft. above the sea. It is broken by the Big Ridge, Riding, and Pembina Mountains, ancient beaches of that vast lake which is supposed at one time to have covered this entire region. The important lakes are Winnipeg, Winnipegosis, and Manitoba, from the latter of which the province derives its name. *Lake Winnipeg* is of irregular shape, being about 260 miles in length and from 6 to 60 miles wide. It is 720 ft. above the sea, contains many islands, and does not exceed 12 fathoms in depth. Ice forms frequently to a thickness of 5 ft., and does not leave the upper part of the lake before the 10th of June. The name Winnipeg in Cree signifies "dirty water." *Lake Manitoba* lies about 60 miles W. of Lake Winnipeg, into which it discharges through the Little Saskatchewan or Dauphin River, and is 120 miles long and 25 wide at the widest part. The name signifies "supernatural strait," the Indians attributing the peculiar agitation of the water in a portion of the lake to the presence of a spirit. The lakes abound in fish. The principal stream in Manitoba is the Red River of the North, which, rising in Minnesota, flows for 120 miles through the province and empties into Lake Winnipeg. Its chief affluent, the Assiniboin, joins it about 45 miles above Lake Winnipeg.

The climate is healthy, but exhibits great extremes of temperature, the thermometer falling in winter to 40° below zero, and in summer rising as high as 100°. Owing to the dryness of the atmosphere, however, the cold is not severely felt, and the snow is seldom very deep. The rainfall in summer is ample for agricultural purposes, and vegetation comes rapidly to maturity. Winter sets in early in November, and lasts until the middle of April. Frosts are liable to occur until the end of May, and cold nights begin toward the end of August. The soil is very fertile.

To the sportsman, Manitoba, being a comparatively virgin field, offers unrivaled attractions. The rivers and lakes abound in white-fish, sturgeon, trout, cat-fish, pike, perch, and gold-eyes. Ducks, geese, cranes, swans, snipe, prairie-hens, and other birds swarm in countless numbers; and among the wild animals are elks, black bears, rabbits, squirrels, and badgers. The great buffalo-ranges, formerly visited by the Indian hunters, lie to the W. and S. W. of the province, but the buffaloes are almost extinct.

The capital of Manitoba is **Winnipeg** (hotels: *Clarendon, Manitoba, Queen's*), at the confluence of the Red River of the North with the Assiniboin, 45 miles S. of Lake Winnipeg. It covers an area of 20 square miles, is regularly laid out, and contains 25,642 inhabitants. The chief public buildings are the *Governor's Residence, Parliament Buildings, St. John's College, Manitoba College, General Hospital*, and several handsome churches; the *Court-House*, the *City Hall*, the *Post-Office*, the *Custom-House*, and the *Bank of Montreal, Imperial Bank, Commercial Bank of Manitoba, Merchants' Bank, Ontario Bank, Hudson Bay Company's Office*, and many warehouses, are large and handsome structures, mostly of white brick manufactured in the vicinity. Winnipeg is the headquarters of the Dominion bureaus relating to the Northwest Territories, except that of the Indian Department, which is in Regina, and in America of the Hudson Bay Co. It has broad and well laid-out streets, street-railroads, a fire-brigade, and gas and electric lights. Opposite, on the E. bank of the river, is *St. Boniface*. The trade of Winnipeg is important, and consists in supplying the retail trades of the province and the territories beyond, besides the retail trade of the city and the country around. The exports consist chiefly of wheat, oats, barley, flax, potatoes, flour, fish, cheese, butter, furs, and cattle for the European market. The principal towns in Manitoba besides Winnipeg are Brandon, Portage, La Prairie, Selkirk, Emerson, and Moosmin. Besides these there are many smaller towns and villages; but the province is rapidly filling up by immigration, and each year sees marked changes.

"Fort Garry," the site of the old fort, is now within the city limits, and nothing remains of it except the gateway, which will be preserved by the city to mark the spot of some important events in its history. (See APPLETONS' CANADIAN GUIDE-BOOK, Part II, Western Canada.)

Manitoba forms part of the territory granted in 1670 by Charles II. to the Hudson Bay Co., which in 1811 sold a tract, including what is now the province, to Thomas Douglas, Earl of Selkirk. Under his auspices a colony was established, which was sometimes called the Selkirk Settlement, but more commonly the Red River Settlement. In 1835 the Hudson Bay Co. bought back this tract, and in 1870 Manitoba became a province of Canada, upon the annexation of the Hudson Bay territory to the Dominion. A previous attempt of the Dominion authorities to take possession of the country led to organized resistance on the part of the French half-breeds under the lead of Louis Riel, who formed a provisional government, adopted a bill of rights, and held possession of the province from about Oct. 20, 1869, to Aug. 24, 1870, when a force under Col. (now Lord) Wolseley entered Winnipeg and reinstated the regular authorities, Riel having previously vacated the place.

64. Toronto.

From Montreal, Toronto is reached by steamer on the St. Lawrence River and Lake Ontario, or by Grand Trunk Railway (333 miles), or by the Canadian Pacific Railway short direct line, or *via* Ottawa. (Montreal to Ottawa, 120 miles, thence to Toronto, 254 miles.) From New York by Route 40 or Route 42 to Lewiston, and thence by steamer on Lake Ontario ; or *via* Grand Trunk Railway, passing through Hamilton. From Boston by Route 26 or 27 to Montreal, and thence as above.

Hotels, etc.—The *Queen's Hotel*, in Front St.; the *Rossin House*, cor. King and York Sts. ; *Walker House*, Front and York Sts.; and *Revere House*, in King St. The range of prices at these hotels is from $1.50 to $4 per day. *Street-cars* (fare, 5c.) render all parts of the city easily accessible.

TORONTO, the capital of the Province of Ontario, is situated on a beautiful circular bay on the N. W. shore of Lake Ontario, between the Don and Humber Rivers. The site of the city is low, but rises gently from the water's edge. The streets are regular and in general well paved, crossing each other at right angles. *King, Yonge,* and *Queen Sts.* are the leading thoroughfares, and contain the principal retail shops. The greater part of the wholesale trade is centered in *Front* and *Wellington Sts.* Yonge St. extends back to Lake Simcoe. Many of the houses and business structures are built of light-colored brick, of a soft, pleasing tint. The growth of Toronto has been more rapid than that of any other Canadian city. It was founded in 1794, by Governor Simcoe, on the site of the old French fort called Fort Toronto or Rouillée, who gave it the name of York, changed, when it was incorporated as a city, in 1834, to Toronto—meaning, in the Indian tongue, "The place of meeting." In 1813 it was twice captured by the Americans, who destroyed the fortifications and burned the public buildings. In 1817 the population was only 1,200 ; in 1852 it was 30,763 ; in 1861, 44,821 ; and in 1891 it was 181,220. The commerce of the city is very extensive. Its manufactories include iron and other foundries, flour-mills, distilleries, breweries, paper-hangings, furniture, agricultural machinery, pianos and organs, etc.

The finest buildings in the city, and among the finest of the kind in America, were those of the * **University of Toronto,** standing in a large park, and approached by College Ave., which is ½ mile long and lined with double rows of noble trees. The buildings form three sides of a large quadrangle. They are of gray rubble-stone, trimmed with Ohio and Caen stone, and are admirable specimens of pure Norman architecture. The University library numbers 29,000 volumes, and there is a fine Museum of Natural History. With it are affiliated *University College, McMaster Hall* (Baptist), *St. Michael's* (Catholic), *Victoria* (Methodist), *Wycliffe* (Episcopal), and *Knox College* (Presbyterian). In front of the University is the *Meteorological Observatory*, the *School of Practical Science*, and the * **Queen's Park,** comprising about 50 acres, skillfully laid out and pleasantly shaded. In the Park is a monument to the memory of the Canadians who fell in repelling the Fenian invasion of 1866, and here stand the *Parliament Buildings*. The *Post-Office*, a handsome stone building in the Italian style, stands at the head of Toronto St., and near it is the *Free Public Library*. The *City Hall*, in Front St., near the lake-shore, is an unpretentious structure in the Italian style. Near by is the *St. Lawrence Market*. The *Custom-House* is a large and imposing cut-stone building, extending from Front St. to the Esplanade; and the *Court-House* is in Church St. **Osgoode Hall,** in Queen St., is an imposing building of the Grecian-Ionic order, containing the Provincial law courts and an excellent law library of 30,000 volumes. The *St. Lawrence Hall*, in King St., is a stately stone structure in the Italian style, surmounted by a dome, and containing a public hall, news-room, etc. The *Masonic Hall* is in Toronto St., and the *Sons of England Hall* in Queen St. The Young Men's Christian Association has a fine edifice in Yonge St., with the largest hall in the city. The Grand Opera House seats about 2,300, and the Pavilion, in the Horticultural Gardens, 3,000.

The *** Church of St. James** (Episcopal), cor. King and Church Sts., is a spacious edifice in the Gothic style of the thirteenth century, with a lofty tower and spire (306 ft. high), a clear-story, chancel, and elaborate open roof, of the perpendicular style. It is 200 by 115 ft., and is surrounded by shady grounds. The *Cathedral of St. Michael* (Roman Catholic), in Church St. near Queen, is a lofty and spacious edifice in the decorated Gothic style, with stained-glass windows and a spire 250 ft. high. The **Wesleyan Methodist Church,** on McGill Square, is the finest church of the denomination in Canada. It has a massive tower, surmounted by graceful pinnacles, and a rich and tasteful interior. *Trinity* and *St. George's* (both Episcopal) are neat examples of the perpendicular Gothic style. The *Jarvis Street Baptist Church* is in the decorated Gothic style, and one of the finest church edifices in the Dominion. *St. Andrew's* (Presbyterian) is a massive stone structure in the Norman style.

The *Normal School*, the *Model Schools*, and the *Educational Museum* are plain buildings in the Italian style, grouped so as to produce a picturesque effect, standing amid park-like grounds in Church St. The Museum contains a complete supply of educational apparatus and some valuable paintings. **Trinity University,** in Queen St. West, is a picturesque building 250 ft. long, surrounded by extensive grounds. *Upper Canada College* is a plain red-brick building in King St. near John, immediately opposite the official residence of the Lieutenant-Governor of the Province. The *Provincial Lunatic Asylum* is a large building with 200 acres of grounds W. of the city. Immediately W. are situated the *Central Prison*, the *Mercer Reformatory for Women*, the *Orphans' Home*, and the *Home for Incurables*. E. of the city (Don St., near Sumach) is the fine structure of the *General Hospital*. The *Crystal Palace*, in which are held annual exhibitions, is an extensive building near the Lunatic Asylum. The *Loretto Abbey*, in Wellington Place, is the principal nunnery in the city. The *Public Library*, in Church St., contains very large and pleasant reading-rooms, and well-selected reference and circulating libraries, numbering 55,000 volumes. Branch libraries are also situated in different parts of the city.

65. The Maritime Provinces of Canada.

To describe these Provinces in detail would require a volume of itself, and, furthermore, would be beyond the purposes of this book. Those who desire fuller information on the subject should consult THE CANADIAN GUIDE-BOOK, Part I, Eastern Canada, published by D. Appleton & Co. All we shall attempt will be to give the outlines of a round trip, which, with short side trips or excursions, will include the principal points of interest in New Brunswick, Nova Scotia, Cape Breton, and Prince Edward Island. This round trip *can* be made in two weeks, but at least three weeks should be assigned to it in order to make it thoroughly enjoyable. The traveler should go warmly clad. As to money, U. S. notes will be found as serviceable as anything else, but at each stopping-point they should be taken to a banker's and exchanged for as much local currency as will be needed during the sojourn.

New Brunswick.

New Brunswick, the third Province of the Dominion of Canada, lies upon the eastern boundary of the State of Maine, and is 180 miles long by 150 wide, containing an area of 27,322 square miles. Its population, according to the census of 1891, was 321,294. The landscape is of great variety and of most picturesque beauty, the whole Province (excepting the dozen miles lying directly on the sea) being broken into attractive valleys and hills, which northward assume a very rugged character. Much of its area is covered with magnificent forests, which, as in the neighboring State of Maine, constitute its chief source of industry and wealth. Like the neighboring Province of Nova Scotia, New Brunswick so abounds in lakes and rivers that ready water access may be had, with the help of a short portage now and then, over its entire area. Thus a canoe may easily be floated from the interior to the Bay of Chaleur, the Gulf of St. Lawrence, and the ocean on the N., or to the St. John River, and thence to the Bay of Fundy, on the S. All the waters of New Brunswick abound with fish of almost every variety. The fisheries of the Bay of Fundy are of immense value, and employ vast numbers of the population.

St. John (*Dufferin, Royal*, and *Victoria*) is the principal city of New Brunswick, and is the starting-point for our tour of the Maritime Provinces. A pleasant way to reach it is by the steamers of the *International Steamship Co.*, from either Boston or Portland. They leave Commercial Wharf, Boston, on Mondays, Wednesdays, and Fridays during the summer months (fare from Boston to St. John, $4.50). The best railway approach is by Route 17 or Route 18 to Portland, thence by Route 24 to Bangor, and from Bangor to Vanceboro by the Maine Central R. R., and thence to St. John by the Atlantic Div. of the Canadian Pacific Railway. These systems and the *Intercolonial Railway* enter the city by a cantilever bridge 810 ft. long, which spans the river St. John a few rods above the Suspension Bridge, to which reference is made below. During the summer several trains a day run from

Boston to St. John, one of which goes through in 14 hours (fare, $8), with buffet drawing-room and sleeping cars on two of the trains. The short line of the Canadian Pacific Railway connects the city with Montreal; time, about 15 hours. The city (containing 39,179 inhabitants) is superbly situated upon a bold, rocky peninsula at the mouth of the St. John River. The scenery of this river is very striking in the passage immediately preceding its entrance into the harbor, and for 1¼ mile above the city. It makes its impetuous way here in a chain of grand rapids, through a rugged gap 270 ft. wide and 1,200 ft. long. The passage is navigable only during the very brief time of high and equal tides in the harbor and river; for at low water the river is about 12 ft. higher than the harbor, while at high water the harbor is 5 ft. above the river. The streets are wide and laid out at right angles; some of them are very steep, and cut through the solid rock to a depth of 30 or 40 ft. On June 20, 1877, a most disastrous conflagration reduced the entire business portion of the city to ashes, but since the fire many very handsome public buildings have been erected; among them, the *Masonic Hall*, at a cost of $80,000, *Post-Office, Custom-House, City Buildings, Odd-Fellows' Hall*, and various banks and churches. One mile and a half distant from the city is the *Rural Cemetery*, containing 110 acres. The *Owens Art-Gallery*, which was opened about 7 years since, is a superb structure, and contains one of the finest collections of paintings by European and American artists to be found in the Dominion. On the W. side of the river is a portion of the city called Carleton. The principal points of interest in the vicinity of St. John are *Lily Lake* and *Mt. Pleasant*, about a mile distant; *Rothesay*, a pretty village on Kennebecasis Bay, much resorted to in summer; *Loch Lomond*, 11 miles N. E., also a favorite resort; and the *Suspension Bridge*, 640 ft. long, and 100 ft. above the river. The favorite drives are on the *Marsh Road* and the *Mahogany Road*.

Several interesting excursions may be made from St. John: 1. A trip up the St. John River to Fredericton, the capital, may be made by steamer in 7 hours (fare, $1.00), or by the Canadian Pacific Railway in 3 hours. The St. John River is about 600 miles long, and from Grand Falls to the sea (225 miles) its course is within British territory. A great part of its course is through wild forest-land, but at some points the banks rise in grand rocky hills. 2. To St. Stephen and St. Andrews; by the Canadian Pacific Railway, or steamer *via* Eastport, Me. Fare to St. Stephen, $1.75. This trip gives the tourist a sight of the turbulent Bay of Fundy, and of the picturesque scenery of Passamaquoddy Bay; it may be made in 7 hours. From *Calais* (opposite St. Stephen) a railway runs 21 miles to the lovely and fish-abounding **Schoodic Lakes.** 3. To the Basin of Minas by steamer, to Annapolis, and thence by Windsor & Annapolis R. R. to Wolfville, whence a small steamer runs to Parrsboro, Kingsport, and Windsor. The **Basin of Minas,** the E. arm of the Bay of Fundy, penetrates 60 miles into Nova Scotia, and is remarkable for its tremendous tides, which rise sometimes to the height of 60 or 70 ft. *Parrsboro* (*Dominion House*) is a pretty little town at the entrance of the Basin, and may be made the center for many agreeable minor excursions.

Across the Basin from Parrsboro is **Grand Pré,** the land of Long-fellow's Evangeline. The picturesque *Gaspereaux Valley* may be visited from Wolfville, and also from Halifax *via* Windsor.

The next stage in our regular round trip is from St. John to Halifax. This may be made without change of cars *via* the Intercolonial Railway (distance, 276 miles; fare, $6 for 1st class, and $4 for 2d class); but the pleasantest route in summer is by steamer to *Annapolis,* and thence by rail (fare, limited, $4.50; unlimited, $5.80). By this route the tourist obtains fine views of the picturesque scenery of **Digby Gut,** and the lovely **Annapolis Basin.**

Nova Scotia.

The Province of Nova Scotia, the ancient Acadia, lies S. E. of New Brunswick, and, besides the peninsula proper, comprises the island of Cape Breton, from which it is separated by the narrow Gut of Canso. Its area is 21,731 square miles, including the 4,775 of Cape Breton, and the total population in 1891 was 450,523, of whom 86,789 resided on Cape Breton. The surface of the peninsula is undulating, and though there are no mountains, there are several ranges of hills, most of which traverse the country in an E. and W. direction. The shores are indented with a great number of excellent bays and harbors, and there are numerous small rivers, mostly navigable by coasting vessels for short distances. The surface is dotted with many lakes and ponds, the largest being Lake Rossignol in the S. W., 15 miles long by about 5 wide. In the N. E. part of the Province, in the vicinity of the St. Mary's River, moose or elk abound, and are hunted successfully in the autumn and early winter. The black bear is also occasionally found, while partridge, plover, and wild fowl are shot in enormous numbers. In the St. Mary's and other rivers large numbers of salmon are taken.

Halifax (*Halifax, Queen's*), the capital of Nova Scotia, is situated near the middle of the S. E. coast of the Province, on the W. side of a deep inlet of the Atlantic, called Halifax Harbor. Besides the routes mentioned above, it is reached direct from Boston by steamer, and from Norfolk or Baltimore. The city is built on the declivity of a hill rising 236 ft. above the level of the harbor, and had a population in 1891 of 36,556. Its plan is regular, most of the streets crossing each other at right angles; many of them are spacious and handsome. The lower part of the city is occupied by wharves and warehouses, above which rise the dwelling-houses and public buildings, while the summit of the eminence is crowned by the granite bastions of the Citadel. The * **Provincial Building,** in which are the Government offices, is in Hollis St., and is 140 ft. long by 70 broad, with an Ionic colonnade. On the third floor is the Provincial Museum, containing specimens of the various natural products of the Province and a number of curiosities. West of the Provincial Building is the *Parliament Building,* a plain gray-stone edifice, surrounded by pleasantly shaded grounds, containing the *Free Library.* In the Legislative Chamber are some fine portraits. Near by is the handsome building of the *Young Men's Christian Association,* containing a free reading-room. The *Court-*.

House is a spacious free-stone structure, on the Spring-Garden Road. Just below it is the fine Roman Catholic *Cathedral of St. Mary.* The *Government House*, in Pleasant St., is a solid but gloomy structure, and is the official residence of the Lieutenant-Governor of Nova Scotia. The *Wellington Barracks*, which comprises two long ranges of substantial stone and brick buildings, is the most extensive and costly establishment of the kind in America. The *Admiralty House, Dalhousie College, Military Hospital, Lunatic Asylum* (in Dartmouth), *Workhouse, Jail, Penitentiary,* the *Academy of Music,* and some of the public schools, are among the most prominent buildings. The *Citadel occupies the summit of the heights commanding the town, and is a mile in circumference. It is a costly work, and, after that of Quebec, is the strongest fortress in British North America. The *Queen's Dockyard* covers 14 acres in the northern portion of the city, and is said to be inferior in equipment to few except those of England. The harbor is over a mile wide opposite the city, but about a mile above it narrows to ¼ of a mile, and then expands into *Bedford Basin,* which has a surface of 10 square miles and is completely land-locked. The road to *Point Pleasant* is a favorite promenade. The *Dartmouth Lakes,* entered on the opposite side of the harbor, afford a pleasing excursion. The best views of the city of Halifax are from the summit of the Citadel or from the Dartmouth side of the harbor.

Three interesting minor excursions may be made from Halifax: 1. To the Basin of Minas and Grand Pré *via* Windsor & Annapolis Ry. to Windsor, and thence by steamer to Parrsboro. This has the same objective as Excursion 3 from St. John (see page 284). 2. To Yarmouth and the Tusket Lakes. The trip from Halifax to Yarmouth may be made by W. & A. Ry. to Annapolis (219 miles), steamer or rail to Digby, and thence by the Western Counties Ry. to Yarmouth (70 miles); or by steamer leaving Halifax, and running all the way to Yarmouth. This latter gives the tourist an opportunity of seeing the richly beautiful scenery of the Atlantic coast of Nova Scotia. **Yarmouth** (*Lorne, Queen*) is a flourishing seaport on the southwest coast of Nova Scotia, containing 6,087 inhabitants. The picturesque **Tusket Lakes** are entered by way of *Tusket* (10 miles from Yarmouth) or *Lake George* (12 miles from Yarmouth). They afford excellent fishing, and the surrounding forests are full of game. 3. To the **Liverpool Lakes,** by stage from Annapolis to *Greenfield* (50 miles), or by preceding steamer route to *Liverpool,* and thence by stage to Greenfield, or by stage *via* Mahone Bay (109 miles) to Liverpool. A road through the forest leads from Greenfield to the Indian village on *Ponhook Lake,* where guides may be procured. From Ponhook 12 lakes may be entered without making a single portage, including *Lake Rossignol,* the largest and finest in Nova Scotia. These lakes and the region around them are the paradise of sportsmen.

From Halifax, the next and final stage in our regular round trip is to Cape Breton and the famous Bras d'Or Lakes. There are three principal routes by which this excursion may be made: 1. From Halifax *via* the Pictou Branch R. R., which diverges from the Intercolonial Railway

at Truro, to *New Glasgow ;* thence by Eastern Extension of the Intercolonial Railway to Port Mulgrave, between which and Port Hawkesbury,
directly opposite, a steam ferry-boat plies across the Strait; and from
Port Hawkesbury by Cape Breton Railway to *Sydney.* Sydney and Louisburg are now connected by a short line called the New Glasgow & Cape
Breton Railway. 2. By steamer on alternate Tuesdays and Saturdays,
direct to Sydney. The fare by Saturday steamers is $10 (with meals); by
the Tuesday steamers, $8 (without meals). 3. By Pictou Branch of the
Intercolonial Ry. to *Pictou,* thence by steamer to Port Hawkesbury,
thence by stage to West Bay, and thence by steamer on the Bras d'Or to
Sydney. Fare, $8. The best way to make the round trip is to take
route 1 or 2 to Sydney, and route 3 for the return. In this way the
sail on the lakes will be made during the day.

Cape Breton.

The Island of Cape Breton is separated from Nova Scotia by the Gut
of Canso, a narrow strait from 1 to 1½ mile wide. Its greatest length is
100 miles, and its greatest breadth 85 miles, with an area of 4,775 square
miles and a population of 86,789. The island is very irregular in shape,
and is nearly divided into two parts by the **Bras d'Or,** which is not a
lake, but a great inland sea with a narrow outlet. At the entrance lies
Boularderie Island, between which and the main island on the S. is
Little Bras d'Or. The Bras d'Or is 55 miles long and 20 miles wide,
and varies in depth from 70 to 300 ft. The coast is for the most part
rocky and elevated, and indented by numerous bays and inlets. There
are several fresh-water lakes, the principal of which are *Lake Margarie,*
in the N. W. division, 40 miles in circumference, and *Grand Lake* and
Mira Lake in the S. division. Mira Lake receives the Salmon River,
which flows from the W.

The chief town on the island is **Sydney,** which is reached from
Halifax or St. John, as explained above. It has 2,426 inhabitants, and
one of the finest harbors on the Atlantic coast. An interesting excursion from Sydney is a stage-ride of 24 miles to the ruins of the once
famous fortress of **Louisburg,** now a small fishing-hamlet. The
steamer which leaves Sydney twice a week (Tuesdays and Thursdays)
for West Bay traverses the entire length of the *Bras d'Or,* and affords
the best opportunity for seeing that remarkable water. It stops at
Baddeck (whose name Charles Dudley Warner has rendered familiar),
and at *West Bay* connects with stages and wagons, which convey passengers 13 miles to *Port Hawkesbury,* where they may take stages or
steamers to Halifax and St. John, or to Prince Edward Island.

Prince Edward Island.

Prince Edward Island lies in the Gulf of St. Lawrence, 9 miles from
New Brunswick, 15 miles from Nova Scotia, and 30 miles from Cape
Breton Island. Its extreme length is 150 miles and greatest breadth
34 miles, and it has an area of 2,173 square miles and a population of
100,988. The surface is generally flat, but rises here and there to a

moderate height, without being anywhere too broken for agriculture. The coasts are bold, and are lined with red cliffs ranging from 20 to 100 ft. in height, and deeply indented by bays, with numerous projecting headlands. The climate is salubrious, and is milder than that of the adjacent continent. The winters are long and cold; the summers are warm, but not oppressive.

Charlottetown (*Davis, Osborne*) is the capital, chief commercial center, and only city. It has 11,374 inhabitants, is regularly laid out, and fronts on a good harbor. The only handsome buildings in the city are the *Colonial Building*, containing the offices and Legislative Chambers of the Provincial Government, and the *Post-Office*. During the season of navigation a line of steamers runs 4 times a week from Charlottetown to Pictou, Nova Scotia (fare, $2), where connection is made with railway to Halifax; and daily to Shediac, New Brunswick, where connection is made with railway to St. John. Weekly lines connect with Quebec, and with Halifax and Boston. The Prince Edward Island Ry. traverses the entire length of the island, connecting Charlottetown with *Summerside* (49 miles), with *Tignish* (117 miles), with *Georgetown* (46 miles), and with *Souris* (60 miles). This railroad affords access to any part of the island.

work devoted to the New England and ——————————————————
19

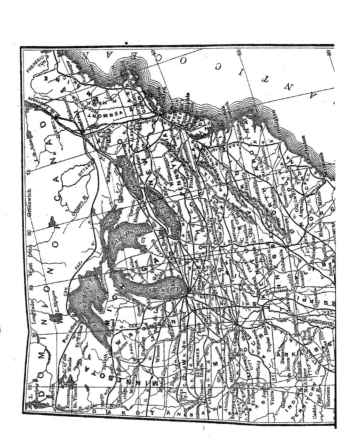

WESTERN AND SOUTHERN STATES.

66. New York to Chicago via Niagara Falls and Detroit.

a.

By the New York Central and Hudson River R. R. and the Michigan Central R. R. The distance from New York to Chicago by this route is 978 miles, and the time of the ordinary fast express trains is about 34 hours. Drawing-room cars are attached to all the day trains, and palace sleeping-cars to all the night trains, and there is no change of cars between New York and Chicago. The North Shore Limited Express, with complete drawing-room car service, leaves the Grand Central Station daily at 4.50 A. M., and reaches Chicago in 25 hours. Distances : New York to Albany, 143 miles ; to Utica, 238 ; to Rome, 252 ; to Syracuse, 291 ; to Rochester, 371 ; to Suspension Bridge, 446 ; to Hamilton, 491 ; to London, 576 ; to Detroit, 677 ; to Chicago, 978.

LEAVING the Grand Central Station at 42d St. and 4th Avenue in New York, the train passes to Albany amid the picturesque scenery of the Hudson River.[1] The Hudson River Div. of this system runs along the E. bank of the river all the way to Albany, and, owing to the fine view of the opposite side of the river, which is obtained from the cars all the way, the journey is a most agreeable one. Going N., the traveler should secure a seat on the left-hand side of the car, and going S. on the right-hand side. The lower Hudson, emptying into New York Bay, is like a huge arm of the sea, and, as we ascend, preserves its noble width, occasionally expanding into lakes, while at several places among the Highlands the mountains approach so close on either side as to reduce the river to a contracted and tortuous channel. The railroad runs close along the bank of the river, in sight of its waters almost continuously, making occasional short cuts from point to point, and ever and anon crossing wide bays and the mouths of occasional tributary streams.

Passing beneath the upper part of New York City through long tunnels, the train crosses the Harlem River, and then, turning to the left, follows the Spuyten Duyvil Creek to the Hudson. On reaching the river, the traveler's attention is at once caught by the * **Palisades,** a series of

---

[1] Our description of this route as far as Suspension Bridge is a mere outline or summary, designed to furnish such cursory information about the places and scenery *en route* as may meet the wants of through passengers to the West. Those who desire a more detailed description will find it in the section of the work devoted to the New England and Middle States (Routes 9 and 40).

19

grand precipices rising in many places to the height of 300 ft. and stretching in an unbroken line along the W. river-bank for more than 20 miles. The rock is trap, columnar in formation, and the summit is thickly wooded. In striking contrast with the desolate and lonely appearance of these cliffs, the E. bank presents a continuous succession of beautiful villas standing amid picturesque and exquisitely-kept grounds, with a frequent sprinkling of villages and hamlets. *Yonkers* (17 miles) is a fashionable suburban town, beautifully situated at the mouth of the Neperan or Saw-Mill River. *Piermont* (22 miles) is on the opposite side of the river at the end of the Palisades, and takes its name from a mile-long pier which extends from the shore to deep water. Here begins the * **Tappan Zee,** a lake-like expansion of the river, 10 miles long and 3 miles wide at the widest part, surrounded by beautiful scenery. *Tarrytown*, immortalized by Washington Irving, and *Sing Sing*, the site of one of the most important of the State Prisons, are on the E. shore of the Tappan Zee; and the pretty little town of *Nyack* is on the W. shore. Croton Point divides the Tappan Zee from *Haverstraw Bay*, another lake-like widening of the river, at the upper end of which stands *Peekskill* (43 miles), at the gate of the *Highlands,** as the mountains through which the Hudson forces its way are called. The scenery for the next 16 miles is unsurpassed in the world; but a very imperfect idea of it is obtained from the cars. The first seen of the Highland group is *Dunderberg Mt.*, which looms up grandly across the river. Nearly opposite is *Anthony's Nose*, whose base is tunneled by the railway a length of 200 ft. In the river, under Dunderberg, is the pretty *Iona Island*, noted as a picnic resort. In the heart of the Highland Pass, a beautiful view is obtained of * **West Point,** the seat of the U. S. Military Academy, with fine buildings on a broad terrace 157 ft. above the river. *Garrison's* (51 miles) is a station on the R. R. nearly opposite West Point. Just above West Point, on the same side, is *Cro' Nest*, one of the loftiest of the Highland group, and then comes *Storm King*, the last of the range on the W. On the E. side, scarcely visible from the cars, are *Mt. Taurus, Breakneck*, and *Beacon Hill*, which are among the most commanding features of the river scenery. At the end of the Highlands the river again expands into the broad *Newburg Bay*, on the W. shore of which is **Newburg,** a beautifully-situated city of 23,087 inhabitants. In this vicinity occurred many of the important events of the Revolutionary War, and Washington's headquarters, an old gray-stone mansion south of the town, is still preserved under the ownership of the State. Connected with Newburg by ferry, and on the line of the railroad, is *Fishkill*, a place with important manufacturing interests. The *Verplanck House*, 2 miles from the landing, was once the headquarters of Baron Steuben, of Revolutionary fame. **Poughkeepsie** (75 miles) is the largest city between New York and Albany on the railway. It contains 22,206 inhabitants, and is the site of Vassar College and other famous educational institutions. Crossing the Hudson at this point is a cantilever bridge 1¼ miles long. It rests on four pyramidal steel towers 100 ft. high; and there are three cantilevers of 548 ft. each, and two connecting spans of 525 ft. each. The

Central New England & Western R. R. uses this bridge, and affords unbroken communication between the Pennsylvania coal-fields and the New England cities. Above Poughkeepsie, on either bank, are many pleasant towns and fine country-seats, but the river-banks are for the most part low and uninteresting. Just before reaching Hudson the noble range of the Catskill Mountains is seen along the W. horizon. *Catskill*, whence these are visited, offers a pleasant view across the river, with the spacious *Prospect Park Hotel* on an elevated plateau above the landing. **Hudson** (115 miles) is a flourishing city of 9,970 inhabitants, at the head of ship-navigation on the river. The heights back of the city command majestic views of the Catskills. Between Hudson and Albany is *Rhinecliff*, which is connected with Rondout and Kingston on the W. bank of the river, and where the trains of the Ulster & Delaware R. R. may be taken for Catskill Mountain resorts. **Albany** (143 miles) is the capital of New York State, and is a city of 94,923 inhabitants, beautifully situated on the W. bank of the Hudson near the head of tide-water. It contains many features of interest, and the tourist who has time to stop over will find them all described on page 72.

The through trains make but a short pause at Albany, and then run on over the N. Y. Central R. R., which traverses the entire length of New York State from Albany to Buffalo, passing through the rich and populous midland counties. The scenery along this portion of the route is mostly of a pastoral character, with nothing bold or striking, but with much that is pleasing. The famous valley of the Mohawk is first traversed. The river, now quiet, now rushing along its rocky bed, is continually in sight, the hills bounding the valley adding to the picturesqueness of the view, and the many villages clustering along the line giving evidence of solid prosperity. The Erie Canal traverses the State from Albany to Buffalo, nearly on the same line with the railroad. *Schenectady* (160 miles) is one of the oldest towns in the State, and is distinguished as the site of Union College. Just beyond Schenectady the train crosses the Mohawk and the Erie Canal on a bridge 1,000 ft. long. *Little Falls* (217 miles) is remarkable for a bold passage of the river and canal through a wild and most picturesque defile. The next station is *Herkimer* (226 miles), the starting-point of the Adirondack & St. Lawrence R. R., which pierces the Adirondack wilderness and makes direct connection at Coteau Junction for Montreal, etc. **Utica** (238 miles) is a manufacturing city of 44,007 inhabitants, on the S. bank of the Mohawk, nearly in the center of New York State, and from where connection is made with the Rome, Ogdensburg & Watertown R. R. for the Adirondack Region and the Thousand Isles. The *State Lunatic Asylum* here holds high rank among the institutions of the kind. **Rome** (252 miles) is a city of 14,991 inhabitants, with fine buildings, and is one of the chief lumber markets of the State. It was formerly the site of Fort Stanwix, and the battle of Oriskany was fought near here. **Syracuse** (291 miles) is the next important city on the line of the road, and is famous for its salt-springs, the chemical works of the Solvay Process Co., and other manufactories. Next comes **Roches-ter** (371 miles), the metropolis of Central New York, with a population of 133,896, and the site of the celebrated Genesee Falls. The train passes about 100 rods S. of the most southerly fall, so that passengers in crossing lose the view. At *Lockport* (428 miles) the wonderful system of locks by which the Erie Canal descends from the level of Lake Erie to the Genesee level is visible from the windows of the cars. The Michigan Central R. R. connects with the New York Central at Niagara Falls and Buffalo. By the former route the trains pass over a cantilever bridge 895 ft. in length, and, in crossing, the passenger has a fine view of the Falls on one side and the Whirlpool Rapids on the other. At *Falls'.View* all express trains stop five minutes to allow a view of the Falls from the verge of the embankment. The point of view is directly above the Horseshoe Falls, and the eye takes in the entire extent of the Horseshoe and American Falls, with Goat Island between. **Buffalo** is reached in 436 miles. This is the third city in size of New York State (population, 255,664), situated at the E. end of Lake Erie, at the head of Niagara River. It is the terminus of the N. Y. Central & Hudson River R. R., the Erie R. R., and of other less important lines; also the terminus of the Erie Canal, which extends E. to the Hudson River at Albany, giving Buffalo a commerce which surpasses that of many maritime cities. The tourist will find it worth his while to stop a day or two at this interesting city, in which case he should consult the detailed description of the city and its environs given in Route 40.

b.

Another route is as described above, except the interval between Buffalo and Detroit, to the new suspension bridge, thence by the Grand Trunk line.

At * **Suspension Bridge** (446 miles) the train crosses the Niagara River, in view of the Falls and of the rapids rushing to the whirlpool below. The bridge was long regarded as one of the achievements of modern engineering. It is 800 ft. long from tower to tower, is 258 ft. above the water, and was finished in 1855 at a cost of $500,000.

From Suspension Bridge to Detroit the route runs through Canada, and is most uninteresting, though the road (the Grand Trunk R. R.) is an admirable one. In the neighborhood of *St. Catherine's* (458 miles), noted for its mineral springs, and *Hamilton* (491 miles) there is some attractive scenery; but with these exceptions everything is dull, flat, and monotonous, and the traveler will be glad when, emerging from a deep cut he suddenly comes upon the bank of the Detroit River at *Windsor* (676 miles), opposite Detroit. Here the train is transferred to the other side of the river on a steam ferry-boat, and the route is resumed on the line of the Michigan Central R. R.

For a detailed description of this route, see THE CANADIAN GUIDE BOOK, Part I, *Eastern Canada.*

Detroit.

Hotels, etc.—The leading hotels are the *Griffen House*, the *Griswold House*, the *Hotel Cadillac*, the *Hotel Normandie*, the *Russell House*, and the *Wayne.* Nine lines of *street-cars* intersect the city, and four *ferry-boats* ply across the river to Windsor, on the Canadian side. There are numerous steamboat lines with boats running frequently to various points on the lakes.

Detroit, the chief city of Michigan, is situated on the N. bank of
the Detroit River, a stream 20 miles long, with a depth of water sufficient for the largest vessels, connecting Lakes Erie and St. Clair. The
city extends along the bank for about 7 miles, and is built up for about
2¼ miles from the water. For at least 6 miles the river-front is lined
with mills, dry-docks, ship-yards, foundries, grain-elevators, railway-
depots, warehouses, lumber-yards, and rolling-mills. For a short distance
from the river-bank the ground rises gradually, and then becomes perfectly level, furnishing an admirable site for a large city. Detroit is laid
out upon two plans: the one, that of a circle with avenues radiating from
the Grand Circus as a center; the other, that of streets crossing each
other at right angles. The result is a slight degree of intricacy in certain
localities, which inconvenience is more than compensated by a number
of little semicircular and triangular parks which diversify and ornament
the place. The avenues are from 100 to 200 ft. wide; the streets vary
in width from 50 to 100 ft., and are generally shaded by an abundance
of trees.

The site of Detroit was visited by the French as early as 1610 ; but no permanent settlement was made until 1701, when Fort Pontchartrain was built. In
1760 it passed into the hands of the English, and in 1763 was besieged for 11
months by Pontiac, in his attempt to expel the whites from that region. In
1788 Detroit was ceded to the United States, but the Americans did not take
possession of it till 1796. During the War of 1812 it fell into the hands of the
British, but was recaptured in 1813. It was incorporated as a city in 1824, when
its population was less than 2,000, and in 1890 it had 205,876 inhabitants. The
manufactures of the city are important, including extensive iron-works and
machine-shops, railroad-car factories, drugs and pharmaceutical preparations,
flour-mills, breweries, and tobacco and cigar factories, tanneries, boot and shoe
and stove factories, and potteries, etc. The shipping interests are also large,
while pork and fish packing employ numerous hands.

The principal streets of the city are *Jefferson Ave.*, parallel with the
river; **Woodward Ave.,** which crosses the former at right angles,
and divides the city into two nearly equal parts; and *Fort St.*, *Cadillac
Square*, *Grand River Ave.*, and *Gratiot*, *Washington*, *Madison*, *Michigan*,
and *Monroe Aves.*, at angles with Woodward Ave. *Griswold St.* is the
financial thoroughfare of the city. Adelaide and Alfred Sts., Edmund
Place and Stimson Place, and Davenport St. are pleasant residence-
streets. Cass Ave. is the fashionable drive. The *Grand Circus Park*,
half a mile back from the river, is semicircular in form, and is divided
by Woodward Ave. into two quadrants, each containing a fountain.
About half-way between the river and the Grand Circus is the **Campus
Martius,** an open space 600 ft. long and 250 ft. wide, which is crossed
by Woodward and Michigan Avenues, and from which radiate Monroe
Ave. and Fort St. Facing the Campus Martius on the W. is the * **City
Hall,** a handsome structure in the Italian style, 200 ft. long, 90 ft.
wide, and 180 ft. high to the top of the tower, completed in 1871, at
a cost of $600,000. From the top of the tower a bird's-eye-view is
afforded of the city and surrounding country which is worth seeing. In
front of the City Hall is a fine * *Soldiers' Monument,* designed by Randolph Rogers, and erected in memory of the Michigan soldiers who fell
in the civil war. Near by is the artistic drinking-fountain presented to

the city by John J. Bagley, a former Governor of the State. Facing the Campus Martius on the N. is the *Detroit Opera-House*, and facing it on the E. is the *Market*. The *Custom-House*, which also contains the *Post-Office*, is a large stone building in Griswold St. The *Board of Trade Building*, cor. Jefferson Ave. and Griswold St., is spacious and ornate. The * *Freight Depot* of the Michigan Central R. R. stands on the wharf, and consists of a single room, 1,250 ft. long and 102 ft. wide, covered by a self-sustaining roof of corrugated iron. The passenger station of this road at the foot of 3d St. is one of the finest in the State. In the vicinity are the great *Wheat-Elevators* of different railway and other corporate companies. Besides the Detroit Opera-House, there are the *Grand* and *Whitney Grand Opera-Houses*, and several large public halls. A fine *Y. M. C. A.* building has been erected on the cor. of Griswold St. and Grand River Ave., with library, gymnasium, public hall, etc. The *Police Headquarters Building* is a spacious structure. The *Museum of Art* building on Jefferson Ave. contains an excellent exhibit, including the Scripps collection of old masters and the Frederick Stearns collection of Japanese, Chinese, and East Indian curios, numbering some 15,000 pieces.

The churches of Detroit are noted for their number and beauty. The * **Cathedral of St. Peter and St. Paul** (Roman Catholic), Jefferson Ave. cor. of St. Antoine St., is the largest church-edifice in the State, though very plain in outward appearance. *St. Joseph's* (Roman Catholic) is a handsome building. * **St. Paul's** (Episcopal), cor. Congress and Shelby Sts., is the parent church of the diocese, and is famous for its beautiful roof, which is self-sustaining. Other handsome Episcopal churches are *Christ's*, on Jefferson Ave. above Hastings St., *St. John's*, on Woodward Ave., and *Grace*, on Fort St. The *First Congregational, Unitarian*, and the *First Presbyterian Churches* are fine structures of large size and of different styles of renaissance architecture, and are on opposite corners of Woodward Ave. and Edmund Pl. The **Fort St. Presbyterian** (Fort St., cor. 3d) has a handsome front and a beautiful interior. The *Baptist Church*, in Woodward Ave., cor. of Winder St., of Ionic stone, is perhaps even finer. The **Central Church** (Methodist), in Woodward Ave., has a richly-decorated interior, and is the oldest Protestant church in the city. The *Jefferson Ave. Presbyterian*, above Rivard St., the *Westminster Presbyterian*, cor. Woodward and Parsons Sts., the *First Baptist*, on Cass Ave., and the *Woodward Ave. Congregational*, are all fine edifices. There are several libraries in the city, of which the principal is the *Public Library*, containing 100,000 volumes. The * **Convent of the Sacred Heart**, in Jefferson Ave. near St. Antoine St., is a large and beautiful building. The *House of Correction*, in the N. portion of the city, is used principally for the confinement of petty criminals. Directly opposite is a home for discharged female prisoners, who are received here and furnished with work until places can be found for them out of the reach of the influences previously surrounding them. The *U. S. Marine Hospital*, on the bank of the river in the eastern part of the city, commands a fine view of the Canada shore. *Harper* and *Grace Hospitals* in John R. St., and *St.*

University of Michigan, Ann Arbor.

Mary's in St. Antoine St., are also fine edifices. *Elmwood Cemetery* is a beautiful burying-ground within the city limits (reached by street-cars). *Woodmere Cemetery*, on high ground, 4 miles W. of the city, is of recent origin. *Fort Wayne* is a bastioned fortification inclosing 65 acres, about 3 miles below the city, standing upon the bank of the river and completely commanding the channel. It is the largest and strongest fortress in the lake region. The Fort St. and Elmwood street-cars run to the fort gate, and it is also a favorite point to which rides and drives are taken. Below Fort Wayne, at the mouth of the Rouge River, is the International Fair and Exposition building, with extensive grounds. Annual exhibits are held here. *Belle Isle*, an island containing 700 acres, lying near the head of Detroit River, was laid out as a park by Frederick L. Olmsted. It is connected by a bridge with the mainland, and contains a casino, drives and walks, a broad canal with pleasure-boats, and other artificial features. *Grosse Point*, projecting into Lake St. Clair, 9 miles above the city, is at the end of a beautiful drive, and contains many summer residences. Across Lake St. Clair, and passing up the U. S. Ship Canal, the St. Clair Flats are reached, which is well known as a resort for sportsmen. Passing Marine City, with its extensive ship-yards, we reach *St. Clair Springs*, famous for its mineral springs and its curative baths. The *Oakland House* is the principal hotel.

From Detroit the route is *via* the Michigan Central R. R., which traverses a fine agricultural country, the general aspect of which is pleasing, especially in spring and summer, but which is not of a striking or picturesque character. In many places it passes through dense virgin woods, and in others across and along the winding rivers which abundantly water this section of Michican. *Ypsilanti* (30 miles from Detroit) is a thriving city of 6,129 inhabitants, on the Huron River, which furnishes water-power for several flour-mills, paper-mills, and other factories. The State Normal School is located here. Beyond Ypsilanti the train follows the Huron River and passes in 8 miles to **Ann Arbor** (*Cook Hotel, Franklin House, Germania*, and *New Arlington*), a city of 9,431 inhabitants, on both sides of the river, known as the seat of the *** University of Michigan,** one of the leading institutions of learning in the West. With fees little more than nominal, and a high standard of scholarship, the University attracts students from every part of the country, and is open to both sexes. The University buildings stand in the midst of grounds comprising 44½ acres, and thickly planted with trees. *University Hall* is 347 ft. long and 140 ft. deep, and is devoted to the uses of the department of literature, science, and art. There are also buildings for the departments of law, medicine, pharmacy, and dentistry, two hospitals, a chemical laboratory, and a residence for the president, but no dormitories. The *Observatory* is on a hill about a mile from the other buildings. The fire-proof library of the University contains 60,000 volumes, and has capacity for 115,000. The museums are large and valuable, and include collections in natural history, the industrial arts, archæology and ethnology, and the Chinese exhibit. The number of students will average not less than 2,400. The building of the *Union School* at Ann Arbor is one of the finest in the State, accommodating 1,000 pupils. There are five mineral springs in the city, an opera-house, and several fine churches. There is a *Y. M. C. A. Building*, and *Harris Hall* fills a similar function for the Episcopal Church. There are also several guilds connected with the different churches at which lecture-courses are given. **Jackson** (*Hibbard House, Hurd House*) is 753 miles from New York, and is a busy manufacturing city of 20,798 inhabitants, on the Grand River, at the intersection of six railroads. It lies on the edge of the coaldeposits of the State, and the mines can be seen from the cars. The city is regularly laid out and substantially built. Several of the churches and the two Union school-houses are handsome edifices. The *Michigan State Penitentiary*, with spacious stone buildings, is located here, and the *Passenger Depot* of the Michigan Central R. R. is one of the finest in the State. The manufactures are extensive and various. At *Parma*, 11 miles beyond Jackson, the road reaches the Kalamazoo River, which it follows to Kalamazoo, passing through a fertile country noted for its wheat. *Marshall* (785 miles) is a very pretty town of 3,968 inhabitants, noted for its flour; and *Battle Creek* (798 miles) is a milling city of 13,197 inhabitants, at the confluence of Battle Creek and the Kalamazoo River. Battle Creek College is the chief school of the Seventh-day Baptists, who likewise have their publishing headquarters here. Twenty-three miles beyond Battle Creek is **Kalamazoo** (*Burdick House, Kalamazoo*), with a population of 17,853. It is regularly laid out, with broad, well-shaded streets, and contains many fine business structures and costly residences. The buildings of the *State Lunatic Asylum* are spacious and imposing; and *Kalamazoo College* (Baptist) and the *Michigan Female Seminary* are flourishing institutions. The manufactures are numerous and varied. *Niles* (868 miles) is a handsome and well-built city of 4,197 inhabitants, on the St. Joseph River, in the midst of a rich agricultural region. The remaining stations are unimportant, being chiefly junctions with connecting railways. **Chicago** (see Route 71).

c.

' Another route to Chicago and the West is *via* the West Shore R. R., Great Western Div. of the Grand Trunk R. R. to Detroit, thence by the Michigan Central R. R. Distances: Newburg, 56 miles; Catskill, 110 miles; Coeyman's Junction, 128 miles; Schenectady, 152 miles; Utica, 231 miles; Syracuse, 278 miles; Rochester, 366 miles; Buffalo, 426 miles; Suspension Bridge, 444 miles; Hamilton, 488 miles; London, 573 miles; Detroit, 673 miles; Chicago, 975 miles. Complete drawing-room service is attached to the day trains, and sleepers to the night trains.

The West Shore R. R.[1] runs from Jersey City (Penn. R. R. depot), or by ferry foot of 42d St. to Weehawken, along the W. Shore of the Hudson River to *Coeyman's Junction*, where the main line diverges to the

[1] The description of the route to Buffalo is a mere summary, designed to meet the wants of west-bound passengers. Fuller details will be found in the section of the route devoted to the New England and the Middle States (Route 42).

West, the branch continuing to Albany. It runs parallel with and close to the N. Y. Central R. R. from Coeyman's Junction to Buffalo, a distance of 310 miles, touching many of the same towns and cities. The road from Weehawken passes in the rear of the Palisades through several small towns in New Jersey, and reaches *Hackensack* (9 miles), a thriving and beautiful country town, full of quaint old houses. The adjacent region is full of fine dairy farms, and, though not specially picturesque, is highly cultivated and interesting. A number of small towns intervene, and, just before reaching *Tappan*, the town where Major André was tried and executed as a spy in 1778, the train enters New York State. *Nyack* (24 miles), a town of 4,111 inhabitants, is on Tappan Bay, and lies at the foot of and on the side of a range of low, picturesque hills. It is noted for its great number of beautiful and costly villas amidst shaded and park-like grounds, and it has a number of summer hotels, which accommodate many visitors. The railroad runs somewhat W. of the town through *Nyack West*, and passes through a half-dozen stations, among which are *Congers* or *Rockland Lake*, one of the most beautiful suburban resorts on the line, and *Haverstraw*, a brisk and thriving town on Haverstraw Bay, and lying on the slope of the Ramapo Hills, before reaching **West Point** (47 miles). This beautiful spot is the seat of the United States Military Academy, and is one of the most noted places on the Hudson, and a very attractive summer resort. *Newburg* is 14 miles beyond, a thriving city, which, as well as other places just mentioned, will be found referred to in the present Route *a*. *Kingston* (88 miles), *Catskill* (110 miles), the points of departure for the Catskill Mts., and *Coxsackie* (120 miles), are the only places of interest before reaching *Coeyman's Junction* (128 miles), where the road begins its curve to the west, and diverges from the Albany route. In 24 miles *S. Schenectady* is reached, and from this point to Buffalo the road passes close to the line of the N. Y. C. & H. R. R. R. through the whole of its length. As the two roads pass through the same principal towns, the reader is referred to the present Route *a* for a description.

The west-bound through train at Buffalo takes the track of the N. Y. Central R. R. to *Suspension Bridge*. Here the train takes the track of the Southern Div. of the Grand Trunk Railway. Hence the route is as described in present Route *b*.

67. New York to Chicago via Buffalo and Cleveland.

By the New York Central & Hudson River R. R. to Buffalo, and thence by the Lake Shore & Michigan Southern R. R. Through trains, with complete drawing-room and sleeping-car service attached, run through without change, making the journey in about 30 hours. The "New York & Chicago Limited" is composed exclusively of sleeping, drawing-room, smoking, buffet, library, and dining cars, leaving New York at 10 A. M., and reaching Chicago 9.45 A. M. next day. Its schedule-time between New York and Chicago is less than 25 hours. The cost of travel is somewhat more, but this is compensated for by added comfort and gain in time. Distances: New York to Albany, 143 miles; to Utica, 238; to Rome, 252; to Syracuse, 291; to Rochester, 371; to Buffalo, 440; to Dunkirk, 480; to Erie, 528; to Cleveland, 623; to Toledo, 736; to Elkhart, 869; to Chicago, 979. The Erie R. R. also sells tickets by the L. S. & M. S. R. R., with which it connects (with change of cars) at Buffalo and Dunkirk.

FROM New York to Rochester this route is the same as Route 66. Leaving Rochester, the train runs W. to Buffalo, passing several small towns, of which the only one requiring notice is *Batavia*, which contains 7,221 inhabitants, and is laid out in broad streets, beautifully shaded. The N. Y. State Institution for the Blind, one of the finest structures of the kind in the country, is located here. A monument to William Morgan, erected by the Anti-Masons, consisting of a shaft of Vermont granite surmounted by a full-length statue, may be seen from the train. (For **Buffalo,** see Route 66 *a.*)

The through train makes a short stop in the Union Depot at Buffalo, and then passes out on the tracks of the Lake Shore & Michigan Southern R. R., with which line it forms an unbroken all-rail route from New York to Chicago, then skirting the S. shore of Lake Erie, and in 40 miles reaching *Dunkirk* (480 miles from New York), where close connection is made with Erie R. R. Just beyond Dunkirk the road leaves New York State and crosses the upper corner of Pennsylvania to **Erie** (528 miles), an old, pleasant, and important lake city, with 40,634 inhabitants and extensive commerce and manufactures (fully described in Route 53). *Conneaut* (556 miles) is the first station in Ohio, and is noted as the landing-place of the party who first settled N. W. Ohio. *Painesville* (595 miles) is charmingly situated on Grand River, 3 miles from and about 100 ft. above Lake Erie. The valley through which the river runs is deep and picturesque, and the railroad crosses it on a stone bridge more than 800 ft. long. Twenty-eight miles beyond Painesville the train reaches

Cleveland.

Hotels, etc.—The *American House*, 128 Superior St.; the *Forest City House*, cor. Superior St. and Public Square; the *Hawley House*, in St. Clair St.; the *Hollenden*, cor. Superior and Bond Sts.; the *Kennard House*, cor. St. Clair and Bank Sts.; the *Stillman*, Euclid Ave. near Erie St.; and the *Weddell House*, cor. Superior and Bank Sts. There are numerous smaller houses. Several *bridges* cross the Cuyahoga, connecting the different portions of the city. There are 2 cable and a number of electric car lines that intersect the city in all directions and extend into the suburbs. *Reading-rooms* at the Public Library (75,000 vols.), the Case Library (30,000 vols.), and at the Y. M. C. A. rooms, cor. Prospect and Erie Sts.

Cleveland, the second city in size and importance in Ohio, is situated on the S. shore of Lake Erie, at the mouth of the Cuyahoga River. Originally the town was confined to the E. bank of the river, but subsequently Brooklyn, or Ohio City, sprang up on the opposite side, and both parts are now united under one corporation. The greater portion of the city stands on a gravelly plain, elevated about 100 ft. above the lake. The river passes through it in a winding course, affording an excellent harbor, which has been improved by dredging out a commodious ship-channel (branching from the river near its mouth), and by the erection of 2 piers 200 ft. apart, stretching several hundred feet into the lake. On each pier is a lighthouse. The *Breakwater* just W. of the river's mouth, erected at an expense to the United States of $1,200,000, incloses 180 acres of water, and affords a safe harbor. Another *Breakwater* is under construction on the east side. The city is laid out with much taste, chiefly in squares, the streets being remark-ably wide and well paved. The abundance of shade-trees, chiefly elms, has given it the title of the "Forest City." The great stone *Via-duct which spans the river-valley between two divisions of the city, on a level with the plateau, was completed in 1878, and is justly reck-oned among the triumphs of American engineering. It extends from the foot of Superior St. to the junction of Pearl and Detroit Sts., and is 3,211 ft. long. A second viaduct, the *Central*, connects the central part of the city with the south side. A third viaduct crosses Wal-worth Run between the south and west sides; and a fourth Kings-bury Run. These viaducts form a belt elevated roadway connecting the three parts of the city.

The growth of Cleveland has been very rapid. It was laid out in 1796, but in 1830 contained only 1,000 inhabitants. It received its first impetus from the completion, in 1834, of the Ohio Canal, which connects Lake Erie at this point with the Ohio River at Portsmouth A further stimulus was given after 1850 by the development of the railroad system ; and since 1860 its prosperity has been greatly increased by the rapid extension of manufacturing industry. In 1870 the population was 92,829, and according to the census of 1880 was 160,142. The census for 1890 gave 261,353 inhabitants. The commerce of the city is very large, especially with Canada and the mining regions of Lake Superior. Its iron and steel manufactories are numerous and extensive. It is noted for prominence in coal-oil refining and industries connected with it. Other important products are sulphuric acid, wooden-ware, agricultural implements, marble and stone, railroad-cars, sewing-machines, and white lead. It is the principal ship-build-ing port on the Great Lakes.

The chief business thoroughfare of the city is *Superior St.* Other important business streets are *Ontario, Water, Bank, Seneca, St. Clair, Euclid Ave., Merwin,* and *River,* on the E. side, and *Detroit, Pearl,* and *Lorain,* on the W. side. After leaving the business portion, which extends from the Park to Erie St., **Euclid Ave.** is lined with costly residences, each surrounded by ample grounds, and is con-sidered the handsomest street in the country. *Prospect St.*, parallel to the avenue, ranks next in beauty. *** Monumental Park** is a square of 10 acres in the center of the city, at the intersection of Ontario and Superior Sts., which divide it into 4 smaller squares. It is shaded with fine trees, and is well kept. In the S. E. quarter stands a stately monument to the soldiers of Cuyahoga County, erected in 1892–'93 by taxation of the county. It occupies the site of the statue of Commodore Perry, removed to Gordon Park. In the N. W. corner of the Park there is a handsome fountain; and in the S. W. a pool and cascade, and a bronze statue of Moses Cleaveland, the founder of the city. W. of the river is another park, called the *Circle,* which has a fountain in the center, and is finely adorned with shade-trees. The United States building, fronting on the Park, containing the *Cus-tom-House, Post-Office,* and Federal courts, is a fine stone structure, as are also the two *County Court-Houses,* the one on Seneca St., the other at the cor. of the Park. The *City Hall,* on Superior St., E. of Monu-mental Park, is a magnificent six-story building—200 × 100 ft.—with stores underneath. *** Case Hall,** a beautiful edifice near the Park, contains, besides the rooms of the Case Library, which has a collection of 30,000 volumes and a reading-room, a fine hall capable of seating 1,240 persons, and used for lectures, concerts, etc. The principal places for dramatic entertainments are the **Euclid Ave. Opera-House,** and the *Lyceum Theatre,* on the N. side of the Public Square; besides these, there are the *H. R. Jacobs Theatre,* in St. Clair St., and *Star Theatre,* in Euclid Ave.; the *Music Hall,* seating 5,000, cor. Superior and Erie Sts., a Bohemian theatre, a German theatre, and several public halls for lectures. The **Union Depot,** built in 1866, is a massive stone structure, one of the largest of its kind in the world. On the keystone over the main entrance is a bas-relief portrait of Amasa Stone. There are similar portraits of Presidents Grant and Lincoln, and various symbolical designs upon keystones at either end of the building. The *Water-Works* stand near the lake, W. of the river. By means of a tunnel, extending 6,600 ft. under the lake, pure water is obtained, which is forced by two powerful engines through the city. Two reservoirs, the *Fairmont* (80,000,000 gallons) and the *Woodland Hills* (40,000,000 gallons), complete the water-works system.

Of the many churches in the city, among the more noteworthy are *St. Paul's* (Episcopal), cor. Case and Euclid Aves., the *Woodland Avenue* (Presbyterian), with the largest Sunday-school in the city, the *Old Stone Church* (Presbyterian), cor. Ontario St. and the Park, the *Second Presbyterian,* cor. Prospect St. and Perry St., *Calvary Presbyterian Church,* cor. Euclid and Madison Aves., the *First Methodist Church,* cor. Erie St. and Euclid Ave., *Plymouth Congregational,* cor. Prospect and Perry Sts., the *Euclid Ave. Baptist,* cor. Huntington St., and the *First Congregational,* cor. Franklin Ave. and Taylor Sts., all of stone in the Gothic style. The *Roman Catholic Cathedral* is a large and handsome building; and *Trinity Church* (Episcopal) is an imposing edifice, also in the Gothic style. The *First* and *Third Presbyterian* churches are fine structures. Among the educational institutions is *Western Reserve University,* removed to Cleveland on the conditional gift of $500,000 from the late Amasa Stone, in 1882. It is located on Euclid Ave., 4 miles west of the Square, and has two fine buildings. Besides the Adelbert College it includes a medical department, cor. Eric and St. Clair Sts., and the Cleveland College for Women. The *Case School of Applied Science* is located on the same ground, and has an endowment of $1,250,000, the gift of the late Leonard Case. The *Medical Department* of the University of Wooster, in Brownell St., and the Homœopathic College Hospital, in Huron St., are flourishing institutions. The *University School,* Hough Ave. near Giddings, is a fine building; and the two High-School buildings are handsome edifices of brick and stone. The *Public Library,* opened in 1869, contains about 75,000 vols. It is free, and is supported by an annual tax upon the citizens. The other libraries are the *Case Library,* 30,000 vols.; the *Cleveland Law Library,* 11,000 vols.; and the *Western Reserve Historical Society's Library,* 22,000 vols. The *Masonic Temple* is in Superior St.

On the shore of the lake stands the extensive building of the *U. S. Marine Hospital.* The *Charity Hospital,* in Perry St., was established partly by the city and partly by private subscriptions, and is attended by the Sisters of Charity. The *Homœopathic Hospital* has a large and handsome building on Huron St. The *Work-House*, on the E. outskirts of the city, is a large and handsome structure, for the confinement and utilizing of city offenders. The *City Infirmary*, to which the sick and homeless poor are taken, has attached to it a good farm, which is worked by the inmates of the institution.

Cleveland has several beautiful cemeteries. *City Cemetery*, in Erie St., is laid out with rectangular walks shaded with trees, and contains many fine monuments. *Woodland Cemetery* is in the E. part of the city. It is prettily laid out with paths winding amid noble trees and abundant shrubbery, and is rich in monuments and statuary. *Lake View Cemetery*, containing 300 acres, on which $500,000 has been expended, is in Euclid Ave., about 5 miles from the center of the city. It is 250 ft. above the level of the lake, commands extensive views, and has been greatly beautified and adorned. Here President James A. Garfield is buried. Two and a half acres on the highest point of the cemetery are reserved for the beautiful monument, beneath which his remains are placed. Three lines of electric railroad run from the Square to the cemetery to accommodate visitors to his grave. *Riverside Cemetery*, on the S. side, has a picturesque location. There are also well-cared-for cemeteries belonging to the Catholics and Jews.

Wade Park, opposite Adelbert College, consists of 83 acres, on which $500,000 have been expended, and eventually it will be one of the finest parks in the West. **Gordon Park,** of about 120 acres, on the lake-shore, in the eastern part of the city, and on which W. J. Gordon had expended large amounts of money, was bequeathed by him to the city in 1892. Other public parks are Lake View Park, on the lake-shore, and Pelton Park, on the S. side.

Leaving Cleveland, the train passes the village of *Berea* and the town of *Elyria*, and in 34 miles reaches *Oberlin* (653 miles from New York), noted as the seat of Oberlin College, from which no person is excluded on account of sex or color. This college, founded in 1834, combines manual labor with study, inculcates entire social equality between whites and blacks, and has had a prosperous career. The next important station is **Toledo** (*Boody House, Burnett House, Hotel Madison, Jefferson House, Merchants' Hotel*, and *Oliver House*), which within a few years has developed from an inconsiderable village into a large and rapidly-growing city. In 1850 the population was 3,820; in 1870 it was 31,693, and in 1890 had reached 81,434. It is situated on the Maumee River, 5 miles from a broad and beautiful bay, and 8 miles from Lake Erie, of which it is regarded as one of the ports. Its commerce is very large, consisting chiefly of the handling of grain; and its manufactures include car-factories, iron-works, locomotive-shops, furniture-factories, flour-mills, and breweries. The city is regularly laid out, having wide streets, that give an easy ascent from the harbor to the table-land on which most of the houses are built, and natural gas from the largest gas-well yet bored is being introduced. It has large and handsome public buildings, several neat parks, street railroads, and costly water-works. Toledo communicates with Cincinnati by the Miami & Erie Canal, and is the converging-point of 17 railroad lines. Six of these lines concentrate at the *Union Depot.* The *Public Library* contains 16,000 volumes, and there are several handsome churches. The principal charitable institutions are the City Hospital, St. Vincent's Hospital, House of Refuge and Correction, Home for Friendless Women, and 3 orphan asylums. Among the public buildings are the *Post-Office* and *Produce Exchange.* The first lunatic asylum on the cottage system erected in Ohio has 34 buildings in Toledo.

At Toledo the road branches, one branch running through Indiana, and known as the Air-Line Division, and the other running through southern Michigan, and known as the Michigan Division. Through trains run by way of both lines, and the same rich agricultural country, numerously sprinkled with small towns, is traversed by both. **Adrian** (769 miles) is the largest city in southern Michigan, with a population, in 1890, of 8,756. It is well built, and has prosperous manufactures. There is a fine Soldiers' Monument to the 77 citizens of Adrian who fell in the civil war, and the Central Union School-building is one of the finest in the West. At *Elkhart* (869 miles) the two divisions of the road unite again; and the route from there to Chicago is through a level prairie-country, which has been well described as having "a face but no features." **South Bend** (894 miles) is a busy manufacturing city of 21,819 inhabitants, one of the chief places in northern Indiana, situated in a great bend of the St. Joseph River, which is navigable to this point and affords a good water-power. The Court-House here is one of the finest buildings in the State, and the University of Notre Dame is a Roman Catholic institution of some note. **La Porte** (921 miles) is a city of 7,126 inhabitants, situated on the edge of the prairie of the same name, and surrounded by a very rich agricultural country. A chain of several beautiful lakes runs N. of the city, which, from their facilities for boating and bathing, are a favorite summer resort. **Chicago** (see Route 71).

68. New York to Chicago via Erie Lines.

By the Erie Lines. Through trains, with complete palace and sleeping car service, run without change of cars from New York to Chicago in about 28 hours. Distances : to Paterson, 17 miles ; to Turner's, 48 ; to Port Jervis, 88 ; to Susquehanna, 193 ; to Binghamton, 215 ; to Elmira, 274 ; to Hornellsville, 331 (to Buffalo, 423 ; to Dunkirk, 460) ; to Salamanca, 413 ; to Corry, 474 ; to Meadville, 515 (to Cleveland, 626) ; to Akron, 615 ; to Mansfield, 682 ; to Chicago, 986.

The Erie R. R. also sells through tickets to Chicago *via* the Lake Shore & Michigan Southern R. R. (with which it connects at Buffalo and Dunkirk), *via* the Canada division of the Michigan Central R. R. (with which it connects at Buffalo), and *via* the Southern Division of the Grand Trunk Railway (with which it connects at Suspension Bridge). All these roads are described in Routes 66 and 67.

THE Erie Railway, which now runs solid trains over its own tracks from New York to Chicago, is one of the greatest achievements of engineering skill in this or any other country, and affords some of the grandest and most varied scenery to be found E. of the Rocky Mountains. Portions of the line were considered impassable to any other than a winged creature, yet mountains were scaled or pierced, and river-cañons passed, by blasting a path from the face of stupendous precipices; gorges of fearful depth were spanned by bridges swung into the air; and broad, deep valleys crossed by massive viaducts. When first completed in 1851, the road, except at a few points, lay through an almost unknown country—a country which was looked upon then pretty much as the Adirondack wilderness is now, but numerous towns and villages have since grown up along the line. The great charm of the Erie Route lies in its romantic and picturesque scenery.[1]

For the first 31 miles the road traverses the State of New Jersey, passing through the great manufacturing city of *Paterson* (17 miles), famed for the beautiful Falls of the Passaic. Just this side of *Suffern Station*, it crosses the line and enters the State of New York, commencing the ascent of the famous Ramapo Valley, which is followed for 18 miles. At *Sloatsburg* (36 miles) the road passes near Greenwood Lake, a noted summer resort, around which are a number of pretty little lakes. *Turner's* (48 miles) is the most picturesque station on this portion of the line. The view from the hill N. of the station is superb, the Hudson River, with Fishkill and Newburg, being in sight. On approaching *Otisville* (76 miles), the eye is attracted by the bold flanks of the Shawangunk Mountain, the passage of which great barrier (once deemed insurmountable) is a great feat of engineering skill. A mile beyond Otisville, after traversing an ascending grade of 40 ft. to the mile, the road runs through a rock-cutting 50 ft. deep and 2,500 ft. long. This passed, the summit of the ascent is reached, and thence we go down the mountain's side many sloping miles to the valley beneath, through the midst of grand and picturesque scenery. Onward the way increases in interest, until it opens in a glimpse, away over the valley, of the mountain-spur known as the *Cuddeback;* and at its base the glittering water is seen, now for the first time, of the Delaware & Hudson Canal. Eight miles beyond Otisville we are imprisoned in a deep cut for nearly a mile, and, on emerging from it, there lie spread before us (on the right) the rich and lovely valley and waters of the *Neversink*. Beyond sweeps a chain of blue hills, and at their feet, terraced high, gleam the roofs and spires of the town of *Port Jervis* (88 miles); while to the S. the eye rests upon the waters of the Delaware, along the banks of which the line runs for the next 90 miles. Three miles beyond Port Jervis the train crosses the Delaware into the State of Pennsylvania, which it traverses for 26 miles to Delaware Bridge, where it again enters New York. Near *Shohola* (107 miles) some of the greatest obstacles of the entire route were encountered, and for several miles the roadway was hewed out of the solid cliff-side at a cost of $100,000 a mile. *Lackawaxen* (111 miles) is a pretty village at the confluence of the Lackawaxen

---

[1] Our description of this route as far as Salamanca is a mere outline or summary, designed to furnish such cursory information about the places and scenery *en route* as may meet the wants of through passengers to the West. Those who desire a more detailed description will find it in the section of the work devoted to the New England and Middle States (Route 41).

Creek and Delaware River. Here the Delaware is spanned by an iron suspension bridge supporting the aqueduct by which the D. & H. Canal crosses the river. The country around *Narrowsburg* (123 miles) was the theatre of the stirring incidents of Cooper's novel, " The Last of the Mohicans." Beyond Narrowsburg for some miles the scenery is uninteresting and the stations unimportant.

At *Deposit* (177 miles) the valley of the Delaware is left, and we begin the ascent of the high mountain-ridge which separates it from the lovely valley of the Susquehanna. As the train descends into the latter valley, there opens suddenly on the right a picture of bewitching beauty. This first glimpse of the *Susquehanna* is esteemed one of the finest points of the varied scenery of the Erie Route. A short distance below we cross the great *Starucca Viaduct*, 1,200 ft. long and 110 ft. high, constructed at a cost of $320,000. From the vicinity of *Susquehanna*, the next station (193 miles), the viaduct itself makes a most effective feature of the valley views. For a few miles beyond Susquehanna the route still lies amid mountain-ridges; but these are soon left behind, and we enter upon a beautiful hilly and rolling country, thickly dotted with villages and towns. **Binghamton** (215 miles) is a flourishing city of 35,005 inhabitants, a railroad center, and the site of the Asylum for Chronic Insane. Twenty-two miles farther is *Owego*, a large and prosperous manufacturing town; and then comes **Elmira** (274 miles), the most important city on the road, with a population of 29,708. At *Hornellsville* (332 miles; 10,996 population) we reach the last and least interesting division of the road, and soon after descend to the Lake Erie level, passing through a wild and desolate region, with few marks of human habitation. At *Salamanca* (413 miles) the train takes the track of the old New York, Pennsylvania & Ohio R. R., now part of the Erie System.

Passengers holding through tickets *via* Buffalo take the Buffalo Div. of the Erie R. R. at Hornellsville, and pass in 92 miles to Buffalo. The scenery at **Portage** on this division is considered by many the finest on the entire road : but the traveler must leave the cars and visit the Falls in order to enjoy it. The famous * Portage Bridge, by which the train crosses the Genesee River, is worthy of attention. At Buffalo the passenger for Chicago takes either of the routes mentioned at the head of this route. Passengers holding tickets *via* Dunkirk continue on the Erie main line from Salamanca, traversing an uninteresting region. At Dunkirk, connection is made with the Lake Shore and Michigan Southern R. R. (see Route 66).

From Salamanca the New York, Pennsylvania & Ohio Division runs along the forest-clad valley of the Alleghany River, enters the Conewango Valley, and in 34 miles reaches **Jamestown,** a city of about 16,038 population (*Everett, Humphrey,* and *Sherman*), a popular summer resort on the Chautauqua Outlet. Here an *Opera-House* has been recently built. **Chautauqua Lake** is the farthest W. of all the New York lakes, being bounded on two sides by Pennsylvania. It is 18 miles long and 1 to 3 wide, and is the highest navigable water east of the Rocky Mountains, being 730 ft. above Lake Erie and 1,291 ft. above the sea. A steamer runs twice a day from Jamestown in 22 miles to **Mayville** (*Chautauqua House, Mayville House*), another popular summer resort at the N. end of the lake. **Chautauqua** (*Athenæum Hotel*), the seat of the Chautauqua Assembly, which holds its sessions during July and August on the shores of Chautauqua Lake, is a handsomely-built summer city, with a number of public buildings, and more than 500 tastefully-designed cottages. The lake and the country in the vicinity afford many delightful excursions. As a resort, Chautauqua holds a high place. But Chautauqua has a broader field than this: it utilizes the summer months in the interest of higher forms of enjoyment, and, in furtherance of this plan, the best lecturers, musicians, and entertainers of the country are engaged. Besides, there are the most complete schools in the world—summer schools, schools of language, literature, art, science; schools of methods for teachers; instruction in practical matters, such as short-hand, type-writing, penmanship, book-keeping; classes for ladies in artistic decorative work, china-painting, wood-carving, and the like; in short, opportunities for work suited to the most varied tastes and requirements. Chautauqua is conducted on a broad, liberal, undenominational basis, and is founded on the idea that recreation means, not idleness, but a change of occupation; and the soundness of this principle is proved by the fact that thousands yearly visit the great assembly. **Point Chautauqua** (*Grand Hotel*), headquarters of the *Baptist Association*, are also reached by steamers from Jamestown. The lake is surrounded by hills 500 to 600 ft. high, and affords some attractive scenery. Passing S. W. from Jamestown, the train soon crosses the line and enters Pennsylvania. **Corry** (474 miles) is a city of 5,677 inhabitants, which has sprung up since 1861 as the product of the oil business. It lies at the entrance of the Pennsylvania Oil Regions (see Route 57), and is at the intersection of several important railways which have given it its prosperity. Beyond Corry the road descends the valley of French Creek, along the banks of which are several of the principal wells in the Oil Region. *Venango* (505 miles) is in this valley. Ten miles beyond Venango is **Meadville** (*Budd House, Commercial Hotel*), a city of 9,520 inhabitants, with important manufactures and an extensive trade with the Oil Regions. It lies on the E. bank of French Creek, and is one of the oldest towns W. of the Alleghanies. The business portion of the city is compactly built, and there are a handsome Court-House, a State Arsenal, an Opera-House, and a Public Library with 3,000 volumes. *Allegheny College* (Methodist) occupies 3 buildings on a hill N. of the city. It was founded in 1817, and has libraries with 12,000 volumes. The *Meadville Theological School* (Unitarian) was established in 1844, and has a library of 12,000 volumes. *Greendale Cemetery*, in the suburbs, is well laid out, and tastefully adorned.

A short distance beyond Meadville the road leaves the French Creek, and, passing several small stations, enters the State of Ohio near *Orangeville* (554 miles), which is the first station in Ohio. From *Leavittsburg* (578 miles) the Mahoning Division diverges, and runs in 49 miles to **Cleveland** (see Route 67). *Ravenna* (598 miles) is a flourishing manufacturing town on the Pennsylvania & Ohio Canal, which affords a good water-power. It is also the point of shipment for large quantities of cheese, butter, grain, and wool. Seventeen miles beyond Ravenna is **Akron** (*Empire Hotel, Hotel Buchtel*), a city of 27,601 inhabitants, at the intersection of the Pennsylvania & Ohio and Ohio & Erie Canals. The canals and the Little Cuyahoga River furnish ample water-power for numerous mills, factories, etc. The chief articles of manufacture are flour and woolen goods. The city is 400 ft. above Lake Erie, being the highest ground on the line of the canal between the lake and the Ohio River. In the vicinity are immense beds of mineral paint, which is exported to all parts of the country. Beyond Akron the road traverses a rich agricultural country, passing 6 or 8 small towns, and soon reaches **Mansfield** (*The Von Hof,* and *Tremont House*), a city of 13,473 inhabitants, compactly built on a beautiful and commanding elevation in the midst of a fertile and populous region. It has a number of handsome public buildings, including several of the churches and school-houses, and the *Court-House.* Many of the residences are costly, and surrounded by spacious ornamental grounds. The principal manufactures are of threshing-machines, machinery, woolens, paper, furniture, and flour.

From Marion, one of the most flourishing towns in Ohio, the Chicago & Erie R. R. extends to Chicago, completing the Erie system between the East and West. The Chicago & Erie was built expressly for the Erie's Chicago connection and is almost an air-line, there being sections of 60 to 70 miles without a perceptible curve or grade.

69. New York to Chicago via Philadelphia and Pittsburgh.

By the Pennsylvania R. R. and the Pennsylvania Lines. The Pennsylvania R. R., formerly a merely local line between Philadelphia and Pittsburgh, is now a vast corporation, including upward of 2,400 miles of track under a single management. It is one of the great highways of traffic and travel between the Atlantic coast and the Western States, and through trains, with complete palace and sleeping car service attached, run through, without change of cars, from New York *via* Philadelphia to Chicago, Cincinnati, St. Louis, and Louisville. The ordinary time from New York to Chicago is 24 to 36 hours. The Chicago Limited Express is a special feature of this route. It consists entirely of vestibule cars, leaves New York at 10 A. M., and runs through in 24 hours, making but eight regular stops on the way. The price of tickets is somewhat higher than by the regular trains. Distances : to Newark, 9 miles ; to New Brunswick, 31 ; to Trenton, 57 ; to Philadelphia, 90 ; to Lancaster, 158 ; to Harrisburg, 195 ; to Altoona, 327 ; to Cresson, 342 ; to Johnstown, 365 ; to Pittsburgh, 444 ; to Mansfield, 620 ; to Fort Wayne, 764 ; to Chicago, 913.

THE station in Jersey City is reached by ferries from the foot of Desbrosses and Cortlandt Sts. The route across New Jersey is through a flat and featureless country, which would be monotonous but for the numerous cities and towns along the line.[1] **Newark** (9 miles), contained 181,831 inhabitants in 1890, and is the largest city and chief manufacturing center of New Jersey, but offers few attractions to the tourist. The Passaic Flour-Mills and the works of the Celluloid Co. are large concerns.

[1] Those who desire a more detailed description will find it in the section of this work devoted to the New England and the Middle States (see Routes 3 and 49).

There are large manufactories of India-rubber goods, boots and shoes, carriages, paper, varnish, and jewelry. *Broad St.* is the principal ·thoroughfare, and the *U. S. Custom-House* and *Post-Office*, the *City-Hall*, the *Court-House*, the *Newark Academy*, and several of the churches, are fine buildings. **Elizabeth** (15 miles) is one of the handsomest cities in New Jersey, and contains many fine residences of New York business men, a few of which are visible from the cars. **New Brunswick** (31 miles) is a city of 18,603 inhabitants, at the head of navigation on the Raritan River, and is noted for possessing one of the most extensive India-rubber factories in the United States, and as the site of *Rutgers College*, an ancient and flourishing institution. *Princeton* (48 miles), is chiefly known as the seat of **Princeton College,** one of the most famous institutions of learning in America, founded in 1746. Several of the college buildings, of which there are fifteen, are handsome structures, standing amid ample, well-shaded grounds. **Trenton** (57 miles; 57,458 population), the capital of New Jersey, is situated at the head of navigation on the Delaware River, and contains some fine public buildings.. The *State-House* is a picturesque old building, occupying a commanding site near the river. The *U. S. Post-Office*, the *State Lunatic Asylum*, the *State Penitentiary*, and the *State Arsenal*, are among the other edifices worthy of notice. The battle which was fought here Jan. 3, 1777, was a turning-point in the Revolution. On leaving Trenton, the train crosses the Delaware on a bridge 1,100 ft. long, and follows the right bank of the river to *Frankford*, where it turns W. and swings round the N. portion of the great city of **Philadelphia,** the third ·in population in the United States, to· the Broad St. station, cor. Broad and Market Sts. The city of Philadelphia is fully described in Route 4.

Leaving the station, the train passes in sight of Fairmount Park, traverses a pleasant suburban region, and enters one of the richest agricultural districts in America, which is traversed for nearly 100 miles. The size and solidity of the houses and barns, and the perfection of the cultivation, will be apt to remind the tourist rather of the best farming districts of England than of what he usually sees in the United States. *Paoli* (109 miles) was the scene of a battle fought Sept. 20, 1777, in which the British under General Gray surprised and defeated the Americans under Gen. Wayne. The battle is commonly called the "Paoli Massacre," because a large number of the Americans were killed after they had laid down ·their arms. A marble monument, erected in 1817, marks the site of the battle-field. Beyond Paoli the scenery grows more picturesque, and fine views are had of the beautiful Chester Valley. *Downington* (122 miles) is the terminus of the Chester Valley R. R. (branch Philadelphia & Reading R. R.), and is near the marble quarries which supplied the marble from which Girard College in Philadelphia was built. At *Coatesville* (128 miles) the W. branch of the Brandywine is crossed on a bridge 850 ft. long and 75 ft. high. *Gap* (141 miles) is so named because it lies in the gap by which the road passes from the Chester Valley to the Paquea Valley. The scenery in the vicinity is attractive. · **Lancaster** (158 miles) is pleasantly situated near the

Conestoga Creek, which is crossed in entering the city. It was incorporated in 1818, and was at one time the principal inland town of Pennsylvania, being the seat of the State government from 1799 to 1812. It is now a prosperous manufacturing city of 32,011 inhabitants, containing many fine public buildings, among which are the *Court-House*, the *County Prison, Fulton Hall*, and *Franklin and Marshall College* (Dutch Reformed). Lancaster has extensive manufactures of locomotives, axes, carriages, etc., and has navigable communication by canal and river with Baltimore. The only station between Lancaster and Harrisburg which requires mention is *Middletown* (186 miles), an important shipping-point on the Susquehanna River at the mouth of Swatara Creek. It has extensive iron-works and machine-shops, and is the terminus of the Union Canal. Nine miles beyond is **Harrisburg** (195 miles; 39,385 population), the capital of the State of Pennsylvania, situated on the E. bank of the Susquehanna River, which is here a mile wide and spanned by two bridges. The city is handsomely built, and is surrounded by beautiful scenery. The *State-House*, finely situated on an eminence near the center, is a spacious brick building in the classic style, and is plainly visible from the cars. The other important public buildings are the *State Arsenal*, the *Court-House*, the *State Lunatic Asylum*, the *County Prison*, the market-houses, the school-houses, and several handsome churches. The iron-manufactures of Harrisburg are extensive, and six important railways converge here.

About 5 miles above Harrisburg the railroad crosses the Susquehanna on a splendid bridge, 3,670 ft. long; the * view from the center of this bridge is one of the finest on the line. Near *Cove Station* (10 miles from Harrisburg) the Cove Mt. and Peter's Mt. are seen; and from this point to within a short distance of Pittsburgh the scenery is superb, and in places grand beyond description. *Duncannon* (210 miles) is at the entrance of the beautiful Juniata Valley, which is followed for about 100 miles to the base of the Alleghany Mts. The landscape of the Juniata is in the highest degree picturesque; the mountain background, as continuously seen across the river from the cars, being often strikingly bold and majestic. The passage of the river through the Great Tuscarora Mt., 1 mile W. of *Millerstown* (228 miles), is especially fine. Four miles beyond *Mifflin* (244 miles) the train enters the wild and romantic gorge known as the * **Long Narrows,** which is traversed by the railway, highway, river, and canal. *Mount Union* (281 miles) is at the entrance of the gap of Jack's Mountain; and 3 miles beyond is the famous Sidling Hill, and, still further W., the Broad Top Mountain. *Huntingdon* (293 miles) is a flourishing village on the Juniata, finely situated, and surrounded by beautiful scenery.

At *Petersburg* (300 miles) the railroad parts company with the canal and follows the Little Juniata, which it again leaves at *Tyrone* (313 miles) to enter the Tuckahoe Valley, famous for its iron-ore. At the head of the Tuckahoe Valley and at the foot of the Alleghanies is **Altoona** (327 miles), a handsome city of 30,337 inhabitants, built up since 1850, when it was a primitive forest, by being selected as the site of the vast machine-shops of the Pennsylvania R. R. The trains stop here for meals, and many travelers arriving in the evening remain over night in order to cross the Alleghanies by daylight. Just beyond Altoona the ascent of the Alleghanies begins, and in the course of the next 11 miles some of the finest scenery and the greatest feats of engineering on the entire line are to be seen. Within this distance the road mounts to the tunnel at the summit by so steep a grade that, while in the ascent double power is required to move the train, the entire 11 miles of descent are run without steam, the speed of the train being regulated by the " brakes." At one point there is a curve as short as the letter U, and that, too, where the grade is so steep that in looking across from side to side it seems that, were the tracks laid contiguous to each other, they would form a letter X. The road hugs the sides of the mountains, and from the windows next to the valley the traveler can look down on houses and trees dwarfed to toys, while men and animals appear like ants from the great elevation. Going W., the left-hand, and coming E., the right-hand, side of the cars is most favorable for enjoying the scenery. The summit of the mountain is pierced by a tunnel 3,670 ft. long, through which the train passes before commencing to descend the W. slope. The much-visited **Cresson Springs** (*Mountain House*) are 2½ miles beyond the tunnel, 3,000 ft. above the sea. There are 7 mineral springs here, and the hotels and cottages are apt to be thronged in summer. In descending the mountains from Cresson, the remains of another railroad are constantly seen, sometimes above and sometimes below the track followed by the trains. This was the old Portage R. R., by which, in the ante-locomotive days, loaded canal-boats were carried over the mountains in sections by inclined planes and joined together at the foot. The stream, which is almost continuously in sight during the descent, is the Conemaugh Creek, which is crossed by a stone viaduct near *Conemaugh Station* (363 miles), the terminus of the mountain division of the road.

From the foot of the mountains to Pittsburgh the road traverses a rich farming region, the scenery of which, though pleasing, will be apt to seem somewhat tame after the magnificent panorama of the Alleghanies. *Johnstown* (365 miles) is a busy manufacturing borough at the confluence of the Conemaugh with Stony Creek. The Cambria Iron-Works, to the right of the road, are among the largest in America. At *Blairsville Intersection* (390 miles) the road branches, the main line running to Pittsburgh by *Latrobe* (403 miles) and *Greensburg* (413 miles); while the Western Pennsylvania Div. runs to Allegheny City by *Blairsville* (393 miles). The former is the route followed by the through trains. **Pittsburgh** (444 miles) is the second city of Pennsylvania in population and importance, and one of the chief manufacturing cities of the United States. It occupies the delta at the confluence of the Alleghany and Monongahela Rivers, with several populous suburbs on the opposite banks, and in 1890 had a total population of 238,617. **Allegheny City,** with a population of 105,287, lies just across the Alleghany River, and contains many costly residences of Pittsburgh merchants. In both cities are numerous places of interest, including the great Westinghouse electric plant and the iron-works of Carnegie, Phipps & Co., in seeing which several days may be pleasantly and profitably spent, and such tourists as can spare the requisite time should consult the detailed description of the two cities given in Route 50. The extent of its steel, glass, and iron manufactures has given it the appellation of the "Iron City," while the heavy pall of smoke that formerly overhung it, before the introduction of natural gas, caused it to be styled the "Smoky City."

After a short stop in the Union Depot at Pittsburg, the train passes out on the tracks of the Pittsburg, Fort Wayne, & Chicago Div., crosses the Alleghany River in full view of several handsome bridges, runs through the heart of Allegheny City, and sweeps past a number of small suburban villages to *Rochester* (26 miles from Pittsburgh, 470 from New York), at the confluence of the Ohio and Beaver Rivers. From Rochester the train runs N. up the Beaver River, passing the busy manufacturing towns of *New Brighton* and *Beaver Falls*, and at *Home-wood* (479 miles) turns W., and in about 15 miles enters the State of Ohio. *Salem* (514 miles) is the first important station in Ohio, and is a neat manufacturing town, surrounded by a highly-cultivated farming country. At *Alliance* (528 miles) the through cars for Cleveland take the track of the Cleveland & Pittsburg R. R., and· run in 3½ hours to **Cleveland** (see Route 67). The Chicago train passes on to **Canton** (546 miles, *Melbourne Hotel*), a city of 26,189 inhabitants, beautifully situated on Nimishillen Creek, and surrounded by a fertile farming country, which enjoys the distinction of sending more wheat to market than any other portion of the State. Bituminous coal and limestone are found in the vicinity, and considerable manufacturing is carried on. **Massillon** (554 miles, *Hotel Conrad*) is a flourishing manu-facturing city of 10,092 inhabitants, situated on the Tuscarawas River and the Ohio Canal, by which it has water communication with Lake Erie. It is regularly laid out, is substantially and compactly built, and contains many handsome residences and an Opera-House costing $100,000. It is surrounded by one of the most productive coal-fields of the State, and the coal obtained here has a wide reputation. The Massillon white sand-stone, which is largely quarried, is shipped to all parts of the country. Large shipments of iron-ore, wool, flour, and grain are also made, and the manufactures are varied and important. Several small stations are now passed, of which the principal is *Wooster* (579 miles), and then comes **Mansfield** (620 miles), which has already been described (see Route 68). From this point to Chicago the route is the same as in Route 68.

70. New York to Chicago and Cincinnati, via Baltimore and Washington.

By the Baltimore and Ohio R. R., which forms one of the great through routes between the Atlantic seaboard and the Western States. With its various branch lines it controls over 2,000 miles of road, and has for its western termini the principal cities of the interior. Through trains, with drawing-room and sleeping car service attached, run without change from New York to Chicago, Columbus, Cincinnati, and St. Louis. The time to Chicago is 30 hours ; to Cincinnati, about 23 hours ; to St. Louis, 32 hours. Distance from New York to Baltimore, 186 miles. Distances from Baltimore : To Relay Station, 9 miles ; to Washington, 40 ; to Point of Rocks, 69 ; to Harper's Ferry, 95 ; to Martinsburg, 118 ; to Cumberland, 192 ; to Grafton, 294 (to Parkersburg, 398 ; to Chillicothe, 495 ; to Cincinnati, 593 ; to Louisville, 721 ; to St. Louis, 934) ; to Bellaire, 390 (to Wheeling, 393) ; to Zanesville, 468 ; to Newark, 494 (to Columbus, 526) ; to Mansfield, 548 ; to Chicago Junction, 584 ; to Chicago, 853.

THE first section of this route is *via* the Bound Brook route to *Philadelphia* (fully described in Part I, Route 3 *b*). From Philadelphia to Baltimore the route is by way of the Baltimore & Ohio R. R., whose route was completed in 1886. The terminal station in Baltimore is in Camden St., near Howard, but the through trains to the West are transferred by ferry-boat across the river at Camden, thus avoiding the circuit through the city.

The grandeur of the scenery along the line of the Baltimore & Ohio R. R. makes it one of the most attractive routes that tourists can take. After leaving Baltimore, the first object of interest is the *Carrollton Viaduct*, a fine bridge of dressed granite, with an arch of 80 ft. span, over Gwinn's Falls, beyond which the road soon enters the long and deep excavation under the Washington turnpike. Less than a mile farther is the "deep cut," famous for its difficulties in the early history of the road. It is 76 ft. deep, and nearly ½ mile long, and just beyond is the deep ravine of Robert's Run. At *Relay Station* (9 miles) the road branches, the main line striking westward through *Ellicott City* (14 miles), *Elysville* (20 miles), *Mount Airy* (42 miles), and *Monocacy* (58 miles) ; while the Washington Branch diverges to Washington City. The latter route is the one taken by the through trains. Just beyond Relay Station the famous * *Washington Viaduct* is crossed, a magnificent piece of masonry whose arches rest on seven lofty piers. The scenery in this vicinity is very attractive, and a fine summer hotel has been erected on the E. side of the river. *Elk Ridge* (10 miles) is a small manufacturing village on the Patapsco; and *Hanover* (12 miles) is near the iron-mines which supply the Avalon Furnaces. At *Annapolis Junction* (19¼ miles) the Annapolis, Washington & Baltimore R. R. diverges to Annapolis, the capital of Maryland, which is worth a visit if the traveler have time. From *Alexandria Junction* (34 miles) a branch road diverges to Shepherd, opposite Alexandria, and 5 miles beyond the train enters **Washington City,** the capital of the Republic. The first view of the Capitol, in approaching the city from this direction, is exceedingly fine, and should not be lost—the dome presents "such majesty and whiteness as you never saw elsewhere." Owing to the number and magnificence of its public buildings, Washington is one of the most interesting cities in America, and no tourist should pass through without stopping at least long enough to visit its principal places of interest. A detailed description of the city and its environs will be found in the portion of the book devoted to the New England and Middle States (Route 8).

From Washington to Point of Rocks (where the main line is again reached) the road traverses a beautiful champaign country, extending to the Catoctin Mountains, a continuation of the Blue Ridge. *Point of*

Rocks (69 miles) takes its name from a bold promontory, which is formed by the profile of the Catoctin Mountain, against the base of which the Potomac River runs on the Maryland side, the mountain towering up on the opposite (Virginia) shore forming the other barrier to the pass. The railroad passes the Point by a tunnel, 1,500 ft. long, cut through the solid rock. Beyond, the ground becomes comparatively smooth, and the railroad, leaving the immediate margin of the river to the Chesapeake & Ohio Canal, runs along the base of gently-sloping hills, passing the villages of Berlin and Knoxville, and reaching the Weverton factories, in the pass to the *South Mountain*, near which was fought the désperate battle of South Mountain (Sept. 14, 1862). From South Mountain to Harper's Ferry the road lies along the foot of a precipice for the greater part of the distance of 3 miles, the last of which is immediately under the rocky cliffs of Elk Mountain, forming the N. side of this noted pass. The Shenandoah River enters the Potomac just below the bridge over the latter, and their united currents rush rapidly over the broad ledges of rock which stretch across their bed. The length of the bridge, over river and canal, is about 900 ft., and at its W. end it bifurcates, the left-hand branch connecting with the Valley Branch of the B. & O. R. R., which passes directly up the Shenandoah, and the right-hand carrying the main road, by a strong curve in that direction, up the Potomac. **Harper's Ferry** (81 miles) is delightfully situated in Jefferson Co., W. Virginia, at the confluence of the Potomac and Shenandoah Rivers, the town itself being compactly but irregularly built around the base of a hill. Before the civil war it was the seat of an extensive and important United States armory and arsenal; but these were destroyed during the war, and have not been rebuilt. The scenery around Harper's Ferry is wonderfully picturesque. Thomas Jefferson pronounced the passage of the Potomac through the Blue Ridge "one of the most stupendous scenes in nature, and well worth a voyage across the Atlantic to witness." The tourist should stop here for at least one day, and climb either Maryland Heights (across the Potomac) or Bolivar Heights (above the town). Apart from its scenery, the chief interest pertaining to Harper's Ferry (which is now a decadent village of only 958 inhabitants) is historical. It was the scene of the exploits which in October, 1859, rendered the name of John Brown, of Ossawattomie-Kansas notoriety, still more notorious; and Charlestown, the county-seat where Brown and his followers were tried and executed, is only 7 miles distant on the road to Winchester. During the civil war Harper's Ferry was alternately in the hands of the Federals and Confederates, and a detailed narrative of its changing fortunes would reflect with fidelity the vicissitudes of the war itself.

A short distance beyond Harper's Ferry the road leaves the Potomac and passes up the ravine of Elk Branch, which, at first narrow and serpentine, widens gradually until it almost loses itself in the rolling table-land which characterizes the "Valley of Virginia." The head of Elk Branch is reached in about 9 miles, and thence the line descends gradually over an undulating country to the crossing of Opequan Creek. Beyond the crossing, the road enters the open valley of Tuscarora Creek, which it crosses twice, and follows to **Martinsburg** (100 miles), where the railroad company have built extensive shops. The town contains some 7,226 inhabitants, and is pleasantly situated on an elevated plateau above Tuscarora Creek, which affords a fine water-power. Much fighting occurred in this vicinity during the civil war, and in June, 1861, the Confederates destroyed 87 locomotives and 400 cars belonging to the B. & O. R. R. The Cumberland Valley R. R. (Route 58) runs from Winchester to Harrisburg in 116 miles, passing through Martinsburg. Seven miles beyond Martinsburg the road crosses North Mountain by a long excavation, and enters a poor and thinly-settled district covered chiefly with a forest in which stunted pine prevails. The Potomac is again reached at a point opposite the ruins of Fort Frederick, on the Maryland side. *Sir John's Run* (128 miles) is but a few miles from **Berkeley Springs** (see Route 130), and just beyond the station the track sweeps around the Cacapon Mountain, opposite the remarkable insulated hill called "Round Top." The next point of interest is the *Doe Gulley Tunnel* (1,200 ft. long), the approaches to which are very imposing. The *Paw-Paw Tunnel* is next reached, and, after passing through some 20 miles of rugged and impressive scenery, the train crosses the N. branch of the Potomac by a viaduct 700 ft. long and enters Maryland. **Cumberland** (*Queen City Hotel*) is in the mountain-region of the narrow strip which forms the W. part of Maryland, and in point of population and commerce is its second city. The entrance to the city is beautiful, and displays the noble amphitheatre in which it lies to great advantage. The city itself has a population of some 12,729, and is the site of the great rolling-mills of the R. R. Co., for the manufacture of steel rails. A few miles W. of Cumberland, upon the summit of the Alleghanies, begins the district known as the Cumberland Coal Region, which extends W. to the Ohio River. Vast quantities of this coal are sent E. by the railroad and by the Chesapeake & Ohio Canal, which has its W. terminus at Cumberland and runs to tide-water at Georgetown. At Cumberland the Pittsburg Div. diverges, and runs in 150 miles to **Pittsburg** (see Route 69). From Cumberland to Piedmont (28 miles) the scenery is remarkably picturesque. For the first 22 miles to the mouth of New Creek, the Knobly Mountain bounds the valleys of the N. branch of the Potomac on the left, and Will's and Dan's Mountains on the right; thence to Piedmont, the river lies in the gap which it has cut through the latter mountains. The crossing of the Potomac from Maryland to W. Virginia is 21 miles from Cumberland, and the view from the bridge, both up and down the river, is very fine. At *Piedmont* (206 miles) the ascent of the Alleghanies is commenced, and *Altamont* (223 miles) is upon the extreme summit of the range, 2,720 ft. above the sea. From Altamont westward for nearly 20 miles are beautiful natural meadows (known as the "Glades") lying along the upper waters of the Youghiogheny River and its numerous tributaries, divided by ridges of moderate elevation and gentle slope, with fine ranges of mountains in the background. Three miles beyond Altamont is the *Deer Park Hotel*, a first-class summer hotel, built and managed by the railroad company. It is 2,800 ft. above the sea, and is surrounded by grand scenery. At *Oakland* (6 miles beyond Deer Park, *Glades*

Hotel, Oakland Hotel). are some excellent trout-streams, and game in the adjacent forests. The descent from the summit plateau to Cheat River presents a succession of very heavy excavations, embankments, and tunnels; and at the foot the famous **Cheat River Valley** is crossed, with fine views on either side. For several miles on this part of the line the road runs along the steep mountain-side, presenting a succession of magnificent landscapes. Descending from Cassidy's Ridge, which forms the W. boundary of Cheat River Valley, the train soon reaches the *Kingwood Tunnel*, which is 4,100 ft. long and cost $1,000,000; and, 2 miles beyond, *Murray's Tunnel*, 250 ft. long. **Grafton** (280 miles) is at the end of the mountain section of the road, and is a village of 3,156 inhabitants, picturesquely situated on the Tygart's Valley River.

At Grafton the Parkersburg Branch diverges to Parkersburg and Cincinnati (see page 315). The Chicago train runs N. W. down the Tygart's River Valley, amid a variety of pleasing scenery, and in 20 miles reaches *Fairmont*, at the head of navigation on the Monongahela River, which is here spanned by a fine suspension-bridge 1,000 ft. in length, connecting Fairmont with the village of Palatine. Just beyond Fairmont the road leaves the valley of the Monongahela, and ascends the winding and picturesque ravine of Buffalo Creek. At the head of the valley, 23 miles from Fairmont, the road passes the ridge by deep cuts and a tunnel 350 ft. long, and descends the other side by the valley of Church's Fork of Fish Creek, through many windings and tunnels. Just beyond *Littleton* (337 miles) the road passes through *Broad-Tree Tunnel* under a great hill, which was originally crossed on a zigzag track with seven angles representing seven V's, and enters the Pan-Handle of West Virginia. *Moundsville* (368 miles) is one of two villages on the Ohio at the mouth of Grave Creek, the other being *Elizabethtown*. The approach to the Ohio at this point is very beautiful. The line, emerging from the defile of Grave Creek, passes straight over the "flats" that border the river, forming a vast rolling plain, in the middle of which looms up the great *Indian Mound*, a relic of the prehistoric inhabitants of America, 80 ft. high and 200 ft. broad at the base. About 3 miles up the river from Moundsville the "flats" terminate, and the road passes for a mile along rocky narrows washed by the river, after which it runs over wide, rich, and beautiful bottom-lands all the way to *Benwood* (375 miles), where the river is crossed and connection made with the Central Ohio Division. Four miles from Benwood, on the same side, is **Wheeling** (*Hotel Windsor, McClure House, Stamm House*), the former capital of West Virginia and a city of 35,013 inhabitants. It has a large commerce on the Ohio River, and its manufacturing interests are extensive, including iron and nail mills, glass-works, etc. The National Road crosses the Ohio here by a graceful suspension bridge, 1,010 ft. long, and the railroad bridge (below the city) is one of the finest in the country. The *Custom-House*, of stone, also contains the Post-Office and the U. S. Court-room. On the transfer of the capital of the State to Charleston, the *Capitol* building was converted into county and city offices. There are an Odd-Fellows' Hall, a Public Library with 5,000 volumes, and an Opera-House. There are also several costly and ornate school-buildings. Near the city is an extensive Fair Ground, with a trotting-course.

At Benwood the Chicago train crosses the Ohio River to *Bellaire* (376 miles), whence the Central Ohio Div. runs in about 100 miles to Newark, through a productive and populous country. The principal station on this portion of the line is **Zanesville** (*Clarendon Hotel, Kirk House*), a city of 21,009 inhabitants, situated on both sides of the Muskingum River at the mouth of Licking River. The Muskingum is here crossed by an iron railroad-bridge 538 ft. long, and by 3 other bridges. The city is well built, with wide, regular streets, lighted with gas, and has water-works costing over $500,000, street railroads, and a stone *Court-House* costing $300,000. Several of the school-buildings are remarkably handsome, and the *Zanesville Athenæum* has a reading-room and a library of 5,500 volumes. The country around Zanesville is fertile, and is the source of a profitable trade; but the chief interest is manufacturing, for which facilities are afforded by the water-power in the rivers, and the bituminous coal, iron-ore, limestone, and clays of the adjacent region. The Muskingum River is navigable to *Dresden,* 17 miles above the city. **Newark** (480 miles) is at the crossing of the Pittsburg, Cincinnati & St. Louis R. R. (Route 73), and is there described. From Newark the Central Ohio Div. passes W. in 33 miles to **Columbus** (see Route 73). The Lake Erie Div. of the B. & O. R. R. runs N. W. through a rich agricultural region, by numerous small towns to **Mansfield** (542 miles), which is at the junction of Routes 67 and 68, and which has already been described in Route 68. *Shelby Junction* (554 miles) is at the crossing of Route 72. From *Chicago Junction* (568 miles) the Lake Erie Div. continues N. to **Sandusky** (596 miles), on Lake Erie, while the Chicago Div. diverges and runs W. across northern Ohio and Indiana. Among the towns *en route* are such important ones as Tiffen, Fostoria, Defiance, and Garrett. The country traversed, though extremely fertile and productive, offers few picturesque features. *Defiance* (656 miles) is at the crossing of the Wabash R. R. (Route 79). At *Garrett* (696 miles) most of the trains stop for meals, and between this place and Chicago there is little to attract attention except the numerous railways that are intersected. **Chicago** (see Route 71).

From Grafton to Cincinnati.

At Grafton, as already mentioned (see p. 314), the Parkersburg Div. diverges from the main line and runs W. to the Ohio River. It passes through a country which is well wooded, and rich in coal and petroleum, but without interest for the tourist, though some rugged mountain scenery is occasionally seen from the cars. *Clarksburg* (302 miles from Baltimore) is the first station of any consequence, and is situated on a high table-land on the W. bank of the Monongahela River, surrounded by hills. It has some 3,008 inhabitants, and in the vicinity there are valuable mines of bituminous coal. *Petroleum* (362 miles) is in the rich Oil Regions of West Virginia, and from *Laurel Fork Junction* (364 miles) a branch road leads N. to *Volcano,* the most important place in the Oil Region. At *Claysville* (377 miles) the Little Kanawha River is reached, and the train follows it for 7 miles to **Parkersburg** (*Central Hotel, Hill's*), a city of 8,408 inhabitants, with a large trade in petroleum. Here the train crosses the Ohio River to *Belpré* on a splendid bridge, 1¼ mile long, with 6 spans over the river and 43 approaching spans, completed in 1871 at a cost of over $1,000,000. At Belpré the train passes on to the tracks of the Baltimore & Ohio Southwestern R. R., and in 37 miles reaches **Athens,** one of the largest towns of S. Ohio, with a population of 2,620, and considerable trade with the surrounding country. It is pleasantly situated on the Hocking River, and is the seat of the *Ohio University*, founded in 1804, and the oldest college in the State. One of the *State Lunatic Asylums* is also located here, and in the vicinity are several Indian mounds, similar to the one at Moundsville (see p. 314). Several small stations are now passed, and then comes the flourishing city of **Chillicothe,** beautifully situated on a hill-environed plateau, through which flows the Scioto River. Chillicothe was settled in 1796, and from 1800 to 1810 was the seat of the State government, which was afterward removed to Zanesville and then to Columbus. It had in 1890 a population of 11,288, and is the center of nearly all the trade of the rich farming country bordering on the Scioto, one of the finest agricultural districts in the United States. Its manufactures are also important, including carriage and wagon factories, flour-mills, machine-shops, a paper-mill, shoe-factory, etc. The city is regularly laid out, the principal avenues following the general course of the river, and being intersected at right angles by others, all lighted by electricity. The two main streets, which cross each other in the center of the city, are each 99 ft. wide; Water St., facing the river, is 81½ ft. wide, and the width of the others is 66 ft. There are many handsome public buildings, including 14 churches, 5 brick school-houses, a *Public Library* of 12,000 volumes, and a *Court-House*, built of stone, at a cost of over $100,000. The Ohio & Erie Canal passes through the city. Between Chillicothe and Cincinnati there is no place requiring mention, though the traveler through this portion of Ohio can not but be struck with the neatness of the villages, the fertility of the land, and the high state of cultivation to which it has been brought. **Cincinnati** (see Route 75).

71. Chicago.

Hotels.—Among the hotels of Chicago are some of the largest in the world. The *Auditorium*, Congress St., Wabash and Michigan Aves.; the *Great Northern*, Dearborn, Quincy, and Jackson Sts.; the *Grand Pacific*, La Salle, Jackson, and Clark Sts.; the *Leland*, Michigan Ave. and Jackson St.; the *Metropole*, Michigan Ave. and 23d St; the *Palmer House*, State and Monroe Sts.; the *Sherman House*, Clark and Randolph Sts.; the *Richelieu*, Michigan Ave., between Jackson and Van Buren Sts.; the *Tremont House*, Lake and Dearborn Sts.; and the *Wellington*, Wabash Ave. and Jackson St., are all first-class houses, and vary in prices from $2.50 a day upward. Good hotels on a more modest scale are the *Atlantic*, Van Buren and Sherman Sts.; the *Briggs House*, Randolph St. and 5th Ave.; the *Clifton House*, Wabash Ave. and Monroe St.; and the *Gault House*, W. Madison and Clinton Sts. Good hotels on the European plan are the *Brevoort*, in Madison St., between La Salle and Clark Sts.; *Gore's Hotel*, at 266 S. Clark St.; *Grace*, on Clark and Jackson Sts.; *McCoy's*, Clark and Van Buren Sts.; and the *Saratoga*, at 155 Dearborn St.; while the *Drexel*, at the entrance to Washington Park; the *Woodruff*, Wabash Ave. and 21st St.; *Hyde*

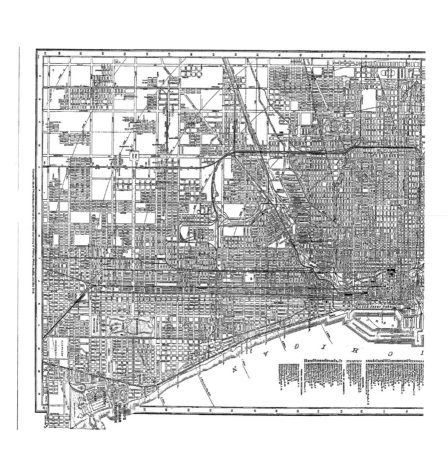

Park. Lake Ave. and 51st St.; the *Lexington,* Michigan Ave. and 22d St.; and the *Virginia,* at 78 Rush St., are more strictly family hotels. Many new hotels have sprung up in the "World's Fair district," some of which are first-class both in accommodations and prices. (See APPENDIX.)

Restaurants.—Besides the restaurants connected with the hotels, first-class' eating-houses will be found conveniently located along the principal thoroughfares. Among these may be mentioned *Kinsley's,* 105 Adams St.; *Rector's* and *Race's,* at cors. Monroe and Clark Sts.; *The American,* at Adams and State Sts.; *Thomson's,* Tribune block, cor. Madison and Clark Sts.; *Werner's,* in Madison St.; the *Boston Oyster-House,* cor. Madison and Clark Sts.; the *Chicago Oyster-House,* near by; *Col. Wilson's* shell-fish house, in Washington St. near Wabash Ave.; and *Vienna Bakeries,* in Madison St. near 5th Ave. and at 36 Washington St. Men may get first-class meals at *Chapin & Gore's,* in Monroe St.; at *Kern's,* La Salle and Washington Sts.; and at *Billy Boyle's Chop-House,* in alley off Dearborn, between Madison and Washington Sts.

Modes of Conveyance.—The street-car service includes cable, electric, elevated, and horse-car systems (fare, 5c.). *Parmelee's omnibuses* are in waiting at the stations, and carry passengers to hotels or stations with baggage (fare, 50c.). *Carriages* are in waiting at the stations and steamboat-landings, as well as at hotels, and at the Court-House and City Hall. The legal rates of fare are as follows : For cabs and other one-horse conveyances, one mile or fraction thereof, for each passenger, for the first mile, 25c.; for each succeeding mile or fraction thereof, for one or more passengers, 25c. By the hour, cab-charges will be : For the first hour, 75c.; for each succeeding quarter-hour, 20c. For service outside the city limits or in the parks : For the first hour, $1 ; for each succeeding quarter-hour, 25c. For hacks and other vehicles drawn by two horses, rates will be as follows : One or two passengers from one railroad depot to another, $1 ; one or two passengers not exceeding one mile, $1 ; one or two passengers any distance over one mile and less than two miles, $1.50 ; each additional passenger of the same party, 50c.; conveying one or two passengers any distance exceeding two miles, $2 ; each additional passenger of the same party, 50c. Children between 5 and 14 years of age will be carried at one-half the rates named for hacks and two-horse vehicles ; and children under 5 years will be carried free not exceeding one mile. Hackney-coach or other two-horse vehicle, $8 a day ; by the hour, including stops, for the first hour, $2; for each additional hour or fraction thereof, $1. At the principal depots and street-corners are the Gurney cabs and the Hansom cabs (25c. for each person from one point to another, or 75c. per hour for two persons).

Railroad Stations.—There are 6 *Union Railway Stations* in the city. The Pittsburg, Ft. Wayne & Chicago R. R., Chicago, Milwaukee & St. Paul R. R., Chicago & Alton R. R., Chicago, Burlington & Quincy R. R., and Pittsburg, Cincinnati, Chicago, & St. Louis R. R. (Pennsylvania system) depart from the station cor. *Canal and Adams Sts.* The Chicago & Northwestern R. R., with all its divisions, occupies a large station of its own, on the cor. of N. Wells and Kinzie Sts. The Chicago, Rock Island & Pacific R. R., Lake Shore & Michigan Southern R. R. (New York Central connection), and the New York, Chicago & St. Louis R. R. depart from the station cor. *Van Buren and Sherman Sts.* The Wabash R. R., Chicago & Grand Trunk R. R., Chicago & E. Illinois R. R., the Santa Fé system, and the Chicago & Atlantic R. R. (Erie system) arrive and depart from the Louisville, New Albany & Chicago (Monon. route) station at the *foot of Dearborn St.,* cor. Polk St. and 3d Ave. The Michigan Central R. R., Illinois Central R. R., and Cleveland, Cincinnati, Chicago & St. Louis R. R. (the "Big Four") depart from the *Central Station, foot of Lake and Randolph Sts.* The Baltimore & Ohio R. R., the Chicago, Detroit & Niagara Falls R. R., and the Wisconsin Central (Northern Pacific System) have their station at *Harrison St. and 5th Ave.*

Theatres and Amusements.—*Alhambra,* cor. State St. and Archer Ave.; *Auditorium,* in Congress St., between Wabash and Michigan Aves.; *Chicago Opera-House,* Clark and Washington Sts.; *Columbia Theatre,* in Monroe, between Clark and Dearborn Sts.; *Criterion,* cor. Sedgwick and Division Sts.; *Germania Theatre,* in Randolph, near Clark St.; *Grand Opera-House,* in Clark, between Washington and Randolph Sts.; *Halsted Street Op'ra-House,* Halsted and Harrison Sts.; *Harlin's Theatre,* in Wabash Ave., between 18th and 20th Sts.; *Haymarket Theatre,* in W. Madison St., between Halsted and Union Sts.; *Hooley's Theatre*, in Randolph, between La Salle and Clark Sts.; *Jacob's Academy*, in S. Halsted St., near W. Madison St.; *Jacob's Clark Street Theatre*, in N. Clark St., near the bri¹ge; *Lyceum Theatre*, in Desplaines St., between Madison and Washington Sts.; *Madison Street Theatre*, in Madison St., between Dearborn and State Sts.; *McVicker's Theatre*, in Madison, between Dearborn and State Sts.; *Park Theatre*, in State, between Congress and Harrison Sts.; *People's Theatre*, in State, between Congress and Harrison Sts.; *Schiller Theatre*, Washington St, between Dearborn and Clark Sts.; *Standard Theatre*, cor. Halsted and Jackson Sts.; *Timmerman's Opera-House*, cor. Stewart Ave. and 63d St.; *Waverly Theatre*, in W. Madison St., between Throop and Loomis St.; and *Windsor Theatre*, cor. N. Clark and Division Sts., are the leading places of amusement. There are 2 *Cyclorama Buildings* on the cor. of Wabash Ave. and Panorama Pl. The *Chicago Fire Cyclorama* is at Michigan Ave. and Monroe St. The *Casino*, in Wabash Ave. near Adams St., contains an exhibition of wax-works. The *Central Music Hall*, cor. Randolph and State Sts., and the *Methodist Church Block Hall*, are used for lectures and concerts, and generally for religious services on Sunday. Besides several cheaper museums, the *Libby Prison Museum*, in Wabash Ave., between 14th and 16th Sts., is worthy of a visit, as it contains many relics of the civil war, and other historical curiosities. *Steele MacKaye's Spectatorium* is in Stony Island Ave., N. of the Exposition Grounds.

Reading-Rooms.—At all the leading hotels there are reading-rooms for the use of guests, well supplied with newspapers. The *Public Library*, one of the best in the country, contains about 175,000 volumes, has an excellent reading-room, containing newspapers and magazines from all parts of the civilized world, and is open from 9 A. M. to 10 P. M. This will occupy a building (now in course of erection) at Michigan Ave., Randolph and Washington Sts. The *Chicago Athenæum*, 44 and 54 Dearborn St., has a system of night and day lectures for mechanics, a large library and gymnasium, and is open from 8 A. M. to 9 P. M. The *Young Men's Christian Association* has a library and reading-room at 148 Madison St. (*Farwell Hall*), to which all are welcome (open from 8 A. M. to 10 P. M.). The *Academy of Sciences*, in Wabash Ave. near Van Buren St., has a small library and a museum. The *Chicago Historical Society Library*, at 142 Dearborn St., has a growing collection of historical books. The *Union Catholic Library*, 94 Dearborn St., has a small collection of books. The great *Newberry Free Library*, endowed with property valued at upward of $2,000,000, is temporarily at the corner of Oak and North State Sts., and is a reference library only. Its new building, to occupy the entire block on the S. E. cor. of North Clark and Oak Sts., is in process of erection. The Library already has 50,000 volumes, chiefly of scientific or scholarly character. It contains the largest and most valuable collection of musical works in this country. The late John Crerar bequeathed $2,000,000 for the establishment of *The John Crerar Library* on "the South Side" of Chicago.

Clubs.—The principal social clubs are the *Calumet*, cor. Michigan Ave. and 20th St.; the *Chicago*, in Monroe St., between Sta'e St. and Wabash Ave.; *Douglas Club*, 3518 Ellis Ave.; *Illinois Club*, 154 Ashland Ave.; *Iroquois Club*, 110 Monroe St.; *Kenwood Club*, Lake Ave. and 47th St.; *Lakeside Club*, Indiana Ave., between 31st and 32d Sts.; *La Salle Club*, 252 Monroe St.; *Oakland Club*, cor. Ellis and Oakland Aves.; *Phœnix Club*, Calumet Ave. and 31st St.; *Press Club*, 131 Clark St.; *Standard Club*, Michigan Ave. and 13th St.; *Union Club*, Washington Pl. and Dearborn Ave.; *Union League Club*, cor. Jackson St. and 4th Ave.; *University Club*, cor. Dearborn St. and Calhoun Pl.; and *Whitechapel Club*, at 173 Calhoun Pl. Admission on introduction by a member. Besides the foregoing there are numerous college and State clubs, as well as a *Chicago Woman's Club.*

Post-Office.—The General Post-Office is in the block bounded by Adams, Jackson, Dearborn, and Clark Sts. It is open for business from 8 A. M. to 7 P. M. on week-days, and from 11 A. M. to 1 P. M. on Sundays. There are, besides, 20 sub-postal stations and 11 carrier stations in different portions of the city, besides numerous lamp-post boxes.

CHICAGO, the principal city of Illinois, has within 50 years grown from a small Indian trading station to the position of the metropolis of the West, and the greatest railway center on the continent. It is situated on the W. shore of Lake Michigan, at the mouth of the Chicago and Calumet Rivers, in lat. about 41° 50′ N., and lon. 10° 33′ W. from Washington. The site of the business portion is 14 ft. above the lake; it was originally much lower, but has been built up from 3 to 9 ft. since 1856. It is an inclined plane, rising toward the W. to the

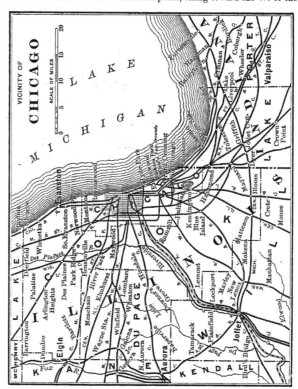

height of 28 ft., giving slow but sufficient drainage. The city stands on the dividing-ridge between the basins of the Mississippi and St. Lawrence, and is surrounded by a prairie stretching several hundred miles S. and W. One eighth of a mile N. of the Court-House the Chicago River extends W. a little more than half a mile, and then divides into the

North and South branches, which run nearly parallel with the lake-shore, about 2 miles in each direction. The river and its branches, with numerous slips, give a water-frontage of 58 miles, while the lake-front affords 22 miles of additional frontage, on which an outer harbor is now in process of construction. Connected with the S. branch is the terminus of the Illinois & Michigan Canal, which extends to the Illinois River at La Salle. The city extends N. and S. along the lake. Its length is 24 miles, and it is 10½ broad at its widest point, embracing an area of nearly 182 sq. miles. Beyond these limits suburbs extend for 10 miles north, west, and south, connected with the city by many local trains on the railways. The rivers divide the city into three distinct parts, known as the North, South, and West Divisions, which are connected by 45 bridges and 2 brick tunnels under the river-bed. The improved water-front is 30 miles long, on which are the lumber and coal yards, elevators and warehouses. It is the greatest corn, cattle, and timber market in the world. The city is regularly laid out, with streets generally 80 ft. wide, and many of them from 3 to 7 miles in length, crossing each other at right angles. State, Halsted, Western Ave., and several other streets extend the whole length of the city (24 miles).

The first white visitors to the site of Chicago were Joliet and Marquette, who arrived in August, 1673. The first permanent settlement was made in 1804, during which year Fort Dearborn was built by the U. S. Government. The fort stood near the head of Michigan Ave., below its intersection with Lake St. It was abandoned in 1812, rebuilt in 1816, and finally demolished in 1856. At the close of 1830, Chicago contained 12 houses and 3 "country" residences in Madison St., with a population (composed of whites, half-breeds, and blacks) of about 100. The town was organized in 1833, and incorporated as a city in 1837. The first frame building was erected in 1832, and the first brick house in 1833. The first vessel entered the harbor June 11, 1834 ; and at the first census, taken July 1, 1837, the entire population was found to be 4,170. In 1850 the population had increased to 29,963 ; in 1860, to 112,172 ; in 1870, to 298,977 ; and in 1880, to 503,304. The population in 1890 was 1,099,850. In October, 1871, Chicago was the scene of one of the most destructive conflagrations of modern times. The fire originated on Sunday evening, October 8th, in a small barn in De Koven St., in the S. part of the West Division, from the upsetting, as is supposed, of a lighted kerosene-lamp. The buildings in that quarter were mostly of wood, and there were several lumber-yards along the margin of the river. Through these the flames swept with resistless fury, and were carried across the South branch by the strong westerly wind then prevailing, and thence spread into the South Division, which was closely built up with stores, warehouses, and public buildings of stone, brick, and iron, many of them supposed to be fire-proof. The fire raged all day Monday, and crossed the main channel of the Chicago River, sweeping all before it in the Northern District, which was occupied mostly by dwelling-houses. The last house was not reached till Tuesday morning, and many of the ruins were still burning several months afterward. The total area burned over, including streets, was nearly 3⅗ sq. m. The number of buildings destroyed was 17,450 ; persons rendered homeless, 98,500 ; persons killed, about 200. Not including depreciation of real estate or loss of business, it is estimated that the total loss occasioned by the fire was $190,000,000, of which about $30,000,000 was recovered on insurance, though one of the first results of the fire was to bankrupt many of the insurance companies all over the country. The business of the city was interrupted but a short time, however. Before winter many of the merchants were doing business in extemporized wooden structures, and the rest in private dwellings. In a year after the fire, a large part of the burned district had been rebuilt; and at present there is scarcely a trace of the terrible disaster save in the improved character of the new buildings over those destroyed. On July 14, 1874, still another great fire swept over the unfortunate city, destroying 18 blocks or 60 acres

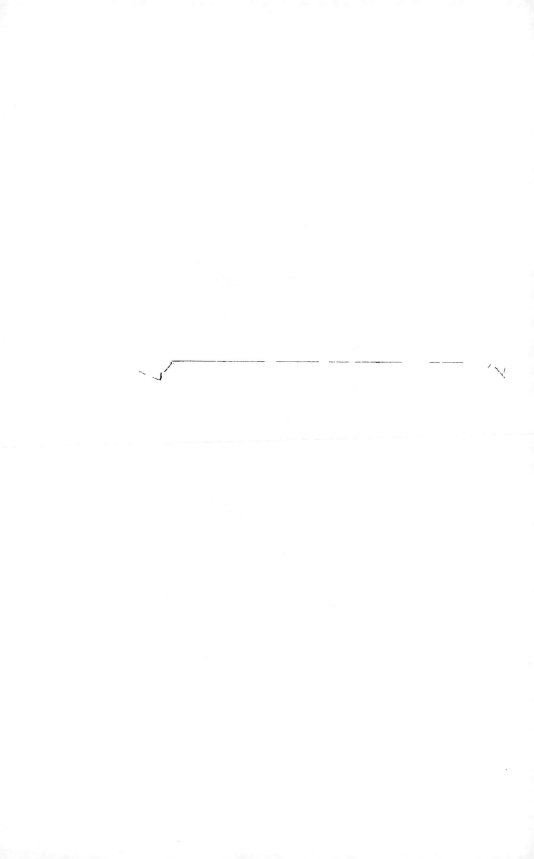

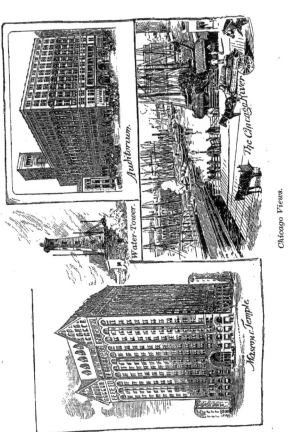

Auditorium.

Water-Tower.

The Chicago River.

Masonic Temple.

Chicago Views.

in the heart of the city, and about $5,000,000 worth of property. Chicago ranks next in commercial importance to New York among the cities of the United States. As early as 1854 it had become the greatest primary depot for grain in the world ; and since then it has also become the greatest grain, live-stock, and lumber market in the world. The manufactures of Chicago are extensive and important, and include iron and steel works, manufactories of car-wheels, cars, and other railroad appliances, flour-mills, furniture factories, boot and shoe factories, and tanneries.

State, Clark, and *Madison Sts.* are the principal retail business thoroughfares. La Salle and Dearborn Sts. are devoted to banking and office business. The S. side avenues that should be seen are *Michigan Boulevard, Calumet, Prairie,* and *Wabash.* On the N. side are *La Salle* and *Dearborn Aves.,* and *Rush, Pine,* and *Cass Sts.,* with the *Lake Shore Drive.* In this quarter are the residences of *Palmer,* the *Farwells, Medill, Archbishop Feehan, McCormick, Tree,* and *Smith.* On the W. side, *Madison* and *Halsted Sts.* are the best shop-streets. *Washington, Drexel,* and *Grand Boulevards* are especially noteworthy. The best drives are southward to the parks, the *Lake Shore Drive,* and *Sheridan Drive.*

The **Board of Trade,** at La Salle, Jackson, and Sherman Sts., is in the heart of the business quarter. In its immediate neighborhood are some of the tallest and finest office-buildings in the world. Among them are * *The Rookery,* 12 stories high, and accommodating 2,500 tenants ; * " *The Woman's Building,*" 16 stories ; the *Home Insurance Building,* 14 stories ; the *Insurance Exchange* and *Rand-McNally Buildings,* each 10 stories ; the *Counselman Building,* 12 stories ; *The Phœnix,* 12 stories ; the *Great Northern Hotel* (just completed), at Jackson and Dearborn Sts., 17 stories ; the *Owings Building,* 14 stories ; the *Manhattan* and *Monadnock,* each 17 stories ; the *Royal,* 12 stories, etc. *The Tacoma,* at La Salle and Madison Sts., and the *Chamber of Commerce,* at La Salle and Washington Sts., are each 14 stories high ; and the *Masonic Temple,* State and Randolph Sts., is 20 stories high. A new *Odd-Fellows' Building* is projected, to be 34 stories high. Other notable structures are *Temple Court,* the *Chicago Opera-House* (Washington and Clark Sts.), the *Monon,* the *Caxton,* the *Rialto, Montauk, Calumet, Pullman, Walker, Goff, Marshall Fields's Wholesale,* the *Leiter, First National Bank,* and *Adams Express Buildings,* and the great * *Auditorium Hotel*—fine view from tower (elevator). The *City Hall and Court-House* cost $5,000,000. The *U. S. Custom-House* and *Post-Office* occupies an entire block. In the *Board of Trade's Chamber* is a free gallery which gives a view of the interior. The tower is 303 ft. high. The " rotundas " of the *Rookery,* the *Royal,* and the *Union Depot,* at Canal and Madison Sts., should be visited.

Out of the 300 churches, there are a few notable structures. The * **Unity Church** (Unitarian), in Dearborn Ave., cor. Lafayette Place, is a light stone structure, in the modern Gothic style, with double spires. The *Second Presbyterian,* cor. Michigan Ave. and 18th St., is a large and imposing stone structure. The *Immanuel Baptist Church,* in Michigan Ave. near 24th St., is of stone, in the Gothic style, with a graceful tower and spire. *Grace Church* (Episcopal), in Wabash Ave., near 14th St., is a

21

handsome stone edifice in the Gothic style, with open timber roof and a richly decorated interior. *St. James's* (Episcopal), cor. Cass and Huron Sts., is large and massive, with a square flanking tower. The *Union Park Congregational*, cor. Ashland Ave. and Washington St., has a lofty spire, and is quite ornate in style. The *Second Baptist*, cor. Monroe and Morgan Sts., is a plain edifice in the Italian style, with a most peculiar spire. The *Church of the Holy Family* (Roman Catholic), in W. 12th St., is pure Gothic in style, and has an extremely rich and noble interior. The Roman Catholic *Cathedral*, N. side, is also a fine building.

Among the literary and educational institutions of Chicago a foremost place must be assigned to the * **Public Library,** the nucleus of which was contributed by English authors and publishers in 1872, and which now numbers 156,000 volumes, including many German, French, Dutch, Norse, Swedish, and Bohemian books. Its rooms are at present on the top floor of the W. wing of the Court-House, but a magnificent fire-proof structure is being erected at Michigan Avenue, Randolph and Washington Sts., which will be the finest library-building in the country. The *Academy of Sciences*, established in 1857, lost a valuable collection of 38,000 specimens in the fire, but has erected a building in Wabash Ave. near Van Buren St., and is slowly gathering a museum and library. The * **Art Institute,** Michigan Ave. and Adams Sts., was established in 1869. It is one of the foremost art museums in the United States, and maintains one of the most successful art schools in the country. It contains a choice collection of modern paintings, several extremely fine examples of the old masters (Rembrandt, Rubens, Van Dyck, Terburg, Ruysdael, Hobbema, etc., from the Demidoff, May, and other collections); a collection of Greek vases and antiquities, ivory carvings, and many interesting art-objects. There are usually loan collections on exhibition. The Art Institute's new building on the lakefront, on the site of the old Inter-State Exposition Building, is 320 ft. long and 170 ft. deep, constructed of granite and marble at a cost of $800,000. The collections of the Art Institute may not be placed therein before next year, owing to the use of the building by the Columbian Exposition for "World's Congresses," etc., during the present year. Three blocks S. is the vast * **Auditorium,** containing a theatre with 7,000 seats and a hotel with 10 floors. From the Tower Observatory a fine view of the city can be obtained (fee, 25c.). The *Chicago Historical Museum* is in Dearborn Ave. near Ontario, and will well repay a visit. The *Baptist Theological Seminary* is located at Morgan Park, a suburb. The *Chicago Theological Seminary* has a fine stone building in the Norman style; and the *Hammond Library* is on the W. side of Union Park, at the intersection of Ashland Ave. and Warren St. The *Presbyterian Theological Seminary* has a fine edifice at the cor. of Fullerton Ave. and Halsted St. It is 5 stories high, and contains a good library. The **St. Ignatius College** (Roman Catholic) has an ornate and costly building, 413 W. 12th St. The **University of Chicago** is an educational institution whose buildings, when completed, will extend from 56th to 59th Sts., between Ellis and Greenwood Aves. There are departments of law, medicine, theology, civil, mechanical, and electrical engineering, pedagogy, fine arts, music, etc., in addition to the regular academic courses. John D. Rockefeller subscribed $2,600,000 toward the establishment of this institution, and citizens of Chicago have subscribed $5,000,000. Instruction began Oct. 1, 1892. Charles T. Yerkes, a wealthy citizen of Chicago, has given $500,000 for a telescope for the university. Considerably over $50,000,000 are invested in the public schools of Chicago.

At the cor. of Harrison and Wood Sts. is the **Cook County Hospital.** Around it stand the *Rush*, the *Homœopathic*, the *Physicians' and Surgeons'*, and the *Woman's Medical Colleges*, and the great *West Side High* and the *Marquette Schools.* The *Chicago Medical College* has a large structure at the cor. of Prairie Ave. and 26th St. The *Hahnemann College* (homœopathic) is at the cor. of Cottage Grove Ave. and 28th St. The large *Woman's and Children's Hospital* is at Adams and Paulina. ***Mercy Hospital** is at the cor. of Calumet Ave. and 26th St. Other important charitable institutions are *St. Luke's,* 1439 Indiana Ave.; the *Home for the Friendless,* 911 Wabash Ave.; the *Protestant Orphan Asylum,* cor. Michigan Ave. and 22d St.; and *St. Joseph's* (male) and *St. Mary's* (female) *Orphan Asylums,* in N. State St. cor. Superior St. The two last named are under the charge of the Sisters of Mercy. The *Michael Reese Hospital,* 29th St., N. E. cor. Groveland Park Ave., recently erected, is maintained by the United Hebrew Relief Association. Other institutions are the *Old People's Home,* in Indiana Ave., the *Foundlings' Home,* on Wood near Madison St., and the *Newsboys' Home,* 146 Quincy St. There are 25 asylums in the city. The ***U. S. Marine Hospital,** situated at Lake View, a little beyond Lincoln Park, is one of the largest and costliest in the country. It is built of Joliet stone, is 340 ft. long, and cost $371,132. The *Washingtonian Home,* for the reformation of inebriates, is in W. Madison St. There are not less than 150 charitable establishments.

Chicago has a system of six exterior parks, joined by boulevards 200 ft. wide, which inclose the city as it is now built. These give 33 miles of drives, besides those around the parks. ***Lincoln Park,** on the lake-shore, in the N. Division, contains about 310 acres, and has several miles of drives and walks, fine trees, artificial hills and mounds, miniature lakes and streams, summer-houses, rustic bridges, shady rambles, and a *Zoölogical Garden.* From the N. end of Lincoln Park a boulevard, 3½ miles long, extends W. to *Humboldt Park,* which contains 225 acres. About 2 miles S. of Humboldt Park, with which it is connected by a similar boulevard, is *Garfield Park,* an irregular tract of land nearly a mile long from N. to S., and containing 185 acres, the middle line of which lies on Madison St., 4 miles from the Court-House. From this park the Douglas Boulevard runs 1½ mile S. E. to *Douglas Park,* which also contains 180 acres. From this another boulevard runs S. 4½ miles, thence E. 4¼ miles to the two *South Parks,* containing 1,055 acres, which are tastefully laid out. The most southerly extends upward of 1½ mile along the shore of the lake. *Washington Park,* S. of the South Parks, contains the Washington Park Race-Track. Three boulevards run thence to the streets that connect with the business portion of the city. *Douglas Monument* is in a small park on the lake-front, the statue at 35th St.; the statues of *Grant, Lincoln, Schiller, Indian Chief*, are in Lincoln Park; *Humboldt*, is in Humboldt Park. The Drexel fountain stands at the end of Drexel Boulevard. *Lake Park*, on the S. side, running about 1 mile on the lake-shore, is ornamented by the elegant Michigan Ave. residences. The *Chicago Base-Ball Park*, on the W. side, occupies a block bounded by Congress, Harrison, Loomis, and Throop Sts., and seats 8,000 spectators. A *League Base-Ball Park* at 35th and Wentworth Sts. has been opened, with a seating capacity for 12,000 spectators. There are several beautiful squares and minor parks—Union, Jefferson, Vernon, Ellis, Washington, and Groveland.

Of the cemeteries, *Graceland, Rose-Hill*, and *Calvary*, in the North Div., are the most interesting. The last two are on the line of the Chicago & Northwestern and Chicago, Milwaukee & St Paul R. Rs., and funeral-trains are run to them daily. *Oakwood*, on the Vincennes road, 3 miles S. of the city limits, is a pretty rural spot. This cemetery can be reached by Cottage Grove cable-line, or by a pleasant drive through the boulevards.

The * **North Water-Works** are situated on the lake-shore (take N. Clark St. cars and get off at Chicago Ave.), and may be freely inspected. They should not be missed by the visitor. They comprise a stone water-tower, 160 ft. high, up which the water is forced by 4 engines, having a pumping capacity of 74,500,000 gallons daily, and flows thence through pipes to every part of the city. A very fine * view of the city, lake, and surrounding country may be obtained from the top of the tower, which is reached by a spiral staircase. From this tower a nearly cylindrical brick tunnel, 62 inches high and 60 wide, extends 2 miles under the lake, lying 66 to 70 ft. below the lake-surface. The water enters the tunnel through a grated cylinder, inclosed in an immense crib, on which are a lighthouse and dwelling. The tunnel was begun in 1864 and finished in 1866, at a cost of $550,000. Another tunnel, 7 ft. in diameter, was completed in 1874, at a cost of $957,622, which also connects with the crib, and, through independent pumping-works, supplies the S. W. section of the city. A new lake-tunnel lately has been completed. The "in-take" is located 4 miles from the shore, and there are 2 great mains, each 6 ft. in diameter. Another abundant source of water-supply has been developed in the *Artesian Wells*, of which there are about 40. The first two sunk are situated at the intersection of Chicago and Western Aves. (reached by W. Randolph St. cars), are respectively 911 and 694 ft. deep, and flow about 1,200,000 gallons daily. The stock-yards, the west-side parks, and numerous manufacturing establishments are supplied from artesian wells.

Intercourse between the three divisions of the city is effected by 45 bridges, which span the river at nearly every street, and swing on central pivots to permit the passage of vessels. A * **Tunnel** was constructed in 1868 under the South Branch at Washington St. It is 1,608 ft. long, with a descent of 45 ft., has a double roadway for vehicles and a separate passage for pedestrians, and cost $512,707.

In 1870 another similar tunnel, with a total length of 1,890 ft., including approaches, was constructed under the main river on the line of La Salle St., connecting the North and South Divisions (cost nearly $500,000).

The * **Union Stock Yards** on S. Halsted St., where the vast live-stock trade of the city is transacted, comprise 400 acres, of which 146 are in pens, and have 32 miles of drainage, 20 miles of streets and alleys, 2,300 gates, and cost $4,000,000. They have capacity for 25,000 cattle, 120,000 hogs, 15,000 sheep, and 1,200 horses. There is a large and handsome brick hotel connected with the yards; also 2 banks, and a Board of Trade. A town of 50,000 inhabitants has sprung up in the immediate vicinity, with post-office, telegraph-offices, churches, schools, etc. Connected therewith are the *Packing and Slaughtering Houses*, whence are shipped annually 10,000,000 lbs. of hog products alone.

The *Grain-Elevators* are a very interesting feature, and should be visited, in order to obtain an idea of the manner in which the immense grain-trade of Chicago is carried on. There are 27 of these buildings, all situated on the banks of the river, and connected with the railroads by side-tracks. They have an aggregate storage capacity of 28,675,000 bushels, and receive and discharge grain with almost incredible dispatch.

NOTE.—An Appendix will be found at the close of this volume, in which full and ample descriptions of all the leading features of the great Columbian World's Fair is given.

Chicago has the World's Fair, which celebrates the 400th anniversary of the discovery of America, and is called the **World's Columbian Exposition.** The Exposition Buildings were dedicated with imposing ceremonies on October 21, 1892, the anniversary of Columbus's landing. Preceded by a grand naval review in New York Harbor in April, the Exposition opened May 1, 1893; it closes October 26th. An Act of Congress, approved April 25, 1890, gave life to the project; selected Chicago as its location; created a supervisory body, the National Commission, composed of eight commissioners-at-large and two from each State and Territory and the District of Columbia; defined its powers; provided for the formation of a Board of Lady Managers; directed the President, when satisfied that a suitable site was provided, and that the Chicago Exposition corporation had $10,000,000, to invite foreign nations to participate; specified that foreign articles intended for exhibition should be admitted free of duty; and required the various departments of the Government to make exhibits, for which $1,500,000 was appropriated. Subsequently Congress appropriated $2,500,000 in support of the enterprise.

Connected with the Exposition, a World's Congress Auxiliary has been organized to promote the holding of congresses, at which shall be discussed arbitration, peace, art, education, sciences, charities, and other subjects.

The following departments, based on an approved classification of exhibits, were established: A, Agriculture, Food and Food Products, Farming Machinery and Appliances; B, Viticulture, Horticulture, and

Floriculture; C, Live-Stock, Domestic and Wild Animals; D, Fish, Fisheries, Fish Products, and Apparatus of Fishing; E, Mines, Mining, and Metallurgy; F, Machinery; G, Transportation Exhibits—Railways, Vessels, Vehicles; H, Manufactures; J, Electricity and Electrical Appliances; K, Fine Arts—Pictorial, Plastic, and Decorative; L, Liberal Arts—Education, Engineering, Public Works, Architecture, Music, and the Drama; M, Ethnology, Archæology, Progress of Labor and Invention—Isolated and Collective Exhibits; N, Forestry and Forest Products; O, Publicity and Promotion; P, Foreign Affairs.

The principal Exposition officials are as follows: Council of Administration, Harlow N. Higginbotham, Chairman; George V. Massey, Charles H. Schwab, J. W. St. Clair, A. W. Sawyer. Director-General, George R. Davis. National Commission, Thomas W.. Palmer, President; J. T. Dickinson, Secretary. Lady Managers, Mrs. Potter Palmer, President; Mrs. Susan G. Cooke, Secretary. Local Directory, Harlow N. Higginbotham, President; H. O. Edmonds, Secretary. World's Congress Auxiliary, C. C. Bonney, President; Benjamin Butterworth, Secretary. Director of Works, D. H. Burnham. President Board of Architects, R. M. Hunt. Secretary of Installation, Joseph Hirst. Following are the chiefs of the respective departments: Agriculture, W. I. Buchanan. Horticulture, J. M. Samuels; John Thorp, Assistant Chief. Live Stock, W. I. Buchanan, Acting Chief. Fish and Fisheries, J. W. Collins. Mines and Mining, F. J. V.. Skiff. Machinery, L. W. Robinson. Transportation Exhibits, Willard A. Smith. Fine Arts, Halsey C. Ives; Charles M. Kurtz, Assistant Chief. Liberal Arts, S. H. Peabody. Manufactures, James Allison; Frank B. Williams, Assistant Chief. Electricity, J. P. Barrett; J. Allen Hornsby, Assistant Chief. Ethnology and Archæology, F. W. Putnam. Forestry, W. I. Buchanan, Acting Chief. Foreign Affairs, Walker Fearn. Publicity and Promotion, M. P. Handy.

The Exposition is held mainly in Jackson Park (which embraces 586 acres, and has nearly 2 miles frontage on Lake Michigan), Washington Park, 371 acres, a mile to the west, and Midway Plaisance, a strip of 80 acres connecting the two, used to accommodate the overflow. Thus the entire site embraces 1,037 acres. The Exposition site is within the city limits, 7 miles southeast of the City Hall. Visitors are able to reach the grounds by steam, electric, and horse-railways, elevated roads, cable-cars, an extensive omnibus and cab service, and steamboat lines on the lake.

The Exposition buildings are magnificent structures. Those erected by the Exposition Company are designated as follows: Administration, Electricity, Agriculture, Machinery, Mines and Mining, Fish and Fisheries, Live-Stock, Manufactures, Horticulture, Forestry, Woman's Building, Fine Arts, Music Hall, and Transportation, including depots. Besides these, the company has constructed ornamental entrances, a pier, peristyle, several power-houses and annexes, fountains, and other additional buildings. It estimates its expenditures as follows: On grounds and buildings, $12,766,890; for administration, $3,808,563; for operating expenses, May to October, $1,550,000; total, $17,625,453.

The several States of the Union made appropriations, and have exhibits of their resources and products. Most of them have erected buildings of their own, which partake of the nature of club-houses.

Foreign nations make extensive exhibits at Chicago. England, France, Germany, Spain, Russia, Turkey, Japan, Mexico, Persia, China, Siam, Jamaica, Hayti, Venezuela, Ecuador, Peru, Chili, Uruguay, the Argentine Republic, Honduras, Bolivia, Nicaragua, Costa Rica, Guatemala, Colombia, Salvador, and Brazil are well represented.

Notched into the S. territory of the city of Chicago is the unique city of **Pullman**, the beginning and growth of which illustrate an important phase in methods of manufacturing enterprise. It is named after the inventor of the "Pullman sleeping-cars." A few years ago Mr. George M. Pullman bought 3,000 acres of land at this point, at a cost of over $1,000,000, and there commenced building a city bearing his name. He has erected vast shops for the manufacture of these cars, putting in the best machinery obtainable, and employs nearly 14,000 workers in wood, iron, glass, painting, upholstering, etc. The city is laid out, graveled, sewered, etc., in the most perfect manner, and the public buildings, churches, free school-houses, and the *Florence Hotel*, are models. There are also a public park, a free public library, and athletic grounds. A prominent feature is the admirable and tasteful style of dwellings built for the workmen. Adjoining Pullman are *S. Chicago* and *Grand Crossing*, which contain rolling-mills, iron and steel mills, and many of the larger manufactures.

Itineraries.

The following series of excursions—each to occupy a day—have been prepared so as to enable the visitor to see as much of the city in as short a space of time as possible. Naturally the great World's Fair will demand longer time than any other feature, and hence in the present instance the itineraries have been condensed to the shortest space possible, giving only suggestions of what may be seen, leaving it to the inclination of the visitor to devote more time to such sights as shall prove most congenial to his tastes. See APPENDIX.

1. *The Business Portion of Chicago.*—Visit the Auditorium: take elevator to top of observatory tower (Congress St. side, adults, 25c.; children, 15c.), and from this point—270 ft. above the street—become acquainted with the enormous extent of Chicago. This view should give the visitor a fair idea of the plan of the city, and should aid him to fix in his mind the location of the principal points of interest. On a clear day, one may see across the end of Lake Michigan to Michigan City. On the S., across miles of streets, and over the acres of the stock-yards, one may see into the State of Indiana. On the N., Evanston, with its University buildings, may be discerned; and, westward, stretch miles on miles of brick and mortar, with bits of green park or boulevard, breaking the monotony here and there. Near at hand are acres of roofs, chimneys belching black smoke, and tall, tower-like structures rising on every side. With a good glass and a good map the visitor may spend an hour here very profitably. From the Auditorium visit the Art Institute (three blocks N., on Michigan Ave.; 25c.; free Saturdays and Sundays). Then walk out Adams St., and take a look up Wabash Ave. (one block W. of Michigan Ave.), mostly devoted to wholesale business. Afterward walk in Adams St. to Dearborn, where is the Post-Office, occupying the block bounded by Adams and Jackson, Dearborn and Clark Sts. In rear of the Post-Office, in Jackson St., is

the handsome Union League Club Building. Continue W. on Adams St. to La Salle. Enter "The Rookery," S. E. cor. Adams and La Salle. See the rotunda. Passing out to La Salle St., a few steps to the S. is the Board of Trade Building. The gallery of the "Exchange Hall" should be visited (free, 10 A. M. to 3 P. M.). Thence back to Adams St., see the beautiful "Women's Temple," La Salle St., near Adams. Walk E. on Adams St. to Clark St., and N. on Clark to Washington. Here are City Hall and Court-House. Visit Public Library in the City Hall (La Salle St. side). Walk E. on Washington St. to Dearborn, and N. on Dearborn to the river. Walk E. to State St. Here are many of the most interesting retail shops. At Randolph St. is the tall Masonic Temple. Walk S. to Congress St., and thence to hotel.

2. *The Southern Residence Region and Parks.*—Drive down Michigan Ave. to 33d St.; thence, by the Grand Boulevard, to Washington Park, and through 69th St. to Jackson Park (where is the World's Columbian Exposition). Return by way of Drexel Boulevard to 39th St., thence to Grand Boulevard and Michigan Boulevard and Ave.

3. *The Northern Residence Region and Parks.*—Drive to the Rush St. bridge, and across to Pine St. Thence to Water Works (fine view from tower). See the great pumping-engines. Go along Lake Shore Drive, where are many fine residences, to Lincoln Park. At extreme northern end of Park, on Sheridan Drive, is a good road-house. Thence continue up Sheridan Drive to Bryn Mawr Ave., and go W. to Rose Hill Cemetery, the handsomest of Chicago's burial-grounds. Return by way of any avenue E. to North Clark St., and thence to Sulzer St., Graceland Cemetery. Continue down North Clark St. to Diversey St., and E. to Lincoln Park. Visit Zoölogical Gardens. Return to the center of the city *via* Dearborn Ave. or State St.

4. *Stock-Yards, Packing-Houses, and "The Crib."*—Among the chief sights of Chicago are the stock-yards and great meat-packing houses. Take State St. cable-car to 39th St. and thence Stock-yards car to the yards. Then walk along main road through yards. Get permit at office of Armour, Swift, Morris or Cudahy, who will send boy to accompany you and exhibit the entire process of slaughtering animals, and converting them into food products. The visit to the slaughtering department is not recommended to persons of nervous temperament. The above may occupy the forenoon, and the afternoon may be devoted to a visit to "The Crib," out in the lake, whence water is pumped into the city water-mains. Take boat at Clark St. bridge (fare 25c.). The Crib is two miles out in the lake, E. of Chicago Ave.

The foregoing itineraries will enable one to obtain a general idea of the city in the shortest possible time. For special visits to the various suburbs, as Evanston, Lake Forest, etc., consult the various railway time-tables. The Rose Hill and Graceland Cemeteries may be visited by the Chicago, Milwaukee & St. Paul, and the Northwestern R. R., and Lincoln Park may be visited by North Clark St. cable-car, or by boat from the Clark St. bridge.

72. New York to Cincinnati via Buffalo and Cleveland.

By the N. Y. Central & Hudson River R. R. (Route 40), or the Erie R. R. (Route 41), or the West Shore R. R. (Route 42) to Buffalo; thence by Lake Shore R. R. (Route 67) to Cleveland; and from Cleveland, by the Cleveland, Columbus, Cincinnati & St. Louis R. R.   The average time from New York to Cincinnati is about 30 hours, but the "Southwestern Limited" of the N. Y. Central R. R. runs through in 22 hours.   Distances : New York to Cleveland. 623 miles; to Crestline, 698; to Delaware, 737 (to Columbus, 761); to Springfield, 787; to Dayton, 811; to Cincinnati, 867.   Through trains with sleepers run without change of cars.

THE portion of this route between New York and Cleveland is described in Routes 66 and 67, or, if the Erie R. R. be taken, in Route 68. Leaving Cleveland by the C. C. C. & St. L. R. R. (better known as the "Big Four "), the train passes in quick succession a number of small towns, which please by their neatness and air of prosperity, but which do not require special mention. *Shelby* (67 miles from Cleveland, 690 from New York) is a busy village at the crossing of the Lake Erie Div. of the Baltimore & Ohio R. R. (Route 70), and *Crestline* (698 miles) is at the intersection of the Pennsylvania System (Route 69).   *Galion* (703 miles), *Gilead* (716 miles), and *Cardington* (720 miles) are small villages. **Delaware** (737 miles ; *Hotel Donovan, St. Charles*) is a thriving town of 8,224 inhabitants, on the right bank of the Olentangy River.   It is pleasantly situated on rolling ground, and is neatly built.   In 1842 the *Ohio Wesleyan University* was founded here, and the *Ohio Wesleyan Female College* in 1863.   Both are prosperous institutions, and the former has a library of 13,000 volumes.   There is also here a medicinal spring which is much resorted to.   At Delaware a branch-line diverges, and runs in 24 miles to **Columbus** (see Route 73).   Beyond Delaware several small stations are passed, and in 50 miles the train reaches **Springfield** (*The Arcade, Lagonda House, St. James*), with a population of 31,895, situated at the confluence of the Lagonda Creek and Mad River, both of which furnish excellent water-power, which is utilized in manufactures.   It is in the heart of one of the most populous agricultural regions in the Union, and has a large trade in wheat, flour, Indiancorn, and other produce.   The city is well laid out and handsomely built, with six large public-school buildings, including a fine edifice for the High-School, several costly churches, the *Court-House*, two Opera-Houses, and many fine residences.   The *Springfield Seminary* is a flourishing institution ; and *Wittenberg College* (Lutheran), founded in 1845, has 209 students and a library of 7,000 volumes.   The *Free-Warder Library* contains 15,000 volumes.   Springfield has an extensive reputation for the manufacture of agricultural implements, 30,000 mowers and reapers being produced annually.   Six lines of railway intersect here.   Twenty-four miles beyond Springfield the train reaches **Dayton** (*Beckel House, Phillips*), a beautiful city of 61,220 inhabitants, on the Great Miami River, at the môuth of Mad River.   It is regularly laid out, with broad, well-shaded streets crossing each other at right angles, and lined with tasteful private residences.   The public buildings are unusually fine. The * *Old County Court-House* is an imposing white-marble edifice, 127

ft. long by 62 ft. wide. The *New County Court-House*, connected with
the old by corridors, is 144 × 94 ft., built of limestone. The beautiful
Soldiers' Monument, at the head of Main St., was erected in 1884. One
of the market-houses, 400 ft. long and paved with blocks of limestone,
has accommodations for the municipal offices in the second story.
There is a large water-power within the city limits, obtained from two
hydraulic canals, and Dayton is a place of great industrial activity. It
is especially noted for its manufactures of agricultural machinery, steam-
engines and boilers, railroad-cars, stoves, paper, and hollow-ware, which
amount annually to over $20,000,000; also glucose, paint, and shirt
manufactories. The public schools are of a high character, and the
Public School Library, in the center of the Public Square, contains
20,000 volumes. There are 50 churches, many of them of much architectural beauty. The principal charitable institutions are the *County
Orphan Asylum*, the *Widows' Home*, and the *Southern Lunatic Asylum
of Ohio*. To the tourist the most interesting feature of Dayton is the
* *Central National Soldiers' Home*, situated on a picturesque elevation
2 miles from the city, and reached by horse-cars and two lines of steam-
cars. The Home is an extensive group of fine buildings, over 40 in
number, including a church, built of white limestone, and a hospital.
The latter is of red brick, with freestone facings and trimmings, and
accommodates 300 patients. The principal other buildings are a brick
dining-hall, capable of seating 2,250 persons, a fine library, a music-hall,
billiard-room, bowling-alley, headquarters building, and several barracks
for the men. There are now 5,000 disabled soldiers in the Home.
The grounds embrace an area of 640 acres, well shaded with natural
forest-trees, and are handsomely laid out, with winding avenues, a deer-
park, three beautiful artificial lakes, an artificial grotto, hot-houses, and
flower-beds. Between Dayton and Cincinnati there are no stations requiring mention, except *Middletown*, but the country *en route* is fertile,
populous, and pleasing. **Cincinnati** (see Route 75).

73. New York to Cincinnati via Philadelphia, Pittsburg and Columbus.

By the "Pan-Handle Route," consisting of the Pennsylvania R. R. to Pittsburg, and the Pittsburg, Cincinnati, Chicago & St. Louis R. R. from Pittsburg to
Cincinnati. Through trains, with palace drawing-room and sleeping cars attached, run through without change of cars in 28 hours. Distances : to Philadelphia, 90 miles ; to Harrisburg, 195 ; to Pittsburgh, 444 ; to Steubenville, 487 ;
to Newark, 604 ; to Columbus, 637 ; to Xenia, 692 ; to Cincinnati, 757.

As far as **Pittsburg** (444 miles) this route is the same as Route
69. Shortly beyond Pittsburg the train enters and crosses that narrow
arm of West Virginia (the "Pan-Handle") which is thrust up between
Pennsylvania and Ohio, and then crosses the Ohio River into the State
of Ohio. The first station of importance in Ohio is **Steubenville**
(*Imperial, United States Hotel*), a city of 13,394 inhabitants, situated on the W. bank of the Ohio River, which is here ½ mile wide.
The city is well laid out and substantially built, is surrounded by a rich
farming and stock-growing country, and is the center of an important

trade. There are also a number of foundries, rolling-mills, machine-shops, flour-mills, etc. Abundance of excellent coal is found in the neighborhood, and there are eight shafts within the city limits. The * *County Court-House* is the finest in E. Ohio, and there are several very handsome churches and school-buildings. Among the educational institutions are an academy for boys and a noted female seminary, the latter delightfully situated on the bank of the river. The scenery in the vicinity of Steubenville is very attractive. Beyond Steubenville a number of small towns are passed, of which the principal is *Coshocton* (568 miles), the capital of the county of the same name, picturesquely built on 4 natural terraces rising above the Muskingum River. The Ohio Canal, connecting the Ohio River with Lake Erie, passes through the village and furnishes a good water-power. *Dresden* (582 miles) is another busy village on the Muskingum River, and 22 miles beyond is **Newark** (*Fulton, Warden*), a flourishing city of some 14,270 inhabitants, situated on a level plain at the confluence of three branches of the Licking River. It is a handsome place, the streets being wide and regular, and the churches, stores, and private residences well built. The surrounding country is very productive, and in the vicinity are quarries of sandstone, an extensive coal-mine, and several coal-oil factories. The Ohio Canal passes through the city. The next important station after leaving Newark is

Columbus.

Hotels, etc.—The leading hotels are the *American House*, cor. High and State Sts.; the *Chittenden; Davidson's* (*European*), S. of Union Depot; *Neil House*, cor. High and Capitol Sts.; *Park Hotel*, cor. High and Goodale Sts.; and the *U. S. Hotel*, cor. High and Town Sts.. These hotels charge from $2 to $3 a day. *Street-cars* (fare, 5c.) reach all parts of the city, also to N. Columbus, a suburb, 6 miles from the Capitol, and there are six bridges across the Scioto River.

Columbus, the capital of Ohio, and one of the largest cities in the State, is situated on the E. bank of the Scioto River, 100 miles N. E. of Cincinnati. It was laid out in 1812, became the seat of the State government in 1816, and was incorporated as a city in 1834, when its population was less than 4,000. The population, according to the census of 1880, was 51,665, and in 1890 was 88,150. The commercial interests of the city are large, and its manufactures numerous and important; but its growth and wealth are chiefly due to the concentration there of the State institutions, and the liberal expenditure of public money. The streets are very wide, and are regularly laid out in squares. **Broad St.** is 120 ft. wide for a distance of more than 2 miles, having beautifully laid asphaltum pavement. It has a double avenue (4 rows) of trees, alternate maple and elm. The finest residences are in the N. and E. portions of the city. The business thoroughfare is *High St.*, which is 100 ft. wide, and paved with the asphalt pavement for 2 miles of its length, and the other 4 miles with stone-block pavement. In the center of the city, occupying the square of 10 acres between High and 3d and Broad and State Sts., is * **Capitol Square**, beautifully laid out. It is proposed to make it a complete *arboretum* of Ohio trees.

The most interesting feature of Columbus to the stranger is its public buildings and institutions, in which it is not excelled by any city in the United States except Washington, and much surpasses most of the Western capitals. The State has concentrated here nearly all the public buildings devoted to its business, benevolence, or justice. The *Capitol, which stands in Capitol Square, is one of the largest and finest in the United States. It is constructed of fine gray limestone, resembling marble, in the Doric style of architecture, of which it is a noble specimen. It is 304 ft. long and 184 ft. wide, and is surmounted by a dome 64 ft. in diameter and 157 ft. high. The interior is elegantly finished. The hall of the House of Representatives is 84 ft. long by 72½ ft. wide, and the Senate Chamber is 56 by 72½ ft. There are also rooms for all the State officers, besides 26 committee-rooms. The *State Penitentiary is another very striking building. It is of hewn limestone, in the castellated style, and with its yards and shops covers 30 acres of ground on the E. bank of the Scioto, just below the mouth of the Olentangy. The Central Ohio Lunatic Asylum has a series of spacious buildings standing amid 300 acres of elevated ground W. of the city. The buildings are in the Franco-Italian style, and have a capacity for 1,000 patients. The *Idiot Asylum*, a plain Gothic structure, occupies grounds 123 acres in extent, adjoining those of the Lunatic Asylum. The *Blind Asylum*, in the E. part of the city, on the grounds of the old one, is a stone structure, in the Gothic style of the Tudor period. The *Deaf and Dumb Asylum, centrally located in extensive and handsome grounds on Town St., cor. Washington Ave., is built in the Franco-Italian style, with Mansard roof. The building is 400 ft. long and 380 deep, and has numerous towers, the central one of which is 140 ft. high. The *U. S. Barracks is located in the midst of spacious grounds, beautifully wooded, in the N. E. suburb of the city. It comprises, besides an immense central structure, numerous other buildings, used for offices, quarters, storehouses, etc. There is a fine drive to the Barracks, and beautiful drives are laid out through and around the grounds. The State has also a large and well-built Arsenal. The *City Hall*, facing Capitol Square on the S. side of State St., is a handsome Gothic structure, 187½ ft. by 80, with a small central tower 138 ft. high. The *Masonic Cathedral* is at 3d and Town Sts. The *High-School* is in Broad St. The *Holly Water-Works* occupy a large building near the junction of the Scioto and Olentangy Rivers. The *Odd-Fellows' Hall*, in High St., near Rich St., is opposite *Opera-House Block*. The *Union Depot* is a spacious structure. The U. S. Government has erected a building, costing $800,000, at the corner of 3d and State Sts. A Court-House, costing over $500,000, has recently been completed. The *Board of Trade Building*, a fire-proof building of stone, is in E. Broad St., opposite the State-House. Northeast of the city are the *Pan-Handle* shops and round-house.

There are more than 50 churches in the city, and some are fine examples of Gothic architecture. Most notable among them are *Trinity Church* (Episcopal), cor. Broad and 3d Sts.; *St. Joseph's Cathedral* (Roman Catholic), cor. Broad and 5th Sts.; the *Second Presbyterian*, cor. 3d and

Chapel Sts.; *St. Paul's* (German Lutheran), cor. High and Mound Sts.; *Broad Street* (M. E. Church), Broad St. and Washington Ave., built of Pennsylvania green-stone; *Wesley Chapel,* cor. Broad and Third Sts.; and the *Third Avenue,* High St. and 3d Ave., and a new Welsh church on E. Long St. The *State Library,* in the Capitol, contains over 46,400 volumes, and the *City Library* 20,000.   *** Starling Medical College** is at the cor. of State and 6th Sts., and the *Columbus Medical College* cor. of Long and Fourth Sts. *Capital University* (Lutheran) is a building in the Italian style, in the E. part of the city. The female seminary of *St. Mary's of the Springs* adjoins the city on the E., and near by is the *Water-Cure.* The Ohio *State University,* endowed with the Congressional land-grant, was opened in 1873. Of the charitable institutions, the *Hare Orphans' Home,* the *Hannah Neil Mission,* and the *Lying-in Hospital* may be mentioned. The *Catholic Asylum* for the reclamation of fallen women is W. of the city, and the Sisters of Mercy have a hospital in the Starling Medical College Building. A convent of the Sisters of the Good Shepherd has been established at West Columbus.

*** Goodale Park,** presented to the city by Dr. Lincoln Goodale, is at the N. end of the city, and comprises about 40 acres of native forest, beautifully improved. *City Park,* at the S. end of the city, is of about the same size as Goodale Park, and resembles it in many respects. The former grounds of the Franklin County Agricultural Society, 83 acres in extent, are on the E. border of the city, and have been made a public park, known as **Franklin Park.** The *Ohio State Fair Grounds,* located directly N. of the city, comprise 100 acres. In the immediate vicinity are the gardens of the *Columbus Horticultural Society,* occupying 10 acres. **Green Lawn** is the most beautiful of the five cemeteries of Columbus.

Leaving Columbus, the train soon reaches *London* (662 miles), a pretty town, capital of Madison County, and containing a fine Union school-house, and then passes on in 30 miles to **Xenia** (*Commercial* and *St. George Hotels*), a city of 7,301 inhabitants, with extensive manufactures. The streets of the city are well paved and beautifully shaded, and there are many substantial business-blocks and costly residences. The chief public buildings are the *Court-House,* one of the finest in the State, in a large and handsome park in the center of the city; the *City Hall,* containing a fine public hall, and the *Jail.* It is known as the center of the *twine* and *cordage* manufacture of the West, having three large and flourishing factories, employing hundreds of hands, and running day and night through most of the year. There are also *paper-mills* of some importance. The *Public Schools* are in fine condition, the High School ranking among the best in the State. *Wilberforce University,* established in 1863, for the higher education of colored youth of both sexes, is a short distance outside the city limits, and has a library of 4,000 volumes. The *Theological Seminary* (Presbyterian) dates from 1794, and has a library of 3,500 volumes. The *Ohio Soldiers' and Sailors' Orphans' Home* has about 30 buildings, accommodating 700 inmates, surrounded by very attractive grounds 200 acres in extent. The country between Xenia and Cincinnati is undulating, fertile, and highly cultivated, but presents nothing calling for special mention. *Morrow* (721 miles) is a thriving village at the junction with the Cincinnati & Muskingum Valley R. R., and *Loveland* (734 miles) is at the crossing of the Baltimore & Ohio Southwestern R. R. *Milford* (743 miles) is a flourishing village on the opposite side of the Little Miami River, and connected with the R. R. station by a bridge. **Cincinnati** (see Route 75).

74. New York to Cincinnati via Erie Railway and Connecting Lines.

By the Erie R. R. and the New York, Pennsylvania & Ohio R. R. Through trains, with palace drawing-room and sleeping-cars attached, run through on this route without change of cars in about 35 hours. (The Erie R. R. also sells through tickets to Cincinnati *via* Buffalo and Cleveland, as explained at the head of Route 72.) Distances : to Port Jervis, 88 miles ; to Susquehanna, 193 ; to Binghamton, 214 ; to Elmira, 273 ; to Hornellsville, 331 ; to Salamanca, 413 ; to Meadville, 515 ; to Akron, 615 ; to Mansfield, 682 ; to Marion, 717 ; to Urbana, 766 ; to Springfield, 781 ; to Dayton, 801 ; to Hamilton, 836 ; to Cincinnati, 861.

As far as Mansfield (682 miles) this route is described in Route 68. **Mansfield** (see the same) is at the junction of the present route with Routes 68, 69, and 70. From Mansfield to Dayton the New York, Pennsylvania & Ohio R. R. closely follows the line of the Cleveland, Columbus, Cin. & St. L. R. R. (Route 72), touching the same places at frequent intervals. *Galion* (697 miles) is a station on both roads, and *Marion* (717 miles) is a prosperous village at the crossing of the Indianapolis Div. of the C. C. C. & St. L. R. R. **Urbana** (766 miles) (*Weaver* and *Sowles Hotels*) is a city of 3,511 inhabitants, capital of Champaign County. The trade with the surrounding country is large, and there are several important manufactories, of which the chief is the U. S. Rolling-Stock Co., which employs 500 hands. *Urbana University* (Swedenborgian) was founded in 1851, and has a library of 5,000 volumes. The High-School building cost $90,000, and accommodates 400 pupils. There is also a free public library. **Springfield** (781 miles) has already been described in Route 72, and **Dayton** (801 miles) in the same. At Dayton the train passes on to the track of the Cincinnati, Hamilton & Dayton R. R., and soon reaches **Hamilton** (*St. Charles* and *St. Clair Hotels*), a city of 17,565 inhabitants, situated on both sides of the Miami River and on the Miami & Erie Canal. Hamilton is surrounded by a rich and populous district, and is extensively engaged in manufactures, of which the most important are machinery, agricultural implements, paper, woolen goods, flour, carriages and wagons, boots and shoes, etc. Abundant water-power is supplied by a hydraulic canal, which gives a fall of 28 ft. There are a number of handsome churches and school-buildings in Hamilton, and a free Public Library.

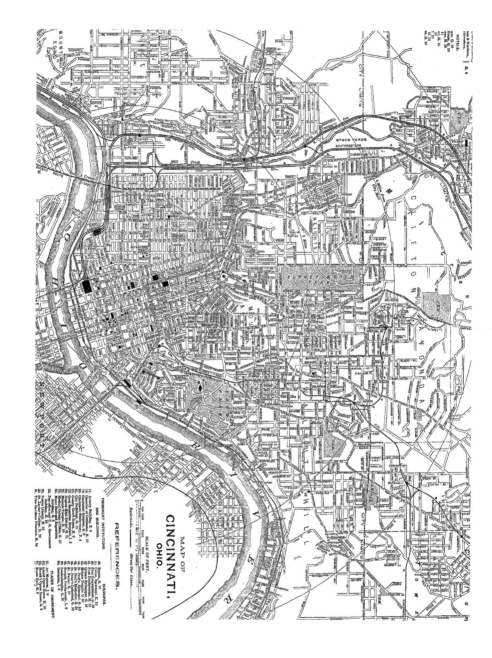

MAP OF
CINCINNATI,
OHIO.

SCALE OF FEET.

REFERENCES.

Railroads.
Street Car Lines.

PROMINENT INSTITUTIONS
AND BUILDINGS.

PLACES OF AMUSEMENT.

CHURCHES.

su
tr
cu
(7
M
of
flo
an
R

Th
th
R.
pl
qu
Sa
71
to

M
R
sy
bu
in
(7
Di
S
Th
po
wh
fo
bu
fr
in
pa
so
17
Mi
dis
im
flo
po
Th
ilt

75. Cincinnati.

Hotels.—The *Grand Hotel* is cor. 4th St. and Central Ave. The *Burnet House*, cor. 3d and Vine Sts. The *Gibson House*, Walnut St., between 4th and 5th. The *Palace Hotel*, cor. 6th and Vine Sts. The *St. Clair Hotel*, cor. Mound and 6th Sts. Rates at these hotels are from $2.50 to $4 a day. Other good hotels are the *St. James*, in E. 4th St., between Main and Sycamore ; *Bristol*, cor. 6th and Walnut Sts. ; the *Dennison Hotel*, 5 N. Main St. Rates at the latter-named houses are from $2 to $2.50 a day. Good hotels on the European plan are the *St. Nicholas*, cor. 4th and Race Sts., and the *Hotel Emery*, in Vine St., between 4th and 5th Sts.

Restaurants.—The best restaurants for ladies and gentlemen are the *St. Nicholas*, cor. 4th and Race Sts. ; the *Woman's Exchange*, with lunch-room attached, cor. Race and Longworth Sts.; the *Vienna Bakery*, cor. 7th and Race Sts.; *Brock's*, in Mound St. near 6th ; *Stewart's*, in Elm, between 4th and 5th, west side ; and the *Hotel Emery*, in the Arcade. The *Glencairn*, in the new Chamber of Commerce, serves lunch from 11.30 to 3.

Modes of Conveyance.—*Horse, electric,* and *cable cars* (fare, 5c.) run to all parts of the city and suburbs, and to Covington and Newport, Ky. *Omnibuses* run from all the stations and steamboat-landings to the hotels (fare, 25c.). *Hacks* are in waiting at the depots and other points in the city. Their legal rates are : For 1 person to any point within the city, 50c. ; 3 or more persons, 50c. each ; large baggage, extra ; by the hour, $1.50 for the first hour and $1 for each additional hour. A Cincinnati *Cab* system charges 25c. a trip for each passenger. *Ferries* to Covington from foot of Central Ave. ; to Newport from foot of Lawrence St. ; to Ludlow from foot of 5th St.

Railroad Stations.—The station of the *Cincinnati, Hamilton & Dayton R. R.* is at the cor. of 5th and Hoadley Sts. The *Grand Central Station* is cor. 3d St. and Central Ave. Here are the Ohio & Mississippi R. R., the Cincinnati Southern R. R., Baltimore & Ohio Southwestern R. R., Kentucky Central R. R., Cleveland, Cincinnati, Chicago & St. Louis R. R., and the Chesapeake and Ohio R. R. The *Pittsburg, Cincinnati, Chicago & St. Louis R. R. Station* is cor. Kilgour and Front Sts. The Louisville & Nashville R. R. also uses this station.

Theatres and Amusements.—Cincinnati is amply supplied with places of amusement : The *Grand Opera-House*, on Vine, between 5th and 6th Sts.; *Havlin's Theatre*, on Central Ave. between 4th and 5th Sts.; *Pike's Opera-House*, on 4th, between Walnut and Vine Sts. *Heuck's Opera-House*, on Vine, between 12th and 13th Sts. ("over the Rhine"), has a seating capacity for over 2,000 people. The *Springer Music Hall*, in Elm St., contains one of the largest organs in the world. It has a seating capacity of 5,900. *Smith and Nixon's Hall* is on W. 4th St., between Main and Walnut Sts. Concerts and lectures are given at *College Hall*, in Walnut St., near 4th ; at the *Odeon*, in Elm St., near Springer Music Hall ; at the *Lyceum*, built for the College of Music, near Music Hall, capacity 450 ; and at *Greenwood Hall*, in the Mechanics' Institute, cor. 6th and Vine Sts. The large German Halls "over the Rhine" are noticed further on. The *Gymnasium* has commodious quarters, cor. Vine and Longworth Sts., over the Grand Opera-House (open from 8 A. M. to 10 P. M.). The *Floating Bath* is moored at the foot of Vine St. (single bath, 15c.). A favorite resort outside of the city is the *Zoölogical Gardens*, N. of the city near Avondale. The buildings are substantial, the grounds beautiful, and include 66 acres, and the collection the best in the country. Admission, 25c. Reached by either Main St. or Elm St. cars, *via* inclined planes, or Vine St. and Sycamore St. cable-cars. The city is one of the best and most substantially paved cities in the country. Owing to the number of cable and electric lines of street-cars, her beautiful suburbs have developed greatly. The tourist will find the time pleasantly spent in taking the Walnut Hills, Sycamore, and Vine St. cable-cars, and Mt. Auburn electric line to the terminus and return (fare, 5 cts.).

Reading-Rooms.—In the leading hotels are reading-rooms for the use of guests, well supplied with newspapers. The *Public Library*, in Vine St., between 6th and 7th, contains 197,484 books and pamphlets, and a well-supplied reading-room (open from 8 A. M. to 10 P. M.). The *Young Men's Mercantile Library* is in the 2d story of the College Building, in Walnut St. between 4th and 5th, and contains 60,000 volumes. The *Philosophical and Historical Society* has a commodious building, at 115 W. 8th St., and its library numbers 12,600 bound volumes and 48,000 pamphlets. The *Law Library*, in the Court-House, has over 18,500 volumes. The *Mechanics' Institute Library*, cor. 6th and Vine Sts., has 7,050 volumes, and a reading-room.

Clubs.—The *Queen City Club*, cor. of 7th and Elm Sts. The *Phœnix Club*, cor. of Central Ave. and Court St. The *Allemania Club*, cor. of 4th St. and Central Ave. The *Cuvier Club* is at 32 and 34 Longworth St. The *University Club*, cor. of 4th St. and Broadway, is for college-bred men. *Political Clubs:* *Lincoln*, cor. Garfield Place and 8th St. ; *Blaine* (Rep.), 66 W. 8th St., near Vine ; *Ohio* (Dem.), 4th St. ; *Young Men's Democratic Club*, 56 W. 4th St. ; *Duckworth* (Dem.), 7th St. ; *Jefferson* (Dem.), Vine St. Introduction by a member.

Post-Office.—The general Post-Office is in 5th St. bet. Main and Walnut, and is open from 6 A. M. to 10 P. M. There are also sub-stations in different parts of the city, and letters may be mailed in the numerous lamp-post boxes.

CINCINNATI, the chief city of Ohio, is situated on the N. bank of the Ohio River, in lat. 39° 6' N. and lon. 84° 27' W. It has a frontage of 10 miles on the river, and extends back about 3 miles, occupying half of a valley bisected by the river, on the opposite side of which are the cities of Covington and Newport, Ky. It is surrounded by hills from 400 to 465 ft. in height, forming one of the most beautiful amphitheatres on the continent, from whose hilltops may be seen the splendid panorama of the cities below, and the winding Ohio. Cincinnati is principally built upon two terraces, the first 60 and the second 112 ft. above the river. The latter has been graded to an easy slope, terminating at the base of the hills. The streets are laid out with great regularity, crossing each other at right angles, are broad and well paved, and for the most part beautifully shaded. The business portion of the city is compactly built, a fine drab freestone being the material chiefly used. The outer highland belt of the city is beautified by costly residences which stand in the midst of extensive and neatly adorned grounds. Here the favorite building material is blue limestone. The names of the suburbs on the hilltops are Clifton, Avondale, Mt. Auburn, Price Hill, and Walnut Hills.

Cincinnati was settled in 1788, but for a number of years a continual series of difficulties with the Indians retarded the progress of the town. In 1800 it had grown to 750 inhabitants, and in 1814 it was incorporated as a city. About 1830 the Miami Canal was built, and during the next 10 years the population increased 85 per cent. In 1840 the Little Miami, the first of the many railroads now centering at Cincinnati, was finished, and in 1850 the population had increased to 115,436, in 1880 to 255,708, and in 1890 to 296,908, exclusive of several populous suburban villages. It is estimated that within a radius of 15 miles of the city there is a population of 500,000. The central position of Cincinnati has rendered it one of the most important commercial centers of the West ; but manufactures constitute its chief interest.

There is no one among the streets of Cincinnati which has the preeminence over the others. Of the business streets, *Pearl St.*, which contains nearly all the wholesale boot and shoe and dry-goods houses, is noted for its splendid row of lofty, uniform stone fronts between Vine and Race Sts. *Third St.*, between Main and Vine, contains the banking, brokerage, and insurance offices. **Fourth St.** is the fashionable promenade and most select retail-business street, and is lined with handsome buildings. In *Pike St.*, in *4th St.* from Pike to Broad-

way, and in *Broadway* between 3d and 5th Sts., are the finest residences
of the "East End"; in *4th St.*, W. of Smith, in *Dayton St.*, and in
Court St., between Freeman and Baymiller Sts., those of the "West
End." The portion of *Freeman St.* lying along the Lincoln Park is a
favorite promenade. *Pike St.*, from 3d to 5th, along the old Longworth

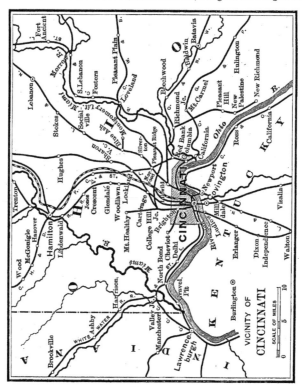

homestead, is known as the "Lover's Walk." Along Front St., at the
foot of Main, lies the *Public Landing*, an open area, paved with bowlders,
1,000 ft. long and 425 feet wide. There are many beautiful drives in
the vicinity of Cincinnati. One of the most attractive is that from the
Brighton House, cor. Central Ave. and Freeman St., to Spring Grove

Cemetery, and thence around Clifton and Avondale, returning to the city
by way of Mount Auburn. This drive affords fine views of the city and
surrounding country.

Of the public buildings, the finest in the city is the new *U. S.
Government Building,* occupying the square bounded by Main and
Walnut and 5th and Patterson Sts., designed to accommodate the Custom-
House, Post-Office, and U. S. Courts. It is of granite, in the Renaissance style, 354 ft. long by 164 ft. deep, and six stories high. The
Chamber of Commerce is at 4th and Vine Sts., and is open on every
business day from 11 A. M. to 2 P. M. It is one of the finest pieces of
architecture in the city, having been built after one of Henry H. Richardson's latest designs. The *County Court-House,* in Main, between N.
and S. Court St., was destroyed by rioters March 29, 1884; the new
Court-House is on the site of the one destroyed; the design is Romanesque. With the *County Jail* in its rear, it occupies an entire square.
The new *City Buildings* occupy the entire square on Plum St., between
8th and 9th. They are just completed, of red sandstone, at a cost of
$1,200,000. The rooms of the *Board of Trade* are in the new Chamber
of Commerce building. The *Masonic Temple,* cor. 3d and Walnut
Sts., is a structure in the Byzantine style, 195 by 100 ft., with 2 towers
140 ft. and a spire 180 ft. high. The *Masonic Scottish Rite Cathedral,*
in Broadway, between 4th and 5th Sts., is an imposing building. The
Odd-Fellows' Hall is cor. 4th and Home Sts. A magnificent new hall is
being finished at the cor. of Elm and 7th Sts. Fine blocks of commercial buildings may be found in Pearl, 3d, 4th, Main, Walnut, and Vine
Sts. The *Music Hall* and *Exposition Buildings,* in Elm St., fronting
Washington Park, cover 3¼ acres, and have 7 acres' space for exhibiting. A part of the Exposition Buildings is the *Springer Music
Hall,* a beautiful building in the modified Gothic style, 178 ft. wide,
293 ft. deep, and 150 ft. high from the sidewalk to the pinnacles of
the front gable. The interior decorations are extremely rich, and the
great organ is one of the largest in the world. In the vestibule is a
statue of the founder, R. R. Springer, by Powers. The carving on the
case of the organ is worth attention. Adjoining is the *College of Music,*
which has a corps of 50 professors, and an average of 850 students.
Pike's Building, in 4th St., between Vine and Walnut, is a very imposing structure. It is of sandstone, in the Elizabethan style, and the
interior is elaborately painted and frescoed. The *Public Library* building, in Vine St., between 6th and 7th, is one of the finest and largest
in the city. It is of stone and brick, in the Romanesque style, is fire-
proof, and will afford shelf-room for 300,000 volumes. The library
now contains 197,484 volumes, including pamphlets, and a well-supplied
reading-room (open from 8 A.M. to 9 P.M.). The *Emery Arcade*
extends from Vine to Race St., between 4th and 5th, a distance of 400
ft. The roof is of glass, and in it are shops of various kinds The
Rookwood Pottery on Mount Adams is famous for its artistic work, and
is well worthy of a visit.

The Cincinnati Museum Association, Eden Park, exhibits paintings, sculpture, and *bric-à-brac,* at their magnificent building, erected

at a cost of over $350,000; open daily (Sundays from 1 to 5 P. M. during the summer months), from 10 A. M. to 4 P. M.; admission, 25 cts.; on Sundays, 10c. The most notable work of art in the city is the **Tyler Davidson Fountain,** in 5th St., between Vine and Walnut. It stands on a free-stone esplanade, 400 ft. long and 60 ft. wide. In the center of a porphyry-rimmed basin 43 ft. in diameter is the quatrefoil Saxon porphyry base supporting the bronze-work, whose base is 12 ft. square and 6 ft. high, with infant figures at each corner representing the delights of children in water. Bas-relief figures around the base represent the various uses of water to mankind. From the upper part of the bronze base extend four great basins, and from the center rises a column, up whose sides vines ascend and branch at the top in palm-like frontage. Around this column are groups of statuary; and on its summit stands a gigantic female figure, with outstretched arms, the water raining down in fine spray from her fingers. The work was cast in Munich, and cost nearly $200,000. It plays during warm days from morning till midnight.

The finest church edifice in the city is * **St. Peter's Cathedral** (Roman Catholic), in Plum St., between 7th and 8th. It is of Dayton limestone, in pure Grecian style, 200 by 80 ft., with a stone spire 224 ft. high, and a portico supported by 10 sandstone columns. The altar, of Carrara marble, was made in Genoa; and the altar-piece, "St. Peter Delivered," by Murillo, is one of the chief glories of art in America. *St. Xavier Church* (Roman Catholic), in Sycamore St., between 6th and 7th, is an exceptionally fine specimen of the pointed Gothic style of architecture. *St. Paul's* (Episcopal), cor. 7th and Plum Sts., is of stone and stuccoed brick, in the Norman style, notable for its square towers, rough ashlar gable, and deep and lofty Norman door, and has fine music. * **St. Paul's** (Methodist), cor. 7th and Smith Sts., of blue limestone, in cruciform style, has a fine interior, and a spire 200 ft. high. The *First Presbyterian,* in 4th St., between Main and Walnut, is noted for its huge tower surmounted by a spire 270 ft. high, terminating in a gilded hand, the finger pointing upward. The *Baptist Church,* in 9th St., between Vine and Race, is a handsome building with massive clock-tower. The *First Congregational* and *Unitarian,* Reading Road above Oak. Some of the German churches "over the Rhine" are very large, and the music excellent. The * **Hebrew Synagogue,** in Plum St., opposite the Cathedral, is of brick, in the Moorish style. The *Hebrew Temple,* cor. 8th and Mound Sts., is in the Gothic style, with double spires.

The * **University of Cincinnati,** founded and endowed by Charles McMicken, has an imposing building at the cor. of Hamilton Road and Elm St. The *Art Academy of Cincinnati,* under the management of the Cincinnati Museum Association, occupies a spacious and recently constructed building near the Art Museum in Eden Park. The instructors and lecturers number thirteen, and the students in day and night classes exceed four hundred. Instruction in all branches of art is practically free. The *Law School* is in Walnut St., between 4th and 5th. * **St. Xavier's College** (Jesuit), cor. Sycamore and 7th Sts., is

a splendid building in the Romanesque style, of brick, profusely ornamented with stone. The *Cincinnati Wesleyan College* has a spacious and
handsome building in Wesley Ave., between Court and Clark Sts. The
Seminary of Mount St. Mary's is a famous Roman Catholic college,
beautifully situated on Price Hill, which commands extensive views.
Lane Theological Seminary (Presbyterian) is situated on Walnut Hills,
and possesses a library of 16,200 volumes. The *Medical College of
Ohio* is one of the most famous in the West, and has a very fine building in 6th St., between Vine and Race. The *Miami Medical College*, in
12th St., near the Hospital, is another famous institution. The *Franklin School*, Walnut Hills, is the leading classical and preparatory school
in the city. The *Hughes High School*, in 5th St., at the foot of Mound
St., is an imposing edifice in the Gothic style, with octagon towers
at the corners. The *Woodward High School* is an ornate building in
Franklin St., between Sycamore and Broadway. The *Technical School*
is temporarily in the N. wing of Music Hall. The *Mechanics' Institute*
is a commodious building, cor. 6th. and Vine Sts., containing a library
of 7,050 volumes and a reading-room.

In 12th St., between Central Ave. and Plum St., occupying a square
of 4 acres, stands the * **Cincinnati Hospital,** said to be the largest
and best-appointed institution of its kind in the country. It consists of
eight distinct buildings arranged *en échelon* round a central court, and
connected by corridors. The central building, through which is the
main entrance, is surmounted by a dome and spire 110 ft. high. The
Good Samaritan Hospital is a fine, large, red-brick building, situated on
a grassy hill at the cor. of 6th and Lock Sts. *St. Mary's Hospital*, cor.
Baymiller and Betts Sts., is also a fine and spacious building. The
* **Longview Asylum for the Insane,** at Carthage, 10 miles N. of
the city, is of brick, in the Italian style, 612 ft. long and 3 and 4 stories
high. Its grounds are laid out in beautiful lawns, walks, and parks,
with greenhouses. There are no bars to the windows, and everything
prison-like is avoided. The *House of Refuge* is situated in Mill Creek
Valley, about a mile N. of Brighton. The buildings are of blue limestone trimmed with white Dayton stone, and are surrounded by 6
acres of ground. The *City Workhouse* is near the House of Refuge.
The main building is 510 ft. long, and is one of the most imposing edifices about the city. The *Cincinnati Orphan Asylum* is a spacious
brick edifice at Mount Auburn, comprising ample grounds which command extensive views.

The chief public park of Cincinnati is * **Eden Park,** situated on a
hill in the E. district, and commanding magnificent views of the city,
the valley of the Ohio, and the surrounding country. It contains 216
acres, beautifully laid out and adorned; and in it are the two city
reservoirs, which look like natural lakes. *Burnet Woods*, on a hill N.
of the city, contains 170 acres, nearly all forest. *Lincoln Park*, in
Freeman St., between Betts and Hopkins, contains only 18 acres, but
is admirably adorned and finely shaded. *Washington Park*, one of the
oldest pleasure-grounds in the city, formerly a cemetery, is situated in
12th St., between Race and Elm Sts. It comprises 10 acres. *Hop-*

kins's Park is a small lawn with shrubbery, on Mount Auburn, N. of the city. *Spring Grove Cemetery, one of the most beautiful in the West, lies 5 miles N. W. of the city, in the valley of Mill Creek, and is approached by an attractive avenue 100 ft. wide. It contains 600 acres, well wooded and picturesquely laid out, and many fine monuments. The entrance-buildings are in the Norman-Gothic style, and cost $50,-000. The chief attractions are the Dexter mausoleum, representing a Gothic chapel, and a *bronze statue of a soldier, cast in Munich, erected in 1864 to the memory of the Ohio volunteers who died during the war.

More than a third of the residents of Cincinnati are Germans or of German parentage. They occupy the large section of the city N. of the Miami Canal, which they have named "the Rhine." The visitor finds himself in an entirely different country "*over the Rhine,*" for he hears no language but German, and many of the signs and placards are in German. The business, dwellings, theatres, halls, churches, and especially the beer-gardens, all remind the European tourist of Germany. Strangers should visit the Great Arbeiter and Turner Halls, in Walnut St., Heuck's Opera House, cor. Vine and 13th Sts., and some one of the vast beer-cellars, which can be found almost anywhere "over the Rhine." The *Suspension-Bridge over the Ohio, connecting the city with Covington, Ky., is the pride of Cincinnati. From tower to tower it is 1,057 ft. long; the entire length is 2,252 ft., and its height over the water 100 ft. The Chesapeake and Ohio Railway bridge across the Ohio River, a magnificent piece of engineering, is built on the principle of the parabolic truss, and has three spans, the middle span 550 feet long (the longest truss-span in the world) and 90 feet high; the other two 490 feet long and 80 feet high each. The total length is 5,716 feet, the total cost of structure and approaches $600,000, and the time of building a year and a half, the first train having passed over December 25, 1888. There is another handsome suspension-bridge over the Licking River, connecting the cities of Covington and Newport. By taking the street-cars at Front St., in an hour's ride one may cross both these bridges, and return to the starting-point, having been in two States and three cities, and having crossed two navigable rivers and a cantilever bridge. There are also four pier railroad-bridges across the Ohio at Cincinnati, connecting Cincinnati with Newport, Covington, and Ludlow, one of which was completed in 1891. The *Water-Works,* in E. Front St., not far from the station on Kilgour St., are well worth a visit. There are 8 pumping-engines with a capacity of 85,000,000 gallons a day. Well worth visiting are the *United Railroads Stock-Yards,* comprising 50 acres on Spring Grove Ave., with accommodations for 25,000 hogs, 20,000 sheep, and 5,000 cattle. There are four *Inclined Planes,* leading from the terrace on which the business portion of the city is built to the top of the surrounding hills. These *Hilltops* form the residential portion of the city, and abound with elegant homes and institutions. No one should miss the views from Price Hill and from the *Lookout House, Mt. Auburn (reached by electric-cars from cor. Main and 5th Sts.).

Itineraries.

The following series of excursions have been prepared so as to enable the visitor whose time is limited to see as much of the city as possible in the least amount of time. Each excursion is planned to occupy a single day, but the visitor can readily spend more time as special features crowd upon his attention.

1. Visit the Arcade, Vine to Race Sts., between 4th and 5th; the Tyler Davidson Fountain, in 5th St., between Vine and Walnut; the Public Library, Vine St., between 6th and 7th; the new City Buildings, occupying the square bounded by 8th and 9th Sts. and Plum St. and Central Ave.; St. Peter's Cathedral, 8th St., opposite the City Buildings; Plum St. Temple (Jewish), opposite St. Peter's; the City Hospital, Central Ave. and 12th St.; the Exposition Buildings and Music-Hall, Elm St., north of 12th St.; the new Court-House, in Main St., between North and South Court Sts., where the terrible riots of 1884 occurred, and when the old Court-House was burned; the Custom-House, which cost the Government $6,000,000; the Chamber of Commerce, which is best visited on any business-day between the hours of 11 A. M. and 2 P. M. This day's excursion does not necessitate the use of a carriage or street-car.

2. Take Covington street-car in Vine St., cross the Suspension Bridge to Covington; see the Covington Custom-House; take street-car to Newport, across the Licking Suspension Bridge, and return to Cincinnati by way of the Pennsylvania R. R. Bridge, or by the Cantilever Bridge. At Fountain Square take an electric-car to Mt. Adams, where the Rookwood Pottery is located; and after a visit to the pottery, continue the electric-car ride to the Art Museum in Eden Park, where the rest of the day can be spent. While in the park, see the reservoirs and the standpipe, just completed.

3. Take a cable-car ride through Walnut Hills; a cable-car ride to Avondale, returning from that suburb on the electric line; a ride on Vine St. cable-road through Clifton, where an hour or more may be spent with satisfaction on any of Clifton's beautiful avenues. From the car, as it crosses Clifton Ave., see the Cincinnati Crematory. Each one of these excursions will consume an hour in transit. The day may be wound up with a horse-car ride to Price Hill, where a magnificent view of the city can be had.

4. A forenoon can be spent at the Zoölogical Garden, reached either by the Mt. Auburn electric-cars or by the Sycamore St. cable. Pass the afternoon by continuing the electric-car ride on to Carthage, where Longview Asylum is located.

5. Spring Grove Cemetery is regarded as among the most beautiful specimens of landscape-gardening in the world. It is located on Spring Grove Ave., and can be reached either by the Colerain Ave. electric-cars or by a steam-car ride either on the Big Four, the Baltimore & Ohio Southwestern, or Cincinnati, Hamilton & Dayton R. Rs. Carriages are always ready in the cemetery to convey a visitor through (round-trip ticket, 15c.). In the cemetery, see the Dexter Chapel and the statue of C. W. West, the donor of the Art Museum. In transit on the Colerain Ave. line the visitor passes the Work-House and the House of Refuge.

6. Visit College Hill by the College Hill R. R., where the Sanitarium is located. While in this suburb, see Belmont College, and Clovernook, once the home of Alice and Phœbe Cary.

7. A 15-mile ride down the Big Four or the Ohio and Mississippi R. R. will take the visitor to North Bend, the home of President William Henry Harrison, where he is now buried. Four miles farther west, on the Ohio & Mississippi R. R., is the residence of John Scott Harrison, the father of President Benjamin Harrison. This is at the mouth of the Big Miami River, near which is located a prehistoric fort, in which may be seen two Indian mounds. The bastions in this ancient fortification are as distinctly defined as if built yesterday.

76. Cincinnati to Louisville.

Besides the routes described below, Louisville may be reached from Cincinnati by steamer on the Ohio River. There are several steamers daily, and in summer the trip is a very pleasant one. The scenery along the river is both varied and attractive (see Route 132).

a.   Via Louisville, Cincinnati & Lexington Div. of the Louisville & Nashville R. R.   Distance, 110 miles.

Leaving Cincinnati by this route, the train at once crosses the Ohio River on a long and lofty pier-bridge to **Newport,** a very handsome city of Kentucky, with a population of 24,918. It is built on an elevated plain commanding a fine view, and ornamented and made attractive by numerous shade-trees. In the city and its suburbs are a large number of fine residences, and the schools are noted for their excellence, and it has mercantile and manufacturing interests of importance. As already mentioned in the description of Cincinnati, a graceful suspension-bridge across the Licking River connects Newport with **Covington,** which in turn is connected with Cincinnati by a suspension-bridge. Covington is a city of 37,321 inhabitants, the largest in Kentucky after Louisville, but is substantially a suburb of Cincinnati, whose business-men have here many costly residences. It is built upon a beautiful plain several miles in extent, and includes within its corporate limits over 1,350 acres. The combined *Court-House* and *City Hall* is a handsome edifice; and the *U. S. Post-Office* and Court building cost $150,000. There are a public library and several flourishing educational institutions. The *Hospital of St. Elizabeth* (Roman Catholic) occupies a commodious building, with ample grounds adorned with shrubbery, in the center of the city, and has a foundling asylum connected with it. Beyond Newport, the Louisville train crosses the Licking River, passes in rear of Covington, and traverses a rich but uninteresting agricultural region. The stations passed are small. From *Walton* (21 miles) stages run to Williamstown, and at *Lagrange* (83 miles) a branch road diverges to Frankfort and Lexington.

b.   Via Ohio & Mississippi R. R.   Distance, 130 miles.

As far as *N. Vernon* (72 miles) this route is described in Route 78. From N. Vernon the road runs W. through one of the most productive and populous sections of southern Indiana, which, however, offers little to attract. Thence the Louisville Div. runs due S. The numerous stations *en route* are mostly small villages, none of which require special mention. At *New Albany* (127 miles), a flourishing city of 21,059 inhabitants, the train crosses the Ohio River on a magnificent bridge, which is described in connection with Louisville. In addition to the routes from New York *via* Cincinnati, Louisville may also be reached *via* Washington. From the latter-named point the Richmond & Danville R. R. connects at Charlottesville, Va., with the great trunk-line of the Chesapeake & Ohio R. R. (see Route 130), which runs through without change of cars to Louisville and Memphis.

77. Louisville.

Hotels.—The *Galt House* has long been celebrated as one of the best hotels in the United States. The *Louisville Hotel*, on Main St., between 6th and 7th; *Willard's*, on Jefferson, between 5th and 6th Sts.; and the *Fifth Avenue Hotel*, on 5th St., between Green and Walnut, are well-kept houses. Prices are from $2.50 to $5 per day.

Modes of Conveyance.—The *street-car* system is excellent, affording easy access to all parts of the city (fare, 5c., with free transfers). The charges for *carriages* are : $1.50 for the first hour, and $1 for each subsequent hour ; from depots and steamboat-landings, 25c. to 50c. for each person. There are two *ferries*, one to Jeffersonville from the foot of 1st St., and one to New Albany from Portland (foot of 34th St.).

Railroad Stations.—The station of the *Louisville & Nashville R. R.* is at Broadway and 10th St. It was completed in October, 1891, and is one of the finest stations in the South. It is also used by the *Louisville, New Albany & Chicago R. R.*, and the *Pennsylvania lines.* The station on the river, between 1st and 2d Sts., is used for local trains between Louisville and Lexington, and by the *Louisville, Harrod's Creek & Westport R. R.* (narrow gauge). The *Newport News & Mississippi Valley R. R.*, the *Chesapeake & Ohio R. R.*, the *Louisville, St. Louis & Texas R. R.*, the *Ohio & Mississippi R. R.*, and the *Louisville Southern R. R.* now enter and leave the Union Depot on river-front, between 7th and 8th Sts.; while the *Louisville, Evansville & St. Louis R. R.* (air-line) uses the station at the cor. of Main and 14th Sts. All the trains of the Pennsylvania lines also stop here.

Theatres and Amusements.—*Macauley's Theatre*, in Walnut St. near 4th, is the most fashionable place of amusement, and is fitted up in handsome style. The *Bijou* and *Harris's* are both in 4th St., between Green and Walnut ; *Liederkranz Hall*, Market St. near 2d, and *Masonic Temple Theatre*, cor. Jefferson and 4th Sts., are tasteful and commodious buildings. *Phœnix Park* is a noted German resort, with concerts in summer.

Reading-Rooms.—In the leading hotels are reading-rooms, provided with newspapers, etc., for the use of guests. The library of the Polytechnic Society of Kentucky, with more than 50,000 volumes, is open to the public daily from 9 A. M. to 10 P. M.; and the reading-room of the Society is open to strangers on introduction. The Y. M. C. A. reading-room is open to members and strangers.

Post-Office.—The Post-Office is at the cor. of Green and 3d Sts. It is open on week days from 7 A. M. to 6 P. M. ; on Sundays from 9 to 10 A. M. Letters may also be mailed in the lamp-post boxes, whence they are collected several times a day.

LOUISVILLE, the chief city of Kentucky, and one of the most important in the country, is situated at the Falls of the Ohio, 400 miles from its mouth, and 130 miles below Cincinnati, where Beargrass Creek enters that river. Its site is one of peculiar excellence. The hills which line the river through the greater part of its course recede just above the city, and do not approach it again for more than 20 miles, leaving an almost level plain about 6 miles wide, and elevated about 70 ft. above low-water mark. The Falls, which are quite picturesque, may be seen from the town. In high stages of the water they disappear almost entirely, and steamboats pass over them; but when the water is low, the whole width of the river has the appearance of a great many broken cascades of foam making their way over the rapids. To obviate the obstruction to navigation caused by the Falls, a canal, 2¼ miles long, has been cut around them to a place called Shippingport. It was a work of vast labor, being for the greater part of its course cut through the solid rock, and cost nearly $1,000,000. The widening of this canal is now in progress. The city extends about 8 miles along the river and about 4 miles inland, embracing an area of 18 square miles.

The first settlement of Louisville was made by 13 families, who accompanied Colonel George Rogers Clarke on his expedition down the Ohio in 1778. The town was established in 1780, and called Louisville, in honor of Louis XVI of France, whose troops were then aiding the Americans in their struggle for independence. It was incorporated as a city in 1828, when its population was about 10,000. In 1860 the population had increased to 68,033; in 1870, to 100,753; and in 1890 was 161,129. Louisville is the largest leaf-tobacco market in the world, and is rapidly becoming one of the most important markets for live-stock in the country. Pork-packing is extensively carried on, and the sugar-curing of hams is a special feature of the business. Louisville is the great distributing market for the fine whiskies made by the Kentucky distilleries. The manufacture of beer has also become a very important interest. Leather, cement, agricultural implements, furniture, and iron pipes for water and gas mains, are the other leading manufactures.

The city is regularly laid out, with wide, well-paved streets, and large squares, which are bisected each way by paved alleys 20 ft. wide. The beauty of the residences is a notable feature of the city; most of them are set back from the street, leaving lawns in front, which are planted with flowers and shrubbery, and the streets are lined with shade-trees. The business portion is compactly built, and contains many fine edifices. *Main, Market, Jefferson,* and *Fourth,* and the cross streets from 1st to 15th inclusive between the river and Walnut St., are the principal streets in this section.

The public buildings of Louisville are not fine architecturally, but are of a solid and substantial character. The * **Court-House,** in Jefferson St., between 5th and 6th, is a large limestone structure, with Doric portico and columns, and cost over $1,000,000. The * **City Hall** is the most ambitious edifice in the city, and is much admired. It is of stone, in the Composite style, with a square clock-tower at one corner, and cost $500,000. The Council-room is very fine. The *Custom-House*, which also contains the *Post-Office*, is an elaborate stone building recently erected at the corner of 4th and Chestnut Sts. The *Masonic Temple*, cor. 4th and Green Sts., is a handsome structure, with tasteful interior decorations. The Board of Trade has a commodious building cor. 3d and Main Sts. The * building of the *Courier-Journal*, cor. 4th and Green Sts., is one of the handsomest in the city.

Of the church edifices, the most noteworthy are the *Warren Memorial Church* (Presbyterian), 4th Ave. and Broadway; *Cathedral* (Roman Catholic), on 5th St. near Walnut; *Christ Church* (Episcopal), on 2d St., between Green and Walnut; the *First Presbyterian*, in 4th near York; the *Calvary* (Episcopal), in 4th near York; the *Second Presbyterian*, cor. Broadway and 2d; *College Street Presbyterian*, cor. College and 2d; the *Church of the Messiah* (Unitarian), cor. 4th and York; the *Temple Adas Israel*, cor. Broadway and 6th; and *Broadway Church* (Baptist), Broadway, between Brook and Floyd. Among the charitable institutions are the *Masonic Widows' and Orphans' Home;* the *Church Home* for females, and infirmary for the sick of both sexes, erected at a cost of $100,000, and the gift of John P. Morton, Esq.; the *John N. Norton Memorial Infirmary*, etc.

The *Polytechnic Society of Kentucky** occupies a commodious edifice in 4th Ave. between Green and Walnut Sts. The library numbers 50,000 volumes, and is opened to the public; and connected with it is a museum and natural-history department, with 100,000 specimens, and an art-gallery with a small but choice collection of paintings. The Gerald Troost collection of minerals, one of the largest in the United States, is included in it. Louisville being the center of one of the finest fossiliferous regions in the world, there are numerous private collections, containing many excellent specimens elsewhere rare. The *University of Louisville*, containing law and medical departments, is a flourishing institution, and has one of the finest buildings in the city, at the corner of 9th and Chestnut Sts. The *Kentucky School of Medicine* and *Hospital College of Medicine* (medical department of The Central University of Kentucky) are prosperous institutions of learning. Other important institutions are the *Southern Baptist Theological Seminary, Kentucky College of Music and Art, Louisville Law School, Louisville Medical College, College of Dentistry, School of Pharmacy for Women,* and *Louisville College of Pharmacy.* The two *High Schools* (male and female) are large and handsome brick structures. The *Colored Normal School,* dedicated in 1873, is one of the finest public-school edifices designed for the instruction of negroes in the country. The entire public-school system for colored people is admirably organized.

The *State Blind School,** on the Lexington Turnpike, E. of the city, is a massive and imposing structure, one of the finest of its kind in the Southwest. In the same building is the *American Printing-House for the Blind,* established in 1858, and endowed by Act of Congress in 1879 with $250,000, the interest of which is to be used in manufacturing embossed books and apparatus for all the schools for the blind in the United States. The *Almshouse* is a large building in the midst of ample grounds near the W. limits of the city (reached by 7th St.). The *City Hospital* is a plain but spacious edifice in Preston St., between Madison and Chestnut. Other important charitable institutions are the *House of Refuge for Boys,* the *House of Refuge for Girls,* the *Eruptive Hospital,* and the *St. Vincent Orphan Asylum* (Roman Catholic), in Jefferson St. near Wenzell. There are over 25 hospitals and asylums, all well conducted. The tornado which struck the city on March 29, 1890, wrecked that portion lying between 7th and 18th, Broadway and Main Sts. So much energy has been shown in rebuilding that the effects of the disaster are nearly obliterated.

By vote of the citizens, Aug. 4, 1890, there was accepted an act of the Legislature providing for the establishment of a system of parks in and adjacent to the city. Already ground for three such parks has been purchased, and the improvements are going on rapidly under the supervision of Frederick Law Olmsted. *Iroquois Park* is a tract of 554 acres, about 5 miles S. of the Court-House, and reached by the 4th Ave. Electric Line. It consists mainly of one great lofty hill, with steep slopes, bearing much crowded, well-grown wood. It is rather a forest than a park. *Cherokee Park* is situated just E. of the city limits, and contains more than 250 acres. Here are found superb umbrageous trees, standing singly and in open groups, distributed naturally upon a gracefully undulating greensward, in higher perfection than is yet to be found in any public park in America. *Shawnee Park* is a much smaller piece of ground in the extreme W. portion of the city, along the river-bank, and affords a beautiful prospect down and across the river to the Indiana knobs. These three parks, differing greatly in character, will give to Louisville one of the most complete park systems of any American city.

Strangers should visit * **Cave Hill Cemetery.** The monument of George D. Prentice, the poet, journalist, and politician, consists of a Grecian canopy, of marble, resting on four columns, with an urn in the center, and on the top a lyre. The cemetery is situated just E. of the city limits, contains other noteworthy monuments, and has appropriately ornamented and carefully preserved grounds. *Silver Creek*, 4 m. below the city, on the Indiana side, is a beautiful, rocky stream, and a favorite fishing and picnic place for the citizens. *Harrod's Creek*, 8 m. up the Ohio, *Riverside, Crescent Hill Reservoir*, and the *Water-Works Grounds*, afford pleasant excursions. The *Lexington* and *Bardstown* turnpikes afford enjoyable drives through a picturesque and well-cultivated country. *Jeffersonville*, a flourishing town on the Indiana shore, opposite Louisville, and connected with it by ferry and bridge, is situated on an elevation from which a fine view of Louisville may be obtained. The great railroad-bridge across the Ohio at this point is 5,219 ft. long, divided into 25 spans, supported by 24 stone piers, and cost $2,016,819. **New Albany,** opposite the W. end of Louisville, is a finely-situated and handsomely-built city of 21,059 inhabitants, with wide and delightfully-shaded streets, fine churches and public buildings, and handsome private residences. The trains of the Ohio & Mississippi R. R. now enter from the north by the bridge connecting New Albany with Louisville. "From the hills back of New Albany," says Mr. Edward King, "one may look down on the huge extent of Louisville, half hidden beneath the foliage which surrounds so many of its houses; can note the steamers slowly winding about the bends in the Ohio." New Albany is now connected with Louisville by the Kentucky and Indiana Bridge, just finished. It is the largest system of cantilevers in the world, and cost about $2,000,000. Total length, 2,453 ft.

The Louisville, Evansville & St. Louis R. R. is a direct line between Louisville and St. Louis (273 miles), and between Louisville and Evansville (122 miles). A double daily line of through vestibule trains are run between these points. Connection is made at Louisville with the Louisville & Nashville R. R. and the Queen & Crescent System for all points East and Southeast. At St. Louis direct connection is made in Union Depot with all diverging lines. Parlor and dining cars on all day-trains, and sleeping-car service on night-trains.

78. Cincinnati to St. Louis.

By the Ohio & Mississippi R. R. Through trains from Baltimore to St. Louis *via* Baltimore & Ohio R. R. (Route 70) run on this line. Close connection is made with the trains of the various routes from New York to Cincinnati. Another way of reaching St. Louis from Cincinnati is by steamer on the Ohio River and Mississippi River. This latter is a pleasant route in summer. *Stations on the Ohio & Mississippi R. R.:* North Bend, 15 miles ; Lawrenceburg, 20 ; Aurora, 24 ; Osgood, 52 ; Nebraska, 62 ; N. Vernon, 73 ; Seymour, 87 ; Mitchell, 127 ; Washington, 173 ; Vincennes, 192 ; Olney, 223 ; Clay City, 238 ; Xenia, 254 ; Salem, 271 ; Odin, 276 ; Sandoval, 280 ; Lebanon, 317 ; Caseyville, 331 ; St. Louis, 340.

THIS route traverses from side to side the great States of Indiana and Illinois, passing through an extremely rich agricultural country which is for the most part under fine cultivation. The numerous towns and villages *en route* are neat and attractive, with that air of busy prosperity about them which is eminently characteristic of the West; but, like the stretches of country between them, they are curiously alike, and few present any features requiring special notice. For 25 miles after leaving Cincinnati the train runs nearly parallel with the Ohio River. *North Bend* (15 miles) is a pretty village on the river, noted as the residence of the late General William Henry Harrison, President of the United States. His tomb, a modest brick structure, stands on a commanding hill, whence there is a fine view, including portions of Ohio, Indiana, and Kentucky. Three miles beyond N. Bend the train crosses the Great Miami River and enters Indiana, speedily reaching *Lawrenceburg* (20 miles), a city of 4,284 inhabitants, on the Ohio River, and a station of the Cleveland, Cincinnati, Chicago & St. Louis R. R. (known as the "Big Four Route"). It is chiefly a manufacturing town (furniture, barrels, stoves, flour, etc.), and is the county-seat. Four miles beyond Lawrenceburg, also on the river, is the beautiful little city of **Aurora** (*Globe House, Kirsch Hotel*), with 3,929 inhabitants, and a large trade derived from the rich farming country of which it is the shipping port. A number of small stations are now passed. From *N. Vernon* (73 miles) the Louisville branch diverges and runs S. in 54 miles to **Louisville** (Route 77). *Seymour* (87 miles) is a thriving town at the intersection of the Pittsburg, Cincinnati, Chicago & St. Louis R. R.; and *Mitchell* (127 miles) is at the crossing of the Louisville, New Albany & Chicago R. R. *Washington* (173 miles) is a small town, capital of Daviess County, where the shops of the Ohio & Mississippi R. R. are located; and 19 miles beyond is **Vincennes** (*Grand Hotel, Laplante Hotel*), a flourishing city of 8,853 inhabitants, on the E. bank of the Wabash River, which is here navigable by steamboats. Vincennes is the oldest town in the State, having been settled by the French Canadians, who established a mission here in 1702, and a few years later built a fort. It became the capital of the Territory of Indiana upon its organization in 1800, and so remained until 1814. The surrounding country is fertile and abounds in coal, and the city enjoys good manufacturing facilities. The leading establishments are the flouring-mills. The public schools are excellent, and there are 10 churches and 4 libraries. *Vincennes University*, chartered in 1807, is now conducted as a high-school.

Leaving Vincennes, the train crosses the Wabash River and enters the State of Illinois, passing at frequent intervals a number of small stations. *Olney* (223 miles) is the capital of Richland County, the general character of which is suggested by its name. It is one of the most prosperous places on the line of the road, and has a population of 3,831. *Clay City* (238 miles), *Xenia* (254 miles), and *Salem* (271 miles) are thriving villages. *Odin* (276 miles) is at the crossing of the Chicago Branch of the Illinois Central R. R. (Route 84), and *Sandoval* (280 miles) is at the crossing of the Main Line of the Illinois Central R. R. Sandoval is the point where the large repair-shops of the Ohio & Mississippi R. R. are located. *Carlyle* (293 miles) is situated on the Kaskaskia River, on the margin of a fine prairie, and is a lumber-market of some importance, logs being floated to this point, where they are made into lumber and sent to St. Louis. *Lebanon* (317 miles) is a beautifully-situated and well-built place of 3,682 inhabitants. It has a handsome Union school-house, and is the seat of McKendree College. At *Caseyville* (331 miles) the train first enters the great American Bottom, or Valley of the Mississippi. The village is built just at the foot of the bluff, and is one of the principal points from which St. Louis is supplied with coal, the bluffs being underlaid for many miles by inexhaustible deposits. At *E. St. Louis* (339 miles) the train crosses the Mississippi on the splendid bridge which is described in connection with St. Louis. **St. Louis** (see Route 81).

79. New York to St. Louis via Cleveland and Indianapolis.

By the N. Y. Central & Hudson River R. R. or the West Shore R. R., and the Lake Shore & Michigan Southern R. R. (Route 65) to Cleveland; and thence by the Cleveland, Columbus, Cincinnati & St. Louis R. R. (commonly called the "Big Four"). Through trains, with drawing-room and sleeping cars attached, run through from New York to St. Louis on the N. Y. Central route in 38 hours. The "Southwestern Limited," composed exclusively of palace-car service, leaves New York at 1.30 P. M., arriving at St. Louis at 7.35 P. M. Distances: New York to Cleveland, 623 m.; to Crestline, 698; to Galion, 703; to Bellefontaine, 764; to Indianapolis, 906; to Terre Haute, 978; to Mattoon, 1,034; to Alton Junction, 1,146; to St. Louis, 1,167.

FROM New York to **Cleveland** (623 miles) this route is described in Route 67. From Cleveland to *Galion* (703 miles) it is described in Route 74. At Galion the Indianapolis Div. of the C., C., C. & St. L. R. R. diverges from the main line, and runs nearly due W. through one of the richest sections of Ohio. *Marion* (724 miles) is at the intersection of the New York, Lake Erie & Western R. R. (see Route 74). *Bellefontaine* (764 miles) is a flourishing town of 4,245 inhabitants, so named from the numerous fine springs in the neighborhood. It is surrounded by a productive and populous agricultural country, and has a large trade. There are also several manufactories, and the County buildings are located here, Bellefontaine being the capital of Logan County. *Sidney* (787 miles) is a neat village, built upon an elevated plateau on the W. bank of the Great Miami River, which affords a fine water-power. A navigable feeder of the Miami Canal also passes through the place. In the center of the village is a neat public square, around which are the principal buildings. *Union* (820 miles) is situated directly on the boundary-line, and is partly in Ohio and partly in Indiana. It is a flourishing place, and an important railroad center. *Winchester* (831 miles) and *Muncie* (853 miles) are pretty towns. *Anderson* (870 miles) is picturesquely situated on a high bluff on the left bank of White River, in the midst of a very fertile region. A few miles above the village is a dam by which a fall of 34 ft. is obtained, the extensive water-power being used in numerous manufacturing establishments. *Pendleton* (878 miles) is a thriving village on Fall Creek, which affords a good water-power. In the vicinity are quarries of limestone. *Fortville* (886 miles) is a small station, 20 miles beyond which the train reaches

Indianapolis.

Hotels, etc.—The leading hotels are the *Bates House*, the *Denison*, the *Grand Hotel*, the *Spencer House*, and the *Union Depot Hotel*. The *Occidental*, the *Browning*, the *Brunswick Hotel*, the *Circle*, the *Circle Park*, the *English*, and the *Sherman House*, are good. Prices are from $2 to $5 per day. *Street-cars* render all parts of the city easily accessible, and there are nine bridges across the river (three of them for railroad purposes). The *Post-Office* is at the cor. of Pennsylvania and Market Sts.

Indianapolis, the capital and largest city of Indiana, is situated near the center of the State, on the W. fork of White River, 110 miles N. W. of Cincinnati, and 195 miles S. E. of Chicago. The city is built in the midst of a fertile plain, chiefly on the E. bank of the river, which is crossed by 9 bridges. The streets are 90 ft. wide (except Washington St., which has a width of 120 ft.), and cross each other at right angles; but there are four long avenues radiating from a central square (the Circle) and traversing the city diagonally. Indianapolis was first settled in 1819, became the seat of the State government in 1825, was incorporated in 1836, and received a city charter in 1847. In 1891 it received a new charter prepared by representative citizens, whose desire is to have a model municipal government. In 1840 it had a population of only 2,692; in 1850, 8,091; in 1860, 18,611; in 1870, 48,244; in 1880, 75,074; and in 1890, 105,436. Its trade has kept pace with the growth of its population, and its manufactures (more than 1,000 in number and employing over 25,000 hands) are varied and important, the principal industries being pork-packing and the manufacture of machinery, agricultural implements, cars, carriages, furniture, and flour. More than 18 completed railways converge here, making it one of the great railway centers of the West.

Washington St. is the principal retail thoroughfare, and the wholesale business houses are in *South Meridian, Pennsylvania, Illinois,
Maryland,* and *Georgia Sts.* The most prominent public building is
the *State-House,* completed in 1887. It occupies two squares, and
cost about $2,000,000. The *Court-House,* completed in 1876,
at a cost of $1,200,000, is an imposing structure. The *State Institute
for the Blind,* in North St., between Illinois and Meridian, was built in
1847, at a cost of $300,000, and is surrounded by 8 acres of grounds.
The main building has a front of 150 ft., and is five stories high, consisting of a center and two wings. The *State Lunatic Asylums,* 1½ mile
W. of the city limits, is a fine group of buildings, surrounded by 160
acres of grounds, a portion of which is handsomely laid out and adorned.
The *State Institute for the Deaf and Dumb,* just E. of the city limits,
was erected in 1848 at a cost of $220,000, and a building costing
$200,000 was added later. The grounds comprise 105 acres, hand-
somely laid out and adorned with trees and shrubbery. The *U. S.
Arsenal,* 1 mile E. of the city, is a handsome building, and is surrounded
by 75 acres of grounds. The **Union Passenger Depot** (in Louisiana
St., between Meridian and Tennessee) has sheds 760 ft. long, and is one
of the most spacious structures of the kind in the country. Other
prominent public buildings are the *Post-Office,* cor. Pennsylvania and
Market Sts.; the *City Hall; Tomlinson Hall* (built from the proceeds
of a bequest left the city by Isaac H. Tomlinson); the building of the
Commercial Club, 8 stories high, cor. Meridian and Pearl Sts.; *County
Jail,* and *City Prison;* the *Masonic Hall,* cor. Washington and Tennessee Sts.; the *Odd-Fellows' Hall,* cor. Washington and Pennsylvania
Sts.; and the *Propylæum,* erected by the women of the city. Of
the churches, the most noteworthy are *Christ* and *St. Paul's,* Episcopal; *Meridian Street* and *Roberts Park,* Methodist; *First* and *Second,*
and *Tabernacle,* Presbyterian; *First,* Baptist; *Plymouth,* Congregational;
the Roman Catholic *Cathedral;* and the Jewish *Synagogue.* The *Butler University,* founded in 1850, occupies a handsome building 4 miles
E. of the city; it admits both sexes, and has a library of 5,000 volumes. The *State Library* contains 15,000 volumes, and there is a *Free
City Library* with 40,000 volumes. The principal charitable institu-
tions are three *Asylums for Orphans,* the *German Orphan Asylum,* the
State Female Reformatory and Asylum, the Catholic *Infirmary,* and a
City Hospital. Among the principal manufacturing industries are the
Atlas Engine-Works, the *Indianapolis Car-Works,* the *Brown-Ketcham
Iron-Works,* the *National Malleable Castings Co.,* the *Kingan Pork-
Packing Houses,* the *Encaustic Tile-Works,* the *Parry Cart Co.,* the
Woodburn-Sarven Wheel-Works, the *Udell Woodware Works,* the *Terra-
cotta Works,* the *National Card-Works,* the *Premier Steel-Works,* etc.
The *Stock-Yards* are also worth notice.

There are nine public parks in the city, viz.: the *Circle,* in the
center, containing 4 acres, the site of the Soldiers' and Sailors' Monument, built by the State at a cost of $400,000, and a statue of *Oliver
P. Morton;* the *Military Park,* 18 acres; *University Park,* 4 acres,
with a statue of *Vice-President Schuyler Colfax;* the *Trotting Park,*

with a course of one mile, 86 acres; *Garfield Park,* S. of the city, about 100 acres; and *Woodruff Place,* E. of the city, next to the arsenal; a park in the N. portion of the city, embracing 100 acres; *Athletic Park,* with 80 acres; *Fairview Park,* 160 acres, a beautiful shaded retreat along the river and canal, N. of the city; and the *State Fair Grounds,* with Exposition Building, containing 40 acres. It is the home of President Benjamin Harrison. *Greenlawn Cemetery* is within the city limits, and 2 miles N. of the city is **Crown Hill,** which is handsomely laid out; and the *Catholic Cemetery* is just S. of the city limits.

At Indianapolis the train takes the track of the St. Louis Div. to St. Louis. *Danville* (20 miles from Indianapolis, 926 from New York) is a pretty village, with county buildings which cost $180,000. **Greencastle** (944 miles; reached also by Vandalia R. R.) is a little academic city of 4,390 inhabitants, in the midst of a farming and stock-raising region. It contains a Court-House, a Jail, a large rolling-mill and nail-factory, seven public schools, including a High-School, and several churches. The *Indiana Asbury University* (Methodist), founded in 1835, is open to both sexes, and has nearly 500 students. The Whitcomb and the college circulating libraries contain 9,000 volumes. There is also in the city a flourishing Presbyterian female college. The Vandalia Line to St. Louis (see Route 80) touches Greencastle, and runs nearly parallel with the present line to **Terre Haute** (*National, Terre Haute House*). Terre Haute is a city of 30,217 inhabitants, on the E. bank of the Wabash River, which is here spanned by three bridges. It contains a Market-House and City Hall, two Opera-Houses, two Orphan Asylums, a new high-school building, costing $75,000, ten public-school buildings, several private schools, and *Coates's Female College,* and 24 churches. The *State Normal School* was erected at a cost of $230,000. The *Rose Polytechnic School,* endowed with $600,000, and *Rose Orphan Home* with $500,000, were founded by Chauncey Rose. *St. Anthony's Hospital* (the building given by H. Hulman) is conducted by the Sisters of St. Francis. A *Court-House* and a *Federal Building* are now erecting. Terre Haute is the center of trade for a populous region, and has extensive manufactures, blast-furnaces, carriage and wagon works, machine-shops, nail-works, car-works, rolling-mills, woolen-mills, etc., and 7 flour-mills with a daily capacity of 3,300 barrels. An artesian well, 2,000 ft. deep, is celebrated for its medicinal value. *Collett Park,* named for its donor, is a handsome woodland of 25 acres, at the city's edge. It is also the point of intersection of 9 railroad lines, and the Wabash River is navigable for steamboats during most of the year. From this point the passenger reaches all points S. by the Evansville & Terre Haute R. R., connecting with the Louisville & Nashville R. R.; St. Louis, by the Vandalia and "Big Four." Lines; Chicago, by the Chicago & Eastern Illinois R. R.; St. Joseph and Michigan, by the Vandalia R. R.; Cincinnati, Washington, and New York, by the Vandalia and "Big Four" Lines; and Southern Indiana by the Evansville & Terre Haute R. R.

Leaving Terre Haute, the train crosses the Wabash River into the State of Illinois, passes several small stations, of which *Paris* (997 miles) and *Charleston* (1,023 miles) are the principal, and soon reaches *Mattoon* (1,034 miles), one of the principal towns between Terre Haute and St. Louis. The Chicago Branch of the Illinois Central R. R. (Route 82) crosses here, and here are the machine-shops, round-house, and car-works of this division of the road. *Pana* (1,073 miles) is a prosperous little city at the crossing of the Northern Div. of the Illinois Central R. R., surrounded by a rich agricultural region. *Litchfield* (1,112 miles) is another busy little city, with coal-mines in the neighborhood. There are several grain-elevators here; and besides several steam-mills it contains the R. R. construction and repair shops. At *Bethalto* (1,142 miles) the road leaves the prairie and enters the "American Bottom," as the strip of rich alluvial land between the Mississippi River and the bluffs is called; scattered over it in all directions are numerous lakes, bayous, and sloughs. From *Alton Junction* (1,146 miles) a branch line diverges to **Alton** (see Route 85). At *E. St. Louis* (1,166 miles) the train crosses the Mississippi on the noble bridge described in connection with St. Louis. **St. Louis** (see Route 81).

Wabash Line.—Another favorite route from New York to St. Louis is *via* the Wabash Railway, which runs from Detroit, Mich., and *Toledo, Ohio*, across northern Ohio, northern Indiana, and central Illinois, to *St. Louis*. Toledo is reached from New York *via* the N. Y. Central and Michigan Central R. R. (Route 67); also *via* Erie R. R. (Route 68). Close connection is made at Toledo, and there is no change of cars between New York and St. Louis. At *Fort Wayne*, the Wabash R. R. connects with Route 68 from New York; so that the "Wabash Line" may be combined with either of the great routes from the seaboard to the far West. The Wabash R. R. runs nearly parallel to and a little N. of the route described above, and through a very similar section of country. The principal cities and towns on the line are *Napoleon, Defiance, Fort Wayne, Wabash, Peru, Logansport,* and *Lafayette,* in Indiana ; and *Danville* and *Decatur,* in Illinois. It intersects Route 79 at Litchfield (55 miles from St. Louis). The time from New York to St. Louis by the "Wabash Line" is 36 hours.

80. New York to St. Louis via Philadelphia, Pittsburg, and Indianapolis.

By the Pennsylvania R. R., the Pittsburg, Cincinnati, Chicago & St. Louis R. R., and the Terre Haute & Indianapolis R. R. This is commonly called the "Pan-Handle Route" and "Vandalia Line." Through trains, with palace drawing-room and sleeping car service attached, run through from New York to St. Louis, without change of cars, in about 31 hours. Distances : to Columbus, 637 miles : to Urbana, 684 ; to Piqua, 710 ; to Richmond, 757 ; to Indianapolis, 825 ; to Terre Haute, 898 ; to Vandalia, 996 ; to St. Louis, 1,065.

As far as **Columbus** (637 miles) this route is the same as Route 73. *Milford* (665 miles) is at the crossing of the Springfield Branch of the Cleveland, Cincinnati, Chicago & St. Louis R. R. (Route 73), and *Urbana* (684 miles) is at the crossing of the Sandusky Div. of the Cleveland, Cincinnati, Chicago and St. Louis R. R. (Route 74). Urbana is described in Route 74. Twenty-six miles beyond Urbana the train reaches **Piqua,** a city of 9,090 inhabitants, situated on the W. bank of the Great Miami River, just at a bend which leaves a level plateau between the city and the water's edge, while on the opposite side the bank rises boldly. The city is regularly laid out, with wide streets, and is sub-stantially built. The surrounding country is rich in agricultural products. Water-power is supplied by the Miami Canal, and considerable manu-facturing is carried on, the principal establishments being car-shops, woolen-mills, foundries, etc. At *Bradford Junction* (720 miles) the road branches ; one division running N. W. to Chicago *via* Logansport, while the present route continues W., enters Indiana a little beyond *Greenville* (721 miles), and soon reaches **Richmond,** a flourishing city of 16,608 in-habitants, situated on the E. fork of the Whitewater River, in the center of a fertile agricultural district, from which it derives an important trade. It has an abundant water-power, and is the seat of numerous mills and factories, the chief articles of manufacture being agricultural machinery and implements. The city is handsomely built, contains many costly residences, and has two theatres, a Public Library of 10,000 volumes, and 20 churches. The Quakers form a large element in the population of Richmond, and they have here two educational institutions : the *Friends' Academy* and *Earlham College*, which was founded in 1859, admits both sexes, and has a library of 3,500 volumes. The college buildings are about ¼ mile W. of the city. In the N. E. corner of the city are Fair Grounds. Four railroads intersect at Richmond, and street-cars traverse the principal streets. *Cambridge City* (772 miles) and *Knightstown* (791 miles) are thriving towns. Near the latter is a Sol-diers' Home, for the disabled soldiers and for the indigent widows and orphans of the soldiers from Indiana who fell in the civil war. The next important station on the line is **Indianapolis,** the capital of In-diana, which has already been described in Route 79.

From Indianapolis to Terre Haute the present route and Route 79 run close beside each other, touching at *Greencastle* (see Route 79) and at **Terre Haute** (see Route 79). Between Terre Haute and St. Louis the present route makes a gain in distance of 24 miles, but trav-erses a newer and more thinly settled region, though the stations along either route are not of much importance. *Effingham* (965 miles) is at the intersection of the Chicago Branch of the Illinois Central R. R. (Route 84). It is situated near the Little Wabash River, and has con-siderable trade and manufactures, with a population of about 3,260. *Vandalia* (996 miles) is a town of 2,144 inhabitants, on the W. bank of the Kaskaskia River. From 1818 to 1836 it was the capital of Illinois, and was then a prosperous place. After the removal of the capital to Springfield it became rapidly decadent, but is reviving now under its railroad advantages, and promises to become an important manu-facturing center. *Greenville* (1,014 miles) is the highest point on the line between Terre Haute and St. Louis, and is a flourishing town of 1,868 inhabitants, on the E. bank of Shoal Creek. To the S. is a fine prairie. *Highland* (1,034 miles) is a busy manufacturing town with 1,857 inhabitants, mainly Germans. It is pleasantly situated and well built. Between here and Collinsville (1,052 miles) are numerous coal mines, whence the city of St. Louis derives its chief supply. At *E. St. Louis* (1,062 miles) the train crosses the Mississippi on the mag-nificent bridge which is described in connection with St. Louis.

MAP OF
ST. LOUIS
MO.

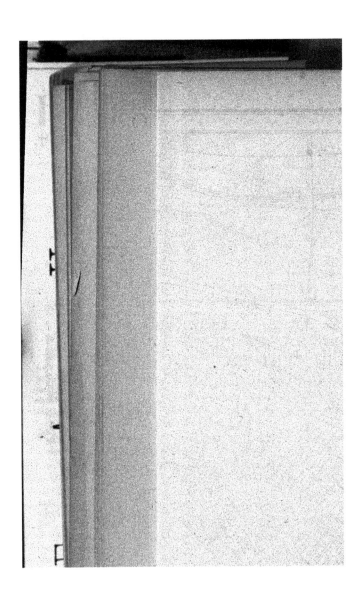

81. St. Louis.

Hotels.—The *Southern Hotel*, between Walnut and Elm and 4th and 5th Sts.; the *Lindell Hotel*, cor. Washington Ave. and 6th St.; the *Laclede*, cor. 6th and Chestnut Sts.; *Hotel Richelieu*, cor. Washington Ave. and 14th St.; and *St. James*, cor. Broadway and Walnut St., are among the best hotels. *Hotel Moser*, in Pine, between 8th and 9th Sts.; the *Belvedere Hotel*, cor. Washington Ave. and 13th St., are on the European plan. The rates are from $2 to $5 a day. Other very comfortable and well-kept houses, on a smaller scale, are *Hurst's*, Broadway and Chestnut St., and *Hotel Rozier*, 13th and Olive Sts., on the European plan (rooms, $1 to $1.50 a day). The *Hotel Beers* (private), cor. Grand Ave. and Olive St., is a fine specimen of architecture, and has an excellent restaurant; while on the opposite corner is the *Grand Avenue Hotel*, a family hotel. The *West End Hotel*, on the cor. of Vandeventer and Bell Aves., is a new and desirable family hotel.

Restaurants.—*Faust's*, cor. South Broadway and Elm, has a high reputation. *Moser's Restaurant*, 811-813 Pine, is the largest restaurant in the city, having a capacity for serving 350 persons at one time. Numerous other restaurants of lesser note are to be found in all parts of the city.

Modes of Conveyance.—*Electric* or *cable cars* traverse the city in every direction, and render all parts accessible (5c.); all lines start within a radius of 5 blocks from Broadway and Olive St. The cars on 5th St. run nearly the entire length of the city from N. to S.; those on Market, Pine, Olive, Locust, Washington Ave. and Franklin Ave., run E. and W. *Carriages* are in waiting at the depots and steamboat-landings, and at stands in different parts of the city. The rates established by law are: For conveying 1 or more persons a distance of 1 m. or less, $1; more than 1 m. and less than 2 m., $1.50, and 50c. for each additional mile. By the hour, $2 for the first hour, and $1.50 for each additional hour. In case of disagreement as to distance or fare, call a policeman, or complain at the City Hall. *Ferries* run to *East St. Louis, Ill.*, from the foot of *N. Market St., Carr St., Spruce St., Poplar St., Sidney St.*, and *Davis St.*; also, electric-cars run across the St. Louis Bridge.

Railroad Stations.—All passenger trains run into the Union Station in Poplar St., between 11th and 12th Sts. (accessible by the Pine St. cars), except some suburban trains on the *Wabash R. R.*, which use the station at the foot of Vine St., and those of the *Iron Mountain R. R.*, which use the station at the cor. of 4th. St. and Chouteau Ave. A very large and handsome Union Station is now being erected in Market St., from 18th to 20th Sts., which will replace the present one at 12th and Poplar Sts. (to be finished in 1894).

Theatres and Amusements.—All places of amusement are usually open on Sunday. The *Olympic*, cor. South Broadway and Walnut; *Grand Opera-House*, Market, between Broadway and 6th; *Pope's*, cor. 9th and Olive, and the *Hagan*, 10th and Pine Sts., are the leading theatres; all of them have large stages and beautiful interiors. The *Standard*, cor. 7th and Walnut, and *Havlin's*, cor. 6th and Walnut, afford first-class attractions at a lesser price. A new German theatre has just been erected at 13th and Locust Sts. The *London*, cor. 4th and Walnut, is devoted to good variety shows. The *Pickwick* is a beautiful little theatre at the West End (2809 Washington Ave.). *Uhrig's Cave*, cor. Washington and Jefferson Aves., is a summer theatre, where light operas are given. There are numerous German beer-gardens. *Schnaider's*, Choteau and Mississippi Aves., the leading one, is a favorite summer-evening resort for many of the best people in the city. During the summer months instrumental concerts, by the best bands in the country, and comic operas are given nightly. *Sportsmans' Park*, in N. Grand Ave., is where the famous St. Louis team play during the base-ball season. *Amateur Park*, Russell and Missouri Aves. (on the Southside) is used for base-ball and athletic games.

Reading-Rooms.—At all the principal hotels there are reading-rooms for the use of guests, well supplied with newspapers, etc. The city has just cause to be proud of its public libraries, which are large and carefully selected. The *Mercantile Library*, cor. Broadway and Locust Sts., has a library of 80,000 volumes and a reading-room, both of which are free to strangers (open from 9 A. M. to 10 P. M.). Besides the library, the hall contains paintings, statuary, coins, etc. The *Public Library*, cor. Locust and 9th Sts., contains 80,000 volumes, and a large reading-room, which are open to the public (from 10 A. M. to 10 P. M.). The *Academy of Science*, founded in 1856, has a large museum and a library of 12,000 volumes in the Washington University. The *Missouri Historical Society*, founded in 1865, has a large historical collection, and is located in Lucas Place, at the cor. of 16th St.

Clubs.—The *University Club*, Pine and Beaumont Sts., occupies an old-time Southern mansion, with handsome grounds. The *St. Louis*, Locust St. and Ewing Ave., has a fine building. The *Jockey Club*, at the Fair-Grounds, has very attractive and commodious quarters that are kept open the entire year, though used mainly during the racing season. The *Marquette*, Grand Ave. and Pine St., has a beautiful house, and its membership is chiefly among Catholics. The *Mercantile*, 7th and Locust Sts., is a business-men's club, and has a handsome building. All the preceding clubs have restaurants. The *Germania*, 8th and Gratiot Sts., is a popular German club, as is also the *Liederkranz*, 13th St. and Choteau Ave.; while the *Harmonie*, 18th and Olive Sts., is a Hebrew club. The *Elks* have comfortable quarters in the Hagan Building, 10th and Pine Sts.; while the *Commercial Club*, composed of the most influential business-men of the city, usually meets in the St. Louis Club-House. The *Pastime* is an athletic club, with a fine gymnasium and club-house at 811 N. Vandeventer Ave. The *Union*, Park and Jefferson Aves., is a social club that has just been organized by the "South-siders," as the other leading clubs are in the "West-End" of the city. The *St. Louis Cycle Club* has a house at 307 N. Ewing Ave.; while the *Camera Club* is located in the Pastime Building, 811 N. Vandeventer Ave.; the *Engineers' Club*, in the Odd-Fellows', at 9th and Olive Sts.; and the *Artists' Guild*, in the Museum of Fine Arts, 19th and Locust Sts. The *Modocs*, *Westerns*, *North-End*, and *St. Louis* are rowing-clubs that have boat-houses along the river-front. The privileges of these clubs may be obtained by non-residents by an introduction from a member.

Post-Office.—The general Post-Office occupies a block between Olive and Locust Sts. and 8th to 9th Sts. It is open on week-days from 7½ A. M. to 8 P. M.; on Sundays from 10 to 11 A. M. There are also sub-stations in different parts of the city, and letters may be mailed in the lamp-post boxes, whence they are collected at frequent intervals by the carriers.

St. Louis is situated geographically almost in the center of the great valley of the Mississippi, or basin of the continent, on the W. bank of the Mississippi River, 20 miles below the entrance of the Missouri, about 175 miles above the mouth of the Ohio, and 1,170 miles above New Orleans, in lat. 38° 37' N. and lon. 90° 15' W. The city is well elevated above the surface of the river, being built on a series of gentle terraces, the first of which rises from the river-bank for about 1 mile to 17th St., where the elevation is 150 ft. above the stream; the ground then gently declines, rises in a second terrace to 29th St.; again falls, and subsequently rises in a third terrace to a height of 200 ft. at Côte Brillante or Wilson's Hill, 4 miles W. of the river; the surface here spreads out into a broad and beautiful plain. The corporate limits embrace 61.37 square miles, extending 19 miles on the river-front, 21 miles on the W. line, and 6 miles back from the river. The densely-built portion is comprised in a district of about 6 miles along the river and 3 miles in width.

In 1762 a grant was made by the Governor-General of Louisiana, then a French province, to Pierre Liguest Laclede and his partners, comprising the "Louisiana Fur Company," to establish trading-posts on the Mississippi; and on February 15, 1764, the principal one was established where the city now stands, and named St. Louis. In 1803 all the territory then known as Louisiana was ceded to the United States. In 1812 that portion lying N. of the 33d degree of latitude was organized as Missouri Territory. In 1822 St. Louis was incorporated as a city. The first census was taken in 1764, and the population was then 120. In 1810 it was only 1,600; in 1850 it had increased to 77,860; in

1860 to 160,773 ; and in 1870 to 310,864. According to the U. S. Census of 1880, the population was 350,522, and in 1890 the U. S. Census gave St. Louis a population of 451,770, with the rank of five. As the natural commercial entrepot of the vast Mississippi Valley, the commerce of St. Louis is immense ; the chief articles of receipt and shipment being breadstuffs, live-stock, provisions, cotton,

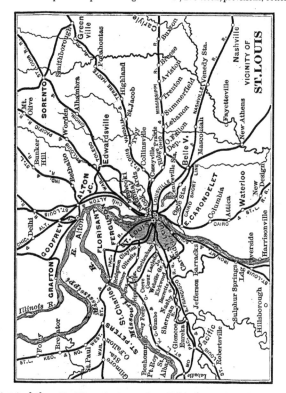

zinc, lead (from the Missouri mines, and smelted in the great metallurgical establishments in the immediate vicinity of St. Louis), hay, salt, wool, hides and pelts, lumber, tobacco, beer, and groceries. It is the greatest mule market in the world. Vast as are its commercial interests, however, the prosperity of the city is chiefly due to its manufactures. According to the census returns of 1890, there were 6,148 industrial establishments, employing 93,610 hands. The capital invested was $140,775,392, the cost of materials used, $122,010,805, while the value at factory of goods manufactured amounted to $228,714,317.

The city is, for the most part, regularly laid out, the streets near the river running parallel with its curve, while farther back they are generally at right angles with those running W. from the river-bank. From the Levee, or river-front, the streets running N. and S. are named numerically, beginning with Main or 1st St., 2d St., 3d St., etc., up to 25th St., with the single exception of 5th St., which is now called Broadway. Streets running E. and W., and those W. of Jefferson Ave., or 26th St., are named arbitrarily or from some historical association. The houses are numbered on the "Philadelphia system"—i. e., each block starts with a new hundred, all streets running parallel to the river being numbered N. and S. from Market St.; while on all streets running E. and W. the numbering begins at the Levee. *Front St.*, which is 100 ft. wide, extends along the Levee, and is built up with massive warehouses. This street, with *Main, Second,* and *Third,* is the principal location of the wholesale trade. **Broadway** and **Olive Sts.** are the fashionable promenades, and contain the leading retail stores. *Grand Ave.,* 12 miles long, runs nearly parallel with the river, on what was once the W. boundary of the city, which is now, however, extended to the W. extremity of Forest Park. *Washington Ave.* is one of the widest and handsomest in the city, its lower end being occupied by the wholesale dry-goods trade, and the upper end by fine residences. The city is divided by an E. and W. valley, in which are all the railroad-tracks and the Union Depot, into the "North Side" and the "South Side," bridges carrying the streets over the valley at 12th, 14th, 18th, 26th, 27th, and 36th Sts. The residences are large, comfortable, and have more or less ground around them, the solid blocks usually seen in large cities being exceptional. The fashionable portion of the city is at the *West-End,* which may be said to have its center at Grand and Washington Aves., where the most recent and finest residences are to be seen. The central portion of the city is occupied by the Americans, while the northern and southern sections are mainly composed of the large German element that forms so great a proportion of the population of St. Louis. St. Louis has many of the characteristics of a Southern city, in its social life, as shown in its famous hospitality, and the manner in which the people live out-doors after sunset during its long summer season. The city is smoky, on account of its many factories and the use of soft coal that is mined in the immediate vicinity, some coal-mines having formerly been worked within the city limits. A very efficient system of street-sprinkling that is kept up throughout the entire year, and the replacement of macadam pavement, made from the limestone on which the city is built, by Belgian granite, Nicholson, and asphalt pavements, have almost entirely done away with the dust-nuisance. The water-supply is derived from the Mississippi River, which, in spite of its muddy character from the suspended sediment that it always contains and which is but imperfectly settled out at the water-works, is an exceptionally wholesome and safe water to drink, especially after filtering.

The city is remarkably well built, stone and brick being the chief materials used, and the architecture being more substantial than showy. One of the finest public buildings in the city is the **Court-House,** occupying the square bounded by 4th, 5th, Chestnut, and Market Sts. It is built of St. Genevieve limestone, in the form of a Greek cross, with a lofty iron dome surmounting its center, and cost $1,200,000. In the dome are frescoes by Karl Weimer. The fronts are adorned with beautiful porticoes, and from the cupola of the dome (which is accessible to visitors) there is a fine view of the city and its surroundings. The present *City Hall*, cor. Market and 10th Sts., is a plain brick structure occupying half a square; a handsome building is now being erected, with brick and stone, on the square bounded by Market St., Clark Ave., 12th, and 13th Sts. The ***Four Courts** is a spacious and handsome limestone building, in Clark Ave. between 11th and 12th Sts., erected at a cost of $1,000,000. In the rear is an iron jail, semicircular in form, and so constructed that all the cells are under the observation of a single watchman at once. (Visitors admitted on Mondays, Wednesdays, and Fridays, from 8 to 9 A. M. and from 3 to 4 P. M.) The **Custom-House,** including Post-Office and United States Sub-Treasury, is on the block between Olive, Locust, 8th, and 9th Sts. It is built of Maine granite with rose-colored granite trimmings, is three stories high, with a French roof and Louvre dome, and the building was erected at a cost of $8,000,000. The *U. S. Arsenal*, situated in the extreme S. limits of the city, immediately on the river, is a beautiful spot (reached by 5th St. cars). The ***Chamber of Commerce,** in 3d St., between Pine and Chestnut, is the great commercial mart of the city. It is 223 ft. long by 187 ft. deep, is solidly built of gray limestone, is six stories high, and cost $1,500,000. The main hall or "Exchange" is a magnificent room, 222 ft. long, 65 ft. wide, and 60 ft. high. The sessions of the Exchange are from 10 A. M. to 1 P. M. Strangers are admitted to the floor on introduction by a member; the galleries are free to all. The *Cotton Exchange*, cor. Main and Walnut Sts., is five stories high, and cost $150,-000. The **Equitable Life-Insurance Building*, cor. 6th and Locust Sts., is a very ornate and showy edifice. It is of rose-colored granite, in the Renaissance style, ten stories high, with a massive cornice on the roof. From the roof (reached by elevators) a fine view is obtained. The *Republic Building*, cor. 3d and Chestnut Sts., and the *Globe-Democrat Building*, 6th and Pine Sts., are among the most complete and admirably-appointed newspaper offices in the world. Other fine buildings are the *Roe Building*, at Broadway and Pine St., constructed of pressed brick, granite, and red sandstone; the *Gould Building*, 7th and Chestnut Sts.; the *Commercial Building*, in Olive and 6th Sts.; the *Houser Building*, at Broadway and Chestnut Sts.; and the *Liggett and Myers Building*, covering the whole square bounded by Washington Ave., St. Charles, 10th, and 11th Sts. Other important structures are the *Mercantile Library Building*, at Broadway and Locust St., the *Bank of Commerce Building*, at Broadway and Olive St., *Laclede Bank Building*, at 4th and Olive Sts.; *Security Building*, 4th and Locust Sts.; *Chemical Bank Building*, 6th and Locust Sts.; *Ely-Walker Building*, 8th St. and Washington Ave; *Public Library Building*, 9th and Locust Sts.; *Turner Building*, 8th and Locust Sts.; and the *Odd-Fellows' Hall*, at 9th and Olive Sts. The *Union Market* occupies the square bounded by 5th,

6th, Christy, and Morgan Sts., and is well worth a visit. So is the *St. Louis Elevator*, on the Levee at the foot of Ashley St. It has a capacity of 2,000,000 bushels, and is one of the largest in the country. The *Levee* should also be visited. At 6th and Poplar Sts., near the entrance of the railroad-tunnel, are grouped the brick warehouses of the *Cupples Woodenware Co.*, the largest concern of its kind in the country. Here the goods are received, stored, and shipped direct from the railroad-cars, which enter the basement of the building, with specially arranged freight-elevators and hand-cars for rapidly transferring goods from the upper floors to the loading-platforms in the basement. Several of the largest wholesale houses also have space in the buildings, which are really large freight clearing-houses.

The most imposing church edifice in the city is probably the *Grand Avenue Presbyterian*, in Grand Ave., at the head of Washington Ave. * *Christ Church* (Episcopal), cor. 13th and Locust Sts., is of stone, in cathedral-Gothic style, with stained-glass windows and lofty nave. The * *Cathedral* (Roman Catholic), in Walnut St. between 2d and 3d, is a splendid edifice, with a front of polished freestone, ornamented by a Doric portico. It is surmounted by a lofty spire in which is a fine chime of bells. The *St. Xavier's*, now building, will be the finest Catholic church in the city, and is on the corner of Grand and Lindell Aves. Other fine Roman Catholic churches are *St. Alphonsus*, on N. Grand and Easton Aves., and *Sts. Peter and Paul*, cor. 7th St. and Allen Ave. The *Church of the Messiah* (Unitarian), cor. Garrison Ave. and Locust St., is a fine Gothic structure; and the *Second Presbyterian*, cor. 17th St. and Lucas Place, is another noble specimen of the Gothic style. The * *First Presbyterian*, cor. Washington Ave. and Sarah St., is a large and costly structure, with a graceful and elegant spire. The *Union Church* (Methodist), and the *Central Presbyterian*, both on corners of Garrison and Lucas Aves., are large and commodious stone structures, as is also the *First Congregational Church*, at 3606 Delmar Ave.; the *Cumberland Presbyterian Church*, at Channing and Lucas Aves.; and the *Lafayette Presbyterian Church*, at Lafayette Park. The building occupied for many years by the Y. M. C. A., cor. 11th and Locust Sts., is a good model of an old Lombard church, believed to be the only structure of the kind in the country. The *Baptist Church*, cor. Beaumont and Locust Sts., is a stone structure of handsome design. The *Pilgrim Congregational Church*, a beautiful stone edifice, cor. Ewing and Washington Aves., is surmounted by a handsome belfry containing a fine set of chimes. The * *Jewish Temple*, cor. 17th and Pine Sts., is one of the finest ecclesiastical structures in the city, and the *Temple Israel*, in Pine and 28th Sts., is a fine structure.

Of the literary and educational institutions the most interesting is the * **Mercantile Library,** which has a large and handsome brick building at the cor. of 5th and Locust Sts. The library and reading-room are in the 5th story, and both are free to strangers (open from 9 A. M. to 10 P. M.). The library numbers 80,000 volumes, and the hall contains paintings, coins, and statuary, among which may be mentioned a series of Indian portraits by George Catlin; Harriet Hosmer's life-size statue of Beatrice Cenci; a life-size statue of Daniel Webster, by Ver Hagen; the West Wind, by Thomas R. Gould; a bronze copy of the Venus de Medici; marble busts of Thomas H. Benton, Robert Burns, Walter Scott, Henry Clay, and Columbus; and a sculptured slab from the ruins of Nineveh. The reading-room is tastefully fitted up, and supplied with 400 newspapers and magazines. The *Public Library* possesses 80,000 volumes, and has an excellent free reading-room, where 235 periodicals are accessible. It occupies the 6th and 7th floors of a fine fire-proof building on 9th and Locust Sts., where it has a model reading, special reference, and technological rooms, and a department for teachers. The library is open every day of the year, from 2 P. M. to 9 P. M. on Sundays and holidays, and 10 A. M. to 10 P. M. on week-days; it is free to every one for reference, but a small fee is charged for books that are taken out. The *St. Louis University* (Jesuit), cor. Grand Ave. and Pine Sts., is the oldest educational institution in St. Louis, having been founded in 1829. It has a valuable museum, several large buildings, very complete philosophical and chemical apparatus, and a library of 25,000 volumes, among which are some rare specimens of early printing. * **Washington University,** 17th St. and Washington Ave., occupies a large building with excellent museums, laboratories, and a library of 6,000 volumes. It was organized in 1853, has a total endowment of $1,800,000, and has an attendance of over 700 students in the University, which includes the College, the *Polytechnic School*, the *Henry Shaw School of Botany*, the *St. Louis School of Fine Arts*, the *St. Louis Law School*, and the *St. Louis Medical School.* The University is non-sectarian, is co-educational, and maintains a very high standard. Connected with it as preparatory schools are the *Smith Academy*, for boys; the *Manual Training School*, the pioneer institution of this kind in the United States; and the *Mary Institute*, an excellent academy for girls, with a total attendance of 900 pupils. The *College of the Christian Brothers* (Roman Catholic), 4 miles west of the Court-House, in Franklin Ave. extension, is a flourishing institution, and has a library of 12,000 volumes. *Concordia College* (German Lutheran) was established in 1839, and has a library of 4,500 volumes. The public-school system is one of the best in the country, and the school-houses are exceptionally fine. The educational development of St. Louis is largely due to the ability of Hon. William T. Harris, who was Superintendent of City Schools in 1868–'80, and now holds the office of U. S. Commissioner of Education. The * *High School* has a handsome building of red sandstone in N. Grand Ave. The Roman Catholics have about 100 parochial, private, and conventual schools.

The * **County Insane Asylum,** on the Arsenal road, 4½ miles from the Court-House, is an immense brick and stone structure, occupying about 40 acres of ground, beautifully laid out. On the premises is an artesian well 3,843 ft. deep. The Asylum is open to visitors from 10 A. M. to 12 M. and from 2 to 5 P. M. The *Poor-House* and the *House of Industry* are just beyond, on the Arsenal road, and are spacious brick buildings. The *Workhouse* and the *House of Refuge* are 4 miles S. of the Court-House (reached by the 5th St. line of cars). The *City Hospi-*

tal, cor. Lafayette Ave. and Linn St., is a handsome building, situated in the midst of pleasant grounds (reached by Pine St. cars ; open to visitors from 2 to 3 P. M.). The *U. S. Marine Hospital* is in Carondelet Ave., 3 miles from the Court-House. *St. Luke's,* under charge of an Episcopal Sisterhood, is cor. 19th St. and Washington Ave. The *Convent of the Good Shepherd,* for the reformation of fallen women, is at the cor. of Chestnut and 17th Sts. The *Deaf and Dumb Asylum* (Roman Catholic) is at the cor. of Beaumont and Lucas Sts. St. Louis is famous for the number of its charitable institutions, of which we have found space to enumerate only a few, and probably no city in the country is better equipped in this way. The *Exposition Building,* at the foot of Lucas Place, between Olive, St. Charles, 13th, and 14th Sts., on what was formerly Missouri Park, is a noble structure. The area of ground covered is 322 ft. by 455 ft. It is 3 stories high, and the exterior is richly decorated with terra-cotta statuary, tracings, and figures in high relief. The interior, in addition to the Exhibition Hall, contains a music-hall, a lecture-hall, and an art-gallery. An exceptionally good industrial exhibition, with daily concerts by famous musicians, is given every year from September 1 to October 16, while the two halls are used for operas, concerts, etc., during the winter season. The *Museum of Fine Arts,* 19th and Locust Sts., is the home of the Art department of Washington University, with 275 students in its day and evening classes. It is a beautiful gray limestone building, with a pleasing lecture-hall, and contains collections of casts from the Grecian and Egyptian antique, statuary, paintings, ceramic-ware, wood-carving, and industrial art. The *Shrine of St. Siebald* is one of the casts, while Harriet Hosmer's *Œnone, Puck,* and *Zenobia* are among the statuary. Several of *Weimar's* valuable studies of frontier life are in the picture-gallery, as also Washington Allston's *Paul and Silas ;* Dupré's masterpiece, *In the Pasture ;* the *Parting Kiss,* by Beyle ; Harrison's much-admired *Le Crepuscule ;* Vely's *Love and Riches ;* Thompson's magnificent landscape, *The Shepherdess ; Sand-Dunes in France,* by Pélouse ; *End of Autumn,* by Luigi Loir ; and *Ecce Homo,* by Naudin.

The public squares and parks of St. Louis embrace in the aggregate some 2,268 acres, and weekly concerts are given in all the parks during the summer. The most beautiful is *Lafayette Park,* which embraces about 30 acres in the S. portion of the city (reached by Chouteau Ave. cars, running on 4th St., and Pine St. cars). It is for pedestrians only, is admirably laid out and adorned, and is surrounded by costly residences. In it is a bronze statue of Thomas H. Benton, by Harriet Hosmer, and one of Washington. *St. Louis Place,* and *Hyde Park,* in the N. part of the city, are attractive places of resort, the former containing 11 and the latter 12 acres. *O'Fallon Park* (160 acres), on the bluffs in the N. portion, is noted for its fine trees. *Carondelet Park,* in the S. end of the city, contains 180 acres, and has some beautiful bits of natural scenery. *Forest Park* is the largest and finest park in the city. It is in the W. portion, with two fine boulevards (the *St. Louis* and the *Lindell*) leading to it ; on the latter are some of the finest residences in St. Louis. The park contains 1,371

acres; is heavily covered with fine old oaks; has numerous lakes and
the Des Peres River meandering through it; has a driving-park, a
zoölogical garden, tennis-courts, and ball-grounds. Bronze statues of
Frank P. Blair and *Edward Bates* occupy prominent sites at the E.
end. The Olive St., Washington Ave., Laclede Ave., Choteau Ave., and
suburban cars go direct to the park. ***Tower Grove Park,** 276
acres, lies in the S. W. part of the city (reached by 4th St. cars and
Gravois Railway lines, from 4th and Pine Sts.). It is beautifully laid
out, with green lawns and shrubbery, and offers the pleasantest drives
of any park in the city. It contains statues of * *Shakespeare;* * *Columbus,* and * *Humboldt,* which are masterpieces, by Von Müller.

Adjoining Tower Grove Park is ***Shaw's Garden,** formerly owned
by Henry Shaw, whose will has made it the property of the city, and
endowed it with a very large fund for maintenance. The garden contains 109 acres, and is divided into three sections. The Floretum or
Flower Garden, embracing 10 acres, contains almost every flower that
can be grown in this latitude; and there are several greenhouses, in
which are thousands of exotic and tropical plants. In the Fruticetum,
comprising 6 acres, are fruits of all kinds. The Arboretum is 25 acres
in extent, and contains all kinds of ornamental and fruit trees that will
grow in this climate. The Labyrinth is an intricate, hedge-bordered
pathway, leading to a summer-house in the center. A brick building
near Mr. Shaw's late residence contains a botanical library and one of
the largest Herbariums in the country. The ***Fair Grounds** of the
St. Louis Agricultural and Mechanical Association embrace 137 acres,
3 miles N. W. of the Court-House, are handsomely laid out and ornamented, contain extensive buildings, and one of the best Zoölogical
Gardens in the country. The Amphitheatre will seat 40,000 persons.
There is a fine one-mile race-track, with grand stand and club-house.
"Fair-week," which is usually the first week in October, is the gala-
season in St. Louis, as the huge annual Fair is open by day, while the
Veiled Prophets and other processions are given by night. The grounds
are reached by cars on Washington Ave., Franklin Ave., Cass Ave. line,
Mound City line, and Jefferson Ave. line.

*Bellefontaine Cemetery, the most beautiful in the West, is
situated in the N. part of the city, about 4½ miles from the Court-
House (reached by 5th St. cars or the Vine St. branch of the Wabash
R. R.). It embraces 350 acres, is tastefully decorated with trees and
shrubbery, and contains some fine monuments. *Calvary Cemetery* (225
acres) lies a short distance N. of Bellefontaine, and is little inferior,
either in size or beauty.

The great ***St. Louis Bridge** across the Mississippi, from the
foot of Washington Ave. to a corresponding point in East St. Louis, is
regarded as one of the greatest triumphs of American engineering. It
was designed by James B. Eads, and was begun in 1869 and completed
in 1874. It consists of three spans resting on four piers. The piers
are composed of granite and limestone, and rest on the bed-rock of the
river, to which they were sunk through the sand from 90 to 120 ft.
by the use of wrought-iron caissons and compressed air. The cen-

ter span is 520 ft., and the side ones are each 500 ft. in the clear; each of them is formed of four ribbed arches, made of cast steel. The rise of the arches is 60 ft., sufficiently high to permit the passage of steamboats at all stages of the water. The bridge is built in two stories, the lower one containing a double car-track, and the upper one two 8-feet foot-ways, and a carriage-way of 30 ft. in width. It passes over a viaduct of five arches (27 ft. span each) into Washington Ave., where the lower roadway runs into a tunnel 4,800 ft. long, which passes under the principal part of the city, terminating at 8th and Poplar Sts. The total cost of bridge and tunnel was over $10,000,000. The *Merchants' Bridge*, a magnificent structure of similar magnitude and completeness, in the N. part of the city, now rivals it in usefulness. The *Grand Ave. Bridge*, across the railroad-tracks, is a handsome suspension bridge that is well worth visiting. The city *Water-Works* are situated at Bissell's Point, on the bank of the river, 3½ miles N. of the Court-House (reached by 5th St. cars). The buildings are substantial, and the seven pumping-engines, with a capacity of 92,000,000 gallons a day, are worth seeing. The engine-rooms are open to visitors at all times. New water-works are now in course of construction at the Chain of Rocks, 7 miles farther up the river.

Itineraries.

The following series of excursions have been prepared so as to enable the visitor whose time is limited to see as much of the city as possible in the least amount of time. Each excursion is planned to occupy a single day, but the visitor can readily spend more time, as special features crowd upon his attention.

1. Visit the Chamber of Commerce, 3d and Pine Sts. (visitors' gallery open from 10 A. M. to 1 P. M.). Then pass down Pine St. to the Levee, alongside of which are the wharf-boats, connected by bridges, by which trucks drive on and off the steamboats with their loads. As the river is subject to a rise of 40 ft. between high and low water, it is necessary to use this wharf-boat system instead of docks. Go on board some of the famous Mississippi steamboats, preferably of the Anchor Line or New Orleans boats (at the foot of Olive St.), which are the largest on the river. Walk up the Levee to the St. Louis Bridge, a good view of which is obtained from the Levee, and cross it by the electric-cars from the 3d St. and Washington Ave. entrance. Transfer, at the East St. Louis terminus, to the Stock-Yards electric-car for the large *National Stock-Yards*, in East St. Louis, and adjoining which are several large packing-houses. Return to St. Louis by same route, and visit the Boatman's Bank, at 4th St. and Washington Ave., which has the finest offices in the city. Then visit the Mercantile Library, 5th and Locust Sts., with its statuary, reading and reference rooms. Then visit the Equitable Building, 6th and Locust Sts., one of the largest office-buildings, from the top of which there is an excellent view. Walk up Locust St. to 9th St., past the Mercantile Club, in 7th St., and the Post-Office, at 8th St., to the Public Library. Walk up two blocks to the old Lombard Church, in 11th St., and go S. one block to Olive St., and walk down town or E. on Olive St., passing the Masonic Temple, in 10th

St.; Pope's Theatre and the Odd-Fellows' Building, in 9th St.; the Turner Building, in 8th St.; Barr's retail store and the Commercial Building, in 6th St.; the Bank of Commerce, in 5th St.; and the Laclede Building, in 4th St. Then walk N. one block on 4th St. to the Security Building.

2. Visit the Union Market, at 5th and Morgan Sts. Then take Washington Ave. cars to Washington University, at 17th St., passing the Lindell Hotel, in 6th St.; the Ely-Walker Building, in 7th St.; the Simmons Hardware Co.'s Building, in 9th St., the largest hardware house in this country; the Liggett & Meyer Building, in 10th St.; the Belvedere Hotel, at 13th St.; and the Richelieu Hotel, at 14th St. At Washington University are museums, laboratories, and libraries. Then walk one block to the Manual Training School, with its shops and machinery. Visit the Art Museum, 19th and Locust Sts., with its paintings, statuary, and industrial art exhibits. Then visit the plant of the Missouri Electric Light Co., in 20th St., one of the largest and finest in the country. Then walk E. four blocks to the Historical Society Building, 16th and Locust Sts., which has a very fine and valuable collection of prehistoric implements and local portraits. Walk down Locust St. to 14th, to the Exposition Building, which be sure to visit if open; and then the large tobacco-factory of Liggett & Meyers, in 13th and St. Charles Sts.

3. Visit the huge freight clearing-houses of the Cupples Woodenware Co., and other wholesale concerns, at 6th and Poplar Sts. Walk W. six blocks to the St. Louis Sampling & Testing Works, 1225 Spruce St., where visitors are welcome to see the interesting gold and silver milling processes. Thence take 14th St. cars to the Electric Lighting Station, in 18th and Papin Sts., which furnishes the electricity for lighting the streets of the city. Then take the blue cars to Schnaider's Garden, Mississippi and Choteau Aves., the best of the beer-gardens and where light operas are given during the summer season; and walk three blocks S. to Lafayette Park, with lovely walks, and statues of Benton and Washington. Take the blue cars at Lafayette and Mississippi Aves. to the City Hospital, Linn St. and Lafayette Ave. Visit the famous Anheuser-Busch Brewery, by the Broadway or 6th St. cars, the largest in the world. Visitors are escorted through its interesting plant by special guides.

4. Take Broadway cars going N. for the Water-Works, passing the Belcher Sugar-Refinery, the Niedringhaus Granite-Ironware Mill, the Horse and Mule Markets, numerous factories, and the St. Louis Stock-Yards. At the Water-Works a fine view is to be had of the Merchants' Bridge; visit the imposing pumping-engines, settling-basins, and finally the Water-Tower, a few blocks up the hill. The Old Water-Tower is a Corinthian column 181 ft. high, from the top of which an extensive view is to be had. Deep quarries in the vicinity of the towers expose the gray limestone which underlies the entire city, and which is largely used for building and paving purposes. Then take Broadway and Baden cars to O'Fallon Park, a beautiful natural park on the bluff. Again take Baden cars to Bellefontaine (Protestant, 350 acres) and Calvary (Catho-

lic, 225 acres) Cemeteries. Return by the Wabash R. R. from the local
cemetery station.

5. Take Mound City cars (red, electric) to St. Louis Park, a pretty
park of 12 acres in the northern part of the city. Then, by same cars,
continue to the Fair-Grounds, in Grand Ave., where there is a zoölogical
garden, race-track, amphitheatre, and exhibition buildings for the huge
annual agricultural fair; the tasteful quarters of the Jockey Club are
here; also a restaurant at the House of Public Comfort. Take Franklin Ave. cable-car to Easton and Grand Aves., passing Sportsman's Park,
the home of the St. Louis Base-Ball Team; walk S. on Grand Ave. from
Easton to Vandeventer Place, passing St. Alphonsus's or the "Rock"
Church, and the High School. Stroll through Vandeventer Place, which
is a select residence park for the aristocracy, with many beautiful homes,
at the foot of which is the house of the Pastime Club. Then take
Washington Ave. cars (electric) for down town, passing many fine residences in the West End, and Cumberland Church, at Channing and
Lucas Aves.; the Central and Union, at Garrison and Lucas Aves.; Pilgrim Church, at Ewing and Washington Aves.; and Washington University, 17th St. and Washington Ave.

6. Drive out on Lucas Place to Beaumont St., passing the Exposition
Building, in 14th St.; the St. Louis Law School, at 15th St.; the Second Presbyterian Church, at 18th St.; Museum of Fine Arts, at 19th
St.; Robert H. Brooking's residence, at 26th St.; and Mary Institute, at
27th St. Then go S. two blocks to Pine St., and drive W. to Grand
Ave., passing the University Club, at Beaumont St.; the Jewish Temple, in 28th St.; and the St. Louis University, on Grand Ave. Then
drive W. to Kingshighway by the Lindell Boulevard, passing many
fine residences. Drive through Westmoreland and Portland Places,
from Kingshighway, which are exclusive parks for residences for the
wealthy. Drive through Forest Park, the largest park in St. Louis,
with many beautiful drives, statues of Blair and Bates, music-pavilion,
and race-track. Drive back to Grand Ave. by way of St. Louis Boulevard and West Pine St., on which latter is the Cupples mansion,
the finest in the city; also many other residences. Then drive S. on
Grand Ave. to Compton Hill Park (passing over the Grand Ave. Suspension Bridge), which contains attractive flower-gardens and the city
reservoir. Then drive to Tower Grove Park, taking the central drive,
with the statues of Columbus, Humboldt, and Shakespeare. Thence to
Shaw's Botanical Garden, where time should be taken to see what
promises to rival the Kew Gardens of London; the Floretum and
greenhouses, with their wealth of native and tropical flowers; the Arboretum, with its variety of forest-trees; and the Fruiticetum, with its
collection of fruit-trees. The Botanical Museum, with its valuable Herbarium, and the tomb of Henry Shaw, the founder and endower of
Tower Grove Park and the Botanical Gardens, should also be visited.
Return down town by way of Tower Grove Park, Grand Ave., and
Lafayette Ave., passing by Lafayette Park.

7. Take the Iron Mountain R. R. to Jefferson Barracks (from 4th
St. and Choteau Ave. depot; 12 miles), the headquarters of the St.

Louis Station of the U. S. Army.· The quarters of the officers and troops are quite interesting, and the cavalry parades well worth seeing, the grand parade being on Sunday, at noon. Concerts are also given by the post band. The grounds are extensive, and beautifully situated on bluffs overlooking the Mississippi. Return by Iron Mountain R. R. (Missouri Pacific System) to Carondelet, and visit the picturesque *Carondelet Park*. Then take the Oak Hill R. R. to the Insane Asylum, an imposing building on the highest piece of ground in the city, and from the dome of which a very extensive panorama is to be seen. The County Poor-House and the Crematory are in the immediate neighborhood.   ·

8. One of the most attractive suburban trips is to take the Missouri Pacific R. R. to Kirkwood (14 miles), the largest and finest of the suburbs. Thence drive over to the Meramec Highlands (about 3 miles) by way of the Springs, where the Meramec River makes a very pretty bit of scenery with its bold bluffs. Drive thence to Old Orchard and Webster, which are younger adjoining suburbs, and return to town by either the St. Louis & San Francisco R. R. or the Missouri Pacific R. R.

.· 9. An interesting day's excursion 35 miles down the river is to Crystal City by boat. The steamboat lines will land on signaling. An excellent general view can be had of St. Louis from the boat; of the manufacturing industries of Carondelet (now part of St. Louis), 5 miles lower down; while Jefferson Barracks are passed at 12 miles, Quarantine at 13 miles, and the mouth of the Meramec River at 20 miles below the city. .The little town of Kimswick, with its sulphur springs, is passed at 22 miles, on the Missouri side of the river, and the hamlet of Sulphur Springs, 23 miles lower down, on the same side. The eastern or Illinois bank is a continuous stretch of low bottom-land as far as Crystal City, famous for its remarkable fertility. The western or Missouri bank consists of bold and highly picturesque limestone bluffs that rise abruptly to a height of 100 to 300 ft. Numerous examples of the attempts to improve the channel of the river are to be seen on this trip. At Crystal City, about one mile back from the river, are the extensive works of the Crystal City Plate-Glass Co. The different operations of casting, rolling, annealing, and polishing the large sheets of glass are interesting. The works are accessible to visitors on application to the superintendent. Return to St. Louis by train over the private road of the glass company to Silica (2 miles), and then the Iron Mountain R. R. to St. Louis (32 miles).

10. To those interested in industrial matters an enjoyable excursion can be made in a day to the establishment of the St. Joe Lead Co., at Bonne Terre, which is one of the᠆ largest lead-producers in this country. (Take the Iron Mountain R. R. to Riverside station (26 miles), and then change cars, taking the Bonne Terre & Mississippi R. R. for Bonne Terre, about 33 miles beyond). At Bonne Terre are huge underground workings in a magnesian limestone at a depth of 100 to 300 ft., the ore yielding only 5 to 10 per cent. of lead. ·The roof of the mine is supported on large pillars of ore. The ore is crushed and concentrated into a rich grade in an imposing dressing-mill, after being hoisted out of the mine through a vertical shaft. It is then roasted, to

free it from sulphur, in a series of long, low reverberatory furnaces, and finally smelted into pig-lead in circular water-jacket furnaces. The pig-lead is remelted and reworked in refining furnaces before it is finally shipped to market.

82. Chicago to Cincinnati.

a. Via Pittsburg, Cincinnati, Chicago & St. Louis R. R.
Distance, 300 miles.

LEAVING Chicago by this route, the train runs S. E. by the small stations of *Lansing* (20 miles), *Crown Point* (41 miles), and *Hebron* (51 miles), to *La Crosse* (67 miles), a small village at the intersection of the Louisville, New Albany & Chicago R. R. *Winamac* (91 miles) is the capital of Pulaski County, Ind., and is pleasantly situated on the Tippe-canoe River. Twenty-six miles beyond Winamac the train reaches **Logansport** (*Murdock Hotel, Barnett, Johnson, Windsor*), on the Wabash River at its confluence with Eel River, and on the Wabash and Erie Canal. It has a population of 13,328, and is at the inter-section of four important railroads, including the Wabash Line, de-scribed in Route 79. The iron bridge by which this road crosses the Wabash at Logansport is a noteworthy structure. The city is sur-rounded by a rich agricultural country, and has an important trade, con-siderable quantities of poplar and black-walnut lumber being shipped. Water-power is abundant, and is used to some extent in manufactures. The principal industrial establishment is the car-works of the Pitts-burgh, Cincinnati, Chicago & St. Louis R. R., which cover 25 acres and employ 600 workmen. Three cars a day can be turned out at these shops. The *Court-House,* one of the finest in the State, is built of cut stone; and several of the churches and other buildings are also of stone. The *Northern Indiana Hospital for the Insane* consists of thir-teen large buildings, in a farm of 281 acres, on the S. bank of the Wa-bash. Beyond Logansport the train traverses a rich agricultural dis-trict, and soon reaches *Kokomo* (139 miles), a pretty village on Wild-Cat Creek. Connection is made here with the Lake Erie & Western R. R. Beginning at Kokomo and extending through *Elwood* (161 miles), *Anderson* (175 miles), to *New Castle* (197 miles), is the Indiana gas-field, in which is an abundant and undiminishing supply of natural gas, making this section peculiarly adapted for manufacturing purposes. *Richmond* is described in Route 80, and *Hamilton* in Route 74.

b. Via Cleveland, Cincinnati, Chicago & St. Louis R. R.
Distance, 306 miles.

This route, popularly known as the "Big Four," is the shortest route. Three trains daily run through both ways without change of cars. As far as *Kankakee* (56 miles) this route is described in Route 84. From Kankakee the train runs S. E. by a number of small sta-tions to **Lafayette** (*Lahr House, St. Nicholas*), one of the principal cities of Indiana, with a population of 16,243, a flourishing trade with the surrounding country, and a number of important manufactories, embracing foundries and machine-shops, marble-works, flouring-mills, woolen-mills, breweries, etc. The city is situated at the head of navi-gation on the Wabash River, is on the line of the Wabash & Erie Canal, and is the point of intersection of four lines of railway, in-cluding the Wabash Line to the West (see Route 79). It is built on rising ground, inclosed in the rear by hills of easy ascent, com-manding a fine view of the river valley. The streets are paved, and lighted with gas and electric lights, and there are many handsome build-ings, among them the *County Jail*, erected in 1869, an *Opera-House*, and *Court-House*. Lafayette is the seat of *Purdue University*, a richly-endowed institution, which is the State College of Agriculture and the Mechanic Arts. The University Building proper is a fine edifice, and there are nine other buildings, with grounds 184 acres in extent. Of the five public-school buildings, *Ford's School-house*, erected in 1869 at a cost of $85,000, is the finest. A *High-School* and *Public Library Building* is just completed, costing $60,000. Near the center of the city is a public square containing an artesian well 236 ft. deep, from which issues sulphur-water possessing curative properties. To the N. and N. E. are Greenbush and Springvale Cemeteries, handsomely situated and adorned with trees; and just S. of the city limits are the Agricultural Fair Grounds of the county. The battle-ground of Tippecanoe, where Gen. William H. Harrison defeated the Indians Nov. 7, 1811, is 7 miles N. of the city.

Beyond Lafayette, the train passes the small villages of *Colfax*, *Thorntown*, and *Lebanon*, and in 64 miles reaches the city of **Indian-apolis** (195 miles from Chicago), which has been described in Route 79. Between Indianapolis and Cincinnati there are many pretty towns and villages, but few that present any noteworthy features. *Shelbyville* (232 miles) is situated on the left bank of Blue River, and is the seat of a large seminary. *Greensburg* (251 miles) attracts attention by its air of neatness and busy thrift. At *Lawrenceburg* (see Route 78) the road turns E. and follows the bank of the Ohio River to Cincinnati.

83. Chicago to Louisville.

By the Pittsburg, Cincinnati, Chicago & St. Louis R. R. Distances : to Kokomo, 139 miles ; to Seymour, 176 ; to Indianapolis, 194 ; to Columbus, 235 ; to Jeffersonville, 302 ; to New Albany, 304 ; to Louisville, 304.

FROM Chicago to Logansport the line is the same as that described in Route 82. At *Kokomo* (see p. 368) the road branches southward to *Indianapolis*, passing through *Tipton*, a thriving village of 2,677 inhab-itants, also on the line of the Lake Erie & Western R. R. The next place of importance is *Noblesville*, on the crossing of the Chicago & Southeast-ern R. R. of Indiana. It is the seat of Hamilton County, and has a popu-lation of 3,054 inhabitants. Natural gas is abundant here, and is used for fuel. *Indianapolis* is then reached, and has already been described on p. 350. Some 41 miles S. is *Columbus*, a city of 6,719 inhabitants. It is noted for its handsome residences and clean and well-shaded streets. It has two large flouring-mills, the largest cerealine-mill in the country,
24

and the plant of the American Starch Factory. *Seymour*, with 5,337 inhabitants, is the only place of importance before Louisville (p. 344) is reached. It is on the junction of the Ohio & Mississippi R. R.

84. Chicago to Cairo.

By the Chicago Division of the Illinois Central R. R. This road traverses Illinois from end to end, nearly in the center of the State. It passes through one of the most productive and populous sections of the Great West, but, important as it is from a commercial point of view, it offers very little by the way to challenge the attention of the tourist. Distances : Chicago to Kankakee, 56 miles ; to Gilman, 81 ; to Paxton, 103 ; to Mattoon, 173 ; to Effingham, 199 ; to Centralia, 253 ; to Du Quoin, 289 ; to Carbondale, 308 ; to Jonesboro, 329 ; to Cairo, 365.

LEAVING Chicago by this route, the train passes several pretty suburban villages, and in 14 miles reaches *Kensington*, at the crossing of the Michigan Central R. R. (Route 65). *Monee* (84 miles) is the highest point on the entire line, being upon the dividing ridge between Lake Michigan and the Mississippi. **Kankakee** (56 miles) is upon the river of the same name, which is one of the principal tributaries of the Illinois (population, 9,025). It is a manufacturing town of considerable importance, having iron-foundries, machine-shops, tool-works, woolen-mills, planing-mills, etc., and here is the diverging-point of the Kankakee & Seneca branch of the Cleveland, Cincinnati, Chicago & St. Louis R. R., and of what is known as the "Kankakee Short Lines" (see Route 82 *b*). When the railroad was begun, a forest stood upon the site. In the immediate neighborhood of Kankakee are quarries of a superior kind of limestone. *Clifton* (69 miles) is supplied with water by artesian wells, a constant supply being obtained at a depth of 80 to 100 ft. The streets of the village are regularly laid out and planted with shade-trees. At *Gilman* (81 miles) the Springfield Division of the Illinois Central R. R. diverges, and runs S. W. in 111 miles to **Springfield** (see Route 85). *Onarga* (85 miles) lies in the midst of a famous fruit-growing region. It is the seat of the Onarga Institution and the Grand Prairie Seminary, both of which are flourishing institutions. *Loda* (99 miles), beautifully situated on undulating ground in the center of Grand Prairie, is the market for the agricultural products of the surrounding country. *Paxton* (103 miles) is the seat of a Swedish college named the *Augustana College of North America*, which has in its library 5,000 volumes presented by the King of Sweden. The public schools of Paxton are noted for their excellence. Twenty-five miles beyond Paxton is **Champaign** (*Doane House, Carter*), a rapidly growing city of 5,839 inhabitants, at the intersection of the Cleveland, Cincinnati, Chicago & St. Louis R. R. It has a female academy, a public library, and 3 newspapers. Here is situated the University of Illinois. *Tolono* (137 miles) is a thriving village, at the crossing of the Wabash R. R. (see Route 79). *Tuscola* (150 miles) and *Arcola* (158 miles) are prosperous and rapidly growing towns. **Mattoon** (173 miles) is at the crossing of Route 79, and is described in the same. *Effingham* (199 miles) is at the crossing of Route 80, and is there described.

We have now entered the great fruit-growing region of central Illinois, and for many miles the road traverses a country of wide-spreading and prolific orchards. *Kinmunday* (229 miles) is noted for the particularly fine fruit raised in its neighborhood, and in which it does a large trade. *Odin* (244 miles) is a very prosperous place, at the crossing of Route 77. Nine miles beyond Odin is **Centralia** (*Centralia House*), a busy little city of 4,763 inhabitants, with a coal-mine and various manufactories. The cultivation of fruit is extensively carried on in the neighborhood, and vast quantities of peaches are shipped annually to Chicago. Centralia is the point of junction of the Chicago Div. and the Northern Div. of the Illinois Central R. R., which continues thence in a single line to Cairo. *Ashley* (266 miles) is a pretty village, attractively situated on a rolling and well watered prairie. *Tamaroa* (280 miles) is another place which derives great prosperity from being the market of a rich fruit-growing region. It also has a large coal-shipping trade, coal of a superior quality being found in the vicinity. Nine miles beyond Tamaroa is **Du Quoin** (*St. Nicholas*), a thriving city of some 4,052 inhabitants, surrounded by highly productive prairie-land. Fruit-raising, tobacco and cotton growing, and general agriculture, are important sources of the city's prosperity; but the principal business is coal-mining, about a dozen companies being in active operation. At Du Quoin connection is made with the St. Louis & Cairo Short Line R. R. (see Route 127). *Carbondale* (308 miles) is a busy town, with a number of cotton-gins, mills, etc., the leading productions of the adjacent plantations being cotton and tobacco. About one fourth of all the tobacco grown in Illinois is sent to market from this place. **Jonesboro** (329 miles) is the principal town of the fruit-region of southern Illinois, and is also the mart of large crops of cotton. It is pleasantly situated in a hilly country, about 4 miles from the Mississippi River. Limestone crops out among the hills, fine building-stone abounds, and iron-ore is found in the vicinity. The *Southern State Insane Asylum* is located here, and is a handsome stone structure. Near the village are some remarkable springs and caves, and 5 miles N. is *Bold Knob*, the highest point of land in the State. *Villa Ridge* (353 miles) is at the commencement of a series of ridges or terraces, rising from the Mississippi River and extending to and along the Ohio. Twelve miles beyond, the terminus of the road is reached at **Cairo** (*Arlington, Halliday House, Planter's*), a city of 10,324 inhabitants, built on a low point of land at the confluence of the Ohio and Mississippi Rivers, forming the southernmost point of the State. It is connected by steam-ferry with Columbus, Ky., where it meets the Mobile & Ohio R. R.; and is the point of connection with the Southern Div. of the Illinois Central R. R., which forms the "Great Jackson Route" from Chicago and St. Louis to New Orleans (see Route 127). Steamers upon the Ohio and Mississippi make this one of their stopping-points. Cairo was founded with the expectation that it would become a great commercial city, and large sums of money were expended in improvements, chiefly in the construction of levees to protect it from inundation. During the civil war it was an important depot of supplies, and enjoyed great prosperity, but is now somewhat decadent. The *County Buildings* are large and handsome; the *U. S. Custom-House* is of cut stone, and cost $200,000.

The *Northern Division* or Main Line of the Illinois Central R. R. runs N. from Centralia in 345 miles to Dubuque (see Route 133), on the Mississippi River; and from Dubuque the *Iowa Division* runs W. to Cherokee, where a branch runs N. W. to Sioux Falls. Another branch runs S. W. to Onawa, while the main line continues W. to Sioux City, on the Missouri River. The principal places on the Northern Division are Vandalia, Pana, Decatur, Bloomington, Mendota, Dixon, Freeport, and Galena. Most of these are described in connection with other routes (see Index). **Dubuque** is one of the chief cities of Iowa (see Route 133), and **Sioux City** is an important railway center.

Besides the foregoing, the Chicago & Eastern Illinois R. R. runs direct to Nashville by way of Danville, Terre Haute, Vincennes, and Evansville. From Nashville connection is made with the Louisville & Nashville R. R., and thence southward to Florida on the east and New Orleans on the west. This is known as the Evansville Route.

85. Chicago to St. Louis.

By the Chicago & Alton R. R. Distances : Chicago to Lockport, 33 miles ; to Joliet, 37 ; to Normal, 124 ; to Bloomington, 127 ; to Springfield, 185 ; to Alton, 257 ; to St. Louis, 283.

THIS road runs S. W. through the rich prairie-lands of central Illinois, which roll off as far as the eye can reach on either hand. The scenery is somewhat monotonous, and, since the country has become thickly settled, has lost the distinctive prairie character which is now seen to perfection only in the W. part of Iowa and on the plains beyond the Missouri. In leaving Chicago, a number of pretty suburban villages are passed in quick succession, and in 33 miles the train reaches *Lockport*, a prosperous town on the Des Plaines River and on the Illinois & Michigan Canal, from which it derives a fine water-power. In the vicinity are some valuable stone-quarries. Four miles beyond Lockport is **Joliet** (*Munroe, St. Nicholas*), a city of 23,264 inhabitants, situated on both sides of the Des Plaines River, and on the Illinois & Michigan Canal, at the intersection of the present route and the Chicago, Rock Island & Pacific R. R. (Route 89 *b*), and Chicago & Alton R. R., also the Chicago, Santa Fé & California R. R. of the Santa Fé System. It is well built, and lighted with gas. The ** State Penitentiaries* and the *Court-House* are particularly fine and imposing. The surrounding country is extremely productive, and Joliet is its principal mart and shipping-point. The canal and river furnish good water-power, and there are several flour-mills, manufactories of agricultural implements, etc. Near the city are extensive quarries of a fine blue and white limestone which is much used for building purposes throughout the Northwest. Beyond Joliet numerous small stations are passed, of which the principal are *Wilmington* (52 miles), *Pontiac* (92 miles), and *Chenoa* (102 miles). *Normal* (124 miles) is a prosperous place, at the crossing of the Northern Div. of the Illinois Central R. R. It is surrounded by the largest nurseries in the State, and by farms devoted to the cultivation of hedge-plants. Coal-mines are also worked in the vicinity. The State Normal School and the Soldiers' Orphans' Home are located in the village. Two miles beyond is **Bloomington** (*Folsom*

House, Phœnix Hotel), one of the principal cities of Illinois, an important railway center, and the seat of large shipping and manufacturing interests. The city has a population of 20,048, is handsomely built, has street railways and steam fire-engines, and is the seat of several important educational institutions. *Durley Hall*, the *Opera-House*, and the *Court-House* are large and handsome buildings, and several of the churches and school-houses are fine edifices. The *Illinois Wesleyan University* (Methodist) is a flourishing institution, with 200 students and a library of 15,000 volumes. The *Major Female College* has a high reputation, and there is a female seminary. The construction and repair shops of the Chicago & Alton R. R. are built of stone, and with the yards attached cover 13 acres of ground.

The *Jacksonville Division* diverges at Bloomington, and is looped up to the main line again at Godfrey. The distance from Chicago to St. Louis by this route is 283 miles. Numerous small towns and villages are passed *en route*, but the only important place on the line is **Jacksonville** (*Dunlap Hote.* and *Pacific Hotel*), a busy city of 10,740 inhabitants, attractively situated in the midst of an undulating and fertile prairie, at the intersection of several railroads, of which the Wabash Line (Route 79) is one. The streets are wide and adorned with shade-trees; the houses are for the most part well built, and surrounded with flower-gardens and shrubbery. Jacksonville is the seat of the State Institution for the Education of the Deaf and Dumb; of the State Institution for the Blind; of a State Hospital for the Insane; of the State Institution for the Education of Feeble-Minded Children; and of a private Asylum for the Insane. All these have handsome buildings. The *Illinois College* (Congregational) and the *Illinois Female College* (Methodist) are flourishing institutions. The former has a library of 10,000 volumes, and the latter of 2,000, and there is a free public library of 1,600 volumes.

Beyond Bloomington, on the main line, six or eight small stations are passed, and in 58 miles the train reaches **Springfield** (*Leland House, St. Nicholas*), the capital of the State, a city of 24,963 inhabitants, built on a beautiful prairie, 5 miles S. of the Sangamon River. Its streets are broad, paved with red and white cedar blocks, and are adorned with shade-trees. From the beauty of the place and its surroundings, Springfield has been called the "Flower City." The * *State Capitol* is a remarkably fine building. Other noteworthy buildings are the *U. S. Building* (containing the Court-House, Custom-House, and Post-Office), the *County Court-House*, the *State Arsenal*, the *High-School*, the *Opera-House*, and *St. John's Hospital*, the *German Lutheran College*, a fine Y. M. C. A. building, and numerous churches. There are a theatre, a commodious concert hall, and a lecture hall. Two miles N. of the city is *Oak Ridge Cemetery*, a picturesque and well-kept burying-ground of 72 acres, containing the remains of President Lincoln and the noble * monument erected to his memory by the Lincoln Monument Association. The monument cost $206,550, and was dedicated on Oct. 15, 1874. There are vast coal-mines in the vicinity of Springfield, the surrounding country is very productive, and the trade of the city is extensive. The principal manufacturing establishments are flouring-mills, foundries and machine-shops, rolling-mills, woolen-mills, breweries, and a watch-factory. The extensive shops of the Wabash R. R. are worth visiting. *Godfrey* (251 miles) is at the junction of the main line with the Jacksonville Div., described above; and 6 miles beyond is **Alton** (*Madison*

Hotel), a prosperous city of 10,294 inhabitants, built upon a high lim_e-stone bluff, overlooking the Mississippi River. It is the center of a rich farming country, and, besides the river navigation, several railroads connect it with all parts of the country. The manufactures are varied and extensive, and lime and building stone are largely exported. There are 14 churches, among them a large Roman Catholic Cathedral, Alton having been made a bishopric in 1868. The State Penitentiary, estab-lished here in 1827, was removed several years since to Joliet. The buildings are still standing, and were used during the civil war as a gov-ernment prison. At Upper Alton, 1½ mile E. of the city, is *Shurtleff College*, an important Baptist institution. Three miles below Alton is the confluence of the Missouri and Mississippi Rivers. At *E. St. Louis* (281 miles) the train crosses the Mississippi on the magnificent St. Louis Bridge. **St. Louis** (see Route 81).

86. Chicago to Milwaukee.

a. Via Milwaukee Div. of Chicago & Northwestern R. R.   85 miles.

This road runs along the W. shore of Lake Michigan through a rich farming region, well cultivated and populous. The first eight or ten stations after leaving Chicago are neat suburban villages. *Wauke-gan* (36 miles) is a flourishing town, with a large export business in grain, wool, and butter. Its site is high, and it is becoming a summer resort. A few miles beyond Waukegan the train crosses the boundary-line and enters Wisconsin, soon reaching **Kenosha** (*Grant House*), a city of 6,532 inhabitants, built on a bluff, and possessed of a good har-bor, with piers extending into the lake. The manufactures are important, and the city has an extensive trade. Eleven miles beyond Kenosha is the academic city of **Racine** (*Commercial Hotel, Merchants' Hotel, Wag-ner House*), which is the fourth city of Wisconsin in population and com-merce. It is situated at the mouth of Root River, on a plateau project-ing about 5 miles into Lake Michigan and elevated about 40 ft. above its level. Its harbor is one of the best on the lake, its commerce is large, and its manufactures varied and extensive. The city is regu-larly laid out, with wide, well-shaded streets. *Main St.* is the business thoroughfare, and its upper portion is lined with fine residences. *Racine College* (Episcopal) is one of the most prominent educational institutions in the West, and has commodious buildings in grounds ten acres in ex-tent at the upper end of Main St. The public schools are excellent, and the Roman Catholics have a flourishing academy. Of the 24 churches, several are handsome edifices. Racine was settled in 1834, was incor-porated as a city in 1848, and in 1890 had a population of 21,014. Between Racine and Milwaukee there are no important stations.

b. Via Chicago Div. of the Chicago, Milwaukee & St. Paul R. R.
85 miles.

This route runs nearly parallel with the preceding, but somewhat further inland.

Milwaukee.

Hotels, etc.—The *Plankinton House*, fronting on Grand Ave., between W. Water and 2d Sts.; the *Schlitz Hotel* (European), at the cor. of Grand Ave. and 3d Sts.; and the *Republican House*, Cedar and 3d Sts., are first class. The *Grand Avenue Hotel*, in Washington Square, is patronized by families. Three main lines of *Street-cars* (fare, 5c.), with numerous branches, converge on either side of Grand Ave. bridge. The *Union Station*, in Everett St., is the point of departure for the Chicago, Milwaukee & St. Paul and the Wisconsin Central R. Rs. The depot of the Chicago & Northwestern and of the Milwaukee, Lake Shore & Western R. Rs. is at the foot of Wisconsin St. The *Post-Office* is at the cor. of Wisconsin and Milwaukee Sts. The Detroit, Grand Haven & Milwaukee and the Flint & Pere Marquette R. Rs. run regular lines of steamers to the city, the docks of which are on the river, between Grand Ave. and E. Water St. bridges.

Milwaukee, the commercial capital of Wisconsin, and, next to Chicago, the largest city in the Northwest, is situated on the W. shore of Lake Michigan, at the mouth of Milwaukee River. This river flows through the city, and with the Menomonee, with which it forms a junction, divides it into three districts, which are severally known as the East, West, and South sides. The river has been rendered navigable to the heart of the city by vessels of any tonnage used on the lakes, and, with Milwaukee Bay, in which the Government is constructing breakwaters on a large scale, forms the best harbor on the S. or W. shore of the lake. A tunnel about half a mile in length was constructed in 1888 between the lake and the river north of the city, through which, by means of the largest single pump in the world, enough lake water is forced to change the entire volume of water in the river every twenty-four hours, clearing it of all objectionable matter. The city embraces an area of 18 square miles, and is regularly laid out. The center, near the Milwaukee River, is the business quarter; and the E. and W. parts, the former of which is built upon a high bluff overlooking the lake, while the latter is still more elevated, are occupied by residences. The low land about the Menomonee River is occupied by manufacturing establishments of all kinds. The peculiar cream-color of the "Milwaukee brick," of which many of the buildings are constructed, has earned for it the name of the "Cream City of the Lakes."

Milwaukee was settled in 1835, and incorporated as a city in 1846. Its population in 1840 was 1,712; in 1860, 45,246; in 1870, 71,440; in 1880, 115,587; and in 1885 it amounted to 162,526, including the former village of Bay View, lately added to the city. Its growth during the last five years has been very rapid, and the population in 1890 was 204,468. The Germans constitute nearly one half the entire population, and their influence upon the social life of the inhabitants is everywhere seen. The commerce of Milwaukee is very large, wheat, flour, and lumber being the most important items. The storage accommodations for grain comprise ten elevators, with a combined capacity of 5,730,000 bushels; and the flour-mills are on an immense scale. Butter, wool, hides, and coal are also important articles of trade. The manufactures are very extensive, and embrace lager-beer (which is widely exported), pig-iron and iron castings, leather, machinery, agricultural implements, steam-boilers, car-wheels, furniture, and tobacco and cigars. Pork-packing is extensively carried on.

The streets of Milwaukee, except those in the commercial quarter, are generally well shaded. *East Water St.*, *Wisconsin St.*, and *Grand Ave.* are very wide and handsome thoroughfares, and on them are the principal hotels and retail stores. In Washington Square, in Grand Ave., is a *Statue of Washington*, with an allegorical group in bronze, modeled by R. H. Park. Among the public buildings is the *U. S. Custom-House*, which also contains the *Post-Office* and the U. S. Courts, which stands at the cor. of Wisconsin and Milwaukee Sts. The * **County Court-House** is a large and handsome edifice, and from the dome a fine view of the city is obtained. The * **Chamber of Commerce** occupies a large building cor. of Broadway and Michigan St. West of it is the **Mitchell Building.** The *Davidson Opera-House*, on 3d St., and the *Bijou Opera-House*, on 2d St., are new and elegant structures. The *New Academy* is also a good theatre, and the *Opera-House* is a fine building, which is now used for theatrical performances in German by an excellent stock company. * **Schlitz Park,** in Walnut St., is a fine garden in the European style, with a theatre, that is regularly occupied during the summer by an English opera company for high-class performances, and has a tower which affords a beautiful view of the city. The finest church-edifice in the city is *St. Paul's* (Episcopal). The *St. James* (Episcopal) *Church*, the *Roman Catholic Cathedral of St. John*, and the *Trinity Lutheran*, are handsome. Of the literary institutions the most prominent are the *Milwaukee College* (for young ladies); the *Marquette College*, under the charge of the Jesuit fathers; and one of the five State normal schools. Four miles S. of the city is the large *Catholic Seminary of St. Francis of Sales*. The *Public Library*, in Grand Ave., has a collection of 50,000 volumes, and a well-supplied reading-room. The * **Exposition Building** contains also the *Public Museum*, which is open free, on Sundays from 9 A. M. to 12 M., other days from 1 to 5.30 P. M. The * **Layton Art Gallery** is at the cor. of Jefferson and Mason Sts., from a design by G. A. Audsley. The * **Northwestern National Asylum** (for disabled soldiers) is an immense brick building, about 3 miles from the city, having accommodations for 2,000 inmates. The institution has a reading-room, and a library of 5,000 volumes. The grounds embrace 425 acres, more than half of which is under cultivation, the residue being laid out as a park. In the city there are five orphan asylums, a Home for the Friendless, three Homes for the Aged, an industrial school for girls, and six hospitals. The large and well-appointed county hospital and insane asylum are at Wauwatosa, 5 miles W. of the city. Several of the industrial establishments are well worth a visit, especially one of the * **Grain-Elevators** in the Menomonee Valley. The seven flour-mills have a capacity of 8,350 barrels of flour daily. The breweries are large and numerous, and in 1889 the combined product reached the amount of 1,413,447 barrels. The rolling-mills of the Illinois Steel Co., at Bay View, and the Reliance Iron-Works, in Clinton St., are very extensive, the latter being the largest engine-manufacturers and mill-furnishers in the world.

The bluff on the lake-shore has been laid out as a park, and named *Juneau Park*. A delightful drive is over the *White-Fish Bay* road, extending 5 miles along the lake, N. of the city.

87. Chicago to St. Paul.

a. Via Milwaukee and La Crosse, and Chicago, Milwaukee & St. Paul
R. R.   Distance, 420 miles.   Time, about 13 hours.

THREE through trains run daily each way on this route. *Brookfield*
(13 miles) is at the junction with the Prairie du Chien Div. *Water-*
town (43 miles) is a small village on the Rock River. Connection is
made here with the Wisconsin Division of the Chicago & Northwestern
R. R. From *Watertown Junction* (44 miles) a branch road runs W. in
37 miles to **Madison** (see present Route), while the present route continues N. W. and soon reaches *Columbus* (63 miles), a pleasant village of
some 1,977 inhabitants, on the Crawfish River. Twenty-eight miles beyond Columbus is **Portage City** (*Corning House, Emder's*), situated at
the head of navigation on the Wisconsin River, and on the canal connecting the Fox and the Wisconsin, at the junction of three divisions of
the Milwaukee & St. Paul R. R. It has a population of 5,143, does a
large trade with the surrounding country, and the water-power furnished
by the canal is extensively used in manufactures. The R. R. Co. has
repair-shops here, and there are 8 churches, a fine Court-House and Jail,
and a handsome High-School building. *Tomah* (153 miles) is a growing
place at the crossing of the Chicago, Milwaukee & St. Paul R. R.; and
Sparta (170 miles) is situated on the La Crosse River, in a very fertile
valley. Twenty-five miles beyond Sparta the train reaches **La Crosse**
(*Cameron, Law*), a city of 25,090 inhabitants on the E. bank of the Mississippi River, at the mouth of the Black and La Crosse Rivers. It is
finely situated on a level prairie, and has many handsome buildings, including the Court-House, which cost $40,000, the Post-Office, an Opera-
House, and the High-School building. There are flourishing graded
schools, a Young Men's Library of 2,400 volumes, and 17 churches.
The city has an extensive trade in lumber, and contains 9 saw-mills, 3
foundries and machine-shops, a large manufactory of saddlery and
harness, and various other establishments.

At La Crosse the train crosses the Mississippi and follows its W.
bank all the way to St. Paul, amid remarkably picturesque scenery.
On the bank of the river, 28 miles from La Crosse, is the prosperous
little city of **Winona** (*The Winona, Jewell House*), charmingly situated
on a plain which commands a fine view of the river for several miles.
Being somewhat sheltered by the high bluffs which line the river above
and below, it is thought to offer conditions favorable to consumptives,
and has some reputation as a winter resort. The streets of the city are
wide, and the business portion is compactly built of brick and stone.
The *First State Normal School* is located here, and has a fine building
which cost $145,000. The High-School building cost $55,000, and there
are several handsome churches. Winona is one of the most important
lumber-distributing points on the Upper Mississippi, and as a grain-
shipping-point it ranks among the first in the Northwest. Two railroads converge here, and manufacturing is extensively carried on. The
population of the city is 18,208. *Wabasha* (256 miles) does a large

grain-shipping business with the productive Chippewa Valley. *Reed's Landing* (262 miles) is at the foot of the beautiful expansion of the river known as Lake Pepin. *Lake City* (268 miles) stands upon a level plain at the foot of high bluffs, and is the port of a rich farming district. It has a population of 2,128, and is growing rapidly. **Frontenac** (279 miles) lies in the center of the lake region, and is a favorite resort in summer on account of its fine scenery, and the hunting, bathing, fishing, and sailing which it affords. Besides the sport furnished by Lake Pepin, there are fine trout-fishing in the streams and deer-hunting in the woods of Wisconsin, on the opposite side of the river, while prairie-chickens are found in abundance in the country back of the village. At the head of Lake Pepin, 6 miles beyond Frontenac, is **Red Wing** (*St. James Hotel*), a well-built city of 6,277 inhabitants, beautifully situated on a broad level plain, which extends to the foot of some majestic bluffs. It is a favorite summer resort, and, being thoroughly protected by high hills, is also a desirable winter residence for consumptives. It is the port and market of a fertile region, and considerable manufacturing is done here. Twenty-one miles beyond Red Wing is the thriving city of **Hastings** (*American House, Bailey House*), situated at the mouth of the Vermilion River, which here falls 110 ft. in ½ mile and furnishes abundant water-power. The population is 3,705, and the principal manufactories are 4 flour-mills, a saw-mill, and a shingle-mill. The Central School-House is a fine building; there are 2 Catholic schools, and 8 churches. The train again crosses the river at Hastings, and passes in 20 miles to St. Paul.

b. Via the Prairie du Chien Division of the Chicago, Milwaukee & St. Paul R. R. Distance, 410 miles. Time, 16 hours.

Two through trains run daily each way on this route. *Brookfield* (14 miles) is at the junction with the La Crosse Div. described above. *Waukesha* (21 miles) is a thriving village on the Fox River, built on the edge of a beautiful prairie. The Court-House and Jail are constructed of a superior quality of limestone, found in abundance in the immediate vicinity. *Whitewater* (51 miles) is another busy village, situated in the midst of a rich farming region, and actively engaged in manufactures. At *Milton Junction* (64 miles) a branch line diverges to Monroe, while the St. Paul train passes on in 32 miles to

Madison.

Hotels, etc.—The *Park Hotel*, near the State Capitol, is a first-class house. The *Ton-ya-wa-tha Hotel, Capitol House, Ogden House,* and *Simons Hotel* are also good. Prices are from $1.50 to $3 a day. There are also several large summer boarding-houses.

Madison, with a population of some 13,426, is at once a State capital, a flourishing commercial center, and a popular summer resort. It lies in the very heart of the "Four-Lake Country," so called from a chain of beautiful lakes which extend over a distance of 16 miles, and discharge their surplus waters into Yahara or Catfish River, a tributary of Rock River. *Mendota* or *Fourth Lake*, the uppermost and largest, is

7 miles long, 5 miles wide, and from 50 to 80 ft. deep in some places. It is fed chiefly by springs, and has beautiful white gravelly shores and pure cold water. *Monona* or *Third Lake* is 3½ miles long and 2 miles wide; and Lakes *Waubesa* and *Kegonsa* are each about 3 miles long by 2 in width. The city lies between Lakes Mendota and Monona. It is about 3 miles in length by 1 mile in breadth, and has wide, straight, and regular streets, with many fine buildings. The * **State Capitol** stands in a square park of 14 acres, and has recently been enlarged and improved at a cost of about $550,000. The *Court-House* (a handsome building, costing $200,000) and *Jail* are situated near the S. cor. of the park; and in an adjacent street is a United States *Post-Office* and *Court-House*, which cost about $400,000. The *University of Wisconsin* stands on a picturesque eminence called College Hill, about a mile W. of the Capitol, and 125 ft. above the lakes. There are five faculties, and an Agricultural Experiment Station. It has 900 students, and a library of 25,000 volumes. *Washburn Observatory* contains a telescope of 15½ inches aperture, and the Woodman astronomical library. The views from this point are extremely fine. The *State Hospital for the Insane*, on the shore of Lake Mendota, 4 miles N. of the Capitol, is a vast and massive building, surrounded by grounds containing 393 acres. The *Wisconsin Historical Society* has an interesting collection of relics in a wing of the Capitol, and a valuable library of 140,000 volumes. The *State Library* contains 7,500 volumes, and there is a fine *City Library* containing 12,000 volumes. The new *Fuller Opera-House* cost $60,000, and will seat 1,500 people. There are several handsome churches in the city, and some fine villa residences in the outskirts. Small steamers ply on Lakes Mendota and Monona, and afford agreeable excursions. *Lake Mendota* is the most beautiful of the lakes, and from Monona the finest views of the city are obtained. On the shores of the latter are the *Monona Lake Assembly Grounds*, and a well-patronized summer hotel. The climate of Madison is delightfully cool and invigorating in summer. The Freeport Div. of the Illinois Central R. R. connects Madison by way of Freeport with the south.

Beyond Madison the St. Paul train passes a number of small stations, but none requiring mention until **Prairie du Chien** (194 miles; *Commercial* and *Depot Hotels*) is reached. Prairie du Chien is a town of about 3,131 inhabitants, situated on the E. bank of the Mississippi River, 2 miles above the mouth of the Wisconsin, on a beautiful prairie which is 9 miles long and 1 mile wide, bordered on the E. by high bluffs. It is an important local shipping-point, and has varied and important manufactures. *St. John's College* and *St. Mary's Female Institute* are under the control of the Roman Catholics. The public schools are well conducted. Leaving Prairie du Chien, the train crosses the river to *McGregor, Iowa*, a flourishing town, and runs W. by several small villages. *Calmar* (238 miles) is a village of 2,074 inhabitants at the junction with the Iowa & Dakota Division. Turning now to the N., the road soon enters Minnesota and reaches *Austin* (306 miles), a prosperous village, pleasantly situated on Red Cedar River. *Ramsey* (309 miles) is at the junction with the Southern Minnesota Div., and *Owatonna* (339 miles) is at the crossing of the Chicago & Northwestern R. R. Fifteen miles beyond Owatonna is **Faribault** (*Arlington* and *Brunswick*), one of the most populous and prosperous interior towns in the State. In 1853 it was the site of Alexander Faribault's trading-post; since 1857 its growth has been rapid, and the population in 1890 was 6,524. It is the seat of the State Asylum for the Deaf and Dumb and Blind, and of an Episcopal Academy, and contains several other schools, 6 or 8 churches, 2 weekly newspapers, 2 national banks, and several flour-mills, saw-mills, foundries, etc. Between Faribault and St. Paul the only important station is *Northfield*, where are located Carlton College (Congregational) and St. Olaf's College (Lutheran).

St. Paul.

Hotels.—The leading hotels are the *Ryan* and *Aberdeen*, $3 to $6 a day ; the *Merchants'*, *Windsor*, and *Metropolitan*, $2.50 to $4 a day.
Opera-Houses.—The *Metropolitan* and *Grand*, both built in 1890, with all the modern improvements.

St. Paul, the capital of Minnesota, is a beautiful city, with 133,156 inhabitants in 1890, situated on both banks of the Mississippi River, 2,200 miles from its mouth, and the head of navigation. It was formerly confined to the east bank, the site embracing four distinct terraces, forming a natural amphitheatre with a southern exposure, and conforming to the curve of the river. The city is built principally upon the second and third terraces, which widen into level, semicircular plains, the last, about 90 ft. above the river, being underlaid with a stratum of blue limestone from 12 to 20 ft. thick, of which many of the buildings are constructed.

The first recorded visit to the site of St. Paul was made by Father Hennepin, a Jesuit missionary, in 1680. Eighty-six years afterward, Jonathan Carver came there and made a treaty with the Dakota Indians, in what is now known as Carver's Cave. The first treaty of the United States with the Sioux, throwing their lands open to settlement, was made in 1837, and the first claim was entered by Pierre Parent, a Canadian *voyageur*, who sold it in 1839 for $30. It is the present site of the principal part of the city. The first building was erected in 1838, and for several years thereafter it was simply an Indian trading-post. It was laid out into village streets in 1847, and a city government was obtained in 1854, when the place contained about 3,000 inhabitants. It derives its name from that of a log chapel dedicated to St. Paul by a Jesuit missionary in 1841.

In 1890 St. Paul had 351 miles of well-graded—including 41 miles of block and asphalt paved—streets, and 123 miles of the best system of sewerage. The streets are lighted with gas and electricity. The street railway system is one of the finest in the world, with 15 miles of cable and 85 miles of electric service ; fare, 5 cents, including transfer. All the lines are owned by one company, and include an interurban line, St. Paul to Minneapolis, distance 10 miles ; fare, 10 cents, including transfer in each city. The principal public buildings are the *State Capitol* and the *U. S. Custom-House*, which also contains the Post-Office. A fine *Court-House* and *City Hall* has just been completed, at a cost of $1,014,000. There are 138 churches of all denominations in the city,

There are four public and as many private circulating libraries, the former including the State Law Library and those of the Historical Society and Library Association, comprising together about 24,000 volumes. The *Academy of Sciences* contains about 126,000 specimens in natural history. The public and private schools are noted for their excellence, the latter including several colleges and female seminaries of a high grade. There are three free hospitals, managed by the county and church organizations, and a Protestant and a Roman Catholic orphan asylum. The *Minnesota Club* has an attractive building. Among the fine business buildings may be mentioned the *Pioneer Press*, 13 stories high ; *Globe*, 10 stories ; the *New York Life*, *Germania Life*, *Germania Bank*, *National German-American Bank*, *Merchants' National Bank*, *Endicott Arcade*, *Grand Arcade*, and the *Manhattan Building ;* also, the *General Office-Buildings* of the Great Northern R. R., Northern Pacific R. R., and the Chicago, St. Paul, Minneapolis & Omaha R. R. Many of the mercantile houses have new and costly buildings. The jobbing business of St. Paul during the decade of the eighties increased from $46,500,000 to $122,250,000, and the manufacturing from $15,500,000 to $61,750,000. St. Paul has 21 banks, with capital and surplus of $10,500,000. St. Paul is the railway center for the Northwest, 28 railroads radiating from this city ; several hundred passenger-trains arrive and depart daily. There are 69 newspapers published in St. Paul, of which 9 are dailies and 40 weekly. The city has 32 public parks and squares, and owns its water-works, which supplies 8,000,000 gallons of water daily from a chain of spring-fed lakes. The average temperature is 43·5°, and average precipitation 28·99.

There are many beautiful drives in and around St. Paul, including the justly renowned Summit Ave. and Boulevard, and many places in the neighborhood of the city which can be reached either by carriage or by rail. Of these the most popular is *White Bear Lake*, 10 miles distant, on the St. Paul & Duluth R. R. It is about 9 miles in circumference, with picturesque shores, and an island in its center. *Bald-Eagle Lake*, a mile beyond White Bear Lake, is noted for its fishing and picturesque scenery, and is a popular resort for picnic-parties. *Minnehaha Falls, immortalized by Longfellow, are reached by the cars of the Chicago, Milwaukee & St. Paul R. R. The Falls are picturesquely situated, but they hardly merit the prominence that Longfellow's poem has obtained for them. *Lake Como* is reached by a drive of 3 miles from the center of the city. The boating here is excellent, and *Como Park*, comprising several hundred acres, is located on its shores. Adjoining the park are the *Fair Grounds* of the State Agricultural Society, and the State experimental farm of 300 acres. *Fort Snelling*, at the western limits of the city, is well worthy of a visit, and is easily reached by carriage or electric cars on West 7th St.

Thirty years ago, Indian tepees stood on the site of the Union Depot in St. Paul, from where hundreds of passenger-trains now go and come every day. Then, the bell of the little Chapel of St. Paul, occupying a spot on a bluff high above the Father of Waters, called a few hundred people to worship. Now more than a hundred church-spires rise above the homes of a large and flourishing population. The pioneer railroad was the St. Paul & Pacific, now the Great Northern. With 4 tracks, it is the main thoroughfare between St. Paul and Minneapolis, passing *en route* through the suburban towns of **Hamline,** the seat of a prosperous Methodist University, and adjoining the extensive grounds of the State Agricultural Society; **St. Anthony Park,** with many pretty residences, is next; and then **University,** containing the spacious buildings and grounds of the State University, Agricultural College, and Experimental Station, a public educational institution having the largest State endowment in the United States. A mile distant lies **East Minneapolis,** with its Exposition Building; and then the train glides across the Mississippi River on a stone-arch curved bridge in full view of St. Anthony Falls and the largest flouring-mills in the world, which have a daily output of 100,000 barrels.

Minneapolis (*West House* and *Nicollet House*) is situated on both sides of the river, and the city limits join those of St. Paul, with which it is connected by 3 lines of railway and electric cars. It is built on a broad esplanade overlooking the Falls of St. Anthony and the river, which is bordered at various points by picturesque bluffs. The city is regularly laid out, with avenues running E. and W., and streets crossing them N. and S. They are generally 80 ft. wide, with 20-ft. sidewalks, and 2 rows of trees on each side. There are many substantial business blocks and elegant residences. The *Court-House, City Hall, Opera-House, Chamber of Commerce, Guaranty Loan, Lumber Exchange, Temple Court* and *Syndicate Block,* and the *Minneapolis Exposition,* are noticeable structures. The *Public Library,* with its building cor. Hennepin Ave. and 10th St., contains more than 42,000 vols., and that of the University of Minnesota upward of 25,000 vols. Besides the University, there are several other important educational institutions; the public schools are numerous and good, and the *High-School* and *Public-School* buildings are worthy of note. The number of churches is about 125, including all the denominations. The business prosperity of Minneapolis and St. Anthony is owing to the * *Falls of St. Anthony,* which afford abundant water-power for manufacturing purposes. The fall is 18 ft. perpendicular, with a rapid descent of 82 ft. within 2 miles. The rapids above the cataract are very fine—in fact much finer than the fall itself, the picturesqueness of which has been destroyed by the wooden "curtain" erected to prevent the wearing away of the ledge. The falls can be seen with about equal advantage from either shore, but the best view is from the center of the suspension-bridge above the falls or the bridge of the Great Northern R. R. Minneapolis is the center of immense lumber and flouring interests, and had a population of 164,738 in 1890. It is the largest flour-manufacturing place in the world. Twenty-four mills have a capacity of 44,100 barrels a day. Wheat receipts in 1890, 45,271,910 bushels. The output of the flouring-mill products in 1890 was 6,988,830 barrels, valued at upward of $30,000,-000. The city has a splendid system of electric street-cars. The streets are well lighted with gas and electricity. The city owns the water-works, and draws the supply from the Mississippi River. Water for family use is largely supplied from springs a few miles out of the city, and is a private enterprise, the water being delivered by wagons, and sold by the gallon.

88. Chicago to St. Paul.

By the Chicago & St. Paul Div. of the Chicago & Northwestern R. R. Two through trains daily, with palace and dining cars attached, run on this line, making the journey in 12¾ hours. Distances : Chicago to Montrose, 8 miles ; to Crystal Lake, 43 ; to Beloit, 91 ; to Madison, 139 ; to Elroy, 213 ; to Black River Falls, 266 ; to Eau Claire, 322 ; to Menomonee, 348 ; to Hudson, 390 ; to St. Paul, 409.

LEAVING the Chicago station (cor. Wells and Kinzie Sts.), the train passes in 8 miles to the pretty suburban village of *Montrose*, and soon reaches *Crystal Lake* (43 miles), a neat village picturesquely situated on a small lake of the same name. The first important station on the line is **Beloit** (90 miles ; *Goodwin House, Grand*), a flourishing city of 6,315 inhabitants, situated on both sides of Rock River, at the mouth of Turtle Creek. It is built on a beautiful plain, from which the ground rises abruptly to a height of 50 to 60 ft., affording excellent sites for residences. The city is noted for its broad, beautifully-shaded streets, and for its fine churches ; the *First Congregational Church*, constructed of gray limestone, is one of the largest and handsomest in the State. *Beloit College* (Congregational), founded in 1847, is a flourishing institution, with about 250 students and a library of 7,200 volumes. Beloit is surrounded by a fine prairie country, which is dotted with numerous groves of timber. The city is well supplied with water-power, and has several flouring-mills, several manufactories of woolen goods, reapers, scales, carriages, etc. The Chicago, Milwaukee & St. Paul R. R. intersects here. The small stations of *Hanover* (104 miles) and *Evansville* (116 miles) are now passed, and the train speedily reaches **Madison** (138 miles), the capital of Wisconsin, which has already been described (see Route 87 b).

Beyond Madison the train runs N. W. by a number of unimportant villages to *Elroy* (213 miles). From Elroy the Madison Division runs W. to La Crosse and Winona (both described in Route 87 a), while the present route traverses the pine-covered portion of Wisconsin. *Black River Falls* (266 miles) and *Augusta* (299 miles) are rapidly-growing villages, near extensive pine-forests. *Eau Claire* (322 miles), the capital of Eau Claire County, is a township of 17,415 inhabitants, on the Chippewa River. . It has an important trade in lumber, and several large saw-mills are in operation. *Menomonee* (348 miles) is another busy lumbering village on the Menomonee River, down which are floated immense numbers of logs from the vast forests above. Forty-six miles beyond Menomonee is **Hudson** (*Commercial Hotel*), the most important place on this section of the road, a flourishing village of 2,885 inhabitants, on the E. shore of Lake St. Croix. Twenty miles beyond Hudson the train reaches **St. Paul** (see Route 87 b).

Another desirable route from Chicago to St. Paul is known as the "Albert Lea" route. It consists of the Chicago, Rock Island & Pacific R. R. to West Liberty ; of the Burlington, Cedar Rapids & Northern R. R. from West Liberty

to Albert Lea ; and of the Minneapolis & St. Louis R. R. from Albert Lea to
Minneapolis and St. Paul. The distances are : from Chicago to West Liberty,
222 miles ; West Liberty to Albert Lea, 191 ; Albert Lea to St. Paul, 100 ; making a total distance of 513 miles.

89. Chicago to Omaha.

a. Via Chicago & Northwestern R. R.   Distance, 492 miles.

FOUR through trains, with palace sleeping and "Northwestern dining" cars attached, run each way on this route.   The road traverses for
the larger portion of the way the great prairie-region of the West, which
fifty years ago was uninhabited, save by the Indian and the trapper, but
which now teems with an industrious and thriving population.   Many of
the towns and cities *en route* exhibit the unmistakable symptoms of
wealth and prosperity, but there are very few which possess any features
of special interest to the tourist.   *Geneva* (35 miles) and *Dixon* (97
miles) are pleasant villages, with a large trade and important manufactures.   From Dixon the train follows the Rock River for 10 miles to
Sterling (*Galt House*), a city of 5,824 inhabitants, attractively situated
on the N. bank.   The river at this point is spanned by a dam of solid
masonry 1,100 ft. long and 7 ft. high, which with the 9 ft. natural
fall of the rapids above affords an immense water-power.   The city is
chiefly devoted to manufacturing, and the articles produced are remarkably varied and valuable.   The St. Louis Div. of the Chicago, Burlington & Quincy R. R. begins here.   *Fulton* (135 miles) is the last station
in Illinois ; and here the train crosses the Mississippi River on a magnificent iron * bridge 4,100 ft. long, with a draw 300 ft. long.   From the
center of the bridge, looking up the river, there is a fine view, taking in
three towns.   At the Iowa end of the bridge is the prosperous city of
Clinton (*Revere House, Windsor*), with a population of 13,619, the extensive repair-shops of the C. & N. R. R., and a large number of saw-mills,
one of which is capable of producing 200,000 ft. of lumber a day.   From
Clinton to Cedar Rapids the road traverses a rolling prairie, dotted with
a succession of small but thriving towns, and relieved from monotony by
numerous plantations of trees.   **Cedar Rapids**, Iowa (219 miles ; *Clifton House, Grand Hotel,* and *Pullman Hotel*), is a rapidly growing city of
18,020 inhabitants, on the Red Cedar River, at the intersection of several
important railways.   Its trade with the surrounding country is large, and
there are a number of manufactories and pork-packing establishments.
The city is regularly laid out and well built, and promises to become one
of the most important in Iowa.

Beyond Cedar Rapids, a fertile but more thinly peopled agricultural
region is traversed, with a number of small stations at frequent intervals
along the line.   At *Ames*, a branch of this road diverges and runs to
Des Moines, the capital of Iowa (see Sub-Route *b*).   *Boone* (340 miles) is
a thriving village, surrounded by a rich and productive farming country.
Soon after leaving Boone the train begins the descent into the valley of
the Des Moines River, amid extremely rugged and picturesque scenery,
and with very heavy grades, in some places of 80 ft. to the mile.   The

Des Moines River, which is the largest river in Iowa, is crossed on a fine bridge. For many miles after leaving the Des Moines Valley the road traverses a superb prairie with many thriving small towns. *Arcadia* (405 miles) is the highest point in Iowa, being 870 ft. above the level of Lake Michigan. In spring and summer the surrounding prairie is rich in long grass and beautiful flowers. *Denison* (423 miles) is a promising young town. At this point the train enters the Boyer Valley, the scenery of which furnishes a pleasing contrast to that of the prairie. *Dunlap* (441 miles) is a growing town, containing one of the R. R. engine-houses. *Missouri Valley Junction* (467 miles) is at the junction of the Sioux City and Pacific R. R. Here the descent into the Missouri Valley begins, and a full view of the "bluffs" is obtained for the first time. The road, turning S. W., almost skirts those on the Iowa side, while those of Nebraska loom up on the opposite side of the broad river-bottom. At the foot of the bluffs, which are here high and precipitous, 3 miles E. of the Missouri River, is the important city of **Council Bluffs** (*Park Hotel, Ogden, Pacific,* and *Grand Houses*), with a population of 21,474. It is the converging-point of all the railroads from the East which connect with the Union Pacific, and is connected with Omaha, on the opposite river-bank, 4 miles distant, by a street car and wagon bridge. The great * **Missouri River Bridges,** which connect the two cities, are 2,750 and 2,920 ft. long. The Union Pacific R. R. Bridge cost over $1,000,000. Council Bluffs is well laid out, with streets crossing each other at right angles, and the principal edifices are of brick. The most important public buildings are the *County Court-House;* the *U. S. Court-House* and *Post-Office;* the *Bloomer School;* the *High-School;* the *Masonic Temple;* and the *Union Depot. Dohany's Opera-House* and the *Y. M. C. A. Building* are also fine structures. The *State Institute for Deaf-Mutes* is just outside the city limits. There are in the city seventeen churches, a Library Association, and a Young Men's Christian Association, with reading-room and gymnasium. The views from Fairmount Park are very fine. It has been decided by the U. S. Supreme Court that Council Bluffs is the E. terminus of the Union Pacific R. R. One and a half mile from the up-town depots of the railroads running into Council Bluffs, and near the bank of the Missouri River, is the great transfer depot, used by the Union Pacific R. R. and connecting lines. All Denver and overland trains start from this city. **Omaha** (Neb.), (see Sub-Route *d*).

b. Via Chicago, Rock Island & Pacific R. R.   Distance, 501 miles.

Two through trains daily, with palace dining and sleeping cars attached, run each way on this route. The country traversed is very similar in character to that along the preceding route, and might be described in the same general terms (see above). The first important place on the line is **Joliet** (40 miles), which has already been described in Route 85. *Morris* (62 miles) is a busy little city of 3,353 inhabitants on the Illinois & Michigan Canal, with an important trade in grain, and a Roman Catholic female seminary of some note. Twenty-three miles beyond Morris is **Ottawa,** a flourishing city of about 9,985 inhabitants,

25

on the Illinois River, just below the mouth of the Fox, and on the Illinois & Michigan Canal. It is lighted with gas, and contains many handsome residences. The chief public buildings are the Court-House, in which the Supreme Court for the N. division of the State is held, and the County Court-House and Jail. The surrounding country is fertile, and abounds in coal. The Fox River has here a fall of 29 ft., affording an immense water-power which is extensively used in manufactures. There are several grain-elevators, and large quantities of wheat are shipped from this point. **La Salle** (99 miles) is a busy manufacturing city of 9,855 inhabitants, on the Illinois River, at the terminus of the Illinois & Michigan Canal, 100 miles long, which connects it with Chicago. It also connects with the Illinois Central R. R., and with steamer to St. Louis. It is the center of extensive mines of bituminous coal, of which large quantities are shipped. *Pond Creek* (128 miles) is at the intersection of the Chicago, Burlington & Quincy R. R. *Geneseo* (159 miles) is in the heart of one of the finest agricultural districts in the State; and 20 miles beyond is **Moline** (*Keator House*), a city of 12,000 inhabitants, on the east bank of the Mississippi River, 3 miles above **Rock Island** (*Harper House, Rock Island*), which is another flourishing city on the E. bank of the river, with a population of 13,634. The river is here divided by the island of Rock Island, which is 3 miles long; and from 16 miles above Moline to 3 miles below are the Upper Rapids. By means of a dam at Moline an immense water-power is obtained, and employed in various manufactories. The scenery about Moline is highly picturesque, and the surrounding country is rich in coal. The city of Rock Island is at the foot of the rapids, opposite the W. extremity of Rock Island, from which it takes its name, and at the confluence of Rock River with the Mississippi. It is an important railroad center, is the shipping-point for the productive country adjacent, and has numerous and varied manufactures. Here are located *Augustana College and Theological Institute.* Rock Island is the terminus of the prospective *Michigan & Mississippi Canal.* The island of * **Rock Island** (960 acres in extent) is the property of the U. S. Government, and the site of the great Rock Island Arsenal and Armory, intended to be the central United States armory. The design embraces 10 vast stone workshops, with a storehouse in the rear of each, besides officers' quarters, magazines, offices, etc. Nearly all of the shops are now completed. The shops are supplied with motive-power by the Moline water-power, three-fourths of which is owned by the Government. There are 20 miles of splendid roadways running in every direction; drives, walks, promenades, and paths; delightful shade, and magnificent prospects from various points of view. Railway and wagon bridges, owned by the Government, conneet the island with the three cities of Moline, Rock Island, and Davenport. A point of interest is *Black Hawk's Watch-Tower,* which overlooks the junction of the Mississippi and Rock Rivers and the country for miles around.

Opposite Rock Island, on the Iowa side of the river, is the city of **Davenport** (*Kimball House, Lindell*), and the train crosses the river between them on the magnificent railroad and wagon * bridge built by the Government in connection with the armory at a cost of $1,000,000. Davenport is the fourth city of Iowa in size, has 26,872 inhabitants, and is the great grain depot of the upper Mississippi. It is also an important manufacturing center, and is situated in the heart of extensive bituminous coal-fields. The city is built at the foot and along the slope and summit of a bluff 3½ miles long, rising gradually from the river, and inclosed on the land side by an amphitheatre of hills half a mile in the rear. It is regularly laid out and handsomely built, and street-cars traverse the principal streets. The County Buildings are substantial structures, the *City Hall* is an imposing edifice, and the *Opera-House* is one of the finest in the West. Several of the churches and school-houses are handsome buildings. *Griswold College* (Episcopal) is a flourishing institution, with a library of 4,000 volumes, and the *Academy of the Immaculate Conception* (Roman Catholic) is of high standing. The *Library Association* has a library of about 5,000 volumes, and there are an *Academy of Natural Sciences,* two medical societies, and the Iowa Orphans' Home.

At Davenport, the S. W. Division of the Chicago, Rock Island & Pacific R. R. diverges and runs to Kansas City, and Leavenworth and Atchison on the Missouri River. On the main line, 40 miles beyond Davenport, is West Liberty, the junction of the Burlington, Cedar Rapids & Northern R. R. **Iowa City** (227 miles) was formerly the State capital, and is now the seat of the *State University*, which has an attendance of 950 students, an extensive laboratory, and a library of 30,000 volumes. The University occupies four buildings, of which the largest, formerly the Capitol, is a fine edifice in the Doric style, 120 by 60 ft. The County Offices and the Court-House are the other principal public buildings. The *State Historical Society* has a library of 3,500 volumes. The Iowa River furnishes water-power for various factories and flour-mills. The city contains 7,016 inhabitants, and is built upon the highest of three plateaus, 150 ft. above the river. *Grinnell* (302 miles) is the seat of *Iowa College*, which was removed here from Davenport. *Colfax* has become quite a noted resort on account of its mineral springs. **Des Moines** (358 miles; *Aborn House, Kirkwood, Morgan*) is the present capital of Iowa, and is situated at the head of navigation on the Des Moines River, at its confluence with the Raccoon. The city, which contains 50,093 inhabitants, is laid out in quadrilateral form, extending 4 miles E. and W. and 2 miles N. and S., and is intersected by both rivers, which are spanned by 8 bridges. The business quarter lies near the rivers, and the finest residences are on the higher ground beyond. The old Capitol is a plain building erected in 1856 at a cost of $60,000. A splendid *Capitol,* costing $3,000,000, has been recently finished. The *Post-Office,* which also accommodates the U. S. Courts and other Federal offices, cost over $200,000. There are 15 churches, 9 public-school houses, and a Baptist college with a spacious building on an eminence commanding a fine view. There are also *Drake University, Calinan College,* and two medical colleges. The *State Library* contains 15,000 volumes, and there is a *Public Library* with about 3,000 volumes. Among other fine buildings are the *Grand Opera-House* and the *City*

Hall. In the N. W. part of the city is a public park of 40 acres, and, in a bend of the Raccoon River, spacious Fair Grounds, with a race-course. A park of some 40 acres has just been laid out in the N. part of the city. Des Moines is an important railroad center, as more than twelve roads pass through the city.

Beyond Des Moines the road passes through the flourishing cities of Avoca and Atlantic, descends the bluffs into the Missouri Bottom, and soon reaches **Council Bluffs** (498 miles), which has been described in Sub-Route *a.*

  c. Via Chicago, Burlington & Quincy R. R.   Distance, 502 miles.

Two through trains daily, with palace sleeping and dining car service attached, run each way on this route. The Chicago, Burlington & Quincy R. R. passes through some of the most fertile farming lands of Illinois, crossing the State diagonally from Lake Michigan to the Mississippi River. It then crosses southern Iowa, a section teeming with agricultural wealth, and better cultivated than some other portions of the State. The country as a whole does not differ greatly from that traversed by the two preceding routes, but there are fewer important cities along the line. The first place requiring mention is **Aurora** (39 miles; *Bishop House, Tremont*), a city of 19,688 inhabitants, situated upon Fox River, which furnishes the power for numerous important manufactures. It contains a handsome City Hall, a college, 14 churches, and many fine stores and dwellings. The construction and repair shops of the R. R., situated here, employ 700 men. **Mendota** (84 miles; *Union Depot Hotel, Warner*) is a growing city of 5,542 inhabitants, at the intersection of the Northern Div. of the Illinois Central R. R. (Route 84). It is surrounded by a rich farming region, and, coal being abundant, manufactures are extensive and varied. *Mendota College* and a *Wesleyan Seminary* are located here, and some of the churches are handsome edifices. **Galesburg** (164 miles; *Brown's, Union Hotel*) is a city of 15,264 inhabitants, noted for its educational advantages, being the seat of *Knox College* (Congregational), with 650 students and a library of 5,000 volumes, and of *Lombard University* (Universalist), with 180 students and a library of 4,000 volumes. Both institutions admit woman-students. The *City Library* contains 7,000 volumes, and that of the *Young Men's Library Association* 4,000 volumes. Galesburg is surrounded by a rich farming country, and has several manufactories, including the machine-shops of the R. R. Co. Thirteen miles beyond Galesburg is **Monmouth,** a city of 5,936 inhabitants, situated on a rich and beautiful prairie. It is the seat of *Monmouth College*, established in 1856, and of the *Theological Seminary of the Northwest*, established in 1839, both under the control of the United Presbyterians. At *E. Burlington* (206 miles) the train crosses the Mississippi to **Burlington** (*Duncan, Union*), the fifth largest city in Iowa, with a population of 22,565, and a place of great commercial importance. The business portion of the city is built upon low ground along the river, while the residences upon the high bluffs command extended views of the fine river scenery. The electric street-railway system is in use, and steam-heat is carried through the business portion of the city. The river at this point is a broad, deep, and beautiful stream, and upon the bluffs between which it passes are extensive orchards and vine-yards. The city is regularly laid out and well built, the houses being chiefly of brick. Several miles of streets in the business portion are paved with hard brick, made near the city. It contains the *Burlington Institute* (Baptist), a business college, a *Free Public Library*, with 20,000 volumes, many handsome churches, and the *Des Moines County Court-House.* Ten railroads converge at Burlington, and it is connected with all the river-ports by regular lines of steamers. A railroad and wagon bridge has recently been built across the Mississippi at this point.

From Burlington to Council Bluffs the road traverses wide-stretching prairie-lands, which rise gradually to *Creston* (397 miles), and then descend more rapidly to the Missouri Bottom. **Mount Pleasant** (235 miles) is a city of 3,977 inhabitants, built on an elevated prairie, nearly inclosed in a bend of Big Creek. It contains *Iowa Wesleyan University* and *German College*, both under the control of the Methodists. The former has 200 students and a library of 3,000 volumes. Near the village, and in full view from the cars, is the spacious building of the *State Hospital for the Insane.* The next important station is *Fairfield* (257 miles), picturesquely situated on Big Cedar Creek. The sur-rounding country is rolling prairie, diversified with forests of hard wood. **Ottumwa** (Iowa) (285 miles) is the largest city on this line between the Mississippi and the Missouri, and has a population of 14,001. It is situated on the Des Moines River, which is here spanned by a bridge, is surrounded by a fertile country, has a good water-power which is extensively used in manufactures, and does a trade amounting to $6,000,000 annually. *Albia* (307 miles), *Chariton* (337 miles), and *Osceola* (363 miles) are small but prosperous places. *Creston* (397 miles) is on the dividing-ridge between the Mississippi and Missouri Rivers, 800 ft. above their level. The engine-houses and repair-shops of this division of the road are located here. The principal stations between Creston and Council Bluffs are *Red Oak*, where the Nebraska City Branch joins the main line, and *Pacific Junction*, the junction with the Burlington & Missouri R. R. in Nebraska. **Council Bluffs** (see Sub-Route *a*).

d. Via Chicago, Milwaukee & St. Paul R. R. 490 miles.

This line, known as the Chicago, Council Bluffs & Omaha Short Line, offers in many ways a desirable route. The through trains are equipped with sleeping and dining-room car service, which are of the best. The country, though not specially interesting, is a fine rolling prairie covered with prosperous farms. The first station of importance out of Chicago is **Elgin** (36 miles), a busy city of 17,823 population, and an important manufacturing place, for which the Fox River gives ex-tensive water-power. It is specially noted for the *National Watch-Works*, which employ some 600 hands and turn out about $800,000 worth of watches. There are also manufactories of carriages and agri-cultural machinery. There are 6 newspapers, 13 churches, 3 banks, and many prosperous business houses in the city. Passing a number of small places, we reach **Rockford** (93 miles), a city of 23,584 people, beautifully situated on both sides of the Rock River. It is the terminus of the Kenosha Div. of the Chicago & Northwestern R. R. The city contains 22 churches, 6 banks, and is notable among the smaller Western cities for its thrift and energy. There are many important manufactories, woolen-mills, iron-foundries, machine-shops, agricultural-implement works, breweries, etc. *Savanna* (138 miles) is a pretty town, and in 194 miles we arrive at **Rock Island** (see Sub-Route *b*). *Marion* (228 miles) is at the junction with the St. Louis Coal R. R., a thriving place of 3,094 inhabitants. **Cedar Rapids** (233 miles) is described in Sub-Route *a*. Passing through *Tama City* (282 miles) and *Pickering* (296 miles), we reach **Des Moines** in 374 miles (see Sub-Route *b*). There is no other station of importance before reaching **Council Bluffs.**

Omaha.

Hotels, etc.—First-class houses are the *Paxton House,* the *Millard,* and *Murray.* Other hotels are the *Dellone, Metropolitan,* and *Windsor, Motor* and *cable cars* traverse the city in various directions, centering at the Union Pacific Depot. The *Post-Office* is at the cor. of 15th and Dodge Sts.

Omaha, the largest city of Nebraska and of the Missouri River Valley, is situated on the Missouri River, opposite Council Bluffs, with which it is connected by the steel railroad-bridge and a wagon-bridge, and is practically the E. terminus of the Union Pacific R. R. It occupies a beautiful plateau, rising gradually into bluffs, and in 1890 had a population of 140,452. The streets are broad, cross each other at right angles, and are lighted with gas and electric lights. The level portion is chiefly devoted to business purposes, and contains many substantial commercial blocks and buildings. The bluffs are occupied by handsome residences with ornamental grounds. The * *U. S. Post-Office and Court-House* is a fine building of Cincinnati freestone, 122 by 66 ft. and 4 stories high, costing $350,000. The * *High-School Building* cost $250,-000, and is one of the finest of the kind in the country. It crowns a far-viewing hill, and has a spire 185 ft. high, from which there is a noble outlook. The *Douglas County Court-House,* costing $225,000, crowns another eminence. *Boyd's Opera-House,* one of the finest theatres in the West, stands at the cor. of Harney and 17th Sts. The *Union Pacific Headquarters Building* and the *C., B. & Q. R. R. General Offices* are handsome and spacious brick buildings, in which are employed some 800 people. Several of the churches are costly and handsome structures. The *Union Pacific R. R. Depot* is a spacious edifice. The *Exposition Building,* on Capitol Ave., between 14th and 15th Sts., will seat 6,000 in its main building, and contains the *Opera-House.* Among the large buildings are those erected by the Omaha *Bee* at a cost of $750,000, the *New York Life-Insurance Building,* costing $500,000, and the *City Hall.*

The prosperity of Omaha is due to its location, being in reality the gateway to the entire West, and as such it is called the "Gate" city; it now has an immense trade and numerous important manufactures. Of the latter, the principal are the *Omaha Smelting-Works,* one of

MAP OF THE PACIFIC RAILWAYS.

F

the largest in America, several large breweries and distilleries, extensive linseed-oil works, steam-engine works, four brick-yards, extensive stock-yards, and the vast machine-shops, car-works, and foundry of the Union Pacific R. R. It is now the third "packing" center in the U. S. The city has an excellent system of water-works. Four miles N. is *Fort Omaha*, a large and handsome post, but the headquarters of the Department of the Platte are in the city. The four great Iowa roads, the Chicago, Burlington & Quincy, the Chicago, Rock Island & Pacific, the Chicago, Milwaukee & St. Paul, and the Chicago & Northwestern, virtually terminate in Omaha. The city communicates with St. Louis and the Southeast by means of the Wabash, with Kansas City and Texas with the Kansas City, St.-Joseph & Council Bluffs and the Missouri Pacific R. R., and with Minnesota and the Northwest by means of the Chicago, St. Paul, Minneapolis & Omaha R. R. system, while the product of the great West is brought in over the Union Pacific and the Burlington & Missouri River R. R. Thirteen railroads converge here with their termini. Á Belt-Line R. R. encircles the city, and, besides accommodating local trade, enables the larger corporations to reach the business center of the city.

90. Omaha to San Francisco.

By the Union Pacific and Central Pacific Railways. Distance, 1,865 miles. Time, less than 3 days. The Pacific Railroads occupy so peculiar a position among achievements of the kind that a brief outline of their history will perhaps prove interesting. The project of a railway across the continent was publicly advocated as early as 1846, by Asa Whitney, and in 1853 Congress passed an act providing for surveys by the corps of topographical engineers. Further acts were passed in 1862 and 1864 providing for a subsidy in United States 6 per cent gold bonds at specified rates per mile. The same acts also gave to the companies undertaking the work 20 sections (12,800 acres) of land for each mile of railroad built, or about 25,000,000 acres in all. The railroad was built from Omaha, Neb., to Ogden, Utah, 1,031 miles, by the Union Pacific Company, and from San Francisco to Ogden, 883 miles, by the Central Pacific Company. Work was begun in 1863 ; the first 40 miles from Omaha to Fremont were completed in 1865 ; and on May 12, 1869, the railroad communication from the Atlantic to the Pacific ocean was opened. The route crosses 9 mountain-ranges, the highest being the Rocky Mountains, at an elevation of 8,247 ft. above the sea, and the lowest, Promontory Mountain, W. of Great Salt Lake, 4,889 ft. The aggregate length of the tunnels, of which there are 15, all occurring in the Sierra Nevada or its spurs, is 6,600 ft. The gradients do not often exceed 80 ft. to the mile, though in one instance they reach 90 ft. and in another 116 ft. to the mile. The cost of the Union Pacific road was reported to the Secretary of the Interior at $112,259,360 ; but the liabilities of the company at the date of the completion of the road were $116,730,052. In 1868 Jesse L. Williams, a civil engineer and one of the government directors, reported the approximate cost of the Union Pacific road in cash at $38,824,821 ; and this was probably not far from correct. The cost of the Central Pacific road and branches (1,222 miles) in liabilities of every sort was reported in 1874 at $139,746,311.

Stations.—*Union Pacific Ry.* : Omaha to Gilmore, 10 miles ; Papillon, 15 ; Millard, 21 ; Elkhorn, 29 ; Waterloo, 31 ; Valley, 35 ; Fremont, 47 ; North Bend, 62 ; Schuyler, 76 ; Benton, 84 ; Columbus, 92 ; Duncan, 99 ; Silver Creek, 109 ; Clark, 121 ; Central City, 132 ; Chapman, 142 ; Grand Island, 154 ; Alda, 162 ; Wood River, 170 ; Shelton, 178 ; Gibbon, 183 ; Kearney, 195 ; Elm Creek, 212 ; Overton, 221 ; Lexington, 231 ; Cozad, 245 ; Willow Island, 250 ; Warren, 260 ; Brady Island, 268 ; North Platte, 291 ; O'Fallon's, 308 ; Dexter, 315 ; Roscoe, 332 ; Ogalalla, 342 ; Big Spring, 361 ; Barton, 369 ; Julesburg, 372 ; Chappell, 387 ; Lodge Pole, 397 ; Sidney, 414 ; Brownson, 423 ; Potter, 433 ; Pine Bluff, 473 ; Egbert, 484 ; Hillsdale, 496 ; Cheyenne, 516 ; Granite Cañon, 535 ; Sher-

man, 549 ; Laramie, 573 ; Lookout, 606 ; Medicine Bow, 645 ; Carbon, 657 ;
Percy, 668 ; Fort Steele, 695 ; Rawlins, 709 ; Creston, 737 ; Rock Springs, 832 ;
Green River, 845 ; Bryan, 860 ; Carter, 905 ; Piedmont, 929 ; Evanston, 957 ;
Wahsatch, 966 ; Castle Rock, 975 ; Echo, 993 ; Weber, 1,009 ; Peterson, 1,017 ;
Uintah, 1,026 ; Ogden, 1,031. *Central Pacific R. R.*—Corinne, 1,053 ; Promon-
tory Point, 1,082 ; Kelton, 1,123 ; Terrace, 1,153 ; Toano, 1,214 ; Wells, 1,250 ;
Elko, 1,307 ; Carlin, 1,330 ; Palisade, 1,339 ; Winnemucca, 1,451 ; Humboldt,
1,493 ; Hot Springs, 1,569 ; Wadsworth, 1,588 ; Reno, 1,622 ; Truckee, 1,656 ;
Summit, 1,671 ; Emigrant Gap, 1,692 ; Blue Cañon, 1,698 ; Dutch Flat, 1,709 ;
Colfax, 1,722 ; Auburn, 1,740 ; Rocklin, 1,754 ; Sacramento, 1,777 ; Elmira, 1,805 ;
Suisun, 1,816 ; Benecia, 1,832 ; Port Costa, 1,833 ; Oakland Pier, 1,862 ; San
Francisco, 1,865.

THE journey from Omaha to San Francisco, by reason of its great
length and the time which it takes, will be in many respects a new experience to the traveler, no matter how extended his previous journeyings
may have been.   It is more like a sea-voyage than the ordinary rushing
from point to point by rail, and, as on a sea-voyage, one ceases to care
about time-tables and connections, and makes himself comfortable.   The
slow-running time of the earlier period has changed to a speed which some
Eastern lines might emulate.   The time consumed from Omaha to San
Francisco, *via* the Union and Central Pacific Rys., is 2 days, 21 hours
and 25 minutes.   Trains do not stop at meal-stations, but dining-cars
are run on this line.   The buffet sleeping-car service is also run, and
the traveler can always find good cheer at the sideboard.   Mr. Charles
Nordhoff, in speaking of the comforts of the long journey, says: " You
may pursue all the sedentary avocations and amusements of a parlor at
home ; and as your housekeeping is done—and admirably done—for you by
alert and experienced servants ; as you may lie down at full length, or sit
up, sleep or wake, at your choice ; as your dinner is sure to be abundant,
very tolerably cooked, and not hurried ; as you are pretty certain to
make acquaintances on the car ; and, as the country through which you
pass is strange, and abounds in curious and interesting sights, and the
air is fresh and exhilarating—you soon fall into the ways of the voyage ;
and if you are a tired business man, or a wearied housekeeper, your care-
less ease will be such a rest as certainly most busy and overworked
Americans know how to enjoy."

Passengers may secure berths in the sleeping-cars for the continuous trip to San Francisco in Chicago, at the ticket-offices of any of
the routes mentioned in Route 87, or may arrange for berths from point
to point.   The price of a berth from Chicago to Omaha is $2.50 ; Omaha
to Ogden or Salt Lake City, $8 ; Ogden to San Francisco, $6.

As there are nearly 300 stations on the line, only the more important are included in the list given above, and the information there
conveyed (names and distances) is all that the traveler will care to have
about most of them.   In such a case as this the only method of
description not likely to prove tedious will be to direct attention in
a general way to the characteristic features of the different sections
of the route.   Those who wish for more detailed information can secure
it from the R. R. folders of the Union Pacific Ry.

Two trains leave Omaha daily for the west, one at 2.15 P. M. and
the other at 6.40 P. M.   Both trains connect for San Francisco, Salt

Lake City, Portland and Utah, Idaho, Montana, and Oregon points. Both trains have through coaches and sleepers for Denver also. Assuming that the passenger takes the early train, he will pull slowly out through a valley in which are located several important manufacturing establishments, and make his first stop at *South Omaha*, the most important suburb of the city of Omaha. A syndicate of capitalists has purchased 1,800 acres of ground here, laid it out into city lots, built large stock-yards, slaughter-houses, and packing establishments, and proposes to conduct an extensive business, giving employment to several hundred men. Westward during the afternoon, the road traverses vast prairies dotted over thickly with farms and farm-houses. On the left is the Platte River, through whose valley. entered at *Elkhorn* (29 miles), the road runs for nearly 400 miles, and whose North Fork is crossed at *Fort Steele* (694 miles). At *Valley* (35 miles) a branch leaves the main line for *Wahoo, Lincoln, Beatrice, Marysville, David City, Osceola*, and *Stromsburg*. At *Fremont* (47 miles), a city of 6,747 people, the Sioux City and Pacific R. R. connects with the Union Pacific Ry. *North Bend* (62 miles) and *Schuyler* (76 miles) are passed, and the train arrives at *Columbus* (92 miles), from which point a branch, known as the Omaha & Republican Valley Branch, leaves the main line for *Albion, Fullerton*, and *Norfolk*. The Burlington & Missouri R. R. has a branch coming from the south and joining the Union Pacific line, with another at *Central City* (132 miles), and a third at *Grand Island* (154 miles). The last-named city is a railway center of some importance. A branch of the Union Pacific extends northward to *St. Paul* and *Ord*, and the St. Joseph & Grand Island R. R., a part of the Union Pacific system, also meets the main line at this point. The train moves westward, continuing through a beautiful agricultural section, passing *Kearny* (195 miles), *Lexington* (231 miles), and reaching *North Platte* (291 miles), the end of the first passenger division, soon after midnight. North Platte has extensive railroad-shops, round-houses, and a large railroad population. It is a city of 3,055 people, and lies on the border between the farming and grazing sections of Nebraska. In the past two years farmers have moved in great numbers upon this section, and have met with good success in cultivating its soil, even as far west as *Sidney* (414 miles). *Julesburg* (372 miles) is the junction of the Julesburg Branch, the short line to Denver, over which through cars pass between Omaha and Denver. Passing through the great stock region in the night, the train arrives at *Cheyenne* (516 miles), where junction is made with the Denver Pacific Branch of the Union Pacific, and the Denver & Kansas City sleeping-cars will be found waiting to deliver their passengers to the main line sleeping-cars.

At Cheyenne a fine glimpse is had of the Rocky Mountains, whose snow-clad tops are taken for clouds. *Long's Peak (14,271 ft. high) soon becomes plainly visible, and the Spanish Peaks are seen in the dim distance ; while away to the N., as far as the eye can reach, the dark line of the Black Hills leans against the horizon.

Cheyenne (Wyoming) (516 miles ; *Interocean Hotel*) is one of the largest towns on the road, though settled only in 1867. It has a population of 11,690, representing chiefly the stock and mining interests, is the point of junction with the Colorado Central Branch (see Route 99), and has an extensive round-house and shops, also a fine depot. The town is substantially built, largely of brick, and contains a fine Court-House and Jail, a neat City Hall, a large public-school building, and a brick opera-house. The military post of Fort D. A. Russell is located here. Cheyenne is another point of departure for the Black Hills, and daily stages run to Deadwood in 48 hours.

A few miles beyond Cheyenne the ascent of the Rocky Mountains is begun, though the train has climbed about one mile in altitude since leaving Omaha, and for thirty miles the road climbs rugged granite hills, winding in and out of interminable snow-sheds. *Sherman* (549 miles) is the highest R. R. station on this transcontinental line (8,247 ft.), and affords grand views. Here commences the descent to the Laramie Plains, which are about 40 miles wide on the average and 100 miles long, bounded by the Black Hills and the Medicine Bow Mountains. They are overrun by enormous flocks of sheep, and are said to afford the best grazing in the United States. In the adjacent hills there is abundance of game, such as mountain-sheep, antelopes, and bears. **Laramie City** (573 miles) is situated on the Laramie River, in the midst of the Laramie Plains, and has a population of some 6,388, which is rapidly increasing. It is the end of a division of the R. R., and has large machine and repair shops, and the rolling-mills of the company. The streets are regularly laid out at right angles with the railway, and there are many handsome buildings of brick and stone. Within 30 miles of Laramie there are deposits of antimony, cinnabar, gold, silver, and lead ores, graphite, and several other minerals; and it is expected that the place will become an important manufacturing center. Beyond Laramie the road traverses the Plains for many miles, crosses a region of rugged hills, and descends once more into the valley of the North Platte. Near *Miser* (616 miles) there are fine views from the cars of Laramie Peak on the right and Elk Mountain on the left. *Rock Creek* (625 miles) is a regular eating-station, and ranks among the best on the line. The North Platte is reached at *Fort Fred Steele* (695 miles), and then another steep ascent is begun. *Creston* (737 miles) is upon the dividing ridge of the continent, from which water flows each way, E. to the Atlantic and W. to the Pacific. At *Green River Station* (845 miles) the train emerges from the desolate plains, and enters a mountain-region, which affords some fine views. Utah Territory is entered W. of *Wahsatch* (966 miles). Within this region, between Green River and Salt Lake Valley, five tunnels are traversed, aggregating nearly 2,000 ft., and cut through solid rock, which never crumbles, and consequently does not require to be arched with brick. *Castle Rock* (975 miles) is a station at the head of Echo Cañon, and we there enter a region whose grand and beautiful scenery has been often described. * *Echo Cañon* and the celebrated * *Weber Cañon* offer the most magnificent sights on the whole Pacific route, and the tourist will be fortunate if he passes them by daylight. The road winds through all the devious turns of these cañons, while rock-ribbed mountains, bare of foliage except a stunted pine, and snow-capped, rise to an awful height on either hand. Emerging from these grim battlements of rock, the train enters the Salt Lake Valley and soon reaches **Ogden** (Utah) (1,031 miles), the junction between the Union Pacific and Central Pacific Railways. Here the road runs S. to Salt Lake City, and N. to Butte, Helena, and Anaconda. Ogden is a flourishing city of 14,889 inhabitants, situated on a high mountain-environed plateau, and remarkably well built. Its streets are broad, with running streams of water in nearly all of them, and it contains a brick Court-House, 3 churches and a Mormon tabernacle, many tasteful residences, and 2 hotels, the *Broome* and the *Depot*, the latter being at the station. The machine and repair shops of the Central Pacific R. R. are located here. Ogden is the regular supper and breakfast station of both Pacific railroads, and here cars are changed. Passengers are allowed one hour in which to get their meals, look about, and secure new berths in the palace-cars.

The *détour* to Salt Lake City (37 miles from Ogden; fare, $1.50) may be made in one day. The country between Ogden and Salt Lake City is quite thickly settled, except within the first 7 miles, and stoppages are made at several Mormon villages, which are attractive on account of the rich and well-cultivated farms.

Salt Lake City.

Hotels, etc.—The best hotels are the *Continental*, the *Cullen*, *Knutsford*, *Metropolitan*, *Templeton*, and *Walker*. *Electric-cars* run on the principal streets and render all parts of the city easily accessible. The population of Salt Lake City is 44,843, of whom about one third are Gentiles and apostate Mormons.

Salt Lake City, the capital of Utah Territory, is situated at the W. base of a spur of the Wahsatch Mts., about 12 miles from the S. E. extremity of the Great Salt Lake. It lies in a great valley, extending close up to the base of the mountains on the N., with an expansive view to the S. of more than 100 miles of plains, beyond which in the distance rise, clear cut and grand in the extreme, the gray and rugged mountains whose peaks are covered with perpetual snow. Great care was displayed in selecting the site and in laying out the city. The streets are 128 ft. wide, and cross each other at right angles. There are 260 blocks, each ½ of a mile square and containing 10 acres. Each block is divided into 8 lots, 10 by 20 rods, and containing 1¼ acre. Several of the blocks in the business quarter have been cut by cross-streets laid out since the founding of the city. Shade-trees and ditches filled with running water line both sides of every street, while almost every lot has an orchard of pear, apricot, plum, peach, and apple trees. The city is divided into 21 wards, nearly every one of which has a public square.

The chief business thoroughfares are *Main St.* and *Temple St.*, and 1st and 2d South Sts. On the former is the great * **Tabernacle,** which is the first object to attract the eye as one approaches the city, although destitute of any architectural beauty. It is of wood, except the 46 huge sandstone pillars which support the immense dome-like roof, is oval in shape inside and out, and will seat 13,000 persons. Its length is 250 ft., width 150 ft., and height 70 ft. It is used for worship, lectures, and debates. The Tabernacle organ is one of the largest in America. A little E. of the Tabernacle, and inclosed within the same high wall, is the completed granite building of the *Temple*, which cost $5,000,000. It is 200 ft. long by 100 ft. wide; the walls are 100 ft. high and the spires 200 ft. high. Within the same walls is the famous *Endowment House*, in which the various Mormon rites are performed, and in which the Mormons are married and receive their "endowments." It is an inferior-looking adobe building. On S. Temple St., E. of Temple Block, is *Brigham's Block*, inclosed by a high stone wall, and containing the Tithing House, the Beehive House, the Lion House, the Assembly Rooms, the office of the "Deseret News," and various other offices, shops, dwellings, etc. Here was the residence of Brigham Young, and 18 or 20 of his wives lived in the Beehive and the Lion House. Nearly opposite is a large and handsome house supposed to belong to the Prophet's favorite wife, and formerly known as *Amelia Palace*, now known as the *Gardo House*. On S. Temple St., opposite the Tabernacle, is *The Museum*, where may be seen the various products of Mormon industry, specimen ores from the mines, and Indian relics and curiosities. The *Theatre* is a vast, plain-looking building, but with a very ornate interior. Walker's *Opera-House* is a fine structure. The *Masonic* and the *Deseret Libraries* are open to the public free. The *City Hall* cost $60,000, and is used as the Territorial Capitol; while the *County Court-House*, a handsome brick building, is on 2d South St. The *U. S. Court* is held in the Wahsatch Building, on Main St. There are several fine churches in the city. Among the educational institutions are the *Deseret University*, *Hammond Hall*, *Collegiate Institute*, *All-Hallows College* (Catholic), and 21 public schools, purely secular, and *St. Mary's Academy*. The hospitals are, *Holy Cross*, *St. Mark's*, and the *Deseret*. The *City Prison* is in the rear of the City Hall. The *Utah Penitentiary* is a curious old structure, 1½ mile S. W. of the city. A new penitentiary is now in construction by the U. S. Government. Other handsome buildings are those of the *Deseret National Bank*, at the cor. of E. Temple and S. 1st Sts.; the *Co-operative Store*, in E. Temple St.; and the new Hooper and Eldredge block. About 2 miles E. of the city is *❋Fort Douglas*, overlooking the city, and garrisoned by a full regiment. The drive around *Liberty Park*, and on the *Boulevard*, 10 to 12 miles long, passes through very fine scenery. From the *Tower*, on the North Bench of the city, a beautiful view is obtained. Park City, Bingham, and Alta mining camps are worthy of a visit.

Most visitors to Salt Lake City will, as a matter of course, wish to see the *❋Great Salt Lake* from which it takes its name, and which is one of the greatest natural curiosities of the West. It is most easily reached *via* the Utah & Nevada, a branch of the Union Pacific, to *Garfield Beach* (18 miles), where there is a large hotel, a fine pavilion, and numerous bath-houses (round trip rate, 50 cents). *Salt-Air Beach*, *Crescent Beach, and Buffalo Park* are resorts that were opened in 1891. Great Salt Lake is 75 miles long and about 30 miles broad, is 4,200 ft. above the sea, and contains 7 islands, of which Antelope and Stansbury are the two largest. Several rivers flow into it, but it has no out-

Mormon Tabernacle, Salt Lake City.

let. The water is shallow, the depth in many extensive parts being not more than 2 or 3 ft. Its water is transparent, but excessively salt, and so buoyant that a man may float in it at full length upon his back, having his head and neck, his legs to the knee, and both arms to the elbow entirely out of the water. Swimming, however, is difficult, from the tendency of the lower extremities to rise above the surface; and the brine is so strong that it can not be swallowed without danger of strangulation, while a particle of it in the eye causes intense pain. A bath in it is refreshing and invigorating, though the body requires to be washed afterward in fresh water.

Leaving Ogden, the westward-bound train passes two small stations, and in 25 miles reaches **Corinne** (*Central Hotel*), the largest Gentile town in Utah, having a large trade with the mining-regions of eastern Idaho and Montana. Beyond Corinne the train winds among the Promontory Mts., and skirts the N. shore of the Great Salt Lake, while the Mormon city lies near the S. end of it. *Promontory Point* (1,082 miles) is interesting as the spot where the two companies building the Pacific Railroad joined their tracks on May 10, 1869. The last tie was made of California laurel trimmed with silver, and the last four spikes were of solid silver and gold. Beyond this the road enters upon an extended plateau, about 60 miles long and of the same width, known as the *Great American Desert*. Its whole surface is covered with a sapless weed 5 or 6 inches high, and never grows any green thing that could sustain animal life. The only living things found upon it are lizards and jackass-rabbits; and the only landscape feature is dry, brown, and bare mountains. "The earth is alkaline and fine, and is whirled up by the least wind in blinding clouds of dust. Rivers disappear in it, and it yields no lovelier vegetation in return than the pallid artemisia or sage-brush. It seems to have been desolated by a fire which has left it red and crisp; the blight which oppresses it is indescribable. The towns along the railway do not enliven the prospect." Yet the process of irrigation, which is beginning to be used in places, shows that even this unpromising soil can be made to yield rich returns. At *Humboldt Wells* (1,250 miles) are some 30 springs in a low basin about ½ mile W. of the station. Some of these springs have been sounded to a depth of 1,700 feet without revealing a bottom, and it is supposed that the whole series form the outlets of a subterranean lake. The most important station on this portion of the line is **Elko** (1,307 miles), which is the county-seat, and has a large brick court-house and jail, a church and a public school, and the State University, founded in 1875. Several important mining districts are tributary to Elko, and secure it a large trade. About 1½ mile W. is a group of mineral springs which are achieving a good deal of local reputation. *Winnemucca* (1,451 miles) is another prosperous town with a large mining trade; and *Humboldt* (1,493 miles) affords a grateful if momentary relief to the now wearied eye of the tourist. "The desert extends from Humboldt in every direction—a pallid, lifeless waste, that gives emphasis to the word desolation; mountains break the level, and from the foot to the crest they are devoid of vegetation and other color than a maroon or leaden gray; the earth is loose and sandy; Sahara itself could not surpass the landscape in its woe-begone infertility; but here at Humboldt a little intelligence, expenditure, and taste have, by the magic of irrigation, compelled the soil to yield flowers, grass, fruit, and shrubbery."

At *Wadsworth* (1,588 miles) the ascent of the Sierra Nevada is begun. The wearying sight of plains covered with alkali and sage-brush is exchanged for picturesque views of mountain-slopes, adorned with branching pine-trees, and diversified with foaming torrents. The ascent soon becomes so steep that two locomotives are required to draw the train. At short intervals there are strong wooden snow-sheds, erected to guard the line against destruction by snow-slides. These sheds, which are very much like tunnels, interrupt the views of some of the most romantic scenery on the line.

Reno (1,622 miles; *Arcade, Golden Eagle, Palace*) is a busy town on the Truckee River, about 5 miles from the base of the Sierra. It has a large trade with the mining districts, is in the heart of an agricultural and grazing valley, and contains the grounds of the State Agricultural Society, the *State University*, and *State Insane Asylum, Bishop Whitaker's School* for girls, and *St. Mary's* (Catholic) *School*, and several flouring-mills and smelting-works.

The Virginia & Truckee R. R. runs from Reno to Carson (31 miles) and Virginia City (52 miles), in the great Nevada mining-region. **Carson** (*Ormsby*) is the capital of Nevada, and is a thriving city of 3,950 inhabitants, containing the Capitol, the U. S. Mint, a Court-House, four churches, the best school-house in the State, and many handsome residences. The State Prison, 2½ miles distant, is a massive building. From Carson daily stages run in 15 miles (fare, $2) to * Lake Tahoe, as far as *Glenbrook* (*Adams's Spring House*), whence a steamer runs to *Tahoe City*. Numerous summer resorts are located on the lake. It is about 22 miles long and 10 miles wide, is 6,247 ft. above the sea, is surrounded by snow-capped mountain-peaks, and has marvelously clear water which has been sounded to a depth of over 1,600 ft. Small steamers circumnavigate the lake, and enable its exquisite scenery to be viewed to great advantage. Twenty-one miles beyond Carson is Virginia City, built half-way up Mount Davidson, completely environed by mountains, and containing 8,511 inhabitants, about one fifth of whom are usually under ground.

Virginia City is built over the famous Comstock Lode, which, since the date of its discovery in 1859, has produced in gold and silver about $500,000,000, and is now yielding about $4,000,000 per annum. The ore body discovered in 1872 in the "Consolidated" ground was of such value as to obtain the name of the "Big Bonanza."

Tourists may leave Reno in the morning, spend the day at Virginia City examining the mines, and return to Reno in the evening, connecting with the overland passenger-train (fare, $4 round trip). An excellent way to make the excursion described above will be to go direct from Reno to Virginia City, then return to Carson and take the stage for Lake Tahoe, cross the lake to Tahoe City, and take the stage thence to Truckee on the S. P. R. R.

Truckee (1,656 miles) is the first important station in California, and is a handsomely-built town of 1,350 inhabitants, perched high up amid the Sierras. Three miles from the town is the lovely * *Donner Lake*, embosomed in the lap of towering hills; and daily stages (fare, $2) run to Tahoe City on Lake Tahoe (see above). *Summit* (1,671 miles) is the highest point on the Central Pacific road (7,042 ft.), and the scenery around the station is indescribably beautiful and impressive. "A grander or more exhilarating ride than that from Summit to Colfax," says Mr. Nordhoff, "you can not find in the world. The scenery is various, novel, magnificent. You sit in an open car at the end of the train, and the roar of the wind, the rush and vehement impetus of the train, and the whirl around curves, past the edge of deep chasms, among forests of magnificent trees, fill you with excitement, wonder, and delight. . . . The entrance to California is as wonderful and charming as though it were the gate to a veritable fairy land. All its sights are peculiar and striking: as you pass down from Summit the very color of the soil seems different from and richer than that you are accustomed to at home; the farm-houses, with their broad piazzas, speak of a summer climate; the flowers, brilliant at the roadside, are new to Eastern eyes; and at every turn of the road new surprises await you." From Summit to Sacramento is a distance of 106 miles, and between these places the descent from a height nearly half as great as that of Mont Blanc to 56 ft. above the sea-level has to be made. The line is carried along the edge of precipices plunging downward for 2,000 or 3,000 ft., and in some parts upon a narrow ledge excavated from the mountain-side by men swung down in baskets. It is thus at * *Cape Horn*, a point grand and imposing in the extreme, which is passed just before *Colfax* (1,722 miles). **Sacramento** (*Golden Eagle, Capitol, Western, State-House*) is the capital of California, the fourth city of the State in size, having a population of 26,386, and second in commercial importance. It is built on an extensive plain on the E. bank of the Sacramento River, immediately S. of the present mouth of the American River. Its site is very low, and the city formerly suffered from inundations; but the business portion has been artificially raised, and the whole city surrounded by a strong levee. The streets are straight and wide, and cross each other at right angles; the shops and stores are mostly of brick; the dwellings are handsome, and surrounded by gardens. Shade-trees are abundant, and a luxuriant growth of flowers and shrubs may be seen in the open air at all seasons of the year. The important public buildings are the * **State Capitol,** one of the finest structures of the kind in the United States (containing a marble statue, by L. G. Meade, of "Columbus appealing to Isabella," presented by D. O. Mills, and two large paintings, "Crossing the Plains" and "Miner's Camp in Early Day," by Charles G. Nahl), the *Court-House*, and the *City Hall*. The Capitol is situated almost in the heart of the city, and the grounds cover eighteen blocks, beautifully laid out with trees, shrubs, and flowers. The *State Library*, in the Capitol, has upward of 85,000 volumes, and the *Free Public Library*, in a fine building belonging to the city, about 16,000 volumes. The Masons, the Odd-Fellows, and the Knights of Pythias have handsome, large buildings, and the national Government is constructing a building for which an appropriation of $500,000 has been recommended. The *State Agricultural Society* has ample accommodations for the exhibition of stock, one of the finest race-courses in the world, and a pavilion that cost over $100,000, for the display of the resources of the State. It holds a fair annually, about the middle of September. There are a number of fine churches in the city, notably the *Catholic Cathedral*, costing nearly $250,000, many schools, charitable institutions, a convent, and several manufactories and machine-shops. The *Crocker Art-Gallery* was presented to the city by Mrs. Margaret E. Crocker, and in the building is a flourishing *Art-School*. Besides several hundred paintings there is also a fine collection of California minerals. *Oak Park* and *Highland Park* are suburban additions to Sacramento, lying S. E. on the line of the Southern Pacific, and connected with the city by electric and other street railways.

The through trains from Sacramento to San Francisco pursue a very pleasant route, being for the most part through the valleys of the Sacramento and San Joaquin. **Benicia,** 57 miles from Sacramento, is a thriving town of 2,591 inhabitants, on the N. side of the *Straits of Carquinez*, and is at the head of navigation for the largest ships. Here are the *large depot and machine-shops* of the Pacific Mail Steamship Co., the *United States Arsenal*, and several noted educational institutions, Catholic and Protestant. The trains cross the straits on a mammoth ferry-boat, and in 27 miles reach **Oakland** (*Albany, Central, Crellin, Galindo, Windsor*), a beautiful city of 48,682 inhabitants, situated on the E. shore of San Francisco Bay, nearly opposite San Francisco, of which it is practically a suburb. It is a favorite residence of persons doing business in San Francisco. The *Fabiola Hospital* is at the cor. of Moss Ave. and Broadway. Oakland is luxuriantly shaded, live-oak being the favorite tree, is remarkably well built, and has a delightful climate. At Berkeley, 4 miles N., is the *State University*, which is open to both sexes, and where tuition is free. The train passes around the city to Oakland Point, where the company has built an immense pier 2¼ miles into the bay. From this pier (which is well worth notice) a ferry-boat conveys the passengers and freight to San Francisco, 3 miles distant.

The old route from Sacramento to San Francisco, and now used only for local travel, is *via* **Stockton,** 54 miles from Sacramento. It is by this route that most of the passengers by the Southern Pacific and other routes coming into the State on the south enter and leave Sacramento. Stockton contains 14,424 inhabitants, is situated at the head of tide-navigation on the San Joaquin River, and the principal hotels are the *Yosemite, Mansion,* and *Grand Central*. It occupies a level site, and is substantially and compactly built, with handsome, wide streets, and public buildings that indicate enterprise and taste. The *Court-House* and *City Hall*, near the center of the city, are surrounded with choice shade-trees and shrubbery, as are also many of the residences. The business blocks are principally of brick, and there are several handsome churches and school-houses. The *State Lunatic Asylum* is located here, and its spacious buildings are seen just before the train enters the city.

The *** Calaveras Grove of Big Trees** is best visited by rail from San Francisco to Lodi, thence by Narrow-Gauge R. R. to Valley Springs, and thence by stage 25 miles to *Murphy's*. It is usual to stay overnight at Murphy's, and take the stage in the morning for the Big Trees, 16 miles distant. There is a good hotel at the grove. The grove occupies a belt 3,200 ft. long by 700 ft. broad,

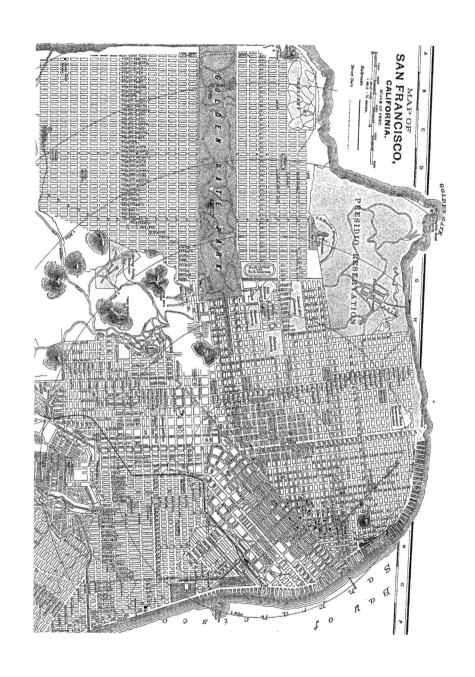

MAP OF
SAN FRANCISCO,
CALIFORNIA.

SCALE OF FEET.

Railroads
Street Cars

GOLDEN GATE

PRESIDIO RESERVATION

GOLDEN GATE PARK

Bay of San Francisco

in a depression between two slopes, through which meanders a small brook that dries up in summer. There are 93 trees of large size in the grove, and a considerable number of smaller ones, chiefly on the outskirts. Several have fallen since the grove was first discovered, one has been cut down, and one has had the bark stripped from it up to the height of 116 ft. above the ground. The tallest now standing is the *Keystone State,* which is 325 ft. high and 45 ft. in circumference; and the largest and finest is the *Empire State.* There are 4 trees over 300 ft. high and from 40 to 61 ft. in circumference. Their age is supposed to be about 1,500 years. The tree which was cut down occupied 5 men 22 days, pump-augers being used for boring through the tree. After the trunk was severed from the stump, it took the 5 men 3 days, with ponderous wedges, to topple it over. The bark was 18 inches thick.—The **Stanislaus Grove** (or *South Grove*) is situated on Beaver Creek, 5 miles S. E. of the Calaveras Grove. There are 700 or 800 trees in this grove, several of them being very fine specimens and in excellent condition. The grove is often visited by tourists, who ride over from the hotel in the other grove, where horses and guides are furnished.

At *Lathrop* (1,833 miles) the Visalia Div. of the Central Pacific R. R. diverges, and constitutes one of the routes to the **Yosemite Valley** (see Route 92). Beyond Lathrop, on the main line, a number of small stations are passed, and the train soon reaches Oakland (see page 400).

91. San Francisco.

Hotels.—*Baldwin,* cor. Market and Powell Sts.; *Beresford,* cor. Bush and Stockton Sts.; *California,* in Bush St. near Grant Ave.; *Grand,* cor. Market and Montgomery Sts.; *Langham,* cor. Ellis and Mason Sts.; *Lick,* cor. Montgomery and Sutter Sts.; *Occidental,* cor. Montgomery and Bush Sts.; *Palace,* cor. Market and Montgomery Sts.; and *Russ,* in Montgomery St., between Bush and Pine Sts. There are also a number of excellent family hotels.

Restaurants. — Restaurants, chop-houses, *rôtisseries,* abound in every quarter of San Francisco. Chop-houses and *rôtisseries* differ from restaurants, in that the cooking-furnaces are arranged on one side of the room, and each person can select the raw food and have it cooked right before his eyes. There are also numerous *tables-d'hôte,* where, by paying from 50c. to $1, one can sit at the table and call for anything he likes, provided it is on the bill of fare, including wines. The *Maison Dorée, Tortoni, California Hotel Café, Maison Riche,* and the *Palace Hotel Café,* are noted for the excellence of their cuisines. At the *Viticultural Café and Restaurant,* 317 Pine St., pure California wines are procurable.

Modes of Conveyance.—*Street-cars* (fare, 5c.) intersect the city in every direction, and most of them run to the foot of Market St. *Steam-cars* run out to the Cliff House. *Hackney-carriages* are in waiting at the steamer-landings and at various stands in the city (they may be found at all hours at the Plaza, opposite the City Hall, Kearny St., and on Sutter St. just above Kearny). The legal charges are : For a carriage drawn by more than 1 horse, for 1 person, not exceeding 1 mile, $1.50 ; for more than 1 person, not exceeding 1 mile, $2.50 ; for each additional mile, for each passenger, 50c. By the hour, $2 for the first hour, and $1.50 for each subsequent hour. For a cab drawn by 1 horse, for 1 person, not exceeding 1 mile, 50c. ; for more than 1 person, not exceeding 1 mile, $1 ; for each passenger, for each additional mile, 25c. By the hour (for 2 persons), $1.50 for the first hour, and $1 for each subsequent hour. No extra charge is allowed for ordinary baggage.

Ferries.—All the ferries, viz., to Oakland, Alameda, Saucelito, San Quentin, Berkeley, San Rafael, and Tiburon, run from the foot of Market St.

Theatres and Amusements.—The *Grand Opera-House,* which is the largest theatre in the city (in Mission St. near 3d), the *California Theatre* (414 Bush St.), and the *Baldwin Theatre* (936 Market St.), are the important theatres, devoted to the higher class of amusements. The *Bush St. Theatre,* in Bush St., between Montgomery and Kearny Sts., is devoted to varieties and negro minstrelsy. *Stockwell's Theatre* (in Powell St. near Market), the *Casino,* and

26

the *Alcazar* (at 114 O'Farrell St.), are cheaper theatres. Cheap opera is a feature of San Francisco; and at the *Tivoli* (in Eddy St.) an excellent performance in opera is given; beer is sold, and smoking in certain parts of the house permitted. There are two *Chinese Theatres*, one at 629 Jackson St., and the other at 816 Washington St., which attract considerable patronage from tourists and visitors on account of the unique character of the performances. The principal *Race-course* is near Golden Gate Park, which is much frequented; and there is a speed race-course in Golden Gate Park. The *Olympic Club Grounds* for athletic sports include the place where the California League games are played.

Reading-Rooms.—In all the leading hotels there are reading-rooms for the use of the guests, supplied with newspapers, etc. The *Free Library*, in the new City Hall, was opened in 1879, and is supported by a percentage of the taxes on public property, giving it an income varying from $40,000 to $50,000 a year. The library contains 70,000 volumes, and the average number of visitors is 1,200 a day. It has branches at convenient points in the city. The *Mercantile Library*, at Van Ness and Golden Gate Aves., has 63,000 volumes, a well-supplied reading-room, and extensive chess-rooms (open from 8 A. M. to 7 P. M.). The *Mechanics' Institute Library*, 31 Post St., has a library of 45,000 volumes, and a reading-room (open from 9 A. M. to 10 P. M.). The *Law Library*, in the new City Hall, has 30,000 volumes, and is open from 9 A. M. to 10 P. M. The *Young Men's Christian Association*, 232 Sutter St., has a library of 4,000 volumes, and a reading-room (open from 9 A. M. to 10 P. M.).

Clubs.—The *Pacific-Union Club* has rooms at the N. W. corner of Post and Stockton Sts. The *Olympic Club* has a fine house in Post St. near Mason St. The *Cosmos Club*, 317 Powell St., has a membership of military and naval officers. The *Bohemian Club*, cor. Grant Ave. and Post St., is the leading literary association in California, made up largely of authors, journalists, actors, musicians, and professional men. The *University Club* has a commodious house on Sutter St., and its membership includes the alumni of educational institutions. The *San Francisco Press-Club* has fine rooms at 430 Pine St. The *Cercle Français*, 421 Post St., has its membership among the French residents. The *Ligue Nationale Française*, 305 Larkin St., is a prosperous club, with a French library of 14,000 volumes. The *San Francisco Verein*, 219 Sutter St., has a library of 6,000 volumes and a reading-room. The *Pacific Turner Bund*, for the cultivation of gymnastic exercises, has rooms at 323 Turk St. The *Concordia Club* is a Jewish club, with a fine house cor. Van Ness Ave. and Post St. Introduction by a member secures the privileges of these clubs.

Post-Office.—The Post-Office is at the cor. of Washington and Battery Sts. Open from 8 A. M. to 8 P. M. on secular days, and on Sundays from 9 to 10 A. M. Letters are collected 6 times daily from the street-boxes.

SAN FRANCISCO, the chief city of California and commercial metropolis of the Pacific coast, is situated at the N. end of a peninsula which is 30 miles long and 6 miles across at the city, and separates San Francisco Bay from the Pacific Ocean, in lat. 37° 46' N. and lon. 122° 46' W. The city stands on the E. or inner slope of the peninsula and at the base of high hills. In 1846 these hills were steep and cut up by numerous gullies, and the low ground at their base was narrow, save in what is now the S. part of the city, where there was a succession of ridges of loose, barren sand, impassable for loaded wagons. The sand-ridges have been leveled, the gullies and hollows filled up, and the hills cut down; and where large ships rode at anchor in 1849 there are now paved streets. The greater part of the peninsula is hilly, bare of trees, and unfit for cultivation; the roads from the city are the *San Bruno*, the *San Miguel*, and the old *San José* roads. The business streets are built up densely, and with the residential portion the settled part of the city may be said to cover an area of 9 square miles. In the N. E. corner of the city is Telegraph Hill, 294 ft. high; in the

S. E. corner Rincon Hill, 120 ft. high; and on the W. side Russian Hill, 360 ft. high. The densely-populated quarters are in the amphitheatre formed by the three hills.

The history of San Francisco is interesting on account of the rapid growth of the place. The first house was built in 1835, when the village was called Yerba Buena, which in Spanish means "good herb" (wild mint), so named from a medicinal plant growing in abundance in the vicinity. In 1847 this was changed to San Francisco; and in 1848, the year that gold was first discovered in California by the white settlers, the population had increased to 1,000. The influx from the East then commenced, and in December, 1850, the population was about 25,000. In 1860 it was 56,802; in 1870, 149,473; and in 1880, according to the census returns, it amounted to 233,956. In 1890 the census returns showed a population of 298,997. The city was incorporated in 1850, and the city and county were consolidated in 1856. In 1851 and 1856, in consequence of bad municipal government and corrupt administration of the criminal laws, the people organized Vigilance Committees, and summarily executed several criminals and banished others. This rough but wholesome discipline had its effect, and the city is now one of the most orderly in the country. The commerce of San Francisco is very large, the chief articles of export being the precious metals, breadstuffs, wines, wool, and fruits; and of import, lumber, coal, coffee, tea, rice, and sugar. The manufactures are important, including woolen and silk mills, and manufactories of watches, carriages, boots, furniture, candles, acids, wire-work, castings of iron and brass, and silver-ware.

The city is regularly laid out, though not on a uniform plan. The streets are broad, and cross each other at right angles. The business streets are generally paved with Belgian blocks or cobble-stones, and many of the residence streets have sidewalks of stone or asphaltum. The leading thoroughfare is *Market St.*, which is broad and lined with handsome buildings. At its S. W. end it formerly extended to the top of a hill too precipitous for the ascent of carriages. But recent improvements have removed this hill, and a cable-road runs through the street from the ferry to Castro St. Market St. is the great center of a cable system which has branches running in Valencia, Haight, Hayes, Turk, Geary, Castro, Oak, and McAllister Sts. All these lines converge in Market St., and all have one common terminus at the ferry. Cable-roads have been widely extended during the last few years. Besides the Market St. system, with its branches, there are the California St., Sutter St., Ferries and Cliff House road, with branches. Omnibus cable line in Post, Leavenworth, Ellis, Oak, and Howard Sts., and cable lines in Geary and Union Sts., with transfers to all parts of the city. There are three cross-town lines, one in Larkin and Polk Sts., one in Powell St., and the other in Jones St. These cable lines afford the best means of seeing the city, as one may ride on the dummy in the open air, and thus obtain as good a view as from an open carriage. There is completed an electric road with about 18 miles of double track running to the cemeteries in San Mateo. It will be extended ultimately to San José, a distance of 50 miles. *Kearny, Montgomery*, and *Market Sts.* are the fashionable promenades, and contain some of the largest retail shops. The principal banks and brokers' and insurance offices are located in *Montgomery, California*, and *Pine Sts.* The importers and jobbers are in *Front, Sansome, Montgomery, Battery Sts.*, and lower end of *Market St.* The handsomest private residences are in California St. Hill (Nob Hill), Van

Ness Ave., Clay St. Hill, Pine St. Hill, Pacific Avenue Heights, and
Washington, Jackson, Franklin, Taylor, Bush, Sutter, and Leavenworth
Sts. Especially worth seeing are the Mark Hopkins Mansion, cor. of
Mason and California Sts.; the Charles Crocker Mansion, cor. Califor-
nia and Taylor Sts.; the James C. Flood Mansion, cor. California and
Jones Sts.; the residence of Leland Stanford, cor. Powell and Cali-
fornia Sts.; and that of Lloyd Trevis, cor. of Taylor and Jack-
son Sts.

A stranger's first impression of San Francisco is that there are no
public buildings, though the new **City Hall,** now partially completed,
in Yerba Buena Park, bounded by Park Ave., MacAllister, and Larkin
Sts., is a fine structure, surpassed by few in the United States. The
U. S. Appraiser's Building is a spacious four-story structure in Sansome
St., extending from Jackson to Washington Sts. The *Custom-House,*
which also contains the *Post-Office,* is at the cor. of Battery and Washington Sts. A new Post-Office will be built at the cor. of Mission and
7th Sts. The *U. S. Branch Mint* is a massive stone structure in
the Doric-Ionic style, at the cor. of 5th and Mission Sts. The machinery
here is believed to be unapproachable in perfection and efficiency (visitors are admitted from 10 A. M. to 12 M.). The *U. S. Treasury* is located
in Commercial St. (office-hours, 10 A. M. to 3 P. M.). The *San Francisco
Stock Exchange* is a splendid six-story granite and marble edifice in Pine
St., surmounted by a handsome tower. The *Merchants' Exchange,*
on the S. side of California St., between Montgomery and Sansome Sts.,
is one of the most costly and spacious buildings in the city. The Exchange is a splendid room in the first story, with lofty ceiling, and is
well supplied with the leading papers and magazines, home and foreign.
In the tower over the building is a fine clock. Other notable commercial
buildings are those of the *Bank of California,* the *Safe-Deposit
Bank,* in California St., the *Nevada Bank,* cor. Montgomery and Pine,
the *Anglo-Californian Bank,* in Sansome St., at the N. E. cor. of Pine;
and the *First National Bank,* at the N. W. cor. of Bush and Sansome
Sts., is the handsomest bank building in the city. The *Palace Hotel*
is a vast and ornate building, at the cor. of Market and New Montgomery
Sts., 275 by 350 ft., 9 stories, erected at a cost (including furniture) of
$7,000,000. It is entered by a grand court-yard surrounded by col-
onnades, and from the roof (reached by elevator) a bird's-eye view of
the whole city can be obtained. It is connected by a bridge with the
Grand Hotel, both being under one management. Another palatial
structure is the *Baldwin Hotel,* at the cor. of Market and Powell
Sts.; though smaller than the Palace, it cost $3,500,000. The *Flood
Building,* at 4th and Market Sts., cost $1,500,000, and is rented
for offices. The building of the *Mercantile Library,* at the cor.
of Van Ness and Golden Gate Aves., is large and fine, of brick and
stone, three stories high. The library contains 62,000 volumes, and
there are several reading-rooms, chess-rooms, and an unusually fine
collection of pictures. The building is of recent erection and contains late library improvements. The *Odd-Fellows' Hall* is a splendid
and commodious building, at the cor. of Market and 7th Sts., lately

completed and occupied by the order. The *Mechanics' Institute* is a substantial building in Post St., between Montgomery and Kearny, with a library of 45,000 volumes. The building of the San Francisco *Chronicle* is a striking edifice, being nine stories high, with a tower 210 ft. high, located at the junction of Market, Kearny, and Geary Sts. The *Mark Hopkins* and *D. O. Mills Buildings* are conspicuous office-structures. The * **California Market,** for fruits, vegetables, meat, and produce of all kinds, is one of the sights of San Francisco. It is between Kearny and Montgomery Sts., extending through from Pine to California. The *Center Market*, at the cor. of Sutter and Dupont Sts., is well worth visiting ; and the old *Washington Market*, in the street of that name, is specially worth a visit in the early morning. The *California Academy of Sciences* occupies a handsome five-story structure in Market St. near 4th, which cost $400,000. The *Mining Bureau* and *Pioneers' Building*, in 4th St. near Market, contain both a fine collection of minerals and relics of early experiences in California. The striking features of recent edifices is in the height of office-buildings, constructed of steel, stone, brick and terra-cotta. Notable among these are the *Crocker Building* (11 stories), cor. Market and Post Sts., which cost $1,000,000 ; and the *Mills Building*, cor. Bush and Montgomery Sts., which cost $1,500,000.

One of the largest and finest church-edifices on the Pacific coast is that of * **St. Ignatius Church and College** (Roman Catholic), in Van Ness Ave. It will accommodate 3,500 persons. *St. Mary's Cathedral*, at Van Ness Ave. and O'Farrell St., is Romanesque, will seat 4,000 persons, and has a magnificent altar of marble and onyx, imported from Munich, with the finest stained-glass windows on the coast. The finest interior is that of old * **St. Patrick's** (Roman Catholic), in Mission St. near 3d. *St. Mary's Cathedral* (Roman Catholic), cor. California and Dupont Sts., is a noble building in the Gothic style, with a spire 200 ft. high. The church of *Nuestra Señora de Guadalupe*, in Broadway near Mason St., is maintained by Spanish and Portuguese residents, and the services are in those languages. * *Grace Church* (Episcopal), cor. California and Stockton Sts., is a stone building with stained-glass windows. Other Episcopal churches are *St. John's*, cor. Valencia and 13th Sts. ; *Church of the Advent*, in 11th St. near Market ; and *Trinity*, cor. Bush and Gough Sts. The temple *Beth Israel*, in Geary St. near Octavia, is a handsome structure, and is attended by wealthy Hebrews. The *Calvary Presbyterian*, cor. Geary and Powell Sts., is a large and costly edifice, in the Composite style. The *First Methodist*, in Powell St., between Washington and Jackson, was founded in 1849, and is the oldest of the denomination in the city. The *First Baptist* is in Eddy St., between Jones and Leavenworth Sts. ; the *Columbia Square Baptist*, in Russ St., between Howard and Folsom ; and the * *First Congregational*, cor. Post and Mason Sts. The Jewish Synagogue of * **Emanu-El,** in Sutter St., between Stockton and Powell, is a large structure in the Byzantine style of architecture, with two lofty towers, and richly-decorated interior. That of the *Sherith-Israel*, cor. of Post and Taylor Sts., is an imposing structure ; the lofty ceiling, arched and frescoed in imitation of the sky at night, is much admired. The *Chinese Mission House*, cor. of Stockton and Sacramento Sts., and the *Joss-Houses*, at 751 Clay St., 230 Montgomery Ave., and 512 Pine St., will prove interesting to strangers.

The most important educational institutions near San Francisco are the **University of California,** at Berkeley (see p. 400), and the **Leland Stanford, Jr., University,** at Palo Alto. In the city are an excellent *School of Design* and *Art Association* combined, two Medical Colleges, and several Academies. Among the charitable institutions the principal are the *United States Marine Hospital*, in extensive and handsome new buildings, on the Presidio Reservation, W. of the city; the *New City Hospital*, in the S. part of the city; *St. Mary's Hospital* (Roman Catholic), cor. of Bryant and 1st Sts.; the *State Woman's Hospital*, cor. 12th and Howard Sts.; the *Children's Hospital*, in California St.; the *Almshouse*, on the San Miguel Road, in the suburbs; the Protestant *Orphan Asylum*, in Laguna St., near Haight; and the *Roman Catholic Orphan Asylum*, in S. San Francisco. The *Alameda Park Asylum for the Insane* is situated on the Encinal, Alameda. The fine building of the *State Asylum for the Deaf, Dumb, and Blind*, near Oakland, was burned in 1875, and now rebuilt.

The * **Golden Gate Park,** W. of the city, comprises 1,043 acres, about one half of which is beautifully laid out in walks, drives, lawns, etc. One of the features of the Park is a magnificent conservatory, in which, at the proper season, the only specimen of the Victoria Regia lily in America can be seen; the building is modeled after the Royal Conservatories at Kew, England, and stands facing the main drive. To the left of the Pavilion is laid out an extensive promenade, in the midst of which the Garfield Monument stands. The music-stand is on the S. side of the Park, and facing it is the fine statue, by W. W. Story, of Francis Scott Key, author of "The Star-Spangled Banner," donated by James Lick. Near by is the Children's Play-House and Grounds, the gift of Senator William Sharon. The Park is reached by several lines of cable-cars. *Portsmouth Square*, commonly called the *Plaza* (W. side of Kearny St., from Washington to Clay Sts.), is inclosed with a handsome iron railing, is tastefully improved with gravel-walks, trees, shrubs, and grass-plots, and has a fountain in the center. *Washington Square, Union Square*, and *Columbia Square*, have also been neatly laid out and planted with trees and shrubbery. * **Laurel Hill Cemetery** is in many respects unsurpassed. It lies 2¼ miles W. of the principal hotels (reached by California St., Sutter St., Geary St., and ferries and Cliff House cable-cars). In the vicinity of the cemetery is Lone Mountain, of conical shape, which rises up singly and alone to a considerable height above the surrounding country, which is tolerably level. On its summit is a large wooden cross; and both mountain and cross are very conspicuous, and may be seen from almost any part of the city. There are several fine monuments in the cemetery, that of Senator David C. Broderick and William C. Ralston (modeled after the Pantheon at Rome) being especially noteworthy; but the great feature is Lone Mountain, with its unrivaled outlook, embracing views of the city, bay, ocean, Mount Diablo, and the Coast Range. There are several other cemeteries, among which are the *Calvary* (Roman Catholic), which contains many costly tombs, among which may be mentioned those of O'Brien, Flood (the "Bonanza Kings"), and Donahue, the *Masonic*, and the *Odd-Fellows*.

There are about 22,000 Chinese in San Francisco, and the "Chinese Quarter" comprises portions of Sacramento, Commercial, Dupont, Pacific, and Jackson Sts. Here they hold undisputed possession of several blocks, and the houses are crammed from sub-cellar to attic. No stranger in San Francisco, who has leisure, should fail to visit one of the two *Chinese Theatres*. He will find the entire audience, not excepting the women, who have a gallery to themselves, smoking energetically, and the performance is carried on amid the clashing of cymbals, the beating of drums and gongs, the blowing of trumpets, and other kinds of noise. A visit to the *Gambling-houses* and *Opium-cellars* will repay the curious tourist; but it had better be made in company with a guide. The Chinese are probably the most inveterate gamblers in the world, and they nearly all gamble. In a cellar, greasy and dirty and filled with smoke, eighty or a hundred will be found sitting around tables, betting. Their mode of gambling is simple: some one throws a handful of copper coins on the table, and, after putting up stakes, they bet whether the number of coins is odd or even; then they count them and declare the result. The opium-cellars are fitted up with benches or shelves, on each of which will be found a couple of Chinamen lying on the boards with a wooden box for a pillow. They smoke in pairs; while one smokes and prepares the opium, the other is dozing in a half-drunken sleep. There are three *Temples*, and at all times the visitor will find them open and joss-sticks smoking in front of the favorite gods. The *Chinese Merchants' Exchange* and their famous *restaurant*, in Grant Ave. near Clay St., should be visited.

One of the chief points of interest in the vicinity is the famous * **Cliff House**, on Point Lobos, or the South Head of the Golden Gate, on the edge of cliffs rising from the ocean and facing west; while behind it rise *Sutro Heights*, the gardens of Adolph Sutro, to which the public are admitted free. The conservatory is well worth a visit, while the grounds include the largest and choicest collection of flowering plants in the State. Mr. Sutro has established in the city a free consulting library; his collection now numbers about 110,000 volumes, including many rare works not to be found in any other American library. The *Cliff House* is 6 miles from the ferries, and is reached by a drive through Golden Gate Park, or through Geary St. and Point Lobos Ave., or by Haight St. cablecars to Park and thence by Park and Ocean steam-cars, or by the ferries and Cliff House cable-road, and steam-cars. The restaurant attached to the house is famous for its excellence. *Seal Rock* is close by the hotel, and the greatest charm of the place is to lounge on the wide, shady piazza and watch the seals (more properly sea-lions) basking in the sun or wriggling over the rocks, barking so noisily as to be heard above the roar of the breakers. Northward lies the *Golden Gate*, the beautiful entrance to San Francisco Bay. Southward is the beach beyond, a rocky shore whose outlines melt in the blue distance. In front is the ocean, on whose horizon on a clear day the peaks of the *Farallone Islands* are visible. The road passes beyond the hotel to a beach several miles long, over which at low tide one can drive to its extreme end, and return to the city by a road behind the Mission Hills. Another popular drive is through Golden Gate Park to the beach near the Cliff House. *Presidio Reservation*, the property of the Federal Government, lies on the narrowest portion of the Golden Gate, and has miles of beautiful drives. It is reached by California St., Jackson St., and Union St. cars. At *Hunter's Point*, 4½ miles S. E. of the City Hall, is a Dry Dock, cut out of the solid rock, and said to be one of the finest in the world. The drive to it is across an arm of the bay, and affords varied and pleasant views. The * **Mission Dolores,** the old mission of San Francisco, lies in the S. W. part of the city (reached by Valencia St. cars). It is an adobe building of the old Spanish style, erected in 1778, and a modern church has been built beside it. Adjoining it is the cemetery, with its well-worn paths and fantastic monuments. *Alameda, Saucelito, Berkeley*, and *Oakland*, across the Bay (reached by ferry), are beautiful towns with fine public gardens.

The ship-yard and machine-shops of the Union Iron-Works at the *Potrero* are said to be as complete as any in the world. Besides the cruiser Charleston and the battle-ship Oregon, other vessels of the new navy were built there for the U. S. Government.

Itineraries.

The following series of excursions has been prepared so as to enable the visitor whose time is limited to see as much of the city as possible in the least amount of time. Each excursion is planned to occupy a single day, but the visitor can readily spend more time as special features crowd upon his attention.

1. Take California St. or ferries and Cliff House cable-cars,[1] both of which lines pass through the finest residence-quarters of the city to Central Ave., where transfer is made to a rapid water-line to the Cliff House. First visit the beautiful park surrounding the residence of Adolph Sutro, called Sutro Heights, from which may be seen a long stretch of coast-line and the far-famed Golden Gate. Descending by an easy slope to the Cliff House, visitors may enjoy from a shaded veranda a closer view of the Seal Rocks, just below. To the right of the Cliff House are the public baths, being established by Adolph Sutro at a cost of $300,000. Water is drawn direct from the ocean into great swimming-tanks, which are to be inclosed by a glass roof. Returning, take water-line from the beach just below the Cliff House and visit the conservatory, containing tropical and other plants in profusion. Close by is the Children's Playground and the Music Area, where,

[1] A system of transfers between various cable and horse-car lines, perfected in 1892, is probably the most extensive in the world, and enables passengers to reach almost any part of the city for one fare. Cards showing all transfer points, and fully explaining the intricate system, may be obtained at the leading hotel offices.

on Thursday, Saturday, and Sunday afternoons, concerts are given by a large band of performers selected by the Park Commissioners. Thence the city is reached by half a dozen cable-lines. The only cost attached to this trip is for car-fare. The rate to the Club House is 10c., and elsewhere, 5c.

2. In the forenoon visit the United States Branch Mint, in Fifth St., near Market, from 10 A. M. to 12 M. In addition to interesting machinery, etc., there is here a very large and valuable collection of antique coins and relics. In the afternoon tourists may profitably see Chinatown by daylight, when no guide is necessary. Along Dupont, Sacramento, Clay, and Washington Sts. are many stores devoted to Oriental goods and curios, as well as workshops, where Chinese jewelers and artisans are engaged. For a night visit to the Chinese quarter guides are necessary. They may be found at any of the leading hotels, and for a moderate fee will escort visitors to the Chinese theatres (admission, 50c.), to the joss-houses and gorgeously equipped restaurants, and into the living quarters and the resorts of the Chinese.

3. An excursion to the military posts on and about the bay is well worth the time. The Government steamer McDowell runs daily from the foot of Clay St. to Alcatraz and Angel Islands, to Fort Mason, and the Presidio. On Alcatraz Island is a large military prison, besides a lighthouse and extensive fortifications. Here, too, is the central firing station of an elaborate system of harbor defense, consisting of submarine torpedoes. Returning, visitors may leave the boat at the Presidio, a military station where the commander of the post has his headquarters. Here may be seen adobe buildings and ancient cannon, relics of the days when the Presidio was the military station of the Spanish occupants. Motor and cable cars afford an easy return to the city. There is no fare charged on the McDowell, and permits for the trip may be obtained at Army Headquarters, in the Phelan Building, at Market St. and Grant Ave.

4. Visit the City Hall at McAllister and Larkin Sts., where are located the municipal offices, the Superior Court in twelve departments, the Law Library, and the Free Public Library. Thence by the Valencia St. cable-cars to Sixteenth St., and walk two blocks to the old Mission Dolores, an adobe church built more than a century ago by the early Fathers. Thence take the San Francisco and San Mateo electric-cars at Guerrero St., and ride through a populous residence district, known as the Mission, through the suburbs across the county line. From this road on unequaled view of the southern portion of the city and its outskirts is afforded. Returning, reach Market St. by the Fourth St. horse-cars, and visit the excellent museum of the Academy of Sciences in Market St., near Fourth. Near by, in Fourth St., is the handsome building of the California Society of Pioneers, the rear portion of which is occupied by the Pioneers as a place of resort. In it are many interesting mementoes of the days of the gold-seekers. In the same building is the State Mining Bureau, a comprehensive museum, containing specimens of ores, minerals, and objects of kindred interest.

5. Take either broad or narrow gauge ferry at the foot of Market St., connecting on the opposite side of the bay with half-hourly trains to Alameda, Oakland, and Berkeley, where reside many city merchants and professional men.   Between these beautiful suburbs communication is effected by horse-cars and electric roads, as well as by the local trains.   At Berkeley is located the University of California, a State-aided institution, with more than 900 students on its registers.   It is surrounded by spacious, well-kept grounds, and has a large library and museum well worth inspecting.   The round-trip rate to any of the three points named is 25c., with an additional charge of 10c. for transportation on local trains between any two of the cities not on the direct route. Street-car fares are the same as in San Francisco.

6. Take Southern Pacific train at Third and Townsend Sts. and ride 32 miles south to Menlo Park, where carriages may be procured for a drive through a beautiful country, about which are the elegant country homes of wealthy San Franciscans, and to the Leland Stanford, Jr., University at Palo Alto.   Adjacent is Senator Stanford's country residence and his stud-farm, justly celebrated for its fine horses.   The fare for the round trip is $1.75.

7. Take Central Pacific train, connecting with broad-gauge ferry from foot of Market St., to Vallejo Junction, and by ferry to Vallejo and the Mare Island Navy-Yard.   A day may be pleasantly spent in examining the dry docks, extensive workshops, and features of interest of the Navy-Yard.   Cruisers home from foreign stations on the Pacific are anchored here for repairs and outfitting ; and here may be seen the old and historic vessel, the Independence, now used as a receiving-ship. The railroad fare each way is 90c.

From San Francisco the steamers of the Pacific Coast S. S. Co. leave regularly from Broadway, Wharf No. 2, stopping at *Santa Cruz, Monterey, San Simeon, Cayucos,* and *Port Harford.*   The latter, sometimes called Port San Luis Obispo, is the most important stop.   It is the port for the different sections of San Luis Obispo County, including the principal city of the county.   Avila is a small seaside place on the shores of San Luis Bay.   It is much frequented in summer on account of the excellent facilities it affords for sea-bathing.   Near here, and within easy reach of the railroad, are located the celebrated **Avila Hot Sulphur Springs,** where there is a comfortable hotel, neat bath-house, swimming-bath, beautiful shady nooks for camping, and a fine stream of fresh water, and on the whole it is one of the most charming and picturesque spots in the country.   The town of *Arroyo Grande* is situated on the bank of the creek of that name, about 12 miles south of San Luis Obispo.   It is in the midst of an exceedingly fertile district.   San Miguel is named after the old mission, situated in the town.   About 9 miles south of San Miguel is the famous **Paso Robles Springs,** around which is growing one of the prettiest towns of the county.   Continuing south for a distance of 10 miles, one comes to the present terminus of the Southern Pacific R. R., **Santa Marga-**

rita, already a bustling and lively village. The city of **San Luis Obispo** is situated near the center of the coast section of the country. It is built on the site of the old mission, which was called San Luis Obispo de Toloso. It is partially surrounded by hills of singular and diversified beauty; the commercial outlook for this city is good. With railroad connections both north and south, the port of the best harbor of the county, surrounded by a fertile and well-tilled agricultural region, the distributing point for many smaller towns, the center of a vast farming country, it is sure to become a flourishing city.

Stops are also made at *Gaviota, Santa Barbara, San Buenaventura, Hueneme, San Pedro,* and *San Diego.* Also, a steamer sails on the 1st of each month for Mexico, landing at *Enseñada, Magdalena Bay, San José del Cabo, Mazatlan, La Paz,* and *Guaymas.*

Going north, the steamers of the San Francisco and Mendocino route stop at *Point Arena, Cuffeys Cove, Mendocino City,* and *Fort Bragg.* Also, a steamer sails from Broadway, Wharf No. 1, San Francisco, every Wednesday at 9 A. M., for *Eureka, Arcata,* and *Field's Landing* (Humboldt Bay), reaching *Eureka* every Thursday at high tide. Returning, it leaves Eureka on Saturday at high tide and reaches San Francisco every Sunday in the forenoon. **Eureka,** the county-seat of Humboldt, is situated on the east side of Humboldt Bay, occupying a pleasant site overlooking the waters of the bay. The city is one of the most tastefully laid out on the coast. Many handsome business blocks and public buildings adorn the thoroughfares. The streets are rectangular and kept in good condition, while the facilities for numerous vessels along the wharves are quite adequate. Eureka is lighted with gas and electricity, and possesses many of the useful institutions of a metropolis.

The Pacific Coast Steamship Co. also runs, in connection with the Union Pacific system, regular steamers to Portland, Ore., which stop at Astoria, and then continue north to Puget Sound, with Tacoma as a terminus. Also, there is a direct line leaving San Francisco every five days for Puget Sound, connecting at Tacoma with the Northern Pacific R. R., and at Vancouver with the Canadian Pacific Railway.

92. The Yosemite Valley.

THERE are two good routes to the valley: 1. By the Southern Pacific R. R. (Visalia Div.) to *Lathrop* (Route 90); then to Berenda; thence by branch-road to Raymond (199 miles from San Francisco); and thence by Yosemite Stage & Turnpike Co., a stage route of 35 miles, to *Wawona Hotel.* Passengers leaving San Francisco at 4 P. M., reach Wawona next day at 6 P. M. The Big Trees are visited on the return trip, and lunch is taken in the afternoon at Wawona, near the Grove. The * **Mariposa Grove of Big Trees** is only 6 miles from Wawona Hotel. The Mariposá Grove is part of a grant made by Congress to be set apart for "public use, resort, and recreation" forever. The area covered by the grant is two miles square, and embraces two distinct groves which are about ½ mile apart. The Upper Grove contains 365 trees, of which 154 are over 15 ft. in diameter, and

many over 300 ft. in height.   The average height of the Mariposa trees
is less than that of the Calaveras (see Route 90), the highest of the
former (272 ft.) being 53 ft. less than the tallest of the latter; but
their average size is greater.   The largest tree in the grove is the
Grizzly Giant (Lower Grove), which is still 94 ft. in circumference
and 31 in diameter, though much decreased in size by burning.   The
first branch is nearly 200 ft. from the ground, and is 6 ft. in diameter.   The remains of a prostrate tree, now nearly consumed by fire,
indicate that it must have reached a diameter of about 40 ft. and a
height of 400.   The trunk is hollow, and will admit of the passage of
three horsemen riding abreast.   There are about 125 trees over 40 ft.
in circumference.   The *Fresno Grove* is also directly on the line of
this route, and contains: over 800 trees spread over an area $2\frac{1}{2}$ miles
long and 1 to 2 broad.   The largest is 95 ft. in circumference at 3 ft.
from the ground.

2. The other route is from Stockton on the S. P. R. R. (see. Route
90) *via* Stockton & Copperopolis Branch to *Milton* (133 miles from San
Francisco), and thence by stage.   There are 147 miles of staging on this
route, and it is only taken by those who wish to visit the *Calaveras
Grove of Big Trees* (see Route 90).

The Yosemite Valley was known to many miners in the summer of
1849, who explored the head-waters of all the rivers in search of gold.
In 1851 Captain Bowling entered the valley, and took the Indians to
the Fresno River, where a treaty was made.

Yosemite Valley.

Hotels, etc.—There are two hotels in the valley—the *Stoneman House* and
Yosemite Falls.   The sleeping accommodations are good, and the tables fairly
provided, considering the distance of the locality from the ordinary markets.
Guides, including their horses, will usually cost $4 a day.

The Yosemite Valley is situated on the Merced River, in the S. portion of the county of Mariposa, California, 140 miles a little S. of E.
from San Francisco, but over 260 miles from that city by any of the
usually-traveled routes.   It is on the western slope of the Sierra Nevada,
midway between its E. and W. base, and nearly in the center of the
State, measuring N. and S.   The valley is a nearly level area, 4,000 ft.
above the sea, about 9 miles in length, and from $\frac{3}{4}$ to $1\frac{1}{4}$ mile in
width, and almost a mile in depth below the level of the adjacent
region, and inclosed in granite walls rising with almost unbroken and
perpendicular faces to the height of from 3,000 to 6,000 ft.   From
the brow of the precipices in several places spring streams of water
which, in seasons of rains and melted snow, form cataracts of a beauty
and magnificence surpassing anything known in mountain scenery.
"The principal features of the Yosemite," says Prof. Josiah D. Whitney, in his excellent "Yosemite Guide-Book," "and those by which it is
distinguished from all other known valleys, are: 1. The near approach
to verticality of its walls; 2. Their great height, not only absolutely, but
as compared with the width of the valley itself; and, 3. The very small
amount of *débris* at the base of these gigantic cliffs.   These are the

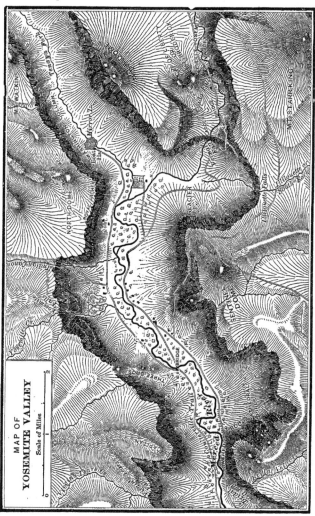

MAP OF
YOSEMITE VALLEY

Scale of Miles

great characteristics of the Yosemite region, throughout its whole length ; but, besides these, there are many other striking peculiarities and features, both of sublimity and beauty, which can hardly be surpassed, if equaled, by those of any mountain valleys in the world. Either the domes or the waterfalls of the Yosemite, or any single one of them even, would be sufficient, in any European country, to attract travelers from far and wide in all directions. Waterfalls in the vicinity of the Yosemite, surpassing in beauty many of the best known and most visited in Europe, are actually left entirely unnoticed by travelers, because there are so many other objects of interest to be visited that it is impossible to find time for them all." The valley is almost one vast flower-garden. Plants, shrubs, and flowers of every hue cover the ground like a carpet ; the eye is dazzled by the brilliancy of the color, and the air is heavy with the fragrance of a million blossoms. Trees of several centuries' growth raise their tall heads heavenward, yet, beside and in comparison with the vast perpendicular clefts of rocks, they look like daisies beside a tall pine. On every side are seen the beautiful and many-colored manzanita and madrone, and trees of such shape and variety as are never seen in the Atlantic States. The Yosemite was discovered in the spring of 1851 by a party under the command of Captain Boling, in pursuit of a band of predatory Indians, who made it their stronghold, considering it inaccessible to the whites. By an act of Congress passed in 1864, the Yosemite Valley and the Mariposa Grove of Big Trees were granted to the State of California upon the express condition that they shall be kept " for public use, resort, and recreation," and shall be "inalienable for all time." The Indian residents of the valley had a name for each of the prominent cliffs and waterfalls, but these are difficult of pronunciation, and have all been discarded except the name of the valley itself (which means " Large Grizzly Bear ").

The most striking feature of the valley scenery is *** El Capitan.** Although not so high by several thousand feet as some of its giant neighbors, yet its isolation, its breadth, its perpendicular sides, its prominence as it projects like a great rock promontory into the valley, make it, as its name indicates, the Great Chief of the Valley. It is 3,300 ft. high, and the sides or walls of the mass are bare, smooth, and entirely destitute of vegetation. " It is doubtful," says Prof. Whitney, " if anywhere in the world there is presented so squarely-cut, so lofty, and so imposing a face of rock." On the opposite side, on the right of the view, is *** Bridal-Veil Fall,** where the creek of the same name leaps over a cliff 900 ft. high into the valley below. The water, long ere it reaches its rocky bed, is converted into mist, and descends in a white sheet of spray. The *Virgin's Tears Creek*, on the other side of the valley, directly opposite the Bridal-Veil, makes a fine fall over 1,000 ft. high, inclosed in a deep recess of the rock near the lower corner of El Capitan. This is a beautiful fall while it lasts, but the stream which produces it dries up early in the season. On the same side as the Bridal-Veil, and a little above it, is *Cathedral Rock*, a massively sculptured pile of granite, 2,660 ft. high, with nearly vertical sides, bare of vegetation. Just beyond are *The Spires*, two graceful columns of gran-

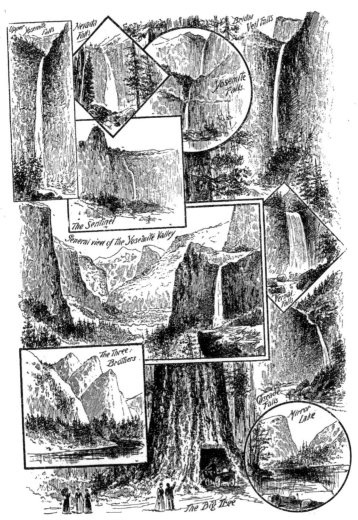

Yosemite Views.

ite standing out from, but connected at the base with, the walls of the valley. From one point of view these spires appear symmetrical and of equal height, and rise above the edge of the cliff exactly like the towers of a Gothic cathedral. Farther up the valley, on the opposite side, is the triple group of rocks known as the *Three Brothers.* The peculiar outline of these rocks, as seen from below, resembling three frogs sitting with their heads turned in one direction, is supposed to have suggested the Indian name Pompompasus, which means "Leaping-Frog Rocks." The highest of the peaks is 3,830 ft. high, and from its summit there is a superb view of the valley and its surroundings. Nearly opposite the Three Brothers is a point of rocks projecting into the valley, the termination of which is a slender obelisk of granite, which, from its peculiar position, or from its resemblance to a gigantic watch-tower, is called * **Sentinel Rock** (3,043 ft. high). This is one of the grandest masses of rock in the Yosemite. Directly across the valley are the * **Yosemite Falls,** which are justly regarded as the most wonderful feature of the Yosemite scenery. The fall has a total height of 2,600 ft., which, however, is not all perpendicular. There is first a vertical leap of 1,500 ft., then a series of cascades down a descent equal to 626 ft. perpendicular, and then a final plunge of 400 ft. to the rocks at the base of the precipice. The rumble and roar of the falls are heard at all times, but in the quiet of the evening they are so great that it seems as if the very earth were shaking. No falls in the known world can be compared with these in height and romantic grandeur. The renowned Staubbach of Switzerland is greatly inferior, both in height and volume. The best time to see the falls is in May, June, and July, as by August or September both the Yosemite and Bridal-Veil have shrunk almost to nothing, but they can be visited up to December. The cliff a little to the east of the Yosemite Fall rises in a bold peak to the height of 3,030 ft. above the valley, and affords a magnificent view of the entire region. Its summit is easily reached by a trail leading up Indian Cañon.

About 2 miles above the Yosemite Falls the main valley ends, and branches out in three distinct but much narrower cañons. Through the middle one of these the Merced River comes down; in the left-hand or N. W. one the Tenaya Fork of the Merced flows in; and in the right-hand or S. W. one, the South Fork or Illilouette. At the angle where the Yosemite branches is the rounded columnar mass called *Washington Column,* and immediately to the left of it the immense arched cavity known as the *Royal Arches.* Above these the symmetrical form of the *North Dome* looms up to the height of 3,568 ft. The * **Half Dome,** on the opposite side of the Tenaya Cañon, is the loftiest and most imposing mountain of those considered as part of the Yosemite. It is a crest of granite, rising to the height of 4,737 ft. above the valley, and was long considered perfectly inaccessible, but in 1879 certain improvements were made by which tourists were enabled and will in future be enabled to reach this commanding point. Lying in perfect quiet and seclusion at the foot of and between the North and Half Domes is the exquisite little * **Mirror Lake,** an expansion of the Tenaya Fork. It is frequently visited (and best early in the morning) for the purpose of getting the reflection upon its mirror-like surface of an overhanging mass of rock to which the name of *Mt. Watkins* has been given. In the middle cañon the Merced River comes down from the plateau above in a series of noble cascades and two grand cataracts, which are among the chief attractions of the Yosemite. The first fall reached in ascending the cañon is the **Vernal Fall,** which has a vertical height of about 400 ft. The ledge over which the fall descends is surmounted by a steep but not difficult path, and the view down the cañon from the summit is extremely fine. "From the Vernal Fall up-stream," to quote Prof. Whitney again, "for the distance of about a mile, the river may be followed, and it presents a succession of cascades and rapids of great beauty. As we approach the Nevada Fall, the last great one of the Merced, we have at every step something new and impressive. On the left hand, or N. side of the river, is the **Cap of Liberty,** a stupendous mass of rock, isolated and nearly perpendicular on all sides, rising perhaps 2,000 ft. above its base, and little inferior to the Half Dome in grandeur. It has been frequently climbed, and without difficulty, although appearing so inaccessible from the cañon of the Merced. The * **Nevada Fall** is in every respect one of the grandest waterfalls in the world, whether we consider its vertical height, the purity and volume of the river which forms it, or the stupendous scenery by which it is environed. The fall is not quite perpendicular, as there is near the summit a ledge of rock which receives a portion of the water and throws it off with a peculiar twist, adding considerably to the general picturesque effect." The height of the fall is about 600 ft. In the cañon of the South Fork, or Illilouette, there is a fine fall estimated at 600 ft. high. It is visible from a point on the trail from the hotel to Mirror Lake, but is seldom visited by travelers, as the cañon is rough and difficult to climb.

Several small encampments of Digger Indians are generally to be found in the valley; and, if not delighted, the visitor will certainly be amused by the primitive mode of living of these "children of Nature." Professor Whitney warmly recommends tourists visiting the Yosemite to make an excursion round the valley on the outside. Such an excursion can be made mostly on beaten trails without the slightest difficulty or danger, will occupy but a few days, and will afford as grand panoramic views of mountain and valley as can be found in Switzerland itself. Those who can not make this tour should at least make excursions to *Inspiration Point*, on the Mariposa trail, and to *Glacier Point* (*Glacier Point Hotel;* 3½ miles), on the McCauley trail. The view from either is indescribably grand.

93. California Resorts.

To the North, San Francisco to Cazadero.

FROM the foot of Market Street the steamers of the North Pacific Coast R. R. Co. carry their passengers along the city front and within a short distance of the Military Reservations (Black Point and the Presidio); thence toward Alcatraz and Angel Islands, with a view of the Golden Gate in the distance to **Saucelito,** a popular bay-side resort .

famous for boating, bathing, and fishing, with pleasant country hotels and attractive residences. It is thirty minutes' ride on the steamers. Thence the train may be taken for **Mill Valley,** which is 5 miles farther. This beautiful valley and its surroundings have been tastefully laid out into avenues and streets by experienced engineers. Another train may be taken for **Larkspur,** which has one of the best-appointed hotels in the county, thence to **San Anselmo** and **Sunnyside.** At the latter place the Presbyterians have built a brown-stone seminary. Still another train may be taken for **San Rafael,** a remarkably pretty town near the W. shore of San Pablo Bay, built on the site of the old Jesuit mission of San Rafael. It is sheltered on the N. and W. by mountains, and is something of a *sanitarium* for those who find the ocean-winds and fogs that prevail at San Francisco too trying. The scenery in the vicinity of San Rafael is extremely picturesque, and there are many charming drives, but the chief attraction is the ascent of *****Mt. Tamalpais** (12 miles distant). The W. summit of the mountain is 2,606 ft. high, and the view from it embraces the cities of San Francisco and Oakland, numerous towns and villages, the bay and the Golden Gate, and the illimitable ocean beyond. Returning to San Anselmo, or the Junction, as it is familiarly called, the traveler can reach **Fairfax,** 1½ miles north, which is famed for its beauties and attractions, and was long the recreation-grounds for Sunday-schools when out for their annual excursions. Thence climbing over White's Hill you reach San Geronimo Valley, and wind along the Lagunitas Creek until you arrive at **Camp Taylor,** where there is a good hotel, surrounded with little cottages and tents, affording every comfort to those who wish to pass a few months in the country with their families, as it is only two hours' ride from the city, and trains leave three times a day. Thence 3 miles farther is **Tocaloma,** the entrance to Bear Valley. The drive through this valley is considered very fine, as immediately after leaving the station the traveler is carried for miles through a perfect labyrinth of wild flowers and evergreens, and upon emerging from this enchanted bower a grand view of the Pacific Ocean bursts upon him and is momentarily enjoyed, until attention is drawn to the beach, strewn with all kinds of shells thrown there by the waves. A good hotel has just been finished here, with all the modern improvements. Thence 5 miles farther is **Point Reyes,** 38 miles from San Francisco. From this station the members of the Country Club drive to their private "shooting-boxes," or the Club House, which latter is a building 150 feet long, and fitted up with all modern ideas of the requirements of true sportsmen. The club has leased 76,000 acres of land, and has determined to make this the largest and best private preserve in the world. They are filling the various lakes with black bass, speckled trout, etc., and the woods with Japanese pheasants, Oregon geese, and European song-birds, so that before long this will be one of the features of the State. Thence 4 miles farther is **Millerton,** the point from which the Presbyterians intend to build a bridge of about 4,000 feet across Tomales Bay to **Inverness,** the site of their new college. This is a lovely, sheltered place, but sufficiently

27

elevated to command a fine view of the beach running along the shores of Tomales Bay. For nearly 2 miles the waters of this bay on that side are quite warm, and form a charming and safe bathing-place. Thence for nearly 20 miles the train carries the traveler through grain and potato fields to **Howards,** 70 miles from San Francisco and 600 feet above the level of the sea. Here one first scents the aroma emitted by the "Sequoia" or redwood-trees, and soon feels the increased ease with which he can breathe, and is anxious for the train to proceed as quickly as possible and carry him through the remaining 17 miles of forest that he has to travel before he can reach **Cazadero,** which is situated on the banks of Austin Creek and a few miles by land from the Gualala River, in both waters of which splendid fishing is to be had; it is also surrounded by the giant redwoods, in whose forests all kinds of game (including an occasional bear) are to be found. This is the paradise of the camper-out, the sportsman, the tired business man, or the exhausted invalid; as pleasure and sport can be found for the one, or rest and quiet for the other, and good food and accommodation in the hotel (*Cazadero*) and its cottages for all.

To the South, San Francisco to San Luis Obispo.

Another favorite excursion is to **Pescadero** (*Swanton House*), which is reached by stage from San Mateo or Redwood City on the Southern Pacific R. R. The stage-ride of 30 miles over the Contra Costa Range affords some noble views. Pescadero is a thriving town, beautifully situated in a remarkably productive valley, on both sides of Pescadero Creek, near its confluence with the Butano, about a mile from the sea-shore. The San Francisco Water Company takes its supply from the head of the creek. Near the town is the famous *Pebble Beach*, where agates, opals, jaspers, and carnelians, of almost every conceivable variety of color, are found in great abundance, with a natural polish imparted by the action of the waves.—Tri-weekly stages run along the coast from Pescadero to **Santa Cruz** (*Pacific Ocean House, Pope House, Sea Beach Hotel*), one of the principal watering-places of California. (Santa Cruz is also reached by narrow-gauge railroad from San Francisco, or *via* Southern Pacific R. R. to *Pajaro*, and thence by the South Pacific Coast R. R.) The *Mission de la Santa Cruz* is situated here. Santa Cruz is attractively situated on the N. side of Monterey Bay, and near by are *Aptos* and *Soquel*, popular seaside resorts. Bathing, fishing, and hunting may be enjoyed here, and in the vicinity there are charming drives. A few miles beyond Santa Cruz is a grove of mammoth trees which go far to rival those of the Mariposa Grove.—Opposite Santa Cruz, at the S. extremity of the bay, is the historic city of **Monterey** (reached from San Francisco by steamer, or by Southern Pacific R. R.). Until 1847 this town was the seat of government and principal port on the California coast; but since the rise of San Francisco its commerce and business have dwindled away, and it is now one of the quietest places in the State. It has attracted much attention as a health-resort. Its climate is warm in winter, cool in summer, and dry all the year round. The Southern Pacific R. R. Co.

have pushed improvements with the design of making Monterey a great health and pleasure resort, and have built the *Hotel del Monte* in a grove of oaks and pine-trees. Among the points of interest are the *Carmel Mission*, established by Father Junipero Serra, and the celebrated 18-mile drive around the shore to *Cypress Point.* A monument to Father Junipero Serra, in granite, has been erected on a hill near where the padre first landed. They have also built a narrow-gauge road (3 miles) to *Pacific Grove*, a pleasant resort under the pines on the shore of the bay. It has several hundred cottages, a fine hotel, *El Carmelo*, and is the place of the annual summer meetings of the Chautauquan Assemblies and religious and literary associations.

Still another favorite excursion from San Francisco is to Calistoga and the Geysers, by way of the Western Div. of the California Pacific R. R. **Napa City** (46 miles from San Francisco) is a thrifty place of 4,387 inhabitants, surrounded by a highly productive agricultural region, rich in fruits of all kinds, and in vast fields of grain that stretch away in every direction. There are many beautiful drives in the vicinity, one of the most attractive of which is that to Santa Rosa, taking in the famous wine-cellars of Sonoma. The highly esteemed *Napa Soda Springs* are situated in the foot-hills about 6 miles N. E. of the town. **Calistoga** (*American Hotel, Magnolia Hotel*), situated 73 miles from San Francisco by the California Pacific R. R., is a pretty town, lying in a valley a mile in width, and encircled by forest-clad hills and mountains. It is supplied with pure water from a reservoir on the adjacent mountain-side, and there are several bath-houses, supplied with water from neighboring springs. The public warm swimming-bath, 40 ft. square, is one of the features of the place. The scenery is exceedingly picturesque, the well-cultivated fields, green lawns, sunny slopes, and shaded villas contrasting pleasantly with the wild grandeur of the rugged mountains. There are numerous mineral springs in the vicinity, the most noted of which are *Harbin's* (20 miles N. of Calistoga), and the *White Sulphur Springs*, situated in a deep and picturesque gorge of the mountains, which rise on either side to a height of about 1,000 ft. About 5 miles S. E. of Calistoga is the * **Petrified Forest,** which is justly regarded as one of the great natural wonders of California. Portions of nearly 100 distinct trees, of great size, scattered over a tract 3 or 4 miles in extent, have been found, the largest being 11 ft. in diameter at the base and 60 ft. long. They are supposed to have been silicified by an eruption of the neighboring Mount St. Helena, which discharged hot alkaline waters containing silica in solution. Daily stages run from Calistoga to the famous * **Geyser Springs,** which are situated in Sonoma County, in a lateral gorge of the Napa Valley, called the "Devil's Cañon," near the Pluton River. The approaches to the springs are very impressive, the scenery being finer, according to Bayard Taylor, than anything in the Lower Alps. The narrow Geyser ravine, which is always filled with vapor, is shut in by steep hills, the sides of which, marked with evidences of volcanic action, are smoking with heat and bare of vegetation. A multitude of springs gush out at the base of these rocks. Hot and cold springs, boiling springs, and quiet springs lie within a few feet of each other. They differ also in color, smell, and taste. Some are clear and transparent, others white, yellow, or red with ochre, while still others are of an inky blackness. Some are sulphurous and fetid in odor, and some are charged with alum and salt. The surface of the ground about the springs, which is too hot to walk upon with thin shoes, is covered with the minerals deposited by the waters, among which are sulphur, magnesium sulphate, aluminum sulphate, and various iron-salts. A properly directed course of these waters is said to afford an almost certain cure for rheumatism, gout, and skin-diseases; but persons suffering from throat or pulmonary affections should not reside in the neighborhood.

The Geysers may also be reached from San Francisco *via* steamer to *Tiburon* (6 miles), or to *Donahue* (84 miles), and thence by the San Francisco & Northern Pacific R. R. to *Cloverdale* on this road (84 miles from San Francisco), where stages run in 16 miles to the Geysers over an excellent road; or by the Southern Pacific R. R. to *Calistoga*, and thence by stages, a distance of 26 miles. A good plan for the tourist is to go by one route and return by the other.

One of the excursions most frequently recommended to the stranger in San Francisco is that to San José and the Santa Clara Valley (*via* Southern Pacific R. R. and South Pacific Coast R. R.). The *Santa Clara Valley* lies between the Coast and the Santa Cruz Mountains, and is about 100 miles in length; it is watered by the Coyote and Guadalupe Rivers and by artesian wells, and claims to be the most fertile in the world. Vineyards covering hundreds of acres, vast wheat-fields one and two miles in length, stately trees, forests of live-oak, and finely cultivated farms, are to be seen on every hand; and the vegetation is of tropical luxuriance and beauty. In the heart of the valley, 40 miles S. E. of San Francisco and 8 miles from the head of San Francisco Bay, is the city of **San José** (*Hotel Vendome, Lick House, St. James*), with a population of 18,060. The main portion of the city occupies a plateau between the Coyote and Guadalupe Rivers, here 1¼ mile apart, with suburbs extending beyond them. The principal public buildings are the *Court-House, a massive Corinthian structure, costing $200,000, with a dome commanding a fine view; the *Jail*, adjoining it, the finest in the State, costing $80,000; the *City Hall;* two *markets*, costing more than $40,000 each; 8 public-school buildings; and 19 churches, of which the largest and most expensive is an edifice belonging to the Roman Catholics. The city is noted for its educational institutions. Besides the public schools, there are the *College of Notre Dame* (Roman Catholic), a day and boarding school for girls; the *San José Institute*, a day and boarding school for both sexes; the *University of the Pacific* (Methodist), connected with which is a young ladies' seminary; and the *State Normal School*, whose building, erected at a cost of $275,000, is the finest of the kind on the Pacific coast. The library of the *San José Library Association* contains 8,000 volumes. There is an *Opera-House*, seating 1,200 persons, and an elegant and commodious *Music-Hall*. The city has three public parks, containing 2, 8, and 30 acres respectively, and owns a tract of 400 acres in Penitencia Cañon, 7 miles E., reserved for a public park, containing a wild, rocky gorge with a mountain-stream

·and a variety of mineral springs. The surrounding country yields grain and fruits abundantly, and in the vicinity are some of the finest vineyards in California. There are many fine drives in the neighborhood, notably one to the *Lick Observatory*, with the largest telescope in the world, on Mt. Hamilton, 26 miles distant. This mountain is 4,443 ft. high, and affords a magnificent view of the Santa Clara Valley. The famous *New Almaden Quicksilver Mines* are about 14 miles from San José, and may be reached by a pleasant two-hours' ride in a stage-coach. They are well worth a visit. Three miles W. of San José is the picturesque village of **Santa Clara,** with a population of 2,887. Electric cars connect the two, running along the **Alameda*, a beautiful avenue bordered by fine residences, and rows of superb trees planted by the Jesuit fathers in 1777. Santa Clara contains several fine churches, and is, the site of the Santa Clara College (Jesuit), which occupies a number of handsome buildings in an inclosure of about 12 acres. Included in this institution is the Old Mission, founded by the Spanish missionaries in early times, and the orchards planted by them may still be seen. Stages run from the depot at Santa Clara to the **Pacific Congress Springs** (10 miles S. W.). These waters contain sodium carbonate and sulphate, sodium chloride, lime, iron, aluminum silicate, and magnesia, and are recommended for rheumatism.

No tourist who has the leisure should fail to stop at *Menlo Park*, on the broad-gauge Southern Pacific R. R., nearly midway between San Francisco and San José. It is the seat of the handsomest suburban villas on the coast, and boasts many beautiful drives. Four miles distant is Senator Leland Stanford's great Palo Alto ranch, the site of the *Leland Stanford, Jr., University*, founded in 1885, several buildings of which are now occupied and courses of instruction in progress. On this ranch are also his stables of famous thoroughbreds.

Paso de Robles hot and cold sulphur springs are reached from San Francisco by Northern Div. of Southern Pacific Co. (215 miles), and lie in the beautiful valley of the Salinas River. The baths are taken at the natural temperature, and are considered efficacious for rheumatism, gout, and skin-diseases. The climate is good and salubrious, and the accommodations for visitors excellent. Other health-resorts on the line of the Northern Div. are *Paraiso Springs*, reached by rail to Soledad (143 miles), thence by stage route (7 miles), and the *Gilroy Hot Mineral Springs*, reached by rail to Gilroy (80 miles), and thence by stage (14 miles).

San Luis Obispo (reached by the Southern Pacific R. R. to Santa Margarita and then 10 miles by stage, or by steamer to Port Hartford and thence by railroad) is a city of 2,982 inhabitants, built on the site of the old mission, in a beautiful country. The *Hotel Ramona* is one of the largest on the coast, and commands a fine view.

On the way to Southern California by the S. P. R. R. the tourist will not regret a day spent at *Fresno* (*Hughes House*), the chief city of the San Joaquin Valley and the center of the raisin industry. For 20 miles around the city one may drive through vineyards of raisin-grapes over roads as level and smooth as a barn-floor. Fresno's wealth is due to the colony system, which has resulted in making it the market for nine tenths of the raisins produced in this State. In 1890 over 15,000,000 pounds of raisins were shipped from this place. Its population in 1890 was 10,818.

Southern California.

Among the health-resorts of Southern California,[1] one of the most popular is **Santa Barbara** (*Arlington Hotel, San Marcus*), lying in a sheltered nook of the shore of the Pacific, 275 miles S. S. E. of San Francisco (from which it is reached by steamer, and also *via* Southern Pacific R. R. to *Saugus* (449 miles), and thence by branch railroad). It is completely protected on the N. by several ranges of mountains, and its climate is extremely equable and mild, the mean temperature for summer being 69·58°, and for winter 53·33°, while the variations are very slight. The air, too, is not only warm, but remarkably *dry ;* and the days are nearly always brilliantly bright and sunny. The town has grown out of an old Spanish mission which was founded in 1780, and which gradually drew around it the native cultivators of the adjacent lands. Its population in 1890 was 5,849, most of whom are Americans that have come here in search of health from the New England and Middle States ; and as most of these latter belong to what are called the "better classes," the society of the place is exceptionally pleasant and refined. There are 2 banks, a college, good public schools, 3 daily and 2 weekly newspapers, and 7 churches. The town contains a "Spanish quarter" and a "Chinese quarter," both of which will prove interesting to strangers by their tumble-down picturesqueness ; but the new or American part of the town and the suburbs are handsomely built. Vines of every sort flourish luxuriantly, heliotrope climbs 20 ft. high, cacti of the rarest and most curious sort grow freely, and a little shoot of the Australian blue-gum (*Eucalyptus globulus*) becomes in 2 years a shade-tree 15 or 20 ft. high. *Montecito* is a suburb 2 or 3 miles from the town, near which are the *Hot Sulphur Springs*, some containing sulphur and hydrogen sulphide, and others containing iron, alumina, and potash. Horseback-riding, surf-bathing, and driving among the cañons are the recreations at Santa Barbara.

Los Angeles (*Hollenbeck, Hoffman, Nadeau, Westminster*), the largest city in Southern California, is situated on the W. bank of the Los Angeles River, a small stream, 30 miles above its entrance into the Pacific, and 350 miles S. S. E. of San Francisco. It is connected by rail with Redondo and San Pedro, whence it has connection with San Francisco by steamer ; also *via* the Southern Pacific R. R. (482 miles). The city was settled by the Spaniards in 1780, and was called Pueblo de la Reina de los Angeles (the Town of the Queen of the Angels). Its population by the census of 1880 was 11,311, but in 1890 was 50,395, and the adobe buildings of which it was originally composed are fast giving way to larger and more imposing structures. With the exception of

---

[1] Full particulars concerning these health-resorts, with details as to climate, changes of temperature, relative dryness, etc., etc., will be found in "Appletons' Illustrated Hand-Book of American Winter-Resorts."

San Diego, its growth has been more rapid since 1880 than that of any city in the State. Its public buildings and private residences are very handsome. During recent years it has built 90 miles of street railways. Its *City Hall* cost $210,000, and the *County Court-House* cost $500,000. There have lately been built the *Post-Office* and *United States Courts Building*, which will cost $500,000. The *Catholic Orphan Asylum*, now building, will cost $175,000. The *State Reform School*, at Whittier, 21 miles from Los Angeles, which cost $200,000, was recently dedicated. In the N. W. portion is a hill 60 ft. high, commanding a fine view of the city, which lies in a sheltered valley, bounded on the W. by low hills, that extend from the Santa Monica Mountains, 40 miles distant, and on the E. by the San Gabriel plateau. The climate of Los Angeles is almost as mild as that of San Diego, and some invalids prefer it. The nights, however, are chilly, and it is not considered a desirable residence for persons affected with throat-diseases. Along both banks of the river below the city extends a fertile plain, planted with vineyards and orange-groves, and there are also large vineyards within the city limits. Los Angeles is the center of the petroleum district of South California, of vast asphaltum deposits, and of the orange and other citrus fruit-growing business of California. It is also the railway-center of California, being a terminal point of the Southern Pacific, the Southern California Railway, and also of several local lines. Los Angeles is the center of the fruit-industry, which has now attained enormous proportions. The large valleys adjacent are a continuous succession of orchards and vineyards in the highest state of cultivation, and are in themselves a source of great beauty and interest to the tourist. *Redondo Beach*, a seaside resort and the seaport of Los Angeles, is 17 miles S. W. Its fine beach, superior hotel (*Hotel Redondo*), and bathing facilities, make it a favorite resort. Ten miles N. E. of Los Angeles is **Pasadena,** first settled in 1874 by citizens of Indianapolis, and soon celebrated for its horticultural enterprise. Health-seekers soon flocked to it, and the population in 1890 was 11,879. The streets are lighted with gas; there are lines of horse-cars, three banks, two daily newspapers, and a public library; and the hotel, the *Raymond*, is large, well appointed, and beautifully situated. The *Carleton* is also a good hotel, centrally situated. *Wilson's Peak*, near Pasadena, is a peculiar vantage-point for sight-seers. Its summit (6,000 ft. high), formerly reached by a burro-train, over a good trail, affords a view of the surrounding country within a radius of 100 miles, the clear, dry atmosphere rendering this possible. It was used during several seasons as a site for astronomical observations conducted under the auspices of Harvard University. An electric railway has been built from Pasadena to the summit.

About 60 miles E. of Los Angeles (by the Southern California and Southern Pacific R. R.) is **San Bernardino** (*Hotel Stewart, Southern Hotel*), in a beautiful valley, with picturesque mountains on three sides of it, and contains some 4,005 inhabitants. The view of Mt. San Bernardino, the loftiest peak of the Coast Range, is exceedingly grand. The air of San Bernardino is drier than that of points nearer the coast, malaria is unknown, and the climate is a perpetual invitation to an open-air life. Invalids find a residence in *Old San Bernardino* (which lies higher than the new town), or in **Riverside** (*Glenwood*), more beneficial than one in the town proper. Riverside, 12 miles S. of San Bernardino, is the finest type of colony town in Southern California. It has a main avenue 12 miles long, with a double drive shaded by magnolia and pepper trees and fan palms. It is the chief seat of navel-orange culture. It has ten churches, eight schools, several large hotels, and many costly residences. Its population is 4,678. Another fine colony town on the Southern Pacific road, 33 miles S. E. of Los Angeles, is *Pomona*, largely settled by English families of means; it is the center of orange-groves and vineyards. The *Hotel Palomares* cost $100,000, and is finely fitted up. About 4 miles distant, near Mt. San Bernardino, are some *hot springs*. Horses may be bought at from $20 to $50 each at San Bernardino; their keep costs very little, and many attractive excursions may be made—to the San Gorgonio Pass, the Great Yuma Desert, the San Jacinto tin-mines, or the placer gold diggings. A few miles from Seven Palms, on the edge of the great desert, is *Palm Valley*, with a good hotel and famous hot springs. The place is worth visiting, because near by in a mountain cañon is the only natural grove of date-palm trees in California.

San Diego (*Brewster, Florence Hotel, Horton House, St. James*), another resort, the Pacific terminus of the Atchison, Topeka & Santa Fé R. R., lies on the N. E. shore of a bay of the same name, about 460 miles S. E. of San Francisco and 15 miles N. of the Mexican border, and is the second city of Southern California. Its harbor is, next to that of San Francisco, the best on the California coast. The town is more than 100 years old, having been founded by the Roman Catholic missionaries in 1769. Its growth during the last few years has been rapid, and it now has 20 churches, 2 academies, several daily and weekly newspapers, 6 banks, a fine court-house, excellent hotels, numerous boarding-houses, and a population in 1890 of 16,159. The climate of San Diego is remarkably equable and salubrious, the thermometer seldom rising to 80°, or sinking to the freezing-point, and the usual mean being 62°. The winter days are as sunny and inviting as those of June in the Eastern States, and an out-door life is possible to all save the feeblest invalids. There is no fog, and very little moisture in the air. San Diego is a healthful place of residence, and, like all this section of California, it is exempt from diseases of a malarious character. The points of interest are the *Sweetwater Dam*, the *San Diego Mission*, the *La Jolla Cave*, the *Mussel-beds, Point Loma Lighthouse, Tia Juana*, the *Monument* on the Mexican boundary-line, and El Cajou Valley. San Diego is connected with San Francisco by steamer along the coast, or by the Southern California Railway (Santa Fé System) to Los Angeles, thence by the Southern Pacific R. R.

On a peninsula directly opposite San Diego is **Coronado Beach.** It is connected with the mainland by a steam-ferry, and thence to the hotel is a short railway on which trains run every ten minutes. This resort is of comparatively recent origin, and a few years ago it was a waste of sand. The *Hotel del Coronado*, which is one of the largest hotels in the world, is situated directly on the edge of the beach overlooking the ocean. Architecturally it is of mixed character, partaking of the Queen Anne style, and having also much that is peculiar to the Elizabethan age. The hotel itself covers more than 7½ acres, and is built around a quadrangular court of 250 by 150 feet, in which is a garden containing many varieties of ornamental shrubs and fruit-trees, and where at night, when illuminated with electric lights, and with the fountain playing amid the music from the orchestra, the scene resembles fairy-land. In the building there are 750 rooms, and the dining-room has a seating capacity for 1,000 people. The peculiar charm of Coronado is its equable temperature. During the summer it is 10° cooler than at Naples, Mentone, Rome, Nice, or Florence, and is 8° warmer during the winter months. The mean temperature of the summer and winter months varies only about 12°. Charles Dudley Warner, who spent part of the winter of 1889–'90 at this beach, writes of the vicinity: "It lies there, our Mediterranean region, on a blue ocean protected by barriers of granite from the northern influences, an infinite variety of plain, cañon, hills, valleys, sea-coast—our new Italy, without malaria and with every sort of fruit which we desire (except the tropical), which will be grown in perfection when our knowledge equals our ambition ; and if you can not find a winter home there, or pass some contented weeks in the months of Northern inclemency, you are weighing social advantages against those of the least objectionable climate within the Union. It is not yet proved that this equability and the daily outdoor life possible there will change character, but they are likely to improve the disposition and soften the asperities of common life. At any rate, there is a land where, from November to April, one has not to make a continual fight with the elements to keep alive." The excellence of the Coronado Springs water, which has a recognized medicinal value, is one of the charms of the resort. Along the beach are hundreds of cottages, with their gardens brilliant with flowers ; and, for those who want other allurements, there are a race-track, a museum, an ostrich-farm, a botanical garden, a labyrinth, good roads for driving, besides many other attractions.

The visitor has choice of various excursions, any one of which will make a pleasant outing for the day. Among these are the ride over the National City & Otay R. R. to the famous Sweetwater Dam and the quaint old town of Tia Juana, across the Mexican border. This great dam, which was begun in 1886 and finished in 1888, is built of solid granite and Portland cement, and has a capacity of six billion gallons. It is 90 feet high, 46 feet thick at the base, and 396 feet long at the top. Its cost was $200,000. From this immense reservoir of 700 acres San Diego, National City, and Chula Vista obtain their water-supply. The trip from Coronado to the dam is a charming one, through orange and olive groves, and in full view of the bay and ocean.

Chula Vista, which the railway traverses from north to south, is a tract of 5,000 acres subdivided into five and ten acre lots, which are sold only to those who will build homes upon them which shall not cost less than $2,000. Thirty-five miles of streets have been graded and lined with trees. The site of Chula Vista seems to have been selected, as its name would indicate, for its picturesque surroundings.

After leaving Chula Vista, the train soon reaches Otay City and enters the valley of the same name, with its acres of fields and orchards, and shortly afterward crosses the Otay River to the valley of Tia Juana. Here he who has never set foot on other than American soil may enjoy the sensation of feeling that he is in foreign territory. Here he also may have his first taste of *frijoles* or *tortillas*—beans or pancakes.

94. San Francisco to Portland, Oregon.

Besides the overland route described below, there are several lines of coast-steamers from San Francisco to Portland, Victoria, and intervening ports. The Pacific Coast Steamship Co. runs semi-weekly steamers, and the voyage from San Francisco to Portland occupies about 2 days. The steamers of the Union Pacific system ply between San Francisco and Portland, leaving each of these places every few days.

THE Shasta Route, consisting of the main line of the Central Pacific R. R. to Sacramento, and thence over the Southern Pacific R. R. to Portland, affords an excursion route of unsurpassed scenic beauty. Trains with sleeping and buffet car service carry the tourist through without change, and give the benefit of day travel to the most attractive sections of the journey.

(From SAN FRANCISCO to SACRAMENTO, see Route 90.) From Sacramento the California and Oregon Div. of the Southern Pacific R. R. diverges from the main line and extends to Ashland, the terminus of the Oregon & California R. R. The total distance from San Francisco to Portland by this route is 772 miles.

Leaving Sacramento, the R. R. follows the Sacramento River in a general northern direction. The country traversed is one of the most productive wheat-growing sections of the State. The first important station is **Marysville** (142 miles), a flourishing town of 3,936 inhabitants at the confluence of the Yuba and Feather Rivers, at the head of navigation on the latter. It is well built, has several foundries and machine-shops, and contains an abundance of choice fruit and shade trees. From Marysville a fine view is obtained of the *Marysville Buttes*, an isolated chain of mountains which rise from the plain of the Sacramento Valley to the height of 1,200 ft. and extend for some 8 miles in length, forming a remarkable feature of the scenery. Beyond Marysville the Feather River is crossed, and the train traverses the upper Sacramento Valley, which was formerly one great wheat-field, but which is now being rapidly converted into small fruit farms. **Chico** (186 miles) is another thriving town of 2,892 inhabitants, situated on the Chico Creek near its junction with the Sacramento River. Here is the branch *Normal School* for Northern California. Just N. of the town is the magnificent estate of Gen. John Bidwell, which comprises 32,000 acres in one tract. The orchard is filled with oranges, lemons, figs, walnuts, almonds, and other choice fruits; and the vegetable and flower gardens are said to be unsurpassed in Northern California. From the plains near Chico can be seen the snowy summits of Mt. Shasta, 216 miles distant. Near *Tehama* (213 miles) the Sacramento River is crossed, and the train passes several small stations to *Redding* (260 miles).

At McCloud, just beyond Redding, the Sacramento Valley is left behind, though the river is ascended for about 80 miles amid the foot-hills. The N. extremity of the Sierra Nevada range is then climbed and crossed, and the road strikes the Pitt and McCloud Rivers, the main affluents of the upper Sacramento. "Near the crossing of the McCloud," says Mr. Williams in his "Pacific Tourist," "is the U. S. fish-hatching establishment. All these rivers abound in trout and salmon, but the best place on them for trout-fishing is the upper waters of the McCloud. *Castle Rocks*, about opposite *Lower Soda Springs* (320 miles), is a startling upheaval of solid granite, with its perpendicular wall rising at a single bound to an altitude of 4,000 ft. above the valley, and is one of the most marvelous scenes in California mountain scenery. The valley of the Sacramento grows narrower as one goes N., and at last is almost a cañon. The *Mossbrae Falls* are seen from the train as it approaches *Upper Soda Springs* (65 miles N. of Redding) ; and, just beyond, the road ascends from the river to an extensive mountain-basin, walled in by yet loftier mountains—a sort of semicircular wall from the Scott Mts. on the N. to Trinity on the W. and Castle Rock on the S. E. On the E. side of the road, and in this great basin, Mt. Shasta rears its lofty head into the dark, deep blue of heaven." The ascent of **＊＊Mt. Shasta** is made from *Sisson's* (78 miles from Redding, 338 from San Francisco), and though tedious is not dangerous. The trip will take about 36 hours, and the cost, including horses, guides, provisions, etc., will be $15 to $20 for each person, according to the size of the party. Shasta from Sisson's is a broad triple mountain, the central summit (14,442 ft. high) being flanked on the W. by a large and quite perfect crater whose rim is 12,000 ft. high. As a whole, Shasta is the cone of an immense extinct volcano, which rises from its base 11,000 ft. in one sweep.

Beyond Sisson's several fine views of Shasta are obtained from various points on the road, the best being from the summit of Scott Mt., which is crossed at an elevation of 5,000 ft. above the sea. Another fine distant view is obtained from a ridge just E. of *Montague* (117 miles from Redding), and beyond which the road climbs up the Siskiyou Mts. over a grade of 220 ft. until an elevation of 4,100 ft. is reached. A mile beyond *Coles* the State line is passed. At *Ashland* (343 miles), the Oregon & California R. R. begins, passing through *Roseburg*, and running, during its whole route of 200 miles, through the beautiful and productive Willamette Valley. Many pretty towns cluster along the railway, but none require special mention until **Salem** (*Willamette Hotel*), the capital of Oregon, is reached, 52 miles from Portland. Salem is a growing city, on the E. bank of the Willamette River, surrounded by a fertile prairie. Mill Creek enters the river at this point, and its rapid fall affords a good water-power, which is extensively used in manufacturing. Here are Willamette University and three State institutions, the Penitentiary, the Deaf-Mute School, and the Institute for the Blind. Thirty-seven miles beyond Salem, at the Falls of the Willamette, is **Oregon City** (*Cliff House*), with a population of 3,062, and several large flouring and woolen mills. The falls have a descent of 38 ft., and constitute one of the finest water-powers in the world. Fifteen miles beyond Oregon City the train reaches **Portland.** (See Route .106.)

95. Portland to Alaska.

By *mail-steamer* leaving Portland twice every month, calling at Fort Wrangel, Sitka, and Juneau ; and excursion-steamers from Port Townsend at more frequent intervals.

THE steamer from **Portland** goes down the *Columbia River*, up the coast of Washington to Port Townsend, Puget Sound, where it receives the mails from San Francisco. *Port Townsend* is the last port of entry in the United States, and here connection is made with steamers from San Francisco.

In recent years the remarkable scenery of Alaska has attracted a large number of visitors. The sea-voyage is one of great fascination, being almost continuously within sight of land, thus affording the tourists a view of the coast and islands from Puget Sound northward. Considerable interest has been attached to Mt. St. Elias, which is believed to be the highest mountain-peak within the territory of the United States. Several expeditions under the auspices of the national Government have endeavored to reach its summit, but thus far without success. The glaciers, notably the *Muir Glacier*, on the steamship route, are unique attractions of Alaska, and they have received much attention from tourists.

Alaska, 580,170 sq. miles in extent, was purchased from Russia, in 1867, for $7,200,000, and by act of Congress, May 17, 1884, was constituted a civil and judicial district, and a governor and other officers appointed.

Sitka, the capital, and residence of the Governor, the seat of the Bishop of the Greek Church, and the site of the *United States Land-Office,* is on **Baranof Island,** in the southern part of the Territory. The *Sitkan* or *Alexander Archipelago* comprises 1,100 islands of considerable fertility. Next to Sitka in importance is **Fort Wrangel,** S. E. from Sitka, at the mouth of the *Stikene River,* which has been aptly described as "a Yosemite 100 miles long." "Here," says John Muir, "Indians may be seen on the platforms of the half a dozen stores, chiefly grim women and cubby-chubby children with wild eyes." **Unalashka,** one of the seventy **Aleutian Islands** that stretch out toward Japan, and **Juneau City,** about 100 miles N. of Sitka, are, with the two places previously named, designated by Congress as the residences of four United States commissioners who have the powers of justices of the peace.

The great river **Yukon,** which Lieut. Frederick Schwatka has made known by his arduous voyages, runs through the center of the Territory, and empties into *Behring Sea,* being navigable in summer for 700 miles.

The **Seal Islands,** discovered by Pribyloff in 1788, lie 240 miles N. of the Aleutian Islands. *St. Paul,* the largest, is 1,491 miles W. of Sitka, and 250 miles from the nearest mainland.

96. St. Louis to Denver.

FROM St. Louis to Kansas City four routes are available: The Wabash R. R. (distance, 275 miles); the Missouri Pacific Railway (distance, 282 miles); the Chicago & Alton R. R. (distance, 323 miles); and the Burlington Route (distance, 335 miles). All four routes traverse a rich and productive section of Missouri. The principal stations on the Wabash R. R. are *St. Charles* (23 miles), where the Missouri River is crossed on a magnificent steel bridge, *Warrenton* (61 miles), *Montgomery* (84 miles), *Mexico* (110 miles), *Centralia* (124 miles), *Moberly* (148 miles), *Salisbury* (169 miles), *Brunswick* (187 miles), *Miami* (197 miles), *Carrollton* (211 miles), and *Missouri City* (255 miles).

The Missouri Pacific Railway has a considerable number of large towns *en route*. *Kirkwood* (13 miles) is a beautiful suburban town with many fine villas of St. Louis merchants. *Pacific* (37 miles), *Washington* (54 miles), and *Hermann* (81 miles) are prosperous and handsome towns. **Jefferson City** (125 miles; *Madison House, Monroe House*) is the capital of the State of Missouri, and is beautifully situated on high bluffs which overlook the Missouri River for many miles. It is well built, and has a population of 6,742. The *State-House* is a handsome stone edifice; the *State Penitentiary* is massive and spacious; there are 8 churches of various denominations; the State Library contains 12,000 volumes; and there are numerous flour-mills and factories. **Sedalia** (188 miles; *Hotel Kaiser, Sichers Hotel*) is a manufacturing town and railroad center, built on one of the highest swells of a rolling prairie, and containing 14,068 inhabitants. The principal street is 120 ft. wide, is finely shaded, and has many handsome buildings. The shops of two R. R. companies are located here, and there are extensive mills, foundries, machine-shops, etc. *Perth Springs*, a famous local resort, is near this place. *Warrensburg* (218 miles), *Holden* (232 miles), *Pleasant Hill* (249 miles), and *Independence* (273 miles) are all neat and thriving towns, with much business activity.

At Kansas City connection is made with all lines to the West, and, in connection with the Union Pacific, through vestibuled sleeping-cars are run from St. Louis to Denver, Cheyenne, Ogden, and as far as Salt Lake City without change, connecting at Ogden with the Southern Pacific for San Francisco.

Also, the route from St. Louis to Texas is over the Iron Mountain Route to that State. This line, in connection with the Texas & Pacific and Southern Pacific R. Rs., run through buffet sleeping-cars from St. Louis, by way of Little Rock, Texarkana, Dallas, Fort Worth, and El Paso to Los Angeles and San Francisco.

The Chicago & Alton R. R. route has already been described in reverse as far as **Alton** (see Route 85). The junction with the Chicago, Burlington & Quincy R. R. is at *Whitehall* (68 miles), and at *Louisiana* (109 miles) connection is made with St. Louis, Keokuk & Northwestern R. R., and with Missouri River steamers. **Mexico** (160 miles; *Windsor Hotel,*

Ring's House) is a city of 4,789 inhabitants, and has considerable trade and manufacturing. It has 9 churches, and is the seat of *Hardin College* (for women). Here also is a point of junction with the Wabash R. R. At **Glasgow** (216 miles; *Glasgow Hotel*), a town of 1,781 population, there are several mills and factories and some excellent educational institutions, among them *Lewis College*. *Marshall* (239 miles; *Minge's Hotel*) is a town of 4,297 population, and has a number of thriving carriage and wagon factories. At this place there are many remarkable salt-springs. *Higginsville* (268 miles) and *Independence* (313 miles) are enterprising towns, both points of junction with branches of the Missouri Pacific R. R.

The Burlington Route has a direct line to Kansas City, passing over the Mississippi Valley and interior Missouri. Among the important places along the route are *Clarksville* (76 miles), noted for its Missouri cider; *Louisiana* (86 miles) has tobacco-works, and ships to the St. Louis market; *Hannibal* (111 miles) is the junction of three railroads, and is a place of some 12,857 inhabitants; also one of the principal shipping points in Missouri. *Shelbina* (158 miles), *Brookfield,* (215 miles), and *Laclede* (220 miles) are among the cities of Missouri through which the Burlington Route passes before reaching Kansas City.

* **Kansas City** (*Midland Hotel, Coates House, Centropolis House, St. James Hotel, Brunswick*) is the second city of Missouri, has a population of 132,716, and is situated on the S. bank of the Missouri River, just below the mouth of the Kansas River, and near the Kansas border. It is well built, chiefly of brick, and contains a fine *Board of Trade Building* and many handsome business blocks and private residences. Among the buildings recently erected are the *Custom-House, New York Life-Insurance Company Building, Midland Hotel, New England Building,* and *Warder Grand Opera-House.* It has an immense and increasing trade, which is brought to it by the 18 important railroads which converge here, and by the steamboat traffic on the river; and some of the largest packing-houses are located here, such as *Armour's, Fowler Brothers,* and *Jacob Dold's.* The stock-yards are next to those of Chicago in size. The first bridge built across the Missouri is located at Kansas City; it is 1,387 ft. long, and cost over $1,000,000. A later railway bridge built by the Chicago, Milwaukee & St. Paul R. R. at the east end of the city is 7,392 ft. long. The *Union Depot* is one of the finest in the West. Several lines of street railroads, seven cable-roads, and one elevated road, run to various parts of the city, and to the suburbs of Kansas City, Kan., and Westport.

Kansas City to Denver via Kansas Div. of Union Pacific System.

On this route there are two through trains daily, which run from Kansas City to Denver without change in 20 hours. They traverse the central portion of Kansas, linking together the principal cities and towns of the State, and affording the opportunity to view its famous wheat and corn fields and immense cattle-ranges. Striking the Kansas (or Kaw) River at Kansas City, the route follows the windings of this beautiful stream for nearly 200 miles amid extremely pleasing scenery, and as it approaches the Rocky Mountains commands some grand views. Leaving Kansas City, the train passes in 38 miles to **Lawrence** (*Eldridge House*), a beautiful city of 9,997 inhabitants situated on both sides of the Kansas River, which is here spanned by 2 bridges. It is built on a rolling slope, and is regularly laid out, with wide streets, partly shaded by trees, and many handsome buildings. Massachusetts St., the principal business thoroughfare, is built up for nearly a mile with blocks of brick and stone. The *State University*, consisting of three large and handsome structures, is located upon a bluff called Mt. Oread in the S. W. part of the city. The *Haskell Institute*, a U. S. Indian school, with over 300 pupils, is located here. The trade of Lawrence is very large. Thirty miles beyond Lawrence the train reaches **Topeka** (*National, Throop, Copeland,* and *Fifth Ave.*), the capital of Kansas, situated on both sides of the Kansas River, which is here spanned by three fine iron bridges. The city contains 31,007 inhabitants, and is well built. The *State-House*, costing $3,000,000, is partially completed. The east and west wings of the *Capitol* are finished, and the rest nearing completion. The *U. S. Federal Building*, the *State Asylum*, and *State Reform School, City Free Library, Washburne, Bethany,* and *Methodist* colleges, and the general offices of the Atchison, Topeka & Santa Fé and of the Chicago, Rock Island & Pacific Railways are among the principal buildings. *Washburne* and *Bethany* are the principal colleges, and there are excellent public schools. The building of the Free Library was completed during 1891 at a cost of $40,000, and contains 11,000 volumes. The river affords a good water-power, and the surrounding country is very fertile and contains deposits of coal. Topeka is provided with street-railroads, electric lights, gas, and all modern improvements. *St. Mary's* (91 miles) is a prosperous town of 1,174 inhabitants, and the seat of St. Mary's College. *Wamego* (103 miles), *St. George* (110 miles), *Manhattan* (118 miles), and *Ogdensburg* (129 miles) are busy and rapidly-growing towns. At *Junction City* (138 miles) connection is made with the Neosho Branch of the Missouri, Kansas & Texas R. R. A highly productive agricultural region is next traversed, with numerous thriving villages *en route*. *Ellsworth* (223 miles) is situated on the Smoky Hill River in a fine stock-raising country. *Fort Hays* (288 miles) is one of the handsomest military posts in the West, situated on a commanding elevation overlooking the plains. Opposite, upon Big Creek, is Hays City, once the center of the buffalo range. *Fort Wallace* (420 miles) is another important military post, situated near the W. boundary-line of Kansas, and just beyond the train enters Colorado. The first noteworthy station in Colorado is *Kit Carson* (487 miles), named after the great " Pathfinder," and situated on Sand Creek, about 20 miles above the spot where Colonel Chivington's Indian massacre took place. Between Kit Carson and Denver there are only " station towns" (*Hugo, Godfrey,* and *Byers* being the most important), but the country along the line is rapidly filling up. At *First View (472 miles) the first view is obtained of the Rocky Mountains. " Towering against the western sky, more than 150 miles away, is Pike's Peak, standing out in this rarefied atmosphere with a clearness which deludes the tourist, if it be his first experience, into a belief that he is already in close proximity to the mountains. Henceforth you feel, in the presence of the mighty peaks which disclose themselves one after another, that you have entered a new world—a land of unapproachable beauty and grandeur—and you reach Denver, having before you an unobstructed panorama of mountains, snow-clad peaks, and plain, more than 300 miles in length."

Kansas City or Atchison to Denver via Atchison, Topeka & Santa Fé.

The Atchison, Topeka & Santa Fé System has four eastern termini at the Missouri River: St. Joseph, Atchison, Kansas City, and Leavenworth. Atchison is 47 miles N. of Kansas City; Leavenworth, 24 miles N. of the same point. The Leavenworth branch unites with the Kansas City line 13 miles from Kansas City; and the Atchison branch unites with the Kansas City line at Topeka, 66 miles from Kansas City. Two daily express-trains, with palace sleeping-cars attached, run on this route from Atchison and Kansas City daily to Denver. St. Joseph is located on a branch 20 miles N. of Atchison, on the E. bank of the Missouri River, and is also the terminus of the branch connecting with the main line to Chicago at Lexington Junction. **Atchison** is a bustling city of 13,963 inhabitants, situated beautifully on the right bank of the Missouri River. It is quite an important railroad center, and it is said that on the 8 railroads meeting here there are some 90 trains arriving and departing daily. It has 12 churches, 5 banks, several theatres and public halls, and a large manufacturing industry in flour-mills, machine-shops, engine-works, breweries, furniture and carriage factories, etc. A fine iron bridge across the river connects the city with the railroads which terminate on the E. bank.

At **Topeka** (66 miles from Kansas City, and 50 miles from Atchison) the branches of the road unite. (See page 431.) Beyond Topeka a number of thriving towns are passed, of which *Burlingame* (93 miles), *Osage City* (101 miles), *Emporia* (128 miles), *Strong City* (148 miles), *Florence* (172 miles), *Peabody* (184 miles), *Newton* (201 miles) [at Newton connection is made with branch running south to *Wichita* (228 miles), *Winfield* (267 miles), *Wellington* (269 miles), and *Arkansas City* (279 miles), running directly through the Oklahoma country and Texas to *Galveston* (949 miles)], and *Burrton* (230 miles), are the most important. At Burrton, which is the terminus of the St. Louis & San Francisco Railway, the latter road makes connection for its through transcontinental route from St. Louis to California over the A., T. & S. F. R. R. From Burrton to Pueblo (414 miles) the road follows the fertile valley of the Arkansas River, through one of the finest agricultural and stock-raising regions in America. The principal towns on this portion of the route are *Hutchinson* (234 miles), *Sterling* (253 miles), *Great Bend* (285 miles), *Larned* (307 miles), *Dodge City* (368 miles), *Garden City* (418 miles), *Lamar* (518 miles), and *La Junta* (571 miles). La Junta is the point of junction with the main line extending to all points in New Mexico, Arizona, and California. Just before reaching La Junta the first glimpse is caught of the Rocky Mountains, still 60 miles distant. Soon

Pike's Peak looms up ; then the Spanish Peaks reveal their snowy crowns ; and finally, as Pueblo is neared, the splendid mountain panorama gradually unfolds itself. **Pueblo** (*Fifth Avenue, Grand, St. James*) is one of the chief cities of Southern Colorado, and is situated at the confluence of Arkansas River and Fontaine Creek. It is the center of an agricultural, petroleum, and mining region, does a large trade, and has a population of 24,558. It has large smelting-works and Bessemer-steel works. From this point the route is directly north, passing *Colorado Springs*, connecting with Colorado Midland Railway for *Manitou, Pike's Peak, Cascade Cañon, Leadville, Glenwood Springs*, and other famous Colorado resorts, whose attractions are described in Route 97. The points of interest in Southern Colorado which can be visited from Pueblo are described in Route 98.

Kansas City to Denver via Missouri Pacific Railway.

Leaving Kansas City toward the southwest, it runs through the midst of the rich farming lands of Central Kansas, and is known as the "Colorado Short Line," passing through *Ottawa*, in Eastern Kansas, the famous resort of that State, and the seat of the Chautauqua Assembly ; *Salina*, in Central Kansas, the most important town in that part of the State, is reached. From there the route is almost directly west, passing up the famous Arkansas River Valley to *Pueblo ;* and thence northward, passing through Colorado Springs, within a few miles of Manitou, the Garden of the Gods, and Pike's Peak, to *Denver*.

Denver.

Hotels, etc.—The leading hotels are the *Palace*, the *Windsor*, the *Albany*, the *St. James*, the *Markham*, the *Oxford*, and the *American*.

Denver, the capital and largest city of Colorado, is situated on the S. bank of the South Platte River, at the junction of Cherry Creek, 15 miles from the E. base of the Rocky Mountains, and about 500 miles W. of the Missouri River. It occupies a series of plateaus, facing the mountains, and commanding a grand and beautiful view. Through the clear mountain atmosphere may be seen Pike's and Long's Peaks, and the snow-capped range extending more than 200 miles, its rich purple streaked with dazzling white, and here and there draped in soft, transparent haze. Its trade is very large, and 23 railroads radiate from it, which, with their stage connections, afford access to all parts of the State. There are numerous hotels, many fine commercial buildings, a *U. S. Branch Mint*, 70 churches, 6 daily newspapers, 22 school-houses, 11 banks, a magnificent *Opera-House*, 2 fine theatres, large manufactories and breweries, and many elegant private residences. The Mint is employed in the melting and assaying of bullion. The city is the site of *Denver University, Baptist College*, and other institutions of learning ; and among other prominent buildings are the *Union Depot, City Hall, Chamber of Commerce, Court-House, State Capitol, U. S. Custom-House, Denver Club*, and *Post-Office*. The great works of the Boston and Colorado Smelting and Refining Co., the Globe Smelting and Refining Co., and the Omaha and Grant Smelter, are worth visiting. The population of the city in 1880 was 35,630 within the city limits, and according to the census of 1890 it was 106,713.

97. Colorado and Manitou Springs.

COLORADO SPRINGS (76 miles from Denver *via* the Denver & Rio Grande R. R., the Union Pacific R. R., and the Colorado Midland R. R.) is a flourishing little city, situated on the plains, 6,022 ft. above the sea, with a fine view of Pike's Peak and the mountains, and with pleasantly-shaded streets, and contains all modern improvements, many fine residences, and a pretty *Opera-House.* (Hotels, *The Antlers, Alamo, Alta Vista.*) Boarding-houses and furnished houses are numerous, amusements plentiful, and the rides and drives charming. In summer the days are warm without being uncomfortable, and the nights always cool. In winter there is almost no rain or snow, and little dampness, while the sunshine is nearly uninterrupted. The following table gives the distances to the chief sights : To Manitou Springs, 5 miles; to Garden of the Gods, 4 miles; to Glen Eyrie, 5 miles; to Monument Park, 8 miles; to Cheyenne Cañon, 5 miles; and to the summit of Pike's Peak, 16 miles. Guides, horses, etc., are easily procured.

The * **Manitou Springs,** reached by the Denver & Rio Grande R. R. from Denver to Colorado Springs, thence by branch line to Manitou; and both Colorado and Manitou Springs are reached directly by the main line of the Colorado Midland R. R. and the Union Pacific R. R. Manitou Springs are so much resorted to as to be known as the "Saratoga of the West," and are the center of excursions in the district. They are situated among the foot-hills at the base of Pike's Peak, on the banks of the beautiful Fontaine Creek. The waters contain sulphur, soda, and iron, and are recommended for their tonic effects in all diseases of which general debility is a feature. Asthmatics and consumptives are usually benefited by a residence at Manitou; the former always. There are several hotels, the *Mansions, Cliff House, Manitou House, Barker Hotel,* and the *Iron Springs Hotel ;* and the adjacent grounds are beautifully laid out. Within easy walking-distance of the hotel is the picturesque and romantic *Ute Pass,* with its mammoth hotel, club-house, artificial lake, bath-house, and other attractions, through which the Colorado Midland R. R. runs to the South Park. A short distance above the mouth of the Pass are the *Ute Falls,* where the creek descends in an unbroken sheet over a precipice 50 ft. high. The road runs close to the edge of this precipice, while on the other hand the rocks tower above to an immense height. In this vicinity is the picturesque *Williams Cañon,* 15 miles long, with walls of rock rising 600 or 800 ft. above a very narrow pass below. In the valley of the pass above the cañon are five summer resorts on the line of the Colorado Midland Railway—*Cascade Cañon, Ute Park, Green Mountain Falls, Woodland Park,* and *Manitou Park*—all of which have excellent hotels and unsurpassed scenic attractions, 6, 8, 9, 15, and 21 miles respectively from Manitou. The summit of * **Pike's Peak** is reached by a cog-wheel-railway from Manitou. The round trip (fare, $5) may be made in a few hours on any day during the season. It is also accessible by a fine carriage-road from Cascade. This peak stands on the edge of the great mountain-range, and the view from its summit (14,147 ft. high) embraces many thousand square miles of mountain and plain. Here is a station of the Weather-Signal Bureau, which is occupied winter and summer.

Colorado Springs and *Manitou* are but 5 miles apart, and are now connected by means of an electric railway. Leaving the Springs, the road runs for 2 miles over a plateau that suddenly terminates at an elevation of several hundred feet above a valley. The view hence is magnificent; the valley is inclosed by lofty mountains, among which Pike's Peak towers on high, and in front are Glen Eyrie, the Garden of the Gods, and Manitou. Descending into the valley you first reach **Glen Eyrie,** a natural park with immense fantastically shaped rock-formations. It has been inclosed, and within it is the elegant summer villa built by General Palmer, and the natural attractions of the place have been enhanced by art. Up the rugged *Queen's Cañon* is the Devil's Punch-Bowl, and a succession of picturesque rapids and cascades. A short drive brings you to the Gates of the **Garden of the Gods,** two piles of red sandstone 300 ft. high and three quarters of a mile in circumference, separated only enough to admit the passage of vehicles. The Garden comprises a tract of land less than 500 acres in extent, hemmed in by mountains on the W. and N., bordered by ravines on the S., and by old red sandstone cliffs on the E., which shut it in entirely from the plains. Its features are a number of isolated rocks upheaved into perpendicular positions, some of them rising to a height of 350 ft. The rocks are mainly of a very soft, brilliantly-red sandstone, although several ridges of cliffs are of a white sandstone. The foot-hills in the vicinity are, many of them, capped by similar upheavals, while all about the main cliff in the valley are numerous separate, spire-like columns.

Cheyenne Cañon, 9 miles from Manitou Springs, is a sequestered mountain-gorge, in which are some striking rock-formations and picturesque cascades. A tortuous trail leads from the mouth of the cañon in 3 miles to the first fall, which is 30 ft. high, and extremely fine. From the ledge above the fall there is a view of a succession of falls, 6 in all, rising one above another at almost regular intervals, the remotest and highest being several miles away. On the side of *Cheyenne Mountain* Helen Hunt Jackson ("H. H.") is buried. In a grove of trees a heap of stones, piled up by visitors, marks her tomb, and an old pine-tree— rudely engraved upon its bark are the letters "H. H."—is the living head-stone. Each visitor places a stone upon the heap, and with each stone placed upon the heap there is added another recollection to her fame.

* **Monument Park,** perhaps the most visited spot in Colorado, is 9 miles from Manitou Springs and ⅓ mile from *Edgerton,* a station on the railway above the Springs. The Park is very striking. It is filled with fantastic groups of eroded sandstone (6 to 50 ft. high), perhaps the most unique in the Western country, where there are so many evidences of Nature's curious whims. They are, for the most part, ranged along the low hills on each side of the park.

98. Southern Colorado and New Mexico.

The *Denver & Rio Grande Railway* runs to Ogden, Utah, where it connects with the Central Pacific R. R., traversing southern and southwestern Colorado, and extending into New Mexico. By this system, over its line west from Denver, the traveler passes through Pueblo, Leadville, Glenwood Springs, Grand Junction, Salt Lake City, and Ogden, and is able to see from the car-windows the wonderful scenery of Pike's Peak, the Royal Gorge, Cañon of the Eagle Run, Mount of the Holy Cross, Cañon of the Grand River, the Book Cliffs, Spanish Forks Cañon, Castle Gate, Utah Lake, and Great Salt Lake. Through-cars to San Francisco may be taken at Denver. It comprises over twenty branches, of which the most important are those from Denver to *Pueblo ;* Pueblo to *Ogden*, Utah ; Pueblo to *Trinidad ;* Cuchara to *Silverton ;* Pueblo to *Leadville, Aspen,* and *Glenwood Springs ;* Antonita to *Santa Fé.* Distances : Denver to Monument, 56 miles ; to Colorado Springs, 75 ; to Pueblo, 120 ; to Ogden, 651 ; to Cañon City, 161 ; to Leadville, 279 ; to Cuchara, 170 ; to El Moro, 206 ; to Antonita, 279 ; to Española, 370 ; to Durango, 453. The *Union Pacific*, from *Denver* to *Leadville* (151 miles), is the shortest line, and by its having acquired the Denver, Texas, and Fort Worth R. R., has now a line to Fort Worth, Texas (803 miles from Denver), by way of Colorado Springs, Pueblo, and Trinidad. The main line of the Colorado Midland R. R. also gives connection between Leadville and Aspen and Glenwood Springs.

THE section of the Denver & Rio Grande R. R. between Denver and Pueblo (120 miles) has been described in Routes 96 and 97. From Pueblo the Leadville Division runs (41 miles) to **Cañon City** (*McClure Hotel*), near which is *Talbott Hill,* where Prof. Othniel C. Marsh has excavated some most remarkable fossils. Two miles beyond Cañon the railway enters the * **Grand Cañon of the Arkansas,** where the Arkansas River cuts its way for 8 miles through walls of granite, in some places 3,000 ft. in height. Through the *Royal Gorge* the track runs for 200 ft. along an iron bridge suspended over the river by steel girders mortised into the rock on either side. From *Salida* (215 miles from Denver ; *Hotel Monte Cristo*) the line runs W. through the wonderful **Gunnison Country,** as it is known in mining parlance. This region, which has attracted so much attention on account of its remarkable mining developments, has an area of not less than 10,000 square miles, being 110 miles long by 80 miles wide. The most important part of the territory is the Elk Mountain Range with its many spurs and foot-hills, constituting one of the most picturesque portions of the Rocky Mountains. Great masses of granite have been upheaved through the surface in giant forms, and there are seven peaks in the range rising to a height of 14,000 ft., while many times that number reach the height of 12,000 ft. Vast beds of coal and iron and innumerable fissure veins of silver are scattered through this region. The principal streams are the Gunnison, Uncompahgre, Cochetopa, Sumichel, Taylor, East, Ohio, Eagle, Rock, Roaring Fork, and Slate, with hundreds of smaller tributaries, all emptying into the Colorado River. In addition to its great mineral resources, the Gunnison region is admirably adapted in many parts to stock-raising, as there are large areas of perennial pasturage, where bunch-grass, blue-grass, and other varieties afford the richest grazing. The most important towns and mining camps in this section are Gunnison, Rocky Camp, Pilkin Gothic, Washington Gulch, Irwin, Crested Butte, Hillerton, Virginia City, Red Cliff, Willard, Cochetopa,

A New-Mexican Pueblo.

and Aspen. **Gunnison** (*Hotel La Veta*), the most important place on
the route after leaving *Salida* (564 miles from Denver and 792 miles
from Kansas City), is a town of 1,105 inhabitants.   It has a court-
house that cost $15,000, five school-houses, and a bank represent-
ing $10,000,000 capital.   It is the great outfitting center of the re-
gion.   Here is the connection with the Crested Butte Branch for the
silver-mines and coal-fields of the Elk Mountains and Crested Butte
regions ; also for those of Irwin, and Gothic.   Five miles from South
Arkansas are the *Poncha Hot Springs*, noted for their medicinal properties; near Nathrop, on the line of the road, are the *Heywood Hot
Springs ;* and near Buena Vista are the *Cottonwood Hot Springs.*
Leadville (*Kitchen Hotel, Grand Pacific*), 10,200 ft. above the sea,
the highest town in the world (except Cuzco), is in the heart of the
silver El Dorado discovered in 1878, and is a busy city of 11,212 inhabitants.   It is the most celebrated mining-camp in the West.   Leadville is lighted with gas and electricity, and has all the conveniences of a
large city.   It is reached also by Colorado Midland R. R. from Colorado
Springs.   Fourteen miles from Leadville are the celebrated *Twin Lakes*,
nearly 2 miles above the sea-level.   The Denver & Rio Grande line
runs from Denver to Leadville, Aspen, and Glenwood Springs, on the
Grand River, and through Grand Junction to Ogden,.thence *via* Southern Pacific R. R. to San Francisco.   *Glenwood Springs* (*Colorado Hotel*),
also reached by the Colorado Midland R. R., is one of the largest hot
and mineral springs in the United States, and much money has been
spent in improving them.

The *Denver & Silverton Division* extends S. W. from Pueblo, and at
the distance of about 80 miles crosses the Sangre di Cristo range by the
wonderful * **La Veta Pass,** at an altitude of 9,486 ft., the *Mule-Shoe
Curve* and the passage around the point of *Dump Mountain* being
among the most daring feats of railway engineering ever accomplished.
The view of *Sierra Blanca* (14,464 ft. high), flanked by the serrated
peaks of the Sangre di Cristo range,|as seen from *Alamosa* (250 miles)
and for 70 miles across the San Luis Park, is nowhere surpassed.
Thirty-five miles W. of *Antonita* (279 miles) are the **Los Pinos Cañon** and the * **Toltec Gorge,** the most wonderful scenic attractions
of Colorado.   Here for a distance of 8 miles the railway runs just below
the brow of a precipitous mountain-range at the height of 1,200 ft.
above the stream, following the irregular contour of the mountains,
till at **Phantom Curve** it comes to the end of a mountain-wall that
juts into the cañon, narrowing it to a mere gorge, 1,400 ft. in depth,
with the wall on the farther side rising to a height of 2,100 ft.   A few
rods from this gorge the railway suddenly enters a tunnel in the granite
cliff, and 600 ft. farther on emerges upon a trestle-bridge overlooking
the precipice that extends to the bottom of the gorge.   From the entrance to the tunnel, in fact all along this.“ aërial trip,” an extended
landscape of mountain and valley adds to the grandeur of the view.
*Durang*o is the supply depot for the San Juan mining district, and the
road continues to *Silverton* (495 miles from Denver), in the heart of
the region.   Thirty miles from Durango are the prehistoric cliff-dwell-

ings on the Rio Mancos. At Silverton, connection is made with the Silverton Railway, which crosses the range at Red Mountain, 12,000 ft.
above the level of the sea, and approaches within a few hours' stage-
ride of Ouray.

At Antonita the *New Mexico Div.* branches off to the S., and runs
in 91 miles to *Española*, connecting with the Santa Fé Southern R. R.
for Santa Fé, which is only 23 miles from Española. Near Española
are six ancient pueblos, inhabited by the Pueblo Indians, whom the
Spaniards found there only forty-eight years after the discovery of
America; and in the neighboring cliffs are numerous cave-dwellings,
prehistoric in their origin.

99. Colorado Central Division of the Union Pacific R. R.

THIS road extends from Denver to Fort Collins (71 miles), and from
Denver to Graymount (58 miles), and a branch from Forks Creek to
Central City (40 miles from Denver). Both routes traverse exceedingly
picturesque regions and afford some of the finest scenery to be enjoyed
in all Colorado. *Golden* (16 miles from Denver) is situated between
two picturesque hills and the North and South Table Mountains. The
State School of Mines is in Golden. It is the center of an extensive
mining-region, and is the point of departure for *Bear Creek Cañon.*

At Golden the Central City branch diverges from the main line,
and passing up * **Clear Creek Cañon,** follows the windings of the
Creek through one of the wildest and most picturesque localities on the
continent. *Black Hawk* (40 miles from Denver) is built irregularly
along the gulches and mountain-sides, and is one of the busiest mining
towns in the State. A mile beyond Black Hawk, reached by a zigzag
course up the mountain-side, is *Central City,* a flourishing mining town
of 2,480 inhabitants, picturesquely situated on the mountain slopes, at an
elevation of 8,300 ft. There are a number of quartz-mills here, and the
town has a U. S. land office and an assay office. Being in the center
of an exceedingly rich gold-mining region, it is at once a depot of supply
and a point of shipment, and business is very active. *James Peak* may
be ascended from Central City, and affords a wide-extended view. The
Georgetown Branch diverges at *Forks Creek* (29 miles from Golden)
and runs in 9 miles to **Idaho Springs,** a quiet little village, beautifully situated in a lovely valley nestling among lofty mountain-ranges
at an elevation of 7,543 ft. above the sea. The air is remarkably dry,
pure, and invigorating, and the surrounding scenery is charming; but
the chief attraction of the place is its hot and cold mineral springs. The
waters contain soda, magnesia, iron, and lime, have fine tonic properties,
and are considered remedial in rheumatism and paralysis. They are
used chiefly for bathing, and there are extensive bathing establishments
and swimming-baths, in which baths may be had at the natural heat of
the water as it bubbles from the ground, or at a lower temperature.
During the summer the little town is thronged with tourists, and its
sheltered position makes it a desirable resort in winter. It is a favorite
gathering-place for excursion parties, and full outfits of carriages, horses,

and guides are here furnished to those desiring to visit Middle Park, the. Chicago Lakes, Green Lake, the Old Chief, or the mining regions. The most popular excursions are to *Fall River* (2¼ miles), and to the lofty-lying *** Chicago Lakes** (15 miles by trail). These lakes are the most picturesque sheets of water in Colorado, and are embosomed on the slopes of Mt. Rosalie at a height of 11,995 ft. above the sea. Georgetown and Idaho Springs are equidistant from them, and though the trail by which they are approached is rough, they are visited by many tourists during the summer months. Twelve miles beyond Idaho Springs is **Georgetown,** an important mining town with a population of 1,927, situated on S. Clear Creek, at an altitude of 8,514 ft., being 5,000 ft. higher than the glacier-walled valley of Chamounix, and but some 2,000 ft. lower than the elevation of Leadville. It is inclosed in a perfect amphitheatre of hills and mountains and cliffs, is laid out with broad streets, and is divided by the creek which winds through it in a silvery current. There are many romantic spots in the neighborhood. Just above the town is the *Devil's Gate*, a profound chasm through which a branch of Clear Creek foams and leaps. About 2½ miles distant is *Green Lake*, with clear waters of a bright green color produced by a coppery sediment on the rocks at the bottom. From Georgetown the railroad passes the "loop," where it crosses its own track on a high bridge, and reaches *** Graymount** (elevation 9,500 ft.), at the foot of Gray's Peak (elevation 14,251 ft.). Here horses and guides can be obtained for an ascent, which can be made by ladies with difficulty. A favorite excursion from *Gray's Peak Hotel* is to ascend and witness the sun rise over this mountainous region.

The Colorado Central Broad Branch runs N. W. from Golden, and in 24 miles reaches *Boulder*, a mining town whose proximity to the famous *** Boulder Cañon** makes it interesting to tourists. A wagon-road leads up the cañon, which is a stupendous mountain-gorge, 17 miles long, with walls of solid rock that rise precipitously to a height of 3,000 ft. in many places. A brawling stream rushes down the center of the ravine, broken in its course by clumsy-looking rocks and the fallen trunks of trees that have been wrenched from the sparse soil and moss in the crevices. About 8 miles from Boulder are the *Falls of Boulder Creek*, and at the head of the cañon is a mining settlement. *Longmont* (13 miles beyond Boulder) is the starting-point for a delightful excursion through the lovely **Estes Park** to the summit of *Long's Peak* (36 miles). Estes Park affords some beautiful views and excellent trout-fishing. Long's Peak is 14,088 ft. high, and affords one of the grandest views to be obtained in Colorado. The ascent is tedious, but not difficult.

The *Denver Pacific Branch of the Union Pacific R. R.* runs from Denver to Cheyenne, and there connecting with the main line of the same road for all points west. The distance by this route is 107 miles, and the country traversed is for the most part a vast level plain covered only with the short gray buffalo grass. The road runs nearly parallel with the principal range of the Rocky Mountains, and 20 to 30 miles from their E. base. The only noteworthy town *en route* is **Greeley** (51 miles from Denver), which is a flourishing place of 2,395 inhabitants, situated on the banks of the Cache.la Poudre River, and named after the founder of the *N. Y. Tribune*. It is watered by an excellent system of irrigation, and is well wooded. No intoxicating liquors are sold within its limits.

100. The Great Natural Parks.

THE surface of Colorado is generally mountainous, but in the E. and N. W. portions are elevated plains, and the spurs or branches of the Rocky Mountains inclose large fertile valleys. These valleys are known as the North Park, Middle Park, South Park, and San Luis Park, and are perhaps the most characteristic feature of Colorado. **North Park,** lying in the extreme northern part of the State, has been less explored and settled than the rest, owing to its remote situation and colder climate. It offers, for these reasons, the greatest attractions for the sportsman and adventurer. The park embraces an area of about 2,500 square miles, and has an elevation of about 9,000 ft. above the sea. Recent discoveries of gold and silver are attracting attention. It is best reached by stage from Fort Collins on the Colorado Central Div. of the Union Pacific system to *Mason City* (80 miles) and *Tyner* (125 miles).

Middle Park lies directly S. of North Park, from which it is separated by one of the cross-chains of the great mountain labyrinth. The snow-range, or continental divide, sweeps around on its E. side, and it is completely encircled by majestic mountains. Long's Peak, Gray's Peak, and Mount Lincoln, from 13,000 to 14,500 ft. high, stand sentinels around it. It embraces an area of about 3,000 square miles, extending about 65 miles N. and S. and 45 miles E. and W., and is about 7,500 ft. above the sea. It is drained by Blue River and the headwaters of Grand River, flowing westward to the Colorado. The portions of the park not covered by forest expand into broad, open meadows, the grasses of which are interspersed with wild-flowers of every hue. There is game in abundance, including deer, mountain-sheep, elk, bears, and antelopes, and the waters teem with fish. The climate, notwithstanding the great elevation, is remarkably mild and equable, with cool nights in summer and warm days in winter. No one, of course, should attempt to winter here who can not safely be cut off from many of the comforts and conveniences of life; but those who are able and willing to "rough it" will hardly find a place where they can do so under more favorable conditions. The usual objective point of tourists who go to the Middle Park is the **Hot Sulphur Springs,** which may be reached from Georgetown by the Berthoud Pass (45 miles); from Central City by the James's Peak trail (60 miles); and from South Boulder. The Colorado Company's fine stages leave the Barton House, Georgetown, every other day for the Springs. A pleasant way of making the journey is on horseback *via* the first-mentioned route. The Springs are situated on a tributary of Grand River, about 12 miles from the S. boundary of the park. The waters are used chiefly in the form of baths, and have been found highly beneficial in cases of rheumatism, neuralgia, chronic diseases of the skin, and general debility. The accommodations for invalids are not first-rate as yet, but sufficient, perhaps, for those who ought to venture upon the journey thither over the mountains. A small town is gradually growing up in the vicinity. One of the pleasantest excursions in Middle Park is up the valley, 27 miles from the Springs, by a good road to *Grand Lake*, the source of the main fork of Grand River. The lake nestles close to the base of the mountains, precipitous cliffs hang frowning over its waters on three sides, tall pines come almost down to the white sand-beach, and its translucent depths are thronged with trout and other fish.

South Park, the best known and most beautiful of all the parks, lies next below Middle Park, from which it is separated by a branch of the Park range. It is 60 miles long and 30 wide, with an area of about 2,200 square miles, and, like the Middle Park, is surrounded on all sides by gigantic ranges of mountains, whose culminating crests tower above the region of perpetual snow. One of the most noted mountains is the *Mount of the Holy Cross*, which can be seen from Robinson's Station, a few miles from Leadville. It is one of Colorado's wonders, and is 14,176 ft. high. The maximum elevation of the park above the sea is 10,000 ft., while the average elevation is about 9,000 ft., and nearly all the land which it contains is well adapted to agriculture. The streams, which are supplied by melting snows from the surrounding mountains, are tributaries of the South Platte, and flow E. through the park to the plains. The climate of the South Park is milder than that of either North or Middle Park, and its greater accessibility gives it peculiar advantages for such tourists and invalids as can not endure much fatigue. *Fairplay* is the chief town of the region, and a good center for excursions. The scenery afforded by any or all of these routes is of incomparable grandeur and beauty, especially at the cañon of the Platte and Kenosha Summit. The visitor to Fairplay in summer should not fail to ascend * **Mount Lincoln,** which is one of the highest of the Colorado peaks (14,296 ft.), and affords a view that Prof. Josiah D. Whitney declares to be unequaled by any in Switzerland for its reach or the magnificence of the included heights. The ascent may be made nearly all the way by wagon or carriage, and presents no difficulty. Another pleasant excursion from Fairplay is to the beautiful *Twin Lakes* (35 miles). The *South Park Div. of Union Pacific System* runs from Denver S. W. through South Park to *Leadville* (see Route 98), and the Colorado Midland R. R. traverses it from E. to W., by way of Colorado Springs, taking in Florrisant, Lidderdale, Spinney, and in its course touching on the *Hartsel Hot Springs* in the S. W. portion.

San Luis Park is larger than the other three combined, embracing an area of nearly 18,000 square miles—about twice the size of New Hampshire. It lies S. of South Park, from which it is separated by the main range, which forms its N. and E. boundary, while its W. boundary is formed by the Sierra San Juan. It is watered by 35 streams descending from the encircling snow-crests. Nineteen of these streams flow into *San Luis Lake*, a beautiful sheet of water near the center of the parks, and the others discharge their waters into the Rio del Norte, in its course to the Gulf of Mexico. On the flanks of the great mountain, dense forests of pine, spruce, fir, aspen, hemlock, oak, cedar, and piñon alternate with broad, natural meadows, producing a luxuriant growth of nutritious grasses, upon which cattle subsist throughout the year without any other food, and requiring no shelter. The highest elevation in the park does not exceed 7,000 ft. above the sea, and this, together with its southern and sheltered location, gives it a wonderfully mild, genial, and equable climate. Thermal springs abound here, as in other parts of Colorado, generally charged with medicinal properties. The Denver & Silverton Branch of the Denver & Rio Grande R. R. (see Route 98) is rendering this vast and attractive region more accessible.

101. Kansas City to San Francisco via Atchison, Topeka & Santa Fé R. R.

The great system of the Atchison, Topeka & Santa Fé R. R. now has two routes to San Francisco and the Pacific coast, and also a route to the city of Mexico. All these routes connecting with the principal routes from New York and other Eastern cities have their point of departure from Chicago, St. Louis, and Kansas City. These routes we shall now successively describe.

a. Via Atlantic & Pacific Short Line.

This route consists of the Atchison, Topeka & Santa Fé R. R. to Albuquerque, N. M. (918 m.); thence of the Atlantic & Pacific R. R. to Mojave (1,736 m.); thence of the Southern Pacific R. R. to San Francisco (2,118 m.). The principal stations on the Atchison, Topeka & Santa Fé R. R. as far as La Junta, Col. (571 m.), have already been described (see Route 96). After leaving La Junta, the most important stations are Trinidad (652 m.), Las Vegas (786 m.), Lamy (851 m.; junction for Santa Fé), and Albuquerque (918 m.). The Atlantic & Pacific R. R. junction is 17 m. S., and thence the notable stations are Wingate (1,064 m.), Manuelito (1,092 m.), Holbrook (1,171 m.), Winslow (1,204 m.), Ash Fork (1,319 m.), Peach Springs (1,384 m.), The Needles (1,493 m.), Daggett (1,653 m.), Mojave (1,736 m.), Caliente (1,768 m.), Tulare (1,874 m.), Madera (1,930 m.), Merced (1,963 m.), Lathrop (2,071 m.), San Francisco (2,115 m.). Complete palace and sleeping car service with dining-cars is used on this line, and there are excellent meal stations along the route under the general supervision of an official of the company. This service runs through from Chicago and St. Louis to San Francisco without change.

THE route from Kansas City to La Junta (571 miles) has already been described in Route 96. The scenery in Kansas, Colorado, New Mexico, and Arizona is very picturesque, and the line is built through a country where there are no extremes of heat or cold. The Atchison, Topeka & Santa Fé R. R. at this point bends southward, skirting the Raton Range of the Rocky Mts. on the W., and affording a succession of beautiful mountain landscapes to the eye of the traveler. **Trinidad** (652 miles), lying at the foot of the Raton Mts., has a population of 5,523, and was one of the most important points between Santa Fé and the Missouri River in the days of the "Old Santa Fé Trail." This, the first typical Mexican town met with on the middle route across the continent, and with its mixture of wooden, brick, and adobe houses, is always an object of interest to travelers on first journeys to this region. Trinidad has become a modern city now, since the railroad arrived, and has water-works, gas, daily papers, graded schools, banks, etc. It is the center of a large mining business, and wool and cattle trade. At Mor-

ley (662 miles) the road climbs the mountains through *Raton Pass*, on a grade 185 feet to the mile. The ascent is attended with many charming "views," not in the least marred by the name which attaches to the pass—Devil's Cañon. The view afforded from the pass of the Spanish Peaks as they rise across the plains, nearly 100 miles to the north, affords an excellent illustration of the vast reach of vision which is possible in these mountain-heights. Five miles farther up the mountain, at an elevation of 7,688 feet, the train suddenly plunges into a tunnel nearly half a mile long, running under the crest of the Raton Range. The light of Colorado quickly vanishes, and that which flashes upon us again in a few minutes is the warm brightness of sunny New Mexico, for we have crossed the border while coming through the tunnel. *Las Vegas* (786 miles) is situated on a branch of the Pecos River, is a place of 2,385 population, and is the trade-center of the great sheep-ranches of New Mexico. About 5,000,000 lbs. of wool are annually exported from this place. The Territorial Insane Asylum is located here. A short branch (6 miles) connects the town with Las Vegas Hot Springs.

Las Vegas Hot Springs is rapidly becoming a rival of the celebrated Arkansas Hot Springs. This attractive sanitarium is at the mouth of a beautiful cañon which opens on the plains 6 m. above Las Vegas, and from that point winds romantically into the Spanish Range of the Rocky Mountains. The springs, some forty in number, varying from very warm to entirely cold, have an altitude of 6,767 ft. The character of the waters closely resembles that of the Arkansas Hot Springs. The excellence of the waters for a wide range of diseases, and the delightful climate, averaging 41° during the winter months, have combined to make this a favorite resort. There are ample hotel accommodations, and the railroad company has erected the *Hotel Montezuma*. Rates, $2 to $3 per day, and $12 upward per week.

At *Lamy* (851 miles), named after the first Archbishop of Santa Fé, the Santa Fé Branch diverges in 18 miles to the ancient and interesting city of *Santa Fé (Palace Hotel)*, which has an altitude of 7,019 ft. This place, the oldest town in the United States, has a population of 6,185 inhabitants. Among the important buildings are the *Territorial Capitol*, a tasteful edifice of cream-colored sandstone; the *U. S. Building*, used for the U. S. courts and other purposes; and the *County Court-House*. There are several schools, including three for the Indians, one of which is called *Ramona School* as a memorial of "H. H."; also *St. Michael's College*, an institution conducted by the Christian Brothers. *St. Vincent's Hospital* is a large building near the Cathedral, and is partly supported by the Territory. Santa Fé is the seat of a Roman Catholic archiepiscopal diocese, and contains the *Cathedral of San Francisco*, the largest and most expensive church in the Territory. It is of light brown-stone, with two towers in front, and is built around the old cathedral, parts of which date back to 1622. Adjoining the church is a museum containing numerous paintings brought to this country by the early Spanish fathers, as well as many interesting historical manuscripts, several of which are very old. The *Chapel of San Miguel* is the oldest church in use in New Mexico. It was originally built between the years 1638 and 1680, but was destroyed by the Indians, and restored about 1710. Across a narrow street stands the "old home," the only remnant of the Pueblo town which preceded the

Spanish city. The mining interests which center here are large and of growing value, and the increasing trade will be likely to make Santa Fé one of the most important cities of the great Southwest. But its interest now is rather historical than actual. Among the relics of its past greatness is the ancient *Governor's Palace*, extending along one whole side of the *Plaza*, a long, low structure built of adobe. At one end of the Palace are the collections of the Historical Society, which are well worth seeing. It was erected in 1598, and of its history Governor Prince wrote in 1890 : " Without disparaging the importance of any of the cherished historical localities of the East, it may be truthfully said that this ancient palace surpasses in historic interest and value any other place or object in the United States. It antedates the settlement of Jamestown by 9 years, and that of Plymouth by 22, and has stood during the 292 years since its erection, not as a cold rock or monument, with no claim upon the interest of humanity except the bare fact of its continued existence, but as the living center of everything of historic importance in the Southwest. Through all that long period, whether under Spanish, Pueblo, Mexican, or American control, it has been the seat of power and authority. Whether the ruler was called viceroy, captain-general, political chief, department commander, or governor, and whether he presided over a kingdom, a province, a department, or a Territory, this has been his official residence. From here Oñate started in 1599 on his adventurous expedition to the Eastern plains; here, 7 years later, 800 Indians came from far-off Quivira to ask aid in their war with the Axtaos; from here, in 1618, Vincente de Salivar set forth to the Moqui country, only to be turned back by rumors of the giants to be encountered; and from here Peñalosa and his brilliant troop started, on the 6th of March, 1662, on their marvelous expedition to the Missouri; in one of its strong-rooms the commissary-general of the Inquisition was imprisoned a few years later by the same Peñalosa; within its walls, fortified as for a siege, the bravest of the Spaniards were massed in the revolution of 1680; here, on the 19th of August of that year, was given the order to execute 47 Pueblo prisoners in the plaza which faces the building; here, but a day later, was the sad war-council held which determined on the evacuation of the city; here was the scene of triumph of the Pueblo chieftains as they ordered the destruction of the Spanish archives and the church ornaments in one grand conflagration; here De Vargas, on September 14, 1692, after the 11 hours' combat of the preceding day, gave thanks to the Virgin Mary, to whose aid he attributed his triumphant capture of the city; here, more than a century later, on March 3, 1807, Lieutenant Pike was brought before Governor Alencaster as an invader of Spanish soil; here, in 1822, the Mexican standard, with its eagle and cactus, was raised in token that New Mexico was no longer a dependency of Spain; from here, on the 6th of August, 1837, Governor Perez started to subdue the insurrection in the north, only to return two days later and to meet his death on the 9th, near Agua Fria; here, on the succeeding day, Jose Gonzales, a Pueblo Indian of Taos, was installed as Governor of New Mexico, soon after to be executed by order of Armijo; here, in the principal reception-room, on August 12, 1846, Captain Cooke, the American envoy, was received by Governor Armijo and sent back with a message of defiance; and here, five days later, General Kearny formally took possession of the city, and slept, after his long and weary march, on the carpeted earthen floor of the palace. From every point of view it is the most important historical building in the country, and its ultimate use should be as the home of the wonderfully varied collections of antiquities which New Mexico will furnish. Coming down to more modern times, it may be added that here General Lew. Wallace wrote 'Ben-Hur,' while Governor, in 1879 and 1880." In the plaza stands the *Soldiers' Monument*, built in honor of those who fell in the Indian and the civil wars, and on the N. E. outskirts of the city stand the remains of the military post of *Fort Marcy*. It overlooks the city and the surrounding country for a distance of over 90 miles. A visit to the makers of the Mexican silver filigree-work is worth making. It is a characteristic industry of this place.

Santa Fé is mostly built of adobe, and its streets present a picturesque commingling of Americans, Mexicans, and Indians. The *Cerillos Mines*, whence are derived the ancient turquoises, are within 20 miles of Santa Fé. The points in New Mexico of most interest to the antiquarian or tourist are the Pueblo Indian villages, as here is to be seen the aboriginal civilization as found by Cortes and Coronado, absolutely unchanged through the centuries that have passed. These Indians are called "Pueblos" on account of their living in permanent towns (a *pueblo* being a town), and, when first discovered, all of their houses were built of stone, and were usually large community houses, accommodating from 100 to 400 persons, built in terrace form, and entered only by means of ladders. The present towns are 19 in number, extending down the Rio Grande Valley from Taos, in the north, to Isleta, a short distance south of Albuquerque, and including Laguna, Ácoma, and Zuñi to the west. Laguna is on the line of the railroad and easily visited. Ácoma is built on a high, flat hill (*mesa*), which was impregnable before cannon were introduced. Taos is composed of two great buildings five stories high; and its festival, on September 30th, draws thousands of visitors. Tesuque is near Santa Fé, and consequently the most accessible to tourists. The Annual Festivals at these pueblos are gorgeous spectacles, entirely unique in character, and well worth a journey from the Atlantic to attend. The principal ones are at Santo Domingo (Wallace Station), August 4th; San Juan, June 25th; Taos, September 30th; Santa Clara, August 12th, etc.

Resuming our journey from Lamy, we arrive in 77 miles at **Albuquerque** (*San Felipe*) (918 miles), a town of 3,785 population, situated on the Rio Grande River, at an elevation of some 5,000 ft. above the sea. The place has an extensive trade in wool and hides. The junction with the Atlantic & Pacific R. R. is 13 miles S. of this point, though officially Albuquerque is the E. terminus of the road. *Laguna*, 66 miles beyond Albuquerque, is a Pueblo Indian village, built upon a rounded elevation of rock, from the foot of which quite an extended view may be had from the train of the clusters of little, square, flat-roofed houses, which are arranged irregularly in terraces, without much regard to streets or alleys, and apparently without any special place of entrance or approach. Closer inspection shows that all the houses are built of adobe, with very thick walls, and that the interior is reached by the aid of scaling-ladders, through apertures in the roof. These buildings are, perhaps wrongly, called houses; they are more properly rooms, arranged in terraces one above the other. The Pueblo Indians are the most ancient race on this continent, and are law-abiding, peaceable, self-supporting citizens. As a rule they are fairly good-looking, especially the young girls, some of whom are quite pretty—even the older members of the tribe are not as repulsive as the average Indian is pictured. Their habits are much the same as those of all other dwellers in arid lands. They depend for their subsistence upon the sale of their pottery to strangers, and to scant cultivation by irrigation. Some of the younger members have been educated by the Government at the Carlisle School, and speak English with a pleasing accent. They have returned to their tribes, dropped into the old ways, and appear to be content. They are friendly to the tourist, and display with every evidence of pride, and with no little ability as beginners, their fantastic and gaudily-colored pieces of pot-tery. Some of the designs, though rude in shape, are quite artistic. Although not extremely religious, many are devout Catholics, and not a few are Protestants. Their church, which is perhaps two centuries old, is an interesting feature of the village. It is quite large, and has the general appearance of all old missions. The interior decorations are worthy of reproduction, and, on the whole, are above the average of such things. The floor is of earth, and is said to cover the remains of many of the celebrities of their tribes. The interior is well arranged and scrupulously clean. The road runs 130 miles W. through New Mexico, through a region full of fine sheep and cattle ranches, inter-spersed with mountainous tracts, till it reaches at the station known as the *Continental Divide* the great mother ridge of the Rocky Mts. Many parts of the region traversed have recently opened rich mining developments. *Wingate* (1,064 miles) is a busy little town, 3 miles from which is Fort Wingate. From this point stages run to the Indian village of Zuni, 45 miles N., about which so much interest has been aroused by the researches of Frank H. Cushing, who claims to have found among these Indians the relics of a high and mysterious civilization. Passing *Manuelito* (1,092 miles), where a stage-line runs to Fort Defiance, the headquarters of the Navajo Agency, *Holbrook* (1,171 miles) is reached after a number of small stations. A stage runs several times a week to the *Moqui Indian Village*, 70 miles away. The towns of these Indians are singularly interesting and well worth a visit. They are generally built on an eminence commanding a view of the surrounding country, so situated that they can only be approached through a narrow defile. The houses are 2 or 3 stories high, built of mud and stone, and ranged in the form of hollow squares. Access can only be had by ladders to the second stories, the first being built solid without any opening. There are seven of these Moqui Pueblos, or Dying Cities, as they have been called (of which Zuni is the chief), and the inhabitants, by their skill in pottery, weaving, and mural decoration, and by their strange religious rites, have deeply excited the curiosity of archæologists. Among a succession of small stations *Winslow* (1,204 miles) is important as being the diverging-point for stage-routes and supply-trains. About 26 miles farther on we reach *Cañon Diablo*, where the scenery is of the most somber and impressive nature, and the railroad spans the mighty chasm by a bridge 500 ft. long and 225 ft. high. Passing Ash-Fork (1,319 miles) and a number of small stations, we reach *Flagstaff* (1,263 miles), to which the tourist will look with greater interest than to any other station *en route*, as it is the point of departure for one of the greatest wonders of nature in the world, the Grand Cañon of the Colorado. A stage-route of 70 miles conveys the tourist to the most interesting portion of the cañon.

The * **Grand Cañon of the Colorado** was made known to the world by the adventurous voyage of Major John W. Powell down the river. The Colorado River is formed by the Grand and Green Rivers, which unite in Utah, and flows southward into Arizona. It passes through a succession of extraordinary cañons, remarkable in themselves ; but all of these preliminary wonders sink into insignificance before the Grand Cañon, which is more than 300 miles long. This cañon opens all the series of geological strata down to the granite foundation. The walls are from 3,000 to 7,000 ft. in height. The plateau adjacent to the cañon is said to be about 7,000 ft. above the sea-level. Major Powell, who has made the most satisfactory study of this great wonder, writes as follows : " To a person studying the physical geography of this country without a knowledge of its geology, it would seem very strange that the river should cut through the mountains, when apparently it might have passed around them to the east through valleys, for there are such along the north side of the Uintahs, extending to the east, where the mountains are degraded to hills. Then why did the river run through these mountains ? The first explanation suggested is, that it followed a previously-formed fissure through the range ; but a very little examination will show that this is unsatisfactory. Then why did not the river turn around this obstruction, rather than pass through it ? The answer is, that the river had the right of way ; in other words, it was running ere the mountains were formed ; not before the rocks of which the mountains are composed were deposited, but before the formations were folded so as to make a mountain-range. The contracting or shrinking of the earth causes the rocks near the surface to wrinkle or fold, and such a fold was started athwart the course of the river. Had it been suddenly formed, it would have been an obstruction sufficient to turn the water into a new course to the east beyond the extension of the wrinkle, but the emergence of the fold above the general surface of the country was little or no faster than the progress of the corrosion of the channel. We may say, then, that the river did not cut its way down through the mountains from a height of many thousand feet above its present site, but having an elevation differing but little, perhaps, from what it is now, it cleared away an obstruction by cutting a cañon, and the walls were thus elevated on either side. The river preserved its level, but the mountains were lifted up—as the saw revolves on a fixed pivot while the log through which it cuts is moved along. . . . The upheaval was not marked by a great convulsion, for the lifting of the rocks was so slow that the rains removed the sandstones almost as fast as they came up. The mountains were not thrust up as peaks, but a great block was slowly lifted up, and from this the mountain was carved by the clouds—patient artists, who take time to do their work. Mountains are often spoken of as forming clouds about their tops : the clouds have formed the mountains. Lift a district of granite or marble into their region, and they gather about and hurl their storms against it, beating the rocks into sand ; and then they carry them out into the sea, carving cañons, gulches, and valleys, and leaving plateaus and mountains embossed on the surface."

Thomas Moran, the artist, who visited the Grand Cañon shortly after Major Powell, gives us the following interesting description : " Our first journey was to the Toroweap Valley. By following down this valley, we passed through the-upper line of cliffs to the edge of a chasm cut in red sandstone and vermilion-colored limestone or marble, 2,800 feet deep and about 1,000 feet wide. Creeping out carefully to the edge of the precipice, we could look down directly on the river, 15 times as far away as the waters of Niagara are below the bridge. Mr. Hillers, who passed through the cañon with Major Powell, was with us, and he informed us that the river below was a raging torrent ; yet it looked from the top of the cliff like a small, smooth, sluggish river. The river, looking up the cañon, is magnificent, and beyond the most extravagant conception of the imagination. In the foreground lies a profound gorge, with a mile or two of the river seen in its deep bed. The eye looks twenty miles or more through what appears like a narrow valley formed by the upper line of the cliffs. The many-colored rocks in which the valley is curved project into it in vast headlands 2,000 feet high, wrought with beautiful but gigantic architectural forms. Within an hour of the time of sunset the effect is strange, weird, and dazzling. Every moment, until light is gone, the scene shifts, as one monumental pile passes into shade, and another, before unobserved, comes into view. Our next visit was to the Karbal Plateau, the highest plateau through which the river cuts. It was only after much hard labor, and possibly a little danger, that we could reach a point where we could see the river, which we did from the edge of Powell Plateau, a small plain severed from the mainland by a precipitous gorge 2,000 feet deep, across which we succeeded in making a passage. Here we beheld one of the most awful scenes on the gulch. While on the highest point of the plateau, a terrible thunder-storm burst on the cañon. The lightning flashed from crag to crag. A thousand streams gathered on the surrounding plain, and dashed down into the depths of the cañon in waterfalls many times the height of Niagara. The vast chasm which we saw before us, stretching away forty miles in one direction and twenty miles in another, was nearly 7,000 feet deep. Into it all the domes of the Yosemite, if plucked from the level of that valley, might be cast, together with all the mass of the White Mountains in New Hampshire, and still the chasm would not be filled."

The Needles (1,497 miles), so named from the curious shape of the huge pinnacles of rock which greet the eye, are on the W. bank of the Colorado River. The bridge spanning the Colorado River at this point is one of the most notable engineering feats of the times. It is the longest cantilever bridge and has the longest single span in America. Its total length is 1,110 ft. ; and while a massive and striking structure, is still of graceful design, and its airy lightness contrasts strongly with its dimensions. Brown, desolate bluffs rise abruptly from the river's side, and to the south the ragged Needles Mountains, with their fantastic shapes and spires, form a grand and towering background for an inspiring picture of a picturesque, romantic, and desolate solitude. This bridge is what is known as a "through" cantilever bridge, and is the second of the kind in America. It is 75 ft. high from the bed of the river, and is entirely of steel, 3,400,000 pounds having been used in its construction. The principal dimensions are as follows : Main suspended span, 660 ft. ; cantilever arms (each 165 ft.), total, 330 ft. ; viaduct, 120 ft. ; and its cost was $480,000. *Barstow* (1,662 miles) is the point where connection is made with the Southern California R. R. for Los Angeles, Pasadena, and San Diego, all of which are described on pages 422 and 423. For 240 miles to *Mojave* (1,733 miles) the route presents no special features which it is necessary to describe. The town is a brisk place of several hundred inhabitants, and an important point of distribution for mining supplies. It is at this point that the trains of the "Santa Fé Route" pass on to the track of the Southern Pacific Railway. *Caliente* (1,768 miles) and Tulare

(1,874 miles) need only passing mention. At *Lathrop* the road proceeds duly W. 94 miles to **San Francisco** (see Route 90). From here also the train is taken to *Raymond*, which is a starting-point for the **Yo-semite Valley** (see Route 92).

b. Northern Route via Colorado Springs, Col., and Ogden, Utah.

The northern route to the Pacific coast is by the Atchison, Topeka & Santa Fé R. R. to Colorado Springs, Col. ; thence by the Colorado Midland Div. of the Santa Fé system to Salt Lake City and Ogden, Utah ; thence by the Central Pacific R. R. to San Francisco. For principal stations as far as Pueblo (635 m.), see Route 96. For stations from Ogden to San Francisco (2,180 m.), see Route 89. This route runs complete sleeping and buffet car service, and there are meal-stations along the whole line.

The route from Kansas City to Pueblo, Col., has already been de-scribed in Route 96. From the latter point to Colorado Springs the road is laid on a prairie track in sight of the Rocky Mountains, and quite near the foot-hills at times. At *Colorado Springs*—an all-the-year-round health resort, a clean, pleasant, beautiful town, with all the con-veniences and luxuries of a metropolitan city—the train takes the track of the Colorado Midland Railway ("Pike's Peak route"), which is the regular Denver & Ogden line, and runs by way of *Manitou*, the famous, the romantic, the beautiful, with its innumerable novelties and attrac-tions, and its famous mineral waters ; thence under the base of Pike's Peak, following the pioneer Ute Pass, reaching Cascade Cañon, a resort built up amid fascinating scenery at the base of the historical peak ; Ute Park, in the center of the Ute Pass, overlooking a lovely valley hemmed in by lofty mountains ; *Green Mountain Falls,* a cosmopolitan resort which became famous in a season, and whose fame is rapidly spreading from day to day ; *Woodland Park,* at the head of the Ute Pass, with a view of Pike's Peak, which, once seen, is never forgotten ; and then *Manitou Park,* to complete the list of resorts in the *Ute Pass,* one of the most beautiful parks in all Colorado. A coach conveys the traveler from the station to the park, 7 miles distant. From Ute Pass the road extends W. over the Hayden Divide, down to and through Granite Cañon, across South Park, over Trout Creek Pass, and down to Buena Vista, in the Arkansas Valley, and thence to Leadville. The route from Leadville W. is over the Saguache Range or Continental Divide, passing through the Hagerman Tunnel at an altitude of 11,528 ft., the highest in the United States. From the summit of the Snowy Range the road passes down the Pacific Slope, circling Hell Gate, on the Frying-Pan River, passing the White Sulphur Springs, and through the Red Rock Cañon to the Roaring Fork of the Grand River. Trav-ersing the Roaring Fork Valley, it reaches Aspen and Glenwood Springs, the famous hot mineral water baths of Colorado, from which latter place it follows the Grand River Valley to Grand Junction, which is the point of departure for pack-trains for the White River country. At Grand Junction the trains take the track of the Rio Grande Western Railway, and traverses a region which is notable for its grand and pict-uresque scenery. *Green River* is on the Green River, across which there is a fine railway-bridge. Pleasant Valley Junction (1,145 miles)

29

is at the junction with the Pleasant Valley Branch. **Provo** (1,269 miles), a prosperous town in Utah of 5,159 population, is on the Provo River, about 3 miles E. of Utah Lake, and lies near the W. base of the Wahsatch Range. It contains a town-hall, a theatre, many flouring-mills and tanneries, reduction-works, etc., and is the center of a region rich in products of wheat, hay, wool, and cattle. Here connection is made with the Utah Central R. R. for important points in southern Utah. At *Bingham Junction* (1,299 miles) connection is made with the Little Cottonwood and Bingham branches; and in 11 miles farther we reach **Salt Lake City** (see Route 90). At **Ogden** (1,346 miles) junction is made with the Central Pacific Div. of the Southern Pacific R. R., by which the passenger continues his route to **San Francisco** (see Routes 90, 91). Ogden is also the point of junction with the Union Pacific system, going W. to Portland, N. to Helena, or S. to San Francisco.

c. Via Deming and Benson to Guaymas over the Atchison, Topeka & Santa Fé R. R., Southern Pacific Co., New Mexico & Arizona Railway, and Sonora Railway; also to Yuma via the Atchison, Topeka & Santa Fé R. R. and Southern Pacific Railway.

As far as Deming this route is by the Atchison, Topeka & Santa Fé R. R. (1,148 m.), thence by the Southern Pacific R. R. to Benson (1,322 m.); then by the New Mexico and Arizona Railway to Nogales (1,410 m.), and Sonora Railway to Guaymas, Mex. (1,587 m.). The fares from Kansas City and New York are the same as by Sub-Routes *a* and *b*, and the same equipment and provision are made for the comfort of the traveler. The principal stations as far as Albuquerque, N. M. (918 m.), have already been described in Routes 96 and 101 *a*. After this station the main places of interest are *Socorro* (994 m.), San Marcial (1,021 m.), Rincon (1,094 m.), and Deming (1,149 m.), the point of junction with the Southern Pacific Co. The main point on this line is Benson (1,322 m.). Through sleeping-car service may be had from Chicago to Albuquerque, from St. Louis to El Paso, and from Deming to Yuma; also tourist-sleeper from Benson to Guaymas.

This route has been described in Route 96 (from Kansas City to La Junta) and in Sub-Route *a* (from La Junta) as far as Albuquerque (918 miles). Following the main line of the A., T. & S. F. R. R., we soon reach one of the most interesting mining regions of New Mexico, which covers several thousand square miles, and extends between the Black Range Mts. on the W. and the San Andres Mts. on the E., the valley of the Rio Grande running between. The mineral belts lying in the foot-hills of the Black Range have proved to be peculiarly rich. **Socorro** (994 miles) has a population of 2,591, and is beautifully situated in the Rio Grande Valley. Mining, grazing, and fruit-growing are the principal industries. Socorro has a stamp-mill and smelting-works, and includes in the mining districts tributary to it many of the best-known mines in New Mexico. Within three miles are the famous Torrence and Merritt mines. The ores are mostly carbonates of lead, carrying silver, some of which run as high as $28,000 to the ton. The development thus far has been sufficient to warrant the belief that the mountains around Socorro contain many millions of gold, silver, and lead. Socorro contains the Territorial School of Mines, and is the center of a fine agricultural, fruit-growing, and stock-raising region, and there are many fine ranches in the vicinity. *San Marcial* (1,119 miles), consisting of the old and new towns, has a population of 611, and is a thriving place, where the railroad company has repair-shops. The battle of Valverde, named after a little Mexican village across the river, was fought here in 1862 between the Federals under Gen. Canby and the Confederates under Gen. Sibley. The railway runs on the W. side of the Rio Grande as far as San Marcial, where it crosses the stream, and in 75 miles reaches *Rincon*, whence the El Paso Branch diverges to the S. to *El Paso* (77 miles), and connects with the Mexican Central R. R. for the city of Mexico and intermediate points (see Route 102). At *Deming* (1,149 miles) is the proper terminus of the A., T. & S. F. R. R.; and here the train takes the track of the Southern Pacific, which runs through southern Arizona.

Arizona is a region comparatively unknown to the tourist, but the country has been now made so accessible by the Southern Pacific and Atlantic & Pacific R. Rs., running in connection with the Atchison, Topeka & Santa Fé R. R., that its beautiful scenery and many attractions are likely to excite much curiosity and interest. It is bounded on the N. by Utah, on the E. by New Mexico, on the S. by Mexico, on the W. by California. It occupies an area of 113,020 square miles, being about twice as large as the State of New York. It is essentially a mountainous and wooded country, though there are vast stretches of sandy plains, which, however, only need irrigation to become fertile. It is exceedingly rich in mineral products, and has many large sheep and cattle ranches. The land dips toward the S. W. from lofty plateaus nearly 6,000 ft. high to plains only a few hundred feet above the sea. This slope is one grand network of mountains, which contains some of the most noble and picturesque scenery in America. The Gila River, which has its sources in the mountains of N. E. Arizona, flows for a portion of its course parallel with and near the S. P. R. R., emptying into the Colorado near Yuma (see present route). Of the Colorado, a river without a parallel in many of its features, we have already spoken (see Sub-Route *a*). The winter climate of S. Arizona can not be surpassed. The climate near Yuma is finer than that of Italy. In the mountains of W. Arizona, for the greater part of the season white with snow on the upper peaks, the air is pure and dry and deliciously cool. An immense variety of climate can be found within a range of 200 m. from N. to S. The winter climate of S. Arizona is both warm and dry. A writer familiar with the region thus describes the country:

"As soon as this great sanitarium is fully known it will become for winter what Colorado now is in summer—a great resort for invalids. From the middle of June to October, however, the heat is intense; but travelers say that, even with the thermometer at 120 degrees, sunstrokes are of rare occurrence. This is due to the rarity of the atmosphere. The average rainfall at Fort Mojave is but little over five inches, distributed through August, December, February, and June. At Camp Grant, which is said to be in all respects a medium climate, the diurnal variations of temperature are from 15 to 30 degrees; the monthly range being about 27 degrees, and the yearly extremes of heat and cold 34 and 96 degrees respectively. There are, annually, about 65 days of rain and hail, and 3 of snow. At Camp Verde the temperature ranges from 5 degrees to 113 degrees, and the average rainfall is 8 inches. At Camp Lowell, 7 miles east of Tucson, the diurnal range is sometimes 70 degrees. Persons afflicted with pulmonary complaints experience speedy relief in this warm atmosphere, and many wonderful and well-authenticated cures of this nature are reported. The scenery is truly charming. It is not so rugged, perhaps, as Colorado, but it is, if possible, more pleasing. Instead of having a continuous mountain-chain running in a given direction, it has isolated peaks and detached sections coming up out of the plain apparently at random. Yet, while her landscapes are thus beautiful to a degree that admits of no rivalry, Arizona has her towering peaks and deep cañons surpassing those of any other locality. The cañons on the Colorado River are some of them 6,000 feet, or more than a mile, in depth. Mention should here be made of the valleys of Arizona. They are numerous and fertile. In the valley of the Verde settlements have been made to a considerable extent. Williamson's Valley, near Prescott, contains not less than 500,000 acres, together with 300,000 acres of adjacent foot-hills, well furnished with bunch-grass. Around Mount Hope, in Yavapai County, there are scores of beautiful valleys containing from 40 to 400 acres of land each; in fact, wherever a river runs, there, at some portion of its course, may be found as lovely depressions as exist anywhere in the United States. It is estimated that there are about 2,800,000 acres of land in the Territory, of the very best quality, with sufficient surface water near at hand to properly irrigate. At least 10,000,000 acres more, it is said, can be reclaimed by the use of artesian wells."

Let us now resume our railroad journey across the southern part of Arizona. *Benson* (1,322 miles) is the point of junction with the Sonora Div., which runs in 265 miles to *Guaymas*, Mexico, on the Gulf of California, through Sonora, passing through *Hermosillo*, the capital of Sonora, noted as the center of a rich agricultural valley, with fine grazing and mineral lands. The ancient city of **Tucson** (1,369 miles), which has 5,150 population, was until recently the capital of Arizona (*Phœnix* now enjoying that honor), was founded in 1560 by the Jesuits, and is almost as quaint in its buildings and social characteristics as Santa Fé. It contains two churches (one Roman Catholic), several excellent schools, two banks, a *Court-House*, and a *U. S. Depository*. It does a large business in exporting gold-dust, wool, and hides. Near *Casa Grande* (1,434 miles) is a remarkable ruin of an ancient **Pueblo City,** these interesting remains being preserved in a very perfect state and extending 2¼ miles by 1¼ miles, showing that it must have had a population in olden times of at least 100,000 people. The city of **Yuma** (1,616 miles) is near the junction of the Gila and Colorado Rivers, and here the train crosses the latter river on a fine bridge. The city has a population of some 1,773. It is approached by steamer up the Colorado River, and is the W. terminus of the Arizona Branch of the S. P. R. R. Here is the location of the *Territorial Penitentiary*, and there are several other notable public buildings, besides a convent and several churches. Crossing the river here we find ourselves in the State of California.

102. Kansas City to the City of Mexico via Atchison, Topeka & Santa Fé R. R. and the Mexican Central R. R.

The trains run sleeping-cars daily between St. Louis, Kansas City, and the City of Mexico *via* El Paso without change. The distance from Kansas City to El Paso is 1,173 miles; to the City of Mexico, 2,398 miles.

The "International Route" to Mexico is *via* Southern Pacific from New Orleans to Spofford, then by branch line to Eagle Pass, where the transfer is made to the International R. R., which connects with the Mexican Central at Torreon. (See page 510, Route 129.)

FROM Kansas City to La Junta this route has been described in Route 96; from La Junta, Col., to Albuquerque, N. M., in Route 101 *a*; from Albuquerque to Rincon, N. M., in Route 101 *c*. At *Rincon* (1,096 miles) the El Paso Branch runs to El Paso on the Rio Grande. *Las Cruces* (1,129 miles) is known as the "Vineyard City," from the surrounding orchards and vineyards, and has a population of 2,472. The Territorial Agricultural College and Experiment Station have been located here. The church, built in 1854, contains a number of interesting paintings. It is the center of a rich mining region, one mine, the "Stephenson," having produced $3,000,000 in five years.

El Paso, in Texas, is 77 miles S. of Rincon, 1,173 miles from Kansas City. The Atchison, Topeka & Santa Fé R. R. connects here in its own depot with the Mexican Central R. R. The population of El Paso is 10,338, and the city is growing very rapidly; a large retail and wholesale trade is done here (the fruit trade is of large importance), and its superior railroad facilities give El Paso merchants many advantages. The Southern Pacific Co. and the Texas & Pacific R. R. center here. This place is of considerable importance as a port of entry from Mexico, and 75 per cent of the exports from that country to the United States pass through the custom-house here. It is the chief point of entry for the Mexican bullion and ores that are sent North for smelting. There are hotels (*Grand Central, Vendome*), banks, 2 street-railways to the old town of Juarez, the only international street-railways in the world, and three newspapers; Methodist, Episcopal, Catholic, Baptist, and Presbyterian churches, and good public schools. The military post about to be established in El Paso will be the most important one on the Southern frontier.

El Paso del Norte,[1] now called Juarez, the first city of Old Mexico which we reach, is just across the river from El Paso. It is built almost entirely of adobe, and the homes of its 7,000 people are scattered along a narrow, rambling, adobe-walled street running several miles down the river. The old church is worth a visit, and the Custom-House is a fine building. The ride to the city of Chihuahua, made during the day, introduces the traveler to the wide expanse of that high table-land which forms the greater portion of the interior of Mexico, but for variety it also includes a view of the beautiful valley of the Rio Carmen; while beyond, on the W., lie the Sierra Madre Mts., which form a natural and effectual boundary between the States of Chihuahua and Sonora. **Chihuahua** (1,398 miles), the capital of the State of the same name, is a beautiful city of 13,128 people. There is a great deal of wealth and refinement in Chihuahua, being the center of a rich mining, agricultural, and stock-growing country, and its magnificent cathedral is one of the most imposing edifices on the continent. (*American House, Robinson House.*) From Chihuahua the road runs along the central plateau, through mountain-passes and among fertile valleys. Immense herds and flocks graze beside the track, while the agriculture is of that diversified character only possible where the products of the tropical and temperate climates may be grown in the same field. Passing *Sombrerete*, with its colleges and mines, a city of 7,000 population, and *Fresnillo*, a city of 13,021 people, we arrive at **Zacatecas,** with a population of 20,722 (1,960 miles). Telephonic and telegraphic service and the electric light—which illuminates the plaza—are evidences that this interior city in the heart of Mexico has begun to feel the influence of the great tide of immigration (*Hotel*

[1] A very brief description is given of all cities mentioned in Mexico. The reader is referred for information to "Appletons' Illustrated Guide to Mexico."

Zacatecano). The mines of Zacatecas have produced, with the primitive method of mining not yet entirely discarded in Mexico, about $1,000,000,000. The city is built on a vein of silver, and contains many fine residences and public buildings, among them being the *Mint, Theatre, Hospital*, and several convents. *Aguas Calientes* (2,035 miles, population 32,355) is noted for its *Hot Springs*, which give name to the city and State. **San Luis Potosi** (37,314 population), the capital of a State of the same name, which lies a little E. of the main line on the route to Tampico, is considered by many to be the most beautiful city of Mexico. **Guadalajara** (*Hotel Hidalgo*), another city not on the main line, has a population of 83,122, and is one of the most important manufacturing centers of the republic. On the Mexican Central R. R. we successively pass *Lagos* (2,103 miles, 14,297 population), *Leon* (2,140 miles, 47,739 population), and *Celaya* (2,207 miles, 26,670 population), till we reach the historic city of **Querétaro** (2,246 miles, 23,520 population), which was the scene of the downfall and execution of the ill-fated Emperor Maximilian in 1867. The city contains many important woolen-mills, and is chiefly noticeable for its numerous ecclesiastical and religious structures, among the latter being the *Franciscan Monastery*, with its noble gardens and grounds (*Hotels, Diligencias, Del Ferro-carril*). Passing *San Juan del Rio* (2,280 miles, 10,000 population), *Polatitlan* (2,304 miles), *Tula*, on the banks of the river of the same name, famous as the ancient capital of the Toltecs (2,349 miles, 10,000 population), and several unimportant places, we reach the * **City of Mexico** in 2,398 miles. (*Hotels, Gillair, Iturbide, San Carlos.*) This ancient capital had a population of 351,804 in 1521, and was a seat of art, science, and commerce long before the Spanish conquest. The present population is 337,600. It is situated in the center of the great valley of Mexico, which measures 45 miles long and 31 miles wide. Its elevation above the sea is 7,420 ft., which gives it a climate of remarkable uniformity, the range of the thermometer being from 50° to 70° Fahr. The rainy season begins early in June and continues until September, showers occurring usually in the afternoons and nights. The city is built on a part of the old bed of Lake Texcoco, and tradition gives it a more romantic origin than it ascribes to the founding of Rome. Science and art have done much to make it a beautiful city, and there seems to be a disposition on the part of the people and government to make their nation's capital compare favorably with the capitals of other countries. The city is encircled by walls and entered by gates. The residences are mostly of stone, one and two stories high, and built around court-yards. The public edifices are numerous and substantial. Chief among the objects of interest is the * **Cathedral,** 500 ft. in length by 420 ft. in breadth, the largest ecclesiastical edifice in the western hemisphere. It is of mixed Gothic and Indian architecture, and is on the site of the chief temple of the Aztecs. The walls are gorgeously decorated, and the high altar is a marvel of magnificence. The dress on the statue of the Virgin is incrusted with gems, the diamonds alone, it is claimed, being worth $3,000,000. The Cathedral is on one side of the *Grand Plaza*, the other sides being occupied by the **National Palace,** comprising the government offices, mint, and prison, the *National Museum*, with an unrivaled Aztec collection, and the *Market-Place*.   Other fine old buildings are the *University of Mexico, Academy of Arts, Public Library*, containing 105,000 volumes, several theatres, and numerous churches and convents, of which there are 60 of the former and 40 of the latter.   Objects of interest are found in the fine *Botanic Garden* and the two aqueducts.   The city is noted industrially for its manufacture of gold and silver lace, and of silversmiths' work.

103. St. Louis to Texas.

a. Via Missouri, Kansas & Texas R. R.

THE Missouri Pacific R. R. makes close connection at St. Louis with all the important Eastern and Northern routes to that city, and runs four daily express-trains with sleeping-car service to principal points in Texas.   For the route to Sedalia (188 miles) see Route 96.   At this point connection for Southern points is made with the Missouri, Kansas & Texas R. R., which has its northern terminus at Hannibal, Mo. (143 miles from Sedalia), where close connection is made with the Chicago, Burlington & Quincy R. R., and other trunk-lines from the East.   Eastern passengers may take a through car at either Chicago, St. Louis, Kansas City, Hannibal, or Sedalia for places in Texas.   The Missouri, Kansas & Texas R. R. brings the rich agricultural and cattle-raising districts of Texas within 3 or 4 days' time of the Northern markets.   It is this which gives the road its importance, and except for this it presents little of interest to the traveler, traversing as it does a region which is for the most part uninteresting.   Sedalia is described in Route 93.   At *Fort Scott* (298 miles from St. Louis) it enters Kansas, and just beyond *Chetopa* (370 miles) it enters the Indian Territory, which it crosses in a nearly straight line from N. to S., passing through some picturesque scenery.   If the tourist desires to visit the great Indian Reservations, this is the road which he should take, as it carries him directly into their midst.

The principal stations on the northern section of the line are Fort Scott (298 miles) and Parsons (347 miles), both in Kansas.   **Fort Scott** is a city of 11,946 inhabitants, situated on the Marmiton River, a branch of the Osage.   It was established as a military post in 1842, and incorporated as a town in 1855, and is now rapidly growing.   Bituminous coal is abundant in the surrounding country, and the manufacturing interests promise to become important.   **Parsons** is a flourishing little city of 6,736 inhabitants, at the junction of the Neosho Div. of the M., K. & T. R. R. with the main line.   It is the site of the R. R. construction and repair shops, and is built on a high rolling prairie between and near the confluence of the Big and Little Labette Rivers.   *Vinita* (399 miles) is the first station in the Indian Territory, which is traversed to *Durant* (600 miles).   The first important station in Texas is **Denison** (621 miles), which is becoming an important railroad center.   It dates only from 1872, and had a population in 1890 of 10,958, with

several important flour-mills and factories. Denison is the N. terminus
of the Houston & Texas Central R. R., which traverses some of the
most fertile portions of Texas. Nine miles beyond Denison is **Sherman,** a city of 7,335 inhabitants. This is also reached by the Houston
& Texas Central R. R., and the Texas & Pacific R. R. It is substantially
built, largely of stone, and has a fine stone Court-House, with excellent schools and churches. Its trade with the surrounding country is
large, and its manufactures include cotton-seed oil, flour, and cotton.
Sixty-three miles beyond Sherman is **Dallas** (*Grand Windsor Hotel,
McLeod, St. George Hotel*), the capital of northern Texas, with a population
of 38,067, an extensive trade with the surrounding country, and numerous manufacturing establishments. It is well built for so young a city
(its population in 1872 was but 1,500), and has 22 churches, 29 schools,
electric street-railways, fire-companies, and gas and water works. The
Court-House, now building, will cost over $3,000,000. Prominent build-
ings are the Catholic and Episcopal Churches, and the Dallas Female
College (Methodist) and Male and Female College (Baptist). Dallas is
on the main line of the Texas & Pacific R. R., which runs from New
Orleans to El Paso, New Mexico, 1,157 miles. *Corsicana* (748 miles)
and *Mexia* (778 miles) are thriving towns. At *Bremond* (816 miles
from St. Louis and 143 miles from Houston) a line, known as the Waco
branch of the Houston & Texas Central R. R., diverges and runs in 45
miles to **Waco** (*McClelland House, Pacific Hotel*), a rapidly-growing
city of 14,445 inhabitants, situated nearly in the center of the State, on
both sides of the Brazos River, which is spanned by a handsome suspension-bridge. The city is regularly laid out and remarkably well built,
and contains a substantial stone Court-House, 9 churches, and a number
of flourishing educational institutions, of which *Baylor University* is the
principal. Waco is the commercial center of a rich and fertile country, which is rapidly filling up with immigrants, and has a number of
prosperous manufacturing establishments.

On the main line, 22 miles S. of Bremond, is *Hearne*, where the International & Great Northern Div. of the M. P. R. R. intersects the
present route. From *Hempstead* (51 miles from Houston) the Western
Div. diverges and runs in 115 miles to **Austin** (*Avenue House, Brunswick Hotel, Driskill*), the capital of Texas, a city of 14,476 inhabitants,
situated on the N. bank of the Colorado River, 160 miles from its mouth.
Just above the city a dam is being built across the Colorado. It is to
be 60 ft. in height and 1,100 ft. in length, constructed of limestone
(which abounds at the spot), and capped with red granite. It will cost
$468,000 together with the canal and appurtenances. Towering nearly
over the dam is a small mountain, on the brow of which the reservoir
is to be constructed. The reservoir, water-motors, and buildings at the
dam, and the mains and service-pipes and electric-light plant, lands,
wires, etc., will cost nearly $900,000 more. The city is built on an am-
phitheatre of hills, and overlooks the valley of the Colorado. *Capitol
Square* contains 20 acres on a gentle elevation in the center of the
city, upon the summit of which the *Capitol* (said to be the seventh
largest building in the world), of Burnet granite (costing $3,600,000), is

situated with the Supreme Court, while on the square is the Treasury building, and on the E. side is the *General Land-Office*, which is a handsome edifice. Other noteworthy buildings are the *University of Texas, Tillotson Institute, County Court-House*, the *County Jail*, the *Deaf and Dumb, Blind*, and *Lunatic Asylums*, and the *Market-House*, in the second story of which are the municipal offices. The city is connected with the region south of the Colorado by an iron bridge (free) that cost $75,000. Austin is also reached by the International & Great Northern R. R., by the Missouri, Kansas & Texas R. R., by the Texas Central R. R., and the Austin & Northwestern R. R.

Houston (*Capitol Hotel, Dissen House, Hotel Boyles*, and *Tremont House*), the third city of Texas in population and commerce and the first in manufactures, is situated at the head of tide-water on Buffalo Bayou, 30 miles above its mouth in Galveston Bay, and 819 miles from St. Louis. It is built on both banks of the bayou, which is spanned by several bridges, embraces an area of 9 square miles, and had by the census of 1890 a population of 27,557. Its manufactures are varied and extensive; and it is the center of the railroad system of the State, with 12 diverging railways, which bring to it the products of a rich grazing and agricultural region. The principal public building is the *City Hall and Market-House*, constructed of brick. Besides the city offices, it contains a hall, fitted up for public entertainments and capable of seating 1,300 persons. From the top of the main tower, 128 ft. high, there is a fine view. The Texas Geological and Scientific Association has rooms in the *Cotton Exchange*, a handsome brick building. The *Masonic Temple* is a spacious structure, costing $200,000, and the *Post-Office* is an attractive building. The city is lighted with gas and electric light, has abundance of water, an excellent fire department, and horse-cars. It has 5 cotton-presses, of a capacity of 1,000 bales a day each, and 3 cotton-seed oil mills. The bayou is navigable by vessels drawing 13 ft. of water, and the Morgan Line of Steamships affords connections with Galveston and New Orleans. From Houston, the International and Great Northern R. R. and the Gulf, Colorado & Santa Fé R. R. run S. E. in 50 miles to **Galveston** (*Beach House, Girardin, Tremont House*, and the *Washington*), the second largest city and commercial metropolis of Texas, situated at the N. E. extremity of Galveston Island, at the mouth of the bay of the same name. The city is laid out with wide and straight streets, bordered by numerous flower-gardens, and in 1890 contained 29,084 inhabitants. Besides the churches, of which several are handsome edifices, the public buildings include the Custom-House and Post-Office, U. S. Court-House, County Court-House, City Hall, Supreme Court-House, Cotton-Exchange, Masonic Temple, Opera-House, two theatres, several public halls, the Union and Harmony Club-Houses, and three market-houses. In the business portion of the city are numerous handsome commercial buildings, and there are many fine residences. Of institutions of learning the most noteworthy are the *University of St. Mary* (Roman Catholic), the *Ursuline Convent*, the *Sacred Heart* (a large convent and school), *Ball's High-School, Rosenberg School*, and other free public-school buildings. *Magnolia Grove Cemetery* embraces 100 acres neatly laid out. The *Island of Galveston* is about 29 miles long and 1¼ to 3½ wide, and is bordered throughout its whole length by a smooth hard beach which affords a pleasant drive and promenade. The harbor is the best in the State, and at present is being improved under the supervision of the U. S. engineers. The commerce of the city is very extensive, the chief business being the shipment of cotton. There is a tri-weekly line of steamers (the Mallory Line) to New York (starting-point in New York, Pier 20, East River), and a weekly line (Southern Pacific) from Pier 25, North River.

From Houston, the Galveston, Harrisburg & San Antonio R. R. runs W. 265 miles through a thickly settled country to **San Antonio** (*Maverick, Menger Hotel, Southern*), the chief city of western Texas, with a population in 1890 of 37,673, one third of whom are of German and one third of Mexican origin. It is situated on the San Antonio and San Pedro Rivers, and was formerly divided into three "quarters": San Antonio proper, between the two streams; Alamo, E. of the San Antonio; and Chihuahua, W. of the San Pedro. The former is the business quarter, and has been almost entirely rebuilt since 1860. Besides many handsome business buildings, it contains the *Federal Building* and the *City Hall*. It consists of the *Military Plaza, Main Plaza,* and Commerce, Market, and Houston Sts., and the streets running parallel with each other from the Main Plaza. Separated from the Main Plaza by a fine Catholic church is the *Plaza de las Armas*. Chihuahua is somewhat Mexican in character and population, and still has houses one story high, built partly of stone and partly of upright logs, with cane roofs. Alamo is considerably higher than the other two sections of the city, and is mostly inhabited by Germans. In the N. part, on the Alamo Plaza, is the famous **Fort Alamo*, where, in March, 1836, a garrison of Texans, attacked by an overwhelming Mexican force, perished to a man rather than yield. Missions San José, San Juan, and Concepcion, built by the Spaniards, who founded San Antonio in 1714, are interesting objects; and the market-places and street-scenes will amuse the visitor as being more foreign and queer than those of almost any other American city. San Antonio may also be reached from northern points by the International & Great Northern R. R., connecting with the Texas extension of the Missouri, Kansas & Texas Div. from Denison at *Taylor.*

b. Via "Iron Mountain Route."

This line is nearly 100 miles shorter than the previous one, and extends S. W. through eastern Missouri and central Arkansas, connecting at Texarkana with the Texas & Pacific R. R. Three trains daily from St. Louis, with sleeping-car service attached, run through without change of cars to Houston, Galveston, San Antonio, Dallas, Laredo, El Paso, Los Angeles, and San Francisco. For about 25 miles from St. Louis the W. bank of the Mississippi River is followed, and afterward the road traverses a rich and highly cultivated agricultural region, and the great mineral fields of Missouri, including the famous Iron Mountain and Pilot Knob. At *Bismarck* (75 miles) the road branches: one line running S. E. to Belmont and Columbus, Ky., where connection is made with the rail way system of the Southern States E. of the Mississippi River; at Knobel, the through trains for Memphis use the Helena branch; while the Texas line passes S. W. and crosses the State of Arkansas in a diagonal direction. The only important place on this section of the line is **Little Rock** (*Capitol, Deming, Grand Central,* and *Richelieu*), the capital and chief city of Arkansas, with a population of 25,874, built upon the first bed of rocks that is met with in ascending the Arkansas River. The city is regularly laid out, with wide streets lighted with electricity and traversed by electric-cars. The business blocks are mainly of brick, and the residences are surrounded by gardens adorned with shade-trees and shrubberies, presenting a handsome appearance. The principal public buildings are the *County Court-House, State-House, Custom-House* and *Post-Office,* the *Lunatic Asylum, Little Rock University,* and *State Medical College.* Several of the churches and school-houses are handsome structures. Little Rock is the seat of a U. S. Arsenal and Land Office, of the State Penitentiary, and of the State Institutions for deaf-mutes, the blind, and the insane. The *State Library* contains 20,000 volumes and the *Mercantile Library* 1,800. *Pine Grove, Glenwood Park, West End Park, Baseball Park,* and *Mountain Park* are among the resorts in the immediate vicinity of the city. The Arkansas is navigable to Little Rock at all times by steamers, and several important railways converge here. From *Malvern* (42 miles S. of Little Rock) the Hot Springs R. R. diverges and runs in 20 miles to the famous **Hot Springs** (*Eastman, Park,* and many others), one of the most frequented health-resorts in America. The town, which is simply an appendage of the sanitarium, contains 8,086 inhabitants, and is built principally in the narrow valley of Hot Springs Creek, which runs N. and S. amid the Ozark Mts. The valley is about 1½ mile long, is 1,500 ft. above the sea, and is very rugged and picturesque. The springs (73 in number) issue from the W. slope of Hot Springs Mountain, vary in temperature from 76° to 157° Fahr., and discharge into the creek about 500,000 gallons a day. The waters are used both internally and externally (but chiefly in the form of baths), and are remedial in rheumatism, rheumatic gout, malarial fevers, scrofula, and diseases of the skin. (For a full description of this place, see APPLETONS' HAND-BOOK OF WINTER RESORTS.) At *Texarkana* (491 miles from St. Louis) Texas is entered, and connections are made with the Texas & Pacific R. R., *via* which are run daily three through trains with complete palace-car service to Marshall, Mineola, Dallas, Fort Worth, and El Paso, the time from St. Louis to Dallas being 24¼ hours. At Texarkana connection is also made with the Transcontinental Div. of the Texas & Pacific R. R. to Clarksville, Paris, Honey Grove, Bonham Bells, Sherman, Whitesboro, Denton, and Fort Worth. At *Longview* (587 miles) connection is made with the International & Great Northern R. R., which runs S. to *Palestine* (765 miles), and there branches, the Gulf Div. leading to Houston and Galveston (see Sub-Route *a*). The International runs S. W. from Palestine to Taylor, where it connects with a branch of the Missouri, Kansas & Texas R. R., thence through Austin and San Antonio (see Sub-Route *a*) to **Laredo,** on the Rio Grande (165 miles from San Antonio), with a population in 1890 of 11,319, an important center of the wool and cattle business, as some of the largest and most profitable ranches in the State lie in this vicinity. This place has lately acquired a new importance as the connecting-link between the Missouri Pacific R. R. system in the U. S. and the Mexican R. R. system, including the Texas, Mexican, and Mexican National R. Rs. This new railway, which when completed will give another through route to the city of Mexico and all other important Mexican cities, is projected to run from Nueva Laredo (connected by R. R. bridge with Laredo across the Rio Grande) to Mexico city, with branches to Matamoras, San Luis Potosi, Tampico, Tuxpan, Papantla, Nautla, Vera Cruz, and to the Pacific coast. The total length, including branches, will be about 1,400 miles.

The Mexican National is already finished from *Corpus Christi*, Texas, to Laredo (161 m.), and from Laredo by the Nueva Laredo bridge over the Rio Grande to *Saltillo*, Mexico (328 m.). On the S. it is completed from the city of Mexico in a northerly direction to San Miguel (409 m.), with several small branches. From Saltillo a line is being constructed to connect with the Mexican International at Jaral. At present it is a stage-drive of 42 miles.

The San Antonio & Aransas Pass Railway Co.'s main line is from Kerrville to Houston (308 m.), with a Corpus Christi Div., running from Kenedy to Corpus Christi (88 m.) ; an Austin Div., running from Austin Junction to Lockhart (53 m.) ; a Waco Div., running from Yoakum to Waco (158 m.) ; a Brownsville Div., running from Skidmore to Alice (43 m.) ; and a Rockport section of 21 m.; making a total of 671 miles of operated lines. This line connects the cities of Houston and San Antonio, and the large towns of Corpus Christi, Rockport, Kerrville, Boerne, Yoakum, Flatonia, Gonzales, Eagle Lake, Beeville, Cuero, Luling, Yorktown, and Giddings. Its northern extension pierces the iron, coal, marble, and granite fields of Gillespie and Llano Counties ; its southern terminus connects the future deep-water ports of Rockport, Corpus Christi, and Aransas Pass ; its eastern terminus is in the heart of the great lumber region of Texas ; and the proposed western terminus, at the entry port of Laredo, will insure the shortest route from the East to all the leading Mexican cities. For the most part this route passes through a rich farming territory, which compares favorably with any section of the Union. The principal cities through which it runs have been previously described.

Corpus Christi, the county-seat of Nueces County, is situated on Corpus Christi Bay, 150 miles from San Antonio and 264 miles from Houston. It has a population of 4,387, and is the terminus of the San Antonio & Aransas Pass, of the Mexican National, and of the Corpus Christi & South American Railways (now building). It is surrounded by fine farming and fruit lands, is near to the harbor of Aransas Pass, and is rapidly improving, owing to the public spirit of its citizens. There are two hotels, recently built, with a capacity of 500 guests, and it is well known as a summer-resort. It is called by the people of Texas the "Naples of America."

Rockport (or *Aransas Pass*), seat of Aransas County, is situated on Aransas Bay, 160 miles from San Antonio and 275 miles from Houston. It has a magnificent harbor, which will soon be opened up to the commerce of the world by the completion of the jetty-work at the mouth of the Pass between St. Joseph and Mustang Islands, connecting the bay with the Gulf of Mexico. The finest fruit and garden lands to be found in Texas surround this place. The winter vegetables are marvels of growth. Good hotel accommodations can be had here. It is the pleasure-resort of large numbers of Northern and Western people, who come to enjoy the fine hunting and fishing. The tarpon, or silver-king, is found in large numbers from the 15th of April until the 1st of August. Trout, sheephead, redfish, and croakers are innumerable ; while the principal game are duck, geese, and braut, which make their feeding-grounds in this vicinity.

104. St. Louis to San Francisco.

THE journey from St. Louis to San Francisco may be made by two routes, either one of which may be recommended. That *via* the Missouri Pacific R. R. main line runs four express trains with sleeping-car service daily to Kansas City and to Pueblo and Denver, connecting at Kansas City with Union Pacific R. R., running through cars between St. Louis, Ogden, and Salt Lake City, connecting there with Southern Pacific R. R. for San Francisco (see Route 90).

Another route (the "Frisco line") is *via* the St. Louis & San Francisco R. R. to Burrton, Kan. (538 miles), thence *via* A., T. & St. F. R. R. to Albuquerque, N. M. (1,106 miles; see Route 101 *a* and *b*), thence *via* Atlantic & Pacific R. R. and Southern Pacific R. R. to San Francisco (2,435 miles; see Route 101 *a*). Sleeping and dining car service is run through on this route the whole distance. After leaving St. Louis *via* the St. Louis & San Francisco R. R., the first station of much importance is **Springfield** (238 miles), with a population of 21,850, and beautifully located on high ground. It is the most important town of S. W. Missouri. It is the seat of *Drury College*, and has a fine *Court-House*, 12 churches, 5 banks, and 8 newspapers. Here also are large machine-shops, engine and boiler works, and woolen and cotton mills. Here junction is made with the Kansas City, Fort Scott & Memphis R. R. At *Monett* (282 miles) connection is made with the Texas Div., and in 5 miles we reach *Pierce City*, the point of junction with the Kansas Div. This place has 2,511 population, and is growing rapidly. **Carthage** (314 miles) is an important place ½ mile S. of Spring River, which furnishes fine water-power, driving flour-mills, woolen-mills, machine-shops, and manufactories of plows, carriages, etc. The population is 7,981. Connection is made here with the Missouri Pacific R. R. At *Oswego* (360 miles) connection is made with the Missouri, Kansas & Texas R. R., and at *Cherryvale* (387 miles) with the Southern Kansas and Kansas City, Fort Scott & Gulf R. Rs. *Wichita*, Kan. (505 miles), is a flourishing place of 1,827 people; and at Burrton (538 miles) we reach the point of junction with the Atchison, Topeka & Santa Fé R. R., the St. Louis & San Francisco train taking the track of that railroad for its further journey. For particulars of the route hence to San Francisco, see Route 101 *b* (Burrton to La Junta, Colorado) and Route 101 *a* (La Junta to San Francisco).

The main line of the St. Louis & San Francisco R. R. is now completed to Sapulpa, I. T. The Atlantic & Pacific R. R. is projected to be completed from Albuquerque, N. M., to Sapulpa. When this is accomplished it will give the St. Louis & San Francisco R. R. control of the shortest through-line between New York, Boston, Philadelphia, and other Eastern points and San Francisco.

105. The Great Lakes.

LAKES Ontario, Erie, Huron, Michigan, and Superior are known as the "Great Lakes," and are the largest bodies of fresh water in the world. They are part of one great system of continental drainage, and are connected in such a manner that one and the same boat can traverse them almost from end to end. **Lake Ontario,** however, is cut off from the others by the Falls of Niagara, and, being the least attractive of the five, is seldom included in the regular routes of summer travel. The tour of the lake may be made in connection with the. tour of the St. Lawrence by taking the Royal Mail Steamers of the Canadian Navigation Co. at *Hamilton,* instead of at Kingston (see Route 60). A steamer leaves Hamilton daily at 9 A. M., stopping at Toronto, Port Hope, and Cobourg, and reaching Kingston, at the E. end of the lake, at 5.30 o'clock next morning. Lakes Erie, Huron, and Superior may be included in a single tour, and afford one of the most delightful trips that can be taken in this country during the summer. The steamers of the Erie & Western Transportation Co. are swift, strong, commodious, and handsomely furnished. These steamers leave Buffalo, Erie, Cleveland, and Detroit. They may be taken at any one of these places, or at Port Huron or Sault Ste. Marie. In the following description of the route we shall suppose ourselves to be starting from Buffalo, at the E. end of Lake Erie, which is fully described in Route 40. Excursion steamers leave Buffalo on Mondays, Tuesdays, Thursdays, and Saturdays for Lake Superior, making the round trip in two weeks (fare, $50).

Lake Erie.

"Among the five great lakes of the Western Chain," says a writer in "Picturesque America," "Erie occupies the fourth place as regards size, the last place in point of beauty, and no place at all in romance." For the rest, the lake is 250 miles long, 60 wide, less than 90 ft. in average depth, and 564 ft. above the level of the sea. It is the shallowest and most dangerous of the entire chain of the Great Lakes. It can be avoided at the cost of a 10 or 12 hours' railway journey, but then the tourist loses the pleasure of the Detroit River trip.

After leaving Buffalo, the scenery for a time is uninteresting, as the steamer does not approach near enough to the land to enable us to see anything, except when entering and leaving port, and many of the steamers make no stops until reaching Detroit. For the convenience of the traveler who ·may be upon a boat making all the landings, brief· mention will be made of the principal ones on the S. shore of the lake. *Dunkirk* (42 miles from Buffalo) has à good harbor, and is described in Route 41. **Erie** (90 miles) is situated on "that sturdy little elbow which Pennsylvania has pushed up to the lake-shore, as if determined to have a port somewhere, on fresh water if not on salt." It is the terminus of the Philadelphia & Erie Div. of the Pennsylvania R. R. (Route 53), and has a very large and beautiful harbor, formed by what was once a long, narrow peninsula, but is now an island. The bar at the mouth has been dredged away so as to afford a good channel, and Erie is a United States naval station.. It was here that Commodore Perry built his fleet, and here he brought his prizes after the battle of Lake Erie, in September, 1813. On the bank above, the embankments of the old French fort, Presque Isle, can be traced. For description of the city see Route 53. Dotted along the coast of the lake are numerous lighthouses, standing on lonely islets and rocky ledges, wherever they can command a wide sweep of the horizon. To the traveler they appear both picturesque and friendly. There is almost always one in view; and, a pillar of cloud by day and of fire by night, they greet the voyager as he journeys, one fading astern as the next shines out ahead. The light at Erie is visible for a distance of 20 miles. **Cleveland** (185 miles) is universally considered the most beautiful city on the Great Lakes. It stands upon a high bluff, and a good view of it is had from the water; though it is so embowered in trees that little save the spires of the churches can be seen through the green. Steamers usually make a stay of several hours at Cleveland, and give passengers an opportunity of seeing the city. It is fully described in Route 67. W. of Cleveland the coast grows more picturesque; the shore is high and precipitous, and the streams come rushing down in falls and rapids. Seven miles from the city is *Rocky River*, which flows through a deep gorge between perpendicular cliffs that jut boldly into the lake and command a wide prospect. " Here is the most extensive and unbroken view of Lake Erie; Black River Point is seen on the W., and the spires of Cleveland shine out against the green curve of the E. shore; but far away toward the N. stretches the unbroken expanse of water, and one can see on the horizon-line distant sails, which are still only in mid-lake, with miles of blue waves beyond." W. of Rocky River, the Black, Vermilion, and Huron Rivers flow into the lake through ravines of wild beauty; and then, after a long stretch of dreary coast, the steamer approaches **Sandusky** (*Colton, Sloane House, West House*), with its beautiful bay, which is 20 miles long and 5 or 6 wide. Sandusky has a population of 18,471, and is handsomely built on ground rising gradually from the shore, and commands a fine view of the bay and lake. Beneath its site is an inexhaustible bed of excellent limestone, which is extensively employed in building and in the manufacture of lime. The *Court-House*, built of white and blue limestone, is one of the finest in the State. The city is celebrated for its manufacture of articles of wood, of which handles, spokes, and hubs, " bent work " for carriages, and carpenters' tools are the most important; and fresh and salt fish, ice, and lumber are extensively exported. It is the largest fresh-water fish-market in the United States, and thousands of tons are frozen every fall to supply the demand, in addition to the great quantity salted. The *State Fish Hatchery* is located here, and 3,000,000 young white-fish are annually put into the lake. The city is also the center of one of the most important vine-growing districts in the United States. *Lakeside*, situated on the lake-shore just outside of the mouth of Sandusky Bay, is a pleasant resort, with a fine summer hotel and a hundred or more cottages. This is a great rendezvous for camp-meetings and Sunday-school and educational gatherings. The Lake Erie Div. of the Baltimore & Ohio R. R. (see Route 70) terminates here, and it is also a lake outlet of the Cleveland, Cincinnati, Chicago & St. Louis R. R., which runs to Springfield, O., and thence to Indianapolis and Peoria, Ill.; and of the Lake Erie & Western R. R., by which connections are made at Bloomington, Ill., for the Northwest and Southwest. From New York, Sandusky is reached by the Lake Shore & Michigan

Southern R. R. and its Eastern connections. During the summer. season there is daily steamer connection with Detroit, Point Au, Pelee Island, and the Canada shore.

After leaving Sandusky, the steamer speedily reaches the * **Put-in-Bay Islands,** a beautiful group, 15 or more in number, lying in the S. W. corner of Lake Erie, near the mouth of the Detroit River. Within a few years past these islands have become a favorite summer resort, as they combine all the advantages of pure air, bathing, fishing, boating, and convenience of access from any of the lake-cities. From Detroit there is a daily steamer to *Kelly's Island*, the largest of the group. *Put-in-Bay Island* has several large summer hotels, including the *Beebe House*, the *Hunker House, Park Hotel*, and *Put-in-Bay House*. The islands are noted for their vineyards and the superior quality of the wine produced; but some of them are still wild and uninhabited. Shortly after passing the islands the steamer enters the Detroit River.

The Detroit and St. Clair Rivers.

There are 15 islands within the first 12 miles of the Detroit River. Father Hennepin, who passed up the river in 1679, enthusiastically writes: "The islands are the finest in the world; the strait is finer than Niagara; the banks are vast meadows; and the prospect is terminated with some hills crowned with vineyards, trees bearing good fruit, groves and forests so well disposed that one would think that Nature alone could not have made, without the help of art, so charming a prospect." Since that day, "art" has done something to mar the freshness of the scene; but the strait still affords some of the loveliest river scenery in America. The river is broad, varying from 3 miles at the mouth to a mile in width at the city of Detroit; the Canadian shore rising abruptly from the water to a height of from 20 to 25 ft., the American shore being low, and in some places marshy. The only island calling for special mention is *Grosse Isle*, which is a favorite summer resort for Detroiters, who find here, within 20 miles of their homes, a delightful retreat from the heat and dust of the city. The island divides the river into two channels, which are known as American and Canadian; the latter, being the deepest, is used by the through-boats, none passing on the American side except to touch at *Trenton* or *Gibraltar*, the former of which is a flourishing place noted for its shipbuilding. *Wyandotte*, Mich. (15 miles below Detroit), is the site of extensive rolling-mills, which may be said to have created the town. Three miles below the steamboat-landing at **Detroit,** the river makes a sudden turn, and the city comes into full view. On the right hand is the village of *Windsor*, in Canada, and directly opposite is *Fort Wayne*, a bastioned redoubt, mounted with heavy ordnance. For at least 6 miles above the fort the river-front is lined with mills, dry-docks, ship-yards, foundries, grain-elevators, railway-depots, and warehouses, and, on the level plateau above, the city extends inland for 2½ miles. The steamers generally stop at Detroit several hours, and the tourist should improve this opportunity for seeing the city, which is described in Route 66.

Beyond Detroit, the steamer passes *Belle Isle*, a small island at the head of the river, and enters **Lake St. Clair**, which is 25 miles long and about the same distance from shore to shore. It is shallow, and at the upper end, where the river St. Clair comes in, large deposits of sand have been made, known as "The Flats." These for a long time greatly impeded navigation, but the difficulty has been lately overcome by the construction of a ship-canal which is justly regarded as a triumph of engineering skill. Around the shores of the lake are large fields of wild rice. Here immense flocks of wild ducks swarm, geese are found in the shooting-season, and the waters teem with fish. *Isle la Pêche* (commonly known as "Peach Island"), near the lower end of the lake, belongs to Canada. It was at one time the summer home of the celebrated Indian chief Pontiac. The **St. Clair River** is really a strait through which the waters of Lake Huron take their way toward the Atlantic Ocean. It is 17 miles long, and has a descent in that distance of 15 ft., which gives a current of $3\frac{1}{2}$ to 4 miles an hour. The scenery along the St. Clair is beautiful, the banks on either side being well cultivated or covered with a thick forest-growth. There are several small towns along the river, but none of much importance (except *St. Clair*, Mich.) until we reach **Port Huron** (*Huron House*), a port of entry at the mouth of Black River, which runs through a rich pine-region, and down which is floated the lumber that supplies the numerous saw-mills at this point. The trade in fish is important, and there are 3 ship-yards and 2 dry-docks. During the season of navigation Port Huron is connected by daily lines of steamers with Detroit, Saginaw, and the principal lake and river ports. **Port Sarnia**, a Canadian port of entry opposite Port Huron (connected by ferry), is a place of active business, being the terminus of the main line of the Grand Trunk Railway, and also of the southern branch of that road. Two miles above Port Huron, between *Fort Gratiot* (a United States military post) and *Point Edward*, the river narrows until it is less than 1,000 ft. wide, the increased velocity of the current being so noticeable that the descent of the water can be seen from the wharves on either side. Here the Grand Trunk Railway crosses the St. Clair River on a handsome bridge, passing which the steamer enters

Lake Huron.

Lake Huron lies between the 43d and 46th degrees of N. latitude, adn is 250 miles in length from the head of the St. Clair River to the Straits of Mackinaw, and 100 miles wide. It is 574 ft. above the level of the ocean, and varies in depth from 100 to 750 ft. *Georgian Bay*, at the N. E. side of the lake, is very large, and lies entirely within the Dominion of Canada; *Saginaw Bay*, on the S. W., being within the limits of the State of Michigan. *Tawas Bay* is a good harbor on the S. W. side of Saginaw Bay. *Thunder Bay* is farther N., and has the *Thunder Bay Islands* at its mouth. The stormiest part of the lake is between the Saginaw and Georgian Bays, where the wind often sweeps with terrific violence. But few islands are seen, and the traveler who has never been at sea can form some idea of what the ocean is, for during a portion of the voyage no land can be seen even from the mast-head; the boundless expanse of water, dotted here and there with a distant sail, stretching on every side.

Mackinac Island.

Mackinac is conveniently reached from Detroit by steamer four times a week, and close connection is made with the steamer-service from Cleveland. There is also a daily line of steamers between Colling-wood, on Georgian Bay, one of the termini of the Grand Trunk Railway, and Chicago, touching at the island; and a mail-boat 3 times a week from Port Sarnia, on Lake Ontario. By rail it can be reached from Detroit by the Mackinaw Div. of the Michigan Central R. R.; also by the Detroit, Grand Haven & Michigan R. R. to Grand Haven, Mich., and thence by the Grand Rapids & Indiana R. R. to Mack-inac City. From Chicago it is reached by the Chicago & Northwest-ern R. R. to Negaunee, Mich., and thence by the Duluth, South Shore & Atlantic R. R. to St. Ignace.

Mackinac (also written Mackinaw, and formerly Michilimackinac) is an island in the Strait of Mackinac, which connects Lakes Michigan and Huron, 260 miles N. W. of Detroit, and about 300 N. of Chicago. It is about 3 miles long and 2 wide, is rough and rocky, and has 800 in-habitants. It is an old military post of the United States, as well as a delightful and popular place of summer resort. The waters surround-ing the island are wonderfully clear and pellucid, and teem with fish of delicious flavor. The fisherman sees the fish toying with his bait, and the active little Indian boys on the piers are always ready to dive for any coins the visitor may throw into the water for them. The inhabit-ants of the decayed and antiquated village at the foot of the cliff are mainly dependent on their seines and fishing-nets for support, and upon the money spent every summer by tourists, there being four good hotels and several stores where Indian curiosities, agates, photographs, and other mementos of the place are offered for sale. Boats for pleasure-excursions may always be had; and the usual accessories of a summer resort, such as bowling-alleys, billiard-rooms, etc., are provided at the best hotels (the *Astor House, Grand Hotel*, and the *Mission House*). On the cliff over the village is *Fort Mackinac* (built in 1780), 200 ft. above the level of the lake, and overlooking the village and beautiful harbor. In rear of and about 100 ft. above this fort are the ruins of old *Fort Holmes*, and in their immediate neighborhood, 320 feet above the lake, the highest point on the island, stands a signal-station. The view from this elevation is very fine.

"The natural scenery of Mackinac," says a writer whom we have already several times quoted, "is charming. The geologist finds mys-teries in the masses of calcareous rock dipping at unexpected angles; the antiquarian feasts his eyes on the Druidical circles of the ancient stones; the invalid sits on the cliff's edge in the vivid sunshine and breathes in the buoyant air with delight, or rides slowly over the old military roads, with the spicery of cedars and juniper alternating with the fresh forest-odors of young maples and beeches. The haunted birches abound, and on the crags grow the weird larches beckoning with their long fingers, the most human tree of all. Bluebells on their hair-like stems swing from the rocks, fading at a touch, and in the deep woods are the Indian pipes, but the ordinary wild-flowers are not to be found. Over toward the British landing stand the Gothic spires of the blue-green spruces, and now and then an Indian trail crosses the road, worn deep by the feet of the red men, when the Fairy Island was their favorite and sacred resort." Chief among the curiosities of the island is *Arch Rock, on the E. side, a natural bridge 149 ft. high by less than 3 ft. thick, excavated in a projecting angle of the limestone cliff. The beds forming the summit of the arch are cut off from direct connection with the main rock by a narrow gorge of no great depth. The portion supporting the arch on the N. side, and the curve of the arch itself, are comparatively fragile, and can not long resist the action of rain and frosts, which in this latitude, and on a rock thus constituted, produce great ravages every season. *Fairy Arch* is of similar formation to Arch Rock, and lifts from the sands with a grace and beauty that justify the name bestowed upon it. The *Lover's Leap* is a rock about a mile W. of the village, having a vertical height of 145 ft. The Indian legend to which the rock owes its name is that a young squaw, standing on this point waiting for the return of her lover from battle, saw the warriors carrying his dead body to the island, and in her grief threw herself into the lake. *Robertson's Folly* is a precipitous cliff E. of the village, 120 ft. high. Its name is taken from the adventure of a British officer who persisted in following the ever-fleeing figure of a most beautiful maiden, with whom, after repeated disappointments, at length coming up at the edge of the precipice, where she stood with her back to the water, he sprang toward her and clasped her to his arms, only to be plunged with her, starting suddenly back, upon the rocks below, where the next morning his mangled body was found alone. The *Sugar-Loaf* is a solitary conical rock, rising 134 ft. from the plateau upon which it stands, and 284 ft. above the lake. The *Devil's Kitchen* is a curious cave. The *British Landing* is a favorite resort for picnics, and received its name from being the point where the British landed when they captured the island in 1812. (Mackinac, the post-town in Mackinac Co., must be distinguished carefully from Mackinaw, in Cheboygan Co., on the S. side of the strait.)

The regular lake-steamers pass a considerable distance to the E. of Mackinaw Island, and enter the **St. Mary's River,** a remarkably beautiful stream 62 miles long, and forming the only outlet to Lake Superior. It is a succession of expansions into lakes and contractions into rivers, and is dotted with beautiful forest-clad islands, while a few small towns are scattered along either shore. The *Ste. Marie Rapids* are avoided by a ship-canal, and 6 miles beyond the steamer traverses the picturesque *Waiska Bay*, and, passing between Iroquois Point on the American and Gros Cap on the Canadian side, enters the vast reaches of

Lake Superior.

Lake Superior, the largest body of fresh water in the world, is 360 miles long and 140 miles wide in its widest part, having an average width of 85 miles, a circuit of 1,500 miles, and an estimated area of 32,000 square miles. It is 800 ft. deep in its deepest portion—the bottom there being 200 ft. below the level of the ocean. It receives its waters from about 200 rivers and streams, draining an area of 100,000 square miles. It contains a number of islands in the E. and W. portions, but very few in the central. The most important of these are Ile Royal, The Apostles, and Grand Island, belonging to the United States, and Michipicoten, Ile St. Ignace, and Pie Islands, belonging to Canada. The early French Jesuit fathers, who first explored and described this great lake, and published an account of it in Paris in 1636, speak of its shores as resembling a bended bow, the N. shore being the arc, the S. shore the cord, and Keeweenaw Point, projecting from the S. shore to near the middle of the lake, the arrow. The coast of Lake Superior is mostly formed of rocks of various kinds and of different geological groups. With the exception of sandy bars at the mouth of some of the rivers and small streams, the whole coast of the lake is rock-bound; and in some places, but more particularly on the N. shore, mountain-masses of considerable elevation rear themselves from the water's edge, while mural precipices and beetling crags oppose themselves to the surges of this mighty lake, and threaten the unfortunate mariner who may be caught in a storm upon a lee-shore with almost inevitable destruction. The waters are of surprising clearness, very cold, and are filled with the most delicious fish.

Once having passed *White-Fish Point*, with its "sand dunes" or hills, and its tall lighthouse, the steamer usually takes a course for *Point au Sable*, 50 miles beyond, keeping in sight of the Michigan shore, which here presents a succession of desolate sand-hills, varying from 300 to 500 ft. in height. Twenty miles beyond the Point are the famous **** Pictured Rocks,** a wonderful exhibition of the denuding effect of water, combined with the stains imparted by certain minerals. They extend for a distance of about 5 miles, rising in most places vertically from the water's edge to a height of from 50 to 200 ft., there being no beach whatever. When the weather permits, the steamers run near enough to give passengers a cursory view of these great curiosities; but in order to be able to appreciate their extraordinary character, the tourist should leave the steamer at *Munesing* and visit them in a small boat. As we can not spare the space required for such a detailed description of these rocks as they deserve, we must content ourselves with briefly mentioning the more conspicuous features in order from E. to W. (the visitor from Munesing approaches them in the opposite direction). The *Chapel* is a vaulted apartment in the rock, 30 or 40 ft. above the level of the lake. An arched roof of sandstone rests on 4 columns of rock so as to leave an apartment about 40 ft. in diameter and the same in height. Within are a pulpit and altar, perfect as if fashioned by the hand of man. A little to the west of the Chapel, Chapel River falls into the lake over a rocky ledge 15 ft. high. The * **Grand Portal,** which appears next, is the most imposing feature of the series. It is 100 ft. high by 168 broad at the water-level, and the cliff in which .it is cut rises above the arch, making the whole height 185 ft. The great cave, whose door is the portal, extends back in the shape of a vaulted room, the arches of the roof built of yellow limestone, and the sides fretted into fantastic shapes by the waves driving in during storms, and dashing a hundred feet toward the reverberating roof. Within this cave there is a remarkably clear echo. *Sail Rock* is about a mile W. of the Grand Portal, and consists of a group of detached rocks bearing a resemblance to the jib and mainsail of a sloop when spread; so much so that, when viewed from a distance, with a full glare of light upon it, while the cliff in the rear is left in the shade, the illusion· is perfect. The height of the block is about 40 ft. Passing to the westward, we skirt the cliffs worn into thousands of strange forms, colored deep brown, yellow, and gray, bright blue, and green. They are arranged in vertical and parallel bands, extending to the water's edge, and are brightest when the streams are full of water. *Miner's Castle,* 5 miles W. of the Chapel, and just W. of the mouth of Miner's River, is the western end of the Pictured Rocks. It resembles an old turreted castle with an arched portal. The height of the advanced mass in which the Gothic gateway may be recognized is about 70 ft., that of the main wall forming the background being 140 ft. The coast of Pictures is not yet half explored, nor its beauties half discovered. "In one place there stands a majestic profile looking toward the north—a woman's face, the *Empress of the Lake.* It is the pleasure of her royal highness to visit the rock only by night, a Diana of the New World. In the daytime, search is vain; she will not reveal herself; but when the low-down moon shines across the water, behold, she appears! She looks to the north, not sadly, not sternly, like the Old Man of the White Mountains, but benign of aspect, and so beautiful in her rounded, womanly ·curves, that the late watcher on the beach falls into the dream of Endymion; but when he wakes in the gray dawn he finds her gone, and only a shapeless rock glistens in the rays of the rising sun."

Leaving Munesing and the Pictures, and going westward past the *Temples of Au Train* and the *Laughing-Fish Point,* the city of **Marquette** (*Clifton House, Hotel Marquette*), the entrepot of the Marquette Iron Region, comes into view. This important center is on the Duluth, South Shore & Atlantic R. R., and is by this route easy of access from Mackinac; or from Chicago *via* the Chicago & Northwestern system, which connects at Negaunee with the Duluth, South Shore & Atlantic R. R. It has a large and picturesque harbor, is well built, and has a population of 9,093. It is the chief depot of supplies for the iron-mines of the Upper Peninsula, and the principal point of shipment for the ore. There are many blast-furnaces and rolling-mills within the city limits and in the vicinity. The place has great attractions for the invalid and tourist, in its healthy, invigorating atmosphere, beautiful walks and drives, fine scenery, boating, and fishing. Persons spending several weeks at Marquette can pass the time very agreeably in making excursions to *Grand Island* and the *Pictured Rocks*, to *Carp River*, *Dead River*, and *Chocolat River*, all of which offer fine trout-fishing. Another excursion is by the Duluth, South Shore & Atlantic R. R. to *Champion* on Lake Michigami (32 miles), where there are good boating, hunting, and fishing, but poor accommodations for travelers. A visit may also be made to the iron-regions by the same route, which has a branch to Houghton. Beyond Marquette the steamer makes no stops until it reaches *Portage Lake*, passing on the way *Granite Island* (12 miles from Marquette); *Stanard's Rock*, a very dangerous granite ledge; the *Huron Islands*, a picturesque group; *Huron Bay* and *Point Abbeye;* and crossing *Keweenaw Bay* to *Portage Entry*. The Entry was originally a narrow, crooked channel, leading from Keweenaw Bay into Portage Lake; but the channel has been deepened, and in conjunction with the Portage Lake ship-canal saves the circuit of 120 miles around Keweenaw Point. In digging the canal indubitable evidences were found that Portage Lake was once an arm of Lake Superior, cutting off Keweenaw Point, which was then a large island. The lake is about 20 miles long, and from $\frac{1}{2}$ to 2 miles in width. *Ontonagon* (336 miles from Sault Ste. Marie) is a small village at the mouth of a river of the same name, and is a station on the Chicago & Northwestern Ry. Twenty miles beyond, the *Porcupine Mountain*, 1,300 ft. high, is a conspicuous object; and 70 miles from Ontonagon are the * **Apostles' Islands,** a large and beautiful group, 27 in number. The clay and sandstone cliffs have been worn into strange shapes by the action of the water, and the islands are covered with fine forest-trees. The fishing here is excellent, and trout, white-fish, and siskowit are caught in abundance. At *Bayfield*, a Wisconsin town on the mainland opposite, is a secure and spacious harbor. At the head of Lake Superior the *St. Louis, River* comes ,in, and on the lake-shore near its mouth is **Duluth** (hotels *Cheltingham*i *Spaulding*, *St. Louis*) (1,235 miles from Buffalo), with 33,115 inhabitants. It derives its commercial importance from its situation at the extreme west point of the Great Lakes; it is a terminus of the St. Paul & Duluth R. R.; of the East Minnesota Div. of the Northern Pacific R. R.; the Chicago, St. Paul, Minneapolis & Omaha R. R.; St. Paul, Minneapolis & Manitoba R. R.; and the Duluth & Iron Range R. R., the latter running N. to Lake Vermilion Iron Mines. It is well built, has 16 grain-elevators, with 2 storage-houses with capacity of 21,-000,000 bushels, churches, banks, and newspapers. Large ship-yards are located here, and several steel vessels have already been built. Among the public buildings are 2 *Opera-Houses*, *St. Luke's Hospital*, and 8 school-buildings, and the *Board of Trade* and *Exchange Buildings*. The manufactures, principally in the way of lumber, are extensive.

The **North Shore of Lake Superior** is comparatively an unknown region. The easiest way of seeing it is by taking the steamers of the Erie & Western Transportation Co. to Sault Ste. Marie, whence the Canadian steamers may be taken to the more important points; but if the tourist desires to visit any number of the many places of interest, he must hire a small boat and two or three experienced men as a crew. North of Duluth the shore rises into grand cliffs of greenstone and porphyry,

800 to 1,000 ft. in height. The ***Palisades** (58 miles from Duluth)
are a remarkable rock formation, presenting vertical columns from
60 to 100 ft. high and from 1 to 6 ft. in diameter. Near by, *Bap-
tism River* comes dashing down to the lake in a series of wild waterfalls. *Pigeon River* (113 miles) is the boundary-line between the
United States and Canada; and here begins the "Grand Portage,"
by which, through a series of lakes and streams, the very names
of which have a wild sound (Rainy Lake, Lake of the Woods, and
Winnepeg), the *voyageurs* are enabled, with short portages, to take
their canoes through to the Saskatchewan and Manitoba. The whole
Canadian coast is grandly beautiful in every variety of point, bay,
island, and isolated cliff. At the mouth of the Kaministiquia is *Fort
William*, formerly an important post of the Hudson Bay Company, but
now the lake terminus of the Canadian Pacific Railway and a divisional
point of that road. At Fort William the company has constructed two
enormous grain-elevators (and is erecting a third), a hotel, and other edifices connected with the operation of the road, while private enterprise
is rapidly building a city in the immediate neighborhood of the station.
About 5 miles E. round the bay is *Port Arthur*, the gathering-place of the
Red River Expedition in 1870 under Colonel Wolseley, and subsequently,
during the construction of the Canadian Pacific Railway, a very busy
town, but now depending principally on the rich mineral discoveries of
the neighborhood and the trade arising from them. Across *Thunder
Bay*, about 16 miles distant, Thunder Cape is seen, a basaltic cliff 1,350
ft. high, on the summit of which is the crater of an extinct volcano.
A short distance to the east of Thunder Cape is *Silver Island*, a small
islet, from which an enormous quantity of silver has been taken, the
story of its productiveness reading like pages of Monte Cristo. Following the line of the Canadian Pacific Railway eastward from Port
Arthur, the route keeps Lake Superior in sight at intervals for a considerable distance, crossing numerous streams noted for their trout-fishing,
and passing through country wild and rugged, but much prized by sportsmen. On this portion of the Canadian Pacific Railway some of the most
interesting illustrations of difficulties in railway-building successfully
overcome occur. Seventy miles east of Fort William is *Nepigon Bay*,
a large indentation of Lake Superior (40 miles long by 15 wide), which
with the river and lake of the same name, is far-famed for its trout-
fishing. All through the summer fishermen singly or in parties make
their way (from the west by Fort William and Port Arthur) to *Nepigon,
Steele River*, and other streams falling into *Lake Superior*, and in the
autumn sportsmen from all parts make the Canadian Pacific Railway
stations their bases of operation for hunting caribou, deer, bear, etc.
By this route hunters and fishermen from points west of Lake Huron
reach the hunting and fishing country of the upper Ottawa and its
tributaries. At *Pic River* (276 miles) is a post of the Hudson Bay
Company, and here the shore-line bends to the S. and the lake begins to
narrow toward the Sault. At *Otter Head* (30 miles S. of Pic River) the
cliff rises in a sheer precipice 1,000 ft. from the water, and on its summit stands a rock like a monument, which on one side shows the profile

of a man, and on the other the distinct outline of an otter's head. The Indians never passed this point without stopping to make their offerings to its manitou. Still farther S. is the broad bay of Michipicoten, or the " Bay of Hills "; and here is another post of the Hudson Bay Company. There are many islands in this portion of the lake, among the most important of which are *Isle Royale* (45 miles long and 8 to 12 wide), *Saint Ignace,* and *Michipicoten Island,* the latter of which will probably become a favorite place of summer resort.

106. St. Paul or Duluth to Portland, Ore., and the Pacific Coast.

Via the Northern Pacific R. R.

The Northern Pacific R. R., completed in 1883, is now one of the greatest railroad systems in the United States, and is destined to be still greater, as it runs through an unbroken belt of country in every way fitted to invite immigration and sustain a dense population. The principal E. terminus is St. Paul, and there are 2 termini on Lake Superior, one at Duluth, Minn., the other at Ashland, Wis. The branches unite to form the main line at Brainerd, 136 m. from St. Paul. The main line of the N. P. R. R. is now completed from Pasco Junction to Tacoma, on Puget Sound, the W. terminus of the road. From Tacoma, Ore., *via* the Pacific Div. of Northern Pacific R. R., Seattle, Olympia, Portland, and other important points are reached. By its branch lines and its connections with the Puget Sound and Alaska S. S. Co., and Southern Pacific R. R., convenient access is given to all important points in California, Oregon, Washington, and British Columbia. The route crosses Minnesota, Dakota, Montana, Idaho, Washington, and Oregon, bisecting a country of unsurpassed advantage for agricultural and grazing purposes, while portions of the route hold in tribute also mining regions which promise to be extraordinarily rich. The scenery on the western half of the route displays many wonderful phases to attract the tourist and the pleasure-seeker, and the sportsman will find inexhaustible resources in shooting and fishing. The largest game known to North America, the grizzly, cinnamon, and black bear, the mountain-lion, the elk, and many varieties of deer and antelope, are found in numbers sufficient to excite the most eager lover of field sports, while creatures of the fin and feather invite the sportsman at every turn. Two daily express trains run from St. Paul to Portland, Ore., and Tacoma, Wash., equipped with sleeping-cars of the most improved pattern, and with elegant dining-cars owned by the company; meals, 75c. each. The principal stations *en route* are as follow (distances given are from St. Paul) : St. Cloud (76 m.), Little Falls (108 m.), Brainerd (138 m.), Wadena (159 m.), Detroit (204 m.), Moorhead (250 m.), Fargo (251 m.), Casselton (271 m.), Valley City (309 m.), Sanborn (320 m.), Jamestown (344 m.), Bismarck (445 m.), Mandan (450 m.), Dickinson (560 m.), Glendive (666 m.), Miles City (744 m.), Custer (838 m.), Billings (891 m.), Livingston (1,007 m.), Bozeman (1,031 m.), Helena (1,130 m.), Garrison (1,180 m.), Butte City (1,232 m.), Missoula (1,254 m.), Hope (1,427 m.), Spokane Falls (1,512 m.), Sprague (1,553 m.), Pasco Junction (1,657 m.), Wallula Junction (1,673 m.), Yakima (1,747 m.), Ellensburg (1,784 m.), The Dalles (1,822 m.), Tacoma (1910 m.), Seattle (1,931 m.), Portland (1,887 m.).

The principal Lake Superior terminus of the Northern Pacific R. R. is the enterprising city of **Duluth,** which has been described (see Route 105). The main line extends from *East Superior,* Wis., to *Ashland,* Wis., forming the Wisconsin Div., 71 miles long. These two outlets on the lake insure for the railroad a large interest in the lake business. The St. Paul Div. meets the main line at Brainerd. **St. Paul** has been already described (see Route 87). The road from St. Paul runs along the E. bank of the Mississippi, furnishing delightful views from the car-windows, through *St. Cloud* (76 miles, population 7,686), *Little Falls* (108 miles, population 2,354), and several smaller stations. **Brainerd** (138 miles) is the junction with the main line. The population is 5,703, and here is the seat of the company car and repair shops, which give employment to 1,200 men. The city is in the midst of a forest of pine, and the initial point of the great Minnesota pineries. The city is lighted by electricity, and has many fine buildings, public and private; among which may be mentioned the *Northern Pacific Hospital*, a fine *Opera-House*, and a capacious hotel. There are 3 public parks, and many churches, schools, etc. The shooting and fishing from this point are excellent. Passing a number of small stations, we reach *Wadena* (159 miles), at the junction of the Black Hills Branch. It is the county-seat, has a population of 552, and is the seat of a thriving trade with the rich agricultural region around. As we proceed westward, the forests of northern Minnesota disappear, except in beautiful groves dotting expansive plains, in which are numerous lakes and water-courses. **Detroit** (204 miles) has a population of 1,510, and is situated on a beautiful lake of the same name in the edge of an extensive timber-belt. It is in the center of the "Lake and Park Region," and the vicinity is famous for quiet, picturesque beauty. This region is said to be the resort and breeding-ground of a greater number of game-birds and water-fowl than can be found elsewhere in America. The town has six hotels, several of them fitted for summer visitors, who are attracted by the well-known *Mineral Springs* here. There are also many summer cottages and villas, owned by residents of St. Paul, Minneapolis, etc. We pass the small stations of *Audubon, Lake Park, Winnipeg Junction,* the diverging point of the Manitoba Branch to Grand Forks, Grafton, Pembina, and Winnipeg, *Hawley,* and *Muskoda,* and arrive at *Glyndon* (241 miles), which is the junction-point with the St. Paul, Minneapolis & Manitoba R. R., and an important grain-shipping point for the Red River Valley. **Moorhead** (250 miles) is a town of 2,088 population on Red River. It is the center of an important trade and thriving manufactures. There are 12 hotels, 7 churches, an opera-house, 2 banks, 3 newspapers, car-wheel works, agricultural-machinery works, brick-yards, etc. Here is the seat of an *Episcopal College.* **Fargo** (*Columbia Hotel, Continental Hotel*) is on the opposite side of the Red River, and the first station in North Dakota. Through western Minnesota and northern Dakota the road passes through the finest wheat-lands in the world, and farms where thousands of acres are inclosed within one fence, and wheat-growing is followed on a gigantic scale. Both Moorhead and Fargo base their prosperity on the fact that they are the entrepots of the wheat-growing interests. The Dakotas are twice as large as all the New England States. They produced, in 1882, 22,000,000 bushels of wheat, and it is believed that, with the same increase in population and production in the future as in the past, the production of cereals at no distant date will reach 200,000,000 bushels. Fargo, the county-seat of Cass Co., has 5,664 inhabitants, 28 hotels, 12 churches, 4 banks, 6 newspapers, a U. S. Land Office, an opera-house, a theatre, a court-house, car-

wheel works, street-railway, water-works, high-school, gas and electric
light, 3 elevators with capacity of 250,000 bushels, 1 planing-mill, large
brewery, 1 paper-mill, manufactories, and all the various branches of
trade which make a thrifty and prosperous city. Brick is manufactured
extensively. It is regarded as the future commercial center of North
Dakota. It is the junction of the Dakota and Minnesota Divs. and of the
Fargo & Southwestern Branch, and here the company has car-shops and
round-houses. *Casselton* (271 miles, population 840) is a busy and enterprising town, important as a grain-shipping point. A half-dozen small
stations intervene before we reach **Jamestown** (344 miles), beautifully
situated in the valley of the James River. It has 2,296 population,
nearly all of whom are native-born to the United States. It has excellent
hotels, four churches, four newspapers, four banks, and considerable
flour-production. The North Dakota Presbyterian College and an Insane
Hospital have been erected here. The Jamestown & Northern Branch
and the James River Valley Branch diverge here.

Minnewaukan (91 miles from Jamestown) is situated on *Devil's Lake*, a salt-
water lake about 50 miles long and 30 miles wide, with high wooded banks and
surroundings of great beauty. Beyond Devil's Lake are very attractive regions
of North Dakota, known as the *Mouse River* and *Turtle Mountain* regions,
reached by the Sanborn, Cooperstown & Turtle Mountain R. R. from Sanborn,
50 miles before we come to Jamestown. This portion of North Dakota is specially attractive to sportsmen, as it swarms with game, and the lakes and
streams are full of splendid salmon and brook-trout.

There are no stations of much importance on the Northern Pacific
R. R. after leaving Jamestown before reaching **Bismarck,** the capital
of North Dakota, situated on the E. bank of the Missouri River, and
having a population of 2,186. Four lines of steamers carry on trade with
the region of the Upper Missouri. The principal buildings are the *State
Capitol, Chamber of Commerce, St. Mary's Catholic Seminary,* and *U.
S. Land Office.* There are 12 hotels, 5 churches, 4 banks, a theatre,
court-house, and town-hall. Various stage-lines converge here from different military posts, Indian agencies, and small settlements. **Mandan**
(450 miles) is on the W. bank of the Missouri, and has 1,328 inhabitants. It is the junction-point of the Dakota and Missouri Divs., and the
machine-shops here built by the company cost $150,000. The Missouri
River is spanned by a fine iron *Railroad Bridge,* and another iron wagon
bridge gives access to Fort Abraham Lincoln. Here is one of the finest
hotels (*Interocean*) in the State, built at a cost of $60,000. Many of the
blocks are built of red brick, home-made, and are handsome and substantial. There is nothing to note in the numerous small stations, all of
which are thrifty and growing, till we reach the twin towns of *Medora*
and *Little Missouri* (600 miles), lying on the E. and W. banks of the Missouri, 80 yards apart. In the vicinity are valuable coal-mines. This is
the headquarters of several large stock-raising companies, and also the
central point of **Pyramid Park,** being but 4 miles distant from Cedar
Cañon and 6 miles from the burning coal-mines. Both places abound
in magnificent scenery, full of interest to scientists and wonder to
pleasure-seekers. **Glendive** (666 miles) is on the Yellowstone River,
and is at the junction of the Missouri and Yellowstone Divs., being the

location of machine-shops, round-houses, etc.   It is the county-seat, and there are several hotels, churches, banks, etc.   The town is a favorite outfitting-spot for hunting-parties.   Stages run from here to Fort Buford, 80 miles away. **Miles City** (744 miles, population 956) is on the Yellowstone at the mouth of Tongue River, and has many fine buildings, public and private, among which is the *Court-House,* a stone edifice costing $50,000.   There is a *Government Land Office* here, and valuable lignite mines in the near vicinity. The Tongue River Irrigation Co. have just completed a ditch 14 miles long, for irrigating purposes.   Two miles farther on the railroad is *Fort Keogh,* a military post of 10 companies.   Passing a number of stations, among which is *Custer* (838 miles), deriving its name from Fort Custer, the largest post in Montana, 30 miles S. (reached by stage), we arrive at **Billings,** on the Yellowstone River (891 miles, population 941), where the R. R. Co. has repair-shops.   A thriving business is done here, and there are large shipments hence of cattle, wool, hides, and bullion.   Stages leave daily, except Sunday, for the Maginnis Mines, Fort Benton, and other points of importance in a valuable agricultural, grazing, and mining region.

We have proceeded thus far without giving any general description of the territory, a region full of interest both to the investor and the tourist. Montana covers an area of 143,776 sq. m., two fifths of it being mountainous and the rest arable. It is splendidly watered, small tributaries of the Missouri and the Yellowstone running in every direction through the E. portion, while a great number of small rivers, tributary to Flathead and Missoula Rivers, forming one of the forks of the Columbia, water the W. section of the State. The country is admirably adapted for cattle-raising, and many extensive ranches are scattered over the region. The climate is mild, dry, and exhilarating like that of the Pacific coast. The mountainous region is rich in mineral deposits. The yield of the precious metals is about $50,000,000 annually, and new mining-camps are continually being opened. The scenery in western Montana is remarkably varied, ranging from the picturesque to the sublime, and offering every kind of weird and strange rock-formation.

Resuming our journey on the Northern Pacific R. R., we reach **Livingston** (1,007 miles), with a population of 2,850, located at the foot of the Belt Mountains, about midway between the Great Lakes and the Pacific coast, and at the last crossing of the Yellowstone River.   It has the largest railroad round-house and machine-shops between Brainerd and Tacoma, also 5 hotels, 2 banks, 1 hall, 75 stores, 1 daily and 2 weekly newspapers, 1 school, and 2 churches.   Large deposits of iron, lime and sand stone, silver ore, and bituminous coal exist in close proximity.   Lumber, lime, and brick are manufactured in the town.   The *White Sulphur Springs* are 65 miles to the N. These springs contain remarkable medicinal qualities, and are becoming somewhat renowned.   The Yellowstone Park Branch diverges here and runs to *Cinnabar* (51 miles).   From the end of the railway, stages convey the tourist in 7 miles to the *Mammoth Hot Springs,* in the Wonderland of the United States,

The Yellowstone Park.

At **Mammoth Hot Springs**, 7 miles from Cinnabar Station, the N. entrance of the Park, is the *Association Hotel*, with excellent accommodation for 350 guests, providing every comfort and luxury. The *Cottage Hotel* (75 rooms) is the annex to the large hotel. Competent guides and camp-outfits are procured to reach any portion of the Park. There are good hotel accommodations at Norris, Lower and Upper Geyser Basins, Yellowstone Lake, and Grand Cañon. On account of great improvements in roads, the circuit of the Park may be comfortably made from June 1st to October 15th. The early and late visitors escape the flies, dust, and crowds, that diminish the comfort between July and September. Camping-parties have good grass and clean camps, with plenty of good fishing, and can see more of the large and small game in the months of June and October. While hunting in the Park is prohibited, there is good hunting across the lines; and fishing with hook and line is permitted all over the Park.

THE Yellowstone National Park, which Congress has "dedicated and set apart as a public park or pleasuring ground for the benefit and enjoyment of the people," lies partly in Wyoming and partly in Montana. It is 65 miles N. and S. by 55 miles E. and W., comprises 3,575 square miles, and is all more than 6,000 ft. above the sea. Yellowstone Lake has an altitude of 7,788 ft., and the mountain-ranges that hem in the valleys on every side rise to the height of 10,000 and 12,000 ft., and are covered with perpetual snow. The entire region was at a comparatively recent geological period the scene of remarkable volcanic activity, the last stages of which are still visible in the hot springs and geysers. In these the Park surpasses all the rest of the world. There are probably 50 geysers that throw a column of water to a height of from 50 to 200 ft., and from 5,000 to 10,000 springs, chiefly of 2 kinds, those depositing lime and those depositing silica. There is every variety of color, and the deposits form around their border the most elaborate ornamentation. The temperature of the calcareous springs is from 160° to 170°; that of the others rises to 200° or more. The chief points of interest are, the *Mammoth Hotel Terraces*, the *Norris Geyser Basin*, extending from the Lake of the Woods to Madison River, the *Mammoth Paint-Pots*, at the foot of Mt. Johnson, the *Monument Geyser Basin*, on the top of Mt. Schurz, in which is the *Prismatic Cañon*, the *Ebony Basin*, containing *Walpurgia Lake*, and the *Black Warrior Geyser*. On the N. of the Park are the sources of the Yellowstone; on the W. those of the principal forks of the Missouri; on the S. W. and S. those of Snake River, flowing into the Columbia, and those of Green River, a branch of the great Colorado, which enters into the Gulf of California; while on the S. E. side are the numerous head-waters of Wind River.

The **Yellowstone River,** which is a tributary of the Missouri, is without exception the most extraordinary river on the continent. Its source is near the S. E. corner of the Park, in the * **Yellowstone Lake,** a beautiful sheet of water 22 miles long and 10 to 15 wide, 7,788 ft. above the sea, and nearly inclosed by snow-clad mountains rising 3,000 to 5,000 ft. higher. Its waters are exquisitely clear and cool, are 300 ft. deep at the deepest part, and abound in salmon-trout. Its shores are rugged but extremely picturesque, and on the S. W. arm

MAP OF YELLOWSTONE PARK.

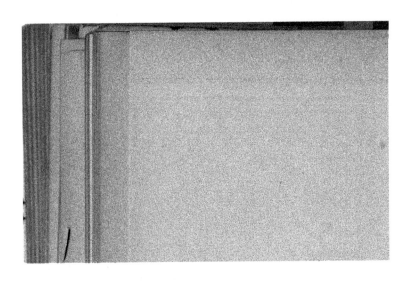

is a belt of hot springs 3 miles long and ½ mile wide, some of which extend into the lake itself. The Upper Yellowstone, the ultimate source of the river, flows into the lake from the S. E. after a course of 25 miles; and from its N. end the Yellowstone River emerges on its course of 1,300 miles to the Missouri. About 15 miles below the lake are the *Upper Falls*, where the river, after passing through a series of rapids, makes an abrupt descent of 140 ft.; and about ¼ mile farther down are the majestic * **Lower Falls,** which are 360 ft. high. Below the Lower Falls the river flows for 20 miles through the ** **Grand Cañon,** whose perpendicular sides, from 200 to 500 yards apart, rise to the height of 1,200 to 1,500 ft.

In Prof. Ferdinand V. Hayden's report to Congress on the explorations which he conducted, to which we are indebted for nearly all the authentic knowledge we have of the Yellowstone region, he says : " No language can do justice to the wonderful grandeur and beauty of the cañon below the Lower Falls ; the very nearly vertical walls, slightly sloping down to the water's edge on either side, so that from the summit the river appears like a thread of silver foaming over its rocky bottom ; the variegated colors of the sides, yellow, red, brown, white, all intermixed and shading into each other ; the Gothic columns of every form, standing out from the sides of the walls with greater variety and more striking colors than ever adorned a work of human art. The margins of the cañon on either side are beautifully fringed with pines. . . . The decomposition and the colors of the rocks must have been due largely to hot water from the springs, which has percolated all through, giving to them their present variegated and unique appearance. Standing near the margin of the Lower Falls, and looking down the cañon, which looks like an immense chasm or cleft in the basalt, with its sides 1,200 to 1,500 ft. high, and decorated with the most brilliant colors that the human eye ever saw, with the rocks weathered into an almost unlimited variety of forms, with here and there a pine sending its roots into the clefts on the sides as if struggling with a sort of uncertain success to maintain an existence—the whole presents a picture that it would be difficult to surpass in Nature. Mr. Thomas Moran, a celebrated artist, and noted for his skill as a colorist, exclaimed, with a kind of regretful enthusiasm, that these beautiful tints were beyond the reach of human art. It is not the depth alone that gives such an impression of grandeur to the mind, but it is also the picturesque forms and coloring. After the waters of the Yellowstone roll over the upper descent, they flow with great rapidity over the apparently flat, rocky bottom, which spreads out to nearly double its width above the falls, and continues thus until near the Lower Falls, when the channel again contracts, and the waters seem, as it were, to gather themselves into one compact mass, and plunge over the descent of 350 ft. In detached drops of foam as white as snow ; some of the large globules of water shoot down like the contents of an exploded rocket. It is a sight far more beautiful than, though not so grand or impressive as, that of Niagara Falls. A heavy mist always rises from the water at the foot of the falls, so dense that one can not approach within 200 or 300 ft., and even then the clothes will be drenched in a few moments. Upon the yellow, nearly vertical wall of the W. side, the mist mostly falls ; and for 300 ft. from the bottom the wall is covered with a thick matting of mosses, sedges, grasses, and other vegetation of the most vivid green, which have sent their small roots into the softened rocks and are nourished by the ever-ascending spray."

Just below the Grand Cañon the river receives Tower Creek, which flows for 10 yards through a deep and gloomy cañon known as the *Devil's Den*. About 200 yards above its mouth the creek pours over an abrupt descent of 156 ft., "forming," as Professor Hayden says, "one of the most beautiful and picturesque falls to be found in any country." Below the mountains the course of the Yellowstone lies through a wide, open valley bounded by high, rolling hills.

As already mentioned, there are immense numbers of hot springs in the Yellowstone Basin, some dead and others evidently dying. A very interesting group is on the E. side of Mt. Washburn, covering an area of 10 or 15 square miles, and there are other extensive groups on both sides of the Yellowstone Lake and also at various points on the river (see map). But the most remarkable group, not only in the Yellowstone region but in the world, is the **Mammoth** or *** Terrace Mountain Hot Springs,** situated on the W. side of Gardiner's River, on the slope of White Mountain, fronting the Grand Terraces. Many of the springs are inactive, but the calcareous deposits from them cover an area of about 2 miles square. The active springs extend from the margin of the river to an elevation nearly 1,000 ft. above.

"After ascending the side of the mountain," says Professor Hayden, "about a mile above the channel of Gardiner's River we suddenly came in full view of one of the finest displays of Nature's architectural skill the world can produce. The snowy whiteness of the deposit at once suggested the name of White (now called Terrace) Mountain Hot Spring. It had the appearance of a frozen cascade. If a group of springs near the summit of a mountain were to distribute their waters down the irregular declivities, and they were slowly congealed, the picture would bear some resemblance in form. We pitched our camp at the foot of the principal mountain, by the side of the stream that contained the aggregated waters of the hot springs above, which, by the time they reached our camp, were sufficiently cooled for our use. Before us was a hill 200 feet high, composed of the calcareous deposit of the hot springs, with a system of step-like terraces, which would defy any description in words. The steep sides of the hills were ornamented with a series of semicircular basins, with margins varying in height from a few inches to 6 or 8 ft., and so beautifully scalloped and adorned with a kind of bead-work that the beholder stands amazed at this marvel of Nature's handiwork. Add to this a snow-white ground, with every variety of shade of scarlet, green, and yellow, as brilliant as the brightest of our aniline dyes. The pools or basins are of all sizes, from a few inches to 6 or 8 ft. in diameter, and from 2 inches to 2 ft. deep. As the water flows from the spring over the mountain-side from one basin to another it loses continually a portion of its heat, and the bather can find any desired temperature. At the top of the hill there is a broad, flat terrace, covered more or less with these basins, 150 to 200 yards in diameter, and many of them going to decay. Here we find the largest, finest, and most active spring of the group at the present time. The largest spring is very near the outer margin of the terrace, and is 25 by 40 ft. in diameter, the water so perfectly transparent that one can look down into the beautiful ultramarine depth to the bottom of the basin. The sides of the basin are ornamented with coral-like forms, with a great variety of shades, from pure white to a bright cream-yellow, and the blue sky, reflected in the transparent waters, gives an azure tint to the whole which surpasses all art. Underneath the sides of many of these pools are rows of stalactites, of all sizes, many of them exquisitely ornamented, formed by the dripping of the water over the margin of the basin."

On the W. side of the Yellowstone River, about 10 miles from the falls, is the *Sulphur Mountain*, rising to a height of 150 ft. from an almost level plain and perforated with numerous fissures and "craters" from which sulphurous vapor pours forth in great abundance. The fissures are lined with sulphur-crystals, and the ground is hot and parched with internal fires. Close by are some boiling *Mud Springs*, and there is another remarkable group of them about 2 miles S. E. on the bank of the river. A few miles above Sulphur Mountain is the *** Mud Volcano,** which has broken out from the side of a well-timbered hill. The crater is 25 ft. across at the top and about 30 ft. deep. The surface of the bottom is in a constant state of ebullition, puffing and throwing up masses of boiling mud, and sending forth dense columns of steam which rise several hundred feet and can be seen for many miles in all directions. Close by are 3 large hot springs, one of which is a geyser having periods of active eruption about every 6 hours.

The great **Geysers** of the Yellowstone region are situated on the Fire-Hole River, the middle fork of the Madison, in the W. portion of the Park. They lie in two large groups, in what are called the Upper and Lower Geyser Basins. The *Lower Basin*, beginning near the junction of the East and Middle Forks of the Madison, comprises an area of about 30 square miles, and contains uncounted numbers of geysers and springs which are distributed in 7 groups. The most interesting of these is the second group, which lies near the center of the basin, and which is said to resemble a factory village, the steam rising in jets from more than 100 orifices. The *Fountain Geyser* is about 20 ft. in diameter, and throws a column to the height of 50 ft. About 100 yards S. of the Fountain is the *Museum* or *Monument Basin*, where Nature has formed what seem images of every conceivable shape. Another, named the *Evangeline Geyser*, is also styled the *Thud*, from the dull, suppressed sound given off as the water rises and recedes. It has a beautiful scalloped rim, with small basins around it. The *Upper Basin* lies in the valley of the same river, about 8 miles S. of the Lower Basin. It is not so large as the latter, covering an area of only 3 square miles, and there are fewer springs ; but the phenomena exhibited are far more remarkable. Close to the Firehole River lies **Hell's Half-Acre,** with the **Excelsior Geyser**, which in 1882 was in eruption for a month, the water rising like a dome 200 ft. high, and rocks and bowlders being ejected. This is the most powerful geyser of the group. The average temperature is over 170°, that of the air being 67°. At the head of the valley, near its S. extremity, stands * **Old Faithful**, a geyser so called for its regularity ; it spouts at intervals of about an hour, throwing a column of water 6 ft. in diameter to a maximum height of 130 ft., and holding it up by a succession of impulses from 4 to 6 minutes. When the action ceases, the water recedes out of sight, and nothing but the occasional hiss of steam is heard until the time approaches for another eruption. On the opposite side of the river is the *Beehive*, which once in 24 hours throws a column of water 3 ft. in diameter to a height of from 100 to 220 ft. The eruption lasts from 5 to 15 minutes. About 200 yards from the Beehive is * **The Giantess,** one of the largest of the geysers. It has an oval aperture 18 by 25 ft. in diameter, the inside of which is corrugated and covered with a whitish silicious deposit. When not in action, no water can be seen in its basin, although its sides are visible to the depth of 100 ft., but a gurgling sound can be heard at a great distance below. When an eruption is about to take place, the water rises in the tube with much sputtering and hissing, sending off vast clouds of steam. When it finally bursts forth, it throws up a column of water the full size of its aperture to the height of 60 ft., and through this rise five or six smaller jets, varying from 6 to 15 inches in diameter, to the height of 250 ft. The eruption, which takes

place at irregular intervals, continues for about twenty minutes. ˙ Farther
down the river, on the same side, is the *Sawmill Geyser,* which throws
a small stream 10 or 15 ft. high almost uninterruptedly.   Near it is the
* **Grand Geyser,** one of the most powerful in the basin.   Its orifice
is 2½ by 4 ft., and when not in eruption the water is quiet and clear
as crystal.   An eruption (which occurs at irregular intervals) is preceded by a rumbling and shaking of the ground, followed by a column
of steam shooting up from the crater, immediately after which the
water bursts forth in a succession of jets, apparently 6 ft. in diameter
at the bottom and tapering to a point at the top, to a height of from
175 to 200 ft., while the steam ascends to 1,000 ft. or more.   This immense body of water is ˙kept up to this height for about 20 minutes,
when it gradually recedes and again becomes quiescent.   Only 20 ft.
from the Grand Geyser, and in the same basin, but apparently having
no connection with it, is *Turban Geyser,* with an orifice 3 by 4 ft.,
which is never wholly quiet, and as often as once in twenty minutes
throws its water to the height of from 15 to 25 ft.   The * **Giant Gey-**
ser has a rugged crater, like a broken horn, 10 ft. in height, 25 ft. in
diameter at the base, and about 8 ft. at the top.   The cone is open on
one side, having a ragged aperture from the ground upward.   Its discharges are irregular and continue for irregular periods.   When Prof.
Hayden saw it in 1871, it played an hour and twenty minutes, throwing
the water 140 ft.; but Lieut. Doane, who visited it the year before,
states that it played 3¼ hours at one time, to a height varying from
90 to 200 ft.   The *Castle,* the *Grotto,* the *Punch-Bowl,* the *Riverside,*
the *Soda,* and the *Fan Geysers,* and numerous others which have not
been named, are worthy of notice.

The **Norris Geyser** Basin is about 20 miles S. of the Mammoth
Hot Springs.   Among its numerous geysers is the *Hurricane,* a new
outbreak caused by the earthquakes of 1886–'87; the water is turbid,
and rises from two craters.

The Yellowstone Park may also be reached by the Utah and Northern Branch
of the Idaho Div. of the Union Pacific R. R., from Ogden (see Route 90), to
Garrison (454 m.), on the N. P. R. R., and thence E. to Livingston (175 m.) or
to Beaver Cañon, and thence by stage to Fire-Hole Basin.

˙Resuming our journey on the Northern Pacific R. R., we reach **Boze-**
man (1,031 miles), after passing a number of small stations.   It has a
population of 2,143.   Among its principal buildings are a fine *Court-*
House, 5 hotels, 6 theatres and public halls, 3 banks, 5 churches, and the
United States Land Office.   There are several flour and planing mills.
Coal, gold, silver, iron, and copper are found near by.   *Fort Ellis,* a
military post, is 3 miles E. on the railroad.   Passing *Gallatin* (1,060
miles), situated at the head of the Missouri River, and *Townsend* (1,096
miles), we arrive at * **Helena** (*Broadwater, Cosmopolitan, Grand*
Central), the capital of Montana (reached also *via* Omaha and the Union
Pacific R. R.), with a population of 13,834, where all routes of transportation converge.   It claims to be the richest city of its size in the United
States.   Among its important buildings are the *State-House, Ming's*

Opera-House, the *U. S. Assay-Office*, *U. S. Land-Office*, 4 national banks, one of which carries deposits exceeding $7,000,000, and 2 fine public halls. The public buildings and the private residences are of a character to attract the eye of the stranger. The city is the center of important manufacturing interests, as well as of trade and commerce. It has telephones, electric lights, an admirable fire-department, and a system of water-works. Helena is situated in the center of a mineral region unsurpassed either in Montana or elsewhere for the number and richness of its gold and silver bearing lodes, there being, within 25 miles, over 3,000 quartz lodes, which have been claimed and recorded, and several hundred patented. The Drum-Lummon. Mine has recently been sold for $1,500,000. Besides the gold and silver lodes, veins of galena, copper, and iron are found in great numbers. Among the attractions of Helena are the noted *Hot Springs*, situated in a romantic glen 4 miles W. of the town, which are much resorted to by persons afflicted with rheumatism and other kindred diseases. The temperature of the water as it bubbles up from the earth varies from 110° to 190° Fahr. Fort Benton, 160 miles N. E., is reached by daily stage.

Helena is the center of a region of remarkable scenic attractions. South of it lies Madison Co., traversed by the Jefferson and Madison Rivers, two of the three streams forming the headwaters of the Missouri River. Near the center of the county is the picturesquely located *Virginia City*, the former capital of Montana (connected with Bozeman by stage), and in the midst of a rich mining region. It is at the very foot of the great chain of the Rocky Mountains. The western boundary of Lewis and Clarke Counties, in which Helena is situated, is the divide of the main range of the Rockies, and the eastern is the Missouri River. From peak to lowland there is a difference of fully 5,000 feet in altitude; the valleys are green and smiling, while the mountains are grand and gloomy, with dark-browed firs and growths of pine. Eighteen miles north of Helena the Benton stage-road enters the noted cañon of the *Little Prickly Pear*. The towering and precipitous walls look down upon a dashing mountain torrent, from 500 to 1,000 feet below, and the rich coloring of the rock-formation blends beautifully with the shades of the foliage which covers every spot where a chink or crevice affords a footing for tree, or shrub, or vine. The most striking scenery, however, next to that of Yellowstone Park, is found along the course of the Upper Missouri. Eighteen miles N. from Helena the tourist finds the great mountain-gate through which the waters of the Missouri plunge between walls 300 ft. wide and 1,000 ft. high. *Atlantic Cañon* is 3 m. farther down, and at the lower end of it is the *Bear's Tooth*, whose tusk-like forms can be seen from Helena, 25 m. away. One hundred miles from Helena are the first of the falls of the Missouri. The principal falls, four in number, are scattered along a distance of 12 miles, where the river flows through a cañon with vertical banks from 200 to 500 ft. in height. First is the *Black Eagle Fall*, where the entire river takes a vertical plunge of 26 feet. Four miles below this, the river, here 1,200 ft. wide, hurls itself over an unbroken rocky rim, forming the beautiful *Rainbow Falls*, with a perpendicular descent of 50 ft. Six miles farther down are the **Great Falls**, whose descent is 90 ft., and whose tremendous roar can be heard a dozen miles away. The river, here possessing a volume three times greater than that of the Ohio, is narrowed to 300 yards and passes between vertical cliffs some 200 ft. high. Nearly half the stream next to the right bank descends with such force as to send into the air clouds of spray 200 ft. high and glowing with all the prismatic hues. The remainder is precipitated over successive ledges, forming a magnificent cataract of fleecy foam, 200 yards in breadth and 90 ft. in perpendicular elevation. In a distance of 10 miles of the river's course there are 12 distinct falls with a total descent of about 400 ft.

After leaving Helena, the next station worthy of notice is *Garrison* (1,180 miles), the northern terminus of the Branch of the Union

Pacific R. R. from Ogden. For a distance of 60 miles the *Deer Lodge Valley*, through which the road now passes, spreads out from 5 to 10 miles wide, and within a short radius are to be found lofty peaks, lovely mountain lakes, glittering cascades, mineral springs, and the Great Geyser Cone, which gives name to the river and valley. *Deer Lodge* (population 1,463), 11 miles S. from Garrison, is a mining center on the Union Pacific R. R. **Butte** (*McDermott, St. Nicholas, Windsor*), population 10,723, is the seat of Silver Bow County, and is the present western terminus of the branch line from Logan. It is also the southern terminus of the Montana Union Railway, and has its western outlet at Garrison. It is lighted by electricity and gas; has a street-railway system operated by steam motors, electric and cable lines, two city water companies, a newly-constructed system of sewerage, court-house and jail, erected at a cost of $200,000, new opera-house, 4 banks, hospitals, first-class schools, 2 telegraph companies, telephone and district telegraph. Butte is said to be the largest and most prosperous mining city in the world. The products are principally copper and silver, with some gold. Some 3,500 miners are employed within a radius of 1¼ miles from the Court-House. There are over 4,000 mines patented in this district. Six smelteries and 6 silver-mills run day and night, reducing ores to copper matte and silver bullion. The largest mining companies are the Anaconda Co., owning the famous "Anaconda" and "St. Lawrence" copper-mines, together with a group of copper and silver mines partly developed, of untold wealth, known as the "Chamber's Syndicate"; the Boston & Montana Co., with a group of splendid copper properties, foremost of which may be mentioned the "Mountain View," "Lloyd & Harris," "Colusa," and many other bonanzas of note. The Butte & Boston Co. ranks third, with a valuable group of silver and copper properties, smeltery, and silver-mills. The Parrot Co. is one of the most prosperous corporations in the city, having copper-mines the product of which is manufactured into pig copper and shipped to the East. The Butte Reduction-Works and Colorado Mining & Smelting Co. also own considerable property, and are large shippers of copper matte to the East. The principal silver mines and mills are the "Blue-Bird," "Silver-Bow," "Lexington," "Alice," and "Moulton." The Anaconda Co. ships 4,000 tons of copper-silver ores daily to their reduction-works at Anaconda, which are by far the largest in the world. The product of copper, silver, and gold at Butte for the year 1889 reached the sum of $21,000,000, and yet the mining industry is still in its infancy in this district. The surrounding country is mountainous, consisting of grazing and timber lands. Shipments are chiefly gold, silver bullion, copper matte, and pig copper. Lumber and wood interests are very large, and an immense capital is invested in furnishing these commodities. **Missoula** (1,254 miles) is an enterprising town, with a population of 3,426, and is beautifully situated near the junction of the Hell Gate and Bitter-Root Rivers, on a broad, high plateau, from which there is a noble outlook. The *Bitter-Root Valley*, noted for its picturesque loveliness, extends S. from this 60 miles. The military post of *Fort Missoula* is 4 miles S. Following the

Jocko River, the road traverses the Flathead Reservation. Here is *Flathead Lake*, 28 miles long by 10 miles wide, lying embosomed in a lovely expanse of country ; a chain of wooded islets stretches across the center, lofty cliffs frown on two sides, while on the ·others lie sunny meadows beyond the sloping shores. In this lake the Pend d'Oreille River takes ·its rise and winds for hundreds·of miles through deep gorges and beautiful valleys before discharging its waters into Lake Pend d'Oreille. About 40 miles from Flathead Lake, near St. Ignatius's Mission, are the *Two Sisters*, cascades of great beauty, which leap down from opposite walls of a great amphitheatre, scooped out of the mountains, a sheer fall of 2,000 ft., like banks of snow against the background of rock. They unite after their descent, and pass on as a single stream. Leaving the Flathead Country, the railroad now follows the charming valley of Clark's Fork of the Columbia River. *Hope* (1,427 miles) is at the junction of the Idaho and Rocky Mts. Divs., and there are railroad-shops at this point. The road skirts Clark's Fork of the Columbia till it reaches the large opening in the river 45 miles long and from 3 to 15 miles in width, known as *Lake Pend d'Oreille*, a sheet of water whose beauty has long made it notable. Both the lake and the surrounding scenery are of the most picturesque ·description. At Sand Point (1,442 miles) the road crosses one end of the lake. *Lake Cœur d'Alene*, 11 miles from Rathdrum, is one of the most beautiful sheets of water in Idaho. Ten miles beyond Rathdrum the road enters the State of Washington.

The States of Washington (69,994 sq. m.) and Oregon (95,274 sq. m.) are so essentially alike in their general geographical, climatic, geological, and physical conditions that the same description will apply in the main to both. This magnificent region is known as the "New Northwest," as it is only within a few years that its great capacities and attractions have been realized by the world. Washington lies within the parallels of 49 degrees and 42 degrees N. latitude, and the population in 1890 was 349,390. The warm Japan current which sweeps down the Pacific coast tempers the climate to a great mildness, giving cool summers and warm winters ; and the soil, which is of great fertility, yields great crops of all the products of temperate climates. The forests of the W. section of this region, consisting of fir, pine, hemlock, spruce, larch, and cedar, and all the hard woods, are immense, surpassing the lumbering wealth of any other part of the country. Trees attain an unusual height, growing so tall and straight as to specially fit them for ships' masts. Yellow fir often attains 250 ft. ; pine, 160 ft. ; silver fir, 150 ft. ; black fir, 150 ft. Cedars have been found 63 ft. in girth and 120 ft. in height. The country is traversed by the Cascade and Coast Ranges. Some of the peaks of the Cascade Range are among the highest in the country, among them *Mount Ranier* (14,444 ft.), *Mount Baker* (10,719 ft.), *St. Helen's* (9,750 ft.), and *Adams* (9,570 ft.), all of them former volcanoes, which still give occasional signs of activity. The Columbia River, which is famous for its beautiful scenery, is navigable for large vessels for 725 m., and for small vessels 150 m. farther. The Snake River is one of the large affluents of the Columbia, and the former has many tributaries traversing Washington and Oregon. Many large streams flow also into the Lower Columbia, the principal one being the Willamette, which passes through a valley of extraordinary beauty and fertility. The largest steamships can come up the Columbia and the Willamette to Portland on the latter river (112 m.) from the Pacific Ocean. All of the coast rivers are navigable from the sea for a long distance. In addition to its great agricultural resources, the mineral wealth of the "New Northwest" is great, including gold, silver, iron, copper, lead, tin, zinc, cinnabar, graphite, etc. The fishing interests are very important ; the Columbia River, Puget Sound, and all the tributaries emptying into them, teeming with salmon and other fish of great commercial value.

Spokane (1,512 miles, pop. 19,922) is beautifully situated on both sides of the Spokane River, on the Idaho Div. of the Northern Pacific R. R., and is a terminal point for the Spokane Falls & Northern R. R. and the Union Pacific system. · It is surrounded by forests of valuable timber, picturesque mountains, and fertile valleys, the latter yielding, with little effort, large returns of grain and fruits. The Falls of the Spokane River, which in the space of half a mile fall 150 feet, are very grand and beautiful. They furnish constant water-power, as the river never freezes. Spokane has 2 electric railways 3½ miles in length, a cable line (water-power) 6 miles long, and a steam-motor line 2½ miles long, also nearly 25 miles of horse-car lines. It has 15 hotels, 9 banks, 3 daily and 4 weekly newspapers, 7 school buildings, 2 colleges, 14 churches, 4 flour-mills, 6 saw-mills, 3 sash and door and 2 furniture factories, 3 iron and machine foundries, all run by water-power. Its two theatres, left after the great fire of August 4, 1889, have a seating capacity of 3,500 people, while the opera-house, built at a cost of $300,000, alone seats 2,500 people. *Wallula Junction* (1,673 miles) is at the junction of the N. P. R. R. with the Oregon Railway & Navigation Co.'s Line. *Castle Rock* (1,750 miles) takes its name from a titanic bowlder on the bank of the river with its turreted and pinnacled sides 500 ft. in the air. A number of small stations intervene before we reach **Dalles City,** with a population of 848 (1,799 miles). This city has fine water-works, 8 hotels, 3 banks, 4 public halls and theatres, 3 newspapers, and 5 churches. It is the center of considerable trade with the surrounding country, and some manufacturing is done. Though there are no important stations between Dalles City and Portland, it is one of the most interesting portions of the route on account of the beauty and grandeur of the river scenery. The Columbia now passes through the very heart of the Cascade Mountains, and the turbulent waves roar through the narrow channel as they flow between the huge cliff-like walls that in many cases frown over the very brink of the river in basaltic masses sometimes 1,200 ft. high. At the *Cascade Locks* (1,865 miles) there are fierce and whirling rapids, where the river falls 40 ft., dashing 20 ft. at a bound. For 5 miles the water is a seething caldron of foam, so that a portage railway conveys the river-passenger around the obstruction. The characteristic feature of the Columbia River in this portion of its course is found in the heavy forests of fir and spruce. Far in the distance loom snowy peaks, and the clouds, trees, and mountains are reflected in the clear water as in a mirror. From time to time mountain streams dash over precipices into the river. Among these the most picturesque is *Multnomah Falls* (1,879 miles), near the railway station of the same name. The water plunges down a distance of 700 ft. in a ribbon of glittering spray. Thence the stream plows its way through moss-lined banks until it makes its leap into the Columbia in a broad, thin sheet of foamy silver.

*Portland, Ore. (1,911 miles) (*Esmond, Gilman, Holton, Merchants' Hotel, The Portland, St. Charles*), is the commercial metropolis of the Pacific Northwest. The city is situated on the Willamette River, 12 miles above its confluence with the Columbia. The population in

The North Puyallup Glacier, Mount Tacoma.

1890 was 46,385. It is a seaport, to which large vessels may come direct. It lies in the very heart of a great producing country, which has no other outlet, and for which it serves as a receiver and distributer of exports and imports. From Portland passenger steamships sail to San Francisco and Puget Sound, British Columbia, Alaska, and Japan, as well as to points on the Willamette, Columbia, and smaller rivers. It is the Pacific terminus of the Union Pacific system, the Southern Pacific R. R., and the Northern Pacific R. R. The streets of Portland are wide, regularly laid out, well paved and well lighted. The buildings of the business thoroughfares would do credit to any city, and the same may be said of many of the churches, the post-office, the custom-house, and other public edifices, as well as private residences. The markets are good and spacious. There are public and other schools of various grades, a large library, well-conducted newspapers, 16 banks, 8 public halls, a good theatre, 12 lines of street-railway, with 76 miles of road, all cable or electric except two miles, water-works owned by the city, gas, manu-facturing establishments, telegraphic communication with all parts, an immense wholesale and retail business, and, in fine, all the features of a flourishing modern city. The wholesale trade of Portland in 1890 amounted to about $131,550,000. The value of building improvements in 1890 amounted to $3,450,000. The factories of the city in 1890 turned out a product of $27,385,000. Within a few years the *Portland Industrial Fair Building* has been erected. For description of the country traversed by the Oregon & California R. R., see Route 94.

Several very agreeable excursions may be made from Portland, of which the easiest and most attractive is that up the Columbia River. Steamers run daily to the *Dalles* (120 miles, fare $3.50), and tri-weekly to *Wallula* (245 miles, fare $12). The scenery all the way is grand and impressive, especially at the Cascades and the Dalles. Another pleasant excursion is to Puget Sound, and may be made in two ways: 1, *via* the Northern Pacific R. R. for various points on the Sound; 2, *via* semi-weekly steamers to Victoria, the capital of British Columbia. *Puget Sound* is a picturesque body of water, and the stopping-places of the steamers are all prosperous towns. A third excursion is to *Astoria*, near the mouth of the Columbia, and thence across the promontory to * *Clatsop Beach*, the great water-ing-place of Oregon, and North Beach, in Washington. All these points, except the beaches, are now reached by railroads, trains daily.

Puget Sound.

Puget Sound is reached direct by Northern Pacific R. R. This sheet of water is one of the loveliest of inland seas, and has not unaptly been called the Mediterranean of the United States. It communicates with the Pacific Ocean by the Strait of Juan de Fuca, in the middle of which is the boundary between the United States and the British pos-sessions. This charming landlocked sea lies in the N. W. portion of Washington, is navigable by the largest ships, and penetrates far into the interior of the State, some parts of the Sound being 200 miles from the Pacific Ocean. Large ships can ride close to the shores without the need of wharves. The lofty hills which encompass the Sound and the interior far back are densely wooded with noble forests, the trees being of great size and straightness. The lumbering interest of Puget Sound has already become enormous. The tourist will be charmed by the great beauty and picturesqueness of this body of water, the ease of traversing it, as steamers plow its waters in every direction, and the general novelty of its surroundings. The principal towns on Puget Sound are Tacoma, Olympia, Seattle, Port Townsend, and Whatcom.

Tacoma (*Fife, Grand Pacific, Sillard House, Tacoma*), the seat of Pierce County, with a population of 36,006, lies at the head of Commencement Bay, the extreme southeastern harbor of Puget Sound. It has an excellent harbor, and is the western terminus of the Northern Pacific R. R. Here are located its extensive shops for over 1,000 miles of road, with a system of wharves for over 2½ miles in length. Water, gas, and electric-light works are in operation. Over 13 miles of electric street-railway are now in operation, and more are in process of construction. It has over a dozen saw-mills; one of them being the largest in the Northwest, with a capacity of 400,000 ft. per day. The output of the lumber-mills of Tacoma is upward of 980,000 ft. daily. Its railroad-shops have a complete plant for building all kinds of cars and machinery, and $6,000,000 have been authorized by the Northern Pacific R. R. Co. to be expended in Tacoma in terminal improvements. Besides the Northern Pacific shops there are several shingle-mills, 4 or 5 iron-foundries, a furniture-factory, box-factory, cornice-works and boiler-works, iron and stove works, cement-works, several breweries, an ice and refrigerator company, electric soap-works, a trunk-manufactory, oatmeal-works, a cracker-factory, a broom-factory, tile and terra-cotta works, etc. There are 2 large flouring-mills in operation, and also a ship-yard. The completion of the Cascade Div. of the Northern Pacific R. R. makes Tacoma one of the largest grain-ports on the Pacific coast, the shipments for 1889 being 18 cargoes containing 1,457,475 bushels, valued at $1,134,525. Several grain-warehouses have already been erected, and a $150,000 elevator, the only one on Puget Sound, is now in operation. The Ryan Smelter, with a capacity of reducing 500 tons of ore a day, has been in operation since May 1, 1890. It derives its support from the rapidly developing mining interests of the Cœur d'Alene, Colville, Salmon River, and other districts in Washington, Idaho, and British Columbia. Thirty miles distant from Tacoma, in the foot-hills of the Cascade Mountains, are immense beds of bituminous coking-coal now being worked. These, taken in connection with large deposits of iron-ore existing in the same locality, will soon be the means of establishing extensive iron-works at Tacoma. The railroad has, at a cost of $150,000, built coal-bunkers at its wharves in Tacoma with a capacity of 4,000 tons. There were shipped from Tacoma, in 1889, 269,400 tons of coal, valued at $1,212,300. There is much good farming-land tributary to Tacoma, lying between the Puyallup and Nesqually Rivers, extending to the Cascade Mountains, including the rich hop-lands of the famous Puyallup Valley. This valley, together with the White River Valley, produces the finest hops in the Northwest, with an average yield of 1,800 pounds to the acre. The country round about Tacoma is adapted to the raising of fruits, hay, hops, and vegetables generally. The chief business streets are well built up with brick edifices. The wholesale trade for the year 1889 ran from $8,000,000 to $10,000,000.

There are 21 banks, with an aggregate capital of $6,755,000; 4 public halls: the Tacoma Theatre, costing $150,000, and seating 1,200 people; 3 daily (2 morning) and 6 weekly newspapers; 7 public-school buildings, valued at $264,480; the Annie Wright Seminary, the Washington College (both of which are Episcopal institutions, endowed with $50,000 each by C. B. Wright, of Philadelphia), and the Methodist University for the Pacific Northwest; 30 churches; 2 hospitals, one entitled the "Fannie Paddock Hospital," costing $50,000. The Chamber of Commerce, organized in 1884, erected its building at a cost of $30,000, and have now under way a larger building costing $250,000. The general offices of the Northern Pacific R. R. on the Pacific coast are located in Tacoma, a fine office-building costing $150,000 having been erected recently for that purpose.

From Tacoma the steamboats of the Puget Sound & Alaska S. S. Co., running in connection with the Northern Pacific R. R., may be taken to all parts of Puget Sound. A trip to Victoria or Bellingham Bay will be found the most enjoyable. By either route the passenger may see the principal attractions of the Sound in a single day. The Olympic Range, snow-capped, rugged, and heavily timbered, so that only recently has it been penetrated and partially explored; also the Cascade Range, its various peaks, from Mt. Tacoma to Mt. Baker, and the islands, both large and small, timbered to the water's edge, are prominent features of the trip. Other trips are to Olympia, to Snohomish, and to Port Townsend, by the inside route, where some of the largest lumber-mills on the Sound are seen at Ports Madison, Gamble, and Ludlow. For the two last-named trips a change of boats is made at Seattle. For Victoria, a steamer leaves every day, except Sunday, at 8 A. M., stopping at Seattle and Port Townsend; arriving at Victoria at 4.30 P. M., and again leaving there at 8.30 P. M. on return trip. This gives the tourist time to see the charming city of Victoria, so different from American cities; and also the Naval Station at Esquimault, with the large dry-dock of granite, while in the harbor are always some of the English men-of-war. At Victoria, connections are made with the steamer for Vancouver and the Canadian Pacific Railway. For Bellingham Bay, stopping at Seattle, Port Townsend, and Anacortes, a steamer leaves every day, except Saturday, at 6 P. M., giving the tourist an opportunity to see Seattle in the evening and all the country by daylight on the return trip, leaving Bellingham Bay at 7.30 A. M.

Olympia (population 4,498), capital of Washington and seat of Thurston County, is on Puget Sound, and is reached by the Port Townsend & Southern R. R. Trains run between Olympia and Tenino, the junction of the Pacific Div. of the Northern Pacific R. R. It has waterworks, gas and electric-light works, street-cars, telephone system, 10 hotels, 1 bank, 4 newspapers, 3 public halls, 6 churches, 2 special religious and educational institutions, 60 stores, a U. S. Land-Office, 6 school-buildings, 5 saw-mills, 2 sash and door factories, 4 shingle-mills, and 1 large water and gas pipe manufactory. Its products are grain, hay, fruit, and vegetables.

Seattle (*Grand, Rainier, Russ*), seat of King County, with a population of 43,914, is situated on Elliott Bay, one of the best harbors on the Pacific. The location is on a peninsula, bounded on the west by Elliott Bay, on the north by Salmon Bay, Lake Union, and the Ocean Canal (now building), connecting Seattle Harbor and Lake Washington, where U. S. engineers have recommended the location of the Government Naval Station and ship-yards for the North Pacific. Lake Washington, 28 miles long and 4 miles wide, and the Falls of Snoqualmin are worth visiting. The great natural resources of King County embrace 60,000 acres of coal-fields, mountains of hematite iron-ore, and valleys of wonderfully productive soil in hops, hay, potatoes, grain, vegetables, and fruits. The surface of the country is covered with fir, spruce, cedar, and hard-wood timber. Seattle is connected almost hourly with Tacoma by both rail and steamer, and with all Sound ports, as well as Pacific coast and foreign ports, by regular lines of steamers. Seattle has 3 standard, besides 1 narrow-gauge, railroads diverging east, north, and south through the coal-fields and valleys, 51 miles of cable and electric railway street-car lines, 13 banks, 4 building and loan associations, 43 churches, 14 school-houses, the University of Washington and a Catholic academy, 2 large hospitals, 5 daily and 10 weekly newspapers, gas-works and 3 electric-light companies, telephone exchange, several large public halls, 2 standard theatres, a paid fire department and very thorough fire and water systems, embracing public water-works costing $1,000,000, and a harbor fire-boat costing $40,000. Its manufacturing interests are important and increasing, there being 12 saw-mills, 41 brick-yards, 2 ship-yards, dry-docks, iron and brass foundries, machine-shops, soap-works, cracker-factory, ice-factory, candy, sash and door, furniture, upholstering and excelsior, carriage and wagon, cigar and box factories, wooden-ware works, salmon-canneries, breweries, tanneries, marble and stone works, plaster and roofing works, soda and bottling works, dye-works, creosoting-works, planing-mills, boiler-works, and boat-yards. Seattle has 33 wholesale houses—dry-goods, groceries, wooden-ware, ship-chandlery, hardware, and liquors. Its shipments are coal, lumber, piles, spars, shingles, laths, hops, salmon, hay, grain, hides, wool, furs, vegetables, fruits, clams, oysters, game (bear, deer, grouse, ducks), and all varieties of fresh and salt water fish, and the products of the various manufactories. On June 6, 1889, nearly the entire business part of the city was destroyed by fire. This, however, has been rebuilt. Old wooden rows of former years have been replaced by massive blocks of solid and beautiful brick buildings; streets have been widened and straightened, and new ones created along the water-front.

Port Townsend, the seat of Jefferson County, is an important port on the west side of Port Townsend Bay, with a population of 4,588. It is the port of entry for the Puget Sound customs district. It has several hotels, 5 national banks, 1 opera-house, 1 saw-mill, 1 foundry and machine-shop, 1 sash and door factory, 2 public halls, 2 daily and weekly newspapers, 1 large school-building, 6 churches, and new water-works. The surrounding country is heavily timbered. Its shipments include oats, barley, potatoes, hay, and pelts. Deer, geese, and ducks are abundant; also halibut, salmon, and bass. *Fort Townsend,* a post of two companies, 2½ miles distant by water and 5 miles by land, is on the west side of the bay. A marine hospital is located here.

107. St. Paul to the Pacific Coast.

Via the Great Northern Railway.

THIS route starts from St. Paul, and between that city and **Minneapolis** (11 miles) the Great Northern Railway has four tracks passing through a well-settled suburban district, with frequent stations, which are described in Route 87. From **Minneapolis** the Great Northern has two lines running to the West, the Montana Pacific route following the Mississippi River to Anoka and St. Cloud, and reaching the Red River at Grand Forks, N. D., while the Manitoba Pacific route leads off to Lake Minnetonka and reaches the Red River at Breckenridge and Wahpeton.

. On the Montana Pacific route, up the Mississippi 65 miles to *St. Cloud*, the Great Northern has tracks on both sides of the river. The outgoing Montana train passes along the east bank and returns on the west side.

a. The Montana-Pacific Route.

Anoka and *Elk River* are the main points on the east side, the former with a population of 4,252, and a number of manufacturing plants with power furnished by Rum River, which here empties into the Mississippi River; while the latter has a population of 1,068, and also good water-power from Elk River. From here a branch line extends to *Milaca*, the nearest point to one of Minnesota's largest inland bodies of water—*Mille Lacs*, located in an Indian Reservation of the same name. Trains on this line run directly through from St. Paul and Minneapolis to *Hinckley*, *Duluth*, and *West Superior*. At various points along the Mississippi to St. Cloud large rafts of logs from the forests of the upper river and tributaries are seen, either in booms along shore or leisurely floating to the mills. The intermediate places are *Itaska, Big Lake, Becker, Clear Lake,* and *Haven.*

. *St. Cloud* is a bustling city of 7,686 population, and the seat of a State Normal School and a State Reformatory; it is also the See city of the Roman Catholic Diocese of Central Minnesota. The granite quarries here are extensive, and large quantities of stone for building and paving purposes are shipped to all parts of the country. On the west side line between Minneapolis and St. Cloud the principal points are *Osseo, St. Michaels, Monticello, Clear Water,* and *Augusta*, modest market-places for the surrounding agricultural districts. Beyond St. Cloud the train plunges into the Park Region, where the track winds and curves around the shores of the lakes, or, as if thwarted in its design of remaining on dry land, pushes through the center of many a lake, giving one the novel sensation of being at sea on a railway train. The water is combined so intricately with the land that one can not tell whether the land is islands or the water lakes.

Seven miles away from St. Cloud is *St. Joseph*, a busy little town, with flour and saw mills driven by water from the Sauk River; and then *Collegeville*, with St. John's University, a popular Catholic school, is reached. In succession, Avon, *Albany*, *Freeport*, and *Melrose* are passed to **Sauk Centre**, a progressive community of 1,695 inhabitants. From here a branch line extends 91 miles to **Park Rapids,** passing on the way through the towns of *Long Prairie* (the seat of Todd County), *Browerville*, *Clarissa*, *Eagle Bend*, *Hewett*, *Wadena* (the seat of Wadena County), *Leaf River*, *Sebeka*, and *Menahga*. Park Rapids is the seat of Hubbard County, in the Itascan basin, where the Mississippi River has its source, around the head-waters of which a State Park has been established. Park Rapids is the outfitting point to the Park, and also to the Leech Lake country, in which there are fine fishing and hunting. Passing the town of *West Union*, on the main line, the train enters the lake-gemmed county of Douglas at *Osakis*, 130 miles from St. Paul. This county contains no less than 150 lakes, none smaller than 50 acres in area. Lake Osakis is 13 miles long and 3 in width, and is well supplied with sail and row boats; but, for this matter, all the principal lakes of the entire region are equipped with sailing craft. Fish in plenty can be lured from the depths of these many waters. Hotels are scattered about, where invalids and rest-seekers can be at ease, and fresh provisions had from farmers and neighboring stores for the campers whose "cotton houses" dot the shores and prairie openings. The summer air is bracing, fogs are unknown, and healthful breezes come from the western plains and the pine forests of the north. Osakis has good hotel facilities. The main summer house, known as *Fairview*, stands on the shores of the lake, with cottages in adjoining groves, occupied during the warm months by city folks. Twelve miles from Osakis lies the town of *Alexandria*, the seat of justice of Douglas County, fairly hemmed in with water. Lakes are in sight in every direction—little gems such as New Englanders would call ponds, teeming with lilies and fringed about with wild rice, the feeding and breeding grounds of wild ducks and geese. A few miles east of Alexandria there is a chain of ten or twelve lakes, connected by channels, affording a variety for fishermen and sportsmen hardly equaled in any similar area in the country. There are numerous club-houses on the shores of these lakes, occupied during the warm season by parties from Eastern and Southern cities. Close to Alexandria is *Geneva Beach*, with a hotel overlooking the water. Continuing westward, *Garfield* and *Brandon* are passed to *Evansville*, where a branch line extends through *Elbow Lake* and *Hereford* to *Tintah*, a junction point on the Manitoba Pacific route.

Melby is the last point in Douglas County, the train passing to **Ashby,** in Grant County. The *Hotel Kittson* is the chief hostelry here, and well deserves its popularity as a cozy country home. Ten miles from Ashby and 176 from St. Paul the train reaches *Dalton*, in Ottertail County, where stages are in waiting to carry tourists to the *McFarland House* on Ten-Mile Lake, where one is offered a plentiful supply of Nature in its most primitive aspects. Ottertail is one of the heart counties of the Park Region. Passing *Parkdale*, the train after an 8-mile run reaches **Fergus Falls**, where the Red River of the North, filled with the tribute of numberless lakes, plunges in a succession of rapids from the highlands to the Red River Valley. This city has 3,772 inhabitants, and is the seat of Ottertail County. The surroundings are very picturesque. Several large factories have already taken advantage of the water-power afforded by the Red River. Continuing to the northwest, the train leaves the rolling uplands of the Park Region and glides into the smoother slopes of the eastern boundary of the Red River Valley. Villages and towns are passed, the clustered buildings of the big farms looking like villages across the dead level of fertile land. *Carlisle, Rothsay, Lawndale, Barnesville, Glyndon, Averill, Felton, Ada, Rollette, Beltrami, Kittson,* and *Carman,* in turn are passed. There is a fascination in the limitless expanse of unfenced fields, of growing wheat and stubble land. The track covers in this distance—100 miles—the counties of Wilkin, Clay, Norman, and Polk. At **Crookston;** the seat of Polk County, the Red Lake River is crossed. This stream has its source in Red Lake, to the eastward, in a vast unexplored region dotted with lakes and clothed with forests, and abounding in minerals. Three counties intervene between this point and Lake Superior, but each one is larger than the State of Connecticut. Here are the sources of the Mississippi, the lakes contributing their supplies to the great river through innumerable streams and rivulets. Along the northern boundary is Rainy River and the great Lake of the Woods. From Crookston, two branches run to the east, one terminating at **Fosston,** a near railway point to Lake Itasca, the head-waters of the Mississippi River, and the other at **St. Hilaire;** and one branch to the north, 99 miles, to **St. Vincent**—a continuous wheat-field interspersed with wooded streams, and farm-houses, villages, school-houses, hay-stacks, and elevators always in sight—through the towns of *Shirley, Euclid, Angus, Warren* (the seat of Marshall County), *Argyle, Stephen, Donaldson, Kennedy, Hallock* (the seat of Kittson County), *Northcote,* and *Humboldt.* Beyond St. Vincent is **Emerson,** just over the line in Manitoba. Turning west from Crookston the railway takes its course through a highly cultivated and beautiful country, passing in the towns of *Fisher* and *Mallory,* to the crossing of the Red River of the North into North Dakota and the city of **Grand Forks.** Well-built residences and business blocks are seen on every side, and wheat-fields encroach upon the city limits. The Red Lake River brings great rafts of logs from the Minnesota forests, to be sawed into lumber at the Grand Forks mills, making this city the chief manufacturing point of North Dakota. The State University is located here; the place is also the crossing of the Manitoba Pacific route from Fargo and Winnipeg. Continuing westward through wheat-fields level as the ocean at rest, the grain elevators stand everywhere on the horizon like ships at sea. A score of towns are passed: *Ojata, Emerado,* and *Arville* to *Larimore* (the crossing of another line of the Great Northern), *Shawnee, Niagara, Petersburg, Michigan City, Mapes, Lakota, Bartlett,* and *Crary,* all passed in turn to *Devils Lake City* on the shores of the largest body of water on the plains east of the Rocky Mountains. The Indian name is Minnewaukon, and many interesting and romantic stories cluster about its history. Until recent years the Indians would not sail or row on its waters, and only after the advent of the white man was the surface navigated. · The Indians believed that a bad spirit lived on the bottom of the lake, and to trouble the water with a canoe would result in destruction of boat and occupants. ·The lake is 50 miles long and from ½ to 8 miles wide, with timbered islands and 300 miles of shore-line. *Fort Totten*, on the south shore, is a Government military post with a small garrison. The trains of the Great Northern Railway line connect at the city of Devils Lake with a steamer for Fort Totten, forming the only direct line to that post. The bathing in the lake is particularly fine. The water has an alkaline flavor, and is most helpful, it is claimed, in cases of rheumatism and skin diseases. The fishing is good and often exciting, on account of the gamy character of the pickerel, this being the only kind of fish in the lake. The Cuthead Sioux Indian Reservation lies along the south shore of the lake. The tribe is well advanced in civilized habits, numbers about 800 persons, with good schools and churches; and yet tourists will see much interesting primitive life. Leaving Devils Lake, the way is westward through **Churches Ferry,** whence a branch line diverges to *Rolla* and *St. John,* in the Turtle Mountains. At **Rugby** another branch leaves the main line, also running to the Turtle Mountains to the towns of *Willow City* and *Bottineau.* Settlements now become less frequent, and bands of horses, cattle, and sheep are seen cropping the buffalo-grass. At every little station heaps of buffalo-bones lie along the tracks, gathered from the surrounding country. Estimating each carcass to weigh fifty pounds, the number of these gigantic animals slain by hide-hunters in two or three years in the territory tributary to the railway must have been over a half-million, for many train-loads have already been hauled away, and the industry of bone-picking is still a profitable one. **Minot,** on the Mouse River, is the last town of consequence in North Dakota. From Minot to *Williston* and **Fort Buford,** a large military post on the upper Missouri, the country is rolling and hilly. At Fort Buford the mouth of the Yellowstone is seen, adding its turbid flood to the waters of the Big Muddy. For many miles the way is along the valley of the Missouri. At *Poplar River* Montana, is a military post, and the largest Indian school on a railway in the country. Leaving the Missouri, the valley of the Milk River is entered near the town of 'Glasgow,' a division headquarters of the railway. For 180 miles now the course is along the Milk River. Here was the scene, in 1887, of the fastest railway building operations ever known, involving the construction of 550 miles of substantial roadway in six months; the unparalleled record of laying $8\frac{1}{10}$ miles of track in one day was accomplished. This is a ranch country, and horses, cattle, and sheep graze the nutritious grasses, wandering about with no seeming ownership, attaining maturity and going to market without ever having a bite of food other than the natural grasses. To the north and south are the outline sentinel hills, warning the traveler that the mighty Continental Range is ahead. At **Chinook** the Bear-Paw Mountains are in plain view, across vast hay-meadows, with more farm-houses in sight than anywhere else so far in Montana. At **Pacific Junction,** 960 miles from St. Paul, the lines divide, one going directly west through the Blackfoot Indian Reservation and over the Rocky Mountains to Kalispel, and through the Flathead Lake country and Kootenai Valley, across the Panhandle of Idaho to Spokane, through the State of Washington to Seattle, Tacoma, and Puget Sound, while the other diverges to the southwest to *Great Falls*, *Helena*, and *Butte*. Continuing on the line a short distance from the Junction, *Fort Assiniboine*, the largest military post in the United States, is reached. The quarters are of brick, and the cost of construction to date is more than $2,500,000. A full regiment of troops is usually stationed here. The location is sightly, on a clear stream in the foot-hills of the Bear-Paw Mountains. Between this post and **Benton** there are a number of trading stations, where, until branch railway lines cut off the business, freight wagon-trains departed for the Judith Basin and the Sweet Grass Hills. These freighting trains were made up of five or six wagons hitched together and drawn by strings of fifteen or twenty spans of mules or yokes of oxen. Benton is a historic point and head of navigation on the Missouri River, 3,000 miles from the Gulf of Mexico. Remains of the old adobe fort are still to be seen. It was formerly a place of great commercial importance, and still lays claim to the distinction of being Montana's seaport. It is the county-seat of Choteau County, which has an area of over 20,000 square miles. A few minor points exist in the 43 miles between Benton and **Great Falls.** This city has its name from the great falls of the Missouri River, first made known through the explorations of Lewis and Clarke early in the present century. The Missouri is a large stream at this point, having a width of 2,800 feet opposite the city front. It narrows to 1,000 feet a half-mile distant, preparatory to the first leap in the series of falls, the aggregate plunge amounting to 520 feet. Close to the first, or Black Eagle Falls, a giant spring bursts from the bank 20 feet above the river, and pours down in fan-shape over a cascade of rocks in volume sufficient to make a stream 200 feet wide and 5 feet deep. Rainbow Falls is perhaps the prettiest of the series, having a drop of full 50 feet, and ranks next to the Great Falls, where the mighty stream leaps 90 feet. From one point of observation three different falls and the giant spring can be seen. Five ranges of mountains are also in sight. A branch line extends from here to *Neihart* and *Barker*, in the Belt Mountain mining region. For 8 miles on this route the railway is through Sluice-Box Cañon, where the solid granite rises fully 2,000 feet above the track, assuming many fantastic forms, and affording the grandest mountain views to be seen from the car-windows in Montana. The Great Falls & Canada Railway runs 200 miles N. to **Lethbridge** from this point. Leaving Great Falls, with its silver and copper smelters and bustling activity, the Sun River is crossed, and the train takes its way along the banks of the Missouri. A high mountain wall stands in front; a mighty rift in the mountainous uplift permits the passage of the river—it is the Gate of the Mountains. The river is navigable for 200 miles above this break in the rocky wall. Leaving the river the train enters Prickly Pear Cañon, and pursues its way in the midst of sublime scenery to **Helena,** the capital of the State. (See

page 480.) This city is surrounded with mountains, forming many
striking views. In a romantic valley close by are the famous *Helena
Hot Springs*, where a fine hotel known as the *Broadwater*, and the most
attractive and commodious plunge-bath building in America, have been
erected. The bath building is of the Moorish style of architecture, 320
by 120 feet, and 100 feet high to the main roof, which is covered with
stained glass. It is 75 miles from Helena to *Butte* over the main range
of the Rocky Mountains, passing *en route* through the important mining
towns of *Jefferson*, *Wickes*, and *Boulder*. The backbone of the continent is pierced at the height of 6,200 feet by a tunnel made at a cost
of over $1,000,000. Passing from the gloom of the great hole through
the rocks the train makes a short turn, and brings into view across a
wide valley the city of Butte, distinguished for its mines of copper and
silver. The mountains around, and the ground under the city, are
honeycombed with shafts, drifts, and tunnels, where thousands of men
work every day in the year and every hour in the day—the men working in 8-hour shifts. Many of the places of business also run night and
day. The liquor-traffic is large, and gambling is public under license.
Close connection is made at Butte with the Union Pacific System for
Anaconda, Pocatello, Ogden, Salt Lake City, Portland,
and **San Francisco.** (See Route 90.)

b. Manitoba Pacific Route.

The trains of this route of the Great Northern use the same tracks
between St. Paul and Minneapolis, but at the latter city take another
direction through the Park Region, touching Lake Minnetonka at *Way-
zata*, whence a branch extends to *Minnetonka Beach* (*Hotel Lafayette*)
and *Spring Park*. Pretty towns are passed in rapid succession, for
this is a well-settled and prosperous agricultural district, diversified
with prairies, forests, and lakes—*Long Lake, Maple Plain, Armstrong*,
in the order to *Delano*, a town of 889 people, located on Crow River,
a water-power stream. Then *Montrose, Waverly, Howard Lake, Smith
Lake, Cokato, Dassel*, and *Darwin*, all small towns, drawing business
and prosperity from the surrounding farm and wood lands. **Litchfield,** 76 miles from St. Paul, is a popular summer resort on account
of Lake Ripley and its charming surroundings. The hotel at Bright-
wood Beach is ample, and has a number of cottages in addition. The
population is 1,899. Near Litchfield occurred a decisive battle during
the Sioux war of 1862, and a monument to the murdered settlers stands
on the spot. *Grove City, Atwater*, and *Kandiyohi* are passed to **Will-
mar,** where the Great Northern branches lead off to **Sioux Falls,**
S. D., and **Sioux City,** Iowa. Then *Pennock, Kerkhoven, Murdock*,
and *Degraff* are passed to *Benson*, where a Great Northern branch starts
for *Watertown* and *Huron*, in South Dakota. Benson is the seat of
Swift County, and is located on the Chippewa River. It has 1,351
inhabitants, and a thrifty farming country all about. *Clontarf* and
Hancock are the intermediate points to *Morris*, the seat of Stevens
County, 157 miles from St. Paul. Several mills and factories make
use of the power furnished by Pomme de Terre River. The popula-

Lake Minnetonka.

tion is 1,503. A branch of the Great Northern runs from here through *Graceville* to *Brown's Valley*, on the boundary of the Sisseton-Wahpeton Indian Reservation in South Dakota. This town has been quite a resort, owing to its proximity to Lake Traverse, one of the sources of the Red River, and of Big Stone Lake, whence the Minnesota River rises. In times of flood the waters of the two mingle, and, dividing, flow north to Hudson Bay and south to the Gulf of Mexico. Leaving Morris, the course is northwesterly through *Donnelly, Moose Island, Herman, Norcross, Tintah,* whence branch lines lead away to North and South Dakota points, including *Hankinson, Lidgerwood, Rutland, Ellendale,* and *Aberdeen*. Then *Campbell* and *Doran* are passed to *Breckenridge,* the seat of Wilkin County. A mile distant, across the Red River, is **Wahpeton,** the seat of Richland County, N. D. It has a population of 1,510. Branch lines of the Great Northern run from here to *Casselton, Hope, Mayville, Portland, Larimore, Dwight,* and other places in North Dakota. Dwight is the seat of the famous Dwight farm, belonging to the New York Congressman of the same name, and managed by Governor John Miller, North Dakota's first chief executive. A number of small towns are scattered along the line to **Moorhead,** a place of 2,088 inhabitants, on the Minnesota side of the Red River, and seat of Clay County. From here a branch extends N. through several towns to **Halstad.** Just opposite Moorhead is **Fargo,** seat of Cass County, N. D. Fargo has a population of 5,664. Continuing northward, the towns of *Harwood, Argusville, Gardner,* and *Grandin,* seat of one of the Dalrymple farms, which produces yearly from 150,000 to 200,000 bushels of wheat, are passed. This farm has averaged 18 bushels to the acre for 12 years, and netted the proprietors over $600,000. *Kelso, Alton, Hillsboro,* seat of Trail County, *Cummings, Buxton, Reynolds, Thompson,* and *Merrifield* are passed to **Grand Forks.** Grain elevators and flouring-mills are prominent objects in every county, for this is the heart of the No. 1 hard-wheat district of the Red River Valley. We continue northward through the towns of *Manvell, Ardock, Minto, Grafton,* seat of Walsh County, whence the Pembina Mountain branch leads off to *Nash, Crystal, Hoople,* and *Cavalier*. Next is *Auburn,* then *St. Thomas, Glasston, Hamilton, Bathgate,* and at last *Neche,* on the forty-ninth parallel, and the boundary between the United States and the Canadian province of Manitoba. The same scenes are repeated—wheat-fields, farm-houses, elevators, and flouring-mills. This is an interesting region. Historically it is the oldest settled part of the Northwest, but actually the youngest in the way of modern growth. The first settlement was a Hudson Bay fur-trading post, established 100 years ago. Then followed the Selkirk colonists, who vacated for more northerly homes after the adjustment of the boundary-line brought them within the United States. There are a good many Canadians and a considerable number of Icelanders in the country, all of whom are thrifty, intelligent, and industrious. An Icelander was a member of a recent Legislature. A striking thing about this part of the country is the long days of summer. It has been said that "the country is so big that it takes the sun all night to go down." One can read the newspapers here till near ten o'clock at night by daylight. Perhaps here is one of the secrets of the rapid maturity of the grain, root, and vegetable crops. This country has over 200 hours more light during May, June, and July than in the Ohio Valley, and light is the great factor in growth. Crossing the boundary the first town under the English flag is **Gretna.** For 66 miles the way is in the midst of settlements of Mennonites, Scotch, and French half-breeds, to **Winnipeg,** the chief city of a vast region. Here the main line of the Canadian Pacific is reached, and the way for 1,800 miles to the Pacific coast is over plains and rivers and across five ranges of mountains, affording scenery not equaled elsewhere in America. (See Route 108.)

108. Canadian Pacific Railway, from Montreal to Vancouver, B. C.

THE city of **Montreal** has been described in Route 60. (See also APPLETONS' CANADIAN GUIDE-BOOK, Part II., Western Canada.) From this city the Canadian Pacific Railway crosses the St. Lawrence above the Lachine Rapids and passes through a portion of the province of Quebec to *Hull*, where it recrosses the St. Lawrence and runs to Ottawa, whence it passes up the banks of the Ottawa River to *Lake Nipissing*, in Ontario, so well known to sportsmen. From Lake Nipissing it passes directly westward through a wild and picturesque region of forest, lake, and rock, until it reaches the north shore of Lake Superior at *Heron Bay.* For nearly 200 miles the line is laid along the shore of this bay through and along enormous cliffs of granite, cutting and tunneling its way to, **Port Arthur,** a prettily situated town on rising ground, overlooking Thunder Bay, a landlocked inlet of Lake Superior; while in front are the cliffs of *Thunder Cape* and *Pie Island.* It can be reached also by the Canadian Pacific Railway Co.'s steamers from Owen Sound, Georgian Bay, on the Toronto branch of the line, or by lake-steamers (Route 104) from Buffalo, Detroit, Sault Ste. Marie, etc. Five miles west of Port Arthur is **Fort William,** with a natural harbor 11 miles in extent, a breadth of 350 ft., and a depth of 10 ft. The city is at the mouth of the Kaministiquia River, a picturesque river with numerous falls, and running through a rich agricultural region. The railway company has erected here two elevators of the capacity of 1,300,000 bushels each, and are building a third of still greater size. Passing through a forest and lake covered region, the traveler reaches **Rat Portage** (293 miles from Fort William), at the head of the *Lake of the Woods.* It is the seat of an extensive lumber-trade, has good water-power, and rich mineral deposits have been discovered in the neighborhood. **Winnipeg** (426 miles from Fort William), the capital of Manitoba, has been described in Route 63. Here is the junction point for traffic *via* St. Paul and Chicago. *Portage la Prairie* (485 miles) is the center of a rich wheat country, and is the E. terminus of the Manitoba & Northwestern Railway, which runs in a N. W. direction to *Prince Albert*, a few miles below the forks of the Saskatchewan, through a

rich grazing and farming country. The next most important town to
Winnipeg is **Brandon** (562 miles), admirably situated at the crossing
of the Assiniboine River, with picturesque hills on both sides, and in
the center of flourishing settlements. It has five elevators, several mills,
and packing-houses. It is the center of trade for the region N. to Min-
nedosa and S. to Turtle Mountain. Passing *Whitewood,* the train
reaches *Broadview* (679 miles), at the head of Weed Lake, the center
of several colonies, and *Indian Head* (743 miles), in a charming situation, with a large and well-appointed brick hotel. The fishing lakes on
the Qu'Appelle River, 8 miles N., and other lakes 6 miles to the N.,
offer special attractions. Here is the celebrated " Bell Farm " of 64,000
acres. The Great Western & Central Railway runs from Brandon to
points N. and N. W. The town of **Qu'Appelle** (753 miles) is situated S. of the fort bearing that name, and is a flourishing town. The
Government has erected commodious Immigration Buildings close to
the railroad-station, and the College Farm of Bishop Anson is about 2
miles N. W. Excellent sport can be had in the valley, and the lakes
abound with fish and water-fowl. To the N. are found wapiti, moose,
and antelope. **Regina** (786 miles) is the capital of Assiniboia, on the
Wascana River. Here reside the Lieutenant-Governor and the Indian
and other officials, and here the Northwest Council meets. A branch
railway runs from Regina by way of Long Lake, a sheet of water 65
miles long, to the Saskatchewan River. At Dunmore the Alberta Railway & Coal Co.'s line branches off S. W. to *Lethbridge* (110 miles), the
center of vast coal-fields, from which large shipments are made to Winnipeg and other points E. At *Medicine Hat* (1,090 miles) the railway
crosses the Saskatchewan River by a substantial iron bridge. The country now becomes a ranch country, of which **Calgary** is the center
(1,269 miles), where cattle are pastured for hundreds of miles on the
rich prairies and along the foot-hills. The town stands about 60 miles
from the foot of the great front range of the Rockies, at the junction
of the Bow and Elbow Rivers, and the great ranches are on the former
river. **Banff** (1,349 miles) is an attractive stopping-place, as it is in
the *Canadian National Park,* and possesses famous *Hot Springs.* This
park is a tract of many square miles, embracing every variety of scenery,
which the Government has made accessible by many carriage-roads and
bridle-paths. In the rivers and lakes trout are plentiful, and in the
hills and forests roam deer, mountain-sheep, and goats. The railway
company has built a very large and well-appointed hotel, capable of
accommodating 800 guests, which ranks among the best in the country. Beyond Banff the traveler sees the splintered and fantastic form
of *Castle Mountain.* About 17 miles W. of Castle Mountain is *Lag-
gan,* the station for the *lakes in the clouds.* These three sheets of
water, Lakes Louise, Mirror, and Agnes, lie above one another, hidden
in the hollows of the mountains, high up above the line of railway.
These lakes are singularly beautiful, and few people visit Banff Hot
Springs without making the trip. Ponies and vehicles are provided for
those who do not care to walk. After surmounting the main range of
the Rockies at Kicking-Horse Pass, the road crosses the upper Colum-

bia, clears the *Selkirk Range*, recrosses the Columbia, and, passing through the *Gold Range*, enters the valley of the Thompson River. The valley of the Thompson is a ranching and farming country. This river empties into the great Fraser River at Lytton (1,754 miles), and thence to the coast, a distance of 150 miles, the railway traverses the depths of the cañons of the Fraser, following all the windings of that mighty stream, between walls of stupendous height. The scenery of this part of the road, by which the Okinagan and Cascade Ranges of mountains are passed, is quite as remarkable in its way as that of the Rocky Mountain division.

A writer, describing the journey from Calgary, says : " Now begins a series of visions and experiences beside which all seen before dwindles into insignificance. Upward, seemingly close at hand, are the naked ledges lifted above the last fringe of vegetation, wide spaces of never-wasting snow, and the wrinkled backs of glaciers whence cataracts come leaping into the concealment of the forest. In some places, where the railway reaches the highest levels, the line is carried almost under the shadow of great fields of perpetual ice, glaciers beside which those of Switzerland would be insignificant, and so near them that the shining green fissures penetrating their mass can be distinctly seen. When finally the Rockies, the Selkirks, and the Gold Range are all crossed, and the pretty lakes of British Columbia have been left behind, there comes the amazing scenery of the Fraser, where a river as large as the Ohio rushes in a mighty torrent between towering cliffs, and the railway follows all its windings. New mountains exhibit themselves, where, above the river-crags, the eye catches glimpses of the Okinagan Range or the snow-peaks of the Cascades ; and it is only after these coast-guarding heights have been traversed that the Pacific shore is reached."

Sleeping-cars run through without change between Montreal and the Pacific coast. Dining-cars accompany all transcontinental trains as far as the Rocky Mountains, where the hotels previously spoken of make them unnecessary. These châlets are not the ordinary stopping-places for meals. They are comfortable and luxurious hotels, where tourists frequently stay for many days, hunting, fishing, or wandering about the glaciers or other localities of scenic interest. The Pacific terminus of the road is at **Vancouver** (*Vancouver Hotel ;* pop., 13,709), on Burrard Inlet—a city with a fine harbor and extensive port facilities. This is the Canadian port of departure of the Canadian Pacific Railway Co.'s royal mail steamships for Japan and China. They sail about once a month for Yokohama and other ports of Japan and China, and are described by travelers as being far beyond any other steamers that have as yet sailed upon the Pacific Ocean. They are called the Empress of India, Empress of Japan, and Empress of China. On one occasion, during the summer of 1891, the Japan mail was delivered in London in 21 days from the time of leaving Yokohama. A daily line of steamers forms a ferry to Victoria, on Vancouver Island, and the capital of British Columbia. From either Victoria or Vancouver daily connection is made by steamer with local ports on Puget Sound, and once a week with San Francisco. There is a regular steamship service from Vancouver for Japan and China, and for Alaska.

Victoria, B. C. (*Driard House*), is also reached by steamer from Portland and by *Puget Sound* steamers, and is situated in the southeastern part of Vancouver Island, on Victoria Harbor, immediately óff the Strait of Juan de Fuca. The Administration Buildings, which include the *Capitol*, are a group of thoroughly English buildings, surrounded by beautiful grounds, on the south side of the bay. An excellent museum, devoted to the products of British Columbia, is worth a visit. Victoria has a large trade during the season of navigation with ports on the Strait of Georgia and Fraser River, being the port of transshipment from ocean to river boats. Population, 16,841. It has waterworks, gas, and electric lights; also a large number of extensive mercantile houses and manufacturing establishments, with several educational and religious institutions. It is garrisoned by British soldiers. The Chinese and Indians are among the features of Victoria. The former have a joss-house, theatre, and several stores devoted to their products; and the latter occupy a reservation opposite the city. At *Esquimault*, 3 miles from Victoria, are the headquarters of the English Pacific squadron, where there is usually a fleet of from three to five ships.

109. Baltimore to Richmond and the South.

Via Steamer on Chesapeake Bay and Connecting Railways.

THE trip down the Chesapeake Bay from Baltimore to Portsmouth or Richmond, if made in pleasant weather, is delightful. The steamers of the *Bay Line* make daily trips, except on Sundays, from Baltimore (Canton Wharf, on arrival of New York train) to Norfolk and Portsmouth, running in about 14 hours, connecting there with the Seaboard & Roanoke R. R. for the South Atlantic States. The principal points of interest seen in the passage of the Bay are the mouth of the Patapsco River and the battle-ground of North Point near Baltimore, referred to in the description of that city (Route 6); the Bodkin, 3 miles distant; and the harbor of Annapolis, 15 miles below, with a distant view of the great dome of the Capitol at Washington. At the lower end of the Bay are the famous fortifications of * **Fortress Monroe** and the *Rip Raps*, protecting the entrance to Hampton Roads and James River. At the head of the steamboat landing at Old Point Comfort, within 100 yards of the Fortress, is the spacious and comfortable *Hygeia Hotel*, accommodating 1,000 guests, and open all the year. Fortress Monroe, the largest in America, is always open to visitors, and presents many features of interest. *Hampton*, 2¼ miles above Old Point Comfort, is the seat of the *National Soldiers' Home* and the *Normal and Agricultural Institute for Colored People and Indians*, one of the most interesting institutions in the country. **Newport News,** 9 miles above Fortress Monroe, on Hampton Roads, is reached by the Chesapeake & Ohio R. R. from Old Point Comfort. This spot has great historic interest in connection with the Revolutionary War and the late civil war. *Hotel Warwick*, at this place, is a popular summer and winter resort. **Norfolk** (*Atlantic Hotel, Purcell House; St. James Hotel*), whose harbor is defended by the above forts, is situated on the N. bank of the Elizabeth River, 8 miles from Hampton Roads and 20 miles from the ocean. After Richmond it is the most populous city of Virginia, with 34,871 inhabitants, and has an extensive trade. Large quantities of oysters and early fruits and vegetables are brought thither by the railways and canals and shipped to Northern ports. It is the third cotton port in the country. The city is irregularly laid out, but the streets are generally wide, and the houses well built of brick and stone. The *Custom-House and Post-Office*, in Main St., is a handsome edifice, erected at a cost of $228,505, and the *City Hall* has a granite front and a cupola 110 ft. high. The *Norfolk Academy*, the *Masonic Temple*, the *Academy of Music*, and the *Norfolk College for Young Ladies*, are handsome structures ; and *St. Mary's* (Roman Catholic), the *First Presbyterian Church*, the *Freemason St. Baptist Church*, and *St. Paul's Episcopal Church* are fine edifices. The grounds of the latter are very lovely, and have some quaint old tombs. There are two cemeteries tastefully laid out and adorned with ornamental trees. Norfolk was founded in 1682, was incorporated in 1705, burned by the British in 1776, severely visited by the yellow fever in 1855, and played a prominent part in the first year of the civil war, when it was captured by the Virginians and became the chief naval depot of the Confederacy. Off Norfolk, on March 8, 1862, was fought the memorable engagement between the Confederate ironclad Merrimac and the Federal iron-clad Monitor, which marks one of the most notable epochs in naval warfare and changed the course of naval construction throughout the world. From Norfolk, Richmond is reached by steamer on the James River or by rail. The boats of the *Va. Steamboat Co.* make the trip in 10 hours, passing amid much pleasing scenery and by many localities of great historical interest. The Norfolk & Western, Norfolk & Southern, New York, Philadelphia & Norfolk, and Chesapeake & Ohio R. Rs., the Old Dominion S. S. Line, the Boston Merchants' and Miners' Transportation Co., and the Baltimore and Washington steamers connect Norfolk with every part of the country. Directly opposite Norfolk, with which it is connected by ferry, is **Portsmouth** (*Ocean House*), a city of 13,268 inhabitants, regularly laid out on level ground, and well built. Its harbor is one of the best on the Atlantic coast, and is accessible by the largest vessels. At Gosport, the S. extremity of the city, is a *U. S. Navy-Yard*, which contains a Dry Dock constructed of granite at a cost of $974,536. In its various departments of construction and repairs it employs a force of over 1,300 men, and the plant is valued at more than $7,000,000. Near by is the *U. S. Naval Hospital*, a spacious brick edifice on the bank of the river, with accommodations for 600 patients. At the time of the secession of Virginia (April 18, 1861) nearly 1,000 men were employed at the Navy-Yard. Two days afterward it was destroyed by fire, with property valued at several million dollars, including 11 vessels of war. At Portsmouth, the Bay steamers connect with the Seaboard & Roanoke R. R., which runs in 79 miles to *Weldon*, where connection is made with through routes to the South.

Daily steamers run from Baltimore (Pier 17, Light St.) to *West Point*, at the head of navigation on York River, whence the Richmond & Danville R. R. runs in 38 miles to Richmond. **Yorktown,** a small village on the right bank of York River, 10 miles above its mouth, is memorable as the scene of that decisive event in the American Revolution, the surrender of the British army under Lord Cornwallis, Oct. 19, 1781. The precise spot where the surrender took place will be pointed out to the inquiring visitor. Remains of the British intrenchments may still be seen, and the country around bears abundant evidences of the operations conducted by Gen. McClellan in 1862. The railway between West Point and Richmond traverses a section of country remarkable as the scene of many important events during the late civil war. A short distance from the point where the railway crosses the Chickahominy River are *Powhite Creek* and *Cold Harbor*, famous as the localities of the great struggles of 1862 and 1864. *Fair Oaks Station* (7 miles from Richmond) was the scene of the bloody but indecisive *Battle of Seven Pines*, fought May 31, 1862, between Generals McClellan and Johnston. **Richmond** (see Route 111).

110. Norfolk and Portsmouth to Atlanta.

Via the Seaboard Air Line.

FROM Norfolk passengers take the ferry from the foot of Market St. to the station in Portsmouth, the point of departure for the Seaboard Air Line. Soon after leaving, the train passes through the famous trucking section tributary to the Western Branch. Here, on account of its great fertility, truck (another name for early vegetables) is produced two weeks earlier than on farms in the surrounding country, and the producers have amassed fortunes. **Suffolk, Va.** (17 miles from Portsmouth), at the head of steam navigation on the Nansemond River, is a lively place of 3,354 inhabitants, situated in the heart of a renowned farming region. Many modern and handsome buildings adorn the town. Besides extensive manufacturing establishments, there are several oyster-packing houses. Passing *Franklin*, a thrifty commercial center at the head of steam navigation on the Blackwater River, where connection is made by steamer for the hunting-grounds and fisheries of the Albemarle Sound, whose praises have been proclaimed far and wide, the town of *Weldon, N. C.* (80 miles from Portsmouth ; *Atlantic Coast Line Hotel*) is reached. A steel railroad bridge spans the Roanoke River, upon the bank of which Weldon is situated (population, 1,286). Here connection is made with the Atlantic Coast Line. The Roanoke River furnishes excellent water-power, though but partially developed.

The course of the Seaboard Air Line, *via* the Raleigh and Gaston R. R., is through a country which has few scenic attractions, though the highly cultivated farms and clustering towns indicate a populous and thrifty section. *Littleton, N. C.*, on account of the Panacea Springs, whose curative properties have undergone the severest tests, and are generally pronounced wonderful, and the Littleton Female College, is fast becoming a popular and attractive resort. *Henderson, N. C. (Henderson House, Martinsburg Hotel,* and *Southern Hotel*), 53 miles from Weldon, at an altitude of 505 ft., is higher than any other point in this part of the State. On account of its great salubrity of climate and good water, it is noted for its absolute freedom from all malarial and mias-

mic influences, and is popular as a summer resort. Population, 4;191.
It has a prominent position in the tobacco-trade which makes it a
money and business center. From Henderson, after a ride of 45 miles
over the Durham & Northern R. R., the town of *Durham*, *N. C.* (*Hotel
Claiborne*) is reached. The word " Durham " and its identity with
the tobacco interest is well known to almost every inhabitant of the
United States, and it would well repay the tourist to stop there for
a few days to visit the extensive manufactories of this staple. To
briefly give an idea of its enormity, a single firm. in Durham made
and shipped in one year 706,810,000 cigarettes ; and they manufactured 914,515 pounds of ·smoking-tobacco. This establishment
is the largest of its kind in the world. Over 100 mammoth brick
structures here are devoted to handling the. leaf. *Wake Forest*, *N. C.*,
the site of the school of that name, 18 miles from Henderson, surrounded by mineral springs, is a well-established resort during the summer months. *Raleigh*, *N. C.*, the " City of Oaks," 97 miles from Weldon, and the capital of North Carolina, is described on page 508. From
Raleigh to Hamlet, N. C., the route is over the Raleigh & Augusta Air
Line R. R., through the heart of the Long-Leaf Pine Belt, whose enormous wealth in lumber and naval stores, and great value from a bygi-
enic standpoint, can scarcely be estimated.

    Southern Pines, N. C., 70 miles from Raleigh, is the highest
point in the long-leaf pine region, and is pronounced by eminent medical authorities to be one of the greatest natural sanitariums. During
the winter months the hotels are frequently unable to accommodate
the applicants, for lack of room. The climate corresponds in dryness
to that of the south of France, due to the fact that the soil is almost
entirely sand, extending to a depth of many yards, thus affording
perfect drainage, and avoiding surface dampness. Hereabout game
abounds, and good hunting and fishing are enjoyed during the winter
by tourists, who come South to avoid the rigors of the North. The
profuse generation of *ozone* is a valuable accessory from a healthful
standpoint.

    Reaching *Hamlet*, *N. C.*, the traveler enters one of the riche·st agri-
cultural districts in the country. The size and solidity of the houses
and barns, and the variety and extent of the cultivation, will remind the
tourist rather of the best farming districts of New England and the
West than what is usually seen South. This is termed the " Great
Pee Dee section."

    From Hamlet the route is *via* the Carolina Central R. R. to Monroe
(distance, 53 miles), where connection is made with the Georgia, Carolina & Northern R. R. for Atlanta, Ga. The G., C. & N. R. R. is
a newly built, well-ballasted, and well-maintained road, and passes
through one of the finest cotton belts in the South. The traveler's
eye is constantly directed to well-cultivated fields,·fine droves of sheep
and cattle, and energetic towns, substantial evidences of vigor and
growth. At *Calhoun Falls*, *S. C.*, a magnificent steel bridge spans the
Savannah River, and the train passes into Georgia. Situated on the
banks of the Oconee River, 199 miles south of Monroe, is **Athens**

(*Commercial House*), the classic city of Georgia (population, 8,639). Its society is cultured and refined, on account of its many institutions of learning, among which is the State University. The Oconee River furnishes water-power for its manufacturing enterprises, and this, together with the easy accessibility to raw material, the general mildness of the climate all the year round, low rate of taxation, and cheapness of labor, makes manufacturing a profitable business. The finely built residences and handsome homes impress the stranger at once. The annual business amounts to upward of $10,000,000. Athens is lighted by electricity, has good water, a paid fire department, good hotels, and the advantages of four railroads, besides a well-regulated system of electric street-cars. Sixty-nine miles from Athens is **Atlanta,** the "Gate City" of the South (see p. 538).

111. Washington to Richmond.

Via Baltimore & Potomac R. R.   Distance, 116 miles.

THE city of Washington is fully described in Route 8. The Richmond train leaves the depot in Washington at the cor. of 6th and B Sts., crosses the Long Bridge into Virginia, and runs down parallel with the Potomac 7 miles to **Alexandria,** which is described in Route 8. Beyond Alexandria it still follows the Potomac for 27 miles to *Quantico,* a small station and steamer-landing, where connection is made with the steamers from Washington. Here the train takes the track of the Richmond, Fredericksburg & Potomac R. R., which runs S. E. across a broken and desolate-looking region, part of which is known as "The Wilderness" and is famous as the scene of the great combats of 1863 and 1864. Twenty-one miles beyond Quantico is **Fredericksburg,** a quaint and venerable old city on the S. bank of the Rappahannock River. It was founded in 1727, contains 4,528 inhabitants, and is notable as the scene of one of the severest battles of the civil war, fought Dec. 13, 1862, in which Gen. Burnside was defeated by Gen. Lee. Many traces of the conflict still remain and may be seen from the cars. In the vicinity are a National and a Confederate cemetery, the latter being adorned with a monument. Eleven miles W. of Fredericksburg, on the E. edge of "The Wilderness," the *Battle of Chancellorsville,* in which "Stonewall" Jackson lost his life, was fought, May 2–4, 1863. Southward from Chancellorsville is *Spottsylvania Court-House,* where, in May, 1864, were fought some of the bloodiest battles of Grant's campaign on his way to Richmond. Just outside the limits of Fredericksburg an unfinished monument, begun in 1833, marks the tomb of the mother of Washington, who died here in 1789. It was in the vicinity of Fredericksburg that Washington himself was born, and here he passed his early years. Leaving Fredericksburg, the train crosses the Rappahannock and passes directly over the ground where Gen. George G. Meade's charge was made in the battle of Fredericksburg. *Guineys* (12 miles beyond Fredericksburg) was the scene of the death of Stonewall Jackson. He was wounded May 2, 1863, and died at the house of William Chandler, May 10th. At *Hanover Junction* (37 miles from Fredericksburg) another battle was fought between Generals Grant and Lee in May, 1864. Remains of the works occupied by the two armies may still be seen. *Ashland* is a favorite residence of many citizens of Richmond, from which it is only 16 miles distant. It is the seat of Randolph-Macon College, and near here Henry Clay was born.

Richmond.

Hotels, etc.—The leading hotels are *Hotel Dodson, Exchange and Ballard House,* facing and connected with each other on Franklin St. The *Davis House, Ford's, Murphy's,* and the *American* are good houses. *Street-cars* (fare, 5c.) traverse the main thoroughfares. The *Electric Street-car* line is the longest now running. *Garber's* omnibuses and hacks are in waiting at the depots and steamboat landings. Fare to any point in the city, 50c. Hacks by the hour, $1.50 for first hour, and $1 for each additional hour. *Post-Office* in Main St., between 10th and 11th. The *Grand Union Station* is at 8th and Byrd Sts.

Richmond, the capital and largest city of Virginia, is situated on the N. bank of the James River (which is navigable), about 127 miles from the ocean. It is built on several eminences, the principal of which are Church and Shockoe Hills, which are separated by Shockoe Creek, and is surrounded by beautiful scenery. It is regularly laid out and well built; the streets, which are lighted with gas and electricity, cross each other at right angles. In the business quarter are many substantial and handsome buildings, and nearly all the residences have grass and flower plots in front.

Richmond was founded in 1737, was incorporated in 1742, and became the State capital in 1779, at which period it was a small village. The city was, in turn, the scene of the conventions of 1788, to ratify the Federal Constitution, those of 1829, 1850, and 1861, and other important political gatherings, which largely shaped the destinies of the Commonwealth. In 1861 still greater prominence was given to it as the capital of the Southern Confederacy; and one of the great aims of the Federal authorities, throughout the war, was to reduce it into their possession. The obstinacy with which the Confederates defended it was a proof of the great importance which they attached to its retention. To effect this, strong lines of earthworks were drawn around the place, and may still be seen as memorials of the great struggle. When General Lee evacuated Petersburg, April 2, 1865, the troops defending Richmond on the E. were withdrawn, and, to prevent the tobacco warehouses and public stores from falling into the hands of the Federal forces, the buildings—together with the bridges over James River—were fired. This resulted in the destruction of a large part of the business section of the city, the number of buildings destroyed having been estimated at 1,000, and the loss at $8,000,000. With the cessation of hostilities, Richmond set to work to rebuild her blackened quarters, which she has now wholly accomplished, and the city is rapidly surpassing its former prosperity. The population in 1870 was 51,038; in 1880, 63,803; and in 1890. 81,388. The commerce is large, the chief articles of export being tobacco and flour. The manufactures include iron-works, machine-shops, foundries, sugar-refineries, cigar-factories, coach and wagon factories, furniture, sheetings and shirtings, and stoneware. Seven lines of railroad intersect at Richmond, and regular lines of steamers run to Norfolk, Baltimore, Philadelphia, and New York.

The most prominent public building of Richmond, and by far the most conspicuous object in the city, or from its approaches, is the ***State Capitol,** together with the City Hall, standing in the center of a park of 8 acres, on the summit of Shockoe Hill. It is a Græco-Composite building, adorned with a portico of Ionic columns, the plan having been furnished by Thomas Jefferson after that of the *Maison* *carrée* at Nismes, in France. The view from the platform on the roof is extensive and beautiful. In the· center of the building is a square hall surmounted by a dome, beneath which stands * Houdon's celebrated statue of ·Washington. It is of marble, of the· size of life, and represents Washington as clad in the uniform worn by an American general during the Revolution. Near by, in a niche in the wall, is a marble bust of Lafayette. The *State Library*, in the Capitol Building, contains 40,000 volumes and many portraits of historical personages, and is said to be one of the best collections in the country. In the Supreme Court-room is the law library of several thousand volumes. The *Historical Society Collection* is in the rooms of the Westmoreland Club and includes a library of 10,000 volumes. On the esplanade leading from the Governor's house to the W. gate of the Capitol Square, and near the latter, is Crawford's equestrian * * **Statue of Washington,** consisting of a bronze horse and rider, of colossal size, rising from a massive granite pedestal, and surrounded by bronze figures of Patrick Henry, Thomas Jefferson, John Marshall, George Mason, Thomas Nelson, and Andrew Lewis. The horse is half thrown upon its haunches, and is thought to be one of the finest bronzes in the world. A life-size marble statue of Henry Clay (near the W.· corner), and Foley's statue of *General " Stonewall " Jackson*, of heroic size, on a granite pedestal (N. of the Capitol), complete the decorations of the Capitol Square. The *Governor's House* is a plain building on the N. E. corner of Capitol Square. The *City Hall*, on Capitol St., is a new and handsome structure. The * **Custom-House,** which also contains the *Post-Office*, is a fine structure of granite, in the Italian style, in Main St., between 10th and 11th. The **Medical College,** in rear of the Monumental Church, is a good specimen of the Egyptian style of architecture. Other educational institutions are the *Richmond College* and *Southern Female Institute*. In the vicinity is the *Brockenbrough House*, which was the residence of Jefferson Davis, President of the Southern Confederacy ; it is now used as a museum for relics of the Southern Confederacy. The *Almshouse* is one of the finest edifices in the city ; the *State Penitentiary* and the *Soldiers' Home*, about 2 miles out on Grove Road, are in the W. suburbs. *Belle Isle* retains some interest as a military prison during ·the civil war. The *Mozart Academy of Music*, in 8th St., is one of the newest edifices.

The churches of Richmond are numerous, and several of them are handsome specimens of architecture. Those with historic associations are St. John's and the Monumental * **St. John's** (Episcopal) is a plain edifice with a modern spire, on Church Hill, cor. Broad and 24th Sts. It is of ante-Revolutionary origin, and in it was held (in 1775) the Virginia Convention to decide the action of the colony, on which occasion Patrick Henry made his celebrated speech containing the words, "Give me liberty or give me death ! " The * **Monumental Church** (Episcopal), cor. Broad and 13th Sts., is a handsome edifice, with a dome, standing on the spot formerly occupied by the Richmond Theatre. In 1811, during the performance of a piece called "The Bleeding Nun," the theatre caught fire, and, in the terror and confusion of the crowd rushing to the doors, 69 persons, including the Governor of Virginia. and some of the most eminent men and beautiful women of the State, were crushed or burned to death. The church was erected as a memorial of the event, the remains of the victims being interred beneath a monument in the vestibule. Of the more modern structures, * **St. Paul's** (Episcopal), cor. Grace and 9th Sts., is the most imposing. In it Jefferson Davis was seated when a messenger brought him the fatal news that Lee was about to evacuate Petersburg.

Of the several cemeteries of Richmond, * **Hollywood** (reached by electric cars) is the principal. It is a spot of great natural beauty, in the W. limits of the city, above James River, and embraces an extensive tract, the whole ornamented with venerable trees, shrubs, and flowers. On the hill at the S. extremity, monuments mark the resting-place of President Monroe and of President Tyler. The remains of Jefferson Davis, the President of the Confederate States, and of General J. E. B. Stuart, the Confederate cavalry leader, are interred here. In the soldiers' section are the graves of thousands of Confederate dead, from the midst of which rises a monumental pyramid of rough stone. *Monroe Park* is near the W. and *Marshall Park* (Libby Hill) near the E. end of the city. From the latter a fine river view may be had. To the N. W. of Monroe Park is the statue of Gen. Robert E. Lee, donated to the State by the Lee Monumental Association. Six bridges connect Richmond with Spring Hill and *Manchester*, the latter a pretty town with 2 fine cotton-mills. The *Tredegar Iron-Works*, which were the great cannon manufactory of the Confederacy, are worth a visit. The buildings cover 15 acres of ground. The *Gallego* and *Haxall Flour-Mills* are among the largest in the world. A carriage may be taken, and within a few hours' ride from the city several battle-fields and National Cemeteries visited.

112. Richmond to Charleston.

a. Via Wilmington and Florence.

The Atlantic Coast, Piedmont Air, and Seaboard Air Lines, with their two fast express-trains daily from New York, with vestibule cars attached, constitute the fast mail and passenger routes to Charleston and Savannah. The route from New York to Philadelphia is *via* Route 3 *a*; from Philadelphia to Baltimore *via* Route 5; from Baltimore to Washington *via* Route 7; from Washington to Richmond *via* Route 111; from Richmond to Charleston as described below; and from Richmond to Savannah *via* Route 113 *a*. The schedule time from New York to Charleston is 33 hours; to Savannah, 39 hours.

LEAVING Richmond, the train crosses the James River on a handsome bridge and runs in 23 miles to **Petersburg** (*St. James Hotel*), a well-built city of 22,680 inhabitants, situated at the head of navigation on the Appomattox River, 12 miles above its entrance into the James. Its trade is large; the handling of tobacco and cotton, with wheat, corn, and general country produce, being the chief business. The principal buildings are the Custom-House and Post-Office, the Court-House, 2 market-houses, and the Theatre. There is a public park called Poplar Lawn. Petersburg was the scene of the last great struggles during the civil war. Since the war the place has prospered, and the marks of the conflict are rapidly disappearing; but the fortifications are still distinctly trace-

A Southern Tobacco Field.

able, and the chief battle-fields, etc., are easily found. *Weldon* (86 miles) is a thriving post-village in North Carolina, at the head of steamboat navigation on Roanoke River. Here the Seaboard & Roanoke R. R. (see Route 106) from Portsmouth and Norfolk connects. Beyond Weldon the country is flat and uninteresting, the road traversing for many miles the great pine belt which extends from Virginia to Florida. **Goldsboro** (164 miles) is a prosperous town of 4,017 inhabitants, near the Neuse River, at the head of navigation, and at the intersection of the Atlantic & North Carolina R. R. Eighty-four miles beyond, passing many small stations *en route*, the train reaches **Wilmington** (*Island Beach Hotel, Orton, Purcell House*), the largest city of North Carolina, situated in the S. E. corner of the State, upon the Cape Fear River, 20 miles from the sea. Wilmington has a population of 20,056, an extensive commerce, both coastwise and foreign, and has long been the leading market for naval stores in the world. There are regular lines of steamers to Baltimore, Philadelphia, and New York. The principal articles of shipment are lumber, turpentine, rosin, tar, pitch, shingles, and cotton. Street-cars run through the principal streets to *Oakdale Cemetery* and to the R. R. stations. *The Sound*, a place of summer resort with railroad connection, is 7 miles distant; and *Fort Fisher*, which played so conspicuous a part in the civil war, is 20 miles below, at the mouth of the river. From Wilmington to Florence (108 miles) the country is of the same featureless and monotonous character, the route now being through South Carolina. **Florence** (356 miles) is a place of considerable commercial importance by reason of its railroad facilities, and is the point of shipment for most of the cotton of the adjacent country. Here the Charleston train takes the track of the Northeastern R. R., which runs to Charleston in 102 miles, through an uninteresting region. **Charleston** (458 miles) is described further on.

b. Via Charlotte and Columbia.

From Richmond to Columbia this route is by the Richmond & Danville Line. Crossing the James River on a substantial covered bridge, the train passes through the populous suburb of *Manchester*, and runs S. W. through a famous tobacco-growing region to *Burkeville* (53 miles), situated at the intersection of the Norfolk & Western R. R., formerly the *South-Side Railway*, which was so prominent in the siege of Petersburg.

In April, 1865, Burkeville became a place of critical importance. General Lee, having evacuated Petersburg on the night of April 2d, retreated up the N. bank of the Appomattox, and, recrossing, reached Amelia Court-House, from which it was his design to advance to Burkeville Junction. General Grant moved more rapidly toward the same point from Petersburg, and, having a shorter distance to pass over, reached the place before Lee, who was forced to halt at Amelia Court-House to obtain rations. The presence of Grant at Burkeville induced Lee to alter his line of march and retire toward Lynchburg, which resulted, April 9, 1865, in the surrender of the Confederate forces at Appomattox Court-House. The scene of the surrender was near *Appomattox*, a station on the Norfolk & Western R. R., 48 miles W. of Burkeville and 28 miles E. of Lynchburg.

Thirty-two miles beyond Burkeville is *Roanoke*, the name of which will recall the famous orator " John Randolph of Roanoke," who passed almost his entire life in this region. *Danville* (140 miles) is a town of 10,305 inhabitants, pleasantly situated at the head of navigation on the Dan River. It is the market-town of the best tobacco-growing section of Virginia, and has an active trade. Connection is made here with the Virginia Midland Div., which forms with the present line a popular through route (known as the Richmond and Danville) from Washington *via* Lynchburg to the South-Atlantic States (see Route 122). Five miles beyond Danville the train enters North Carolina, and soon reaches *Greensboro* (*McAdoo House*), a rapidly growing town, situated in the midst of a rich tobacco-producing region, and near valuable deposits of coal, iron, and copper.

From Greensboro a branch line runs S. E. in 130 miles to *Goldsboro* (see Route 112 *a*), passing **Raleigh** (*Central, Yarborough House*), the capital of North Carolina. Raleigh, with a population in 1890 of 12,678, is pleasantly situated on an elevation 6 miles W. of the Neuse River and a little N. E. of the center of the State. It is regularly laid out, with a park of 6 acres in the center (*Union Square*), from which extend 4 streets, dividing the city into 4 parts, in each of which is a square of 4 acres. In Union Square is the beautiful *State House, built of granite, after the model of the Parthenon, at a cost of \$531,000. Other public buildings are the *U. S. Custom-House and Post-Office*, a fine granite structure, the *Supreme Court* and *State Building*, *State Geological Museum*, the *State Insane Asylum*, the *Institution for the Deaf and Dumb*, and the *Penitentiary*. Raleigh is also reached from *Weldon* (see present route). The chief educational establishments are *St. Mary's Female College*, the *Peace Institute*, the *Leonard Medical School*, the *Shaw University* (colored), the *State Agricultural and Mechanical College*, *Trinity College* (males), and *Baptist State University* for women. There are also many cotton, tobacco, shoe, clothing, car, locomotive, and ice factories.

The next important station S. of Greensboro on the main line is *Salisbury* (238 miles), where connection is made with the Western North Carolina Div., by which the tourist may reach the Mountain Region (see Route 131). Forty-four miles beyond is **Charlotte,** a busy little city of 11,557 inhabitants, on Sugar Creek, at the junction of several important railways. It is situated on the gold range of the Atlantic States, and its prosperity is chiefly owing to the working of the mines in its vicinity. A U. S. Assay Office (formerly a branch mint) is located here. A plank road 120 miles long connects Charlotte with Fayetteville. From Charlotte the route is *via* the Charlotte, Columbia & Augusta R. R., which runs S. through a farming region, and in 110 miles (392 miles from Richmond) reaches **Columbia** (*Grand Central, Hotel Jerome,* and *Wright's Hotel*), the capital of South Carolina. Columbia is a beautiful city, situated on the bluffs of, and 15 ft. above, the Congaree, on an elevated level plateau a few miles below the charming falls of that river. It was famous for its delightfully shaded streets and its wonderful flower-gardens, but the aspect of the city was greatly changed by the unfortunate conflagration which destroyed so large a part of it during its occupation by General Sherman's forces, in Feb.,

1865. The streets, however, are still abundantly shaded, and there are many attractive drives in the vicinity. The view from *Arsenal Hill* is the most beautiful in this portion of South Carolina. The *State House,* when completed, will be one of the handsomest public buildings in the United States; it has cost $3,000,000, and about $1,000,000 more will be required to finish it. It is surrounded by grounds in which are a statue of George Washington, a monument to the Confederate dead, and the Palmetto Monument to those South Carolinians who fell in the Mexican War. The *Executive Mansion* has grounds laid out in walks, gardens, and drives, and commands a picturesque view of the Congaree Valley. The *State Penitentiary* is a vast structure situated in a plot of 20 acres at the junction of the Broad and Saluda Rivers, within the city limits. The *Lunatic Asylum* occupies a group of spacious buildings in the N. E. part of the city. The grounds, 20 acres in extent, are surrounded by an inclosure and beautified with gardens, hot-houses, and walks, and has a farm of 200 acres around it. Other note-worthy public buildings are the U. S. Court-House and Post-Office, the City Hall, and the Market-House. There are several important educational institutions, of which the principal are the *University of South Carolina,* founded in 1801, which has substantial brick buildings in grounds 40 acres in extent, with a library of 35,000 volumes; the *Presbyterian Theological Seminary,* with a library of 18,340 volumes. The car-shops of the Columbia Div. of the Richmond & Danville R. R. occupy 4 acres of ground, and there are other large manufacturing establishments. The * *Fair Grounds* of the South Carolina Agricultural and Mechanical Society, in the N. W. suburbs, contain about 30 acres, with spacious buildings, and are well supplied with fountains, fish-ponds, a race-course, etc. *Sydney Park* contains about 25 acres, tastefully laid out and adorned with trees and shrubbery.

From Columbia to Charleston (*via* Atlantic Coast Line) the journey will give the traveler some inkling of the lowland features of Southern landscape, though not in its most interesting character, since the country is level, and most of the way is through extensive pine-forests. The only station on the line requiring mention is *Summerville* (22 miles from Charleston), a small village situated on a pine-clad ridge which extends across from the Cooper to the Ashley River. Its climate is remarkably agreeable, and the place is attracting attention as a winter resort.

Charleston.

Hotels, etc.—The best hotels are the *Charleston Hotel,* in Meeting St. between Hayne and Pinckney Sts., and the *St. Charles,* cor. Meeting and Hasel Sts. *Street-cars* (fare, 5c.) traverse the city and afford easy access to the chief points of interest. *Omnibuses* are in waiting at the stations and landings on the arrival of trains and steamers, and convey passengers to any portion of the city (fare, 50c.). Besides the rail-routes described above, Charleston is reached from New York by *steamers* (Clyde Steamship Line), leaving Pier 29, East River, at 3 P. M. on Mondays, Wednesdays, and Saturdays. The New York steamers continue to Jacksonville.

Charleston, the chief commercial city of South Carolina, is picturesquely situated at the confluence of the Ashley and Cooper Rivers, in

lat. 32° 45' N., and lon. 79° 57' W.   The rivers run a parallel course
for nearly 6 miles, widening as they approach the sea, and thus gradually narrowing the site of the city to a peninsula.   The corporate limits
of the city extend from Battery or White Point, on the extreme S.
verge of the city, to an arbitrary line on the N. about 3 miles above.
Within this area the city is laid out with tolerable regularity, the streets
generally crossing each other at right angles, and being laid with Belgian pavement.   On August 31, 1886, and for some time afterward,
Charleston experienced a succession of earthquakes, which injured
many of the older buildings, which are now replaced by less picturesque edifices.   The two principal streets are King and Meeting, which
run N. and S., nearly parallel, the whole length of the city, but converge
to intersection near the northern limits.   *King St.* contains the leading
retail stores, and is the fashionable promenade.   The jobbing and wholesale stores are chiefly in *Meeting St.* and *Bay St.;* and the banks, and
brokers' and insurance offices, are in *Broad St.*   The * **Battery** is a
popular promenade, on the water's edge, and commanding a view of
the Bay; it is lined with fine private residences.   Fine residences are
also found in Meeting St. below Broad, in Rutledge St. and Ave., and at
the W. end of Wentworth St.   The roads leading out of the city along
the Ashley and Cooper Rivers are singularly beautiful, and afford interesting drives.   They are all embowered in loveliest foliage; pines, oaks,
magnolias, myrtles, and jasmines vying with each other in tropical luxuriance and splendor.   There is a beach drive on Sullivan's Island
(reached by ferry).   A bridge has been built across the Ashley River,
which will give access to some charming drives on the mainland among
the old plantations.

Charleston was settled in 1679 by an English colony under William Sayle,
who became the first Governor. It played a conspicuous part in the Revolution,
having been the first among the chief places of the South to assert a common
cause with and for the colonies. It was thrice assaulted by the British, and only
yielded to a six weeks' siege by an overwhelming force, May 12, 1780. It was
the leading city both in the nullification movement during Jackson's administration and in the incipient stages of Southern secession. Open hostilities in
the civil war began at Charleston, with the bombardment of Fort Sumter on
April 12, 1861; and for the next four years it was one of the chief points of
Federal attack, without being lost by the Confederates, however, until Sherman's capture of Columbia on February 17, 1865. During the war many buildings were destroyed, and the towers and steeples of churches riddled with shot
and shell. Since its close, rapid progress has been made in the work of rebuild-
ing, and Charleston is now more prosperous than ever. The growth of population has been as follows : In 1800 it was 18,711 ; in 1850, 42,985 ; in 1860, 40,519 :
in 1870, 48,956 ; in 1880, 49,999 ; and in 1890, 54,955. The commerce of the city
is large, the chief exports being cotton (for which it is one of the chief shipping ports), rice, naval stores, and fertilizers. The manufacture of fertilizers from the valuable beds of marl and phosphate, discovered in 1868, is
now one of the principal industries. Ten companies are now engaged in the
business. There are also flour and rice mills, bakeries, carriage and wagon
factories, and machine-shops. Lumber is taking a place among the leading
articles of export.

Of the public buildings of Charleston, several of the most important
are clustered at the intersection of Broad and Meeting Sts.   On the N. E.
corner is the * **City Hall,** an imposing building, entered by a double
flight of marble steps.   It is situated in Washington Park, which con-

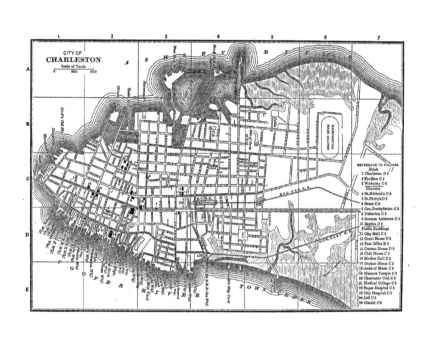

CITY OF
CHARLESTON
Scale of Yards

REFERENCE TO FIGURES
Hotels
1 Charleston D 2
2 Pavilion C 3
3 Waverley C 2
Churches
4 St.Michael's C 3
5 St.Philip's D 2
6 Grace C 3
7 Cen.Presbyterian C 3
8 Unitarian C 3
9 German Lutheran C 3
10 Baptist D 3
Public Buildings
11 City Hall C 2
12 Court House C 2
13 Post Office D 2
14 Custom House D 2
15 Club House C 2
16 Market Hall C 2
17 Orphan House C 2
18 Acad.of Music C 2
19 Masonic Temple C 2
20 Charleston Coll.C 2
21 Medical College C 2
22 Roper Hospital C 2
23 City Hospital C 2
24 Jail C 2
25 Citadel C 2

tains a handsome fountain, and a marble statue of William Pitt erected before the Revolution. . One of the arms was shot off by the British during the siege in 1780. The Council-Chamber is handsomely furnished, and contains some interesting portraits. On the N. W. corner is the *Court-House*, a substantial structure of brick, faced so as to resemble stone. On the S. W. corner, where the *Guardhouse*, or Police Headquarters, which was injured by the earthquake, stood, a handsome new Post-Office is building, to replace the one used for more than a century; and on the S. E. corner stands the venerable * **St. Michael's Church** (Episcopal), built in 1752, it is said, from designs by a pupil of Sir Christopher.Wren. The tower is considered very fine, and the situation of the church makes the spire a conspicuous object far out at sea. Its chimes are celebrated for their age and sweetness. The * view from the belfry is very fine, embracing the far stretch of sea and shore, the fortresses and the shipping. The body of the church was severely injured by the earthquake of 1886. At the foot of Broad St. stands the *Post-Office*, a venerable structure, dating from the colonial period, the original material having been brought from England in 1761. It was much battered during the war, but has since been renovated. It will soon be abandoned, as a new building is in course of construction. The * **U. S. Custom-House,** which has been completed within a few years, at a cost of $3,000,000, is just S. of the Market-wharf, on Cooper River. It is of white marble, in the Roman-Corinthian style, and is now the finest edifice in the city. A noble view is obtained from its graceful Corinthian portico. The *Chamber of Commerce* occupies the 2d and 3d floors of a handsome building at the cor. of Broad and E. Bay Sts.; it has a good reading-room, conveniently arranged for the use of the members. The *Academy of Music*, cor. King and Market Sts., is one of the finest theatres in the South. It is 60 by 231 ft., and cost $160,000. Besides the theatre, with accommodations for 1,200, it contains two large halls for concerts, lectures, etc. The *Grand Opera-House*, a handsome building in Meeting St., has recently been opened to the amusement-loving public. The Charleston Club, the leading club of the city, is handsomely located in Meeting St., near the Battery; and the German Artillery Co. has a fine armory and club-house in Wentworth St., near King. The *Masonic Temple* is a large but fantastic building, cor. King and Wentworth Sts. The old * **Orphan-House,** standing in the midst of spacious grounds, between Calhoun and Vanderhorst Sts., is a most imposing edifice, and a famous institution. John C. Fremont, once a candidate for the presidency, and C. G. Memminger, Confederate Secretary of the Treasury, were educated there. The *South Carolina Military Academy* occupies the building known as the Citadel, on Marion Square. The *College of Charleston*, founded in 1788, has spacious buildings, located in the square bounded by George, Green, College, and St. Philip Sts. It has a library of about 6,000 volumes, and a valuable museum of natural history. The *Medical College*, cor. Queen and Franklin Sts., and *Roper Hospital*, cor. Queen and Logan Sts., are large and handsome buildings, the latter especially so. On the same square with these two is the *County Joil.* The *Charleston Library,* founded in 1748, has a plain but commodious building at the cor. of Broad and Church Sts. It lost heavily in the fire of 1861, but now contains about 17,000 volumes. The *South Carolina Society Hall,* in Meeting St. near St. Michael's Church, is a substantial structure, with colonnade and portico, and a fine interior. The *City Hospital* has been removed to the W. end of Calhoun St., where a handsome building was erected as a memorial to the charity so generously extended to Charleston after the earthquake. * **Market Hall,** in Meeting St. one block S. of Charleston Hotel, is a fine building, in temple form, standing on a high, open basement, having a lofty portico in front, reached by a double flight of stone steps. In rear of this building are the markets, consisting of a row of low sheds supported by brick arches, and extending to E. Bay St. Between 6 and 9 A. M. these markets present one of the most characteristic sights that the stranger can see in Charleston. The refuse is cleaned up by several American buzzards, and is one of the sights of the markets. A fine line of wharves has been recently built by the South Carolina R. R., extending ½ mile on the Cooper River from the foot of Columbus St., and freight tracks extended to the water's edge.

After St. Michael's (already described) the most interesting church edifice in Charleston is * **St. Philip's** (Episcopal), in Church St. near Queen. It was the first church establishment in Charleston; but the present structure, although of venerable age, is yet not quite so old as St. Michael's. The view from the steeple is fine; but there is a keener interest in the graveyard than even in the old church itself, for here lie South Carolina's most illustrious dead. In the portion of the graveyard that lies across the street is the tomb of John C. Calhoun. It consists of a granite sarcophagus, supported by walls of brick, and for inscription has simply the name of "CALHOUN." *St. Finbar's Cathedral* (Roman Catholic), which was destroyed in the great fire of 1861, at the cor. of Broad and Friend Sts., is now partially rebuilt. It was one of the costliest edifices in Charleston. The *Citadel Square Baptist Church,* cor. Meeting and Henrietta Sts., is a fine building, in the Norman style. The *Westminster Presbyterian,* in Meeting St. near Society, has an elegant Corinthian portico with 8 columns. The *Unitarian Church,* in Archdale St. near Queen, is a fine specimen of the perpendicular Gothic style, and has a very rich interior. The *German Lutheran Church,* in King St. opposite the Citadel, is a handsome building, in the Gothic style, with lofty and ornate spire. *Grace Church* (Episcopal), in Wentworth St., is one of the most fashionable in the city. The old *Huguenot Church,* cor. Church and Queen Sts., is worthy of a visit. *St. Patrick's Church,* cor. St. Philip and Radcliffe Sts., is a handsome building in Gothic style, with spire and belfry. *Bethel* and *Trinity Methodist Churches* have recently secured the most powerful organs ever built in this city.

Washington Park, at the junction of Meeting and Broad Sts., is ornamented with the Pitt statue. A promenade, at the W. end of Broad St., on the banks of an inlet of the Ashley, converted into an artificial lake; *Marion Square,* a parade-ground at Calhoun and Meeting Sts., with a bronze statue of John C. Calhoun; and Hampstead Mall, in the N. E. part of the city, are all favorite resorts. *White Point Garden*, or Battery Park, has beautifully shaded promenades, and contains a fine bronze monument erected in honor of Sergeant Jasper, of Revolutionary fame; also a bronze bust of William Gilmore Simms, the novelist, by J. Q. A. Ward. Just outside of the city, on the N. boundary, is *** Magnolia Cemetery** (reached by street-cars). It is embowered in magnolias and live-oaks, is tastefully laid out, and contains some fine monuments, of which the most noteworthy are those to Colonel William Washington, of Revolutionary fame, Hugh S. Legaré, and W. Gilmore Simms, the novelist. Perhaps the most interesting spot in the neighborhood of Charleston is the old *** Church of St. James,** on Goose Creek (reached by carriage, or by Northeastern R. R. to Porcher's Station, 15 miles). It is situated in the very heart of a forest, is approached by a road little better than a bridle-path, and is entirely isolated from habitations of any sort. The church was built in 1711, and was saved from destruction during the Revolutionary War by the royal arms of England that are emblazoned over the pulpit. The floor is of stone, the pews are square and high, the altar, reading-desk, and pulpit are so small as to seem like miniatures of ordinary church-fixtures, and on the walls and altar are tablets in memory of the early members of the congregation. One dates from 1711 and two from 1717.—A short distance from the church, on the other side of the main road, is a farm known as *The Oaks*, from the magnificent avenue of those trees by which it is approached. The trees are believed to be nearly 200 years old; they have attained great size, and for nearly ¼ mile form a continuous arch over the broad road.

The harbor of Charleston is a large estuary, extending about 7 miles to the Atlantic, with an average width of 2 miles. It is landlocked on all sides except an entrance about a mile in width. The passage to the inner harbor is defended by four fortresses. Looking outward on the left, at the entrance, is *Fort Moultrie*, on Sullivan's Island, occupying the site of the fort which, on June 28, 1776, beat off the British fleet of Sir Peter Parker. On the right, raised upon a shoal in the harbor and directly covering the channel, is *** Fort Sumter,** rendered famous by the part which it played in the opening scene of the civil war, and now entirely rebuilt. Immediately in front of the city, and but 1 mile from it, is *Castle Pinckney*, covering the crest of a mud-shoal, and facing the entrance; while to the extreme right is Fort Johnson, now used as a quarantine station. A fine view of the city is obtained in entering the harbor from the sea; and as it is built on low and level land, it seems to rise from the water as we approach, whence it has been called the "American Venice." *Sullivan's Island* is a summer resort of South Carolina, and contains many cottages and a beach drive. A fine hotel, the *New Brighton*, has accommodations for 300 guests. It has a large Casino for amusements, and is open from June to October. A steamboat plies at intervals during the season between the city, Sullivan's Island, and *Mount Pleasant;* the latter being a popular picnic resort.

113. Richmond to Savannah.

a. By "*Atlantic Coast Line.*"

THE Savannah through cars of the "Atlantic Coast Line" run by way of Charleston; and the route from Richmond to Charleston has been described in Route 112 *a*. From Charleston to Savannah the route is *via* Savannah & Charleston R. R. (distance, 115 miles), which runs within a few miles of the Atlantic coast line, though never in sight of the ocean. For miles, near the Savannah River, the rails are laid on piles, passing through marsh and morass, and crossing swift-rushing, muddy streams, dignified by the name of rivers, and called by euphonious Indian names. There are no towns of importance on the line, but the scenery is wild and rich. Extensive pine-forests, lofty cypresses, wreathed in garlands of pendent moss, the bay and the laurel, draped with the vines of the wild grape and of ivy, and huge oaks that have stood the wear and tear of centuries, line the road on either side. Noble avenues are created by these forest giants, and pendent from their stalwart limbs hang long festoons of moss and vine, dimly veiling the vista beyond. At *Yemassee* (60 miles) the Savannah & Charleston R. R. is intersected by the Port Royal & Augusta R. R., which extends from Augusta to *Beaufort* and *Port Royal* (112 miles).

b. *Via Charlotte, Columbia, and Augusta* (609 *miles*).

As far as **Columbia** this route is the same as Route 112 *b*. Beyond Columbia the South Carolina Div. of the Richmond & Danville R. R. continues on through a level, wooded region, unmarked by any striking feature. *Graniteville* (511 miles) is a manufacturing town, with several large granite cotton-mills, giving employment to several hundred operatives who constitute the bulk of the population. Here connection may be made with the South Carolina R. R., and on this railway, 6 miles beyond, is **Aiken** (*Highland Park Hotel, Park Annex Hotel*), one of the most frequented winter-resorts in America.[1] The land upon which it lies is an elevated plateau, some 600 or 700 ft. above the sea. The soil is an almost unmixed sand, covered by a scanty crust of alluvium which is so thin that a carriage-wheel easily breaks through. It bears but little grass and hardly any of the minor natural plants; but the great Southern pine finds here a congenial habitat, and vast forests of it encircle the town on all sides. The streets of the town are remarkably wide, the main avenue being 205 ft. wide, and the cross-streets 150 ft. The houses are generally large and pleasant, and very far apart. Within the town, the natural barrenness of the soil has been overcome by careful culture and a liberal use of fertilizers; and every house has its garden full of trees and Southern plants. Inside the white palings are dense thickets of yellow jasmine, rose-bushes, orange, wild-olive, and fig-trees, bamboo, Spanish bayonet, and numberless sorts of vines and

[1] For full and minute description of Aiken, giving tables of comparative temperature, relative dryness (or humidity), etc., see APPLETONS' ILLUSTRATED HAND-BOOK OF AMERICAN WINTER RESORTS.

creepers, to say nothing of the low bush and surface flowers that are common in the North; but without ·the palings, the sand is as dry and white as it is upon the sea-shore. The air of Aiken is remarkably pure and dry, and the balsamic odors of the pines endow it with a peculiar healing power. The winter climate is wonderfully mild and genial, consisting, as some one has described it, of "four months of June." From observations recorded during the year 1870, it was found that the mean temperature of Aiken in spring is 63·4°; in summer, 79·1°; in autumn, 63·7°; in winter, 46·4°; for the year, 63·1¼°. The average rainfall during the same period was: spring, 11·97 inches; summer, 13·89; autumn, 7·34; winter, 7·16; for the year, 40·36. The climate is as beneficial to rheumatic and gouty patients as to consumptives; and many visit Aiken who, without being sick, desire to escape the rigors of a Northern winter.

Eleven miles beyond Graniteville the train reaches **Augusta** (*Arlington, Bon Air, Planters' Hotel*), the third city of Georgia in population (33,300), and one of the most beautiful in the South. It is situated at the head of navigation on the Savannah River, and embraces an area of about 3 miles in length and a mile and a half in breadth. It is regularly laid out, with broad streets crossing each other at right angles, and many of them beautifully shaded. *Broad St.* is the main thoroughfare of the city, and is 165 ft. wide and 2 miles long. On it are the principal banks, hotels, and shops; and in the center of it is the * *Confederate Monument* (the finest in the South), consisting of an obelisk 80 ft. high, surmounted by a statue of a soldier, and with 4 portrait statues (including Lee and Jackson) on the corner pedestals. * *Greene St.* is 168 ft. wide, and lined with handsome residences; tall, spreading trees not only grace the sidewalks, but a double row, with grassy spaces between, runs down the center of the roadway. Of the public buildings, the * *City Hall*, completed in 1824 at a cost of $100,000, is the most attractive. In front of it stands a granite monument erected by the city in 1849 to the memory of the Georgian signers of the Declaration of Independence. The *Augusta Exchange* is at the cor. of Jackson and Reynolds Sts. The *Odd-Fellows' Hall*, the *Orphan Asylum*, and the *Opera-House*, cor. of Jackson and Green Sts., are handsome edifices. The commerce of Augusta is very prosperous, and the fine water-power of the *Augusta Canal*, 9 miles long, which brings the upper waters of the Savannah River to the city at an elevation of 40 ft., is enriching it with extensive manufactures. Just outside of the city, and E. of the *City Cemetery*, are the * *Fair Grounds* of the Cotton States Mechanics' and Agricultural Fair Association, comprising 47 acres, laid out in attractive walks and drives. A most charming view of Augusta and its environs may be had from **Summerville,** a suburban town of handsome villas situated on high hills about 3 miles from the city (reached by street-cars). Here is located the *Bon Air Hotel*, open all the year round. Among the objects of interest at Summerville are the *U. S. Arsenal*, built in 1827, and the range of workshops, 500 ft. in length, built and used by the Confederates during the war. Across the river from Augusta at Hamburg there are some beautiful wooded and grassy terraces, known as *Schultz's Hill,* and much resorted to as a picnic-ground.

From Augusta to Savannah (132 miles) the route is *via* the Central R. R. of Georgia, which passes through one of the most productive and populous sections of the State. There are no points, however, of special interest on the line, all the towns being small and of merely local importance. At *Millen* (53 miles from Augusta) the road forks, one branch going to Macon and the other to Savannah.

Savannah.

Hotels, etc.—The leading hotels are the *De Soto,* in Bull St.; *Pulaski House,* in Johnson Square; the *Screven House,* in Johnson Square; the *Marshall House,* in Broughton St.; and *Harnet House,* in Barnard St. Besides the routes described above, Savannah is reached from New York by *steamers,* leaving Pier 35, North River, four times a week. Time, about 55 hrs.; fare (cabin), $20. There are also steamers to Savannah from Philadelphia every ten days, from Boston every four days, and Baltimore once a week. It may also be reached *via* Norfolk and the sea-coast (see Route 109) and the Charleston & Savannah R. R. (see above). By the Savannah, Florida & Western R. R. the city is connected with all important points in the Gulf States. By the Central R. R. of Georgia, Asheville may be reached on the N. by way of Augusta, while to the W. branch lines run to Atlanta, Birmingham, and Montgomery.

Savannah, next to Atlanta the chief city and commercial metropolis of Georgia, is situated on the S. bank of the Savannah River, 18 miles from its mouth. The site was selected by General Oglethorpe, the founder of the colony of Georgia, who made his first settlement at this point in February, 1733. The city occupies a bold bluff, about 40 ft. high, extending along the river-bank for a mile, and backward, widening as it recedes, about 6 miles. The river making a gentle curve around Hutchinson's Island, the water-front of the city is in the shape of an elongated crescent about 3 miles in length. The corporate limits extend back on the elevated plateau about 1¼ mile, the total area of the city being 3⅛ sq. m. In its general plan Savannah is universally conceded to be one of the handsomest of American cities. Its streets are broad and beautifully shaded, they cross each other at right angles, and at many of the principal crossings are small public squares or parks from 1½ to 3 acres in extent. These parks, 24 in number, located at equal distances through the city, neatly inclosed, laid out in walks, and planted with the evergreen and ornamental trees of the South, are among the most characteristic features of Savannah; and, in the spring and summer months, when they are carpeted with grass, and the trees and shrubbery are in full foliage, afford delightful shady walks, and playgrounds for the children, while they are not only ornamental, but conducive to the general health by the free ventilation which they afford. Many of the residences are surrounded by flower-gardens, which bloom throughout the year; and among the shrubbery, in which the city is literally embowered, are the orange-tree, the banana, the magnolia, the bay, the laurel, the crape-myrtle, the stately palmetto, the olive, the flowering oleander, and the pomegranate.

Savannah was founded by James E. Oglethorpe in 1733. In 1776 the British attacked it and were repulsed; but on December 29, 1779, they reappeared in

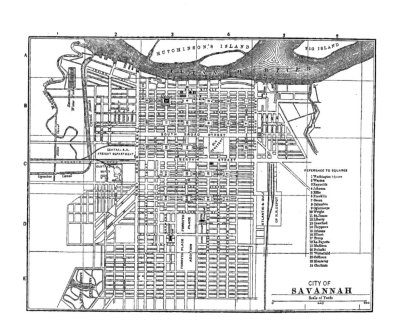

REFERENCE TO SQUARES

1 Washington Square
2 Warren
3 Reynolds
4 Johnson
5 Ellis
6 Franklin
7 Green
8 Columbia
9 Oglethorpe
10 Wright
11 St. James
12 Liberty
13 Crawford
14 Chippewa
15 Orleans
16 Elbert
17 Troup
18 La Fayette
19 Madison
20 Pulaski
21 Whitefield
22 Calhoun
23 Monterey
24 Chatham

CITY OF
SAVANNAH
Scale of Yards

overwhelming force and took possession of the city. In October, 1779, the combined French and Americans attempted to recapture it, but were unsuccessful, and Count Pulaski fell in the engagement. Savannah received a city charter in 1789. In 1850 it had 15,312 inhabitants; in 1860, 22,292; and in 1870, 28,235. According to the U. S. Census of 1880, its population amounted to 30,681, and in 1890 was 43,189. The chief business of the place is in cotton, though the trade in lumber is also considerable. As a cotton port it ranks second in the United States. It recovered rapidly from the effects of the civil war, and its commerce has doubled. The chief manufacturing establishments are planing-mills, foundries, flouring and grist mills, and fertilizer-factories.

The great warehouses of the city are located on a narrow street at the foot of the steep bluff; they open below on the level of the piers, and from the uppermost story on the other side upon a paved road 200 ft. wide and divided by rows of trees. This is called *Bay Street*, and is the great commercial mart of Savannah. The principal business streets are *Congress, Broughton, Whitaker,* and *Barnard* Sts., and the favorite promenade is Bull St. to Forsyth Park. Among the noteworthy buildings are the granite ***Custom-House,** the *Chatham County Court-House,* the *Post-Office* (cor. Whitaker and State Sts.), the *City Exchange* (in front of which General Sherman reviewed his army, January 7, 1865), the *Court-House,* the *Police Barracks,* and the *Artillery Armory.* From the tower of the Exchange the best * view of the city and neighborhood is to be had. The building on the N. E. cor. of Bull and Broughton Sts., formerly the *Masonic Hall,* but now the *Oglethorpe Club,* is interesting as the place where the Ordinance of Secession was passed, Jan. 21, 1861. Four years later (Dec. 28, 1864) a meeting of citizens was held in the same apartment to commemorate the triumph of the Union arms. The *Georgia Historical Society* has a large and beautiful hall, in which are a fine library and some interesting relics; and the *Telfair Academy of Arts,* in Barnard St., contains an admirable collection of casts, many paintings, and other objects of art. There is an excellent system of public schools, of which *Chatham Academy* is the center. The *Independent Presbyterian Church,* at Bull and Broad Sts., has erected its third edifice, which is considered one of the finest churches in the South. *Christ Church,* fronting on Johnson Square, is the mother church of the Episcopal communion in Georgia. On its site the first Sunday-school was established in America by John Wesley. *St. John's Episcopal Church,* fronting Madison Square, is in the English style of Gothic architecture, with rich stained-glass windows. There are other fine churches, including the *Mickva Israel Synagogue,* fronting Monterey Square, that are worth visiting.

The most attractive place of public resort is ***Forsyth Park,** an inclosure of 30 acres in the S. part of the city. It is shaded by some venerable old trees, is laid out in serpentine walks, and ornamented with evergreen and flowering trees and shrubs. In the center is a handsome fountain, after the model of that in the Place de la Concorde, Paris, and a stately *Confederate Monument* stands in the new portion. In Johnson or *Monument Square,* near the center of the city, is a fine Doric obelisk erected to the memory of General Greene, the corner-stone of which was laid by Lafayette during his visit in 1825. The ***Pulaski Monument** stands in Monterey Square, and is one of the most perfect specimens of monumental architecture in the United States. The steps are plinths of granite; the shaft is of purest marble, 55 ft. high, and is surmounted by an exquisitely carved statue of Liberty holding the national banner. The monument appropriately covers the spot where Pulaski fell, during an attack upon the city while it was occupied by the British, in 1779. It was designed by Robert E. Launitz, and cost $22,000. The *Jasper Monument*, in Madison Square, was unveiled on Feb. 22, 1888. The bronze statue of Sergeant Jasper, which surmounts the pedestal, is 15 ft. high, and represents a sturdy specimen of manhood. The left hand clutches at arm's-length a battle-worn banner; while the right hand, holding an upturned saber. is pressed tightly over a bullet-wound in his side. The statue was designed by Alexander Doyle, of New York. In Court-House Square is a *Monument* erected in 1883 by the Central Railroad & Banking Co. in memory of William W. Gordon, its first president.

Though built upon a sandy plain, Savannah is not without suburban attractions, there being several places in its vicinity whose sylvan character and picturesque beauty are in keeping with the "Forest City" itself. Thunderbolt, Isle of Hope, Beaulieu, Montgomery, and White Bluff, are all rural retreats on "The Salts," within short driving-distance of the city, where, in the summer months, bracing sea-breezes and saltwater bathing may be enjoyed. The great drive is to * **Bonaventure Cemetery,** which is situated on Warsaw River, a branch of the Savannah, about 4 miles from the city. The scenery of Bonaventure has long been renowned for its Arcadian beauty; for its broad avenues of live-oaks draped in pendent gray moss. *Laurel Grove*, the municipal cemetery, lies S. W. of the city, near Forsyth Park. *Thunderbolt*, a popular drive and summer resort, is on the Warsaw River, 1 mile beyond Bonaventure. According to local tradition, this place received its name from the fall of a thunderbolt. A spring of water which issued from the spot upon that event has continued to flow ever since. *Jasper Spring*, 2¼ miles W. of the city, is the scene of the famous Revolutionary exploit of Sergeant Jasper, who, with only one companion, successfully assailed a British guard of eight men and released a party of American prisoners. *White Bluff*, 10 miles out, is another favorite resort of the Savannah people, and the road to it is one of the most fashionable of the suburban drives. *Tybee Beach*, on an island of the same name at the mouth of the Savannah River, is the great seaside resort of the city. It is connected by rail, and 5 daily trains run during the summer months. The beach is 5 miles long, and a magnificent and solid roadway commands a view of the ocean. The hotel (*Beach House, Hotel Tybee, Ocean House*) accommodations are excellent, and the surf-bathing unsurpassed.

114. Charleston or Savannah to Jacksonville, Florida.

Steamer Routes.—A steamer of the Florida line leaves Charleston three times a week (on the arrival of the New York steamer) for Jacksonville, connecting by rail and steamer lines with all points in Florida.

THE most direct all-rail route from Savannah to Jacksonville is *via* Savannah, Florida & Western R. R. to *Waycross* (96 miles), and thence *via* the Jacksonville Division of that road (total distance, 172 miles; time, 5 hours; fare, $5.15). The Savannah, Florida & Western Railway is the great connecting-link between the railways from the North (*via* Savannah) and southern Georgia and Florida. The main line runs S. W. from Savannah to *Bainbridge* on the Flint River (236 miles). *Climax,* a few miles E., is the point of connection with Chattahoochee, Fla., the junction of the through Florida line from Jacksonville to Pensacola. Numerous small towns are clustered along the line, but the only one that need be mentioned is **Thomasville** (200 miles from Savannah), which attracts attention as a popular winter health resort. It is a pretty town of about 5,514 inhabitants, situated at the N. verge of the great pine-forest which stretches across Southern Georgia from E. to W. in a belt 75 miles wide. It stands on the highest ground between the Savannah and Flint Rivers, 300 ft. above the sea, and has the dry pure atmosphere, laden only with the odors of pine-forests, which consumptives highly prize. The streets of the town are broad and shady, and in the surrounding country, besides corn and cotton, grapes are produced in abundance. There are several well-conducted hotels, among which are the *Gulf House,* the *Masury Hotel,* the *Mitchell House,* and the *Piney Woods Hotel.* The town is supplied with water from an artesian well 1,900 ft. deep. This region is a great center of the famous "Le Conte pear" culture. There are Episcopal, Methodist, Baptist, Presbyterian, and Roman Catholic churches, and many good schools.

Diverging from the main line at Dupont, the Florida Div. of the Savannah, Florida & Western R. R. runs S. in 49 miles to *Live Oak,* situated at the junction with the Florida Central & Peninsular R. R. Eleven miles beyond Live Oak, on the latter road, is the village of **Wellborn,** and in the neighborhood are *Lake Wellborn* and other lakes, well stocked with fish. Several miles north of Live Oak, on the bend of the Suwanee River as it turns eastward, is the Lower Mineral Spring, sometimes called the Lower Suwanee Spring. It may be reached by the railroad from Dupont, Ga. (42 miles). This spring is a cove of the river, and rises and falls with the same. It is a place of much local resort. The spring is picturesque, boiling up from a dark gorge, and rises and falls, so that the bathing-house is formed by several stories to reach the water conveniently. From Live Oak the connection is continued by rail, by the Savannah, Florida & Western R. R., into southern Florida, *via* Branford (Rowland's Bluff), thence merging into the Florida Southern Railway at Gainesville. Twelve miles E. from Wellborn is **Lake City** (*Central, Gee House*), the most important place in this portion of Florida, with 2,020 inhabitants. Within the city limits are Lakes

Isabella, De Soto, and Hamburg, and Indian or Alligator Lake is only half a mile away. The climate of Lake City is very similar to that of Jacksonville, but the air is thought to be somewhat drier, while the rich balsamic odors from the surrounding forests endow it with exceptional curative and healing power, and render the neighborhood remarkably beneficial to consumptives in the more advanced stages of the disease. *Olustee* (12 miles beyond Lake City) is noted as the site of a battle between the Federal and Confederate forces, fought in February, 1864, in which the former were defeated. *Baldwin* is a small station at the crossing of the Florida Central & Peninsular R. R. (Route 119).

Jacksonville.

Hotels, etc.—The principal hotels are the *St. James*, the *Windsor*, the *Everett House*, the *Carleton*, the *Duval*, the *Tremont*, and the *Hotel Togni*. Prices at these hotels range from $3 to $5 per day. There are a number of boarding-houses, at which the prices range from $8 to $20 a week. Good furnished rooms, including lights, fuel, and attendance, may be had in private houses for from $4 to $10 per week, and board without rooms is $11 per week at the hotels, and less at the boarding-houses. Unfurnished cottages can be hired at from $20 to $30 per month.

Steamer Routes.—Besides the other steamer lines mentioned in Route 114, the Clyde line of steamships runs four first-class steamers three times a week from New York to Jacksonville, stopping at Charleston *en route*, leaving New York Mondays, Wednesdays, and Fridays. The Mallory line also runs four first-class steamers to Fernandina, leaving New York every Tuesday and Friday. The Ocean Steamship Co. run regular steamers from Boston, New York, and Philadelphia (freight only) to Savannah, connecting with the Savannah, Florida, & Western R. R. to Jacksonville.

During the winter season the Atlantic Coast line runs a vestibule train with most luxurious accommodations three times a week each way, leaving New York after breakfast and arriving in Jacksonville and St. Augustine in time for dinner on the following day ; also daily trains having through buffet sleeping-cars.

Jacksonville, the largest city in Florida, is situated on the left bank of the St. John's River, about 25 miles from its mouth. It was named after General Andrew Jackson, was laid out as a town in 1822, had a population of 1,045 in 1850, of 6,912 in 1870, of 7,648 in 1880, and of 17,201 in 1890. Its resident population is largely increased during the winter months by transient visitors. The city is regularly laid out, with streets crossing each other at right angles and shaded with trees. The leading thoroughfares are *Bay, Forsyth, Main,* and *Laura Sts.*, and on these are situated the principal commercial buildings. The suburban villages (East Jacksonville, La Villa, Springfield, Brooklyn, Riverside, Arlington, St. Nicholas, South Jacksonville, and Alexandria, are now included in the city), and those on the other side of the river, are connected with the city by steam ferry and an iron bridge and the Jacksonville, St. Augustine & Indian River R. R. Besides fine public schools, Jacksonville contains Baptist, Catholic, Congregational, Episcopal, Methodist, and Presbyterian churches ; a circulating library and a free reading-room ; 3 daily newspapers and 6 weeklies ; banks, an opera-house, public halls, 6 street-car lines, and telegraphic connections with all parts. The chief business is the sawing and shipment of lumber ; cotton, sugar, fruit, fish, and early vegetables are also shipped to Northern and foreign ports. The *Subtropical Exhibition*, with extensive buildings and beautiful grounds, is open at times. Near by are the Jacksonville waterworks, where the local water supply is obtained from artesian wells. Jacksonville is much resorted to by invalids on account of its mild and salubrious climate; and many prefer remaining here to going farther into the interior, on account of the superior accommodations which it offers, and its social advantages. The mean temperature of Jacksonville, as reported by the chief signal-officer of the United States, is 69·6°; of the coldest month (January) 52·7°; of the hottest month (July) 83·4°. Frost is very variable at different seasons, but is slight in Florida in proportion to its location in latitude. It occurs oftenest between November and March, being most frequent in December and January, and rarely showing itself in October and April as far north as Jacksonville. As a general thing no frost occurs throughout the year below lat. 28° N. Summer being the rainy season in Florida, the winters are usually clear and dry. By observations taken for a period of 22 years at Jacksonville, it was found that January averaged 20 clear days; February, 19; March, 20; April, 25; May, 22; June, 17; July, 18; August, 19; September, 17; October, 19; November, 20; and December, 20. It may be said in general terms that from October to May there are not more than four or five rainy days in a month. *Brooklyn* and *Riverside* are residential suburbs along the river, connected with the city by street-cars. Among the amusements at Jacksonville are excursions on the river and drives on the excellent shell-roads which lead out of the city. A favorite drive is to *Moncrief's Spring* (4 miles), whose waters are said to cure malarial diseases. Other favorite excursions are to *Pablo Beach*, on the Atlantic coast, and *Mayport*, *Burnside Beach*, and *Fort George Island* at the mouth of the St. John's River, both reached by railroads.

Pablo Beach.

The most popular local resort near Jacksonville, next to St. Augustine, bears a name almost new to the general public. It is Pablo Beach. This is a beautiful location on the Atlantic beach, a few miles S. of the mouth of the St. John's. It is connected with Jacksonville by a short but thoroughly equipped railroad, known as the Jacksonville & Atlantic R. R. Both this railroad and Pablo Beach are the result of Jacksonville enterprise. A ride of thirty minutes lands the tourist or pleasure-seeker upon a beach incomparable in hardness, smoothness, and extent. On this beats the Atlantic, and the finest of marine views opens before him. The growth of Pablo Beach has been phenomenal. It is within a few years that the first cottage was erected, and since that time many buildings have been completed and occupied by private citizens. The railroad company have put up extensive buildings, bath-houses, pavilions for dancing, skating, etc., handsome pagodas, and other attractive improvements. Aside from the disbursements of the railroad company, $300,000 have been expended in the development of the place. It seems to have filled a need of Jacksonville. It has certainly attained an unparalleled popularity with the people and sojourners of that city. The attractions and amusements are hunting, fishing, boating, bathing in an unequaled surf, and riding, driving, and bicycle-riding on the magnificent beach.   This beach stretches away in an unbroken line to St. Augustine on the S., and to Mayport at the mouth of the St. John's on the N. (*Murray Hall* and *Ocean House*).

115. Jacksonville to St. Augustine.

St. Augustine is reached from Jacksonville most directly by the Jacksonville, St. Augustine & Indian River R. R., 38 m.; fare, $1.50, or $2.90 for the round trip, running through a region of much historical interest.   Also by Jacksonville, Tampa & Key West R. R. to Palatka, thence by the Jacksonville, St. Augustine & Indian River R. R.

St. Augustine.

Hotels, etc.—The principal hotels are the *Alcazar*, in King St.; the *Cordova*, in King St.; the *Florida House*, in Treasury and St. George Sts.; the *Magnolia*, in St. George St.; the *Ponce de Leon*, in King St.; and the *San Marco*, in San Marco Ave., outside the city gates.   There are also numerous boarding-houses, at which board may be had for from $10 to $15 a week.

St. Augustine is situated on the Atlantic coast of Florida, about 30 miles S. of the mouth of the St. John's River and 33 S. E. of Jacksonville.   It occupies a narrow peninsula formed by the Matanzas River on the E. and the St. Sebastian on the S. and W., the site being a flat, sandy level, encompassed for miles around by a tangled undergrowth of low palmettos and bushes of various descriptions.   Directly in front lies Anastasia Island, forming a natural breakwater, and almost entirely cutting off the sea-view.   On the N. end of the island is a lighthouse with a revolving light, situated in lat. 29° 53' N., and lon. 81° 16' W. The *Alameda* is the thoroughfare of the place, and the other principal streets are *Cordova, St. George, Charlotte*, and *Bay Sts.*   The latter commands a fine view of the harbor, Anastasia Island, and the ocean.   All the streets are extremely narrow, being only twelve or fifteen feet wide, while the cross-streets are narrower still.   An advantage of these narrow streets in this warm climate is that they give shade, and increase the draught of air through them as through a flue.   The principal streets were formerly paved with shell-concrete, portions of which are still to be seen above the shifting sand; and this flooring was so carefully swept that the dark-eyed maidens of Old Castile who once led society here could pass and repass without soiling their satin slippers. No rumbling wheels were permitted to crush the firm road-bed, or to whirl the dust into the airy verandas.   All the old Spanish residences are built of coquina-stone.   Many of them have hanging balconies along their second stories, which in the narrow streets seem almost to touch, and from which their respective occupants can chat confidentially and even shake hands.   It must not be supposed, however, that St. Augustine is built wholly of coquina and in the Spanish style; there are many fine residences there in the American style, and St. Augustine may yet rival Newport in the number of its beautiful villas.   A profusion of oranges, lemons, bananas, figs, date-palms, and all manner of tropical flowers and shrubs, ornament their grounds.   A charming drive is out St. George St., through the City Gate to the beach of the San Sebastian.

The most interesting feature of St. Augustine is the old *Fort of San Marco (now *Fort Marion*), which is built of coquina, a unique conglomerate of fine shells and sand found in large quantities on Anastasia Island, at the entrance of the harbor, and quarried with great ease, though it becomes hard by exposure to the air. The fort stands on the sea-front at the N. E. end of the town. It was 100 years in building, and was completed in 1756, as is attested by the following inscription, which may still be seen over the gateway, together with the arms of Spain, handsomely carved in stone: "Don Fernando being King of Spain, and the Field-Marshal Don Alonzo Fernando Herida being governor and captain-general of this place, St. Augustine of Florida and its provinces, this fort was finished in the year 1756. The works were directed by the Captain-Engineer Don Pedro de Brazos y Gareny." While owned by the British, this was said to be the prettiest fort in the king's dominions. Its castellated battlements; its formidable bastions, with their frowning guns; its lofty and imposing sally-port, surmounted by the royal Spanish arms; its portcullis, moat, and drawbridge; its circular and ornate sentry-boxes at each principal parapet-angle; its commanding lookout tower; and its stained and moss-grown massive walls—impress the external observer as a relic of the distant past; while a ramble through its heavy casemates—its crumbling and dark chapel, with elaborate portico and inner altar and holy-water niches; its dark passages, gloomy vaults, and more recently-discovered dungeons—bring you to ready credence of its many traditions of inquisitorial tortures; of decaying skeletons found in the latest opened chambers, chained to the rusty ring-bolts, and of alleged subterranean passages to the neighboring convent. Next to the fort the great attraction is the *Sea-Wall, which, beginning at the water-battery of the fort, extends S. for nearly a mile, protecting the entire ocean-front of the city. It is built of coquina, with a granite coping 4 ft. wide, and furnishes a delightful promenade of a moonlight evening. Near the S. end of the wall are the *U. S. Barracks*, which was formerly a Franciscan monastery, but has undergone extensive modifications and repairs. The old Spanish wall, which extended across the peninsula from shore to shore and protected the city on the N., has crumbled down or been removed, but the *City Gate, which originally formed a part of it, still stands at the head of St. George St. It is a picturesque structure, with quaint, square towers and loop-holes and sentry-boxes in a fair state of preservation.

In the center of the town is the *Plaza de la Constitucion*, nearly in the center of which stands a monument about 20 ft. high, erected in 1812 in commemoration of the Spanish Liberal Constitution. Another monument erected to the Confederate dead, which was removed from St. George St. in 1879, now stands within the plaza in front of the old Market. A fire on April 12, 1887, destroyed some of the most picturesque buildings in the city, including the old **Catholic Cathedral,** which has been rebuilt and enlarged. The belfry contains a new chime of bells, as well as the old ones which used to hang in the often-sketched belfry. One of these bells bears the date of 1682. The old *Market* was also burned, but has been restored. A small Episcopal church fronts on the Plaza, and there are Methodist and Baptist churches in the city. The *Memorial Presbyterian Church*, erected in 1889, is an elaborate structure in the style of the Venetian renaissance. The old *Convent of St. Mary's* is an interesting building in St. George St., just W. of the Cathedral. In its rear is a more modern structure designated as the Bishop's Palace. The new *Convent of the Sisters of St. Joseph* is a tasteful coquina building in St. George St., S. of the Plaza; the old convent of this sisterhood is in Charlotte St., N. of the Barracks. The nuns are mainly occupied in teaching young girls, but they also manufacture lace of a very fine quality, and excellent palmetto hats. After the Cathedral, the most imposing edifice on the Plaza was the *Governor's Palace*, formerly the residence of the Spanish Governors, but, like the Cathedral, it was destroyed by the conflagration in 1887. It has since been rebuilt, and is used as the Post-Office. The old **Huguenot Burying-Ground,** in King St. near the City Gate, is a spot of much interest; and so is the *Military Burying-Ground* (just S. of the Barracks), where rest the remains of those who fell near here during the prolonged Seminole War.

In January, 1888, was opened a group of buildings which deserve inspection even by the transient visitor. The *Ponce de Leon Hotel* covers 4 acres, and is half a mile around. It is built of coquina, in the Spanish style, inclosing a court 150 yds. square. In the main building is a rotunda 54 by 80 ft., 4 stories high, a dining-room 150 ft. long, and private dining-rooms. The roof is flat, and is intended for an orangery. Besides the usual accommodations for guests, there are ladies' billiard-rooms, an immense children's play-room, artists' studios, ladies' reading and writing rooms, etc., and attached to the hotel are grounds of great beauty. Opposite the Ponce de Leon is the *Alcazar*, an annex to the Ponce de Leon, which contains immense bathing-pools and dancing-rooms. Near both is the *Hotel Cordova*, built in similar style. A unique feature of this is the *Sala del Sol*, or sun-parlor, 108 ft. long, paved with tiles, with a roof of glass. It is designed for the use of invalids on cool days.

There are many fine orange-groves in the environs of St. Augustine, and visits to them are among the unfailing delights of visitors. The harbor affords excellent opportunities for boating, and numerous points of interest attract excursion-parties. Among the most popular of these are those to the *North Beach*, one of the finest on the coast, affording an admirable view of the ocean; to the *South Beach;* to the sand-hills, where General Oglethorpe planted his guns and laid siege to Fort Marion; to *Fish's Island;* and to the light-houses and coquina-quarries on Anastasia Island. A pleasant trip is across *Matanzas Bay*, and thence by railroad across Anastasia Island to the light-house; and *Matanzas Inlet* affords excellent camping-places for hunting and fishing parties. About 2¼ miles off Matanzas an immense *Sulphur Spring* boils up out of the ocean where the water is 132 ft. deep, and is well worth a visit. Salt-water bathing may be indulged in at St. Augustine in suitable bathing-houses, but the sharks render open sea-bathing dangerous.

St. Augustine is the oldest European settlement in the United States, except-ing perhaps Santa Fé, N. M., having been founded by the Spaniards under Menéndez in 1565, more than half a century before the landing of the Pilgrims at Plymouth. It experienced many vicissitudes ; was several times attacked by the French, English, and Indians ; and was twice assailed by expeditions from the neighboring English colonies of South Carolina and Georgia. With the rest of Florida it came into the possession of the English by the treaty of 1763, was ceded to Spain in 1783, and was transferred to the United States in 1819. During the civil war it changed masters three times. The resident popula-tion, according to the census of 1890, was 4,742 ; but this is increased to 10,000 or more by visitors during the winter, and St. Augustine is then one of the gay-est places in the South. The *climate* of St. Augustine is singularly equable both winter and summer, the mean annual temperature being 70°. The mean temperature for winter is 58·08° ; for spring, 68·54° ; for summer, 80·27° ; and for autumn, 71·73°. Frosts seldom occur even in midwinter, and the sea-breezes temper the heats of summer so that they are quite endurable.

116. The St. John's River.

Steamers of the Clyde Line leave Jacksonville daily, except Saturday, at 3.30 P. M. for Sanford and Enterprise. Time to Sanford, 16 hrs. There are numer-ous other steamers on the river, some running through to Palatka and Sanford, and others running only to the lower landings. The following list of principal places on the St. John's may prove useful to the tourist : Riverside, 3 miles from Jacksonville ; Black Point, 10 ; Mulberry Grove, 12 ; Mandarin, 15 ; Fruit Cove, 19 ; Hibernia, 23 ; Remington Park, 25 ; Magnolia, 28 ; Green Cove Springs, 30 ; Hogarth's Landing, 38 ; Picolata, 44 ; Tocoi, 49 ; Federal Point, 58 ; Orange Mills, 63 ; Dancy's Wharf, 66 ; Whitestone, 68 ; Russell's Landing, 69 ; Palatka, 75 ; Rolleston, 78 ; San Mateo, 79 ; Buffalo Bluff, 87 ; Ocklawaha River, 84 ; Wela-ka, 100 ; Beecher, 101 ; Orange Point, 113 ; Mount Royal, 105 ; Fort Gates, 106 ; Georgetown, 113 ; Lake View, 132 ; Volusia, 134 ; Orange Bluff, 140 ; Haw-kinsville, 160 ; De Land's Landing, 162 ; Lake Beresford, 163 ; Blue Spring, 168 ; Emanuel, 185 ; Shell Bank, 193 ; Sanford, 193 ; Mellonville, 195 ; Enterprise, 198 ; Cook's Ferry and King Philip's Town, 224 ; Lake Harney, 225 ; Salt Lake, 270.

THE St. John's River has its sources in a vast elevated savanna mid-way down the peninsula, flows almost directly N. for 300 miles to Jacksonville, and then turning E. empties into the Atlantic. Its whole course, which lies through an extremely level region, is about 300 miles, including the windings of the river above Palatka, and throughout the lower 150 miles it is little more than a succession of lakes, expanding in width from $1\frac{1}{2}$ to 6 miles, and having at no point a width of less than $\frac{1}{2}$ mile. Its banks are lined with a luxuriant tropical vegetation, handsome shade-trees and orange-groves, and here and there are pic-turesque villages. " The banks are low and flat," says Edward King, "but bordered with a wealth of exquisite foliage to be seen nowhere else upon this continent. One passes for hundreds of miles through a grand forest of cypresses robed in moss and mistletoe; of palms tower-ing gracefully far above the surrounding trees ; of palmettos whose rich trunks gleam in the sun; of swamp, white and black ash, of magnolia, of water-oak, of poplar and plane trees ; and, where the hammocks rise a few feet above the water-level, the sweet-bay, the olive, the cotton-tree, the juniper, the red cedar, the sweet-gum, the live-oak, shoot up their splendid stems : while among the shrubbery and inferior growths one may note the azalea, the sumach, the sensitive plant, the agave, the poppy, the mallow, and the nettle. The vines run not in these thickets, but over them.   The fox-grape clambers along the branches, and the woodbine and bignonia escalade the haughtiest forest-monarchs.   When the steamer nears the shore, one can see far through the tangled thickets the gleaming water, out of which rise thousands of 'cypress-knees,' looking exactly like so many champagne-bottles set into the current to cool. The heron and the crane saucily watch the shadow which the approaching boat throws near their retreat.   The wary monster-turtle gazes for an instant, with his black head cocked knowingly on one side, then disappears with a gentle slide and a splash.   An alligator grins familiarly as a dozen revolvers are pointed at him over the boat's side, suddenly 'winks with his tail,' and vanishes! as the bullet meant for his tough hide skims harmlessly over the ripples left above him. . . . For its whole length the river affords glimpses of perfect beauty.   It is not grandeur which one finds on the banks of the great stream : it is Nature run riot.   The very irregularity is delightful, the decay is charming, the solitude is picturesque."

Although the development of railroads has rendered the old regulation-trip up the St. John's a thing of the past, yet the voyage is worth taking by admirers of scenery, and it is traversed by daily steamers to its upper waters.   Noticeable points are *Orange Park*, whose spires can be seen from the steamer, and four miles above it, on the E. bank, *Mandarin*, one of the oldest settlements on the St. John's. It is a village of some 1,739 inhabitants, and is the winter-home of Mrs. Harriet Beecher Stowe, whose cottage is situated near the river, a few rods to the left of the shore-end of the pier.   Seven miles above Mandarin, on an island near the opposite bank, is *Hibernia* (16 miles from Jacksonville).   **Magnolia** (*Magnolia Hotel*) is situated on the W. bank, and is considered one of the most desirable resorts in Florida for consumptives.   It has a sandy soil, covered with beautiful groves of pine and orange trees, and there are no dangerous hummocklands near by.   It is one of the most beautiful points on the river.   In the vicinity is *Magnolia Point*, one of the highest points of land extending into the river between Jacksonville and Palatka.   A little to the N. of the Point, Black Creek, a navigable stream, up which small steamers make weekly trips as far as *Middleburg*, empties into the St. John's.   From the banks alligators are sometimes seen, which are apt to be mistaken for logs which are floated down this stream in large quantities to market. Three miles above Magnolia are the **Green Cove Springs** (*Clarendon Hotel, St. Clair*), one of the favorite resorts on the river.   The place takes its name from a sulphur-spring, situated about 100 yds. from the landing amid a grove of great water-oaks, covered with hanging festoons of gray moss and mistletoe.   The spring discharges about 3,000 gallons a minute, and fills a pool some 30 ft. in diameter with greenish-hued crystal-clear water.   The water has a temperature of 78° Fahr. ; contains magnesium and calcium sulphate, sodium and iron chlorides, and hydrogen sulphide ; is used both for bathing and drinking, and is considered beneficial for rheumatism, gouty affections, and Bright's disease of the kidneys. *Picolata* is the site of an ancient Spanish settlement, of which no traces now remain. On the opposite side of the river are the ruins of a great earthwork fort of the time of the Spanish occupation. Passing *Federal Point*, a wood-station, *Orange Mills* (63 miles), and *Dancy's Wharf* (66 miles)—the two latter noted for their fine orange-groves—the steamer stops at **Palatka** (*Putnam House, Saratoga, The Berkshire*), the largest town on the river above Jacksonville. It has a permanent population of 3,039, and is admirably situated on high ground on the W. bank of the river, where the surface-land is for the most part sandy. The blandness of its climate renders Palatka favorable to consumptives, and it offers all the advantages in the way of postal and telegraphic facilities, etc., possessed by any of the interior resorts. Palatka was the steamboat headquarters for the Upper St. John's and its tributaries, but, since the completion of the Jacksonville, Tampa, & Key West R. R. to *Enterprise*, much of the traffic is diverted to the railroad. It is a terminus of the Jacksonville, St. Augustine & Indian River R. R. Steamers run from Palatka up the Ocklawaha River to Silver Spring, Ocala, and the head of navigation (see Route 117). Another line runs *via* Deep River to *Crescent City*, on Lake Crescent, 25 miles S. of Palatka. All travel formerly went over this route, but now it goes *via* the Jacksonville, Tampa & Key West system to Crescent City Landing, where all trains stop, and direct connections are made with boat for Crescent City.

All the towns on the W. side of the river from Jacksonville to Palatka are also traversed by the Jacksonville, Tampa & Key West R. R., which is continued from Palatka through Seville, Orange City to Sanford. From Sanford this line passes the rising resort of Winter Park (*The Seminole Hotel*), Orlando, and Kissimee City, the headquarters of the Disston Co., to Tampa. Two through trains run daily over this road by the Jacksonville, Tampa & Key West R. R. from Jacksonville to Tampa. Palatka is also the W. terminus of the Florida Southern R. R. The latter road runs to Rochelle, Leesburg, Pemberton, Lakeland, Bartow, and Punta Gorda, on Charlotte Harbor.

Above Palatka the vegetation becomes more characteristically tropical, and the river narrows down to a moderate-sized stream, widening out at last only to be merged in Grand and Little Lake George, Dexter's Lake, Lake Beresford, and Lake Monroe, at Enterprise. The steamers make the run during the night from Palatka to Enterprise in about 10 hours. Five miles above Palatka, on the opposite bank, is *San Mateo*, a thriving settlement situated on a high ridge overlooking the river. **Welaka** (*McClure House*), 25 miles above Palatka, is opposite the mouth of the Ocklawaha River, and is the site of what was originally an Indian village, and afterward a flourishing Spanish settlement. The name is Indian, meaning "river of lakes." Just above Welaka the river widens into *Little Lake George*, 1 mile wide and 2 miles long, and then into **Lake George,** 6 miles wide and 13½ miles long. This is one of the most beautiful sheets of water in the world, being considered by many tourists equal in attractions to its namesake in the State of New York. Among the many lovely islands which dot its surface is one called *Drayton*, which is 1,700 acres in extent, and contains one of the largest orange-groves on the river. All along the lake the eye is delighted and the ear charmed by the brilliant plumage and sweet song of the Southern birds. One finds here the heron, the crane, the white curlew, the pelican, the loon, and the paroquet; and there are many varieties of fish. *Volusia* (5 miles above Lake George) is a landing-station, with a settlement of considerable size back from the river. An ancient Spanish town used to stand here, this formerly being the principal point on the line of travel between St. Augustine and the Mosquito Inlet country. *Orange Grove* and *Hawkinsville* are other landings; and 35 miles above Volusia is *Blue Spring*, one of the curious mineral springs in the State. It is 500 yards from the St. John's, but the stream flowing from it is large enough at its confluence with the river for the steamers to float in it. From Blue Spring a railroad runs in 30 miles to New Smyrna, by way of Lake Helen. Pursuing its voyage to the south, the steamer speedily enters *Lake Monroe*, a sheet of water 4 miles long by 4 miles wide, teeming with fish and wild-fowl. On the south side of the lake is **Sanford** (*Sanford, San Leon*), a young but rapidly growing city of 2,016 inhabitants, the metropolis of S. Florida. It is situated at the head of navigation for large steamers on the St. John's, and is the principal avenue of entrance to Orange County, whither so many of the new settlers are going. The South Florida R. R. extends S. W. to Tampa (the fourth city in size in the State), opening up an excellent country, and passing the growing towns of *Altamonte Springs*, *Winter Park* (*Seminole Hotel*), and *Orlando* (*San Juan Hotel*). Near Sanford and Orlando are a number of fine orange-groves. Opposite Sanford, reached by the railroad bridge, is Enterprise, one of the most popular resorts in Southern Florida for invalids, especially for those suffering from rheumatism. Frederick De Bary, the well-known champagne importer of New York, and the founder of the steamboat line on the St. John's, has his Florida country-seat near here. The climate is rather warmer than that of Jacksonville and Magnolia, but it is said to have special invigorating qualities which speedily convert invalids into successful fishermen and hunters. The *Brock House* is famous among travelers, and board may be had in private families for from $8 to $15 per week. A mile E. of the town is the *Green Spring*, a sulphur-spring, with water of a pale-green hue, but quite transparent. It is nearly 80 ft. in diameter, and about 60 ft. deep.

Although Sanford is the terminus of regular navigation on the St. John's, there is for the sportsman still another hundred miles of narrow river, deep lagoons, gloomy bayous, and wild, untrodden land, where all sorts of game are plentiful, while the waters teem with fish. Small steamers run through Lake Harney to *Salt Lake*, the nearest point to the Indian River from St. John's; and a small steamboat makes frequent excursions to *Lake Jessup* and *Lake Harney*, for the benefit of those who wish to try their hand at the exciting sport of alligator-shooting, or of those who wish simply to enjoy the charming scenery. The trip to Lake Harney and back is made in 12 hours. Lake Jessup is near Lake Harney; it is 17 miles long and 5 miles wide, but is so shallow that it can not be entered by a boat drawing more than 3 ft. of water. The St. John's rises in the elevated savanna before mentioned, fully

120 miles S. of Sanford, but tourists seldom ascend farther than Lake Harney.   About 30 miles S. E. of Enterprise is the ancient town of *New Smyrna* (see Route 118).

117. The Ocklawaha River.

THE Ocklawaha empties into the St. John's about 25 miles S. of Palatka, opposite the small town of Welaka (see Route 116), after flowing for about 250 miles through Putnam, Marion, and Sumter Counties. ·The channel possesses no banks to speak of, being mainly a navigable passage through a succession of small lakes and cypress-swamps; but small steamers ascend it for a distance of nearly 200 miles.   An excursion up the Ocklawaha to Silver Springs (109 miles) is perhaps the most unique experience of the tourist in Florida, and every one who can should make it.   Alligators of immense size are seen, and birds of the most curious forms and brilliant plumage are everywhere conspicuous.   From Palatka steamers run to Silver Springs, which can also be reached by the Southern Division of the Florida Central & Peninsular R. R. *via* Waldo.

The principal landing on the Ocklawaha is **Silver Springs** (*Silver Springs Hotel*), the largest and most beautiful of the springs of Florida, navigable by steamers of several tons' burdén.   This spring is said to be the traditional "fountain of youth" of which Ponce de Leon heard, and for which he so vainly searched.   The clearness of its waters is wonderful; they seem more transparent than air.   "You see on the bottom, 80 ft below, the shadow of your boat, and the exact form of the smallest pebble; the prismatic colors of the rainbow are beautifully reflected, and you can see the fissure in the rocky bottom through which the water pours upward like an inverted cataract."   A deep river, 100 ft. wide, is formed by the water of this spring, which in the course of 9 miles forms a junction with the Ocklawaha.   This is known as the "Run," and a little steamer plies on it and the spring. *Ocala* is only 5 miles distant from the spring, and the *Ocala House* accommodates 400 guests. .

Sixteen miles south, on the Florida C. & P. Co.'s line, is the *Lake Weir Country ;* 10 miles farther is *Wildwood* (whence a branch line of railroad runs to Leesburg, where connection is made with the South Florida R. R. for Plant City).   At *Tavares* connection is made with *Sanford,* on the St. John's River, and Orlando.   Perhaps the best way to visit Silver Springs is by the Florida Southern R. R. (see Route 116) to Ocklawaha and Silver Springs, and then down the Ocklawaha River to Palatka.

34

118. The Indian River Country.

At *Enterprise* the Indian River Division of the Jacksonville, Tampa & Key West R. R. conveys the traveler in parlor-coaches to a hitherto inaccessible region. It is thus, by a pleasant ride through Enterprise, Osteen, Oak Hill, Mims, and La Grange to Titusville, brought within easy reach of Jacksonville.

INDIAN RIVER is a long lagoon or arm of the sea, beginning near the lower end of Mosquito Inlet (with which it is connected by a short canal), and extending S. along the E. side of the peninsula for a distance of nearly 150 miles. It is separated from the Atlantic by a narrow strip of land, through which it communicates with the open water by the Indian River Inlet (latitude 27° 30′ N.) and by Jupiter Inlet; and for more than 30 miles of its northern course the St. John's River flows parallel with it, at an average distance of not more than 10 miles. The water of the lagoon is salt, though it receives a considerable body of fresh water through Santa Lucie River, an outlet of the Everglades; there are no marshes in the vicinity; the adjacent lands are for the most part remarkably fertile, producing abundantly oranges, lemons, limes, bananas, cocoanuts, pineapples, guavas, grapes, sugar-cane, strawberries, blackberries, and all varieties of garden vegetables; and the river itself teems to an almost incredible degree with fish of every kind, including the pompano, the mullet, the sheepshead, tarpon, turtles, and oysters of the most delicious flavor. Along the shore of the lagoon toward the Atlantic is a belt of thick, evergreen woods, which, breaking the force of the chilling east winds that sometimes visit these latitudes in winter, renders the climate of the Indian River country peculiarly favorable to consumptives. "The sportsman who pitches his tent for a few days on the splendid camping-ground of the W. shore will see the pelican, the cormorant, the sea-gull, and gigantic turtles, many of them weighing 500 pounds; may see the bears exploring the nests for turtles' eggs; may 'fire-hunt' the deer in the forests; chase the alligator to his lair; shoot at the 'raft-duck'; and fish from the salt-ponds all the finny monsters that be. Hardly a thousand miles from New York one may find the most delicate and delightful tropical scenery, and may dwell in a climate which neither Hawaii nor southern Italy can excel." Thus wrote Mr. Edward King in 1873. Since then things have greatly changed. *Titusville, City Point,* and *Rockledge* are now flourishing settlements, and the entire region is rapidly filling up with inhabitants. It was by one of the many southern outlets of Indian River that General Breckenridge escaped to Nassau after the collapse of the Confederacy.

At its N. end, as already mentioned, the Indian River connects by Clifton Canal with the Mosquito Lagoon; and at the N. end of Mosquito Lagoon the Halifax River comes in, which begins about 40 miles S. of St. Augustine. The principal settlements are *Ormond, Daytona,* and *Port Orange,* on the Halifax River; *New Smyrna,* on the Hillsboro' River, 3 miles S. of Mosquito Inlet, near the coast; *Titusville* (formerly Sand Point) and *Rockledge,* on the W. bank of the Indian River. Titusville is now the terminus of the Jacksonville, Tampa & Key West R. R., which runs from Jacksonville, *via* Enterprise, to Titusville; thence the Jacksonville, St. Augustine & Indian River R. R. is taken for 20 miles to Rockledge. The Indian River Steamboat Co.'s boats run from Titusville to Jupiter through the Halifax and Indian Rivers, affording tourists ample facilities for visiting this portion of Florida, and also Lake Worth.

119. Fernandina to Cedar Key.

FERNANDINA (*Egmont Hotel*) is an interesting old seaport town, situated on the W. shore of Amelia Island, at the mouth of Amelia River, 50 miles N. of Jacksonville. It is reached by rail from Jacksonville; by steamer, direct from New York; and by the "inside route" steamers from Charleston and Savannah (see Route 114). Fernandina was located by a Spanish grantee early in the present century, and at the present time has a population of 2,803, which is largely increased during the winter season. Its harbor is the finest on the coast S. of Chesapeake Bay, being landlocked and of such capacity that, during the War of 1812, when the town was Spanish and neutral, more than 300 square-rigged vessels rode at anchor in it at one time. It has an important trade in lumber; possesses a large cotton-ginning establishment and a manufactory of cotton-seed oil; and it is in the neighborhood of numerous sugar, cotton, and orange plantations. The *climate* of Fernandina is very similar to that of St. Augustine; mild and equable in winter, and in summer tempered by the cool sea-breezes. The town, which is the headquarters of the Florida Central & Peninsular R. R., contains 7 churches, a large number of business houses, a flourishing young ladies' seminary, and a weekly newspaper. Fernandina possesses other attractions for visitors besides its delightful climate. There is, for instance, a fine shell-road 2 miles long, leading to the ocean-beach, which affords a remarkably hard and level drive of nearly 20 miles. A favorite excursion is to *Dungeness*, the purposed home of the Revolutionary hero, General Nathanael Greene. This estate, of about 10,000 acres of choice land, was the gift of the people of Georgia to the general, in recognition of his services as commander of the Southern provincial army. The grounds were beautifully laid out, and are embellished with flower-gardens and groves of olive-trees, and live-oaks draped with festoons of Spanish moss. General Greene, however, never lived here, having died on the Savannah River. The building was burned during the civil war. In 1884 Cumberland Island, on which Dungeness is situated, was purchased by Thomas M. Carnegie, who built a mansion there. On the beach, about half a mile from the mansion, is the grave of another Revolutionary hero, General Henry Lee, marked by a head-stone erected by his son, General Robert E. Lee.

Beginning at Fernandina, the Florida C. & P. R. R. extends directly across the State to Tampa (212 miles), with a branch to Cedar Key, on the Gulf coast, passing through some of the most picturesque scenery in Florida. There are a number of small stations on the line, but few requiring mention. *Baldwin* is at the crossing of the Western Division of the line (see Route 109). Noteworthy towns are **Lawtey,** famous for its strawberry farms, and **Waldo** (84 miles; *Waldo House*), which is at the junction of the branch running to Cedar Key. The climate here is dry and the air balsamic, and the region is regarded as particularly favorable to invalids suffering from lung-diseases. The woods in the vicinity of the village abound in deer, ducks, quail, etc.; and about 2 miles distant is *Santa Fé Lake*, which is 9 miles long and 4 wide, and affords good facilities for boating and fishing. The streams in the neighborhood are filled with trout and perch. The Santa Fé River disappears underground a few miles from Waldo, and, after running underground for two miles, rises and continues to its discharge into the Suwanee River. **Gainesville** (*Arlington, Brown, Rochemont*) is the principal town on the Cedar Key branch of the road. It has 2,790 inhabitants, four churches, and two newspapers, and, owing to its situation in the center of the peninsula and in the midst of the pine-forests which clothe this portion of Florida, Gainesville is much frequented by consumptives and other invalids. The hotels and a large portion of the town were destroyed by fire on May 3, 1884, but have been substantially rebuilt since the disaster. The *Alachua Sink* teems with fish of various kinds, and with alligators. **Cedar Key** (the *Bettelini Schlemmer, Suwanee House*), the branch terminus of the railway, is a town of some 1,869 inhabitants, pleasantly situated on a large key forming one of a cluster of islands, which affords excellent facilities for bathing, boating, and fishing. The chief commerce of the place is in cedar and pine wood (used in the making of lead-pencils), turtles, fish, and sponges, the sponging-grounds being about 60 miles distant. The climate of Cedar Key is nearly similar to that of Jacksonville, and is beneficial to rheumatism as well as consumption. Sportsmen will find unlimited occupation for both rod and gun. Eighteen miles W. of Cedar Key, the *Suwanee River*, navigable to Ellaville, enters the Gulf; and the *Withlacoochee River*, 18 miles S. Steamers ply between New Orleans and Key West, Tampa, and Manatee, and others connect it with the country on the Suwanee River. A steamer sails twice a week for *Tarpon Springs*, on the Anclote River. The principal town on the southern extension of the road toward Tampa is **Ocala,** whose large bearing orange-groves, wide-spread truck farms, cotton plantations, corn-fields, and other agricultural industries, make it an important commercial center. It is also the center of great phosphate interests. The city is lighted by electricity, has a street-railway system, first-class water-works, fire protection, and paved streets.

120. Middle Florida.

THAT portion of Florida known as "Middle Florida" (in the midst of which Tallahassee lies) differs from the rest of the State in that its surface is more broken and undulating, reaching here and there an elevation of from 300 to 400 ft. The hills are singularly graceful in outline, and the soil is exceedingly fertile, producing all the characteristic products of the Southern States, including tobacco and early garden vegetables. The vegetation is less tropical in character than that of eastern and southern Florida, but it is very profuse and comprises many evergreens. **Tallahassee** (*New Leon, St. James*), the capital of the State and county-seat of Leon County, is situated on the Florida C. & P. R. R. (Western Division), 165 miles W. of Jacksonville and 21 miles N. of the Gulf of Mexico. It is beautifully located on high ground, and is regularly laid out in a plot a mile square, with broad streets and several public squares, shaded with evergreens and oaks. The abundance and variety of the shrubs and flowers give it the appearance of a garden. The business portion of the city is of brick. The public buildings are the *Capitol* (commenced in 1826), a large three-story brick edifice, with pillared entrances opening E. and W.; the *Court-House*, a substantial two-story brick structure; and the *West Florida Seminary*, a large two-story brick building, on a hill commanding a view of the entire city. In the immediate neighborhood of Tallahassee are *Lake Iamonia, Lake Jackson* (17 miles long), and *Lake Lafayette* (6 miles long). During the winter months these lakes swarm with duck and brant; and to the angler Lake Jackson is well stocked with bass. and bream. Quail are also very abundant. The St. Mark's Branch of the Florida C. & P. R. R. runs hence to *Wakulla* (16 miles), the nearest station to the celebrated **＊Wakulla Spring**, which is reckoned among the chief wonders of Florida. It is an immense limestone basin, 106 ft. deep, and with waters so crystalline clear that the fish near its bottom can be seen as plainly as though they were in the air, and so copious that a river is formed at the very start.

Along the line of railway on which Tallahassee is situated (Florida Central & Peninsular R. R.) there are several towns which offer great attractions to invalids, tourists, and sportsmen. **Quincy** (24 miles W. of Tallahassee) is a prosperous village of 681 inhabitants, the county-seat of Gadsden County. This place has several noted tobacco plantations thousands of acres in extent. Its climatic characteristics are the same as those of Tallahassee, and there is a similar abundance of game in the vicinity. Board may be had at the *Metropolitan* and at private boarding-houses. **Monticello** (33 miles E. of Tallahassee) is an important town of some 1,218 inhabitants, and the terminus of a branch road 4¼ miles in length. It contains Baptist, Episcopalian, Methodist, and Presbyterian churches, several schools, and a weekly newspaper. The *St. Elmo House* and *The Oakley* are good houses, and board may be had in private families. In the vicinity of Monticello is *Lake Miccosukie*, whose banks were, according to tradition, the camping-ground of De Soto, and the field of a bloody battle between General Jackson and the Miccosukie Indians. At its S. end the lake contracts to a creek and disappears underground. Near Monticello is the Lipona plantation, where Murat resided for some time while in Florida. The Florida Central & Peninsular R. R. terminates at *River Junction*, on the Chattahoochee River, where it makes connection with the Pensacola & Atlantic R. R. of the Louisville & Nashville System (see Route 121) for Pensacola, Mobile, and other points W.

121. The Gulf Coast and Key West.

MUCH the larger part of the coast-line of Florida is washed by the Gulf of Mexico; but this immense stretch of sea-front is almost inaccessible on account of shallow soundings, and has few good harbors. The principal place in this part of the State is **Pensacola** (*Escambia, Merchants'*), a city of 11,750 inhabitants, situated on the N. W. side of the bay of the same name, 10 miles from the Gulf of Mexico. Its commerce is extensive and its lumber business important. It is the western terminus of the Florida branch of the Louisville & Nashville system, which brings Pensacola into connection with the general railway system of the country. The Pensacola & Perdido R. R. runs in 9 miles to *Millview*, on Perdido Bay, where there are extensive lumbering establishments. The principal public buildings of Pensacola are a Custom-House and several churches. The remains of the old .Spanish forts, San Miguel and St. Bernard, may be seen in rear of the city. A weekly line of steamers was established in 1878 to ply between Pensacola and Tampa, calling at Cedar Key. **Appalachicola,** recently a decadent city of some 2,727 inhabitants, but now springing into new life, is situated at the entrance of the river of the same name into the Gulf of Mexico, through Appalachicola Bay. To the W. of Appalachicola, *Carrabelle*, a village of 482 inhabitants, with a good, deep harbor, promises to be a great lumber port. It is connected with Columbus, Georgia (see Route 122 *b*), by steamers on the Appalachicola and Chattahoochee Rivers. **Cedar Key** has been described in Route 119. Semi-weekly steamers run from Cedar Key to Tampa, Charlotte Harbor, Manatee, Key West, etc.

Tampa (*Almeria, Plant, Tampa Bay, The Inn*), the first noteworthy point below Cedar Key, is situated near the center of the W. coast, at the head of the beautiful Tampa Bay (formerly Espíritu Santo Bay). The bay is about 40 miles long, is dotted with islands, and forms a good harbor. Its waters swarm with fish and turtle, and there is an abundance of sea-fowl, including the beautiful flamingo. Deer swarm on the islands. The surrounding country is sandy, and for miles along the shore there is a luxuriant tropical vegetation. Large groves of , orange, lemon, and pine trees are everywhere to be seen. The town contains 5,532 inhabitants, and is probably destined to become one of the chief health resorts of Florida. Tampa is now directly connected by the Florida Central & Peninsular R. R. and Jacksonville, Tampa & Key West R. R. with Jacksonville and Palatka, and by South Florida R. R. with Sanford and Enterprise. This line now extends to **Port Tampa,** a distance of 9 miles, which is the southern deep-water terminus of the Plant system of railroad and steamship lines. The South Florida R. R., now a portion of the Plant system, connects at Sanford (see Route 116), on the St. John's River, with steamboat lines. The railroad runs S. 40 miles, to *Kissimee City*, near *Kissimee Lake*, a place which has grown into importance within the last two years. The route is through one of the most wild and picturesque portions of the State, where the greatest abundance of fish and game of every description attracts the sportsman to the delights of camping out. At ·Kissimee the road bends to the W., and reaches Tampa in a run of 75 miles, passing *en route* near several beautiful lakes—*Lake Maitland, Lake Hamilton, Lake Parker,* and *Lake Hancock. Manatee* is a small village situated on the Manatee River about 8 miles from its mouth. There are two or three boarding-houses here, where fair accommodations may be had at $2 a day or· $40 a month. *Charlotte Harbor* is about 25 miles long and from 8 to 10 miles wide, and is sheltered from the sea by several islands. The fisheries in and around the harbor are very valuable, the oysters gathered here being remarkably fine and abundant. On one of the islands in Charlotte Harbor there are a number of Indian shell-mounds, from one of which some curious Indian relics have been dug. Opposite is **Punta Gordo,** the southern terminus of the Florida Southern Railway, and a landing-place of the Morgan Steamship Line, which goes thence to Key West and Havana. *Punta Rassa* is a small hamlet near the mouth of the Caloosahatchie River, chiefly noteworthy as the point where the Cuban telegraph-line lands and as a U. S. Signal-Service station. .The thermometrical observations recorded here are interesting as indicating the climate of all this portion of the coast. In 1874 the range was as follows : January, highest 79°, lowest 42° ; February, highest 84°, lowest 50° ; March, highest 85°, lowest 55° ; April, highest 87°, lowest 55° ; May, highest 90°, lowest 59° ; June, highest 91°, lowest 70° ; July, highest 91°, lowest 70° ; August, highest 91°, lowest 70° ; September, highest 91°, lowest 67° ; October, highest 85°, lowest 64°: November, highest 82°, lowest 50° ; December, highest 80°, lowest 49°.

Key West (*Russell House*), the largest city of Florida, next to Jacksonville, is situated upon an island of the same name off. the S. extremity of the peninsula, and occupies the important post of key to the Gulf passage. The island is 7 miles long by from 1 to 2 miles wide, and is 11 ft. above the sea. It is of coral formation, and has a shallow soil, consisting of disintegrated coral, with a slight admixture of decayed vegetable matter. There are no springs, and the inhabitants are dependent on rain or distillation for water. The natural growth is a dense, stunted chaparral, in which various species of cactus are a prominent feature. Tropical fruits are cultivated to some extent, the chief varieties being cocoanuts, bananas, pineapples, guavas,·sapodillas, and a few oranges. The air is pure and the climate healthy. The thermometer seldom rises above 90°, and never falls to freezing-point, rarely standing as low as 50°. The mean temperature, as ascertained by 14 years' observation, is.: for spring, 75·79° ; for summer, 82·51° ; for autumn, 78·23° ; for winter, 69·58°. The city has a population of 18,058, a large portion of whom are Cubans and natives of the Bahama Islands. They are a hardy and adventurous race, remarkable for their skill in diving. The language commonly spoken is Spanish, or a *patois* of that tongue. The streets of the town are broad, and for the most part are laid out at right angles with each other. The residences are shaded with tropical trees, and enbowered in perennial flowers and shrubbery, giving the place a very picturesque appearance. Almost the entire city was recently destroyed by fire. Key West has a fine harbor, and, being the key to the best entrance to the Gulf of Mexico, it is strongly fortified. The principal work of defense is **Fort Taylor,** built on an artificial island within the main entrance to the harbor, and mounting about 200 guns. Among the principal industries of Key West are turtling, sponging, and the catching of mullet and other fish for the Cuban market. Upward of 30 vessels, with an aggregate of 250 men, are engaged in wrecking on the Florida Reef, and the island profits by this industry to the amount of upward of $200,000 annually. The manufacture of cigars employs about 800 hands, chiefly Cubans, and more than 25,000,-000 cigars are turned out yearly. There are a number of charming drives on the island, and the fishing and boating are unsurpassed.

From New York, Key West is reached *via* steamer to Jacksonville, and thence *via* Jacksonville, Tampa & Key West R. R. to Tampa, and from Tampa by steamer thrice weekly for Key West and Havana, or *via* New York and Galveston steamers, leaving Pier 20, East River, every Saturday at 3 P. M. From Baltimore by semi-monthly steamers. Key West is the most important supply-station on the Gulf of Mexico.

122. Washington to New Orleans.

a. Via Piedmont Air-Line, Charlotte, Atlanta, and Montgomery.

THE first section of this route, between Washington and Danville, Va., is over the Virginia Midland Div. of the Richmond & Danville R. R., which traverses a portion of Virginia full of memorials, both of the Revolutionary era and of the late civil war. Leaving *Alexandria* (7 miles), which has been described in Route 8, the trains pass amid the scenes of the earliest struggles of the war, the outposts of the opposing armies occupying this ground for a large part of the time. **Manassas** (34 miles) was the scene of the first great battle of the civil war, fought July 21, 1861, between the Confederates under Beauregard and the Federals under McDowell, in which the latter were routed; and also of another battle, fought August 29 and 30, 1862, between the Confederates under Lee and the Federals under Pope, in which the latter were again defeated. The battle-ground of the "first Manassas" is 3 or 4 miles from the station, and intersected by the Sudley, Brentsville & Warrenton Turnpike, which crosses at Stone Bridge. The battle-ground of the "second Manassas" was nearly identical with the first, with, however, a *change of sides* by the combatants. At Manassas the Manassas Branch diverges and runs in 63 miles to *Strasburg ;* and from *Warrenton Junction* (48 miles) a branch road runs to *Warrenton.* At *Rappahannock* (58 miles) the train crosses the Rappahannock River. **Culpeper Court-House** (69 miles) was an important military point during the war, the place having been occupied and reoccupied time after time by both armies, between whom numerous engagements occurred in the fields surrounding the village. Culpeper County was famous in Revolutionary times for its company of "Culpeper Minute-Men," in which Chief-Justice Marshall was enrolled and fought, and whose flag bore a picture of a coiled rattlesnake with the motto, " Don't tread on me ! " Of this body of men, John Randolph is reported to have said that " they were summoned in a minute, armed in a minute, marched in a minute, fought in a minute, and vanquished in a minute." Twelve miles beyond Culpeper the train crosses the Rapidan River, which was the line of defense frequently held by the Confederates during the war, and soon reaches **Charlottesville** (113 miles), a busy place at the junction with the Chesapeake & Ohio R. R. (*Parrott's Hotel, Wright's Hotel*), famous as the seat of the University of Virginia and for its proximity to Monticello, the home and tomb of Thomas Jefferson. It is an attractive and well-built town of 5,591 inhabitants, situated on Moore's Creek, 2 miles above its entrance into Rivanna River. The * *University of Virginia* is situated 1¼ mile W. of Charlottesville, is built on moderately elevated ground, and forms a striking feature in a beautiful landscape. It was founded in 1819, and its organization, plan of government, and system of instruction are due to Thomas Jefferson, who, in the inscription prepared by himself for his tomb, preferred to be remembered as the " author of the Declaration of Independence and of the statute of Virginia for religious freedom, and father of the University of Virginia." * *Monticello*, once the home and now the burial-place of Jefferson, is about 4 miles W. of Charlottesville. It stands upon an eminence, with many aspen-trees around it, and commands a view of the Blue Ridge for 150 miles on one side, and, on the other, one of the most extensive and beautiful landscapes in the world. The remains of Jefferson lie in a small family cemetery by the side of the winding road leading to Monticello. Congress appropriated $5,000 to erect a suitable monument over them, in place of the ruined granite obelisk which now marks the spot. **Lynchburg** (173 miles ; *Lynch House, Norvell-Arlington*) is a city of 19,709 inhabitants, which derives its importance from the lines of railway which center here, and the extent and character of its manufactures, and especially from its large trade in tobacco, which is the chief article of export. It is situated on the S. bank of James River, and enjoys an inexhaustible water-power. It occupies a steep acclivity, rising gradually from the river-bank, and breaking away into numerous hills, whose terraced walks and ornamental dwellings give a picturesque and romantic appearance to the city. Lynchburg contains 15 churches, five national banks, and has a good public-school system. It has several iron-foundries, rolling-mills, and flouring-mills. The city was founded in 1786, and incorporated in 1805. About 20 miles in the background rises the Blue Ridge, together with the Peaks of Otter, which are in full view, and from here the cars can be taken to Natural Bridge. In the neighborhood of Lynchburg are vast fields of coal and iron-ore, and the celebrated *Botetourt Iron-Works* are not far distant.

From Lynchburg to Danville (66 miles) a number of streams are crossed, among which are the Otter River, Stanton River, Dry Fork, and Fall Creek. From Danville to Charlotte the route is the same as that described in Route 112 *b*. At Charlotte the Atlanta & Charlotte Division of the " Piedmont Air-Line " is taken. This road runs S. W. through South Carolina and Georgia, reaching Atlanta in 648 miles. The country traversed is rolling and hilly, being on the border of the picturesque mountain-region of both States. Numerous small towns are passed *en route*, but most of them are mere railroad-stations, and only three or four worthy of notice. The first of these is *King's Mountain* (33 miles from Charlotte), near an eminence of the same name which was the scene of a battle, Oct. 7, 1780, between the British and the patriot forces, in which the former were defeated and their entire detachment captured. Near *Cowpens* (66 miles) is the memorable Revolutionary *battle-field of the Cowpens*, situated on the hill-range called the Thickety Mountain. The battle was fought Jan. 17, 1781, and resulted in the defeat of the British under Tarleton. In the olden time the cattle were allowed to graze on the scene of the conflict—whence the name. Nine miles beyond Cowpens is **Spartanburg** (*Merchants' Hotel, Windsor Hotel*), the most important town in this portion of South Carolina, with a population of 5,544. It is pleasantly situated in the midst of a region famous for its gold and iron, and is much resorted to in summer by people from Charleston and the lowlands. Near Spartanburg are the *Glenn Springs*, whose waters are strongly impregnated with sulphur, and recommended for rheumatism and dyspepsia; and the *Limestone Spring*, a chalybeate possessing valuable tonic properties. A branch of the Richmond & Danville R. R. connects Spartanburg with Columbia. Thirty-two miles beyond Spartanburg is **Greenville** (*Exchange Hotel, Mansion House*), a city of 8,607 inhabitants, beautifully situated on Reedy River, near its source, and at the foot of Saluda Mountain. It is one of the most popular resorts in the up-country of the State, lying as it does at the threshold of the chief beauties of the mountain region of South Carolina (see Route 131). The Columbia & Greenville Div. runs S. E. in 143 miles to Columbia. At *Seneca City* (148 miles from Charlotte) connection is made with the Columbia & Greenville Div., and, a short distance beyond, the road crosses the Savannah River and enters the State of Georgia. *Toccoa* (176 miles) and *Mount Airy* (189 miles) are convenient entrances to the mountain region of Georgia (see Route 131). From *Lula* (203 miles) a branch runs to the collegiate town of *Athens* (39 miles), and to *Macon* (144 miles). The principal place on this portion of the line is **Gainesville** (216 miles, *Arlington House*), a town of 3,202 inhabitants, which has grown wonderfully since the completion of the railway. One mile from Gainesville (reached by horse-cars) is the *Gower Springs Hotel*, and 2 miles E. are the *New Holland Springs*, a favorite resort. The *Porter Springs* are 28 miles N., attractively situated among the mountains. Just before reaching Atlanta, Stone Mountain comes into view far away on the left, and shortly afterward the train crosses Peach-Tree Creek, the scene (lower down) of the bloody conflict of July 22, 1864. **Atlanta** (*Kimball House, Markham, National, Hotel Weinmeister*) is the capital of Georgia, and the most important commercial city in the State, though Savannah surpasses it in the cotton-trade. The population is 65,533. Atlanta ranks high as a manufacturing city, in proportion to its population. This is the outgrowth of the railroad-system centering there, and in its activity and enterprise reminds one of a Northern rather than of a Southern city. The railroads are the East Tennessee, Virginia & Georgia R. R., Georgia Pacific Railway, Georgia R. R., Atlanta & West Point R. R., Atlanta & Florida R. R., Western & Atlantic R. R., Central R. R., Richmond & Danville R. R., and Georgia, Carolina & Northern R. R., connecting Atlanta with every part of the country. Three new railroads are now being constructed with Atlanta as terminus. The city is picturesquely situated upon hilly ground 1,100 ft. above the sea, and is laid out in the form of a circle 3¼ miles in diameter, the Union Passenger Station occupying the center. The most noteworthy buildings are the *City Hall* (beautifully located), (South), the *Union Passenger Depot*, the *Opera-House*, the *U. S. Custom-House*, built at a cost of $350,000, the *County Court-House*, and the *Chamber of Commerce*. A State Capitol has recently been finished, at a cost of $1,000,000, and contains statues of Senator Benjamin H. Hill and Henry W. Grady. Among the benevolent institutions worthy of note are the *Soldiers' Home for Confederate Veterans* and the *Hebrew Orphans' Home*. The old Kimball House, which was destroyed by fire, is replaced by a fireproof edifice 7 stories high, built and equipped at a cost of $1,000,000. The *State Library* contains about 16,000 volumes; the *Young Men's Library* about 15,000. A ride out *Peachtree St.* on the electric-cars shows one of the finest residence streets in the South, filled with handsome modern dwellings. The chief interest which Atlanta possesses for the tourist is the memorable siege with which it is associated. Its position made it of vital importance to the Southern cause, and with its capture by Gen. Sherman, Sept. 2, 1864, the doom of the Confederacy was sealed. Before abandoning the city, to fall back upon Macon, Gen. Hood set fire to all machinery, stores, and munitions of war which he could not remove, and Gen. Sherman, on leaving it a month later, destroyed the business part of the city. Atlanta became the State capital in 1868.

From Atlanta the route is *via* the Atlanta & West Point R. R., which runs S. W. through a prosperous agricultural region, and in 87 miles reaches West Point, on the Alabama border. The principal towns *en route* are *Newnan* (39 miles), where connection is made with the Central R. R. of Georgia, and *La Grange* (71 miles), which is noted throughout the State for the excellence of its educational establishments. *West Point* is a thriving town of 1,254 inhabitants on both sides of the Chattahoochee River, with an active trade in cotton, and several cotton-factories. At West Point the Western R. R. of Alabama is taken, which runs W. in 88 miles to Montgomery. *Opelika* (22 miles) is a flourishing village at the junction of the branch line from Columbus, Georgia, 29 miles distant (see Sub-Route *b*). **Montgomery** (*Exchange, Houston, Windsor*) is the capital of Alabama, and the second city of the State in size and commercial importance. It is situated on a high bluff on the left bank of the Alabama River, was founded in 1817, named after Gen. Richard Montgomery, who fell at Quebec, has a population of 21,833, and was the first capital of the Confederate States (from February to May, 1861). The principal public building is the *U. S. Court-House and Post-Office*, which cost $125,000. The * *State-House*, now enlarged, is an imposing structure. It is situated on Capitol Hill, at the head of Dexter Avenue, and from its dome there is a fine view. Other noteworthy buildings are the *City Hall*, a fine edifice containing a market and rooms for the fire department, the *Court-House*, several of the churches, and the two theatres. The Alabama River is navigable to Montgomery by steamers at all seasons, and 5 important railroads converge here. From Montgomery to Mobile the route is *via* the Mobile & Montgomery Div. of the Louisville & Nashville R. R., which extends S. W. through one of the most productive portions of Alabama (distance, 180 miles). The most important town *en route* is *Greenville* (44 miles from Montgomery), with a population of 2,806.. **Mobile** (see Sub-Route *b*). The total distance from Washington to Mobile by this route is 1,002 miles.

b. Via Augusta, Macon, and Columbus.

Between Washington and *Augusta* (571 miles) the tourist may take either of the routes described in Route 113. In Route 113 *a* the Wilmington, Columbia & Augusta R. R. is followed to Columbia, whence the route is the same as in *b*. From Augusta to Macon there are two routes: the Georgia R. R. and the Central R. R. of Georgia. The principal towns on the Georgia R. R. are *Camak* (47 miles from Augusta), whence a branch line runs in 124 miles to **Atlanta** (see Sub-Route *a*); *Warrenton* (51 miles) and *Spàrta* (71 miles), both pretty towns; and *Milledgeville* (93 miles), the former capital of the State, and the site of the State Penitentiary and of the Georgia Asylum for the Insane. The most important places on the Central R. R. of Georgia are *Millen* (53 miles from Augusta), where the road forks, one branch going to Savannah (see Route 113 *b*); and *Gordon* (144 miles), whence a branch line runs to Milledgeville (see above) and *Eatonton*, a pleasant town with excellent schools. The former route is the shorter, but, owing to an advantage in connections, the latter is the route usually followed by through travel. — **Macon** (*Brown House, Commercial, Hotel Lanier*) is one of the most populous and prosperous cities of Georgia, and is picturesquely situated on the Ocmulgee River, which is here crossed by a bridge. It contains 22,746 inhabitants, is the site of several important iron-foundries, machine-shops, carriage and cotton manufactures, and flour-mills, is regularly laid out and well built, and is embowered in trees and shrubbery. The * *Central City Park*, combining pleasure and fair grounds, possesses great beauty; and * *Rose Hill Cemetery*, comprising 50 acres on the Ocmulgee, ¼ mile below the city, is one of the most beautiful burial-grounds in the United States. The *Court-House* and the *U. S. Building* are fine structures. Another handsome edifice is the *Academy of Music*. In the Park on Mulberry St. are a monument to the Confederate dead, erected in 1879, and a bronze statue to William M. Wadley. Macon is the seat of the *State Academy for the Blind*, which occupies an imposing brick edifice four stories high, and has a library of 2,000 volumes. *Mercer University* (Baptist) is a prosperous institution, with a library of 9,000 volumes; and the *Wesleyan Female College*, rebuilt and endowed by George I. Seney, of New York, in 1881, with a fund of $125,000, has a wide reputation.

It conferred the first degree on women in 1840. The *Pio Nono College* is Roman Catholic. A belt-line of electric-cars makes a circuit of the city and extends to *Vineville*, a suburban village about a mile from the city. It is the center of the Central R. R. of Georgia system, which has branches to Atlanta on the N., to Savannah on the E., and to Columbia and other points on the S. and W. Several other railroads pass through Macon, and secure it an extensive trade.

From Macon the route is *via* the Southwestern Div. of the Central R. R. of Georgia, which runs in 100 miles to Columbus, through a level, sandy, and unpicturesque region. The most important place on the line is *Fort Valley* (29 miles from Macon), an attractive village of 1,752 inhabitants, at the junction of two important branches of this division. **Columbus** (*Central Hotel, Rankin House*) is situated on the E. bank of the Chattahoochee River, and is the fifth city of Georgia in population (17,303), and the chief manufacturing center in the South. Opposite the city the river rushes over huge, rugged rocks, forming a water-pówer which has been greatly improved by a dam 500 ft. long, and which is extensively utilized in manufactures. There are eight cotton-factories, five run by water-power and three by steam; one of the companies (the Eagle and Phœnix Co.) is the largest established in the South. There are also 14 flour and grist mills, and machine-shops, iron-foundries, saw-mills, planing-mills, etc. The Chattahoochee is navigable from Columbus to the Gulf of Mexico during eight months of the year; and from the end of October to the 1st of July its waters are traversed by numerous steamboats laden with cotton. The city is regularly laid out with streets from 99 to 165 ft. wide, and residences surrounded by ample gardens. The most noteworthy buildings are the *Court-House*, the *Presbyterian Church*, *Temperance Hall*, the *Springer Opera-House*, the *Georgia Home Insurance Co.*, the *Bank of Columbus*, and the *Garrard Building*. Four handsome bridges connect Columbus with its suburbs in Alabama. From Columbus the route is *via* the Columbus & Western Div. of the Central R. R. of Georgia, which connects with the main line at *Opelika* (29 miles). Beyond Opelika the route is the same as that described previously (see page 539). By the present route the total distance from Richmond to Mobile is 1,006 miles.

Mobile.

Hotels, etc.—The *Battle House*, cor. Royal and St. Francis Sts., the *Hotel Royal*, and the *Farley House* are the leading hotels. *Street-cars* traverse the city, and make all points easily accessible. Besides the routes described above, Mobile is reached from the North by Route 122, and from the West by Louisville & Nashville R. R. (Route 126) and by Mobile & Ohio R. R., and by the Mobile & Birmingham Railway, the southern branch of the East Tennessee, Virginia, and Georgia system. The Plant S. S. Line makes direct connection with the Mobile & Ohio R. R., and steamers leave four times a week for Key West and Hávaña. The route from Mobile to New Orleans is described in the present route. *Steamers* ply between Mobile and the interior by way of the Alabama, Tombigbee, and other rivers.

Mobile, the largest city and only seaport of Alabama, is situated on the W. side of Mobile River, immediately above its entrance into Mobile Bay, 30 miles from the Gulf of Mexico, in lat. 30° 42′ N, and lon. 88°

W. Its site is a sandy plain, rising as it recedes from the river, and bounded, at a distance of a few miles, by high and beautiful hills. The corporate limits of the city extend 6 miles N. and S. and 2 or 3 miles W. from the river. The thickly inhabited part extends for about a mile along the river, two miles and a half back toward the hills. It is laid out with considerable regularity, and the streets are generally well paved and delightfully shaded. *Fort Morgan* (formerly Fort Bowyer), on Mobile Point, and *Fort Gaines*, on the E. extremity of Dauphine Island, command the entrance to the harbor, which is about 30 miles below the city. On Mobile Point is also a lighthouse, the lantern of which is 55 ft. above the sea-level. The remains of several batteries erected during the war may be seen in and about the harbor; and on the E. side of Tensas River are the ruins of *Spanish Fort* and *Fort Blakely.*

Mobile was the original seat of French colonization in the Southwest, and for many years the capital of the colony of Louisiana. Historians differ as to the precise date of its foundation, though it is known that a settlement was made a little above the present site of the city at least as early as 1702. Many of the first settlers were Canadians. In 1723 the seat of the colonial government was transferred to New Orleans. In 1763, Mobile, with all that portion of Louisiana lying E. of the Mississippi and N. of Bayou Iberville, Lakes Maurepas and Pontchartrain, passed into the possession of Great Britain. In 1780 England surrendered it to Spain, and that Government made it over to the United States in 1813. It was incorporated as a city in 1819, the population being then about 800. Mobile was one of the last points in the Confederacy occupied by the Union forces during the late war, and was not finally reduced until April 12, 1865, three days after the surrender of General Lee. On August 5, 1864, the harbor fortifications were attacked by Admiral Farragut, who ran his fleet past the forts, and closed the harbor against blockade-runners, though he failed to capture the city itself. The trade of Mobile has of late been much improved by the Government work of deepening the ship-channel from deep water in Mobile River to deep water in the lower bay. The Government's present project is to give this channel a depth of 23 ft. Vessels drawing 18 ft. load and discharge at the wharves. The chief business is the receipt and shipment of cotton, coal, and lumber, and naval stores. The manufactures include carriages, furniture, cypress shingles, staves, barrels, rope and twine, boxes, pressed brick, street cars, cigars, leather and saddlery, cotton-seed oil, foundries and machine-shops. The population in 1890 was 31,076.

Government St. is the finest avenue and favorite promenade of the city. *Bienville Park,* between Dauphin and St. Francis Sts., is adorned with live-oaks and other shade-trees, also a handsome fountain. The **Custom-House,* which also contains the *Post-Office,* at the cor. of Royal and St. Francis Sts., is the finest, largest, and most costly public edifice in the city. It is built of granite, and cost $250,000. The *New Exchange,* occupied by the Cotton Exchange Association and the Chamber of Commerce, is at the corner of St. Francis and Commerce Sts. The *Theatre* and *Market-House,* with rooms in the upper story for the municipal officers, are in Royal St. The *Battle House* presents an imposing façade of painted brick, opposite the Custom-House. The *Court-House,* built in 1888, cor. Government and Royal Sts., and *Temperance Hall,* cor. St. Michael and St. Joseph Sts., are conspicuous buildings. The large public school, **Barton Academy,* in Government St., is a handsome building surmounted by a dome. Of the church edifices the most notable are the *Cathedral of the Immaculate Conception* (Roman Catholic), which has one of the most impressive interiors that can be seen in the South, in Claiborne St., between Dauphin and Conti; *Christ Church* (Episcopal), cor. Church and St. Emanuel Sts.; *Trinity* (Episcopal), with massive campanile; the *St. Francis Street Baptist Church*, in St. Francis, between Claiborne and Franklin Sts.; the *First Presbyterian*, cor. Government and Jackson Sts.; and the *Jewish Synagogue*, in Jackson St., between St. Michael and St. Louis Sts. The principal charitable institutions are the *City Hospital*, the *United States Marine Hospital*, and four Orphan Asylums. The *Medical College* is prosperous.

Spring Hill is a pleasant suburban retreat 6 miles W. of the city (reached by the St. Francis St. steam-cars). The *College of St. Joseph*, a Jesuit institution, is located here. It was founded in 1832 by Bishop Portier, and has a fine building 375 ft. long surmounted by a tower from which noble views may be obtained. The college has a library of 8,000 volumes and a valuable collection of scientific apparatus. A statue of the Virgin Mary, brought from Toulouse, France, stands in rear of the building. The **Gulf Shell-Road** affords a delightful drive, 7 miles in length, along the shore of the bay. *Arlington* and *Frascati* are resorts on the bay shore, and are reached by the drive or by street-cars.

123. Richmond to New Orleans via Mobile.

BETWEEN Richmond and Mobile either of the routes described in Route 122 may be taken. From Mobile the route is *via* the Louisville & Nashville R. R. (distance, 141 miles). There are no important stations on the line, but the journey is one of great interest from a scenic point of view. "Nothing in lowland scenery," says Mr. Edward King, in his "Great South," "could be more picturesque than that afforded by the ride from New Orleans to Mobile, over the Mobile & Texas R. R. [the name of this road before consolidation], which stretches along the Gulf line of Louisiana, Mississippi, and Alabama. It runs through savannas and brakes, skirts the borders of grand forests, offers here a glimpse of a lake and there a peep at the blue waters of the noble Gulf; now clambers over miles of trestle-work, as at *Bay St. Louis*, *Biloxi* (the old fortress of Bienville's time) and *Pascagoula;* and now plunges into the very heart of pine-woods, where the foresters are busily building little towns and felling giant trees, and where the revivifying aroma of the forest is mingled with the fresh breezes from the sea." (See "The Gulf Coast" in HAND-BOOK OF WINTER RESORTS.)

124. Washington to Mobile and New Orleans.
a. Via Great Southern Mail.

THE first section of this route, so historic in its importance from Washington to Lynchburg, is the same as that described in 122 *a*.

Beyond Lynchburg the route is *via* the Norfolk & Western R. R. This road passes through southwestern Virginia, famous for its wild scenery and inexhaustible mineral resources. It intersects or passes between the parallel ramparts of the great range of the Alleghanies, the backbone of the Atlantic slope of the continent, as the Rocky Mountains are the backbone of the Pacific slope; and scenes full of picturesque grandeur meet the eye of the traveler on every side. At *Liberty* (25 miles from Lynch-burg, 203 miles from Washington) the views are very fine. The Blue Ridge runs across the N. W. horizon, and attains its greatest height in the famous *****Peaks of Otter,** about 7 miles distant. These peaks are isolated from the rest of the range, and, with the exception of some peaks in North Carolina, are the loftiest in the Southern States (4,200 ft. above the plain, 5,307 above the sea). The S. peak is easily ascended, and affords a magnificent view. At *Bonsack's* (225 miles from Wash-ington, 47 miles from Lynchburg) are the much-frequented *Coyner's Springs* (see Route 130). From Roanoke direct rail connection can be made to the wonderful *Natural Bridge* (see for other routes Sub-Route *b*).

The Spring region is then passed through, which is described in Route 130. **Bristol** (382 miles; *Hotel Fairmont*) is a lively town of some 6,226 inhabitants, situated on the boundary line between Virginia and Tennessee. Here the train takes the track of the E. Tennessee, Vir-ginia & Georgia R. R., which runs S. W. through a highly picturesque portion of East Tennessee. *Greenville* (438 miles) is a pretty place with 1,779 inhabitants, and seat of a well-known college. **Knoxville** (512 miles; *Vendome Palace*) is a city of 22,535 inhabitants, situated at the head of steamboat navigation on the Holston River, 4 miles below the mouth of the French Broad. It is built on a healthy and elevated site, commanding a beautiful view of the river and surrounding coun-try. It is the principal commercial place in E. Tennessee, and has some important manufactures. The *East Tennessee University*, with which is connected the State Agricultural College, is located here; also the *Knoxville University* (Methodist) and the *Freedmen's Normal School* (Presbyterian). The *State Institution for the Deaf and Dumb* is a prominent edifice.

At *Cleveland* (594 miles) the road branches, one line running W. in 30 miles to **Chattanooga** (see Sub-Route *c*), while the present route continues S. W. to Rome (662 miles). **Dalton** is a mountain-environed town of 3,046 inhabitants, at the junction of three rail-ways. It was the initial point of the famous campaign of 1864, was strongly fortified by Gen. J. E. Johnston, but the position was flanked by Gen. W. T. Sherman, and consequently evacuated by the Confed-erates on May 12, 1864. **Rome** (*Central Hotel*), the most important town of Northern Georgia, is situated on Coosa River, and has a popu-lation of 6,957. It has a growing mercantile and manufacturing busi-ness, ice, cotton, and oil factories, and its agricultural interests are con-siderable.

The *Alabama Division of the E. Tenn., V. & G. R. R.* connects with the main branch at Rome, and affords another through route to Mobile and New Orleans. It extends S. W. through Georgia and Alabama, and the distance from Rome to Selma is 196 miles. The principal places on the line are *Cave Spring*, the seat of the State Asylum for the Deaf and Dumb; *Talladega*, the seat of the Alabama State Asylum for the Deaf and Dumb; *Shelby Springs*, with valuable mineral waters; *Calera*, at the crossing of the Louisville & Nash-ville System; and **Selma** (*St. James Hotel*), a busy manufacturing city of 7,622 inhabitants, on the right bank of the Alabama River, 95 miles below Montgom-ery. From Selma the traveler can reach New Orleans *via* Mobile or *via* Me-ridian, Miss.

From Dalton the route we are describing is *via* Western & Atlantic R. R., which traverses a region interesting as the arena of one of the most obstinate struggles of the civil war—the campaign, namely, between Sherman and Johnston, which culminated in the fall of Atlanta (see Route 122 *a*). This campaign began in the vicinity of Chattanooga, and extended directly down the line of the railway to Atlanta. Mementos of the struggle may be seen by the traveler on the crests of nearly every one of the huge ranges of hills which mark the topography of the country, in the shape of massive breastworks and battlements, which time and the elements are fast obliterating. At *Dalton*, as we have already said, occurred the initial struggle of the campaign. *Resaca* (15 miles beyond Dalton) was the place of the next stand made by Johnston, and was the scene of severe and indecisive fighting between the two armies; it was finally captured by a flank march on the part of Sherman. Retreating from this point, Johnston took a position at *Allatoona* (44 miles below), which was considered impregnable; but it too was successfully flanked, and the Confederates forced back to the Chattahoochee and Atlanta. On the line is *Cartersville* (52 miles from Dalton), which has a population of 3,171. *Marietta* (20 miles from Atlanta) is the most elevated point on the line, has a delightful climate in summer, and is then much resorted to. It contains some 3,384 inhabitants, and is the site of a National Cemetery, in which are buried 10,000 Federal soldiers. *Kennesaw Mountain* (2¼ miles distant) overlooks a vast extent of country, and played an important part in the campaign in this vicinity. **Atlanta** (721 miles from Washington) is described in Route 122 *a*. From Atlanta the route is the same as in the above-mentioned route. (The total distance from Washington to Mobile by this route is 1,076 miles; to New Orleans, 1,217 miles.)

b. Via Shenandoah and Roanoke Div. of the Norfolk & Western R. R.

This route, a little longer than the preceding, is *via* the Baltimore & Ohio R. R. to *Shenandoah Junction* (61 miles); thence *via* the Shenandoah and Roanoke Div. of the Norfolk & Western R. R. to Roanoke (217 miles), an industrial center of 16,159 inhabitants; thence as in Sub-Route *a*. The only important point on the Baltimore & Ohio R. R. before reaching the junction is **Harper's Ferry** (55 miles), which has been described (see Route 70). At Shenandoah Junction we take the old Shenandoah Valley road. *Charlestown*, 6 miles from Shenandoah Junction and 12 miles S. W. from Harper's Ferry, is celebrated as the scene of John Brown's execution, December 2, 1859. The road passes through a very beautiful and fertile region, though the towns are small and not specially notable till we reach **Luray** (127 miles), remarkable for its great subterranean cavern. There is a spacious hotel here, known as the *Luray Inn.* The caverns are situated about one mile from the station, and offer a spectacle not to be surpassed by any similar wonder, in vastness, variety, and beauty. The unsupported spans are vaster than were any of the World's Fair Buildings at Philadelphia. The roof of its highest room is 100 ft. high, and from this is suspended the most enormous stalactite in the world.

35

Every form known to similar subterranean caverns is present at Luray, with peculiar forms known only to this cave. It is asserted that the formation is older than the Tertiary Period. It was accidentally discovered some years ago in digging down through a sink-hole. At *Waynesboro Junction* (175 miles) connection is made with the Chesapeake & Ohio R. R. Passing a number of unimportant stations, we arrive at **Natural Bridge** (230 miles). This wonder of nature is situated in Rockbridge County, Va., at the extremity of a deep chasm in which flows the little stream called Cedar Creek, across the top of which, from brink to brink, there extends an enormous rocky stratum, fashioned into a graceful arch. The bed of the stream is more than 200 ft. below the surface of the plain, and the sides of the chasm, at the bottom of which the water flows, are composed of solid rock, maintaining a position almost perpendicular. The middle of the arch is 40 ft. in perpendicular thickness, which toward the sides regularly increases with a graceful curve, as in an artificial structure. It is 60 ft. wide, and its span is almost 90 ft. Across the top of the Bridge passes a public road, and as it is in the same plane with the neighboring country, one may cross it in a coach without being aware of the interesting pass. The most imposing view is from about 60 yards below the Bridge, close to the edge of the creek; from that position the arch appears thinner, lighter, and loftier. A little above the Bridge, on the W. side of the creek, the wall of rock is broken into buttress-like masses, which rise almost perpendicularly to a height of nearly 250 ft., terminating in separate pinnacles which overlook the Bridge. On the abutments of the Bridge there are many names carved in the rock of persons who have climbed as high as they dared on the face of the precipice. Highest of all for nearly three quarters of a century was that of George Washington, who when a youth ascended to a point never before reached, but which was surpassed in 1818 by James Piper, a student in Washington College, who actually climbed from the foot to the top of the arch. The main line of the Chesapeake & Ohio R. R. crosses here, running to *Clifton Forge*, in W. Va. In 41 miles (271 miles from Washington) we arrive at the village of *Roanoke*, where connection is made with the Norfolk & Western R. R. Thence the route is the same as in Sub-Route *a*.

c. Via East Tennessee, Virginia & Georgia R. R. (Chattanooga and Meridian).

As far as *Cleveland* (594 miles) this route is the same as in Sub-Route *a*. From Cleveland a branch of the East Tennessee, Virginia & Georgia R. R. runs W. in 30 miles to **Chattanooga** (*Merchants' Hotel, Read House, Southern Hotel*, and *Stanton House*), a city of 29,100 inhabitants, situated on the Tennessee River near where the S. boundary of Tennessee touches Alabama and Georgia. Seven railroads converge here, and the river is navigable to this point by steamboats for 8 months of the year, and by small boats at all times. Chattanooga is the shipping-point for most of the surplus productions of East and of a portion of Middle Tennessee, and contains a large number of iron-mills, blast-furnaces, and cotton-factories; a manufactory of railroad cars, sash and blind factories, tanneries, rolling-mills, and the workshops of the Alabama Great Southern R. R. It is the seat of the *Methodist University.* During the war Chattanooga was an important strategic point. Above the city the celebrated * **Lookout Mountain** towers 2,200 ft. above the sea. It was on this mountain that the so-called battle "above the clouds" was fought. The summit of the mountain is reached by a picturesque turnpike-road, which leads through a variety of interesting scenes, and a railroad is near completion. The points on Lookout best worth visiting are Lake Seclusion, Lulah Falls, Rock City, and the battle-field.

From Chattanooga the route is *via* the Alabama Div., which runs S. W. to Meridian, in the State of Mississippi. **Birmingham** (143 miles; *Florence Hotel, Hotel Caldwell, Wilson House*), incorporated in 1879, has a population of 26,178, the chief industry being iron manufactures and mining. It contains five furnaces, one rolling-mill, iron-foundries, chain-works, etc. It has seven public schools, fourteen churches, a female college, and an opera-house. Likewise it is the terminus of the Kansas City, Fort Scott & Memphis R. R. **Tuscaloosa** (198 miles; *Washington House*) is a city of some 4,215 inhabitants, situated on the left bank of the Black Warrior River, at the head of steamboat navigation. It is the commercial center of a district rich in resources, and has a considerable trade in cotton, wheat, coal, etc. The streets of the city are wide and well shaded. A mile distant are the grounds of the *University of Alabama ;* the buildings, with their contents, were burned in 1865, and have been only partially restored. The *Alabama Insane Hospital,* about a mile beyond the University, has a front of 780 ft., with extensive out-buildings and grounds. The city takes its name from the Indian chief Tuscaloosa ("black warrior"), who was defeated by De Soto in the bloody battle of Malvila, Oct. 18, 1540. From 1826 to 1846 it was the capital of the State. *Eutaw* (233 miles) is a pretty town, capital of Greene County, situated 3 miles W. of the Black Warrior River. The adjacent country is one of the most fertile portions of the State, and Eutaw is surrounded by rich plantations. *York* (262 miles) is at the junction with a branch which runs across the State from Selma to Meridian. At *Cuba* (274 miles) the road crosses the boundary-line and enters the State of Mississippi, and 21 miles beyond reaches **Meridian,** whose importance is due chiefly to its position at the junction of several railways. It was captured by Gen. Sherman on Feb. 16, 1864, and, according to his own account, his troops accomplished "the most complete destruction of railways ever beheld." At Meridian the passengers for Mobile take the Mobile & Ohio R. R., which runs S. to Mobile in 135 miles. Those going to New Orleans can go *via* New Orleans & Northeastern Div. (Queen & Crescent System), which runs direct from Meridian to New Orleans, a distance of 196 miles.

125. Charleston or Savannah to Mobile or New Orleans via Savannah, Florida & Western R. R.

ALL northern points are brought in close connection with Mobile and New Orleans *via* Charleston by this route. For the routes to Charleston, see Routes 111 and 112. From Charleston to Savannah, see Route 113. At Savannah the train takes the track of the main line of the Savannah, Florida & Western R. R. as far as *Bainbridge Junction* (342 miles). This route has been described in Route 114. *Chattahoochee*, Fla. (373 miles), on the Appalachicola River near the confluence of the Flint, is the seat of the *State Penitentiary* and *Lunatic Asylum*, and is the present W. terminus of the Savannah, Florida & Western R. R. Crossing the river on a fine bridge, the train takes the track of the Louisville & Nashville R. R., which runs 161 miles across the N. W. portion of Florida to Pensacola. At Careyville, 100 miles from Pensacola, there is connection with local steamboats on the Choctawhatchee River, and at Pensacola (see Route 121) the train takes the track of the main line of the Louisville & Nashville R. R. for Mobile and New Orleans (see Route 123). (For description of New Orleans, see Route 128.)

126. Louisville to Mobile and New Orleans.

By the Louisville & Nashville R. R. This is one of the great highways of travel and traffic between the Northern and Southern States. At Louisville close connections are made with the various routes converging there from the North and West (see Routes 77 and 83). Through palace-cars are run without change from New York, Philadelphia, Baltimore, and Washington to New Orleans; and from Cincinnati, Louisville, and St. Louis to Montgomery, Mobile, and New Orleans. The time from Louisville to Mobile is about 22 hours; to New Orleans, about 25 hours. Distances : to Cave City, 85 miles ; to Memphis Junction, 118 (Memphis, 377) ; to Nashville, 185 ; to Decatur, 308 ; to Birmingham, 395 ; to Calera, 428 ; to Montgomery, 490 ; to Mobile, 670 ; to New Orleans, 811.

LOUISVILLE is described in Route 77. Leaving Louisville, the train runs S. W. across a productive and populous portion of Kentucky, then crosses Tennessee from N. to S., and continues S. through central Alabama. *Bardstown Junction* (22 miles) is the point whence the Bardstown Branch runs to *Bardstown* (17 miles distant). At *Lebanon Junction* (30 miles) the Knoxville Branch diverges. *Mumfordsville* (73 miles) is a pretty village on the right bank of Green River, which is here spanned by a fine bridge. This neighborhood was the scene of numerous encounters between Generals Buell and Bragg in the campaign of 1862. At *Glasgow Junction* (90 miles) all trains make connection with the Mammoth Cave R. R., running to the * * **Mammoth Cave,** 12 miles distant, and allow a "stop over" to visitors. At the *Mammoth Cave Hotel*, near the cave-entrance, guides, boats, etc., can be procured. The mouth of the cave is reached by passing down a wild, rocky ravine through a dense forest ; it is an irregular, funnel-shaped opening, from 50 to 100 ft. in diameter at the top, with steep walls about 50 ft. high. The cave, which is the largest known, extends about 9 miles ; and it is said that to visit the portions already explored requires from 150 to 200 miles of travel. This vast interior contains a succession of marvelous avenues, chambers, domes, abysses, grottoes, lakes, rivers, cataracts, etc., which for size and wonderful appearance are unsurpassed. The rocks present numerous forms and shapes of objects in the external world; while stalactites and stalagmites of gigantic size and fantastic form abound, though not so brilliant and beautiful as are found in some other caves. Two remarkable species of animal life are found in the cave, in the form of an eyeless fish and an eyeless craw-fish, which are nearly white in color. Another species of fish has been found with eyes, but totally blind. Other animals known to exist in the cave are lizards, frogs, crickets, rats, bats; etc., besides ordinary fish and craw-fish washed in from the neighboring Green River. The atmosphere of the cave is pure and healthful; the temperature, which averages 59°, is about the same in winter and summer, not being affected by climatic changes without. To describe the cave in detail would require a volume, and, after all, the visitor would have to intrust himself to the guides. These give him the choice between the *Short Route* (fee, $2) and the *Long Route* (fee, $3). They carry lamps and torches, and impart all the needful information regarding special localities.

"The stars were all in their places as I walked back to the hotel. I had been 12 hours under ground, in which time I had walked about 24 miles. I had lost a day, a day with its joyous morning, its fervid noon, its tempest, and its angry sunset of crimson and gold, but I had gained an age in a strange and hitherto unknown world—an age of wonderful experience, and an exhaustless store of sublime and lovely memories. Before taking a final leave of the Mammoth Cave, however, let me assure those who have followed me through it that no description can do justice to its sublimity, or present a fair picture of its manifold wonders. It is the greatest natural curiosity I have ever visited, Niagara not excepted; and he whose expectations are not satisfied by its marvelous avenues, domes, and starry grottoes, must either be a fool or a demigod."— BAYARD TAYLOR.

Twenty-nine miles beyond Cave City is **Bowling Green** (114 miles), a thriving town of 7,803 inhabitants, at the head of navigation on Barren River. At the beginning of the civil war Bowling Green was regarded as a point of great strategic importance, and was occupied in Sept., 1861, by a large force of Confederates for the purpose of defending the approach to Nashville. After the capture of Fort Henry by the Federals (Feb. 6, 1862) the Confederates found themselves outflanked, and were obliged to evacuate the town. At *Memphis Junction* (118 miles) the Memphis Line diverges from the main line, and runs, in 259 miles, to **Memphis** (see Route 133). At *Edgefield Junction* (175 miles) connection is made with the St. Louis, Evansville & Nashville Div. of the Louisville & Nashville R. R., which forms a short line between St. Louis and points in the Southern States. *Edgefield* (184 miles) is now a portion of Nashville, on the other bank of the river.

Nashville (*Bailey House, Hotel Duncan, Linck's Hotel, Maxwell House, Nicholson House*) is the capital of Tennessee, and the largest city in the State in point of population (76,168), and is situated on both banks of the Cumberland River, 200 miles above its junction with the Ohio. The land on which the city is built is irregular, rising in gradual slopes, with the exception of *Capitol Hill*, which is more abrupt. This eminence is symmetrical, resembling an Indian mound, and overlooks the entire city. Nashville is regularly laid out, with streets crossing each other at right angles. It is generally well built, and there are numerous imposing public and private buildings. Among the former is the * **Capitol**, situated on Capitol Hill, and constructed inside and out of a beautiful variety of fossiliferous limestone. It is three stories high, including the basement, and is surmounted by a tower 206 ft. in height. The dimensions of the whole building are 239 by 138 ft.; it was erected in 1845 at a cost of over $1,000,000. It is approached by four avenues, which rise from terrace to terrace by broad marble steps. The * *Court-House* is a large building on the Public Square, with an eight-columned Corinthian portico at each end, and a four-columned portico at each side. The *Custom-House* is a handsome structure, in the Gothic style, of limestone and granite, costing nearly $1,000,000. The *State Penitentiary* has spacious stone buildings occupying three sides of a hollow square inclosed by a massive stone wall, within which are numerous workshops. The *State Institution for the Blind* is located at Nashville, and the *State Hospital for the Insane* is about 6 miles distant. The educational institutions of the city are numerous and important. The buildings of *Vanderbilt University* (named in honor of the late Cornelius Vanderbilt, of New York) deserve attention. This institution is under the control of the Methodist Episcopal Church, South, and comprises academic, engineering, theological, law, medical, dental, and pharmaceutical departments, and is the largest institution in the South, having over 800 students. The campus covers 76 acres and contains 20 buildings. The *University of Nashville* has over 300 students and a library of 12,000 volumes. The main building is a handsome Gothic edifice of stone. The Medical Department also has a fine building and museum. It operates in connection with the Peabody Normal College, which is maintained by the income derived from the educational fund left by George Peabody. *Fisk University* was established in 1866 by Northern philanthropists for the colored people of the State. *Roger Williams University* is another large institution for colored people, under the patronage of the Baptist Church. The *Tennessee Central College* (Methodist), also for colored people, was established in 1866. The Roman Catholic Church has 3 large academies. The *Watkins Institute* is a fine building, containing the collections of the Tennessee Historical Association and the Nashville Art Association, also the Haward Library, which is one of the most complete in the South. The city is lighted with gas and electricity, is supplied with water by expensive works, and has several lines of street railway—about 50 miles of track entirely " electric system." The railways converging here, and the river, enable the city to command the trade of an extensive region, and its manufactures are varied and important. *The Hermitage*, the celebrated residence of Andrew Jackson, is 10 miles E. of Nashville.

The Battle of Nashville.—In November, 1864, the Confederate General Hood, having lost Atlanta, placed his army in Gen. Sherman's rear and began an invasion of Tennessee. After severe fighting with Gen. Schofield on Nov. 30, he advanced upon Nashville and shut up Gen. Thomas within its fortifications. For two weeks little was done on either side. When Thomas was fully ready, he suddenly sallied out on Hood, and in a terrible two days' battle (Dec. 15 and 16) drove the Confederate forces out of their intrenchments.

Between Nashville and Montgomery there is little to attract the tourist's attention. The country traversed offers few picturesque features, and the towns along the line are for the most part unimportant. The largest of them is **Columbia** (48 miles beyond Nashville), a flourishing town of 5,370 inhabitants, situated on the left bank of Duck River, in the midst of a fertile and productive region. It is the seat of a female Athenæum, a female institute, several schools, and a conference college. At *State Line* (281 miles) the train leaves Tennessee and enters Alabama. *Decatur* (308 miles) is a neat village at the junction of the East Tennessee, Virginia & Georgia R. R. on the route from Memphis to Charleston. *Birmingham* (395 miles) is at the crossing of the Alabama Great Southern R. R. of the Queen & Crescent System (Route 124 *b*), and *Calera* (428 miles) is at the crossing of the Alabama Div. of the E. Tennessee, Virginia & Georgia R. R. (see Route 124 *a*). **Montgomery** (490 miles) has been described in Route 122 *a*. From Montgomery to Mobile and New Orleans the route is the same as in Route 122 *a*.

127. Chicago and St. Louis to New Orleans.

By the Southern Division of the Illinois Central R. R., which is now a through route. This is one of the main trunk lines between the Northern and Southern States, and palace-cars are run through without change from Chicago, Cincinnati, Louisville, and St. Louis, and with but one change from New York, Boston, Philadelphia, and Baltimore. Two trains are run daily, and the time from Chicago to New Orleans is about 50 hours ; from St. Louis to New Orleans, about 38 hours. Distances : Chicago to Cairo, 305 miles ; St. Louis to Cairo, 140 miles ; *Cairo* to Milan, 86 miles ; to Jackson, Tenn., 109 ; to Bolivar, 138 ; to Grand Junction, 156 ; to Grenada, 256 ; to Canton, 344 ; to Jackson, Miss., 368 ; to Magnolia, 453 ; to New Orleans, 550.

FROM Chicago to Cairo this route is *via* the Chicago Division, and has been described in Route 84. From St. Louis the route is *via* the St. Louis & Cairo Short Line, which runs S. E. from St. Louis, and connects at Du Quoin with the main line of the Illinois Central R. R. At **Cairo** (see Route 84) the Ohio River is crossed, and the road runs due S. across portions of Kentucky and Tennessee and through Central Mississippi. The country traversed is for the most part populous and pleasing, but there are no large cities *en route*, and very few important towns. At *Fulton* (44 miles from Cairo) connection is made with the Western Div. of the Newport News & Mississippi Valley R. R., and at *Martin's* (55 miles) with the Nashville, Chattanooga & St. Louis R. R. **Milan** (86 miles) is at the crossing of the Memphis Div. of the Louisville & Nashville R. R., 93 miles from Memphis (see Route 126). The cars from Cincinnati and Louisville going south on the Illinois Central R. R. run through *via* Milan. Twenty-three miles beyond Milan is **Jackson** (*Robinson House*), the largest city in this section of Tennessee, with a population of 10,039. It is pleasantly situated on the Forked Deer River, in the midst of a fertile region, and has a large and growing trade. There are several manufacturing establishments, including the extensive machine-shops of the Louisville & Nashville R. R. Jackson is the seat of the *West Tennessee College*, which is in a prosperous condition, and of a Methodist female institute. *Bolivar* (138 miles) is a handsome and thriving town of 1,100 inhabitants, situated 1 mile S. of the Hatchee River, which is navigable by steamboats for 6 to 9 months of the year. *Grand Junction* (156 miles) is at the crossing of the E. Tennessee, Virginia & Georgia R. R. on the way from Memphis to Charleston. Shortly beyond Grand Junction the train crosses the State line and enters Mississippi, soon reaching the flourishing town of *Holly Springs* (181 miles), which is noted for its educational institutions and the pleasing scenery adjacent. *Grenada* (256 miles) is pleasantly situated on the Yallowbusha River, at the head of steamboat navigation. It contains a U. S. land-office and several churches. Connection is made here with the Memphis Div. of the Ilinois Central R. R. *Canton* (344 miles) is a neat and lively village; and 24 miles beyond the train reaches **Jackson** (*Edwards House*), the capital of the State of Mississippi. It is regularly built upon undulating ground on the W. bank of Pearl River, and has some 5,920 inhabitants. The * *State-House* is a very handsome edifice, erected at a cost of $600,000. The other chief public buildings are the *Executive Mansion*, the *State Lunatic Asylum*, the *State Institutions for the Deaf, Dumb, and Blind*, and the *City Hall*. The *State Penitentiary*, a spacious and handsome edifice, was nearly destroyed during the civil war, but is to be rebuilt. The *State Library* contains 15,000 volumes. Jackson was captured by General Grant on May 14, 1863, after a battle with General Johnston, in which the Confederates were defeated. The railroad depots, bridges, arsenals, workshops, storehouses, and many residences were destroyed. Between Jackson and New Orleans there are numerous small towns, but none requiring mention.

128. New Orleans.

Hotels.—The *St. Charles Hotel*, in St. Charles St., between Gravier and Common, is the largest in the city. The *Hôtel Royal*, bounded by St. Louis, Royal, and Chartres Sts. The *Hôtel Vonderbanck*, in Magazine St., between Gravier and Natchez, is on the European plan. The *Lafayette Hotel*, 130 Camp St.; the *Hôtel de la Louisiane*, 107 and 109 Custom-House St.; the *Hôtel Denéchaud*, cor. of Perdido and Carondelet Sts.; the *Christian Women's Exchange* (for women only), 1 South St.; and the *Hôtel des Étrangers*, 129 Chartres St., are also excellent houses. All the hotels make considerably lower rates to guests remaining a week or more, though most are kept on the European plan. Good board may be obtained in all parts of the city at from $6 to $20 a week.

Restaurants.—Of restaurants, New Orleans has some of the best in America; in many of them is still practiced the famous creole *cuisine* of ante-war times. The most noted are *Moreau's*, in Canal St.; *Bezaudon's*, 107 and 109 Custom-House St.; *Cassidy's*, 174 Gravier St.; the *Cosmopolitan*, 13 and 15 Royal St.; *Fabacher's*, 23 Royal St.; the *Gem*, 17 Royal St.; *Harris's*, 31 Natchez St.; *Lamothe's*, 23 St. Charles St.; *Nicholl's*, 74 Camp St.; the *Acme*, 9 and 11 Royal St.; *Vonderbanck's*, 126 to 130 Common St.; *Antoine's*, 65 St. Louis St.; and *Denechaud's*, 8 Carondelet St. In the French quarter *cafés* are to be found in nearly every block.

Modes of Conveyance.—The *street-car* system of New Orleans is perhaps the most complete in the country. Starting from the central avenue—Canal Street—tracks radiate to all parts of the city and suburbs, and passengers are

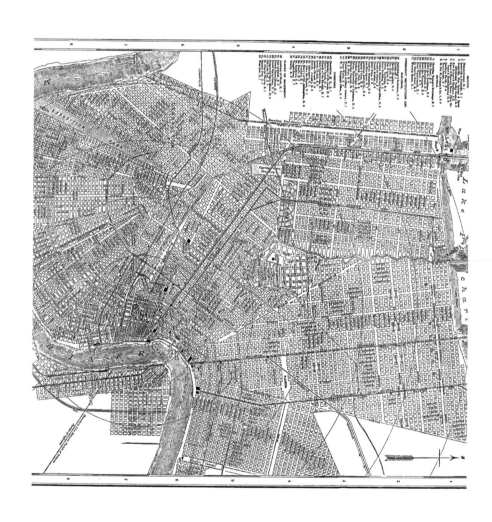

carried to any point within the city limits for 5c., except to Carrollton, whither the fare is 10c. *Omnibuses* attend the arrival of trains and steamers, and convey passengers to the hotels, etc. (fare, 25c.). *Carriages* and *cabs* can be found at the stands in front of the St. Charles and other leading hotels. Fare, $1 an hour ; $5 for the forenoon or afternoon. The best plan for strangers is to hire a suitable conveyance by the hour and discharge at the end of each trip. *Ferries* connect the city with Algiers, McDonoghville, and Gretna, on the opposite side of the river. Three steam-railroads connect the city with Lake Pontchartrain ; fare, 15c. for the round trip.

Theatres and Amusements.—The French *Opera-House*, cor. Bourbon and Toulouse Sts., has seats for 2,000, and is fitted up in the style of the Théâtre Français, Paris. The *Academy of Music*, in St. Charles St., between Poydras St. and Commercial Place, and the *St. Charles Theatre*, in St. Charles St., between Perdido and Poydras, are well appointed. The *Grand Opera-House* is in Canal St. Besides the theatres there are a score or more of halls in which entertainments of various kinds are given. The principal of these are the *Masonic Hall, Odd-Fellows' Hall, Washington Artillery Hall*, and *Grünewald Hall*, in Baronne St. near Canal. *Horse-races* occur at the Fair-Grounds race-track (reached by Shell-Road and 3 lines of horse-cars), and at Audubon Park (reached by the St. Charles Ave. line of horse-cars). Besides the regular sources of amusement which it enjoys in common with other cities, New Orleans is noted for its great displays, during the holiday and carnival season, of troops and processions of masqueraders. Among the many societies which contribute to these displays, the most famous are the *Mystick Crewe of Comus*, the *Knights of Momus*, the *Krewe of Proteus*, and the *Atlanteans*, Momus appearing on preceding Thursday, and Comus and Proteus on the night of * **Mardi Gras**, or Shrove-Tuesday. On the same day (Shrove-Tuesday), Rex, King of the Carnival, arrives with a large retinue, takes formal possession of the city for the nonce, and makes a grand display, followed by his staff, courtiers, and attendants, all mounted and dressed in gorgeous Oriental costumes. The processions are followed by receptions, tableaux, and balls, which are largely attended by the *élite* of the city, and by strangers sojourning there, who are generally the recipients of cards of invitation.

Clubs.—The prominent clubs in the city are the *Boston*, the *Pickwick*, the *Chess, Checkers, and Whist*, the *Liedertafel*, the *Commercial*, the *Harmony*, the *Union*, and the *Jockey Club*. The last-named has a fine house and beautifully decorated and cultivated grounds near the Fair-Grounds. Also the *Southern Athletic Club* and the. *Young Men's Gymnastic Club*. There is also a Woman's Club. The privileges of these clubs are obtained by introduction by a member.

Post-Office.—The Post-Office occupies the basement of the Custom-House, which fronts on Canal St., between Peters and Decatur Sts. It is open day and night for mailing, and from 6.30 A. M. to 6 P. M. for delivery ; Sundays, 9 A. M. to 12 M. There are four sub-stations, and letters may also be mailed in the lamp-post boxes, whence they are collected at frequent intervals.

NEW ORLEANS, the chief city, and commercial metropolis of Louisiana, is situated on both banks (but chiefly on the left) of the Mississippi River, 100 miles above its mouth, in latitude 29° 57′ N. and longitude 90° W. The older portion of the city is built within a great bend of the river, from which circumstance it derives its familiar title of the " Crescent City." In the progress of its growth up-stream, it has now so extended itself as to follow long curves in opposite directions, so that the river-front on the left bank presents an outline somewhat resembling the letter S. The statutory limits of the city embrace an area of some 187 square miles, but the actual city covers an area of about 41 square miles. It is built on land gently descending from the river toward a marshy tract in the rear, and considerably below the level of the river at high-water mark, which is prevented from overflowing by a vast embankment of earth, called the Levee. This Levee

is 15 ft. wide and 14 ft. high, is constructed for a great distance along
the river-bank, and forms a delightful promenade.

The site of New Orleans was surveyed in 1717 by De la Tour ; it was settled
in 1718, but abandoned in consequence of overflows, storms, and sickness ; was
resettled in 1723, held by the French till 1729, then by the Spaniards till 1801, and
by the French again till 1803, when, with the province of Louisiana, it was
ceded to the United States. It was incorporated as a city in 1804. The most
memorable events in the history of New Orleans are the rebellion against the
cession by France to Spain in 1763, the battle of January 8, 1815, in which the
British were defeated by Andrew Jackson, and the capture of the city by Admiral Farragut on April 24, 1862. In 1810, seven years after its cession to the
United States, the population of New Orleans was 17,243. In 1850 it had increased to 116,375 ; in 1860, to 168,675 ; in 1870, to 191,418 ; in 1880, to 216,140 ;
and in 1890, to 242,039. In the value of its exports and its entire foreign
commerce, New Orleans ranks close to New York, though several ports surpass
it in the value of imports. Not unfrequently more than 1,000 steamers and other
vessels from all parts of the world may be seen lying at the Levee. New Orleans is the chief cotton mart of the world ; and, besides cotton, it sends abroad
sugar, rice, tobacco, flour, pork, etc.

The streets of New Orleans, in width and general appearance, are
second to those of no city of its size. As far back as Claiborne St.
those running parallel to the river and to each other present an unbroken line from the lower to the upper limits of the city, a distance of
about 12 miles. Those at right angles to them run from the Mississippi toward the lake with more regularity than might be expected from
the very sinuous course of the river. **Canal St.** is the main business
thoroughfare and promenade. It is nearly 200 ft. wide, and has a grass-
plot 25 ft. wide and bordered with two rows of trees, extending in the
center through its whole length. Claiborne, Rampart, St. Charles, and
Esplanade Sts. are similarly embellished. *Royal, Rampart,* and *Esplanade Sts.* are the principal promenades of the French quarter.—The
favorite drives are out the *Shell-Road* to Lake Pontchartrain and over
an asphalt road to Carrollton ; also out St. Charles Ave., through the
residential quarter to Audubon Park.

Chief among the architectural features of New Orleans is the * **Custom-House.** This noble structure is built of Quincy granite brought
from the Massachusetts quarries. Its main front on Canal St. is 334 ft. ;
that on Custom-House St., 252 ft. ; on Peters St., 310 ft. ; and on Decatur St., 297 ft. Its height is 82 ft. The Marble Hall, or chief business
apartment, is 116 by 90 ft., and is lighted by 50 windows. The building was begun in 1848, and is not yet entirely finished. The *Post-
Office* occupies the basement of the Custom-House, and is one of the
most commodious in the country. The * **U. S. Branch Mint** stands
at the cor. of Esplanade and Decatur Sts. It is built of brick, stuccoed
in imitation of brown-stone, in the Ionic style, and, being 282 ft. long,
180 ft. deep, and 3 stories high, presents an imposing appearance. The
* **City Hall,** at the intersection of St. Charles and Lafayette Sts., is
the most artistic of the public buildings of the city. It is of white
marble, in the Ionic style, with a wide and high flight of granite steps
leading to a beautiful portico supported by 8 columns. The *City Library*
occupies suitable rooms in this building. The *Court-Houses* are on the
right and left of the Cathedral, in Jackson Square. They were con-

structed toward the close of the last century. The *Masonic Hall,* cor. St. Charles and Perdido Sts., is an imposing edifice, 103 by 100 ft. *Odd-Fellows' Hall* is a massive square structure in Camp St. opposite Lafayette Square. *City Courts,* on the site of the old Odd-Fellows' Hall, is one of the conspicuous buildings in the city. *Washington Artillery Hall* is in St. Charles St., between Julia and Girod, in which are given floral displays and other exhibitions. The *Cotton Exchange,* in Carondelet St., cor. Gravier, built at a cost of $500,000, is a magnificent structure. The *Sugar Exchange,* on the levee foot of Bienville St.; the *Produce Exchange,* in Magazine St., between Gravier and Natchez Sts.; the *Pickwick Club,* cor. Canal and Carondelet Sts.; the *Morris Building,* cor. Canal and Camp Sts.; the *Lyons Building,* cor. Gravier and Camp Sts.; the *Baldwin Building,* cor. Camp and Common Sts.; and the *Howard Library,* cor. Howard Ave. and Camp St. (one of the last works of Henry H. Richardson, of Boston), are handsome buildings. Adjoining the library is the *Annex,* occupied by the Louisiana Historical Association.

One of the most interesting church edifices in New Orleans is the old *Cathedral of St. Louis (Roman Catholic), which stands in Chartres St., on the E. side of Jackson Square. The foundation was laid in 1792, and the building completed in 1794 by Don Andre Almonaster, perpetual *regidor* of the province. It was altered and enlarged in 1850, from designs by De Louilly. The paintings on the roof of the building are by Canova and Rossi. The *Church of the Immaculate Conception* (Jesuit), cor. Baronne and Common Sts., is a striking edifice in the Moorish style. *St. Patrick's* (Roman Catholic) is a fine Gothic structure in Camp St., N. of Lafayette Square. Its tower, 190 ft. high, was modeled after that of the famous minster of York, England. The church of * *St. John the Baptist,* in Dryades St. between Clio and Calliope, is a fine building. The Episcopal churches are *Trinity,* cor. Jackson and Coliseum Sts., and *Christ,* cor. St. Charles Ave. and 6th St. The **First Presbyterian,** fronting on Lafayette Square, is a Gothic structure, much admired for its elegant steeple, which is the highest in the city. The *McGhee Church,* in Carondelet St. near Lafayette, is the principal of the Methodist Episcopal churches, South. The *Unitarian Church,* cor. St. Charles and Julia Sts., is a handsome building. The *Temple Sinai (Jewish synagogue), in Carondelet St. near Calliope, is one of the finest places of worship in the city. Party-colored bricks and pointing give its walls a light, airy appearance, and it has a handsome portico, flanked by two towers capped with tinted cupolas. The Gothic windows are filled with beautifully stained glass. One of the most interesting relics of the early church history of New Orleans is the old *Ursuline Convent* in Chartres St. This quaint and venerable building was erected in 1787, during the reign of Carlos III, by Don Andre Almonaster. It is now occupied by the archbishop, and is known as the "Archbishop's Palace." The *Chapelle St. Roch,* in the French quarter, is well worth a visit; and there young ladies who offer prayers on certain occasions are said to be rewarded by speedy marriage.

The **Tulane University of Louisiana,** formerly the University of Louisiana, but now largely endowed by the late Paul Tulane, occupies the south half of the square bounded by Tulane Ave., Baronne, and Dryades Sts. It embraces complete faculties of arts, letters, science, law, and medicine. The *Sophie Newcomb College for Girls* occupies the old Burnside Mansion, on Washington Ave., between Chestnut and Coliseum Sts. *Leland University, Straight University,* and the *Southern University* are exclusively for colored students, and give instruction of good grammar-school grade. The * **Charity Hospital,** in Tulane Ave., is one of the most famous institutions of the kind in the country. It was founded in 1784, has stood on its present site since 1832, and has accommodations for 800 patients. The *Hôtel Dieu,* half a mile farther back from the river, is a very fine hospital established by the Sisters of Charity, and supported entirely by receipts from patients, some of whom are, nevertheless, beneficiary. It occupies a full square, and is surrounded by a well-kept garden of shrubbery and flowers. The *Touro Infirmary,* in Prytania between Aline and Foucher Sts., principally supported by Jewish charity, but open to patients of all creeds, is a model hospital, built upon the pavilion plan, with a large garden in the center. A free clinic is operated in connection with the institution. Other prominent charitable institutions are the *Poydras Female Orphan Asylum* (in Magazine St.), the *Jewish Widows' and Orphans' Home* (cor. St. Charles and Peters Aves.), the *St. Anna's Widows' Asylum,* the *St. Vincent Orphan Asylum,* the *Shakespeare Alms-House,* the *German Protestant Asylum,* and the *Indigent Colored Orphan Asylum.* The *Howard Association* is one of the greatest charitable bodies in the world, its special mission being to labor for the relief of sufferers in epidemics, particularly the yellow fever.

Among the pleasure-grounds of the city is * *Jackson Square* (formerly known as the *Place d'Armes*), covering the center of the river-front of the old Town Plot, now Second District. It is adorned with beautiful trees and shrubbery, and shell-strewn paths, and in the center stands * Mills's equestrian statue of General Jackson. The imposing fronts of the cathedral and courts of justice are seen to great advantage from the river-entrance to the square. *Lafayette Square,* in the First District, bounded by St. Charles and Camp Sts., is another handsome inclosure. The fine marble front of the City Hall, the tapering spire of the Presbyterian Church, and the massive façade of Odd Fellows' Hall present a striking appearance. In the square is a fine white-marble statue of Franklin, by Hiram Powers. In Canal St., between St. Charles and Royal, is a colossal bronze statue of Henry Clay, by Joel T. Hart. *Annunciation Square* and *Lee* (formerly *Tivoli*) *Circle,* on St. Charles St., are worth a visit. The latter now contains the monument of Gen. Robert E. Lee, 65 ft. high, surmounted by a statue in bronze. *Margaret Place* is a triangle at the junction of Prytania and Camp Sts., and contains a memorial statue to Margaret Haughery, who founded an asylum and was distinguished for her charitable work. The *City Park,* near the N. E. boundary (reached by Canal St. and Ridge Road cars), embraces 150 acres. The *Great Exposition* of 1884–1885

was held in Audubon Park, in St. Charles Ave. In *Horticultural Hall*
an exhibition of plants is held, and in the immediate vicinity is a grove
of moss-covered oaks that are well worth visiting.

The *Cemeteries* of New Orleans are noteworthy. From the nature of
the soil, the tombs are above ground. Some of these are very costly and
beautiful structures, of marble, iron, etc.; but the great majority consist of cells, placed one above another, generally to the height of 7 or
8 ft. Each cell is only large enough to receive the coffin, and is her-
metically bricked up at its entrance as soon as the funeral rites are
over. In most instances a marble tablet, appropriately inscribed, is
placed over the brickwork by which the vault (or ",oven," as it is called
here) is closed.   There are 33 cemeteries in and near the city; of these
the *Metairie* (which contains the grave of Albert S. Johnston), *Cypress
Grove*, and *Greenwood* (on the Metairie Ridge, at the N. end of Canal
St.) are best worth visiting.   The *Metairie Cemetery* contains a fine
bronze equestrian statue of Gen. Johnston, erected as a monument to
the Army of Tennessee.   The *Monument to the Confederate Dead* in
Greenwood Cemetery, and the *Monument to the Union Dead* in the
National Cemetery at Chalmette, will attract visitors.

The great "sight" of New Orleans, and perhaps the most picturesque to be seen in America, is the * **French Market,** which
comprises several buildings on the Levee, near Jackson Square. The
best time to visit it is between 8 and 9 o'clock on Sunday morning, or
at 6 A. M. on other days. At break of day the gathering commences,
and it would seem as if all nations and tongues were represented in the
motley crowd which surges in and out until near 10 o'clock. French
is the prevailing language, and it will be heard in every variety, from
the silvery elegance of the polished creole to the childish jargon of the
negroes. Canal St. divides the city into two parts, the old and the new.
That toward the N. E. is the **French Quarter,** and there one meets
"odd little balconies and galleries that jut out from the tall, dingy,
wrinkled houses, peering into each other's faces as if in an eternal con-
fab."   Here are queer little shops—apothecaries', and musty stores
where old furniture, brasses, bronzes, and books are sold, and bird-
stores innumerable, where alligators are to be purchased as well. The
signs hereabout are all in French, and that of "avocat" seems predomi-
nant. Groups of men, chattering over their cigarettes, interfere with
pedestrianism in the alley, and stare with Gallic curiosity and gallantry
after every petticoated individual that passes. A priest in cassock—
and he plump and good-tempered, with face shining like a newly-peeled
onion—leans laughing against the black balustrade of one of the old
French houses. The residents of the poorer localities are great lovers
of potted flowers and singing birds. Some of the streets are rich with
color, owing to the brilliant red masses of geraniums that blossom boldly
in defiance of the hottest sun; and many a tiny bit of iron gallery projecting in curious fashion out of some tall window is transformed into
the coolest of arbors by looped-up cypress-vines. The mover's cart is
never seen in this quarter, and many a young matron lives in the house
her grandmother occupied. In these you will find "cool, red, sanded

floors, quaint spindle-legged dressing-tables, cabinets positively antique, rich with carvings and black with age, mosaic tables, pieced together long before the grand mosaic of these United States was designed, and over the tall, high, and narrow mantel-shelves with their heavy cornices and mimic Corinthian columns, reared about an absurdly small bit of a fireplace, are gigantic vases of Sèvres, odd bits of Bohemian ware, bottles, and absinthe-glasses." The **Levee** affords the visitor one of the most striking and characteristic sights of the Crescent City. For extent and activity it has no equal on the continent. The best points from which to obtain a view of the city and its environs are the roof of the St. Charles Hotel, the roof of the Custom-House, and the tower of St. Patrick's Church.

One of the most interesting spots in the vicinity of New Orleans is the **Battle-Field,** the scene of General Jackson's great victory over the British, Jan. 8, 1815. It lies 4½ miles S. of Canal St., and may be reached either by carriage along the Levee or by horse-cars. It is washed by the waters of the Mississippi, and extends back about a mile to the cypress-swamps. A marble monument, 70 ft. high and yet unfinished, occupies a suitable site overlooking the ground, and serves to commemorate the victory. A National Cemetery occupies the S. W. corner of the field. Between the Battle-Field and the city the *Ursuline Convent*, an imposing building 200 ft. long, overlooks the river. *Lake Pontchartrain*, 5 miles N. of the city, is famous for its fish and game It is 40 miles long and 24 miles wide. It is reached by three lines of railway with cars drawn by steam, and by drive in carriages on a fine shell-road. At the lake-side are extensive pleasure resorts, embracing hotels, public gardens, summer theatres, and concert halls, all illuminated by electric light. *Carrollton*, in the N. suburbs, has many fine public gardens and private residences. *Algiers*, opposite New Orleans (reached by ferry), has extensive dry-docks and ship-yards. *Gretna*, on the same side, is a pretty rural spot, abounding in pleasant, shady walks.

Itineraries.

The following series of excursions has been prepared so as to enable the visitor whose time is limited to see as much of the city as possible in the least amount of time. Each excursion is planned to occupy a single day, but the visitor can readily spend more time, as special features crowd upon his attention.

1. START from Clay Statue, on Canal St., at the junction of Royal and St. Charles Sts., and walk to the Custom-House, cor. of Canal and Peters Sts. Visit the Marble Hall and Post-Office, then go to the top of the building, where a superb view of the city, and especially the Mississippi River, can be obtained. Take the Esplanade and Levee car in front of the Custom-House, which runs along the Levee, to the French Market, a scene of much interest. Visit the market, then cross over to Jackson Square. Directly opposite is the Cathedral of St. Louis. From there take a short walk to the old Courts, which should be visited. After walking around in that interesting quarter, with its old houses, cross over to Royal St. and take the car which passes directly in front of the St. Charles Hotel.

Levee at New Orleans.

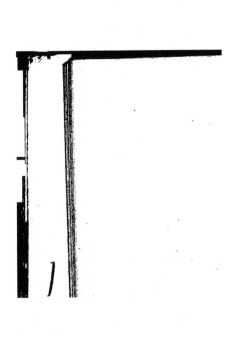

2. Take car at cor. of Canal and Bourbon Sts., going down Bourbon St. to Esplanade Ave. Stop at cor. of Esplanade Ave. and Burgundy St., walk a block to the Mint (visitors are admitted during office hours), then take car in front of the Mint, going along Esplanade Ave., and see the finest residences in the French part of the town. Stop in front of the Jockey Club, which can be visited by applying to the president for an admission card. Near the Jockey Club are the old cemeteries, where some of the monuments of early residents can be seen. From the cemeteries walk a short distance to the Bayou St. John. Picturesquely situated on the other side of the Bayou is the City Park, famous for its beautiful oaks. Take Canal and Rampart St. cars at station in front of the Bayou, which pass along Rampart St. in front of some of the oldest houses, and also pass the Congo Square, where formerly slaves were sold.

3. In front of Clay Statue take Meterarie and Canal St. cars along Canal St. to the Meterarie Ridge, where the famous Howard and other cemeteries are. A day might easily be spent in visiting that cemetery and those adjoining, where the Confederate, the Washington Artillery, and many other beautiful monuments are to be seen. In the Howard Cemetery, prominent among the monuments, is that of Charles Howard, who gave his name to the spot. Jefferson Davis was formerly interred here.

4. Take Canal and Claiborne St. cars at cor. Canal and Bourbon Sts. to the Chapelle St. Roch, a picturesque and beautiful little chapel in an old church-yard. It is said that more than one fair creole girl obtained a husband by going on a "pilgrimage to the little chapelle and praying to the good saint." Taking same car, return to Canal St. and visit Tulane University; then go to the Jesuit Church of the Immaculate Conception, in Baronne St. It is of Moorish architecture and quite unique in its style. The middle altar was made in Rome, and is of gold, studded with precious stones.

5. Take City & West End R. R. in Canal St. and go to Lake Ponchartrain, West End, which is the popular outdoor resort of New Orleans. There are yachts which ply between West End and the Spanish Fort, which must be visited, and where can be seen the old fort, with the cannons that were used during the Spanish war. Return by taking train at Spanish Fort to Canal St.

6. Take St. Charles Ave. car at cor. of Baronne and Canal Sts. and pass up St. Charles Ave., the finest avenue in New Orleans, lined with handsome residences, to Audubon Park, where the Exposition of 1884 and 1885 was held, and which afterward was converted into a park. Within a short walk of the park is Carrollton, where a fine view of the river can be obtained.

7. A day may be well spent in visiting Bay St. Louis, Pass Christian, or any of the numerous summer resorts along the Gulf coast. They are easily accessible by frequent trains which leave the Louisville & Nashville R. R. station. Another trip of considerable interest is to the famous jetties at Port Eads. These are accessible by the river-boats, and there are special excursion-boats that make the trip. The plan-

tations on either side of the river are thus seen, as well as the famous
Forts Jackson and Philip, which took part in the naval engagement
that resulted in the surrender of New Orleans during the civil war.
At proper seasons of the year a visit to the sugar plantations will be
found instructive. Among these, the McCall Plantation in Ascension
Parish, and the Kernochan Plantation in Plaquemine Parish, are the
best. The sugar refineries of New Orleans are very large, and worthy
of visiting if the tourist is interested in technical processes.

129. New Orleans to San Francisco.

*a. Via "Sunset Route" of the Southern Pacific Co.   Distance, 2,492
miles.*

This is a direct and popular route between New Orleans and the Pacific
coast, passing through a region of great variety of scenery and industrial pursuits, much of which is unique and interesting. Tickets are sold through to
San Francisco at all principal Eastern and Northern cities, from which points
to New Orleans there is a through sleeping-car service, connecting with daily
buffet-car service at New Orleans for San Francisco. The line is admirably
equipped, both as to road-bed and rolling-stock, and the cars are of the latest
improved patterns. The low latitude along which this line runs makes the absence of snow in winter a desirable feature.

THE Mississippi River is crossed from New Orleans to *Algiers.*
The tropical character of the region comes prominently into notice at
Shriever, 55 miles from New Orleans. Large trees draped with Spanish moss, bayous with mirror-like surfaces reflecting the luxurious verdure on their banks, and the prevalence of giant magnolia-trees, and, in
season, the perfume of their handsome blossoms, give a distinct character to the scenery. The country is perfectly level, with the exception
of an isolated low hill here and there. Comfortable farm-houses with
wide verandas, placed amid great plantations of sugar-cane, cotton, and
tobacco, make an interesting picture. *Lafayette* (144 miles), on Bayou
Vermilion, which is navigable from the Gulf of Mexico, is an attractive
town. *Lake Charles* (218 miles) is an important town on Lake Charles,
which is connected with the gulf by the Calcasieu River. *Orange,*
Texas (257 miles), the seat of Orange County, is on the Orange River,
which is the dividing-line between Louisiana and Texas. This is the
great lumber region of Texas, contrasting strongly with the plains farther west. *Beaumont* (278 miles) is another town in the lumber country. *Houston* (362 miles; see Route 103, *a*) is an important city; it is
a railroad center, and connects with the Gulf by Buffalo Bayou; it has
several large cotton-compresses, and is the distributing point for southeastern Texas. Beyond this lies a fine stretch of oak-timbered country
and a number of attractive towns. *San Antonio* (571 miles) has a population of 37,673, and is described in Route 103, *a.* The railroad touches
the Rio Grande at *Del Rio* (741 miles), and 15 miles beyond crosses
Devil's River. Some very fine scenery in the grand gorge of the river
is encountered as the road ascends the stream. After leaving the river
the open plains are traversed to El Paso, the otherwise dull aspect of
the country being relieved by mountain-ranges in the distance. At *El*

Paso (see Route 102) the road connects with the Atchison, Topeka & Santa Fé and the Mexican Central R. Rs. (For the further journey to San Francisco, see Route 91.)

b. Via Texas & Pacific R. R.

The Texas & Pacific R. R. extends from New Orleans to El Paso, Texas (1,157 m.), and the route is thence by the Southern Pacific R. R. to San Francisco (2,435 m.). Two trains a day leave New Orleans for San Francisco, and the route is furnished with the most improved conveniences for the comfort of the passenger. Every train is equipped with palace sleeping-coaches.

As far as *Donaldsonville* (64 miles) the route follows closely the course of the river. At *Baton Rouge Junction* (89 miles) a short branch diverges to *West Baton Rouge*, immediately opposite to which, on the E. side of the Mississippi, is *Baton Rouge* (see Route 133). At *Cheney-ville* (170 miles) connection is made with the Southern Pacific R. R. *Alexandria* (194 miles, population 2,861), capital of Rapides Parish, is on the S. bank of the Red River, which gives it an important water traffic. It exports cotton, rice, sugar, and fruits, and has a fine court-house, a bank, and a number of schools and churches. **Shreveport** (326 miles) is an enterprising city of 11,979 population, situated on the W. bank of the Red River, near *Soda Lake*, and is one of the principal points on this important waterway. Steamboats ply regularly to New Orleans. The city contains a handsome court-house, 11 churches, 3 banks and banking-houses, and a number of steam mills, factories, and machine-shops. Cattle and cotton are the chief articles of export. The next place of importance is **Marshall,** Texas (362 miles), where the New Orleans Division unites with the main line of the Texas & Pacific R. R. The city has a population of 7,207 people, and is growing fast. Here are located the headquarters and machine-shops of the road. In addition to the court-house and a number of churches and schools, the city has a *Woman's College* and the *Wiley University*, which was founded in 1873. *Longview* (386 miles) is the place of junction with the International & Great Northern R. R. and with the Texas, Sabine Valley & Northwestern R. R. At *Mineola* (432 miles) intersection is made with a branch of the International & Great Northern R. R. For description of **Dallas** (510 miles) see Route 102 a. Fort Worth (547 miles) has a population of 23,076 people, and is the beginning of the Rio Grande Div. of the line, and here also intersects the Gulf, Colorado & Santa-Fé R. R. *Cisco* (657 miles) is at the junction with the Texas Central R. R. The many stations on the line between this point and El Paso are unimportant. The middle and western portions of Texas tributary to the railroad constitute a pastoral region of unsurpassed attraction to cattle-raisers, and it is on this industry that this section of the State mostly depends. As we approach **El Paso** (1,157 miles) the country becomes mountainous, and the mining industry begins to assume importance, the geological characteristics being identical with those of southern New Mexico and eastern Arizona. El Paso and the further route are described in Route 102. (For description of **San Francisco** (2,489 miles) see Route 91.)

36

130. The Virginia Springs Region.

Hotels, etc.—As a general thing, the hotel, and its cottages, bath-houses, and other buildings, are the only houses in the immediate vicinity of the springs. The charges at the springs are from $2 to $3 a day ; $30 to $70 per month. Other expenses are light. Horses may be hired in the country for $1.50 a day. At the springs, the charges for horses and vehicles are higher, but very moderate in comparison with the liveries of Northern resorts. Carriages seating four may usually be hired for $5 a day. "Let the tourist," says Mr. Pollard, "bring his fishing-rod, and a gun to shoot deer. A common fault at the springs, and which is perhaps prevalent at all watering-places, is the idle and dawdling life ; but the spas of Virginia have this great and peculiar advantage—that instead of the visitor being compelled to walk or ride on a dusty thoroughfare, or take a paltry stroll on the beach, he may lose himself in a few moments in the neighboring forest, where recreation may be sweetened with perfect solitude, or exercise freshened with the mental excitement that makes it alike pleasant and profitable."

THE most important of the Virginia Springs are either directly on the line of the railways which intersect the W. portion of the State— the Baltimore & Ohio, the Chesapeake & Ohio, and the Richmond & Danville R. R. (Virginia Midland Div.)—or are easily accessible from them by stage. A convenient center for the tourist is *Staunton* (which is reached *via* the Virginia Midland and Chesapeake & Ohio R. Rs., or *via* the Baltimore & Ohio R. R. from Harper's Ferry). **Staunton** (*Virginia House*) is 148 miles from Washington, has a population of 6,975, and is situated at the foot of the Blue Ridge Mts. It contains a court-house, ten churches, two banks, and a number of important educational institutions. Among the latter are the *Augusta Female Seminary*, the *Staunton Female Seminary*, the *Virginia Female Institute*, and a *Methodist Female Institute*. Other institutions are the *Western Lunatic Asylum*, and an institution for the deaf, dumb, and blind. Among the manufactories are the *Staunton Iron-Works*, and a number of flouring and planing mills.

Weyer's Cave, 18 m. N. E. from Staunton, and reached thence by stage, is one of the most celebrated and the oldest known stalactite caverns in the United States. It derives its name from Bernard Weyer, a hunter, who discovered it in 1804. It is situated on a spur of a small ridge which branches out from the Blue Ridge. The entrance is about 7 ft. high, and there are many apartments beautifully adorned with stalagmites, stalactites, and other objects of interest. *Washington's Hall,* the largest chamber, is upward of 90 ft. high and 250 ft. long. Near by is *Madison's Cave,* of inferior interest.

Lynchburg is reached by Route 122, *a,* and is described in the same. The best route to reach the springs region is by the Chesapeake & Ohio R. R., which penetrates the central portion of that region. The F. F. V. (Fast Flying Virginian) limited train, leaving New York in the afternoon, reaches the springs or stations (where conveyances are taken for reaching them) without change from New York, Philadelphia, Baltimore, and Washington. The Chesapeake & Ohio R. R. is the only line over which solid trains are run to the Virginia resorts in the Central Spring Region. Those reached from Lynchburg, however, are much visited, and have the advantage of being situated in a more picturesque country than those farther N.

The springs which, owing to the facility with which they are reached, are much resorted to by Northern visitors, are the **Berkeley Springs,** situated in Morgan County, W. Virginia. They are reached by a branch of the Baltimore & Ohio R. R. running from Hancock to the springs. The surrounding scenery is highly picturesque, and the spot possesses historic and social associations as connected with Washington, who frequently visited it. From a remote period it has been the resort of large numbers of people from the lower Valley of Virginia and Maryland; and was a popular watering-place as far back as 1816, when Paulding visited it and described it in his "Letters from the South." The waters flow from five springs at the rate of 2,000 gallons per minute. The temperature is 74° Fahr. The bathing-pools are very large, and rank with the finest in Virginia. The water is not remarkable for its curative properties, and is but slightly impregnated with mineral ingredients, but the bathing is highly invigorating. The main building is a commodious hotel, in which dancing takes place nightly throughout the season.—**Capon Springs** is a highly popular resort near the top of the North Mountain, 16 miles from *Capon Road*, on the Valley Div. of the Baltimore & Ohio R. R. (113 miles from Baltimore), where stages meet the guests. The *Capon Springs Hotel* ($30 to $60 a month) is an excellent hotel, with several cottages attached, furnishing accommodations for about 750 guests. Fronting the *Mountain House* is the bathing establishment, presenting a beautiful colonnade front of 280 ft., with a central building two stories high, 42 by 30 ft., containing parlors, etc., for the use of bathers. The Capon water contains silicic acid, magnesia, soda, bromine, iodine, and carbon dioxide gas; and is recommended for idiopathic and sympathetic affections of the nervous system, various forms of dyspepsia, chronic diarrhœa, irritation of the intestinal canal, and gravel. *Candy's Castle*, the *Tea-Table*, and other curiosities of the region are accessible from this watering-place.—The **Rawley Springs** are situated in Rockingham County, 11 miles by stage from *Harrisonburg* on the Valley Div. of the Baltimore & Ohio R. R. (181 miles from Baltimore). The hotel accommodations are excellent, the grounds are tastefully improved, and the surrounding scenery is very attractive. The Rawley water is a compound chalybeate, is alterative and tonic in its effects, and is held to be remedial in those chronic diseases which are characterized by low and deficient vital action.

The most famous and most frequented of all the West Virginia resorts are the **White Sulphur Springs,** in Greenbrier County, on the line of the Chesapeake & Ohio R. R., 91 miles W. of Staunton and 227 miles from Richmond. The immediate vicinity of the Springs is very beautiful. About 50 acres are occupied by the hotels and cottages and the surrounding lawns and walks, which are admirably kept. The adjacent scenery is unsurpassed in beauty and picturesqueness. Kate's Mountain, which recalls some heroic exploits of an Indian maiden of long ago, is one fine point in the scene southward; while the Greenbrier Hills lie 2 miles away, toward the W., and the lofty Alleghanies tower up majestically on the N. and E. It is not known precisely at what period this spring was discovered. Though the Indians undoubtedly knew its virtues, there is no record of its being used by the whites until 1778. Log-cabins were first erected on the spot in 1784–'86, and the place began to assume something of its present aspect about 1820. Since then it has been yearly improved, until it is capable of pleasantly housing some 1,500 guests. The spring bubbles up from the earth in the lowest part of the valley, and is covered by a pavilion, formed of 12 Ionic columns, supporting a dome, crowned by a statue of Hygeia. Its effect is alterative and stimulant, and it is considered beneficial in cases of dyspepsia, liver-disease, nervous diseases, cutaneous diseases, rheumatism, and gout. The position of the White Sulphur is central to nearly all the prominent springs of the region, which may thus be conveniently visited in turn. The Hot Spring is 38 miles distant, on the N.; the Sweet Spring, 17 miles E.; and the Salt and the Red Springs, 24 and 41 miles respectively, on the S.

The **Old Sweet Springs** are situated in Monroe County, and are reached by stage in 9 miles from *Alleghany*, a station on the Chesapeake & Ohio R. R., 86 miles W. of Staunton. This watering-place is said to be the oldest in Virginia, and to have been frequented for its medicinal properties as early as 1764. The water derives a peculiar briskness from the carbon dioxide which predominates in it, and is prescribed for all the varieties of dyspepsia, for diarrhœa, dysentery, and general disorder of the system. The springs are situated in a lovely valley, between the Alleghany Mts., which bound the northern prospect, and the Sweet Springs Mt., rising on the S. The hotel is large, and there are commodious baths for ladies and gentlemen. The **Sweet Chalybeate Springs** are situated 1 mile from the "Old Sweet," and 8 miles from Alleghany station. The waters are chalybeate and tonic, and the accommodations for visitors ample. The temperature of the water varies from 75° to 79° Fahr., and the 3 springs discharge 250 gallons a minute.

The **Salt Sulphur Springs** connect by stages with *Fort Spring*, on the Chesapeake & Ohio R. R., 108 miles W. of Staunton. This watering-place is near *Union*, the county-seat of Monroe, about 24 miles from the White Sulphur, and is completely shut in by mountains— Swope's Mt., Peters's Mt., and the Alleghanies—the place being near the E. base of the first named. The springs were discovered in 1805 by Irwin Benson while boring for salt-water, which he was induced to hope for from the fact that the spot had been a well-known "lick" for deer and buffalo. The hotel and cottages have accommodations for about 400 guests. There are 3 springs, one of which is styled the "Iodine." The Salt Sulphur water is recommended for chronic affections of the brain; for chronic diseases of the bowels, kidneys, spleen, and bladder; and for neuralgia and the various nervous diseases.

The **Red Sulphur Springs**, in the S. portion of Monroe County, are 41 miles from the White Sulphur, 17 from the Salt, and 39 from the Sweet. They are reached by stage from *Lowell* on the Chesapeake & Ohio R. R., 127 miles W. of Staunton. The approach to these springs is beautifully romantic and picturesque. The springs themselves lie in a verdant glen surrounded on all sides by lofty mountains, and the hotels and cottages afford accommodations for about 450 guests. The water of the spring is collected in 2 white-marble fountains, over which is a tasteful cover. It is clear and cool, with a temperature of 54° Fahr., and is strongly charged with hydrogen sulphide gas, besides containing several of the neutral salts. Its effects are stated to be directly sedative, and indirectly tonic, alterative, diuretic, and diaphoretic; and it is used with advantage in cases of scrofula, jaundice, chronic dysentery, and dyspepsia, and is a specific in consumption and diseases of the throat.

The **Healing, Hot,** and **Warm Springs** of Bath County are grouped together a short distance N. of the Chesapeake & Ohio R. R., and are unrivaled by any others yet discovered, either in Europe or America. They lie within a short distance of each other, and the visitor may pass from one to another in an hour or two, through magnificent scenery. These springs are under the control of the Southern Improvement Co., which has built a 24-mile railway from Covington, a station on the main line of the Chesapeake & Ohio R. R., to Hot Springs, passing through **Healing Springs.** The scenery around this watering-place is extremely agreeable; there is a fine cascade near, and the Springs buildings make a charming little village, shining pleasantly through the green trees. The waters of this spring are stated to be almost identical in their chemical analysis with the famous Schlangenbad and Ems waters of Germany. Their temperature is uniformly 84° Fahr., and the water is regarded as highly beneficial in cases of scrofula, chronic thrush, obstinate cases of cutaneous disease, neuralgia, rheumatism, ulcers of the lower limbs of long standing, and dyspepsia, in some "hopeless cases" of which it is said to have worked cures. The **Hot Springs** are 2½ miles from the Healing Springs, and are said to be the hottest baths in the world, the temperature reaching 110° Fahr. There are 9 springs, and 9 baths attached, all in the grounds of the hotel. The most marked effect of the free use of these waters is in cases of rheumatism and torpid liver, which are promptly and remarkably relieved. A bathing establishment to cost $150,000, and an elegant hotel to accommodate 500 guests, are in course of construction at Hot Springs, which will be conducted in addition to the old hotel, which has been remodeled and newly fitted up. The **Warm Springs** were discovered by the Indians, and have long been a popular resort. The water is very abundant, and is used for bathing as well as drinking, chiefly the former. It contains sulphuric, carbonic, silicic, and organic acids, as the first bases, and potash, ammonia, lime, magnesia, protoxide of iron, and alumina, as the second bases. The diseases for which the baths are beneficial are gout, chronic rheumatism, swellings of the joints and glands, paralysis, chronic cutaneous diseases, and calculous disorders. At the lower end of the Warm Spring Valley is the * *Cataract of the Falling Springs*, where a foaming mountain-brook tumbles over a rocky ledge 200 ft. high. These springs may also be reached by a ride from Millboro, with a magnificent view from the top of Warm Spring Mountain, which the tourist crosses at an elevation of nearly 1,500 ft. above its base (2,250 ft. above the sea). On the summit of the mountain is a spot called *Flat Rock*, from which there is a superb view of the long mountain-ranges, extending as far as the eye can see, "like a dark-blue sea of giant billows, instantly stricken solid by Nature's magic wand." On this route is also seen the curious *Blowing Cave*, situated near the banks of the Cow-Pasture River. The **Bath Alum Springs** are near the E. base of the Warm Springs Mts., 5 miles from the Warm Springs, and 10 from Millboro, with which they connect by stages. The waters issue from a slate-stone cliff, and are received into small reservoirs. The springs differ—one of them being a strong chalybeate, with but little alum; another, a milder chalybeate, with more alumina; while the others are alum of different strength, with traces of iron. The waters are decidedly tonic and astringent, and are recommended for scrofula, dyspepsia, eruptive affections, chronic diarrhœa, nervous debility, and in various uterine diseases.

The **Rockbridge Alum Springs** are situated in Rockbridge County, and are reached by Victoria R. R. from *Goshen*, on the Chesapeake & Ohio R. R., 32 miles W. of Staunton, and are also reached by stage from Millboro. The springs consist of 5 fountains, issuing from beneath irregular slate-stone arches. There are 3 hotels, which, with cottages, have accommodations for 1,200 guests. The waters are regarded as highly beneficial in cases of chronic dyspepsia, diarrhœa, scrofula, gastric irritation, and diseases of the skin. In the immediate vicinity are *Jordan's Alum Springs*. The waters possess qualities similar to those of the other alum springs in this vicinity. **Rockbridge Baths** are reached by the line of the Baltimore & Ohio R. R. to Timber Ridge, and thence by stage 4½ miles to the baths. The springs are within a few feet of the banks of North River, and are surrounded by picturesque scenery. The waters are impregnated with iron, and are strongly charged with carbon dioxide gas. As a tonic bath (adapted to nervous diseases, general debility, especially after the use of alterative mineral waters, and that comprehensive class of cases in which tonic bathing is beneficial) the Rockbridge Baths are highly recommended.

About ¾ of a mile from *Bonsack's*, on the Norfolk & Western R. R. (see Route 124, *a*), are **Coyner's Springs,** a favorite resort with the people of Lynchburg, from which they are only 47 miles distant. The buildings are spacious and comparatively new, and the place has the reputation of being one of the gayest in Virginia. The waters are sulphurous, and, of their class, mild and pleasant. They are recommended in cases of difficult, imperfect, or painful digestion, enfeebled condition of the nervous system, chronic diseases of the bladder or kidneys, saltrheum, tetters, indolent liver, and in some of the affections peculiar to females.—The **Blue Ridge Springs,** in Botetourt County, directly on the line of the Norfolk & Western R. R., have lately become one of the favorites in Virginia. They are situated near the summit of the Blue Ridge Mountains, 1,300 ft. above the sea, in the midst of delightful scenery, and the air is pure and cool. The *Blue Ridge Springs Hotel* is excellent; there are a number of commodious cottages, and the waters have a special reputation for the cure of dyspepsia. From *Shawsville* on the Norfolk & Western R. R. (see Route 124, *a*), stages run in 2 miles to the **Alleghany Springs,** which have long been popular. The large hotel and cottages are situated upon undulating ground, surrounded by wild and picturesque scenery. In the neighborhood (8 miles distant) are the ** Puncheon-Run Falls,* a wonderful series of cascades, where a mountain-brook tumbles for 1,800 ft. down an almost perpendicular ledge. ** Fisher's View* (5 miles from the Springs) is a point on the mountain from which a fine view of the wild and beautiful scenery of the surrounding region may be obtained. The Alleghany water is cathartic, diuretic, and tonic, and is recommended for dyspepsia, depressed biliary secretions, costiveness, scrofula, jaundice, and incipient consumption. From *Big Tunnel,* on the Norfolk & Western R. R. (4 miles from Alleghany), a tramway extends 1 mile to the **Montgomery White Sulphur Springs,** located in Montgomery County. The Springs are beautifully situated in the midst of fine scenery, diversified by rippling streams; and the buildings are unusually handsome and substantial, with accommodations for about 1,000 guests. The waters are of two kinds: one a strong sulphur, resembling that of the Greenbrier White Sulphur; the other a tonic chalybeate. The sulphur is said to be less cathartic and stimulant than other sulphurs, and to act more mildly.—The **Yellow Sulphur Springs** are 5 miles S. W. of the Montgomery White, and 3 miles from *Christiansburg* on the N. & W. R. R., with which they connect by stages. This spring is located high up on the E. side of the Alleghany Mountains, and, "in consequence of this elevation, the air is elastic, pure, and invigorating during the hottest days of summer." The water possesses valuable tonic properties, and is delightfully cool, the temperature in the hottest weather remaining at 55°.

The foregoing springs are the most prominent and popular of the "Springs Region." Among other less frequented watering-places are the *Bedford Alum Springs,* W. of Lynchburg, near the Norfolk & Western R. R.—The *Grayson White Sulphur Springs,* in Carroll County, near the point where New River passes through the Iron Mountain, and connecting with the N. & W. R. R. at Max Meadows.—The *Sharon Alum Springs,* connecting with the N. & W. R. R. at Wytheville, 25 miles by stage.—The *Pulaski Alum Springs,* connecting with the N. & W. R. R. at Newbern, 10 miles by stage.—*Eggleston's Springs,* in Giles County, Va., near the Salt Pond.—The *Fauquier White Sulphur Springs,* in Fauquier County, 7 miles from Warrenton.—*Jordan's White Sulphur Springs,* in Frederick County, 5 miles from Winchester, and 1½ from Stephenson's Station on the Baltimore & Ohio R. R. This is a popular and agreeable summer resort.—The *Orkney Springs,* in Shenandoah County, 12 miles by stage from Mt. Jackson on the Valley Div. of the B. & O. R. R.

The Chesapeake & Ohio R. R., partly described in the above route, forms, with its connections, a great trunk line between the North, West, and Southwest. Its proper E. terminus is **Newport News,** Va. Thence it runs N. W. to **Richmond** (75 m.; see Route 111), and connects at *Gordonsville* (151 m.; with the Richmond & Danville R. R., which gives through-connection from Washington and all Northern points to the West and Southwest. From this point the Chesapeake & Ohio R. R. runs in a nearly direct line W. through-

Virginia and West Virginia to *Huntington,* W. Va. (494 m.), on the Ohio River, just below the mouth of the Guyandotte. Here the Lexington Div. begins, running to *Lexington,* Ky. (644 m.). The Lexington Branch of the Louisville & Nashville R. R. is the connecting-link in the route as far as **Louisville** (788 m.; see Route 77). Hence the route is by the Newport News & Mississippi Valley R. R., which runs from Louisville to Memphis (1,180 m.). From Washington to Memphis (see Route 133) the distance by this route is 1,067 m.

131. Mountain Region of North Carolina, South Carolina, and Georgia.

THE great Appalachian range of mountains, called also the Alleghanies, extends from that part of Canada lying between the New England States and the St. Lawrence River, through the whole length of Vermont, across the W. part of Massachusetts and the middle Atlantic States, to the N. part of Alabama. The White Mts. of New Hampshire and the Adirondack Mts. of New York are really outliers of this range, though separated from it by wide tracts of low elevation. The Catskills form a link of the main range. *Blue Ridge* is the name given to the most eastern of the principal ridges of the chain. It is the continuation S. of the Potomac of the same great ridge which in Pennsylvania and Maryland is known as the South Mountain. It retains the name Blue Ridge till it crosses the James River, from which to the line of North Carolina its continuation is called the Alleghany Mt. Running through North Carolina into Tennessee, it again bears the name of the Blue Ridge. The extreme length of the Appalachian range is 1,300 miles; its greatest width (about 100 miles) is in Pennsylvania and Maryland, about midway of its course. In all their extent the Appalachian Mts. are remarkable, not for their great elevation, nor for their striking peaks, nor for any feature that distinguishes one portion of them from the rest, but for a singular uniformity of outline. While varying little in height, the ridges pursue a remarkably straight course, sometimes hardly diverging from a straight line for a distance of 50 or 60 miles, and one ridge succeeding behind another, all continuing the same general course in parallel lines, like successive waves of the sea.

North Carolina.

The mountain region of North Carolina, where the Appalachian system reaches its loftiest altitude, presents scenes of beauty and sublimity unsurpassed by anything E. of the Rocky Mountains. It consists of an elevated table-land, 250 miles long and about 50 broad, encircled by two great mountain-chains (the Blue Ridge on the E. and the Great Smoky on the W.), and traversed by cross-chains that run directly across the country, and from which spurs of greater or lesser height lead off in all directions. Of these transverse ranges there are four: the Black, the Balsam, the Cullowhee, and the Nantahala. Between each lies a region of valleys, formed by the noble rivers and their minor tributaries. The Blue Ridge is the natural barrier, dividing the waters falling into the Atlantic from those of the Mississippi, and its bold and beautiful heights are better known than the grander steeps of the western chain. This W. rampart, known as the Great Smoky, comprises the groups of the Iron, the Unaka, and the Roan Mountains; and from its massiveness of form and general elevation is the master-chain of the whole Alleghany range. Though its highest summits are a few feet lower than the peaks of the Black Mountain, it presents a continuous series of lofty peaks which nearly approach that altitude, its culminating point, *Clingman's Dome*, rising to the height of 6,660 ft. The most famous of the transverse ranges is that of the Black Mountain, a group of colossal heights, the dominating peak of which—*Mount Mitchell*—is now known to be the loftiest summit E. of the Mississippi. With its two great branches it is over 20 miles long, and its rugged sides are covered with a wilderness of almost impenetrable forest. Above a certain elevation, no trees are found save the balsam-fir, from the dark color of which the mountain takes its name. N. of the Black Mountain stand the two famous heights which Arnold Guyot calls "the two great pillars on both sides of the North Gate to the high mountain region of North Carolina." These are the *Grandfather Mountain* in the Blue Ridge, and *Roan Mountain* in the Smoky. Next to the Black, in the order of transverse chains, comes the Balsam, which in length and general magnitude is chief of the cross-ranges. It is 50 miles long, and its peaks average 6,000 ft. in height, while, like the Blue Ridge, it divides all waters and is pierced by none. From its S. extremity two great spurs run out in a northerly direction; one terminates in the *Cold Mountain*, which is over 6,000 ft. high, and the other in the beautiful peak of *Pisgah*, which is one of the most noted landmarks of the region.

The key of the mountain region, and converging-point of all the roads W. of the Blue Ridge, is **Asheville** (*Battery Park Hotel, Swannanoa, Grand Central Hotel, The Oaks*), situated in the lovely valley of the French Broad River, 2,250 ft. above the sea, surrounded by an amphitheatre of hills, and commanding one of the finest mountain-views in America. Just above its site the beautiful Swannanoa unites with the French Broad, charming natural parks surround it, and within easy excursion-distance is some of the noblest scenery in the State. The town itself is adorned with many handsome private residences, the hotel accommodations are superior, and there are good churches, schools, banks, and newspapers. There are five routes by which Asheville may be reached from the north, west, and south, and, as each of them presents special attractions to tourists by the way, we shall describe them separately.

1st Route (all-rail).—From Salisbury (see Route 112, *b*) by the Salisbury and Point Rock Branch of the Western North Carolina Div. of the Richmond & Danville System to Asheville. **Morganton** (*Mountain Hotel*), 80 miles from Salisbury, is a popular resort, and well worth the attention of all lovers of mountain scenery. It is situated on the slopes of the Blue Ridge, 1,100 ft. above the sea, and a very beautiful view may be obtained from any eminence in the vicinity. About 15 miles W. of Morgantown are the *Glen Alpine Springs,* whose waters are of the lithia class, and are said to possess diuretic, tonic, and alterative properties. In this neighborhood the *Hawk's Bill* and *Table Rock* are situated. The latter is a high, bleak rock rising above the top of a mountain to the height of over 200 ft. It can easily be ascended, and upon the summit there is about an acre of rock with a smooth surface. About 25 miles from Morgantown is the grand * **Linnville Gorge,** where the Linnville River bursts through the massive barrier of the Linnville Mountains.

2d Route.—From Spartanburg, S. C., by the Spartanburg and Asheville Branch of the Western North Carolina Div. of the Richmond & Danville System.

3d Route.—From Charlotte, N. C., to Statesville by the Charlotte and Taylorsville Branch of the South Carolina Div., or to Lincolnton, and by narrow-gauge railway to Hickory on the Western North Carolina Div. of the Richmond & Danville System. Near Shelby are *Wilson's Springs*, somewhat noted as a summer resort. This route lies through the famous * **Hickory-Nut Gap,** the scenery of which has been declared by some European travelers to be equal in beauty and grandeur to any pass in the Alps. The entire length of the Gap is about 9 miles, the last 5 being watered by the Rocky Broad River. The gateway of the gorge on the E. side is not more than ¼ mile wide, and from this point the road winds upward along a narrow pass, hemmed in on all sides by stately heights. The loftiest bluff is on the south side, and, though 1,500 ft. high, is nearly perpendicular. A stream of water tumbles over one portion of this immense cliff, and falls into an apparently inaccessible pool. From the summit of the Gap there is a most impressive view in all directions.

4th Route.—By stage from Greenville, South Carolina (see Route 122, *a*), *via* Saluda Gap, Flat Rock, and Hendersonville, to Asheville (60 miles). This route traverses some of the finest portions of the South Carolina mountain-region (described below), and the entire road lies through the most enchanting and picturesque scenery. **Flat Rock,** once the most frequented of Carolina resorts, has been shorn of its former glories, but the lovely valley still contains some noble mansions, surrounded by beautiful gardens.

5th Route.—By the stage from Greenville, South Carolina (see Route 122, *a*), *via* Jones's Gap and Cæsar's Head, to Asheville (about 75 miles). *Cæsar's Head* is a bold and beautiful headland in South Carolina (see present route). Beyond Cæsar's Head the route passes near * **Cashier's Valley,** a lofty table-land lying on the side of the Blue Ridge, so near the summit that its elevation above the sea can not be less than 3,500 ft., and hemmed in on all sides by noble peaks, among which *Chimney-Top* stands forth conspicuously. On the S. W. edge of the valley is * **Whiteside Mountain,** which is in many respects the most striking peak in North Carolina. Rising to a height of more than 5,000 ft., its S. E. face is an immense precipice of white rock, which, towering up perpendicularly 1,800 ft., is fully 2 miles long, and curved so as to form the arc of a circle. The ascent to the summit can be made partly on horseback and presents no difficulties, and the view is of surpassing grandeur.

"To the N. E., as far as the eye can reach, rise a multitude of sharply defined blue and purple peaks, the valleys between them, vast and filled with frightful ravines, seeming the merest gullies of the earth's surface. Farther off than this line of peaks rise the dim outlines of the Balsam and Smoky ranges. In the distant S. W., looking across into Georgia, we can descry Mount Yonah, lonely and superb, with a cloud-wreath about his brow ; 60 miles away, in South Carolina, a flash of sunlight reveals the roofs of the little German settlement of Walhalla ; and on the S. E., beyond the precipices and ragged projections, towers up Chimney-Top Mountain, while the Hog-Back bends its ugly form against the sky, and Cold Mountain rises on the left. Turning to the N., we behold Yellow Mountain, with its square sides, and Short-Off. Beyond and beyond, peaks and peaks, and ravines and ravines ! It is like looking down on the world from a balloon."—EDWARD KING.

6th Route.—From the north, west, or southwest, Asheville may be reached *via* East Tennessee, Virginia & Georgia system (Route 124, *a*) to Morristown, Tenn. ; thence *via* its North Carolina Div. to Wolf Creek ; and thence by the Paint Rock branch of the Western North Carolina Div. of the Richmond & Danville R. R., which traverses the valley of the French Broad River amid magnificent mountain scenery.

Having reached Asheville (see present route), the tourist may spend days or weeks in visiting the many picturesque spots in the vicinity, or in hunting, fishing, or exploring the caves, mines, and Indian mounds. A few miles from the town are some white sulphur springs, from which a variety of lovely views may be had ; and 9 miles N. are the so-called *Million Springs*, beautifully situated in a cave between two mountain-ranges, where sulphur and chalybeate waters may be had in abundance. But the excursion which above all others he should not fail to make is that down the * **French Broad River,** the supreme beauty of which has long been famous. Below Asheville the river flows through an ever-deepening gorge, narrow as a Western cañon and inexpressibly grand, until it cuts its way through the Smoky Mountains, and reaches Tennessee. For 36 miles its waters well deserve their musical Cherokee name (Tahkeeostee, "the Racing River"), and the splendor of their ceaseless tumult fascinates both eye and ear. The railroad follows its banks, and often trespasses upon the stream, as it is crowded by the overhanging cliffs. About 35 miles from Asheville, on the right of the road, is the famous rock *Lover's Leap ;* and just below it, where the left bank widens out into a level plain, the **Hot Springs** (*Mountain Park Hotel,* open all the year) nestle in a beautiful grove of trees. These springs are also reached directly by Western North Carolina Div. of the Richmond & Danville R. R., and their virtues have been known for nearly a century. An analysis of the water shows that it contains free carbonic acid, free sulphureted hydrogen, carbonic acid, and sulphuric acid, in combination with lime, and a trace of magnesia. Though quite palatable as a beverage, it is taken chiefly in the form of baths, for which there are excellent facilities, and is recommended for dyspepsia, liver-complaint, diseases of the kidneys, rheumatism, rheumatic gout, and chronic cutaneous diseases. Five miles below the springs, on the Tennessee boundary, the road passes beneath the bold precipice of the *Painted Rocks,* a titanic mass over 200 ft. high, whose face is marked with red paint, supposed to be Indian pictures. Near by are the *Chimneys,* lofty cliffs, broken at their summits into detached piles of rock bearing the likeness of colossal chimneys, a fancy greatly improved by the fireplace-like recesses at their base.

Among the mountain-ascents that may be readily made from Asheville, those of Mt. Pisgah and Mt. Mitchell will best repay the trouble. *Pisgah* lies to the S., and commands an extensive view over Tennessee, South Carolina, and Georgia, as well as over the greater part of western North Carolina. The excursion to * **Mt. Mitchell,** including the ascent to the peak and the return to Asheville, can be made in three days, and, though arduous, is entirely free from danger. The summit of Mt. Mitchell is the highest in the United States E. of the Mississippi (6,701 ft.), and affords the visitor a view of unsurpassed extent and grandeur. Another attractive mountain-excursion (less often made, however) is to the **Balsam Range,** lying to the W. The route is to *Brevard*, a pleasant village lying in the matchless valley of the Upper French Broad; and thence along the N. fork of the river into what is called the *Gloucester Settlement*. Here a guide can be secured, and the peaks easily ascended.

South Carolina.

The town of *Greenville* (see Route 122, *a*) lies at the threshold of the chief beauties of the South Carolina mountain region, and affords easy access to all the rest. It is beautifully situated on the Reedy River, near its source, at the foot of Saluda Mountain. About 20 miles from Greenville is * **Table Mountain,** one of the most remarkable of the natural wonders of the State, rising 4,300 ft. above the sea, with a long extent on one side of perpendicular cliffs, 1,000 ft. in height. The view of these grand and lofty rock-ledges is exceedingly fine from the quiet glens of the valley below, and not less imposing is the splendid amphitheatre of hill-tops seen from its crown. Among the sights to be seen from Table Mountain is * **Cæsar's Head,** a lofty peak with one side a precipice of great height, just back of which is a large hotèl. It is the highest point in the vicinity, and well worth a visit. At the base of Table Mountain, in a romantic glen, are the famous * **Falls of Slicking,** a wonderful series of cascades and rapids. They are situated on the two branches of the Slicking River, of which the right-hand branch is the more picturesque. - The **Keowee** is a beautiful mountain-stream in Pickens County, which, with the Tugaloo River, forms the Savannah. The route from Greenville to the valley of Jocasse lies along its banks amid the most lovely scenery, and the entire region is full of romantic memories of the Cherokee wars. **Jocasse Valley,** near the N. boundary-line, is one of the most charmingly secluded nooks in the State, environed as it is on every side, except that through which the Keowee steals out, by grand mountain-ridges. The great charm of Jocasse is that it is small enough to be seen and enjoyed all at once, as its entire area is not too much for one comfortable picture. It is such a nook as painters delight in. **White Water Cataracts** are an hour's brisk walk N. of Jocasse. Their chief beauty is in their picturesque lines, and in the variety and boldness of the mountain-landscape all around. Adjoining this most attractive region of South Carolina, and easily accessible therefrom, are Tallulah, and Toccoa, and Yonah, and Nacoochee, lying in Georgia and described below.

Georgia.

The most convenient point from which to visit the mountain region of Georgia is **Clarksville,** in Habersham County, much resorted to by the people of the "Low Country." It is reached by the Blue Ridge & Atlantic R. R. from Cornelia, on the Richmond & Danville R. R. (Atlanta & Charlotte Div.); or by stage from Walhalla (on the Columbia & Greenville Branch) to *Clayton.* Fair accommodations for travelers may be had at Clarksville, and also horses or wagons for the exploration of the surrounding country. A few miles from Clarksville is the celebrated * **Toccoa Fall,** where a brook "comes babbling down the mountain's side" and plunges over a precipice 180 ft. high. The * **Cataracts of Tallulah** are 12 miles from Clarksville, on the same railroad line. From Toccoa to Tallulah the cut across is only 5 or 6 miles. There is a comfortable hotel near the edge of the gorges traversed by this wild mountain-stream, and hard by its army of waterfalls. The Tallulah, or *Terrora,* as the Indians more appositely called it, is a small stream, which rushes through a chasm in the Blue Ridge, rending it for several miles. The ravine is 1,000 feet in depth, and of an equal width. Its walls are gigantic cliffs of dark granite, whose heavy masses, piled upon each other in the wildest confusion, sometimes shoot out, overhanging the yawning gulf. Along the rocky and uneven bed of this deep abyss the Terrora frets and foams with ever-varying course. The wild grandeur of this mountain-gorge, and the variety, number, and magnificence of its cataracts, give it rank with the most imposing waterfall scenery in the Union. The * **Valley of Nacoochee** (or the Evening Star) is a pleasant day's excursion from Clarksville. The valley is said by tradition to have won its name from the story of the hapless love of a beautiful Indian princess, whose scepter once ruled its solitudes; but with or without these associations, it will be remembered with pleasure by all whose fortune it may be to see it. *Mt. Yonah* looks down into the quiet heart of Nacoochee, lying at its base; and if the tourist should stay overnight in the valley, he ought to take a peep at the mountain panorama from the summit of Yonah. Another interesting peak in this vicinity is *Mt. Currahee,* which is situated S. of Clarksville, a few miles below the Toccoa Cascade. The * **Falls of the Eastatoia** are about 3 miles from *Clayton,* in Rabun, the extreme N. E. county of Georgia. Clayton may be reached easily from Clarksville, or by a ride of 12 miles from the cataract of Tallulah. The falls lie off the road to the right, in the passage of the Rabun Gap, one of the mountain ways from Georgia into North Carolina; they would be a spot of crowded resort were they in a more thickly-peopled country. The scene is a succession of cascades, noble in volume and character, plunging down the ravined flanks of a rugged mountain-height.

Union County, adjoining Habersham on the N. W., is distinguished for natural beauty, and for its objects of antiquarian interest. Among these latter is the *Track Rock*, bearing wonderful impressions of the feet of animals now extinct. *Pilot Mountain*, in Union, is a noble elevation of some 1,200 ft. The *Hiawasse Falls*, on the Hiawasse River, present a series of beautiful cascades, some of them from 60 to 100 ft. in height. The much-visited **Falls of Amicalolah** are in Lumpkin County, 17 miles W. of the village of Dahlonega, near the State road leading to East Tennessee.

132. The Ohio River.

During portions of the summer and in the autumn, when the water is low, the larger steamboats ascend no farther than Wheeling, and even below this point they pass with difficulty. Those who desire only to see the more interesting portions of the river can take the steamer at *Wheeling* (see Route 70), at *Parkersburg* (see Route 70), at *Huntington* (see Route 130), or at *Cincinnati*, the W. terminus of the Chesapeake & Ohio R. R. (see Route 75). Those who wish to see the entire river can take a packet from Pittsburg to Wheeling, whence large and comfortable steamers ply to Cincinnati. From Cincinnati very fine steamers run down the river to Louisville and Cairo.

THE Ohio River is the largest affluent of the Mississippi River from the E., and was known to the early French settlers as *La Belle Rivière*. It is formed by the junction at Pittsburg of the Alleghany and Monongahela Rivers, and has a total length of about 1,000 miles. No other river of equal length has such a uniform, smooth, and placid current. Its average width is about 2,400 ft., and the descent, in its whole course, is about 400 ft. It has no fall, except a rocky rapid of 22¼ ft. descent at Louisville, around which is a ship-canal 2¼ miles long. The course of the Ohio and of all its tributaries is through a region of stratified rocks, little disturbed from the horizontal position in which they were deposited, and nowhere intruded upon by uplifts of the azoic formations, such as in other regions impart grandeur to the scenery. For these reasons the scenery of the Ohio, though often beautiful, is for the most part tame. One interesting feature is the succession of terraces often noticed rising one above another at different elevations. Though they are often 75 ft. or more above the present level of the river, they were evidently formed by fluviatile deposits made in distan periods, when the river flowed at these high levels. Evidence is altogether wanting to fix the date of these periods; but mounds and earthworks, constructed on the lower branches of the river fully 2,000 years ago, show that the river must have flowed at its present level at least so far back.

LANDINGS.	Miles.	LANDINGS.	Miles.
Pittsburg, Pa	0	Racine, Ohio	249
Economy, Pa	19	Guyandotte, W. Va	811
Rochester, Pa	29	Huntington, W. Va	816
Wellsville, Ohio	52	Ashland, Ky	819
Steubenville, Ohio	71	Ironton, Ohio	827
Wheeling, W. Va	94	Greenupsburg, Ky	837
Bellaire, Ohio	98	Portsmouth, Ohio	862
Newport, Ohio	151	Maysville, Ky	415
Marietta, Ohio	170	Cincinnati, Ohio	476
Parkersburg, Ohio	188	Covington, Ky	476

LANDINGS.	Miles.	LANDINGS.	Miles.
Lawrenceburg, Ind	498	Evansville, Ind	813
Madison, Ind	567	Henderson, Ky	821
Jeffersonville, Ind	617	Mount Vernon, Ind	855
Louisville, Ky	618	Shawneetown, Ill	877
New Albany, Ind	621	Elizabethtown, Ill	907
Leavenworth, Ind	680	Smithland, Ky	945
Hawesville, Ky	744	Paducah, Ky	957
Rockport, Ind	769	Mound City, Ill	1,001
Owensboro, Ky	778	Cairo, Ill	1,005

The most important places enumerated in the above list have already been described. *Economy* was settled in 1825 by a German sect called "Harmonists," who hold all property in common. *Beaver* is a busy manufacturing village situated at the mouth of the Beaver River, from which it derives a fine water-power. *Wellsville* is an important wool-shipping point, and contains a number of foundries and machine-shops. Two miles below, near the mouth of Great Yellow Creek, is the locality of the murder of the family of Logan, the Mingo Chief. **Steubenville** (see Route 73). *Wellsburg* is a town of W. Virginia, beautifully situated on the E. bank of the river. **Wheeling** (see Route 70). *Bridgeport,* opposite Wheeling, is connected with it by a magnificent suspension-bridge. *Bellaire* is where the Central Ohio Div. of the Baltimore & Ohio R. R. crosses the river (see Route 70). **Marietta** (*Nation Hotel, Pillsbury*) is a flourishing city of about 8,273 inhabitants, picturesquely situated at the confluence of the Ohio and Muskingum Rivers. It is the E. terminus of the Marietta Div. of the Baltimore & Ohio Southwestern R. R., and the S. terminus of the Cleveland & Marietta R. R., and has a large trade in petroleum, which is obtained in the vicinity. It is the seat of *Marietta College,* which has 4 buildings, surrounded by ample grounds, and a library of 25,000 volumes. On the site of the city is a * group of ancient works which are described by Squier and Davis in their "Ancient Monuments of the Mississippi Valley." *Parkersburg* and *Belpre,* together with the splendid railway bridge uniting them, are described in Route 68. Two miles below Parkersburg is **Blennerhassett's Island,** noted for having been the residence of Harman Blennerhassett, an Irishman of distinction, who improved the island, and built on it a splendid mansion for himself, in 1798. When Aaron Burr was planning his celebrated conspiracy, he induced Blennerhassett to join him, and to embark all his means in the scheme. Although not convicted of treason, Blennerhassett was ruined, his house went to decay, and his beautiful gardens were destroyed. **Pomeroy** (*Grand Dilcher House*) is the fifth place on the river above Cincinnati in trade and commerce, and has a population of 4,726. Its prosperity rests mainly on the mines of bituminous coal within its limits and in the immediate vicinity. It is also the center of the salt basin of the Ohio Valley, and there are 26 salt-furnaces within its limits and in the neighborhood, with an investment of $1,000,000, and yielding about 12,000,000 bushels a year. At *Point Pleasant,* 14 miles below, the Great Kanawha River empties into the Ohio, and at *Guyandotte* the Big Guyandotte River comes in. *Huntington* is an important shipping-point, and the railway connects here with several lines of steamboats. The Big Sandy River, 7 miles below Huntington, is the boundary-line between Kentucky and W. Virginia. **Ironton** is a city of 10,939 inhabitants, built at the foot of lofty hills in the center of the "Hanging Rock" iron-region (embracing a portion of S. Ohio and N. E. Kentucky), of which it is the principal business point. Its iron-trade amounts to about $8,000,000 a year, and it contains a number of blast-furnaces, rolling-mills, machine-shops, etc. *Greenupsburg* is situated at the mouth of Little Sandy River, and 25 miles below is the prosperous Ohio city of **Portsmouth** (*Biggs House*), beautifully situated at the mouth of the Scioto River, and at the terminus of the Lake Erie & Ohio Canal. It is substantially built, and has a population of 12,394. Being the entrepot of the rich mineral regions of S. Ohio and N. E. Kentucky, it has a large trade, besides numerous iron-furnaces, rolling-mills, foundries, etc. The Scioto Valley is a productive agricultural district. A branch of the Baltimore & Ohio Southwestern R. R. terminates at Portsmouth. **Maysville** (*St. Charles, Central*) is the largest place in N. E. Kentucky, and one of the most extensive hemp-markets in the United States. It lies in a bend of the river, and is backed by a range of hills which gives it a very attractive appearance. Its population is 5,358, and it contains several handsome public buildings. **Cincinnati** (see Route 75).

The view from the steamer when opposite Cincinnati is remarkably fine. On the one hand is the densely populated city, its rows of massive buildings rising tier above tier toward the hill-tops, which, crowned with villas and gardens, form a semicircular background. On the opposite bank rise the beautiful Kentucky hills, at whose feet nestle the twin cities of *Covington* and *Newport*, divided only by the Licking River and connected by a graceful suspension-bridge (see Route 76, *a*). There are few places of importance on the river between Cincinnati and Louisville, and they are separated by long stretches of virgin woodland and plain. *North Bend* (see Route 78). The Great Miami River, 4 miles below North Bend, is the boundary between Ohio and Indiana. *Lawrenceburg* and *Aurora* are described in Route ·78· At *Carrollton*, 74 miles from Cincinnati, is the mouth of the Kentucky River, a navigable stream 200 miles long, noted for its beautiful scenery. **Madison** (*Western Hotel, Madison*) is one of the principal cities of Indiana, is beautifully situated and well built, and contains about 8,937 inhabitants. Several pork-packing establishments are located here, the trade in provisions is important, and there are brass and iron foundries, flouring-mills, machine-shops, etc. ·Madison is the terminus of one branch of the Pittsburg, Cincinnati, Chicago & St. Louis R. R. The approach to **Louisville** (see Route 77) is very fine, affording an impressive view of the city and of *Jeffersonville* on the opposite bank. The river is here about a mile wide, and is crossed by one of the finest bridges in the United States (see the same). The Falls of the Ohio just below Louisville descend 23 ft. in 2 miles, and, to avoid this obstruction, a canal 2½ miles long has been constructed around them.

Besides **New Albany** (see Route 77) the only important cities between Louisville and Cairo are Evansville, Ind., and Paducah, Ky.

Evansville (*St. George Hotel, Sherwood House*) is the principal ship-ping-point for the grain and pork of S. W. Indiana, and its manufac-tures are important. It is the terminus of 2 railroads, and of the Wabash & Erie Canal, which extends 462 miles to Toledo (see Route 67). The city contains a handsome *Court-House, City Hall, U. S. Marine Hospital*, an *Opera-House*, and upward of 30 churches. The population is 50,756, and coal and iron are found in the vicinity. *Shawneetown* is a prosperous village. **Paducah** (*Richmond House*) si a city of about 13,076 inhabitants, on the S. bank just below the mouth of the Tennessee River. It is the shipping-point of the surrounding country, the chief productions of which are tobacco, pork, and grain, and contains several tobacco and other factories. The Western Div. of the Newport News and Mississippi Valley R. R. passes through here on its way from Louisville to Memphis. **Cairo** is situated at the conflu-ence of the Ohio and Mississippi Rivers, and has been described in Route 84.

133. The Mississippi River.

The tour of the Mississippi River is usually made in two distinct stages : From St. Paul or Minneapolis to St. Louis, or *vice versa ;* and from St. Louis to New Orleans, or *vice-versa.* A daily line of commodious and comfortable side-wheel passenger packets plies between Minneapolis and St. Paul and St. Louis. The steamers plying between St. Louis and New Orleans are large and fine. That portion of the river above St. Louis is known as the Upper Mississippi ; that below St. Louis as the Lower Mississippi.

THE Mississippi River, "Father of Waters," rises in Minnesota, on the dividing-ridge between the waters which flow into Hudson's Bay and those flowing into the Gulf of Mexico, and so near the source of the Red River of the North that in times of freshet their waters have been known to commingle. It is, at its source, 3,160 miles from its mouth, a rivulet flowing from a small pool fed by springs. Thence it flows through a number of pools or ponds, each larger than the pre-ceding one, until it expands into Itasca Lake, whence it emerges as a stream of some size, and soon becomes a river. It first flows N. through Cass, Sandy, and other lakes, and then, turning toward the S., rolls downward to the Gulf of Mexico, passing over more than 18 degrees of latitude. Between the source and the Falls of St. Anthony are many rapids and waterfalls, but the only one of any magnitude is the Pecagama Rapids, 685 miles above St. Anthony. From these rapids down to the St. Anthony Falls the river is navigable, and much of the scenery is very beautiful. The Falls of St. Anthony form an insuper-able barrier to navigation, and here the St. Louis steamers stop. From St. Paul to Dubuque the river flows between abrupt and lofty bluffs, distant from each other from 2 to 6 miles, and rising from 100 to 600 ft., the valley or bottom being very beautiful, filled with islands, and intersected in every direction by tributaries of the Mississippi, and by the various channels and "sloughs" of the river itself. The bluffs are principally of limestone; they are almost uniformly vertical and rugged, and nearly destitute of vegetation, except at the base and summit. The limestone is generally of grayish white, but is stained and streaked

37

until it is of every hue, from that of iron-rust to that of the white cliffs of St. Paul. There are grandeur and sublimity in every mile of this portion of the river; but it becomes monotonous after a time, the eye becoming surfeited with too much beauty. Below Dubuque the valley continues to preserve the same general characteristics, but the bluffs are lower and more like hills, and the scenery, though still beautiful, is tamer. Below Alton it begins to assume more the appearance of the "Lower River" (as the portion below St. Louis is called); and the waters, turbid and muddy, roll on, a mighty torrent, between banks often low, flat, and sandy, and the vegetation continually more and more tropical in its nature.

Principal Landings on the Mississippi River.

LANDINGS.	Miles.	LANDINGS.	Miles.
Minneapolis, Minn	0	Louisiana, Mo	691
St. Paul, Minn	14	Mouth of Illinois River	762
Hastings, Minn	46	Alton, Ill	780
Prescott, Wis	49	Mouth of Missouri River	785
Red Wing, Minn	79	St. Louis, Mo	805
Winona, Minn	160	Cape Girardeau, Mo	955
La Crosse, Wis	194	Cairo, Ill	1,005
Lansing, Iowa	239	Columbus, Ky	1,025
Prairie du Chien, Wis	269	New Madrid, Mo	1,080
MacGregor, Iowa	272	Memphis, Tenn	1,255
Dunleith, Ill	335	Helena, Ark	1,345
Dubuque, Iowa	335	White River, Ark	1,425
Galena, Ill	355	Napoleon, Ark	1,445
Fulton, Ill	413	Young's Point, La	1,655
Clinton, Iowa	415	Vicksburg, Miss	1,665
Davenport, Iowa	458	Natchez, Miss	1,785
Rock Island, Ill	458	Red River, La	1,855
Muscatine, Iowa	488	Bayou Sara, La	1,895
Burlington, Iowa	550	Port Hudson, La	1,905
Nauvoo, Ill	582	Baton Rouge, La	1,925
Keokuk, Iowa	597	Plaquemine, La	1,955
Quincy, Ill	641	Donaldsonville, La	1,975
Hannibal, Mo	661	New Orleans, La	2,055

Between St. Paul and *Hastings* there are half a dozen small villages, one of them being somewhat noticeable on account of its name, *Red Rock*, which was given by the Indians, who worshiped a large rock at this point, which they painted red, and called Wacon, or Spirit Rock. *Point Douglas* is the last point of Minnesota on the E. bank of the river, as the *St. Croix River*, which empties here, marks the boundary-line of Wisconsin, between which State and Minnesota the Mississippi now forms the boundary-line for many miles. *Red Wing* (see Route 87, *a*) is situated at the head of **Lake Pepin,** an expansion of the river, about 30 miles long, and 3 miles in average width. By many this is considered the most beautiful portion of the Mississippi. The bluffs on either side present peculiar characteristics, which are found in such perfection nowhere else; grim castles seem only to want sentries to be perfect, and all the fantastic forms into which the action of the weather can transform limestone cliffs are to be seen. The forests reach to the river-bank, and the water is so beautifully clear that fish may be seen many feet below the surface. Just below Red Wing is *Barn Bluff*, a well-known landmark, 200 ft. high. *Frontenac* lies in the center of the lake-region, and is a favorite resort in summer on account of its fine scenery, and the hunting, bathing, fishing, and sailing, which it affords (see Route 87, *a*). * **Maiden Rock,** 3 miles below Frontenac, is a promontory 409 ft. high, near the lower end of the lake, on the E. side. Its name is derived from an incident which is reported to have happened about the commencement of the present century. A young Dakota maiden, named Winona, loved a young hunter; but her parents wished her to marry a warrior of the Wabashaw tribe, to which they belonged, and tried to compel her to accede to their wishes. On the day before that appointed for the marriage she went to the verge of this precipice, and commenced chanting her death-song. Her relatives and friends, seeing her on the brink of destruction, called to her that they would yield to her wishes; but she did not believe them, and, before any one could reach her, she·leaped over the precipice, and was dashed to pieces on the rocks below. *Reed's Landing* is at the foot of Lake Pepin, where the river again contracts, and is opposite the mouth of the *Chippewa River*, a navigable lumbering-stream.

Near *Fountain City* (48·miles below Lake Pepin) is the famous * **Chimney Rock,** and between this point and Winona there are 12 miles of remarkably fine scenery, in which are séen bluffs conical in form and covered with verdure, others with precipitous fronts worn by the weather into most fantastic shapes, the river lake-like, and almost filled with islands. **Winona** is described in Route 87, *a*. Below Winona the scenery continues bold and striking; and 20 miles down is * **Trempealeau Island** (sometimes called *Mountain Island*), a rocky island, 300 to 500 ft. in height, and one of the most noted landmarks on the Upper Mississippi. There is a winding path up Trempealeau, and the view from the summit is exquisite. **La Crosse** (see Route 87, *a*). All this portion of the river from·La Crosse to Dubuque is delightful, from the great variety of the scenery, the wooded hills, and the exquisitely pure character of the water, which is clear and limpid as that of·Lake Leman. The bluffs alternate from massive, densely wooded hills to long walls of limestone, which front precipitously on the river, and assume all manner of quaint, fantastic, and striking shapes. Rivers and rivulets come in at intervals, and the rapid succession of the towns indicates a more thickly-settled region. *Prairie du Chien* has already been described in Route 85, *b*. Just above Dubuque one of the landmarks· of the pilots of the upper river is pointed out—* **Eagle Rock,** a splendid bluff, 500 ft. high. **Dubuque** (*Julien House, Lorimier*), the third largest city of Iowa, containing 30,311 inhabitants, is built partly upon a terrace, 20 ft. above the river, and partly upon the bluffs, which rise 200 ft. The lower or business portion is regularly laid· out and compactly built, while in the upper portion the streets rise picturesquely one above another. Among the public buildings worthy of notice are the *U. S. Building*, of marble, 3 stories high, and costing over $200,000; the *Central Market*, and the 4 ward school-houses. The Methodist Episcopal, one of the Presbyterian, the Universalist, the Congregational, and St. Mary's (German Catholic) Churches, and the Cathedral, are imposing structures, the last three being surmounted by lofty spires. Dubuque is the commercial center of the great lead-region of Iowa, N. W. Illinois, and S. W. Wisconsin, some of the mines being within the city limits. Branches of the Illinois Central and Chicago, Milwaukee & St. Paul R. Rs. converge here, and the shipping business is immense.

Below Dubuque the character of the scenery changes, and, though still pleasing, is decidedly tamer. The most noteworthy feature of this portion of the river is the number of important towns and cities that stand on either bank. Twenty miles below Dubuque is the mouth of the Fevre River, 6 miles up which is **Galena,** an important city of 5,635 inhabitants, on the Northern Div. of the Illinois Central R. R. (see Route 84). **Fulton** on the E. bank and **Clinton** on the W. bank, with the great bridge which crosses the river at this point, are described in Route 89. *Le Clair* (25 miles below Clinton) is at the head of the * *Upper Rapids,* which extend for 15 miles to Rock Island. The descent of the rapids is exciting, but seldom dangerous. The cities of **Rock Island** and **Davenport,** on opposite sides of the river, the magnificent * bridge connecting them, and the extensive U. S. arsenals on Rock Island, are described in Route 89. *Muscatine* is a flourishing Iowa city of about 11,454 inhabitants, situated on a rocky bluff at the apex of the Great Bend of the Mississippi. It is the shipping-point of an extensive and fertile country, and its lumber business is large. **Burlington** (see Route 89, *b*). *Nauvoo City* was founded by the Mormons in 1840, and contained about 15,000 inhabitants at the time of their expulsion in 1846 by the neighboring people. It is now a place of small importance. *Montrose* is at the head of the "Lower Rapids," which extend for 12 miles to Keokuk and greatly obstruct navigation. **Keokuk** (see Route 89). **Quincy** (*Tremont House*) is one of the largest cities of Illinois, with a population in 1890 of 31,494. It is picturesquely situated on a limestone bluff 125 ft. above the river, and is regularly laid out and well built, containing many substantial business blocks and handsome residences. The streets are lighted with gas, and the principal ones are traversed by horse-cars. There are 4 small parks and several cemeteries; and about 2 miles from the center of the city are well-appointed Fair-Grounds comprising about 80 acres. Eight lines of railway center at Quincy, and the trade of the city is extensive. The Hannibal & St. Joseph R. R. crosses the river here on a splendid bridge. Twenty miles below Quincy is the flourishing city of **Hannibal** (*Park Hotel*), with a population of about 12,857, important manufactures (including foundries and car-works, flour and saw mills, tobacco-factories and pork-packing houses), and an extensive trade in tobacco, pork, flour, and other produce. After St. Louis, Hannibal is the greatest lumber market W. of the Mississippi, and there are numerous spacious lumber-yards. It is one of the northern termini of the Missouri, Kansas & Texas R. R. (see Route 103, *a*), and several other important railways converge here. **Alton** (see Route 85). Three miles below Alton is the * **Meeting of the Waters** of the Missouri and Mississippi Rivers. This has been pronounced one of the most impressive views of river scenery in the country. The Missouri nominally empties into the Mississippi, but it is really the Mississippi that empties, as any one can see who ever looks upon the scene. **St. Louis** is fully described in Route 81.

The scenery of that portion of the river below St. Louis is very different from that above. "The prevailing character of the Lower Mississippi," says a recent traveler, "is that of solemn gloom." The dreary solitude, and often the absence of all living objects save the huge alligators, which float past apparently asleep on the drift-wood, and an occasional vulture attracted by its impure prey on the surface of the waters; the trees, with a long and melancholy drapery of pendent moss fluttering in the wind; and the gigantic river, rolling onward the vast volume of its dark and turbid waters through the wilderness, form the leading features of one of the most dismal yet impressive landscapes on which the eye of man ever rested. Every now and then a stop is made at a small landing, or at the towns and villages that cluster along the banks; and the clamor of lading and unlading causes a momentary excitement that subsides at once as the steamer resumes her course.

About 125 miles below St. Louis the mouth of the Ohio River is reached (see Route 132), and a somewhat prolonged stay is made at **Cairo** (see Route 84). Cairo is connected by ferry with **Columbus, Ky.,** which lies on the river 18 miles below. Columbus is situated on the slope of a high bluff, commanding the river for about 5 miles, and at the outbreak of the Civil War was strongly fortified by the Confederates, who regarded it as the northern key to the mouth of the Mississippi. They collected in the town and its vicinity an army of 30,000 men; but after the fall of Forts Henry and Donelson, in February, 1862, it was promptly evacuated. *Island No.* 10 (51 miles below Columbus) was the scene of a terrific bombardment by the Mississippi River fleet, extending from March 16 to April 17, 1862, in which the Federals were completely successful. The canal which was cut to assist in the investment of the island, and the remains of some of the earthworks, can still be seen in passing the island. Ten miles below, in Missouri, is *New Madrid,* which was captured at the same time as Island No. 10, both places having formed parts of one position, and mutually dependent upon each other. This was the first battle of the war in which the superiority of gunboats to stationary batteries was clearly demonstrated. New Madrid was settled in 1780, and was the scene of a great earthquake in 1811.

From Columbus to Memphis the river skirts the bluffs of the E. or Kentucky shore, having on its W. the broad, alluvial lands of Missouri and Arkansas. A number of small towns dot either bank, and at intervals spots are pointed out which events of the Civil War have rendered interesting. Conspicuous among these is *Fort Pillow* (148 miles below Columbus), situated on the first Chickasaw Bluff. It was evacuated by the Confederates on June 4, 1862; but on April 12, 1864, was the scene of the shameful butchery by the troops under General Forrest, known in history as the Fort Pillow Massacre, concerning which the testimony is conflicting, and probably exaggerated, on both sides. Below Fort Pillow a journey of about 100 miles brings the voyager to **Memphis** (*Peabody Hotel, Gayoso, Gaston's*), the second city of Tennessee, and the largest on the Mississippi between St. Louis and New Orleans. It is situated on the fourth Chickasaw Bluff, 450 miles below St. Louis, and 800 above New Orleans, and had in 1890 a population of 64,495. The city presents a striking appearance as seen from the water, with a levee several hundred feet in width, sweeping along the bluff. The streets are broad and regular, and lined with handsome buildings; and many of the residences on the avenues leading from the river are surrounded with beautiful lawns. Among the larger buildings are the *Cotton Exchange*, the *Merchants' Exchange*, and the *Appeal-Avalanche Building*, which is regarded as the finest newspaper office in the South. The city extends over 5 square miles. In the center there is a handsome park, filled with trees, and containing a bust of Andrew Jackson. The principal of the six cemeteries is *Elmwood*, on the S. E. border of the city. Memphis is the center of a vast cotton-trade, lumber, grocery, and brick industries, and numerous manufactures. There are a *U. S. Custom-House*, four theatres, fine churches and charitable institutions, excellent public and private schools, and the *Cossett Free Library*. Memphis was captured by the Federals early in the war (June 6, 1862), and was never afterward held by the Confederates. This city is a terminal point for the Memphis & Charleston Div. of the East Tennessee, Virginia & Georgia, the Louisville & Nashville, the Illinois Central, the Kansas City, Fort Scott & Memphis, the Little Rock & Memphis, the Louisville, New Orleans & Texas, the Newport News & Mississippi Valley, the Kansas City, Memphis & Birmingham, the St. Louis & Iron Mountain, and the Tennessee Midland R. Rs.; and there is an all-year river communication with New Orleans by 14 lines of steamboats. There are 5 ferries crossing the Mississippi, and a large railroad bridge.

A short distance below Memphis the Mississippi turns toward the W., and crosses its valley to meet the waters of the Arkansas and White Rivers. The latter enters the Mississippi 161 miles below Memphis, and the former about 15 miles farther down. The Arkansas River is 2,000 miles in length, for 800 of which it is navigable by steamers. It rises in the Rocky Mountains, and, next to the Missouri, is the largest tributary of the Mississippi. The town of *Napoleon* lies at its mouth. Near this point commences the great cotton-growing region, and the banks of the river are an almost continuous succession of plantations. Fifty miles below begins the growth of the Spanish moss, which, covering the trees with its dark and somber drapery, forms one of the most notable features of the river scenery. Having received the waters of the two affluents above mentioned, the Mississippi again crosses its valley to meet the Yazoo near Vicksburg, creating the immense Yazoo reservoir on the E. bank, extending from the vicinity of Memphis to Vicksburg, and the valleys and swamps of the Macon and Tensas on the W. side. **Vicksburg** (*Hotel Piazza, Washington House*) is situated on the Walnut Hills, which extend for about 2 miles along the river, rising to the height of 500 ft., and displaying the finest scenery of the Lower Mississippi. It is a well-built city of 13,373 inhabitants, the largest between New Orleans and Memphis, and about equidistant from both. As at Memphis, the view of the city from the water is in the highest degree picturesque and animated, and the pleasing impression is confirmed by a closer examination of the town. Vicksburg was founded in 1836 by a planter named Vick, members of whose family are still living there. As the chief commercial mart on this portion of the river, it has long been a place of some note, but it is more widely known as the scene of one of the most obstinate and decisive struggles of the Civil War. After the loss successively of Columbus, Memphis, and New Orleans, the Confederates made here their last and most desperate stand for the control of the great river. The place was surrounded by vast fortifications, the hills crowned with batteries, and a large army under General Pemberton placed in it as a garrison. Its capture by General Grant after a protracted siege (July 4, 1863) "broke the backbone of the Confederacy, and cut it in twain." Above Vicksburg, at the point where Sherman made his entrance from the "Valley of Death," is the largest national cemetery in the country, containing the remains of nearly 16,000 soldiers.

From Vicksburg to Baton 'Rouge the river hugs the E. bluffs, with Mississippi on one side and Louisiana on the other. *Grand Gulf*, in Mississippi, is a pretty little town 60 miles below Vicksburg, lying upon some picturesque hills overhanging the river; and **Natchez** (*Bonturas Hotel*), 60 miles nearer New Orleans, is built on a high bluff, 200 ft. above the stream. That portion of the city lying on the narrow strip of land between the foot of the hill and the river is called "Natchez-under-the-Hill," and, though containing some important business houses, can make no claim to beauty. It communicates by broad and well-graded roads with the upper town, called "Natchez-on-the-Hill," which is beautifully shaded, and contains many handsome residences and other buildings. The houses are principally of brick, and the residences are adorned with gardens. The brow of the bluff along the whole front of the city is occupied by a park. The principal buildings are the *Court-House*, in a public square shaded with trees, the *Masonic Temple*, the *Catholic Cathedral*, with a spire 128 ft. high, the Episcopal Church, and the Presbyterian Church. On the bluff adjoining the city there is a *National Cemetery*.

Natchez was founded by D'Iberville, a Frenchman, in 1700, and is replete with historic associations. Here once lived and flourished the noblest tribe of Indians on the continent, and from that tribe it takes it name. Their pathetic story is festooned with the flowers of poetry and romance. Their ceremonies and creed were not unlike those of the Fire-worshipers of Persia. Their priests kept the fire continually burning upon the altar in their Temple of the Sun and the tradition is that they got the fire from heaven. Just before the advent of the white man, it is said, the fire accidentally went out, and that was one reason why they became disheartened in their struggles with the pale-faces. The last remnant of the race were still existing a few years ago in Texas, and they still gloried in their paternity. It is probable that the first explorer of the Lower Mississippi River, the unfortunate La Salle, landed at this spot on his downward trip to the sea. It is a disputed point as to where was the location of the first fort. Some say it lay back of the town, while others say it was established at Ellis's Cliffs. In 1713 Bienville established a fort and trading-post at this spot. The second, Fort Rosalie, or rather the broken profile of it, is still visible. It is gradually sinking, by the earth being undermined by subterranean springs, and in a few years not a vestige of it will be left. Any one now standing at the landing can see the different strata of earth distinctly marked, showing the depth of the artificial earthworks.

The capital of Louisiana, **Baton Rouge,** is a city of 10,478 population, and is pleasantly situated on the last bluff that is seen in descending the Mississippi. The site is 30 to 40 ft. above the highest overflow of the river. The bluff rises by a gentle and gradual swell, and the town, as seen from the water, rising regularly and beautifully from the banks, with its singularly shaped French and Spanish houses, and its queer squares, looks like a finely painted landscape. It contains a *State Prison, Arsenal,* and the *State Institution for the Deaf, Dumb, and Blind,* founded in 1852. From Baton Rouge to New Orleans "the coast," as it is called, is lined with plantations. Every spot susceptible of cultivation is transformed into a beautiful garden, containing specimens of all those choice fruits and flowers which flourish only in tropical climes. **New Orleans** is fully described in Route 128.

Those who, taking an ocean steamer, pursue the journey below New Orleans, traverse a portion of the river not less interesting if less attractive than that left behind. Very soon after leaving the city the phenomena of a "delta-country" become conspicuous, and one can fairly witness the eternal and ever-varying conflict between land and sea. The thick forest vegetation disappears, giving place to isolated and stunted trees; the river-banks grow less and less defined, and finally lose themselves in what appears to be an interminable marsh; and through this marsh the "passes" furnish channels to the Gulf, which are discernible only by the practiced eyes of the pilots. It is impossible, however, for the inexperienced traveler to say where land ends and sea begins; and before he is aware of having reached the "mouth" of the river, he is far out on the Gulf of Mexico, where a muddy surface-current is the only relic of the mighty "Father of Waters."

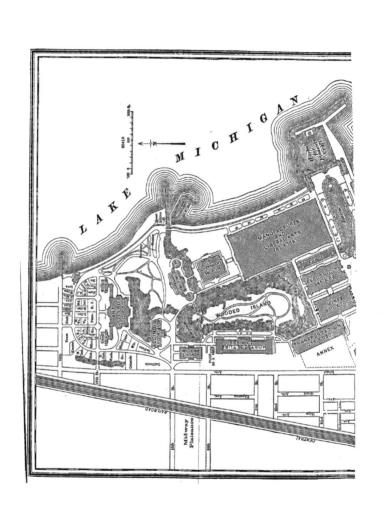

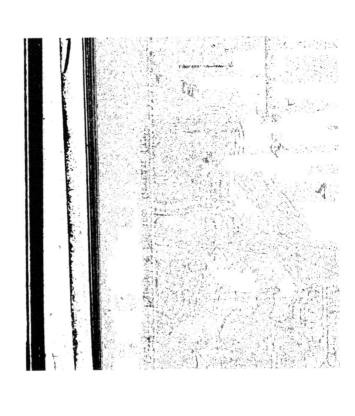

APPENDIX.

THE WORLD'S COLUMBIAN EXPOSITION IN CHICAGO, ILLINOIS.

Location.—The Exposition is held chiefly in Jackson Park, a tract of land embracing 586 acres, with a frontage of nearly 2 miles on Lake Michigan, and in Washington Park, embracing 371 acres; while between the two is a tract of 80 acres called the Midway Plaisance. The entire site covers 1,037 acres.

Admission Fee.—During the Exposition, from May 1st to October 30th, the fee for each admission is 50c.

Administration.—For a complete list of the officers and details of the administration, see the list given on page 325.

Access to the Exposition.—The grounds of the Exposition extend from 56th St. to 67th St., and there are entrance gates along the line of Stony Island Ave., at 57th St., 59th St., 60th St., 62d St., 63d St., and 67th St. Visitors from Chicago, about 7 miles distant, approach the grounds by the Illinois Central R. R., with its South Park Station at 57th St., or its Woodland Park Station at 63d St. (fare, 25c., round-trip ticket ; time, 30 minutes). The Cottage Grove Ave. lake cars run to the South Park entrance at 57th St. (fare, 5c. each way ; time, 45 minutes). The boats of the World's Fair Transportation Co. will leave the docks on the lake front between Monroe and Van Buren Sts., landing at the Exposition pier, opposite the foot of 58th St. (fare, 25c., round trip ticket ; time, 45 minutes). Also by the elevated railway (fare, 5c.).

Hotels.—It will be desirable for all persons who visit the fair to arrange beforehand for their accommodations. On p. 316 will be found a full list of the hotels in Chicago, in addition to which the following houses have been erected in the immediate vicinity of the fair grounds : *Cornell Avenue Hotel*, on Cornell Ave., between 51st and 52d Sts. ; *Park Gate Hotel*, cor. 63d St. and Stony Island Ave., at the terminus of the elevated railroad, lake car lines, and electric car line; *South Shore Hotel*, cor. Bond and 73d Sts., and the *Strickland Hotel*, on Lake Ave., between 38th and 39th Sts.

The World's Fair.

Introduction.—By Act of Congress approved by President Harrison on April 25, 1890, the International Exhibition of Arts, Industries, Manufactures, and the Products of the Soil, Mine, and Sea, to celebrate the 400th anniversary of the discovery of America by Christopher Columbus, was created. Very promptly and with characteristic energy the proper officials effected an organization, and the buildings grew into existence. On page 325 there has already been given a brief account of the development and preliminary events culminating with the opening of the fair by President Cleveland on May 1st. There also may be found a list of the departments under which the exhibits are classified, together with the names of the leading officers of the management. In this Appendix a short description of the more important buildings and other features of the Great Chicago Exposition is given.

Buildings.

In the space at our command there is scarce opportunity to do adequate justice to the magnificent buildings which form so important a feature of the great Columbian World's Fair, but a brief description of each, carefully compiled from official sources, is herewith given in concise language.

Administration Building (architect, Richard M. Hunt; cost, $550,-000).—It is located at the west end of the great court in the southern part of the site, looking eastward, and at its rear are the transportation facilities and depots. The most conspicuous object which will attract the gaze of visitors on reaching the grounds is the gilded dome of this lofty building. It covers an area of 260 ft. square, and consists of four pavilions 84 ft. square, one at each of the four angles of the square, and connected by a great central dome 120 ft. in diameter and 220 ft. in height, leaving at the center of each façade a recess 82 ft. wide, within which are the grand entrances to the building. The general design is in the style of the French renaissance. The first great story is in the Doric order, of heroic proportions, surrounded by a lofty balustrade, and having the great tiers of the angle of each pavilion crowned with sculpture. The second story, with its lofty and spacious colonnade, is of the Ionic order. The four great entrances, one on each side of the building, are 50 ft. wide and 50 ft. high, deeply recessed, and covered by semicircular arched vaults, richly coffered. In the rear of these arches are the entrance doors, and above them great screens of glass, giving light to the central rotunda. Across the face of these screens, at the level of the office floor, are galleries of communication between the different pavilions. The interior features of this building exceed in beauty and splendor those of the exterior. Between every two of the grand entrances, and connecting the intervening pavilion with the great rotunda, is a hall or loggia 30 ft. square, giving access to the offices, and provided with broad, circular stairways and swift-running elevators. Above the balcony is the second story, 50 ft. in height. From the top of the cornice of this story rises the interior dome, 200 ft. from the floor, and in the center is an opening 50 ft. in diameter, transmitting a flow of light from the exterior dome overhead. The under side of the dome is enriched with deep panelings, richly molded, and the panels are filled with sculpture in low relief, and immense paintings representing the arts and sciences.

Agricultural Building (architects, McKim, Mead & White; cost, $100,000).—The style of architecture of this structure is classic renaissance. It is very near the shore of Lake Michigan, and is almost surrounded by the lagoons that lead into the Park from the lake. The building is 500 by 800 ft., its longest dimensions being east and west. For a single-story building the design is bold and heroic. The general cornice-line is 65 ft. above grade. On either side of the main entrance are mammoth Corinthian pillars, 50 ft. high and 5 ft. in diameter. On each corner and from the center of the building pavilions are reared, the center one being 144 ft. square. The corner pavilions are connected by curtains, forming a continuous arcade around the top of the building. The main entrance leads through an opening 64 ft. wide into a vestibule, from which passage is had to the rotunda, 100 ft. in diameter. This is surmounted by a mammoth glass dome 130 ft. high.

The Agricultural Building.

MCKIM, MEAD, AND WHITE, ARCHITECTS.

All through the main vestibule statuary has been designed illustrative of the agricultural industry. Similar designs are grouped about all the grand entrances in the most elaborate manner. The corner pavilions are surmounted ·by domes 96 ft. high, and above these tower groups of statuary. The design for these domes is that of three female figures, of herculean proportions, supporting a mammoth globe. To the southward of the Agricultural building is an annex, devoted chiefly to a live-stock and agricultural assembly hall. This building is conveniently near one of the stations of the elevated railway. On the first floor, near the main entrance of the building, is located a bureau of information. This floor also contains suitable committee and other rooms for the different live-stock associations. On this floor there are also large and handsomely equipped waiting-rooms. Broad stairways lead from the first floor into the assembly-room.

Dairy Building (cost, $30,000).—The dairy building was specially designed to contain a complete exhibit of dairy products and also a dairy school, in connection with which will be conducted a series of tests for determining the relative merits of different breeds of dairy cattle as milk and butter producers. The building stands near the lake-shore, in the southeastern part of the park and close by the general live-stock exhibit. It covers nearly half an acre, measuring 95 by 200 ft., and is two stories high. In design it is of quiet exterior. On the first floor, besides the necessary office headquarters, there is in front a large open space devoted to exhibits of butter, and farther back an operating-room 25 by 100 ft., in which the model dairy will be conducted. On two sides of this room are amphitheatre seats capable of accommodating 400 spectators. Under these seats are refrigerators and cold-storage rooms for the care of the dairy products. The operating-room, which extends to the roof, has on three sides a gallery where the cheese exhibits will be placed. The rest of the second story is devoted to a *café*, which opens on a balcony overlooking the lake. The dairy school will be most instructive and valuable to agriculturists.

Electrical Building (architects, Van Brunt & Howe; cost, $410,000). —The Electrical building is 345 ft. wide and 700 ft. long, the major axis running N. and S. The S. front is on the great quadrangle or court; the N. front faces the lagoon; the E. front is opposite the manufacturers' building; and the W. faces the Mines building. The general scheme of the plan is based upon a longitudinal nave 115 ft. wide and 114 ft. high, crossed in the middle by a transept of the same width and height. The nave and the transept have a pitched roof, with a range of sky-lights at the bottom of the pitch, and clear-story windows.· The rest of the building is covered with a flat roof, averaging 62 ft. in height, and provided with sky-lights. The second story is composed of a series of galleries connected across the nave by two bridges, with access by four grand staircases. The area of the galleries in the second story is 118,546 sq. ft., or 2·7 acres. The exterior walls of this building are composed of a continuous Corinthian order of pilasters, 3 ft. 6 in. wide and 42 ft. high, supporting a full entablature, and resting upon a stylobate 8 ft. 6 in. The total height of the walls from the grade outside is 68 ft. 6 in. At each of the four corners of the building there is a pavilion, above which rises a light open spire or tower 169 ft. high. Intermediate between these corner pavilions and the central pavilions on the E. and W. sides there is a subordinate pavilion bearing a low square dome upon an open lantern. The electricity building has an open portico extending along the whole of the S. façade, the lower or Ionic order forming an open screen in front of it. The various subordinate pavilions are treated with windows and balconies. The details of the exterior orders are richly decorated, and the pediments, friezes, panels, and spandrils have received a decoration of figures in relief, with architectural *motifs*, the general tendency of which is to illustrate the purposes of the building. The appearance of the exterior is that of marble, but the walls of the hemicycle and of the various porticoes and loggia are highly enriched with color, the pilasters in these places being decorated with scagliola and the capitals with metallic effects in bronze.

Fine Arts Building (architect, C. B. Atwood; cost, $670,000).—This structure is of Grecian-Ionic order and is a pure type of the most refined classic architecture. The building is oblong, and is 500 by 320 ft., intersected north, east, south, and west by a great nave and transept 100 ft. wide and 70 ft. high, at the intersection of which is a dome 60 ft. in diameter. The building is 125 ft. to the top of the dome, which is surmounted by a colossal statue of the type of the famous figure of Winged Victory. The transept has a clear space through the center of 60 ft., being lighted entirely from above. On either side are galleries 20 ft. wide and 24 ft. above the floor. The collections of the sculpture are displayed on the main floor of the nave and transept, and on the walls both of the ground floor and of the galleries are ample areas for displaying the paintings and sculptured panels in relief. The corners made by the crossing of the nave and transept are filled with small picture galleries. Around the entire building are galleries 40 ft. wide, forming a continuous promenade around the classic structure. Between the promenade and the naves are the smaller rooms devoted to private collections of paintings and the collections of the various art schools. On either side of the main building, and connected with it by handsome corridors, are very large annexes, which are also utilized by various art exhibits. The main building is entered by four great portals, richly ornamented with architectural sculpture, and approached by broad flights of steps. The walls of the lóggia of the colonnades are highly decorated with mural paintings illustrating the history and progress of the arts. The frieze of the exterior walls and the pediments of the principal entrances are ornamented with sculptures and portraits in bas-relief of the masters of ancient art. The construction, although of a temporary character, is necessarily fire-proof. The main walls are of solid brick, covered with "staff," architecturally ornamented, while the roof, floors, and galleries are of iron. The building is located beautifully in the northern portion of the park, with the south front facing the lagoon. It is separated from the lagoon by beautiful terraces, ornamented with balustrades, with an immense flight of steps leading

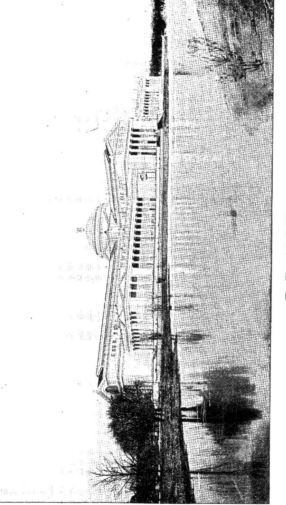

The Fine Arts Building.

C. B. ATWOOD, ARCHITECT.

down from the main portal to the lagoon, where there is a landing for boats. The north front faces the wide lawn and the group of State buildings. The immediate neighborhood of the building is ornamented with groups of statues, replica ornaments of classic art, such as the Choragic monument, the "Cave of the Winds," and other beautiful examples of Grecian art.

Forestry Building (cost, $100,000).—The Forestry building is in appearance the most unique of all the Exposition structures. Its dimensions are 200 by 500 ft. To a remarkable degree its architecture is of the rustic order. On all four sides of the building is a veranda, supporting the roof of which is a colonnade consisting of a series of columns composed of three tree-trunks each 25 ft. in length, one of them from 16 to 20 inches in diameter and the others smaller. All these trunks are left in their natural state, with bark undisturbed. They are contributed by the different States and Territories of the Union and by foreign countries, each furnishing specimens of its most characteristic trees. The sides of the building are constructed of slabs with the bark removed. The window-frames are treated in the same rustic manner as is the rest of the building. The main entrances are elaborately finished in different kinds of wood, the material and workmanship being contributed by several prominent lumber associations. The roof is thatched with tan and other barks. The visitor can make no mistake as to the kinds of tree-trunks which form the colonnade, for he will see upon each a tablet upon which is inscribed the common and scientific name, the State or country from which the trunk was contributed, and other pertinent information, such as the approximate quantity of such timber in the region whence it came. Surmounting the cornice of the veranda and extending all around the building are numerous flagstaffs bearing the colors, coats of arms, etc., of the nations and States represented in the exhibits inside.

Fisheries Building (architect, Henry I. Cobb ; cost, $225,000).—The Fisheries building embraces a large central structure with two smaller polygonal buildings connected with it on either end by arcades. The extreme length of the building is 1,100 ft. and the width 200 ft. It is located to the northward of the United States Government building. In the central portion is the general fisheries exhibit. In one of the polygonal buildings is the angling exhibit, and in the other the aquaria. The exterior of the building is Spanish-Romanesque, which contrasts agreeably in appearance with that of the other buildings. . To the close observer the exterior of the building can not fail to be exceedingly interesting, for the architect exerted all his ingenuity in arranging innumerable forms of capitals, modillions, brackets, cornices, and other ornamental details, using only fish and other sea forms for his *motif* of design. The roof of the building is of old Spanish tile, and the side walls of pleasing color. In the center of the polygonal building is a rotunda 60 ft. in diameter, in the middle of which is a basin or pool 26 ft. wide, from which rises a towering mass of rocks covered with moss and lichens. From clefts and crevices in the rocks crystal streams of water gush and drop to the masses of reeds, rushes, and ornamental semi-aquatic plants in the basin below. In this pool gorgeous gold-fishes, golden ides, golden tench, and other fishes disport. From the rotunda one side of the larger series of aquaria may be viewed. These are ten in number, and have a capacity of 7,000 to 27,000 gallons of water each. Passing out of the rotunda, a great corridor or arcade is reached, where on one hand can be viewed the opposite side of the series of great tanks, and on the other a line of tanks somewhat smaller, ranging from 750 to 1,500 gallons each in capacity. The corridor or arcade is about 15 ft. wide. The glass fronts of the aquaria are in length about 575 ft., and have 3,000 square ft. of surface. The total water capacity of the aquaria, exclusive of reservoirs, is 18,725 cubic ft., or 140,000 gallons. This weighs 1,192,425 pounds, or almost 600 tons. Of this amount about 40,000 gallons is devoted to the marine exhibit. In the entire salt-water circulation, including reservoirs, there are about 80,000 gallons. The pumping and distributing plant for the marine aquaria is constructed of vulcanite. The pumps are in dupli-cate, and each has a capacity of 3,000 gallons per hour. The supply of sea-water was secured by evaporating the necessary quantity at the Wood's Holl station of the United States Fish Commission to about one fifth its bulk, thus reducing both quantity and weight for transporta-tion about 80 per cent. The fresh water required to restore it to its proper density is supplied from Lake Michigan.

Government Building (architect, W. J. Edbrooke ; cost, $400,000).— This is near the lake-shore, S. of the main lagoon and of the area re-served for the foreign nations and the several States, and E. of the Woman's building and of Midway Plaisance. The buildings of England, Germany, and Mexico are to the northward. It is classic in style, and bears a strong resemblance to the U. S. National Museum and other Government buildings at Washington. It covers an area of 350 by 420 ft., and is constructed of iron and glass. Its leading architectural fea-ture is an imposing central dome 120 ft. in diameter and 150 ft. high, the floor of which will be kept free from exhibits. The building fronts to the W., and connects on the N., by a bridge over the lagoon, with the building of the fishing exhibit. The S. half of the Government building is devoted to the exhibits of the Post-Office department, Treasury de-partment, War department, and Department of Agriculture. The N. half is devoted to the exhibits of the Fisheries Commission, Smithsonian Institution, and Interior department. The State department exhibit ex-tends from the rotunda to the E. end, and that of the Department of Justice from the rotunda to the W. end of the building. The allot-ment of space for the several department exhibits is : War department, 23,000 square ft. ; Treasury, 10,500 square ft. ; Agriculture, 23,250 square ft. ; Interior, 24,000 square ft. ; Post-Office, 9,000 square ft. ; Fishery, 20,000 square ft. ; and Smithsonian Institution, balance of space.

Horticultural Building (architect, W. L. B. Jenney ; cost, $300,-000).—This building is immediately south of the entrance to Jackson Park from the Midway Plaisance, and facing east on the lagoon. In front of it is a flower terrace for outside exhibits, including tanks for

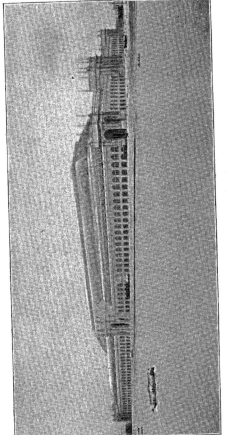

Manufactures and Liberal Arts Building.

nymphæa and the Victoria Regia. The front of the terrace, with its low parapet between large vases, borders the water, and at its center forms a boat-landing. The building is 1,000 ft. long, with an extreme width of 250 ft. The plan is a central pavilion with two end pavilions, each connected with the central one by front and rear curtains, forming two interior courts, each 88 by 270 ft. These courts are beautifully decorated in color, and planted with ornamental shrubs and flowers. The center of the pavilion is roofed by a crystal dome 187 ft. in diame-ter and 113 ft. high, under which are exhibited the tallest palms, bam-boos, and tree-ferns that can be procured. There are galleries in each of the pavilions. The galleries of the end pavilions are designed for *cafés*, the situation and the surroundings being particularly adapted to recreation and refreshment. These *cafés* are surrounded by an arcade on three sides, from which charming views of the grounds can be ob-tained. In this building are exhibited all the varieties of flowers, plants, vines, seeds, horticultural implements, etc. Those exhibits requiring sunshine and light are shown in the rear curtains, where the roof is en-tirely of glass and not too far removed from the plants. The front curtains and space under the galleries are designed for exhibits that re-quire only the ordinary amount of light. Provision is made to heat such parts as require it. The exterior of the building is in "staff," tinted in a soft warm buff, color being reserved for the interior and the courts.

Machinery Hall (architects, Peabody and Stearns; cost, $1,200,-000).—This building measures 850 by 500 ft., and includes the machin-ery annex and power-house. It is located at the extreme south end of the park, midway between the shore of Lake Michigan and the west line of the park. It is just south of the Administration building, and west and across a lagoon from the Agricultural building. The building is spanned by three-arched trusses, and the interior presents the ap-pearance of three railroad train-houses side by side, surrounded on all the four sides by a gallery 50 ft. wide. The trusses are built separately, so that they can be taken down and sold for use as railroad train-houses. In each of the long naves there is an elevated traveling crane running from end to end of the building for the purpose of moving ma-chinery. These platforms are built so that visitors may view from them the exhibits beneath. The power for this building is supplied from a power-house adjoining the south side of the building.

Manufactures and Liberal Arts Building (architect, George B. Post; cost, $1,700,000).—This is the mammoth structure of the Exposition. It measures 1,687 by 787 ft. and covers nearly 31 acres, and is the largest Exposition building ever constructed. Within the building a gallery 50 ft. wide extends around the four sides, and projecting from this are 86 smaller galleries 12 ft. wide, from which the array of exhibits and the scene below may be viewed. The galleries are approached upon the main floor by 30 staircases, the flights of which are 12 ft. wide each. "Columbia Avenue," 50 ft. wide, extends through the mammoth build-ing longitudinally, and an avenue of like width crosses it at right angles at the center. The main roof is of iron and glass, and arches an area

385 by 1,400 ft., and has its ridge 150 ft. from the ground. The build-ing, including its extensive galleries, has about 40 acres of floor-space. The architecture is in the Corinthian style, and in point of being severely classic excels nearly all the other edifices. The long array of columns and arches, which its façades present, is relieved from monoto-ny by very elaborate ornamentation, in which female figures, symbolical of the various arts and sciences, play a conspicuous and very attractive part. The exterior of the building is covered with "staff," treated to represent marble. The huge fluted columns and the immense arches are apparently of this beautiful material. There are four great en-trances, one in the center of each façade. These are designed in the manner of triumphal arches, the central archway of each being 40 ft. wide and 80 ft. high. Surmounting these portals is the attic story or-namented with sculptured eagles 18 ft. high, and on each side above the side arches are great panels with inscriptions, and the spandrels are filled with sculptured figures in bas-relief. At each corner of the main building are pavilions forming arched entrances, which are designed in harmony with the great portals.

Mines and Mining Building (architect, S. S. Beman ; cost, $265,-000).—Located at the southern extremity of the western lagoon or lake, and between the Electricity and Transportation buildings, is the Mines and Mining building. It is 700 ft. long by 350 ft. wide. Its architecture has its inspiration in early Italian renaissance, with which sufficient lib-erty is taken to invest the building with the animation that should characterize a great general Exposition. There is a decided French spirit pervading the exterior design, but it is kept well subordinated. In plan it is simple and straightforward, embracing on the ground floor spacious vestibules, restaurants, etc. On each of the four sides of the building are placed the entrances, those of the north and south fronts being the most spacious and prominent. To the right and left of each entrance, inside, start broad flights of easy stairs leading to the gal-leries. The galleries are 60 ft. wide and 25 ft. high from the ground-floor, and are lighted on the sides by large windows, and from above by a high clear-story extending around the building. The main fronts look southward on the great central court, and northward on the western and middle lakes and an island gorgeous with flowers. These prin-cipal fronts display enormous arched entrances, richly embellished with sculptural decorations emblematic of mining and its allied industries. At each end of these fronts are large square pavilions, surmounted by low domes, which mark the four corners of the building, and are light-ed by large arched windows extending through the galleries. Between the main entrance and the pavilions are richly decorated arcades, form-ing an open loggia on the ground floor, and a deeply recessed prome-nade on the gallery floor level, which commands a fine view of the lakes and islands to the northward and the great central court on the south. These covered promenades are each 25 ft. wide and 230 ft. long, and from them is had access to the building at numerous points. These loggias on the first floor are faced with marbles of different kinds and hues, which will be considered part of the mining exhibit, and so util-

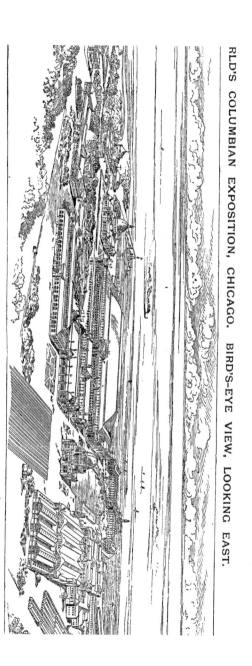

RLD'S COLUMBIAN EXPOSITION, CHICAGO. BIRD'S-EYE VIEW, LOOKING EAST.

Fisheries.

U. S. Government.

Naval.

Casino and Pier.

Manufactures and Liberal Arts.

Electricity.

Agriculture.

Forestry.

Horticulture.

Transportation.

Mining.

Administration.

R. R. Approaches.

Machinery.

in's.

ized as to have marketable value at the close of the Exposition. The loggia ceilings will be heavily coffered, and richly decorated in plaster and color. The ornamentation is massed at the prominent points of the façade. The exterior presents a massive though graceful appearance.

Naval Exhibit (designer, Frank W. Grogan; cost, $100,000).— This imitation battle-ship of 1893 is erected on piling on the lake front in the northeast portion of Jackson Park. It is surrounded by water, and has the appearance of being moored to a wharf. The structure has all the fittings that belong to the actual ship, such as guns, turrets, torpedo-tubes, torpedo-nets and booms, with boats, anchors, chain-cables, davits, awnings, deck fittings, etc., together with all appliances for working the same. Officers, seamen, mechanics, and marines are detailed by the Navy Department during the Exposition, and the discipline and mode of life on our naval vessels are completely shown. The detail of men is not, however, as great as the complement of the actual ship. The crew gives certain drills, especially boat, torpedo, and gun drills, as in a vessel of war. The dimensions of the structure are those of the actual battle-ship, to wit: Length, 348 ft.; width amidships, 69 ft. 3 in.; and from the water-line to the top of the main deck, 12 ft. Centrally placed on this deck is a superstructure 8 ft. high with a hammock berthing on the same 7 ft. high, and above these are the bridge, chart-house, and the boats. At the forward end of the superstructure there is a cone-shaped tower, called the "military mast," near the top of which are placed two circular "tops" as receptacles for sharpshooters. Rapid-firing guns are mounted in each of these tops. The height from the water-line to the summit of this military mast is 76 ft., and above is placed a flagstaff for signaling. The battery mounted comprises four 13-in. breech-loading rifle cannon; eight 8-in. breech-loading rifle cannon; four 6-in. breech-loading rifle cannon; twenty 6-pounder rapid-firing guns; six 1-pound rapid-firing guns; 2 Gatling guns, and 6 torpedo-tubes or torpedo-guns. All these are placed and mounted respectively as in the genuine battle-ship. On the starboard side of the ship is shown the torpedo-protection net, stretching the entire length of the vessel. Steam launches and cutters ride at the booms, and all the outward appearance of a real ship of war is imitated.

Transportation Building (architects, Adler and Sullivan; cost, $370,000).—The Transportation building is at the southern end of the west flank, and lies between the Horticultural and the Mines buildings. Facing eastward, it commands a view of the floral island and an extensive branch of the lagoon. It is exquisitely refined and simple in architectural treatment, although very rich and elaborate in detail. In style it savors much of the Romanesque, although to the initiated the manner in which it is designed on axial lines, and the solicitude shown for fine proportions, and subtle relation of parts to each other, will at once suggest the methods of composition followed at the École des Beaux Arts. The main entrance to the Transportation building consists of an immense single arch enriched to an extraordinary degree with carvings,

38

bas-reliefs, and mural paintings, the entire feature forming a rich and beautiful yet quiet color climax, for it is treated in leaf and is called the golden door. The remainder of the architectural composition falls into a just relation of contrast with the highly wrought entrance, and is duly quiet and modest, though very broad in treatment. It consists of a continuous arcade with subordinated colonnade and entablature. Numerous minor entrances are from time to time pierced in the walls; and with them are grouped terraces, seats, drinking-fountains, and statues. The interior of the building is treated much after the manner of a Roman basilica, with broad nave and aisles. The roof is therefore in three divisions. The middle one rises much higher than the others, and its walls are pierced to form a beautiful arcaded clear-story. The cupola, placed exactly in the center of the building and rising 165 ft. above the ground, is reached by 8 elevators. These elevators of themselves naturally form a part of the transportation exhibit, and as they also carry passengers to galleries at various stages of height, a fine view of the interior of the building may easily be obtained. The main galleries of this building, because of the abundant elevator facilities, prove quite accessible to visitors. The main building of the transportation exhibit measures 960 ft. front by 250 ft. deep. From this extends westward to Stony Island Avenue an enormous annex covering about 9 acres. This is one story only in height. In it may be seen the more bulky exhibits. Along the central avenue or nave the visitor may see facing each other scores of locomotive-engines, highly polished, and rendering the perspective effect of the nave both exceedingly novel and striking. Add to the effect of the exhibits the architectural impression given by a long vista of richly ornamented colonnade, and it may easily be seen that the interior of Transportation building is one of the most impressive of the World's Fair. The transportation exhibits naturally include everything, of whatsoever name or sort, devoted to the purpose of transportation, and range from a baby-carriage to a giant engine, from a cash conveyor to a balloon or carrier pigeon.

Woman's Building (architect, Miss Sophia B. Hayden; cost, $138,000).—This structure is designed on the Italian renaissance. Directly in front of the building the lagoon takes the form of a bay about 400 ft. in width. From the center of this bay a grand landing and staircase leads to a terrace 6 ft. above the water. Crossing this terrace other staircases give access to the ground 4 ft. above, on which, about 100 ft. back, the building is situated. The first terrace is designed in artistic flower-beds and low shrubs. The principal façade has an extreme length of 400 ft., the depth of the building being half this distance. The first story is raised about 10 ft. from the ground line, and a wide staircase leads to the center pavilion. This pavilion, forming the main triple-arched entrance, with an open colonnade in the second story, is finished with a low pediment enriched with a highly elaborate bas-relief. The corner pavilions have each an open colonnade added above the main cornice. Here are located the hanging gardens. A lobby 40 ft. wide leads into the open rotunda, 70 by 65 ft., reaching through the height of the building, and protected by a richly orna-

The Woman's Building.

mented sky-light. This rotunda is surrounded by a two-story open arcade, as delicate and chaste in design as the exterior, the whole having a thoroughly Italian court-yard effect, admitting abundance of light to all rooms facing this interior space. On the first floor are located, on the left hand, a model hospital; on the right, a model kindergarten; each occupying 80 by 60 ft. The whole floor of the south pavilion is devoted to the retrospective exhibit; the one on the north to reform work and charity organization. Each of these floors is 80 by 200 ft. The curtain opposite the main front contains the library, bureau of information, records, etc. In the second story are located ladies' parlors, committee-rooms, and dressing-rooms, all leading to the open balcony in front. The whole second floor of the north pavilion incloses the great assembly-room and club-room. The first of these is provided with an elevated stage for the accommodation of speakers. The south pavilion contains the model kitchen, refreshment-rooms, reception-rooms, etc. The building is incased with "staff," the same material used on the rest of the buildings, and as it stands with its mellow, decorated walls bathed in the bright sunshine, the women of the country are justly proud of the result.

State and Territorial Buildings.

Suitable sites were allotted by the management of the World's Fair to every State and Territory for their special exhibits. Many of them have erected characteristic or special buildings. In several instances, as in Florida, Massachusetts, and Pennsylvania, historical structures have been reproduced. A brief description of the more important State buildings is herewith given:

California.—The architecture of this structure is. in the style of the old California Missions. P. Brown, of San Francisco, was the architect, and the cost was $75,000. State appropriation, $300,000.

Colorado.—The style of architecture chosen was that of the Spanish renaissance. The architect was H. T. E. Wendell, of Denver, and the cost $35,000. State appropriation, $100,000.

Connecticut.—This State building is a type of a Connecticut colonial residence. Warren R. Briggs, of Bridgeport, was the architect, and the cost $12,000.—

Florida.—For this State there has been reproduced a miniature of old Fort Marion, in St. Augustine. The architect was W. Mead Walter, of Chicago, and the cost $20,000. State appropriation, $50,000.

Illinois.—The style of architecture is an adaptation of the Italian renaissance. The architects were Boyington & Co., of Chicago, and it cost $250,000. State appropriation, $800,000.

Indiana.—This building is in the French-Gothic style of architecture, such as is seen in the chateaux of France. The architect was Henry I. Cobb, and the cost $60,000. State appropriation, $75,000.

Iowa.—The building of this State includes the "Shelter" and a wooden structure combined, after the style of a French chateau. Architects, the Josselyn & Taylor Co., of Cedar Rapids. Cost, $35,000. State appropriation, $130,000.

Maine.—The building of this State is octagon in form, and is partially constructed of Maine granite. The architect is Charles S. Frost, of Chicago, and the cost was $20,000   State appropriation, $40,000.

Maryland.—This building is a reproduction of the State-House in Annapolis. It cost $35,000.   State appropriation, $60,000.

Massachusetts.—This building is in the colonial style, and is largely a reproduction of the John Hancock House. It was designed by Peabody & Stearns, of Boston, and cost $50,000.   State appropriation, $150,000.

Minnesota.—The architecture of this State's building is in the Italian rennaissance style. William C. Whitney is the architect, and the cost $30,000.   State appropriation, $50,000.

Montana.—The style of architecture followed in this building is the Romanesque. The architects are Galbraith & Fuller, of Livingston, and cost $16,000.   State appropriation, $50,000.

New York.—For this State, McKim, Mead & White have designed a large summer-house resembling an Italian villa. Its cost was $77,000.   State appropriation, $300,000.

Ohio.—The building from this State is colonial in style, and built largely of native woods. James McLaughlin, of Cincinnati, is the architect, and the cost $30,000.   State appropriation, $125,000.

Pennsylvania.—From this State we have an exact reproduction of the old Independence Hall, of Philadelphia. The architect is R. Lonsdale, of Philadelphia, and the cost $60,000.   State appropriation, $300,000.

South Dakota.—This State has erected a frame structure, the exterior of which is covered with Yankton cement, in imitation of stone work. The architect is W. L. Dow, of Sioux Falls, and the cost $15,000.   State appropriation, $25,000.

Texas.—The building of this State is a good example of the Spanish renaissance architecture, resembling one of the old Spanish Missions.

Washington.—A building entirely of wood from the Puget Sound region has been erected by this State. Warren P. Skillings, of Seattle, is the architect, and the cost was $100,000.   State appropriation, $100,000.

Wyoming.—The building of this State is in the French château style. It cost $20,000.   State appropriation, $30,000.

Midway Plaisance.

This is a strip of land 600 ft. wide and seven eighths of a mile long, between 59th and 60th Sts., containing 80 acres, connecting Jackson and Washington Parks. In it are located all the amusements and other attractions of the Fair, outside the main Exhibition buildings.

The following entertainments may be seen in the Midway Plaisance.

Austrian Village.—A representation of a section of a street in old Vienna, called "Der Graben." The character of this concession is similar to that of the German Village.

Bohemian Glass Factory.—The entire process of making the celebrated Bohemian glassware is shown by the native Bohemian workmen. The building is a reproduction of the native factories.

Captive Balloon.—A balloon with a capacity of carrying from 12 to 20 people to a height of 1,500 ft. The latest machinery known to aërial navigation is introduced in connection with this balloon, and it is also proposed to demonstrate to what practical uses balloons can be put.

Dahomey Village.—This consists of a settlement of from 30 to 60 natives, of both sexes, including a king and several chiefs. These people will execute their various dances, give their war-cries, and perform such rites and ceremonies as are peculiar to them. They have the privilege of selling such native merchandise as they may produce.

Dutch Settlement.—A practical demonstration of the habits and customs of the people of the South Sea islands. The natives will sell their manufactured articles, and give entertainments peculiar to their race.

East India Settlement.—This is similar in character to the Dutch settlement. Natives show their mode of living, sell their wares, and typical jugglers and snake-charmers will perform.

Ferris Wheel.—This attraction is a wheel, 250 ft. in diameter, swung on an axle, which rests upon towers 135 ft. high. The purpose of the wheel is that there shall be hung from it, at different points on the perimeter, cars similar in character to those used in elevators, the lowest car resting on the ground as the people get into it. The wheel is then started in motion, and the occupants make the complete circuit of 250 ft.

German Village.—A group of houses representative of a German village of the present time; also a German town of mediæval times. There are the houses of the upper Bavarian mountains, the houses of the Black Forest, the Hessian and Altenberg house of Silesian peasants, representing the middle Germans, the Westphalien Hof, the Lower Saxons, the Hallighaus, the Friesen, and the house from Spreewald and Niederdeutsche. In the various houses is installed original household furniture, so characteristic as to be readily distinguished as belonging to particular tribes.

Hagenbeck Animal Show.—A trained troupe of from 60 to 90 animals, including lions, tigers, dogs, cattle, horses, elephants, etc., at play about the cage. They go through many athletic performances, which can be believed only after it has been seen.

Ice Railway.—The railway is built on an incline, and is a practical summer toboggan slide. The ice which covers the surface of the incline is made and perpetuated by machinery.

Irish Industries.—An exhibit of the Irish cottage industries, including a reproduction of the ruins of Donegal Castle, making habitable such rooms as possible without destroying the historical beauty of the ruins.

Japanese Bazaars.—These show the Japanese people, their customs, and merchandise. The bazaars are operated under contract with the Imperial Japanese Commission.

Libbey Glass Exhibit.—The company will demonstrate the production of glassware, except plate and window glass. The building is largely constructed of glass, and the exterior set with prisms of cut glass, like great diamonds. The plant includes a sixteen-pit furnace, cutting, etching, engraving, and decorating shops, and a great display of glassware.

Minaret Tower.—A reproduction of a Turkish structure, the concession being operated by Turks. Among the attractions here is a silver bed once owned by a Sultan. It is said to weigh two tons, and to be composed of 2,000 pieces. There is also shown an immense embroidered tent, once owned by the Shah of Persia.

Moorish Palace.—This building is in design after the style of old Moorish temples, the remains of which are still found in some portions of Spain and northern Africa. In this building are various novelties in the line of illusions, camera-obscura, etc. There is also a restaurant which is capable of seating 500 people. One of the great attractions in this building is the exhibit of $1,000,000 in gold coins.

Morocco.—Similar to the other national sections.

Natatorium.—The building is 190 to 250 ft., and has a large swimming-pool. There are a *café* and bakery in connection with the natatorium.

Nursery Exhibit.—This is the final exhibit in the Plaisance, occupying about five acres in the western end of the tract. It is sought here to show the most artistic effects possible in a combination of flowers and shrubbery.

Panorama of the Bernese Alps.—Here is shown the scenery of the Alps, and in connection with this feature is an exhibition of the manufactured products of the country.

Panorama of the Volcano of Kilauea.—This volcano is supposed to have the greatest crater in existence. The visitor is taken to an island in the center of the crater, and, while surrounded by a sea of fire, views the scenery around the volcano.

Pompeiian House.—A reproduction of a typical house of ancient Pompeii. Installed in the house is an exhibit of articles gathered from the excavated ruins of the ancient city.

Sliding Railway.—On the southern edge of the Plaisance and extending its entire length. It is a French invention, and was first given a practical demonstration in the Paris Exposition of 1889. It is an elevated road, the cars having no wheels. The rail is 8 in. wide, the substitute for the wheel being a shoe, which sets over the side of the rail, and is practically water-tight. The speed claimed by the inventors is 120 to 160 miles an hour.

Street in Cairo.—The street is constituted of reproductions of historic buildings in the Egyptian city. Shops, mosques, a theatre, a dancing-hall, etc., are shown in the buildings. The customs of the people are delineated, many attractions peculiar to Arabia and the Soudan are introduced, and curiosities from the museums in Cairo and Alexandria are exhibited.

Tower of Babel.—This structure is 400 ft. in height, diameter at base 100 ft. The ascent of the tower is made by a double-track, circular electric railway, by elevators, and by a broad walk. At the top a chime of bells is installed, and meteorological experiments are conducted.

Tunisian and Algerian Section.—Typical people of northern Africa show here their mode of life, their amusements, and their manufactures. Several tribes are represented, each having its chief or sheik. The minaret tower is in this section.

Turkish Village.—A reproduction of one of the old street squares in Stamboul. The people and the goods of Turkey in Europe and Turkey in Asia are shown. Entertainments peculiar to the people are given.

Other Features.

Besides the foregoing, many features of interest well worthy of notice should be described; but it is impossible in this APPENDIX to do adequate justice to everything, and therefore the minor attractions are left for those works which deal exclusively with the World's Fair.

TABLE OF RAILWAY AND STEAMBOAT FARES

From New York to the Leading Cities and Places of Interest in the United States and Canada.

☞ *The Railway named is that by which the traveler leaves New York.*

☞ The rates given are those which obtain at the time of going to press, but are liable to slight variations. They are both for unlimited and limited tickets. Unlimited tickets are good until used, and permit of stop-over at any place and for any time *en route*. The limited tickets are good for continuous passage only, and will not permit of stop-over. We do not give the price of excursion-tickets (good for passage both ways), as these are so variable at different times and are issued to but few points.

NEW YORK TO	VIA	Unlimited.	Limited.
Aiken, S. C.	Baltimore, Washington, Lynchburg		$23 25
" "	Washington, Richmond, & Wilmington		23 25
Albany, N. Y.	Hudson River *or* West Shore R. R.	$3 10	
" "	Steamboat	2 00	
Atlanta, Ga.	Harrisburg, Luray, and Roanoke		24 00
" "	Washington, Lynchburg, and Charlotte		24 00
" "	Washington, Lynchburg, and Bristol		24 00
" "	Baltimore & Ohio and Shenandoah Junction		24 00
" "	Washington, Richmond & Seaboard Air Line		24 00
Atlantic City, N. J.	Pennsylvania *or* New Jersey Southern R. R.	3 25	
Augusta, Ga.	Baltimore, Norfolk, Weldon, and Columbia		23 00
" "	Washington, Richmond, and Wilmington		23 00
" "	Washington, Lynchburg, and Danville		23 00
Baltimore, Md.	Pennsylvania R. R.	5 30	
" "	Baltimore & Ohio R. R.	5 30	
Boston, Mass.	New York & New Haven R. R.	5 30	
" "	Fall River, *or* Stonington, *or* Providence, *or* Norwich steamers	4 00	
Buffalo, N. Y.	New York Central, Erie, *or* West Shore R. R.	9 25	
Burlington, Iowa	New York Central R. R.	29 40	26 15
" "	Erie *or* West Shore R. R.	29 40	26 15
" "	Pennsylvania R. R.	32 65	26 15
" "	Baltimore & Ohio R. R.		23 15
Burlington, Vt.	New York Central & Hudson River R. R.	8 00	
Cape May, N. J.	New Jersey Southern *or* Pennsylvania R. R.	4 25	
Charleston, S. C.	Baltimore, Norfolk, and Weldon		21 55
" "	Washington and Richmond		21 55
" "	Washington, Lynchburg, Danville, Charlotte, and Columbia		21 55
" "	Steamer (Pier 19, East River)	20 00	
Chattanooga, Tenn.	Washington, Lynchburg, and Bristol		23 00
" "	Washington, Lynchburg, Asheville and Knoxville		23 00

Table of Railway and Steamboat Fares.—(Continued.)

NEW YORK TO	VIA	Unlimited.	Limited.
Chattanooga, Tenn.....	Baltimore & Ohio, and Shenandoah Junction.........................		$23 00
Chicago, Ill............	New York Central R. R.............	$22 25	20 00
" "	Erie *or* West Shore R. R..........	26 50	20 00
Chicago, Ill............	Baltimore & Ohio R. R.............	26 50	17 00
" "	Pennsylvania R. R.	26 50	20 00
Cincinnati, Ohio	New York Central, Erie, *or* West Shore R. R.....................	21 25	18 00
" "	Pennsylvania R. R................	21 50	18 00
" "	Baltimore & Ohio R. R...........	21 50	16 00
Cleveland, Ohio........	New York Central, Erie, *or* West Shore R. R.....................	14 25	13 00
" "	Pennsylvania R. R................	16 50	13 00
Colorado Springs, Col..	New York Central R. R............	52 90	49 90
" " " ..	Erie *or* West Shore R. R..........	52 90	49 90
" " " ..	Pennsylvania R. R................	56 95	49 90
" " " ..	Baltimore & Ohio R. R...........		45 15
Columbus, Ohio........	New York Central, Erie, Baltimore & Ohio, *or* Pennsylvania R. R ...	18 40	15 25
Cooperstown, N. Y.....	New York Central & Hudson River R. R............................	6 15	
Delaware Wat.-Gap, Pa.	Morris & Essex (Delaware, Lackawanna, & Western) R. R.:.......	2 55	
Denver, Col............	New York Central R. R............	52 90	49 90
" "	Erie *or* West Shore R. R..........	52 90	49 90
" "	Pennsylvania R. R	56 95	49 90
" "	Baltimore & Ohio R. R............		45 15
Detroit, Mich..........	New York Central, Erie, Baltimore & Ohio, *or* Pennsylvania R. R. ...	21 00	16 25
Frankfort, Ky..........	New York Central, Erie, *or* Pennsylvania.......................	25 35	20 80
Galveston, Texas.......	New York Central, Erie, Pennsylvania, *or* Baltimore & Ohio R. R. (Western Route)..............		46 30
" "	Washington, Atlanta, and New Orleans........................		46 30
" "	Steamer (Pier 20, East River).....	50 00	
Halifax, Can...........	New York & New Haven R. R......	20 00	18 00
Hartford, Conn........	" " "	2 65	
Hot Springs, Ark.......	Pennsylvania, Erie, Baltimore & Ohio, *or* New York Central R. R.		32 40
Houston, Texas........	New York Central, Erie, Pennsylvania, *or* Baltimore & Ohio R. R. (Western Route)................		44 80
" "	Washington, Atlanta, and New Orleans...........................		44 80
Indianapolis, Ind......	New York Central *or* Erie.........	22 50	19 00
" "	Baltimore and Ohio R. R..........	23 80	17 00
Jacksonville, Fla......	Baltimore, Norfolk, and Weldon ...		29 15
" "	Washington, Richmond, and Wilmington......................		29 15
" "	Washington and Atlanta, *or* Columbia............................		29 15
" "	Steamer (Pier 19, East River)......	25 00	
Kansas City, Mo.......	New York Central R. R............		31 75
" " "	Erie *or* West Shore R. R..........		31 75
" " "	Pennsylvania R. R................		31 75
" " "	Baltimore & Ohio R. R............		27 00
" " "	Washington, Atlanta, and Memphis...........................		31 75
Kingston, Can.........	New York Central & Hudson River R. R............................	9 20	

Table of Railway and Steamboat Fares.—(Continued.)

NEW YORK TO	VIA	Unlimited.	Limited.
Leadville, Col.........	New York Central R. R............	$60 90	$57 90
" "	Erie or West Shore R. R...........	60 90	57 90
" "	Pennsylvania R. R	64 95	57 90
" "	Baltimore & Ohio R. R............		53 15
Little Rock, Ark......	Pennsylvania R. R., Baltimore & Ohio R. R., and St. Louis or Cairo		33 00
" " "	Baltimore and Ohio R. R. and St. Louis or Cairo......		33 00
" " "	Washington, Atlanta & Memphis...		33 00
Long Branch, N. J....	Central R. R. of New Jersey.......	1 35	1 00
" "	New Jersey Southern (Pier 14, North River).......................	1 00	
Los Angeles, Cal......	Pennsylvania or New York Central R. R.............................		91 75
" "	Erie or West Shore..............		91 75
" "	Baltimore & Ohio R. R.		87 00
" "	Washington, Atlanta, New Orleans & So. Pac. R. R..........		91 75
Louisville, Ky.........	New York Central, Erie, or West Shore R. R...................	24 75	21 50
" "	Pennsylvania R. R................	25 00	21 50
" "	Baltimore & Ohio R. R...........	25 00	19 50
Lynchburg, Va........	Pennsyl. or Baltimore & Ohio R. R.	11 70	
Madison, Wis.........	New York Central R. R...........	26 15	23 90
" "	Erie or West Shore R. R...........	26 15	23 90
" "	Pennsylvania R. R...............	30 40	23 90
" "	Baltimore & Ohio R. R...........		20 90
Mauch Chunk, Pa......	Morris & Essex or New Jersey Central R. R......................	3 45	
Memphis, Tenn........	Cincinnati and Louisville (Western Route).......................		29 50
" "	Washington and Lynchburg (Atlanta & Birmingham).............		29 50
Mexico, Mex.........	New York Central, Erie, West Shore, Baltimore & Ohio, and Pennsylvania R. R. via Laredo		85 20
Milwaukee, Wis.......	New York Central R. R...........	24 80	22 55
" "	Erie or West Shore R. R...........	24 80	22 55
" "	Pennsylvania R. R...............	29 05	22 55
" "	Baltimore & Ohio R. R...........		19 55
Minneapolis, Minn.....	New York Central R. R...........	33 75	31 50
" "	Erie or West Shore R. R...........	33 75	21 50
" "	Pennsylvania R. R...............	38 00	31 50
" "	Baltimore & Ohio R. R...........		28 50
Mobile, Ala..........	Cincinnati and Louisville (Western Route).......................		32 00
" "	Baltimore or Washington (Southern Route).......................		32 00
Montgomery, Ala......	(Same routes as to Mobile) (Southern Route)...................		28 50
Montreal, Can........	New York Central or New York & New Haven R. R..............	10 00	
Nashville, Tenn.......	Cincinnati and Louisville (Western Route).......................		23 65
" "	Washington and Lynchburg (Southern Route)..................		25 65
New Haven, Conn.....	New York & New Haven R. R.....	1 75	
New Orleans, La......	Cincinnati direct (Western Route)..		31 00
" "	Harrisburg, Roanoke, and Chattanooga (or B. & O.), and Shen. Jct.		34 00
" "	Washington, Richmond (or B. & O.), and Atlanta		34 00

NEW YORK TO	VIA	Unlimited.	Limited.
New Orleans, La......	Washington, Lynchburg, Charlotte & Atlanta.....................		$34 00
" "	Washington, Richmond, Weldon & Seaboard Air Line................		24 00
" "	Steamer (2 lines)..................	$35 00	
Newport, R. I..........	New York & New Haven R. R.....	5 00	
" "	Fall River steamers...............	3 00	
Niagara Falls..........	N. Y. Central, or West Shore, or Erie R. R......................	9 25	
Norfolk, Va...........	Pennsylvania R. R................	8 30	
" "	Baltimore & Ohio, and Bay Line...	8 30	
" "	Steamer, via Washington..........	8 00	
" "	Steamers direct..................	8 00	
Northampton, Mass....	New York & New Haven R. R......	4 25	
Oil City, Pa............	Erie R. R........................	13 35	10 90
Omaha, Neb......	New York Central R. R............	35 00	32 75
" "	Erie or West Shore R. R..........	35 00	32 75
" "	Pennsylvania R. R...............	39 25	32 75
" "	Baltimore & Ohio R. R...........		29 75
Ottawa, Can..........	N. Y. Central & Hudson River R. R.....	11 35	
Philadelphia, Pa......	Pennsylvania or New Jersey Central R. R........................	2 50	
Pittsburg, Pa..........	Pennsylvania or New Jersey Central R. R........................	12 50	
" "	Baltimore & Ohio R. R...........	15 15	10 50
Pittsfield, Mass........	New York & New Haven R. R.....	3 28	
Plattsburg, N. Y........	New York Central & Hudson River R. R.........................	8 00	
Portland, Me..........	New York & New Haven R. R.....	8 50	
" "	Steamer to Boston, thence by R. R.	7 00	6 50
Portland, Ore.........	Pennsylvania or New York Central R. R.........................	104 00	87 50
" "	Erie or West Shore R. R......	99 75	87 50
" "	Canada Pacific..................	104 00	75 50
" "	Baltimore & Ohio R. R...........		86 50
Portsmouth, N. H.....	(Same routes as to Portland).......	6 20	
Providence, R. I.......	New York & New Haven R. R.....	4 50	4 50
" "	Steamer (Pier 29, North River)	4 00	
Quebec, Can..........	New York & New Haven R. R.....	12 00	
Raleigh, N. C..........	Pennsylvania R. R...............		14 25
" "	Baltimore & Ohio, and Washington.		15 25
Richmond, Va.........	Pennsylvania or Baltimore & Ohio R. R........................	10 00	
" "	Steamers (foot of Beach Street, North River)......................	9 00	
" "	B. & O. R. R.. or Penn. R. R., and York River Line..............	7 80	
Rutland, Vt...........	New York Central & Hudson River R. R.........................	5 64	
Sacramento, Cal.......	New York Central R. R............		91 75
" "	Erie or West Shore R. R..........		91 75
" "	Pennsylvania R. R...............		91 75
" "	Baltimore & Ohio R. R...........		91 75
St. Augustine, Fla.....	Washington, Richmond, and Wilmington		30 65
" "	Washington, Lynchburg, Charlotte & Columbia or Atlanta		30 65
" "	Steamers to Charleston or Savannah		
St. John, N. B........	New York & New Haven R. R.....	15 50	13 50
St. Joseph, Mo........	New York Central R. R............	34 75	31 75
" "	Erie or West Shore R. R...........	34 75	31 75

Table of Railway and Steamboat Fares.—(Continued.)

NEW YORK TO	VIA	Unlimited.	Limited.
St. Joseph, Mo.........	Pennsylvania R. R.................	$38 80	$31 75
" "	Baltimore & Ohio R. R. ...		27 00
St. Louis, Mo.........	New York Central R. R............	39 75	24 25
" "	Erie or West Shore R. R............	39 75	24 25
" "	Pennsylvania R. R	31 80	24 25
" "	Baltimore & Ohio R. R.............	31 80	21 00
St. Paul, Minn.........	New York Central R. R............	33 75	31 50
" "	Erie or West Shore R. R.	33 75	31 50
" "	Pennsylvania R. R.... .	38 00	31 50
" "	Baltimore & Ohio R. R		28 50
Salt Lake City, Utah...	New York Central R. R.............	74 75	71 75
" " " ...	Erie or West Shore R. R...........	74 75	71 75
" " " ...	Pennsylvania R. R.................	78 80	71 75
" " " ...	Baltimore & Ohio R. R.............		67 00
San Francisco, Cal.....	New York Central R. R.		91 75
" " "	Erie or West Shore R. R... ...		91 75
" " "	Pennsylvania R. R. & New Orleans.		91 75
" " "	Baltimore & Ohio R. R............		91 75
" " "	Canada Pacific via Portland, Ore...		81 75
Saratoga Springs, N. Y.	New York Central & Hudson River or West Shore R. R..............	4 20	
Savannah, Ga....... ..	Washington, Richmond, & Charleston........................		24 00
" "	Washington, Lynchburg, and Charlotte........................		24 00
" "	Steamer (Pier 35, North River)	20 00	
Sharon Springs, N. Y..	New York Central & Hudson River R. R............................	4 90	
Springfield, Ill.........	New York Central R. R............	27 00	23 75
" "	Erie or West Shore R. R...........	27 00	23 75
" "	Pennsylvania R. R.................	29 60	23 75
" "	Baltimore & Ohio R. R...	29 60	21 10
Springfield, Mass......	New York & New Haven R. R.	2 75	
Staunton, Va...........	Pennsylvania or Baltimore & Ohio R. R........................		11 05
Toledo, Ohio..........	New York Central or Erie.........	17 35	16 25
" "	Baltimore & Ohio R. R.	19 75	16 25
Toronto, Can..........	New York Central, Erie, or Pennsylvania R. R......................	11 85	
Trenton Falls, N. Y....	New York Central & Hudson River R. R............................	5 54	
Vancouver, B. C........	Canada Pacific....................	104 00	77 50
" "	B. & O., and Northern Pacific R. R.		86 50
" "	B. & O., and Canada Pacific R. R.		76 50
Washington, D. C......	Pennsylvania or Baltimore & Ohio R. R........................	6 50	
Watkins Glen, N. Y....	Erie or New York Central R. R....	7 85	
White Mountains, N. H.	New York & New Haven R. R.....	9 75	
" "	Any steamer route to Boston, thence by R. R.....................	8 00	
White Sul. Springs, Va.	Pennsylvania or Baltimore & Ohio R. R........................	13 80	
Wilkesbarre, Pa.......	Morris & Essex or New Jersey Central R. R.	5 00	
Wilmington, N. C.	Pennsylvania or Baltimore & Ohio R. R........................		16 35
Winnipeg, Manitoba...	New York Central R. R............	47 95	45 70
" " ...	Erie or West Shore R. R...........	47 95	45 70
" " ...	Pennsylvania R. R.............r..	52 20	45 70
" " ...	Baltimore & Ohio R. R.............		42 70
Yosemite Valley, Cal,..	Side excursion from San Francisco.	50 00	

INDEX.

CARDS OF LEADING HOTELS.

By referring to the advertising pages of these GUIDES, the traveler will find advertisements giving full information of many of the leading Hotels, as also Bankers and others.

ALBANY, N. Y.

HOTEL KENMORE.
Leading hotel of Albany, N. Y. Just added, at an outlay of over $100,000, one hundred elegant rooms, grand dining hall (handsomest in the State), lobbies, reading-rooms, etc. This is the only hotel in Albany serving late dinner.

H. J. ROCKWELL, Proprietor.
F. W. ROCKWELL, Manager.

BOSTON, MASS.

THE BRUNSWICK,
Boylston and Clarendon Streets, opposite Trinity (Phillips Brooks's) Church. American plan.

BARNES & DUNKLEE, Proprietors.

THE VICTORIA,
Dartmouth and Newbury Streets, opposite Boston Art Club. European plan.

BARNES & DUNKLEE, Proprietors.

THE VENDOME,
Commonwealth Avenue.

C. H. GREENLEAF & Co., Proprietors.
Amos Barnes,   J. W. Dunklee,
C. H. Greenleaf.

UNITED STATES HOTEL.
Directly opposite the Boston and Albany, and only one block from the Old Colony and Fall River Lines, three blocks only from the New York and New England and Providence and Stonington Stations, and connecting directly by horse-cars with all the Northern and Eastern Railroads and Steamboats, giving guests every possible facility and convenience of rapid and economical transfer from all points.

TILLY HAYNES, Proprietor.

BOSTON, MASS.

COPLEY SQUARE HOTEL.
American and European plans. Huntington and Exeter Street, Boston. Located in the fashionable and beautiful Back Bay District. Containing 300 rooms, single and en suite, richly furnished. It is but six minutes' ride by horse or electric cars to the shopping and amusement centers, and five minutes to different railroad depots. Passengers via Boston & Albany R. R., at Huntington Ave. Station, within one minute's walk of hotel. Hotel porter will be in attendance at trains arriving from New York and the West.

F. S. RISTEEN & Co., Proprietors.

BUFFALO, N. Y.

HOTEL IROQUOIS.
WOOLLEY & GERRANS, Proprietors. The leading hotel in Buffalo. The only absolutely fire-proof house in the city; metropolitan in its structure, arrangements, equipments, and management. American and European plans. Most central location.

CATSKILL MOUNTAINS.

CATSKILL MOUNTAIN HOUSE.
This famous summer Hotel is situated on the Catskill Mountains, eight miles west of the Hudson River and twelve miles from the village of Catskill, N. Y. It has accommodations for 400 guests, and is the largest and leading Hotel of the Catskill region. Open June 20th to September 20th. Great reduction in rates. New Otis elevator from station to house. Send for circular.

C. L. BEACH, Proprietor, Catskill, N. Y.

24

CANADA.

THE QUEEN'S,
Toronto. Celebrated for its home comforts, perfect quiet, good attendance, and the peculiar excellence of its *cuisine*. Delightfully situated near the bay on Front Street, convenient to business center, railroad depot, steamboats, etc.

McGaw & Winnett, Proprietors.

CLIFTON HOUSE.
Niagara Falls, directly fronting the Park Reservations. Sanitary condition perfect. The spray from the Falls keeps the air always pure. This House, situated directly in front of the Fall, possesses superior advantages.

G. M. Colburn, Proprietor,
Niagara Falls.

ST. LAWRENCE HALL,
Montreal. For upward of thirty-five years the name of the St. Lawrence Hall has been familiar to all travelers on this continent. The Hotel is conveniently situated in the heart of the business center of Montreal, and is contiguous to the General Post-Office and other important public buildings. It is handsomely decorated, luxuriously furnished, lighted by the electric light, and fitted with a passenger elevator. The Hotel is under the personal supervision of the proprietor, Mr. Henry Hogan.

THE FLORENCE,
Quebec, is one of the most pleasant, attractive, and comfortable houses for tourists that can be found on this continent. Its location unequaled, and the panoramic view to be had from the Balcony is not even surpassed by the world-renowned Dufferin Terrace, as it commands a full view of the River St. Lawrence, the St. Charles Valley, Montmorency Falls, Laurentian Range of Mountains, and overlooks the largest part of the City. The rooms are large, elegantly furnished, and well ventilated, and the table first class. Benj. Trudel, Proprietor.

HOTEL ROBERVAL,
At Roberval, Lake St. John. An elegant new hotel, accommodates three hundred guests, on a commanding site, affording a magnificent view of the whole expanse of Lake St. John. Almost in front of the hotel is the steamboat wharf, where tourists may embark on the passenger steamers making daily trips and excursions to all points on Lake St. John during the season of navigation.

DEER PARK AND OAKLAND.

ON THE CREST OF THE ALLEGHANIES, 3,000 feet above tide-water. These famous mountain resorts, situated at the summit of the Alleghanies and directly upon the main line of the Baltimore & Ohio R. R. All Baltimore & Ohio trains stop at Deer Park and Oakland during the season. Rates, $60, $75, and $90 a month, according to location.

George D. DeShields, Manager.

EUREKA SPRINGS, ARK.

THE CRESCENT HOTEL.
It is in the midst of the numerous springs, and is conducted for comfort of guests.

LAKE CHAMPLAIN.

HOTEL CHAMPLAIN.
The Superb Summer Hotel of the North. On west shore of Lake Champlain, three miles south of Plattsburg. Delaware and Hudson station in grounds.

O. D. Seavey, Manager.

LEBANON, MO.

HOTEL GASCONADE.
First-class House, with all modern improvements and conveniences.

LENOX, MASS.

CURTIS HOTEL.
This magnificently located house, in the midst of the Berkshire Hills, is open all the year round. All modern improvements and conveniences.

W. O. Curtis, Proprietor.

NEWPORT NEWS, VA.

HOTEL WARWICK,
On Hampton Roads, eight miles above Old Point Comfort and twelve miles from Norfolk. A new brick building, commanding extensive marine, river, and inland views; elegant and complete in its appointment; elevators, steam heat, open fireplaces, artesian well, thorough drainage; natural park and pleasure-ground; pier 300 feet long, with handsome pavilion; separate music and ball-room on the bluff; billiards and bowling-alley; a sloping beach miles in length; interesting drives. Open all the year.

J. R. Swinerton, Manager.

NEW HAMPSHIRE.

HOTEL PONEMAH,
Amherst Station. June to October.
D. S. PLUMER, Proprietor.

NEW YORK.

FIFTH AVENUE HOTEL,
Madison Square. The largest, best appointed, and most liberally managed Hotel in the city, with the most central and delightful location.
HITCHCOCK, DARLING & Co.

PARK AVENUE HOTEL.
Absolutely fire-proof. European plan, $1 per day and upward; American plan, $3.50 per day and upward. Park Avenue, 32d and 33d Streets. WM. H. EARLE & SON, Proprietors, New York. Free baggage to and from Grand Central and Long Island Depots.

THE PLAZA HOTEL.
Located in the heart of New York, at Fifth Ave., 58th and 59th Streets, overlooking Central Park and Plaza Square. The Hotel is absolutely fire-proof. Conducted on American and European plans.
F. A. HAMMOND, Proprietor.

BROADWAY CENTRAL HOTEL,
Opposite Bond Street, under entire new management. One Hundred Thousand Dollars have been spent in thorough refurnishing and alterations, making it now one of the finest and best arranged hotels in the city. Both American and European plans. Rooms, $1 to $2, and board from $2.50 to $3.50.
TILLY HAYNES, Proprietor.

GRAND UNION HOTEL.
Passengers arriving in the City of New York *via* Grand Central Depot save Carriage-Hire and Transfer of Baggage by stopping at the GRAND UNION HOTEL, opposite the depot. Passengers arriving by West Shore Railroad, *via* Weehawken Ferry, by taking the 42d Street horse-cars at ferry entrance reach Grand Union Hotel in ten minutes. 600 rooms, $1 and upward per day. European Plan. Guests' baggage delivered to and from Grand Central Depot free.
FORD & Co., Proprietors.

HOTEL CAMBRIDGE,
Fifth Avenue and 33d Street. A select family hotel, having every comfort and convenience for the accommodation of permanent and transient guests. *Cuisine* and service unexcelled. Convenient to the principal points of interest.
HENRY WALTER, Proprietor,
Late of the Albemarle Hotel, New York city.

26

NIAGARA FALLS.

THE PROSPECT HOUSE
is under the same Owner and Management as the original Prospect House, Canada side, established in 1874. This hotel is fitted with all modern improvements, conveniently and admirably located on high, shady ground, and is a strictly first-class transient hotel, open all the year round. Rates, $3.50 to $5.50.
D. ISAACS, Owner and Manager.

THE CATARACT HOUSE.
One of the best known hotels at Niagara. Improvement in its appointment of *cuisine* and service. Under new and progressive management. Adjoining State Reservation, and directly opposite Goat Island. For terms, address
J. E. DEVEREUX, Manager.

PHILADELPHIA, PA.

COLONNADE HOTEL,
Chestnut Street, corner 15th, Philadelphia. Most desirably located, and adapted in all respects to the requirements of the best class of the traveling public. European and American plans.
H. J. & G. R. CRUMP.

SARATOGA SPRINGS, N. Y.

GRAND UNION HOTEL,
WOOLLEY & GERRANS, Proprietors, is the most magnificent summer hotel in the world. It is in the finest location, and adjacent to the most famous springs. Splendid orchestra. Second season under present management.

WINDSOR HOTEL.
Season of 1893. Opens May 31. Closes October 1. This quiet and elegant hotel will be conducted as a strictly high-class house. Location unrivaled. Appointments, *cuisine*, and service unexcelled. Rooms may be engaged and contracts made by addressing
WILLARD LESTER, Manager,
Saratoga Springs, N. Y.

WASHINGTON, D. C.

RIGGS HOUSE,
Washington, D. C. Reopened under new management; refurnished and redecorated in first-class style; table the best in the city.
RIGGS HOUSE Co., Proprietors,
G. DEWITT, Treasurer.

New York Central!

WHY not take an early opportunity to investigate the unparalleled facili we offer, and unite your verdict with that of the press public of this great country, that one of its greatest and m valuable institutions is the

New York Central?

WHO would not prefer to have the *best*, when it can be had at practica the same cost, taking into consideration time, comfort, and sirability of route and surroundings, viz.,

New York Central?

WHAT can be superior to the magnificent equipment of the most famo railroad in the world, the

New York Central?

WHEN you go to the World's Fair at Chicago, or make your customal trip to the East, go by "America's Greatest Railroad," the

New York Central,

WHICH is the best route between the East and the West. There is no but one answer to this question. Without doubt the answer every experienced traveler is, the

New York Central.

WHERE can you find such varied and delightful scenery, such super trains, or such reliable service generally, as are presented by th

New York Central?

WOULD you have a clearer conception of our great system and its advan tages? Send five two-cent stamps for a copy of "Health an Pleasure on America's Greatest Railroad," to

GEORGE H. DANIELS, General Passenger Agent,

Grand Central Station, New York

28

TABLE OF DISTANCES and TIME OF TRAINS

ON THE

Mauch Chunk,
Summit Hill, &
Switch-Back
RAILROAD.

THE SWITZERLAND OF AMERICA.

Length of Mount Pisgah...........2,322 feet.	Fall from Mount Jefferson to Summit Hill...................... 45 feet.
Height of Mount Pisgah........... 664 feet.	Grade from Summit Hill to Mauch Chunk, to the mile............ 96 feet.
Distance from Mount Pisgah to Mount Jefferson................ 6⅔ miles.	Summit Hill above the Lehigh.... 975 feet.
Fall from Mount Pisgah to Mount Jefferson...................... 302 feet.	Mount Pisgah above the Lehigh... 850 feet.
Length of Mount Jefferson.......2,070 feet.	Mount Pisgah above the tide...... 1,500 feet.
Height of Mount Jefferson........ 462 feet.	Mount Jefferson above the tide.... 1,662 feet.
Distance from Mount Jefferson to Summit Hill.................. 1 mile.	Distance from Mauch Chunk to Summit Hill and return....... { 18 miles circuit.

TIME-TABLE SWITCH-BACK R. R., TAKING EFFECT MAY 18, 1893.

TRAINS.	No. 1	No. 2	No. 3	No. 4	No. 5	No. 6	No. 7*
LEAVE.	A. M.	A. M.	A. M.	P. M.	P. M.	P. M.	P. M.
Mansion House (stage)........	8.00	9.40	11.17	12.40	1.50	3.25	5.20
American " "	8.05	9.45	11.22	12.45	1.55	3.80	5.25
Switch-Back Depot..........	8.80	10.00	11.87	1.00	2.20	3.45	5.35
Mount Pisgah.................	8.45	10.25	11.47	1.10	2.30	3.55	5.40
Mount Jefferson..............	9.10	10.45	12.12	1.85	2.55	4.20	6.05
Summit Hill, arrive..........	9.15	10.50	12.17	1.40	3.00	4.25	6.10
Summit Hill, leave.	9.40	11.10	12.85	1.50	3.20	4.85	6.15
Upper Mauch Chunk.........	10.00	11.80	12.55	2.10	8.40	4.55	6.85
American House..............	10.15	11.45	1.10	2.25	8.55	5.05	6.45
Mansion House...............	10.20	11.50	1.15	2.80	4.00	5.10	6.50

*After September 20th Train No. 7 will be 30 minutes earlier.

The above make connection with trains to and from New York, Phila., Wilkesbarre, Scranton, Elmira, etc. Stages connect with all trains on Central R. R. of N. J. and Lehigh Valley R. R. for "Switch-Back," enabling passengers to enjoy a ride over this road and return the above same day.

SUMMER EXCURSIONS.

SPECIAL CARS FOR PRIVATE PARTIES WILL BE RUN AT ANY TIME DURING THE DAY.

T. L. MUMFORD, Less. and Manager.   H. J. MUMFORD, Supt. and Pass. Agent.

General Office, Mauch Chunk, Pa.

29

81

A SENSIBLE ROAD.

THE

CANADIAN PACIFIC
Railway Company

OPERATE THEIR OWN

DINING-CARS,   TELEGRAPHS,   STEAMSHIPS,
SLEEPING-CARS,   HOTELS, Etc., Etc.

EVERYTHING THEY DO IS DONE WELL. IF YOU ARE GOING TO

- Canada, The Eastern Provinces,
     The Upper Lakes,
         The Western Prairies, Over the
            Rockies, to the Pacific Coast,
To Alaska, To China, Japan, and India,
       To the Sandwich Islands,
          New Zealand, Australia,
            or Around the World,

YOU CAN DO SO WITH MORE COMFORT AND SATISFACTION BY TRAVELING VIA

THE GREATEST OF RAILWAYS AND STEAMSHIPS.

Its Rates are the Lowest and its Equipment the Best.

PUBLICATIONS:

Around the World Folder.   **Banff and the Lakes in the Clouds.**
New Highway to the Orient.   **Summer Tours.**
Westward to the Far East.   **Fishing and Shooting.**
         Annotated Time-table, etc.

Send for a copy to

E. V. SKINNER, 358 Broadway, New York. | R. KERR, Winnipeg.
C. E. McPHERSON, 197 Washington St., Boston, and St. John, N.B. | G. McL. BROWN, Vancouver, B.C.
  | M. M. STERN, 648 Market St., San Francisco.
W. R. CALLAWAY, 1 King St., E. Toronto. | D. E. BROWN, Hong-Kong.
C. SHEEHY, 11 Fort Street, Detroit. | A. BAKER, 67 King William Street, London,
J. F. LEE, 232 South Clark Street, Chicago. | England.

Or to D. McNICOLL, General Passenger Agent, MONTREAL.

85

The Intercolonial Railway of Canada.

CPSIA information can be obtained
at www.ICGtesting.com
Printed in the USA
BVHW091845101218
535228BV00031B/1349/P